Functional Food and Human Health

Vibha Rani • Umesh C. S. Yadav
Editors

Functional Food and Human Health

Springer

Editors
Vibha Rani
Department of Biotechnology
Jaypee Institute of Information Technology
Noida, Uttar Pradesh, India

Umesh C. S. Yadav
School of Life Sciences
Central University of Gujarat
Gandhinagar, Gujarat, India

ISBN 978-981-13-1122-2 ISBN 978-981-13-1123-9 (eBook)
https://doi.org/10.1007/978-981-13-1123-9

Library of Congress Control Number: 2018956794

This Springer imprint is published by the registered company Springer Nature Singapore Pte Ltd.
The registered company address is: 152 Beach Road, #21-01/04 Gateway East, Singapore 189721, Singapore

Foreword

I am pleased to write a foreword to the book entitled *Functional Food and Human Health*, coedited by Dr. Vibha Rani and Dr. Umesh C. S. Yadav and published by Springer India Ltd. Dr. Rani and Dr. Yadav have performed excellent research in their niche areas of oxidative stress biology, inflammatory pathogenesis and herbal intervention and produced good research articles. The editors have expertise in oxidative stress and natural therapeutics for human health that is evident from their reputed publications. This book includes chapters that comprehensively and succinctly discuss a wide range of traditional functional foods and their revived usage for various indications; the sources of the traditional functional foods; their history, functionality and their chemical, physical and physiological properties; health benefits; mechanisms of antioxidant, anticancer, anti-inflammatory and anti-ageing properties; as well as clinical and epidemiological evidences. The chapters in this book have been contributed by eminent scientists working on various human diseases and in the field of functional food. These contributions have provided a comprehensive and updated review of their respective topics. The expert contributors have elucidated theoretical and practical aspects of functional foods, from fundamental concepts of biochemistry, nutrition and physiology to food technology. I am certain that the vast repertoire of readers including students, researchers, scientists and academicians from universities and institutes will find this book extremely informative, interesting and stimulating. This book indeed acts as an updated reference treatise on the functional foods and reinvigorates their usage for better human health in the modern time.

With best wishes,

(Dr. Y. K. Gupta)

Professor and Head & Chief, National Poison
Information Centre National Scientific
Coordinator: PVPI Chairman: National
GLP Technical Committee

Former President: Indian Pharmacological Society
Former President Toxicology Society of India (STOX)
Chairman, INSA – ICSU IUPHAR National Committee

Preface

This book entitled *Functional Food and Human Health* is an effort to bring together the new and merging aspect of alternative therapy by using the functional food to the main stream scientific discourse. Functional foods are defined as the items of regular diet and nutrition that go beyond the fulfilment of basic energy and nutritional requirement. By virtue of possessing bioactive molecules and compounds, the functional food promotes good health and prevents disease development. A number of such food items have been recognised to have the constituents that promote health and also found to be effective against many pathological conditions. Health-promoting biomolecules such as flavonoids, terpenoids, alkaloids and other phenolic compounds including curcumin, resveratrol and numerous others have been demonstrated to possess positive effects against several indications. These molecules have been demonstrated to exhibit anti-oxidant, anti-inflammatory, antibacterial, anticancer and other beneficial activities. In this regard, the identification of plants and plant-derived molecules from time immemorial has played important role. Many traditional systems of medicine were based upon the use of such naturally derived molecules and are in practice even today. Lately, the research on the possible use of these compounds to promote good health and treat various diseases through understanding their molecular interactions in the cells and tissues has gained momentum. This has led to tremendous increase in the number of research and review articles being added to the scientific literature every month. The excitement towards the medicinal plants and functional food derived from such plants entails greater understanding of their role in ameliorating human diseases.

This book encompasses the efforts by the contributors in different area of expertise and includes excellent chapters which provide up-to-date information in the area of functional food and its effect on human health. This book contains a total of 28 chapters which have been divided into 5 parts for enhanced cohesive learning and interpretation. The first section covers the functional food in daily diet that is regularly consumed by human. This section discusses the different compounds that are present in diet, their roles in the health promotion and disease prevention and their current perspective in terms of the latest research findings and prospects to be used as functional food by fortifying our regular diet. The second part of this book provides an in-depth knowledge and mechanistic outlook on different groups of functional food and their health-promoting properties at the molecular as well as genetic levels. The third part of this book includes the discussion on the efficient methods of delivery of the functional food to exploit their medicinal properties better. This section also deals with the aspects of product development that could enhance the efficiency of the functional food and also addresses the toxicity-related concerns by scanning the various clinical trials that have been undertaken using the potential lead molecules from the repertoire of well-studied compounds. The fourth and the last section of this book includes other miscellaneous effects of the functional food, probiotics and their role on human health and an overview of world diet containing the different functional foods.

The area of functional food is currently drawing the attention of researchers, scientists, governments, policymakers and various global agencies to enhance the awareness and public usage of these food items due to their known health-promoting effects and least toxicity concerns. A huge number of publications including books on the topic of functional food are being published; however, this book presents a concerted approach of scientists and experts in this area who have put forth excellent compilation of information which should to its readers. Accordingly, this book will be of extreme interest to the target audience including the researchers and scientists working in the area of herbal plants, nutrition and alternative therapy for disease amelioration, pathologists, students, academicians and the broad scientific community. To enhance and update the understanding of functional food-related research, the authors have compiled diverse data that covers the latest advancement in the area of functional food. It is our sincere hope that the included topics in this book will help a wider section of scientific readers in understanding recent developments in functional food and human health and diseases, addressing their pertinent queries, enhancing inquisitiveness and generating fruitful research ideas and tools that shall direct and advance their research.

We, the editors of this book, would like to extend our immense gratitude to all the contributors of the chapters. We would like to especially acknowledge the encouraging words of Prof. Y. K. Gupta, New Delhi, for inscribing a foreword for this book. We would like to acknowledge and thank our mentors Prof. Shyamal Goswami and Prof. Najma Z. Baquer, School of Life Sciences, JNU, New Delhi, India, and Prof. Satish K. Srivastava, Department of Biochemistry and Molecular Biology, University of Texas Medical Branch (UTMB), Galveston, Texas, USA, for giving their incessant encouragement, guidance and valuable training. We also acknowledge the efforts, time and patience of our colleagues, students and family who helped in the various stages during the preparation of this book. We are grateful to the Department of Biotechnology (DBT) and Department of Science and Technology (DST), Govt. of India, for supporting our research endeavours through various grants and awards. We are grateful to the publisher, Springer India Pvt. Ltd., for agreeing to publish our book and special thanks to Ms. Saanthi Shankhararaman (Project Coordinator) and Madhurima Kahali (Associate Editor, Biomedicine), both at Springer, for the logistical support. We appreciate their professionalism, support and commitment for the successful creation and execution of this book.

Noida, Uttar Pradesh, India Vibha Rani

Gandhinagar, Gujarat, India Umesh C. S. Yadav

Contents

Contributors

Hina Agraval School of Life Sciences, Central University of Gujarat, Gandhinagar, Gujarat, India

Asrar Alam Laboratory of Vaccinology and Applied Immunology, Kanazawa University School of Pharmacy, Kanazawa, Japan

Asmita Arora Department of Biotechnology, Jaypee Institute of Information Technology, Noida, Uttar Pradesh, India

Stuti Awasthy Department of Biotechnology, Jaypee Institute of Information Technology, Noida, India

Dibyendu Banerjee Molecular and Structural Biology Division, CSIR-Central Drug Research Institute, Lucknow, Uttar Pradesh, India

Academy of Scientific and Innovative Research (AcSIR), Chennai, India

Joyita Banerjee Symbiosis School of Biological Sciences, Symbiosis International (Deemed University), Pune, Maharashtra, India

Audesh Bhat Centre for Molecular Biology, Central University of Jammu, Jammu, India

Vivek Bhosale Department of Clinical and Experimental Medicine, CSIR-Central Drug Research Institute, Lucknow, Uttar Pradesh, India

Academy of Scientific and Innovative Research (AcSIR), Chennai, India

Maxine P. Bonham Department of Nutrition, Dietetics and Food, Monash University, Melbourne, VIC, Australia

Sagarika Bose Molecular and Structural Biology Division, CSIR-Central Drug Research Institute, Lucknow, Uttar Pradesh, India

Fenisha Dilipkumar Chahwala School of Life Sciences, Central University of Gujarat, Gandhinagar, Gujarat, India

S. Chennakesavulu Department of Animal Biology, School of Life Sciences, University of Hyderabad, Hyderabad, India

Shabnam Chhabra Department of Home Science, V.M.L.G (PG) College, Ghaziabad, Uttar Pradesh, India

Sayantani Chowdhury Division of Molecular Medicine, Bose Institute, Kolkata, West Bengal, India

Shweta Dang Department of Biotechnology, Jaypee Institute of Information Technology, Noida, Uttar Pradesh, India

Arpita Devi Department of Animal Biology, School of Life Sciences, University of Hyderabad, Hyderabad, India

Yogita Dhas Symbiosis School of Biological Sciences, Symbiosis International (Deemed University), Pune, Maharashtra, India

Monisha Dhiman Department of Biochemistry and Microbial Sciences, School of Basic and Applied Sciences, Central University of Punjab, Bathinda, Punjab, India

Aimee L. Dordevic Department of Nutrition, Dietetics and Food, Monash University, Melbourne, VIC, Australia

Atiya Fatima Department of Applied Chemistry, Delhi Technological University, New Delhi, India

Reema Gabrani Department of Biotechnology, Jaypee Institute of Information Technology, Noida, Uttar Pradesh, India

Gauri Gaur Department of Biotechnology, Jaypee Institute of Information Technology, Noida, Uttar Pradesh, India

Kunj Bihari Gupta Department of Biochemistry and Microbial Sciences, School of Basic and Applied Sciences, Central University of Punjab, Bathinda, Punjab, India

Sanjay Gupta Department of Biotechnology, Jaypee Institute of Information Technology, Noida, Uttar Pradesh, India

Aditi Jain Department of Biotechnology, Jaypee Institute of Information Technology, Noida, Uttar Pradesh, India

Gauransh Jain Department of Biotechnology, Jaypee Institute of Information Technology, Noida, Uttar Pradesh, India

Tushar Jain Molecular and Structural Biology Division, CSIR-Central Drug Research Institute, Lucknow, Uttar Pradesh, India

Academy of Scientific and Innovative Research (AcSIR), Chennai, India

Dhwani Jhala Biomaterials and Biomimetics Laboratory, School of Life Sciences, Central University of Gujarat, Gandhinagar, Gujarat, India

Kushal Kandhari School of Life Sciences, Central University of Gujarat, Gandhinagar, Gujarat, India

Atinderpal Kaur Department of Biochemistry and Microbial Sciences, School of Basic and Applied Sciences, Central University of Punjab, Bathinda, Punjab, India

Sharanjot Kaur Department of Biochemistry and Microbial Sciences, School of Basic and Applied Sciences, Central University of Punjab, Bathinda, Punjab, India

Sukhchain Kaur Department of Biochemistry and Microbial Sciences, School of Basic and Applied Sciences, Central University of Punjab, Bathinda, Punjab, India

Aditi Khare Department of Biotechnology, Jaypee Institute of Information Technology, Noida, Uttar Pradesh, India

Ajay Kumar Department of Zoology, Banaras Hindu University, Varanasi, India

Bhawna Kumari Metabolic Disorders and Inflammatory Pathologies Laboratory, School of Life Sciences, Central University of Gujarat, Gandhinagar, Gujarat, India

Deepak Kumar Department of Botany, Central University of Jammu, Jammu, India

Kunal Kumar School of Life Sciences, Central University of Gujarat, Gandhinagar, Gujarat, India

Sandeep Kumar Department of Biochemistry and Microbial Sciences, School of Basic and Applied Sciences, Central University of Punjab, Bathinda, Punjab, India

Shashank Kumar Centre of Biochemistry and Microbial Sciences, Central University of Punjab, Bathinda, India

Dipeeka Mandaliya Institute of Science, Nirma University, Ahmedabad, Gujarat, India

Anil K. Mantha Department of Animal Sciences, School of Basic and Applied Sciences, Central University of Punjab, Bathinda, Punjab, India

Deeksha Mehtani Department of Biotechnology, Jaypee Institute of Information Technology, Noida, Uttar Pradesh, India

Neetu Mishra Symbiosis School of Biological Sciences, Symbiosis International (Deemed University), Pune, Maharashtra, India

Sujata Mohanty Department of Biotechnology, Jaypee Institute of Information Technology, Noida, Uttar Pradesh, India

Margaret Murray Department of Nutrition, Dietetics and Food, Monash University, Melbourne, VIC, Australia

Vijay Nema Division of Molecular Biology, National AIDS Research Institute, Pune, India

Darshika Nigam Department of Biochemistry, School of Life Sciences, Dr. Bhimrao Ambedkar University, Agra, Uttar Pradesh, India

Krunal Ramanbhai Patel School of Life Sciences, Central University of Gujarat, Gandhinagar, Gujarat, India

Rohit Patel Metabolic Disorders and Inflammatory Pathologies Laboratory, School of Life Sciences, Central University of Gujarat, Gandhinagar, Gujarat, India

Sweta Patel Institute of Science, Nirma University, Ahmedabad, Gujarat, India

Utkrishta L. Raj Department of Biotechnology, Jaypee Institute of Information Technology, Noida, Uttar Pradesh, India

Vibha Rani Department of Biotechnology, Jaypee Institute of Information Technology, Noida, Uttar Pradesh, India

Hilal Rather Biomaterials and Biomimetics Laboratory, School of Life Sciences, Central University of Gujarat, Gandhinagar, Gujarat, India

Aramati B. M. Reddy Department of Animal Biology, School of Life Sciences, University of Hyderabad, Hyderabad, India

Purnam Hoshe Ruba Department of Biotechnology, Jaypee Institute of Information Technology, Noida, Uttar Pradesh, India

Rubal Department of Biochemistry and Microbial Sciences, School of Basic and Applied Sciences, Central University of Punjab, Bathinda, Punjab, India

Lisa Ryan Department of Natural Sciences, Galway-Mayo Institute of Technology, Galway, Ireland

Akash Sabarwal School of Life Sciences, Central University of Gujarat, Gandhinagar, Gujarat, India

Indira P. Sarethy Department of Biotechnology, Jaypee Institute of Information Technology, Noida, India

Sharad Saxena Department of Biotechnology, Jaypee Institute of Information Technology, Noida, Uttar Pradesh, India

Sriram Seshadri Institute of Science, Nirma University, Ahmedabad, Gujarat, India

Akanksha Sharma Metabolic Disorders and Inflammatory Pathologies Laboratory, School of Life Sciences, Central University of Gujarat, Gandhinagar, Gujarat, India

Arpana Sharma School of Life Sciences, Central University of Gujarat, Gandhinagar, Gujarat, India

Ritis Shyanti School of Life Sciences, Central University of Gujarat, Gandhinagar, Gujarat, India

Parames C. Sil Division of Molecular Medicine, Bose Institute, Kolkata, West Bengal, India

Mohit Singh Metabolic Disorders and Inflammatory Pathologies Laboratory, School of Life Sciences, Central University of Gujarat, Gandhinagar, Gujarat, India

Ram Singh Department of Applied Chemistry, Delhi Technological University, New Delhi, India

Rana P. Singh Cancer Biology Laboratory, School of Life Sciences, Jawaharlal Nehru University, New Delhi, India

Kopal Singhal Department of Biotechnology, Jaypee Institute of Information Technology, Noida, Uttar Pradesh, India

Krishnendu Sinha Department of Zoology, Jhargram Raj College, Jhargram, West Bengal, India

Division of Molecular Medicine, Bose Institute, Kolkata, West Bengal, India

Gandreddi V. D. Sirisha Department of Biochemistry and Bioinformatics, Institute of Science, GITAM University, Visakhapatnam, Andhra Pradesh, India

Himanshu Sukhpal Department of Biotechnology, Jaypee Institute of Information Technology, Noida, India

Chava Suresh Department of Animal Biology, School of Life Sciences, University of Hyderabad, Hyderabad, India

Udita Tiwari Department of Biochemistry, School of Life Sciences, Dr. Bhimrao Ambedkar University, Agra, Uttar Pradesh, India

Shishir Upadhyay Department of Animal Sciences, School of Basic and Applied Sciences, Central University of Punjab, Bathinda, Punjab, India

Johnna Francis Varghese Metabolic Disorders and Inflammatory Pathologies Laboratory, School of Life Sciences, Central University of Gujarat, Gandhinagar, Gujarat, India

Rajesh Vasita Biomaterials and Biomimetics Laboratory, School of Life Sciences, Central University of Gujarat, Gandhinagar, Gujarat, India

Kappala Vijaya Rachel Department of Biochemistry and Bioinformatics, Institute of Science, GITAM University, Visakhapatnam, Andhra Pradesh, India

Renu Yadav Department of Biochemistry, School of Life Sciences, Dr. Bhimrao Ambedkar University, Agra, Uttar Pradesh, India

Umesh C. S. Yadav School of Life Sciences, Central University of Gujarat, Gandhinagar, Gujarat, India

About the Editors

Dr. Vibha Rani is working as an Associate Professor in the Department of Biotechnology, Jaypee Institute of Information Technology, Noida, India. After completing her doctoral studies from the School of Life Sciences, Jawaharlal Nehru University, New Delhi, she continued her academic career where she teaches as well as guides graduate, postgraduate and doctorate students in the area of oxidative stress-induced cardiomyopathy and other development-related heart diseases. Dr. Rani is an academician as well as an accomplished young scientist and has been in research for 16 years. Her current research focuses on understanding the mechanism of action of phytomolecules and also developing ideal drug molecule for prevention of non-communicable diseases. She has been focusing on developing natural therapies and microRNA-based therapeutics to address the most severe disorders that affect the human society such as cardiomyopathy, diabetes and breast cancer. Dr. Rani has received many extra-mural funding for her research work from the Department of Science and Technology (DST) and Department of Biotechnology (DBT), Govt. of India. She has been successfully able to communicate her research findings to various international journals. She has been honoured with the prestigious A.R. Rao Memorial Young Scientist Award, S.C. Tyagi Young Faculty Award and Academic Brilliance Young Faculty Award, 2017, which accounts for just few among the much recognitions that she has.

Dr. Umesh C. S. Yadav is currently working as an Associate Professor at the School of Life Sciences, Central University of Gujarat, Gandhinagar, India. His current focus area of research includes understanding the biochemical and molecular mechanisms of metabolic disorder-induced chronic inflammatory diseases including diabetic and cardiovascular disorders, cancer, asthma and COPD. Dr. Yadav has a Ph.D. degree from the School of Life Sciences, Jawaharlal Nehru University, New Delhi, where he investigated the effects of *Trigonella foenum-graecum* on diabetic complications. He has over 13 years of research experience, 6 years as postdoctoral fellow and more than 7 years as faculty member. Dr. Yadav spent 9 years in the University of Texas Medical Branch (UTMB) Galveston, Texas, USA, where he performed excellent research in oxidative stress-induced inflammatory pathologies such as diabetic and cardiovascular complications, lung inflammatory diseases including asthma and chronic obstructive pulmonary disease (COPD) and ocular inflammatory diseases. He has published nearly 50 research articles in high-impact journals including reviews and book chapters which indicate towards his expertise in inflammatory diseases and associated signalling pathways and their modulation using different drug molecules, including natural products. Dr. Yadav has been awarded a prestigious Ramanujan Fellowship from the Department of Science and Technology (DST), New Delhi, India.

Part I

Functional Food Components in Daily Diet

Composition of Functional Food in World Diet

1

Vibha Rani, Asmita Arora, Purnam Hoshe Ruba, and Aditi Jain

Abstract

The interest of researchers and consumers for functional food is rapidly growing along with exhaustive research to find out the properties and applications that are beneficial to human health. The main reason for growth of functional food market is the lifestyle and ever-growing population. With competition at every level, at times health takes the last seat which effects life in the long run. Today in twenty-first century, obesity is recognized as a global issue which has severely affected the United States, followed by India. Genetics play a major role in the development of diseases, but today lifestyle has a dominant effect. This chapter analyzes the current Indian market situation for function foods as compared to the market across the globe. It also covers the major food segments that are now regarded as functional foods. Probiotics, some fruits and vegetables, and beverages like green tea and wine are regarded as functional food. Lastly, the chapter also talks about the challenges and future aspects of functional foods.

Keywords

Functional foods · Probiotics · Functional food market · Antioxidants

1.1 Introduction

There is not much of a difference in western parts of the world and India when it comes to food and health. The Indian food industry is rapidly growing with exhaustive response to the demand of healthy food. The Indian consumer of the

V. Rani (✉) · A. Arora · P. H. Ruba · A. Jain
Department of Biotechnology, Jaypee Institute of Information Technology, Noida, Uttar Pradesh, India
e-mail: Vibha.rani@jiit.ac.in

V. Rani, U. C. S. Yadav (eds.), *Functional Food and Human Health*,
https://doi.org/10.1007/978-981-13-1123-9_1

twenty-first century is looking for food options that are beyond nutrition, options that provide specific health benefits and benefits related to skin and beauty and decrease risks related to diseases. Functional foods for this reason as readily being accepted in the Indian market and the consumers for the same are demanding this type in food categories [36]. But there is still a lot of scope for functional food in the Indian scenario before it is established naturally. This chapter provides in-depth knowledge about benefits of functional foods and its status in the Indian market.

Industrialization and globalization of the country have a major impact on the food habits of the citizens. The bifurcated urbanization of the country has led to two major nutritional issues: excess and deficiency of nutrition. The changes in lifestyle of the consumers have led to a shift to functional foods from ordinary food. The diseases related to lifestyle like diabetes, obesity, and blood pressure are increasing at an alarming rate. This is due to high work pressure and cut throat competition where health and nutrition are gradually taking a back seat. Decrease in field activities and more dependence on fast foods are also major causes for the birth of such diseases in India [46].

Due to such rationales and limited physical activities, consumers are looking for options in food that provide them limited risks of diseases and more health benefits, something that functional food is fulfilling. The children nowadays are more into gadget gaming than playing in the field. Along with this they are taking food with high calories and low nutrition value which leads to lifestyle diseases in early stages of life. Such urbanization has led to change in the eating habits and lifestyles affecting majorly the children and elderly with diseases that stick with them throughout the lifetime [32].

In the course of the most recent decade, interest for nourishments and refreshments that enhance or advantage well-being has expanded in numerous parts of the world, close by increasing expenses of social insurance, increments in future, and longing for a higher nature of life. In such regard, functional foods play a vital part, offering another sort of well-being instrument that guarantees particular impacts identified with specific nourishment components.

Functional foods were first introduced in Japan in the mid-1980s. At that time, it referred to processed foods containing ingredients that beneficially influence specific body functions, in addition to being nutritious. Today, there is no accurate definition of what functional foods exactly are. A functional food can thus be a natural food or it can be a food that has been modified to have a functional influence on the health and well-being of the consumer through the addition, removal, or modification of specific components.

Functional foods are divided into three classes: food items that have increased content of fatty acids, vitamins, minerals, or fiber, allergy or intolerance-causing food with eliminated active components, and food items with nontraditional nutrients having immune boost [17].

The rise in functional food usage can be credited to advances in science and technology, expanding social insurance costs and changes in nourishment laws influencing marketplaces.

Another term homogenously used with functional foods is "nutraceuticals," although it is less supported by the consumers. Nutraceuticals refer to nearly any bioactive component that delivers a health benefit. Functional food carries a different meaning to different countries of the world. The difference lies in the regulatory market and different definitions. The concept of functional food was first promoted in 1984 by Japanese scientists who studied the relationships between nutrition, sensory satisfaction, fortification, and modulation of physiological systems [40]. There it referred to the food products fortified with special constituents that possess advantageous physiological effects [21, 42].

In Europe, functional food is defined as "a food product that can only be considered functional if together with the basic nutritional impact it has beneficial effects on one or more functions of the human organism, thus either improving the general and physical conditions or/and decreasing the risk of the evolution of the diseases. The amount of intake and form of the functional food should be as it is normally expected for dietary purposes. Therefore, it could not be in the form of pill or capsule just as normal food forms" [16].

In India, functional foods are defined as "foods which are specially processed or formulated to satisfy particular dietary requirements which exist because of a particular physical or physiological condition or specific diseases and disorders and which are presented as such, wherein the composition of these foodstuffs must differ significantly from the composition of ordinary foods of comparable nature if such ordinary foods exist, and may contain one or more of the following ingredients, namely plants, minerals or vitamins, substance from animal origin and a dietary substance used to supplement the diet".

This definition is according to the Food Safety and Standards Act, 2006, which is an amalgam of functional foods, nutraceuticals, and health supplements. This includes food present in different forms like powder, capsules, and granules.

1.2 Market of Functional Foods

Food that has health and nutritional benefits has made a significant impact on changing consumers' lifestyle. This provides a great opportunity for the market of functional foods. Today food not only satisfies hunger but also a sense of satiety. Factors like demographic, economic, social, and cultural have a significant impact on attitudes toward consumers for food products. The consumer today understands the relationship between diet and health and demand products that are beyond nutrition.

1.2.1 Status of Functional Food at International Level

The global market of functional foods today is estimated to be at least 33 billion US $. The food market of the United States has almost 50% of the market share. Out of this 50%, 2–3% is owned by the functional foods market [25].

Japan is regarded as the birthplace of functional foods which have more than 1700 types of products available [25]. In 2006, the market was at 5.73 billion US $ with almost 500 products under the umbrella of FOSHU (Foods for Specified Health Uses) [40].

Europe, Germany, France, the United Kingdom, and the Netherlands hold the dominancy in functional food. Euromonitor has predicted that the sales will boom in regions of Poland, Hungary, and Russia [6].

Today, not only the food manufacturers but even pharmaceutical industries are trying to enter the market of functional foods [28]. Some leading pharmaceutical industries that have entered this domain are Novartis Consumer Health, Johnson & Johnson, and Abott Laboratories. These companies organize clinical trials to prove health claims of a particular food product [5].

1.2.2 Status of Functional Foods at Indian Level

The market of functional foods today functions by keeping in mind that the urban lifestyle of India is rapidly damaging the health and food habits of the citizens. The companies and various startups are keen on the experimentation of products even though there is a high risk of long-term losses. Various consultancies present in India estimate that Health & Wellness (H&W) food markets in India reach INR 101 billion in 2012 and forecast it to grow at a CAGR of 33% to INR 550 billion by fiscal year 2015 with advances in product development and government-mandated fortification [22].

There have been various new products launched in the Indian market which come under the category of functional foods. According to a report by Mintel, a UK-based market research company, 116 new products were launched in the Indian market. Eighty products were aimed at benefitting the cardiovascular functioning of the human body, while 36 were promoting strong immunity and health. Products launched were mainly edible oil, baby food, dairy, bakery, and confectionaries.

Since a basic Indian household uses edible oil as a chief cooking medium, it was considered as the main category of functional foods. Consumers now look for edible oil that has low trans-fat and low cholesterol which results in reduced cardiovascular diseases.

Another category which was very popular among the Indian consumers were dairy and baby food categories, where probiotics and prebiotics are increasingly becoming popular due to their abundant health benefits like providing immunity by increasing the population of good bacteria in the gastrointestinal tract.

1.2.3 Suppliers of Functional Food to the Indian Market

The Indian food market is divided into six segments as summarized in Table 1.1 (Adapted and modified from [34]).

Table 1.1 Indian functional food companies and startups

Main segments	Companies and start ups	Functional foods
Broad product range	Nestle, Danone, Unilever, Kelloggs, Pepsico, Yakult, Danone	Probiotic yoghurt/Dahi, snacks, energy drinks, breakfast cereals
Pharmaceuticals	GSK, Amway, Ranbaxy	Malted food, supplements, fortified products
National Category leaders	Amul, Dabur, ITC, Britannia, Parle	Dairy products, fruits & vegetable juices, biscuits
Small and medium sized companies	Heritage Foods, Ruchi Soya, Mother Dairy	Soya milk, dairy, oils
Retail companies	Reliance Wellness, Apollo Pharmacy, Patanjali	Sweetners, cereals, energy drinks, ayurvedic & herbal products
Functional ingredients	Chr Hansen, Orana, Danisco	Cultures, enzymes, phytonutrients, natural colors

The food industry is increasingly becoming "health conscious," where the main players of the industry are focusing more on nutrition, health, and safety to gain a competitive edge. Multinational food companies like Nestle and Pepsico have introduced the concept of functional foods in the market. Nestle has been the market leader of the food industry since long. It has introduced many healthy food product chains like infant food, noodles, condensed milk, and vegetable multigrain Maggi noodles, while Kelloggs holds a 60% share in the INR 4 billion cereal markets in India. The brand is popular for breakfast cereals, Muesli, Kelloggs Special K for weight reduction, and various flavors of cereals.

All functional food products require a strong R&D department. Many of the multinational companies have their own R&D on which they spend about 2–3% of the annual turnover. Domestic companies have a low R&D budget due to which they lag behind, and multinational companies having their own R&D with appropriate budget lead the food industry. Also the multinational companies are more capable to establish the specific health claim related to the functional food products with the scientific verification of the efficacy of functional food based on statistically validated data from different model systems, from retrospective and prospective epidemiological studies, as well as from intervention studies on humans [36].

1.3 Fast-Growing Categories of Functional Foods

Functional foods are regarded as any substance that may be considered food or a part of food that provides health benefits that includes treatment and prevention of diseases. The products range from dietary, genetically engineered food, herbal products, and processed food. The potential of such type of food is often related to the maintenance and improvement of health beyond nutrition [1, 44].

1.3.1 Probiotics as Functional Foods

Probiotics are defined as selected, viable microbial dietary supplements that, when introduced in sufficient quantities, beneficially affect human organism through their effects in the intestinal tract [15, 20].

Probiotics have been extensively consumed in Japan, European countries, and the United States. It is now emerging as an important category of food supplement in India. Probiotics are considered functional foods because they provide benefits apart from nutrition [30].

The health benefits of bacteria in foods and consumption led to the concept of probiotics. Probiotics are live microbial feed that when consumed provide beneficial health effects by maintaining the flora in the intestinal tract. Probiotics, antonym of antibiotics, is now increasingly rising as an alternative approach for the latter.

Probiotic bacteria are present in yogurt, kefir, dark chocolate, microalgae, kimchi, raw cheese, apple cider vinegar, salted gherkin pickles, and miso (Japanese spice).

The human body is a place of residence for microorganisms and has more prokaryotic cells than eukaryotic cells. Some situations alter the ecology maintained by the microorganisms, thus exposing the body to pathogenic bacteria and infections [20]. These situations include misuse of antibiotics, the diet we take, and the environment we live in. Hence, probiotics acts as a rescuer and helps to reflourish the microflora present in our body, thus promoting health.

Benefits of probiotics include reduced lactose intolerance, blood pressure, and respiratory infections and increased resistance to pathogens causing infections [26]. Sources of probiotics include yogurt, kefir, dark chocolate, raw cheese, kimchi, and apple cider vinegar.

1.3.2 Beverages as Functional Foods

Beverages are fortified with vitamins A, C, and E along with antioxidants and other functional ingredients. There are a number of products available under this domain, yet the market for beverages is still small and limited to European countries. Germany is the only country that has a good functional beverage market [40].

1.3.2.1 Green Tea

Tea is the most common drink consumed after water in India. Green tea is a "non-fermented" tea and contains more catechins than black tea or oolong tea. Catechins are strong antioxidants which reduce the formation of free radicals leading to prevention of cell damage. Green tea is proven to reduce cardiovascular diseases by maintaining the blood pressure and lowering the cholesterol [10].

The health benefits are due to its polyphenol content. Flavonols, a class of flavonoids, comprise 30% of the dry weight of the leaf. The benefits are due to the presence of catechins [10].

Green tea in Asia is known to effectively treat diarrhea and typhoid [7]. The catechins present have been shown to have an inhibitory effect on infections caused by *Helicobacter pylori* [43].Green tea has the right amount of caffeine and amino acid L-theanine. When these two work synergistically, they contribute toward better functioning of the brain.

The laboratory trials have proven to show health benefits. Human trials are still limited, and further extensive research is required to analyze the health benefits of green tea along with their mechanism of action. The catechins present reduce the risk of cancer as well by preventing cell damage [45].

1.3.2.2 Dairy-Based Beverages

Under the category of dairy industry belong milk, fermented milk, and yogurt drinks. These beverages are loaded with ω-3 fatty acids, α-linoleic acid (C18:3 n-3, ALA), eicosapentaenoic acid (C20:5 n-3, EPA), and docosahexaenoic acid (C22:6 n-3, DHA) [35]. These have known to prevent and treat epilepsy [8].

Milk contains a protein called casein which is known to act as a precursor of biologically active peptides [12]. These compounds can inhibit the angiotensin-converting enzyme (ACE-I) playing a major role in converting angiotensin-I to angiotensin-II and degrading bradykinin by blocking the active site of the enzyme. The conversion of angiotensin-I to angiotensin-II and the degradation of bradykinin result to increased blood pressure, while the inhibition of the enzyme reduces pressure increase [35].

1.3.2.3 Sports Drinks

Sports drinks are drinks that are consumed before exercise or during exercise. The drink is infused with carbohydrates, vitamins, minerals, and electrolytes. There is no sign of caffeine in the drink due to strict safety regulations [23].

Sports drinks are considered to be the best substitute of water [24]. Various sports drink contains combination of sucrose, glucose, and maltodextrin/glucose polymers. Maltodextrin is less sweet than glucose and provides carbohydrates in larger quantities without making the drink sweet [11].

But consumption of such drinks is advised to be regulated and avoided because the overall intake of calories exceeds the daily limit. This leads to unnecessary weight gain, dental problems, and low nutrition diet [29]. Gatorade from Pepsico, Powerade, and Accelarade are a few examples of sports drink available.

1.3.2.4 Wine

Wine is traditionally regarded as health-promoting beverage due its positive effect on coronary heart disease [36]. It has antioxidant properties due to the presence of phenolic compounds [27]. Of the phenolics, the stilbene group is one of the most important, with resveratrol (3,5,40-trihydroxystilbene) being one of the main stilbenes found in wine [16]. The polyphenols present contribute to the aging process and give the characteristic taste and color. Red wine is preferred over white wine since it has 6 times more phenolic compound due to longer contact with grapes. The concentration of phenolic compounds in red wine is 1800–3000 mg/L. These have healthy effects on the body due to the antioxidant properties [19].

Wine has been a popular beverage for consumption in the European Union, and a study conducted in Spain showed that the origin, price, and quality certification are some leading factors that consumers look into before the purchase [9]. Consumers in Europe have identified wine as a healthy product, while the similar trend is yet to touch India [4]. There have been several reports that wine was considered as a medicine in India and China in the prehistoric time [33]. Wine was used as a tranquilizer, an antiseptic for wounds, an appetite stimulant, and a cooling agent [33].

Benefits of wine consumption were first identified by the observations made by Renaud and de Lorgeril [36]. He along with his team reported there were low mortality rate from heart diseases in French people who were regular wine drinkers in spite of the fact that they consumed saturated fats and smoked a lot.

1.3.3 Dietary Fiber as Functional Foods

Dietary fiber is derived from plants and it is not digested by the stomach or small intestine. It reaches to the large intestine unchanged [39]. These fibers are not being able to get hydrolyzed by endogenous enzymes in the small intestine (indigestibility). Dietary fibers are the polymers of carbohydrate with ten or more monomeric units and belong to one of the three categories of carbohydrates: naturally occurring edible carbohydrate in food; carbohydrate polymers which have been obtained from raw food material by physical, enzymatic, or chemical means; and synthetic carbohydrate polymers. Many of these fibers are the source of food and nourishment for a large number of bacteria that normally reside in the colon. These bacteria play a major role in promoting and maintaining good health. They produce vitamins and enzymes, enhance the immune system, and control cholesterol and triglyceride level. It also helps in the prevention of certain cancers and provides many other health benefits when consumed in an adequate amount (25–35 grams per day) [41]. There are various types of fibers that are categorized. One type is insoluble fiber. This fiber is insoluble in water and is unable to get fermented by the bacteria in the gut. It helps to promote a softer, bulkier stool by retaining water. Thus these fibers may prove to be important in sweeping out certain toxins and cancer-causing carcinogens. Soluble type of fibers is the fibers that are fermented by colon bacteria. These microorganisms require their own support and nourishment source. The medical advantages these bugs give are entirely dependent on the amounts of solvent fiber.

There are some special types of fibers known as prebiotic soluble fiber which have shown to have the most significant medical advantages. Inulin, oligofructose, and galacto-oligosaccharide are the three proven prebiotic soluble fiber. Vegetables like onions, garlic, bananas, leeks, asparagus, chicory root, yams, wheat, and artichokes contain prebiotic fibers in large amount. Consumption of fibers have lots of advantages like increasing the population of colon bacteria, enhancing immune system, controlling appetite and weight, increasing bowel regularity, increasing bone density, and many more.

Plants have both insoluble and soluble fiber present in it, but the amount of the fibers present varies. Fibers present in wheat and corn are 90% insoluble, while in oats the ratio of insoluble fiber to soluble fiber is about 50/50. Soluble fibers are present in large amounts in artichokes. Prebiotin™ is a dietary fiber supplement which can also be used.

1.3.4 Fruits and Vegetables as Functional Foods

Regular consumption of fruits and vegetables has proven to reduce the risk of cancer and heart-related diseases which are prevalent in industrialized countries and developing countries. The antioxidants present in fruits and vegetables are known to reduce the risks of Alzheimer's, stroke, cardiovascular disorders, and cancer [31]. Functional foods contain bioactive compounds that produce desirable health effects beyond the basic requirement of nutrition. They are a part of the daily diet which are easily available in all parts of the world. The benefits of some fruits and vegetables are summarized in Table 1.2.

Table 1.2 Fruits and vegetables as functional food

Food	Function
Berries	Berries are a rich source of phytochemicals and have flavonoids and other phenolics that promote health. Some examples are highbush blueberry, cranberry, and red and black raspberries [3]
Cranberries	Cranberries are known to reduce inflammation and were once used to dress wounds. Other than being a good source of fiber, cranberries serve as an excellent source of anthocyanins, flavonol glycosides, and phenolic acids [2]
Grapes	Phenolics like salicylic, gallic, and cinnamic are present in grapes. While flavonols like q 3-glucoronide and q 3-rutinoside are present, flavononols like astilbin and engeletin are present as well. The stem, grape seed, and skin are a rich source of flavonoids. Antioxidants present in grapes have reported to protect the body from heart and cancer diseases as well [13]
Tomato	Tomato is the world's largest cultivated vegetables consumed raw as well as fresh. They serve as a major antioxidant source containing vitamin C and lycopene. The red color of the tomato is due to the presence of lycopene which is the most effective carotenoid. A diet which includes moderate amount of tomatoes has reportedly reduced the cause of cardiovascular diseases [47]
Garlic	Garlic is a combination of spice, herb, and vegetable and is also known as Russian penicillin. Inclusion of garlic in diet reduces blood platelet clumping, blood clots, LDL cholesterol, as well as fungal infections. Therapeutic effects of garlic are categorized into antimicrobial properties, cardiovascular effects, and anticarcinogenic components [38]

1.4 Challenges and Opportunities of Functional Foods

Functional foods are called so because they provide health benefits from a particular known ingredient [18]. The challenge of researchers right now is to find that ingredient and manipulate it in the best possible way. Vegetables, fish, and fruits provide benefits which are known, but the exact molecular weight of the ingredient is unknown. Vegetables, for example, sometimes fail to give the expected benefits via involvement of minerals and vitamins. Much more knowledge is needed to bridge the gap between the mechanism of action of the molecule and the food that contains the particular molecule.

Functional food is a key concept for the future of nutrition as a science because it results from the implementation in nutrition of all the basic scientific knowledge that has accumulated over the past two or three decades. To the benefit of public health, this progress cannot be ignored; it needs to be recognized fully and used. But, today, functional food is still mainly a scientific concept that serves to stimulate research and the development of new products [18].

While functional foods provide number of health benefits, there are certain concerns that are to be taken care of before we permanently switch to this kind of food supply. Technology-wise speaking today there is number of ways with which we can alter the nutritional supply of food, but functions of many ingredients are not determined [37].

1.5 Conclusion

Functional food research is still in a state of infancy where there is a lot of scope of manipulating the ingredients present in the food. Every food item has some nutritious element in it, which when identified and isolated can make the food item as functional foods. The Japanese were the first to identify and associate with the term "functional foods," which is now popularly followed by European countries. India is still figuring out the term and is rapidly growing in the market of food.

Acknowledgment We acknowledge the Jaypee Institute of Information Technology for providing the infrastructure and literature support for conducting the detailed study presented in the chapter.

References

1. Aluko RE (2012) Functional foods and nutraceuticals. Springer, New York
2. Anhe F, Roy D, Garofalo C et al (2015) A polyphenol-rich cranberry extract protects from diet-induced obesity, insulin resistance and intestinal inflammation in association with increased Akkermansia spp. population in the gut microbiota of mice. Gut Microbiota 64:872–883
3. Anisimoviene N, Jankauskiene J, Jodinskiene M (2013) Phenolics, antioxidative activity and characterization of anthocyanins in berries of blackcurrant interspecific hybrids. Acta Biochem Pol 60:767–772

4. Barreiro-Hurle J, Colombo S, Cantos-Villar E (2008) Is there a market for functional wines? Consumer preferences and willingness to pay for resveratrol-enriched red wine. Food Qual Prefer 19:360–371
5. Bech-Larsen T, Scholderer J (2007) Functional foods in Europe: consumer research, market experiences and regulatory aspects. Trends Food Sci Technol 18:231–234
6. Benkouider C (2005) Dining with the Dutch. Functional foods and nutraceuticals. Agriculture magazines. http://www.ffnmag.com/ASP/articleDisplay.asp?strArticleId=753&strSite=FFNSITE&Screen=CURRENTISSUE. Last accessed on 15 Aug 2017
7. Blumberg J, McKay DL (2002) The role of tea in human health: an update. J Am Coll Nutr 21:1–13
8. Boroski M, Giroux HJ, Sabik H, Petit HV, Visentainer JV, Matumoto-Pintro PT, Britten M (2012) Use of oregano extract and oregano essential oil as antioxidants in functional dairy beverage formulations. Food Sci Technol 47:167–174
9. Burgarolas M, Martinez-Carrasco L, Martinez-Poveda A (2005) Quality wines and wines protected by a designation of origin: identifying their consumption determinants. J Wine Res 16:213–232
10. Cabrera C (2013) Beneficial effects of green tea—a review. J Am Coll Nutr 2:77–99
11. Campbell B (2013) Dietary carbohydrate strategies for performance enhancement. In: Campbell B (ed) Sports nutrition–enhancing athletic performance. Taylor & Francis Group, Florence, pp 75–124
12. Damodar S, Li Y (2017) A two-step enzymatic modification method to reduce immuno-reactivity of milk proteins. Food Chem 237:724–732
13. De Rosso M, Tonidandel L, Larcher R et al (2014) Identification of new flavonols in hybrid grapes by combined liquid chromatography–mass spectrometry approaches. Food Chem 163:244–251
14. Delmas D, Lancon A, Colin D et al (2006) Resveratrol as a chemopreventive agent: a promising molecule for fighting cancer. Curr Drug Targets 7:423–442
15. Dimer C, Gibson GR (1998) An overview of probiotics, prebiotics and synbiotics in the functional food concept: perspectives and future strategies. Int Dairy J 8:473–479
16. Diplock A (1999) Scientific concepts of functional foods in Europe: consensus document. Br J Nutr 1:1–27
17. Faria AMC, Gomes-Santos AS, Gauncalves JN (2013) Food components and the immune system: from tonic agents to allergens. Front Immunol 4:102
18. Gidley MJ (2004) Naturally functional foods—challenges and opportunities. Asia Pac J Clin Nutr 13:531–531
19. Goldberg DM, Tsang E, Karumanchiri A et al (1998) Quercetin and p-coumaric acid concentrations in commercial wines. Am J Enol Vitic 49:142–151
20. Grajek W, Olejnik A, Sip A (2005) Probiotics, prebiotics and antioxidants as functional foods. Acta Biochemica Pol 52:665–671
21. Hardy G (2000) Nutraceuticals and functional foods: introduction and meaning. Nutrition 16:688–689 Elsevier
22. Herald D (2012) Tata Strategic pegs wellness foods market at Rs 55K crore. Available at: http://www.deccanherald.com/content/61120/tata-strategicpegs-wellness-foods.html
23. Heckman M, Sherry K, Gonzalez E (2010) Energy drinks: an assessment of their market size, consumer demographics, ingredient profile, functionality, and regulations in the United States. Compr Rev Food Sci Food Saf 9:303–317
24. Higgins J, Tuttle T, Higgins C (2010) Energy beverages: content and safety. Mayo Clin Proc 85:1033–1041
25. Hilliam M (2000) Functional food—how big is the market? World Food Ingred 12:50–53
26. Kechagia M, Basoulis D, Maria E et al (2013) Health benefits of probiotics: a review. ISRN Nutr 2013:1–7
27. Kirimlioglu V, Ara C, Ozgar D et al (2006) Resveratrol, a red wine constituent polyphenol, protects castric tissue against the oxidative stress in cholestatic rats. Dig Dis Sci 51:298–302

28. Kotilainen L, Rajalahti R, Ragasa C et al (2006) Health enhancing foods: opportunities for strengthening the sector in developing countries. Agric Rural Dev Disc 1:1–95
29. Larson N, DeWolfe J, Story M et al (2014) Adolescent consumption of sports and energy drinks: linkages to higher physical activity, unhealthy beverage patterns, cigarette smoking, and screen media use. J Nutr Educ Behav 46:181–187
30. Lin D (2003) Probiotics as functional foods. Nutr Clin Pract 18:497–506
31. Liu R (2003) Health benefits of fruit and vegetables are from additive and synergistic combinations of phytochemicals. Am J Clin Nutr 78:5175–5205
32. Lobo V, Patil A, Phatak A et al (2010) Free radicals, antioxidants and functional foods: impact on human health. Pharm Rev 4(8):118–126
33. Medic-Saric M, Rastija V, Bojic M (2009) From Functional food to medicinal product: systematic approach in analysis of polyphenolics from propolis and wine. Nutr J 8:33
34. Menrad K (2003) Market and marketing of functional food in Europe. J Food Eng 56:181–188
35. Ozer B, Kirmaci H (2010) Functional milks and dairy beverages. Int J Dairy Technol 63:1–15
36. Renaud S, De Lorgeril M (1992) Wine, alcohol, platelets, and the French paradox for coronary heart-disease. Lancet 339:1523–1526
37. Roberfroid M (2007) Global view on functional foods: European perspectives. Br J Nutr 2:133–138
38. Schafer G, Kaschula C (2014) The immunomodulation and anti-inflammatory effects of garlic organosulfur compounds in cancer chemoprevention. Anti Cancer Agents Med Chem 14:233–240
39. Shewry PR, Hey SJ (2015) The contribution of wheat to human diet and health. Food Energy Secur 4:178–202
40. Siro I, Kapolna E, Kapolna B et al (2008) Functional food. Product development, marketing and consumer acceptance—a review. Appetite 51:456–457
41. Staffalo M, Albertengo L, Bevilacqua A et al (2012) Dietary fiber and availability of nutrients: a case study on yoghurt as a food model. In: The complex world of polysaccharides. InTech Europe, University Campus SteP Ri, Rijeka, pp 455–490
42. Stanton C, Ross RP, Fitzgerald GF et al (2005) Fermented functional foods based on probiotics and their biogenic metabolites. Curr Opin Biotechnol 16(2):198–203
43. Takabayashi F, Harada N, Yamada M et al (2004) Inhibitory effect of green tea catechins in combination with sucralfate on *Helicobacter pylori* infection in Mongolian gerbils. J Gastroenterol 39:61–63
44. Wildman R (2006) Handbook of nutraceuticals and functional foods, 2nd edn. CRC Press, New York
45. Yang C, Wang H (2016) Cancer preventive activities of tea catechins. Molecules 21:318–322
46. Yiridoe EY, Bonti-Ankomah S, Martin RC (2005) Comparison of consumer perceptions and preference toward organic versus conventionally produced foods: a review and update of the literature. Renew Agric Food Syst 20(4):193–205
47. Zuorro A, Fidaleo M, Lavecchia R (2011) Enzyme-assisted extraction of lycopene from tomato processing waste. Enzym Microb Technol 49:567–573

2 Dietary Fibre - Nutrition and Health Benefits

Shabnam Chhabra

Abstract

Numerous researches have suggested a positive impact of fibre-rich diets on human health and its association with decreased prevalence of several non-communicable diseases. Population consuming adequate dietary fibre have shown reduced risk of chronic diseases due to the beneficial effects of its intake on the disease-associated risk factors. Although new references of fibre intake have been suggested for adults and young people, the recent data on dietary assessments of population indicate low intake of fibre-rich foods such as fruits, vegetables, nuts, oilseeds and wholegrain starchy foods in diet. Dietary fibre intake is highly recommended in various gastrointestinal diseases, cardiovascular conditions, type 2 diabetes and colorectal cancer; however the importance of fibre consumption for human health is still not very well acknowledged as compared to other nutrients such as carbohydrates and fats. The message of increasing consumption of fibre-rich foods containing both soluble and insoluble fibres, such as whole grains, legumes, fruits and vegetables, needs to be strongly propagated, and the most suitable advice of consuming dietary fibre from natural foods must be delivered.

Keywords

Cardiovascular diseases · Colorectal cancer · Diabetes · Dietary fibre · Gastrointestinal diseases · Insoluble fibre · Soluble fibre

S. Chhabra (✉)
Department of Home Science, V.M.L.G (PG) College, Ghaziabad, Uttar Pradesh, India

V. Rani, U. C. S. Yadav (eds.), *Functional Food and Human Health*,
https://doi.org/10.1007/978-981-13-1123-9_2

2.1 Introduction

Industrialization, urbanization and technological advancements have, on one hand, drastically improved the standard of living for our people while, on the other hand, escalated the incidence of non-communicable diseases [1]. The association between food and health has resulted in consumers' interest to know the nutritional and therapeutic value of the food they are consuming [2]. Since the work demand has replaced the traditionally cooked foods with commercially processed foods and ready-to-eat foods, more and more people are becoming aware that they are being deprived of some food components, which may be of immense importance to their health. There has been a striking change in the perception related to the intake of dietary fibre over the past few decades, and its importance as a nutrient has received lots of consideration in recent years. The processing methods being used currently, like milling of grain to refined flour and canning of fruits and vegetables, have compromised the availability of fibre from the diet.

Dietary fibre may be explained as the plant derived-food component that is resilient to hydrolysis/digestion by the vital enzyme machinery present in humans. Fibres are basically the residual parts of plant which are safe to consume and include components like polysaccharides, oligosaccharides, lignin and other related plant substances. Organic acids (butyric acid) and polyols (sorbitol) also constitute the fibre. Animal foods, however do not have any fibre components [3]. They are similar carbohydrates that are resilient to digestion and absorption while in the small intestine with full or part fermentation in the large intestine of human beings. They demonstrate support of one or more benefits such as either laxation (faecal bulking and softening, improved frequency and/or regularity), blood cholesterol reduction and/or blood glucose regulation [4].

Traditionally, fibre was known as roughage/bulk and was measured as crude fibre. Crude fibre may be explained as the residue remaining after the treatment with sulphuric acid, alkali and alcohol. The main constituent of crude fibre is a polysaccharide called cellulose, which is a component of dietary fibre. Other components such as pectins, hemicelluloses and lignins are also resistant to digestion, and these together with cellulose form part of the group, commonly known as dietary fibre.

2.2 Constituents of Dietary Fibre

The edible plant foods such as cereals, fruits, vegetables, dried peas, nuts, lentils and grains contain fibre that is grouped by its natural characteristics, mainly its dissolvability in water. Total dietary fibre (TDF) can be classified into two types, namely, soluble dietary fibre (SDF) and insoluble dietary fibre (IDF), both of which are composed of dense indigestible polysaccharides and are described below.

2.3 Soluble Dietary Fibre (SDF)

This edible portion of plant foods does not get digested and absorbed in the small intestine. Colonic bacteria cause part or total fermentation while the food passes through the large intestine (colon). Soluble fibre is gummy and is primarily found in the pulp of fruits, legumes, greens, oats, barley, psyllium, etc. This type of fibre is soluble in water and, eventually, thickens to form a gel-like substance [5]. The advantages of SDF include reduction of total and LDL cholesterol (bad cholesterol), thereby decreasing the proneness to heart disease as well as maintenance of blood glucose levels in patients with diabetes.

2.4 Insoluble Dietary Fibre (IDF)

It is made up of the structural material of plant cell wall consumed as part of diet, controls bowel movement and increases the transit time through the intestinal tract. Some insoluble fibres unite with certain minerals, such as calcium, magnesium, phosphorous and iron. The advantages of IDF are that it encourages bowel regularity and prevents constipation, gets rid of noxious waste from the body and prevents microbes from stagnating and producing cancerous substances. Though most of the plant-based foods contain both soluble and insoluble fibre, the amount of SDF/IDF or both is variable in different plant foods. Therefore, in order to gain maximum health benefits, a wide variety of high-fibre foods must be consumed.

2.5 Dietary Fibre Requirements

The recommended dietary allowances for fibre are 30 g/day or 12 g/1000 kcal for a normal healthy adult. It is also suggested that fibre intake should be from natural food sources and the suggested ratio of soluble and insoluble fibre being 1:2 [5]. Whole-grain products, fruits, vegetables, beans, peas and other legumes, nuts and seeds are considered as excellent sources of fibre in diet. Moreover, foods that undergo various refining processes such as canned fruits and vegetables, clear (pulp-free) juices, white breads and pastas and non-whole-grain cereals possess less fibre. The grain-refining procedure removes the outer coat (bran) from the grain, and they lose their fibre substance. Although fortified foods are usually replenished with some of the B vitamins and minerals including iron and calcium and post-processing, this food remains to be deficit in its fibre content.

The daily fibre intake can be increased by the following ways:

1. Incorporating whole wheat, oats and barley in our daily diet.
2. Including wholemeal/ multigrain breads and rotis (Indian bread) instead of processed/refined cereals.

Fig. 2.1 **Total dietary fibre in common food groups.** All values are expressed in g per 100 g of the edible portion. (**a**). Cereals, 10–15; (**b**). grain legumes, 16–25; (**c**). leafy vegetables, 3–5 (**d**). other vegetables, 4–8; (**e**). roots and tubers, 3–4; (**f**). fruits, 5–8; (**g**). nuts and seeds, 10–16; (**h**). spices (fresh), 4–6; (**i**). spices (dry), 25–48. (Source: Longvah T, Ananthan R, Bhaskarachary K, Venkaiah K (2017). Indian Food Composition Tables, National Institute of Nutrition, ICMR, Hyderabad. 1–19)

3. Adding an extra vegetable serving to every meal.
4. Consuming fresh fruits (with peel, if possible), dried fruits, nuts or wholemeal snacks.

It is believed that a daily intake of more than 30 g can be easily achieved if one eats wholegrain cereal products, fruits, vegetables and legumes and replaces consumption of low-fibre snacks, cakes and biscuits with nuts/oil seeds as a snack between meals. Fibre content range of the commonly consumed foods belonging to various food groups is shown in Fig. 2.1 [6].

2.6 Dietary Fibre: Nutritional and Health Significance

The fibre-rich diets have a positive effect on health as their ingestion has been related to decreased prevalence of several diseases. It has been documented that Indian diets have the possibility of reducing the occurrence of many diseases [7].

Physiological Effects The effect of fibre on the gastrointestinal tract is affected by the characteristics of the fibre itself, the particle size and the interaction between fibre, other dietary components and bacterial flora.

Nutrient Absorption Susceptibility of fibre to disintegration during cooking, processing and mastication into fine particles causes delay in the uptake of nutrients that can be taken up by the epithelial cells lining the mucosa. This further leads to the eventual excretion of nutrients. Even after cooking, processing and other treatments, there remains a barrier to the digestion as the structure around the nutrients envelops it. For instance, legumes take longer to digest, and hence, they have a slow glycaemic response.

Nutrient Binding Several soluble minerals get converted into unabsorbed forms that get excreted. Sometimes very-high-fibre diets have undesirable effects on mineral absorption due to the presence of phytic substances and polysaccharides coupled with strong mineral-binding capacity.

Mobility of Intestinal Contents Presence of soluble dietary fibre increases the viscosity of the intestine, decreases the peristaltic movements of the gut and reduces the chances of movement of nutrients towards the villi for absorption.

Faecal Output Dietary fibre increases the faecal bulk and frequency of stool. Wheat bran, due to its high-fibre content, resists colonic fermentation, thereby increasing the dry faecal matter. In addition, the capacity to retain water contributes to softer stools and easy defaecation.

Faecal Transit After ingesting food, it takes at least half a day for the undigested food material to reach the colon. Complex carbohydrates take much longer to pass through the stomach and small intestine. Accelerated colonic mobility is observed in cases of high dietary fibre intake, as it causes reduced transit time in the colon and also faster bowel movements [5].

2.7 Role of Dietary Fibre in Physiological Well–Being

1. Normalizes bowel movements: Dietary fibre helps in formation of stool, increases its bulk and softens its texture. Such a stool is easier to pass and, thus, reduces the chance of constipation. The addition of bulk by fibre to the stool helps solidify it, in case of loose stool.
2. Helps retain good bowel health: An optimum amount of fibre in diet possibly reduces the risk of developing haemorrhoids and diverticular disease. A small quantity of fibre undergoes fermentation in the large intestine.
3. Decreases cholesterol levels: Soluble fibre advantageously lowers total blood cholesterol by lowering the low-density lipoprotein. Studies have documented heart-health benefits on consumption of high-fibre foods via mechanism of reducing blood pressure and inflammation.
4. Helps regulate blood glucose levels: In known cases of diabetics, fibre – especially the soluble form – can slow the assimilation of glucose and positively impact blood glucose levels. The risk of developing type 2 diabetes can be minimized by consuming a healthy diet with sufficient amount of insoluble fibre.
5. Assists in attaining healthy weight: Fibre-rich diet tends to be more substantial, making one consume less and attain higher satiety value. It takes longer time to eat/chew such foods, and also, these are less energy giving, implying that they contribute lesser calories on consumption of the same food quantity.
6. Prevents occurrence of colorectal cancer: This is another reported benefit attributed to consumption of dietary fibre. However, there still exists a mixed opinion upon the potential of dietary fibre to reduce the incidence of colorectal cancer.

A schematic representation of the role of dietary fibre in human nutrition is shown in Fig. 2.2.

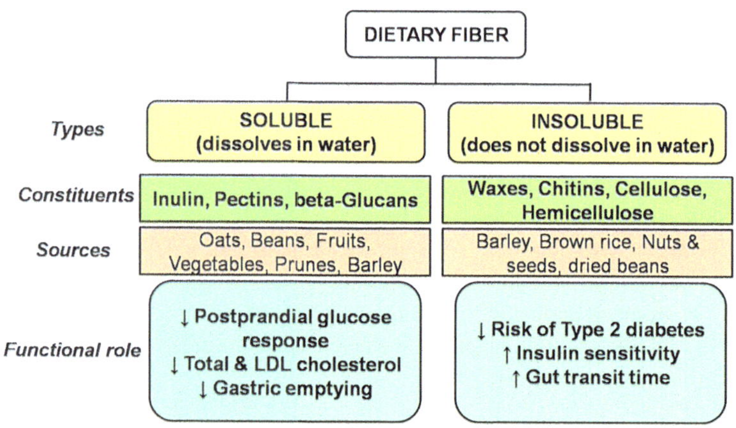

Fig. 2.2 Dietary fibre in human nutrition: an overview

2.8 Conditions Linked to Low-Fibre Diets

Consuming a diet low in fibre can give rise to many disorders like:

1. Constipation – small, hard and dry faecal matter with difficulty in passing.
2. Haemorrhoids – varicose veins of the anus.
3. Diverticulitis – small hernias of the digestive tract caused by continuing constipation.
4. Irritable bowel syndrome – pain, flatulence and bloating of the abdomen.
5. Overweight and obesity – having excess body fat.
6. Coronary artery/heart disease – narrowing of the arteries due to fatty deposits.
7. Diabetes – a condition characterized by raised levels of glucose in the blood.
8. Colorectal cancer – malignant cell growth in the large intestine.

2.9 Dietary Fibre and Cardiovascular Diseases

Several research studies have recommended that improved fibre intake is linked with reduced cardiovascular disease risk. Reviews and meta-analysis suggest that greater intake of total dietary fibre, insoluble fibre and cereal, fruit, or vegetable fibre sources is associated with a lower incidence of cardiovascular disease/coronary artery disease in healthy populace. The dietary fibre intake covers the risk of coronary artery disease (CAD), and many probable systems come into play through which fibre works on particular risk factors [8]. Sources of the soluble viscous fibre solidify to form gels and impact absorption from the small intestine, thereby regulating the postprandial blood glucose and lipid levels [9]. The gel formation also slows gastric emptying, gives a feeling of fullness and assists in checking weight gain. Short-chain fatty acids are produced when the intestinal bacteria ferment soluble fibre and the resistant starch molecules, and this helps reduce the blood cholesterol levels [10].

Diets high in fibre, especially from cereal/vegetable sources, and those that have abundant insoluble fibre are significantly linked with lower risk of coronary artery disease. Fruit fibre, if included in diet in liberal amounts, has shown to be linked with reduced cardiovascular risk. Recommendations related to data advocate whole food consumption and do not support intake of foods specifically fortified with cereal-/vegetable-derived fibre. A UK-based study reported 9% decrease in the risk for coronary artery disease by consuming an extra 7 g/day of total fibre. Current recommendations and the research findings both propagated increased fibre intake and showed a considerable risk factor decline with an attainable amount of daily dietary fibre intake. CAD has been reported as one of the major causes of mortality among population in the UK. The rate of incidence is estimated to be approximately 13–16%; little drop in risk factor by adopting suitable measures can positively impact the population and reduce mortality rate [11]. An extra 7 g of fibre can be gained in diet by including one additional serving of whole grains and another one of beans/lentils/legumes or otherwise two to four servings of fibre-dense fruit/vegetables. CAD risk is reported to decline with higher intake of insoluble, cereal, fruit and vegetable fibre.

2.10 Dietary Fibre and Diabetes

Recommendations from most nutrition and dietetic associations emphasize on a high dietary fibre intake. The viscous and gel-forming nature of soluble dietary fibre obstructs macronutrient absorption, lowers postprandial glucose response and positively impacts the levels of certain blood lipids. The consumption of soluble dietary fibre along with carbohydrate-dense meals decreases postprandial glucose response and lowers the total and LDL cholesterol levels [12]. These consequences can possibly be justified by the viscous and/or gel-forming properties of soluble DF, which slow gastric clearing and macronutrient assimilation from the gut. In large prospective cohort studies, owing to primarily the consumption of insoluble cereal dietary fibre and whole grains, association with reduced risk of type 2 diabetes is quite visible [13]. Recent research has given novel insights and assisted in establishing a strong metabolic association between insoluble DF consumption and decreased diabetes risk. Known cases have shown enhanced insulin sensitivity and modulation of inflammatory markers; and direct/indirect effects on the gut flora too have been reported.

Possible mechanisms proposed to explain as to how dietary fibre intake may change proneness to diabetes include effects on satiety and body weight. Numerous studies have indicated that consumption of fibre-dense diet increases the post-meal satiety or reduces the subsequent hunger, both under situation where calorie intake was the same and also the energy consumption was ad libitum [14]. Conversely, there were other studies that concluded no significant effects [15, 16]. Quite a few studies do not indicate an inverse association between postprandial glucose and insulin response with satiety, and thus, it cannot be concluded that low vs. high glycaemic index meals are an important reason to prompt satiety [17, 18]. Moreover, there is higher likelihood of developing diabetes in population with insulin resistance. Therefore, improved insulin sensitivity could be a pertinent factor favouring reduced diabetes risk in those consuming insoluble DF-rich diets.

2.11 Dietary Fibre and Colorectal Cancer

Fermentation process occurs in the large bowel, and as a result short-chain fatty acids are formed. These plausibly exercise a shielding impact against the development of colon cancer [19–21]. Fibre increases mucin secretion for lubrication purposes, and deficiency of fibre results in mucosal tenderness of colon [9]. In epidemiological studies, high consumption of DF has been associated inversely with a lower incidence of colorectal cancer [19]. Direct interventional studies implicate that soluble fibre in diet could even lower the risk of developing of colon cancer.

2.12 Dietary Fibre and Healthy Ageing

The Journals of Gerontology, Series A: Biological Sciences and Medical Sciences reported a study from the Westmead Institute for Medical Research suggesting that consuming the right amount of fibre from breads, cereals and fruits/vegetables helps delay progression of disease/disability into old age [22]. Relationship between carbohydrate intake and healthy ageing was evaluated in a populationbased cohort study conducted on about 1600 adults aged 50 years and older by assessing their longterm sensory loss risk factors and general diseases. During the study several factors such as subject's total carbohydrate intake, total fibre intake, glycaemic index, glycaemic load and sugar intake were assessed, and it was concluded that fibre intake contributed towards 'successful ageing'. Nonexistence of disability, symptoms of depression, cognitive impairment, respiratory symptoms and other chronic diseases such as malignancy, coronary artery disease and stroke may be explained as features of successful ageing [22]. Although more such studies are required to draw conclusive results and give recommendations, it definitely has paved way for further investigation.

Population consuming liberal quantity of DF, as compared to the ones having low fibre intake, have reduced risk for developing lifestyle diseases such as coronary artery disease, stroke, hypertension, diabetes, obesity and also various gastrointestinal diseases [23]. Including high-fibre foods/high-fibre supplements in daily diets attenuates serum lipoprotein levels, reduces hypertension, improves blood glucose control in diabetics and also assists in checking body weight. Hard data has also indicated towards improved immune function in individuals on inclusion of soluble fibre in human diet [24].

2.13 Conclusion

An appropriate amount of dietary fibre intake demonstrates positive effects showing primary (health-protective) as well as secondary (disease-reversal) benefits. The very hypothesis of linking higher incidence of chronic diseases to inadequate consumption of dietary fibre has helped researchers recognize the importance of dietary fibre intake in our routine diets from community nutrition and health perspective. Although adequate dietary fibre inclusion has been recommended worldwide, it has also been observed that certain types/quantity of fibre may cause adverse symptoms of gas, distension, disturbed gastrointestinal condition and altered bowel system. Studies have shown that very high-fibre intake (> 60 g over a day) can cut down on the affectivity of nutrient assimilation and may cause irritable bowel symptoms and also lead to diarrhoea. Though these conditions are purely due to fibre intake, they certainly cannot be attributed to the fact that they are noxious. Recommendations of

US Agencies propose a minimum intake of 20–35 g of fibre per day for an individual and advise this amount as to be beneficial for continuous good health among general population. Though researches are ongoing to evaluate optimum dietary fibre requirements for Indians, the WHO Committee on chronic degenerative diseases continue to put forward a daily requirement of 30 g dietary fibre through diet. With the increasing trend of consuming processed and refined foods among all age groups, especially in the high/affluent class, the intake of dietary fibre is going alarmingly low. Conclusively, the intake of 40 g/2000 kcal may be rationalized in different groups based on recommended energy intake.

Diet is the cornerstone in prevention and treatment in most of the non-communicable chronic degenerative diseases [25]. A diet that represents a wide selection of foods and nutrients includes optimum measures of fruits, vegetables, legumes, whole grains, nuts and oilseeds and needs to be encouraged in larger relevance in order to combat the unprecedented rise in the chronic degenerative diseases.

References

1. Hakajima (2000) Health Promotion – new challenges for the future. IUHPE-SEARB Publications, pp 15–9
2. Chawla R, Patil GR (2010) Soluble dietary fiber. Compr Rev Food Sci Food Saf 9:178–196
3. Lunn J, Buttris JL (2007) Carbohydrates and dietary fiber. Nutr Bull 32:21–64
4. AACC Report (2001) The definition of dietary fiber: a report of the American Association of Cereal Chemists (AACC). Cereal Foods World 46:113–126
5. Srilakshmi B (2002) Carbohydrates. Nutrition Science, New Age International (P) Ltd, New Delhi, pp 20–39
6. Longvah T, Ananthan R, Bhaskarachary K, Venkaiah K (2017) Indian food composition tables. National Institute of Nutrition, ICMR, Hyderabad, pp 1–19
7. Paul S (2014) Carbohydrates. A textbook of bio-nutrition – curing diseases through diet. CBS Publishers & Distributers Pvt. Ltd, New Delhi, pp 47–67
8. Chhabra S (2009) Diet/Lifestyle related risk factors and the impact of educational intervention – a study among known cases of coronary artery disease. PhD thesis submitted to the University of Delhi, pp 125–129
9. James SL, Muir JG, Curtis SL, Gibson PR (2003) Dietary fiber: a roughage guide. Intern Med J 33:291–296
10. Slavin JL, Martini MC, Jacobs DR Jr, Marquart L (1999) Plausible mechanisms for the protectiveness of whole grains. Am J Clin Nutr 70:459–463S
11. Threapleton DE, Greenwood DC, Evans CE, Cleghorn CL, Nykjaer C, Woodhead C, Cade JE, Gale CP, Burley VJ (2013) Dietary fiber intake and risk of cardiovascular disease: systematic review and meta-analysis. BMJ 347:f6879
12. Weickert MO, Pfeiffer AF (2008) Metabolic effects of dietary fiber consumption and prevention of diabetes. J Nutr 138:3439–3442
13. Schulze MB, Schulz M, Heidemann C, Schienkiewitz A, Hoffmann K, Boeing H (2007) Fiber and magnesium intake and incidence of type 2 diabetes: a prospective study and meta-analysis. Arch Intern Med 167:956–965
14. Howarth NC, Saltzman E, Roberts SB (2001) Dietary fiber and weight regulation. Nutr Rev 59:129–139

15. Weickert MO, Spranger J, Holst JJ, Otto B, Koebnick C, Mohlig M, Pfeiffer AF (2006) Wheat-fiber-induced changes of postprandial peptide YY and ghrelin responses are not associated with acute alterations of satiety. Br J Nutr 96:795–798
16. Howarth NC, Saltzman E, McCrory MA, Greenberg AS, Dwyer J, Ausman L, Kramer DG, Roberts SB (2003) Fermentable and nonfermentable fiber supplements did not alter hunger, satiety or body weight in a pilot study of men and women consuming self-selected diets. J Nutr 133:3141–3144
17. Keogh JB, Lau CW, Noakes M, Bowen J, Clifton PM (2007) Effects of meals with high soluble fiber, high amylose barley variant on glucose, insulin, satiety and thermic effect of food in healthy lean women. Eur J Clin Nutr 61:597–604
18. Raben A (2002) Should obese patients be counselled to follow a low-glycaemic index diet? No. Obes Rev 3:245–256
19. Reddy BS (1999) Role of dietary fiber in colon cancer: an overview. Am J Med 106(1A):16S–19S
20. Annison G, Topping DL (1994) Nutritional role of resistant starch: chemical structure vs physiological function. Annu Rev Nutr 14:297–320
21. Goni I, Garcia-Diz L, Manas E, Calixto FS (1996) Analysis of resistant starch: a method for foods and food products. Food Chem 56:455–459
22. Dietary fiber intake tied to successful aging, research reveals. The Journals of Gerontology, Series A: Biological Sciences and Medical Sciences-Science Daily. Science Daily, 2016
23. Passi SJ, Suri S (1997) Dietary fiber and CHD, the cholesterol facts, Life Publications (1stst ed.), Delhi, 70–73
24. Singh A, Singh SN (2015) Dietary fiber content of Indian diets. Asian J Pharm Clin Res 8:58–61
25. Kris-Etherton PM, Hilpert KE, Krauss RM (2006) Nutrition. Preventive cardiology – a practical approach. Tata McGraw Hill, New Delhi, pp 256–295

Phlorotannins and Macroalgal Polyphenols: Potential As Functional Food Ingredients and Role in Health Promotion

3

Margaret Murray, Aimee L. Dordevic, Lisa Ryan, and Maxine P. Bonham

Abstract

Marine macroalgae are rapidly gaining recognition as a source of functional ingredients that can be used to promote health and prevent disease. There is accumulating evidence from in vitro studies, animal models, and emerging evidence in human trials that phlorotannins, a class of polyphenol that are unique to marine macroalgae, have anti-hyperglycaemic and anti-hyperlipidaemic effects. The ability of phlorotannins to mediate hyperglycaemia and hyperlipidaemia makes them attractive candidates for the development of functional food products to reduce the risk of cardiovascular diseases and type 2 diabetes. This chapter gives an overview of the sources and structure of phlorotannins, as well as how they are identified and quantified in marine algae. This chapter will discuss the dietary intake of macroalgal polyphenols and the current evidence regarding their anti-hyperglycaemic and anti-hyperlipidaemic actions in vitro and in vivo. Lastly, this chapter will examine the potential of marine algae and their polyphenols to be produced into functional food products through investigating safe levels of polyphenol consumption, processing techniques, the benefits of farming marine algae, and the commercial potential of marine functional products.

Keywords

Hyperglycaemia · Hyperlipidaemia · Macroalgae · Phlorotannin · Polyphenol

M. Murray · A. L. Dordevic · M. P. Bonham (✉)
Department of Nutrition, Dietetics and Food, Monash University, Melbourne, VIC, Australia
e-mail: maxine.bonham@monash.edu

L. Ryan
Department of Natural Sciences, Galway-Mayo Institute of Technology, Galway, Ireland

V. Rani, U. C. S. Yadav (eds.), *Functional Food and Human Health*,
https://doi.org/10.1007/978-981-13-1123-9_3

3.1 Introduction

Polyphenols are non-nutrient compounds that are produced in both terrestrial plants [1, 2] and marine macroalgae (seaweeds) [3]. Their natural function is predominantly to act as the defence system of these organisms, protecting against infection [1, 2, 4], ultraviolet radiation [1, 2, 4, 5] and consumption by herbivores [4, 5]. They can also play a role as integral structural components of cell walls [4, 6, 7]. When consumed by humans, polyphenols demonstrate biological activity that may be beneficial to human health and disease prevention [3, 8, 10–12].

Over 8000 structurally different polyphenols have been identified in plants, from simple monomer units to complex polymerised structures [8, 9, 11, 13]. However, only several hundred of those varieties exist in edible plants [1]. Other varieties of polyphenols (e.g. phlorotannins) have been identified in both edible and nonedible forms of marine algae. There is ongoing investigation of successful extraction techniques and the potential for marine algal polyphenols to be used as supplements and functional food ingredients [14, 15–18]. This chapter discusses polyphenols from marine macroalgae, their dietary intake levels, their potential as functional food ingredients and their potential role as mediators of risk factors for cardiovascular disease and type 2 diabetes.

3.2 Macroalgal Sources of Polyphenols

There exist over 10,000 known species of macroalgae, which are classified into three categories based on their pigmentation: Chlorophyta and Charophyta (green algae), Rhodophyta (red algae) and Phaeophyta (brown algae) [19].

Green algae get their characteristic colour from a combination of chlorophyll *a* and *b*, beta-carotene and various xanthophylls. Due to the diversity of green algae, they are classified into two phyla (Chlorophyta and Charophyta) [19]. There are about 4500 species of Chlorophyta, including species found in freshwater and marine habitats, and about 3500 species of Charophyta all of which are freshwater-dwelling. Some applications of green algae include commercial use for the production of beta-carotene and use in nutritional supplements, particularly the *Chlorella* genus, to improve healing and enhance the immune system [19, 20].

Approximately 6500 species of algae are classified as red algae, due to the presence of the pigments phycoerythrin and phycocyanin, which mask the other pigments and provide the red colour [19]. Almost all red algal species are found in marine habitats and reside in the intertidal and subtidal zones. Several types of red algae are eaten as food, the most popular of which is Nori (*Porphyra*) [19, 21, 22], now the most valuable marine crop grown by aquaculture, valued at over US$1 billion [19]. Red algae are also a source of carrageenan, a commonly used food stabiliser, and agar, an ingredient used in food, pharmaceuticals, and cosmetics and as a growth medium for microorganisms [19, 23].

There are about 1800 known species of brown algae, all of which get their brown colour from the dominant pigment fucoxanthin [19]. Brown algae are larger than red and green algae, with the largest kelps growing up to 70 m in length. They are mostly found in marine habitats, commonly in cooler Northern hemisphere waters [19]. Japan, Korea and China grow brown algae for use in food and alginate production. Ireland and Scotland also use brown algae for the production of alginates and as a fertiliser for land. Other places where brown algae are harvested include Atlantic France and the coasts of California [19].

Of the three varieties of macroalgae, brown algae contain the highest levels of polyphenols, in particular phlorotannins which alone can make up 2–30% of the dry weight (DW) of the organism [4, 5], compared with approximately 0.1–4% and 0.2–20% total phenolic content in green and red algae, respectively [24, 25]. Brown algae will therefore be the focus of this chapter. A number of polyphenol classes, including catechins, flavonoids, flavonols and phlorotannins, are all found in marine macroalgae [3, 25]. However phlorotannins are the predominant class of polyphenol found in brown algae [18, 26] and are unique to marine sources [4]. Marine macroalgae thrive in harsh environments, including exposure to varying light intensity, salinity, pressure and temperatures, and therefore produce a variety of potent polyphenolic substances, which are not found in terrestrial plants [28].

3.3 Classification and Structure of Phlorotannins

Polyphenol is an umbrella term for a large group of highly heterogeneous compounds, characterised by the presence of at least one phenol structural unit (aromatic ring) [8]. Phlorotannins are a class of polyphenol that are synthesised in marine macroalgae through the acetate-malonate pathway by the polymerisation of phloroglucinol monomer units (1,3,5-tri hydroxybenzene) [3, 4, 18, 27] (Fig. 3.1). Phlorotannins are highly hydrophilic molecules that contain both phenyl (C_6H_5-) and phenoxy (C_6H_5O-) groups (Fig. 3.1) and range in size from 126 Da to 650 kDa [3], a much broader range than terrestrial polyphenols (up to 30 kDa) [28]. Phlorotannins vary in degree of polymerisation, structure and type of chemical bonds [5], resulting in a complex group of compounds with numerous isomers for any given molecular weight [4]. Phlorotannins are classified into four subclasses based on the chemical bonds they contain [3]. The subclasses are (1) fucols which have a phenyl linkage, (2) fuhalols and phlorethols which contain an ether linkage, (3) fucophloroethols which have a mixture of a phenyl and ether linkage and (4) eckols which contain a dibenzodioxin linkage (Fig. 3.1) [3]. However the literature often defines phlorotannins based on the species of their algal source or by naming the specific type of phlorotannin (e.g. phlorofucofuroeckol A, dieckol) rather than which subclass they belong to.

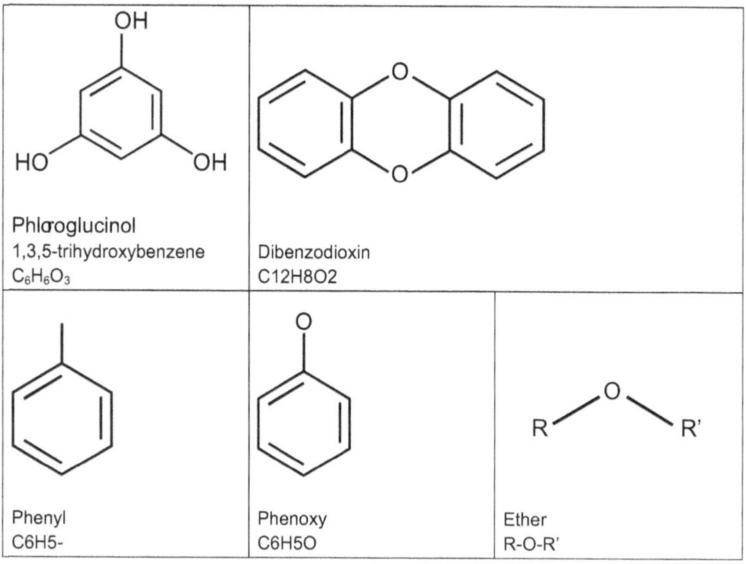

Fig. 3.1 Chemical structures of phloroglucinol, dibenzodioxin, phenyl, phenoxy and ether groups [105]

3.4 Phlorotannin Content of Macroalgae

Different algal species contain varying combinations and concentrations of phlorotannins, and a range of low- and high-molecular-weight phlorotannins can be found within a single species of marine alga [4] (Table 3.1). Within an individual algal body, phlorotannins are generally more concentrated in the outer layers of the organism, where it is exposed to the environment [27]. Phlorotannin content can fluctuate within the same population area and within a single algal body [5]. Environmental factors such as ultraviolet radiation, salinity, light and nutrient availability and herbivore grazing are likely causes for differences in phlorotannin content [5]. Furthermore, since the natural function of phlorotannins is as part of the defence system of the organism, algae that require a greater level of defence against harsh environments or attack from predators are likely to have a higher phlorotannin content [1, 4, 5]. This effect is seen in land-based fruit crops, where crops that are grown organically, without synthetic pesticides, have a higher polyphenol content than those grown the conventional way because they have had more exposure to stressful situations, resulting in an increased natural defence [29].

The location of brown algal species can also impact phlorotannin content. Brown algal species grown in the intertidal zones have the highest phlorotannin content, whereas those grown at lower and upper levels of the shore have lower phlorotannin content [30]. This is likely due to differences in exposure to environmental factors as those in the intertidal zone would experience a more rapidly changing and varied environment with the movements of the tide. The phlorotannin content of brown algae also varies according to season, and the degree of

Table 3.1 Summary of phlorotannins isolated from marine algal species [105]

Seaweed species	Reported phlorotannin content	References
Ascophyllum nodosum	Approx. 5.80% of dry weight	[30]
Bifurcaria bifurcata	3.73 (0.57)% of dry weight	[30]
Cystoseira nodicaulis	89.14 (2.57) g phloroglucinol equivalents (PGE)/mg sample	[4]
Ecklonia cava	3.3% crude phlorotannins:	[27]
	4.7% phloroglucinol	
	0.7% phloroglucinol tetramer	
	6.4% eckol	
	16.6% phlorofucofuroeckol A	
	22.2% dieckol	
	12.4% 8,8′-bieckol	
Ecklonia cava	Dieckol – 1.52 mg/g dry weight	[18]
Ecklonia cava	Phlorofucofuroeckol A – 0.93 mg/g dry weight	[18]
Ecklonia cava	Total phlorotannin content – 3.39 mg PGE/mL in crude phlorotannin extract solution	[107]
Ecklonia kurome	3.0% crude phlorotannins:	[27]
	2.6% phloroglucinol	
	0.3% phloroglucinol tetramer	
	9.2% eckol	
	28.6% phlorofucofuroeckol A	
	24.6% dieckol	
Ecklonia kurome	7.8% 8,8′-bieckol	[27]
Ecklonia stolonifera	Dieckol – 1.52 mg/g dry weight	[18]
Ecklonia stolonifera	Phlorofucofuroeckol A – approx. 1.20 mg/g dry weight	[18]
Eisenia bicyclis	3.1% crude phlorotannins:	[27]
	0.9% phloroglucinol	
	4.4% phloroglucinol tetramer	
	7.5% eckol	
	21.9% phlorofucofuroeckol A	
	23.4% dieckol	
	24.6% 8,8′-bieckol	
Eisenia bicyclis	Phlorofucofuroeckol A – 1.30 mg/g dry weight	[18]
Eisenia bicyclis	Dieckol – 1.33 mg/g dry weight	[18]
Eisenia bicyclis	Contains eckol	[106]
Eisenia bicyclis	Contains 6,6′-bieckol	[106]
Eisenia bicyclis	Contains 8,8′-nieckol	[106]
Eisenia bicyclis	Contains dieckol	[106]
Eisenia bicyclis	Contains phlorofucofuroeckol A	[106]

(continued)

Table 3.1 (continued)

Seaweed species	Reported phlorotannin content	References
Fucus serratus	4.27 (1.12)% of dry weight	[30]
Fucus serratus	180.55 (16.98) μg PGE/mg sample	[4]
Fucus spiralis	3.88 (0.65)% of dry weight	[30]
Fucus vesiculosus	Phlorotannins approx. 5.80% of dry weight	[30]
Fucus vesiculosus	231.95 (8.97) μg PGE/mg sample	[4]
Fucus vesiculosus	Total phlorotannins ranged from 12 to 23 mg/g dry weight	[6]
Himanthalia elongata	2.17 (1.40)% of dry weight	[30]
Himanthalia elongata	198.28 (9.17) μg PGE/mg sample	[4]
Laminaria digitata	0.13 (0.03)% of dry weight	[30]
Pelvetia canaliculata	3.39 (0.64)% of dry weight	[30]
Sargassum aquifolium	6.770 (0.001) mg phlorotannins/g dry weight	[32]
Sargassum denticarpum	0.978 (0.004) mg phlorotannins/g dry weight	[32]
Sargassum mcclurei	2.057 (0.003) mg phlorotannins/g dry weight	[32]
Sargassum oligocystum	2.369 (0.004) mg phlorotannins/g dry weight	[32]
Sargassum serratum	1.305 (0.008) mg phlorotannins/g dry weight	[32]
Sargassum polycystum	0.735 (0.002) mg phlorotannins/g dry weight	[32]

Abbreviations: *PGE* phloroglucinol equivalents

seasonal variation differs among species [30]. During summer the *Fucales* genus, *Pelvetia canaliculata* (3.72% dry weight (DW)) and *Ascophyllum nodosum* (7.18% DW) species exhibit their maximal phenolic content, whereas the *Laminariales* genus has its highest phenolic content in winter. During spring months the *Fucus vesiculosus* (7.84% DW) and *Ecklonia radiata* species have their highest phenolic levels [30, 31]. As well as being influenced by season, time spent in storage can impact phlorotannin content. In a study of six *Sargassum* species of macroalgae, grown in Vietnam, the amount of time spent in storage was negatively associated with phlorotannin content; however the rate at which phlorotannin content decreased varied among species [32].

3.5 Identification and Analysis of Macroalgal Phlorotannins

There is currently limited knowledge of the variety of phlorotannins in macroalgae, and the distribution of phlorotannins within specific algal species, largely due to the structural complexity and polymeric nature of phlorotannins, with variations in the number of monomer units, their positions, and chemical bonds with which they are joined [4]. Historically, only low-molecular-weight phlorotannins (2–8 phloroglucinol units) could be characterised [4]. However, recent technological advancements in chromatographic and mass spectrometric techniques allow for more thorough study of the complex structures of phlorotannins and their distribution in macroalgae, with isomers of up to 16 monomer units successfully detected [4]. One technique by which phlorotannins have been successfully separated from a crude phenolic extract is using high-performance liquid chromatography (HPLC) with UV photodiode array detection [7]. Ultrahigh performance liquid chromatography (UPLC) with mass spectrometry has also been used to profile individual phlorotannins [4, 6]. This technique has identified 61 isomers corresponding to 12 phloroglucinol units, from the brown alga *Fucus vesiculosus*, and determined the isomerisation of phlorotannins ranging from 3 to 16 monomers [4]. With improved technologies we are now able to identify and characterise a greater variety of phlorotannin molecules and accurately determine the phlorotannin content of macroalgae.

Furthermore, depending on the type of extraction solvent used, different quantities of individual phlorotannins have been extracted from the same macroalgae [18]. When phlorotannins were extracted from *Ecklonia cava*, *Ecklonia stolonifera* and *Eisenia bicyclis* using two different solvents, the dieckol yield by boiling water extraction was 86%, 93% and 98% of the organic solvent extraction, respectively. The phlorofucofuroeckol A yield from boiling water extraction was 74%, 86% and 62% of the organic solvent extraction, respectively [18]. This highlights the need for the identification of the most efficient extraction solvent.

Efficient extraction techniques result in a greater yield of phlorotannins from the same biomass, meaning greater potential for biological activity. Organic solvent, such as ethanol and methanol, extraction is the most common method used for extraction of phlorotannins [18]. Boiling water extraction produces a slightly lower yield of phlorotannins than organic solvent extraction (see figures above); however it is a safer and more cost-effective method [18]. It has been suggested that a 70% aqueous acetone solution is most efficient for extraction of phlorotannins due to its ability to inhibit interactions between tannins and proteins and to break hydrogen bonds [7]. However more recent findings indicated that this method was less effective than both boiling water and organic solvent extraction [18]. For the extraction of phlorotannins for use as a functional food ingredient, it appears that either organic solvent extraction or boiling water extraction is the most effective method, depending on the cost and safety considerations. Elucidation of the most efficient extraction technique, along with improved technologies for identifying individual phlorotannins, will enable future researchers to identify the most effective phlorotannin molecules for the promotion of health and prevention of disease in humans.

3.6 Dietary Intake of Macroalgal Phlorotannins

The difficulties associated with quantifying macroalgal phlorotannin content have likely contributed to the lack of information on population intake of macroalgal phlorotannins. However, macroalgae consumption is documented in Asian countries, such as Japan, where it is a traditional part of the diet [21, 33]. With this information it is possible to retrospectively extrapolate phlorotannin intake using existing macroalgal phlorotannin content data (Table 3.1). In 2006, Japanese households consumed 450 g per year of kombu (*Laminaria japonica* – a brown macroalgae), although generally consumption was four times higher in elders than in young adults (<29 years) [22]. Despite a decrease in kombu consumption in Japan, daily macroalgae intake remained relatively stable, 4.3 g/day in 1955 and 5.3 g/day in 1995, with an increase in wakame (*Undaria pinnatifida* – a brown macroalgae) and nori (*Porphyra* genus – a red macroalgae) varieties making up for the decline in kombu [22]. Based on an average dietary intake of 5.3 g of macroalgae per day, the Japanese population would be consuming approximately 160 mg of phlorotannins per day [27, 30]. However this value would vary depending on individual intake, macroalgal variety and other factors that affect phlorotannin content as previously described, such as season and processing. From the red alga family, *Porphyra* is most frequently consumed (nori). From the brown algae, the *Laminaria japonica* (kombu), *Undaria pinnatifida* (wakame) and *Hizikia fusiforme* (hijiki) species are the most commonly consumed [21, 22].

In most western cultures, macroalgae is relatively new to the diet, but consumption has been steadily increasing since the early 1980s [21, 33] due to consumer demand for interesting, natural and sustainable food products and due to globalisation of the food system [33, 34]. There is, however, limited literature available regarding daily intakes of macroalgae among western cultures; the red alga *Palmaria palmata*, common in Atlantic waters, is one of the few algal species that is documented to have been used for human consumption in Europe [33]. However, there are now polyphenol-rich macroalgal extracts that are commercially available as health food products in the United States of America, Canada and Korea. These supplements may significantly increase the average population intake of macroalgal polyphenols in these countries, especially given the health claims these products carry (including antioxidant and anti-inflammatory activity, improvement of lipid balance, weight loss and protection against cardiovascular disease and diabetes) and the likelihood of health claims to positively influence consumer purchase intentions [35]. However, even though these health claims are as yet unsubstantiated in humans, there is some support for their role in these health outcomes based on evidence from in vitro and animal studies.

Accurate estimation of phlorotannin intake based on dietary intake data, like any other dietary component, is difficult [36]. Collection of dietary data is predominantly self-reported and therefore is likely to be inexact and carry bias. Perceived 'unhealthy' foods are often under-reported, while perceived 'healthy' foods are typically over-reported [37], resulting in an overestimation of polyphenol intake. The difficulty of estimating polyphenol intake is further exacerbated as the polyphenol content of foods is not included in most food composition databases. An online European database provides some information on the polyphenol content of 459 common foods (including 500 different polyphenols), but as yet this is limited to polyphenols from terrestrial food sources and does not include phlorotannins [38]. Furthermore, the variation in the phlorotannin content of algal species increases the difficulty of accurate intake estimation, as the recorded phlorotannin content of macroalgae in food databases may not be an accurate representation of all individuals in that species. However, with ongoing improvement (e.g. information on effects of cooking/processing on polyphenol content, polyphenol metabolites formed in the body and additional polyphenol sources), food databases such as Phenol-Explorer may be a useful tool to improve understanding of the impact of polyphenols on health [38]. The use of biomarkers, such as urinary excretion or plasma levels of polyphenols, is becoming more widely used to determine phlorotannin intake and can be used to make conclusions about the potential health effects of phlorotannins [39]. However only 5–10% of dietary polyphenol intake is actually absorbed into the plasma and excreted in urine [40], thus while these are more technically accurate methods of measurement than dietary intake, they may not be better tools for determining total polyphenolic intake and resulting biological activity.

3.6.1 Bioavailability of Polyphenols

Bioavailability is defined as the proportion of a compound that is digested, absorbed and metabolised in the body through normal pathways [9]. In order to properly assess the biological functioning of macroalgal polyphenols, it is important to measure their bioavailability in humans [9]. Most evidence for the biological activity of macroalgal polyphenols to date has been derived from cultured cells or animal models, which do not account for the effects of other dietary components, or human digestion and absorption, and therefore may not represent the biological actions of polyphenols in people [1, 10]. Thus polyphenols that have exhibited strong biological activity in vitro may not have the same effect in the human body [8, 41].

The bioavailability of any compound is affected by its ability to cross biological membranes, withstand pH changes in the gastrointestinal tract and maintain its structural integrity [41, 42]. Factors that affect the absorption and metabolism of polyphenols from food are presented in Table 3.2. During digestion, polyphenols are metabolised in the small intestine, in the large intestine by colonic microflora as well as in the liver and other organs whereby they go through numerous structural modifications [1, 9, 14, 43]. This means that human body tissues are not exposed to polyphenols in their original form [8, 41]. Thus, in vitro studies that examine polyphenol extracts (which have not undergone digestion) are not a true representation of the activity or concentration of the metabolites present in the human body [8, 14, 41]. Studies of polyphenols need to take into account the changes in structure and concentration that occur when the compounds enter the human body [8, 41]. To further complicate the issue, personal variations in intestinal microflora may also impact an individuals' metabolism and absorption of polyphenols; however this

Table 3.2 Factors that affect the absorption and metabolism of food-derived polyphenols

Factors	References
Chemical structure of polyphenols (degree of glycosylation/acylation, basic structure, conjugation with other phenolics, molecular size, degree of polymerisation and solubility)	[1, 9, 13, 28]
Dietary factors (interactions with proteins and polysaccharides, pH of the gut, transit time, intestinal fermentations, biliary excretion, the influence of certain food components on enzymes and carriers)	[1]
Human behavioural factors (such as smoking)	[10]
Person-to-person variation in enzyme activity	[10]
Whether absorption takes place in the small intestine or colon	[1, 41]

area is not yet well understood [41]. These issues highlight the need for more studies investigating the healthful effects of macroalgal polyphenols to be performed in humans.

3.7 Role of Macroalgal Polyphenols in Human Health

As already alluded to in this chapter, polyphenols from terrestrial sources have been linked to positive health effects regarding several risk factors for chronic conditions including obesity, diabetes and cardiovascular diseases [8, 11, 43–47]. Recent research has extended to polyphenols from marine macroalgae, possibly because of epidemiological data from Asia which indicate that a diet rich in macroalgae or seaweed is associated with longevity and a decreased risk for cardiovascular diseases, some cancers and other chronic diseases [48–50]. Another factor that has contributed to increased interest in macroalgae for health promotion is the observation of the lack of photodynamic damages that macroalgae experience during metabolism. Macroalgae grow in harsh environments where they are exposed to light and high oxygen concentrations, factors which lead to the formation of free radicals and other strong oxidising agents [51]. Despite these damaging molecules, the lack of damage that macroalgae experience implies that they have some protective antioxidant mechanisms and/or compounds [51], which may be beneficial to humans as well. The accumulating evidences do indicate that polyphenols from macroalgae can influence glycaemic control, blood cholesterol and lipid levels (Fig. 3.2), which may potentially reduce metabolic abnormalities associated with cardiovascular diseases (CVDs) and diabetic complications [15–17, 43, 52–68].

3.7.1 Anti-hyperglycaemic Effects

Hyperglycaemia is a key characteristic and cause of type 2 diabetes [43]. Both acute and chronic hyperglycaemia cause overloading of the metabolic pathways with glucose, resulting in oxidative stress and free radical formation which lead to diabetic complications such as cardiovascular disorders, nephropathy, retinopathy, neuropathy, foot and leg ulcers and limb amputation [3, 69]. Alpha-amylase,

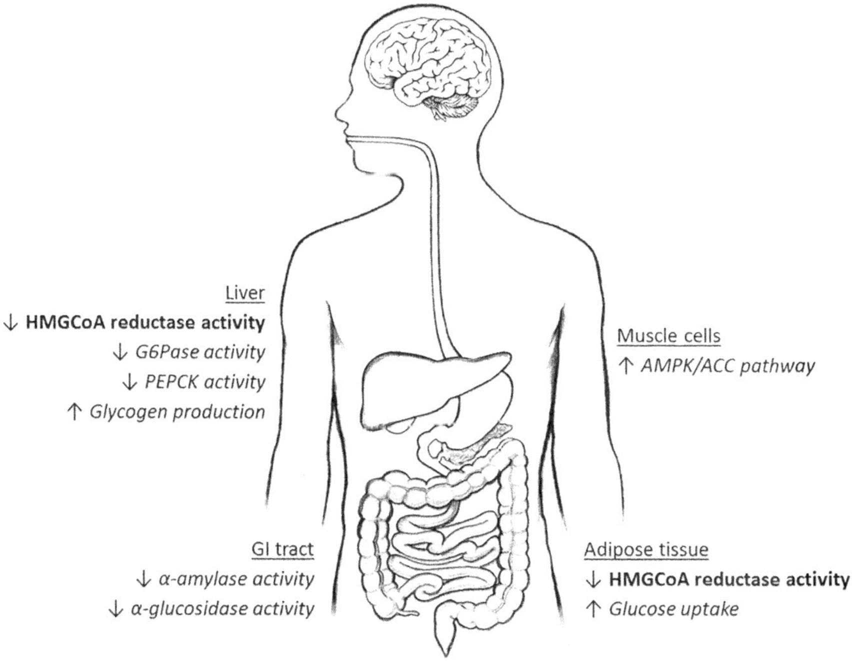

Fig. 3.2 The proposed anti-hyperglycaemic (in *italics*) and anti-hyperlipidaemic (in **bold**) effects of marine polyphenols. (Adapted from [105])
↑, increased, ↓, decreased; AMP-activated protein kinase/acetyl-CoA carboxylase (AMPK/ACC) increases GLUT4 transporters to transport glucose out of the blood and into the tissue. Inhibition of glucose-6-phosphatase (G6Pase) reduces the amount of glucose released from the liver. Inhibition of 3-hydroxy-3-methylglutaryl-coenzyme A (HMGCoA) reductase reduces plasma cholesterol levels. Inhibition of phosphoenolpyruvate carboxykinase (PEPCK), an enzyme involved in gluconeogenesis, reduces the amount of glucose released from the liver. Proliferator activated receptor γ (PPARγ) and CCAAT/enhancer-binding protein α (C/EBPα) are genes involved in adipogenesis and adipocyte differentiation

located in the pancreas, and α-glucosidase, at the brush border of intestinal cells, are key enzymes that break down carbohydrates into monosaccharides to be absorbed into the bloodstream, resulting in increased blood glucose levels following a meal [3, 70, 71]. Inhibition of these enzymes reduces the rate at which glucose is released from carbohydrate foods following a meal and can be an effective strategy for managing postprandial blood glucose [3, 70, 71]. Oral glucosidase inhibitor drugs are the common clinical treatment for type 2 diabetes; however long-term use can result in side effects such as renal tumours, acute hepatitis and serious hepatic injury [3]. A number of studies suggest that marine polyphenols may be safer alternatives [3, 43, 54, 63].

The inhibition of α-amylase and α-glucosidase results in short-term management of blood glucose levels; however there are a number of other ways that marine polyphenols may also contribute to longer-term glycaemic control (Fig. 3.2). Marine polyphenols upregulate the AMP-activated protein kinase/acetyl-CoA carboxylase (AMPK/ACC) signal transduction pathways in muscle cells, which increases the number of GLUT4 glucose transporters at the cell membranes to increase the uptake of glucose into the tissues [58, 72]. They also downregulate glucose-6-phosphatase and phosphoenolpyruvate carboxykinase (PEPCK) enzymes in the liver, which reduces the formation of new glucose through gluconeogenesis. In turn, the rate of glycogen production is increased in the liver, which ultimately promotes the uptake of glucose by the liver and reduces blood glucose levels [61].

3.7.1.1 In Vitro Studies

Polyphenolic-rich extracts from the marine macroalgae *Alaria*, *Pulmaria* and *Ascophyllum* exhibited some α-amylase inhibitory activity in cultured colon cancer cells [62]. The extract from *Ascophyllum* demonstrated strongest α-amylase inhibition (IC_{50} approximately 0.1 μg/mL gallic acid equivalents (GAE)) and was the only extract that also inhibited α-glucosidase activity (IC_{50} approximately 20 μg/mL GAE) [62]. An *Ascophyllum nodosum* extract has also been shown to induce dose-dependent α-amylase and α-glucosidase inhibition [52]. Furthermore, the α-glucosidase inhibition observed from the phlorotannin-rich *Ascophyllum nodosum* extract (0.24 μg phenolics) was greater than that of acarbose (0.37 μg), a current antidiabetic drug [52]. Diphlorethohydroxycarmalol (DPHC), a phlorotannin extracted from *Ishige okamurae*, dose-dependently inhibited α-amylase and α-glucosidase activity (IC_{50} values of 0.53 and 0.16 mM, respectively) in a chemical assay and was more effective than acarbose [54]. Strong α-glucosidase inhibitory effects have also been demonstrated from phlorotannin-rich *Ecklonia stolonifera* extracts [56].

Marine polyphenols act on other enzymes involved in carbohydrate metabolism to reduce hyperglycaemia. Phloroglucinol, eckol, phlorofucofuroeckol A, dieckol, 8,8′-bieckol and an unidentified tetramer phlorotannin were extracted from *Eisenia bicyclis* and examined for enzyme inhibitory action in vitro [65]. Phlorofucofuroeckol A, dieckol and 8,8′-bieckol inhibited α-fucosidase, β-galactosidase and β-mannosidase (extracted from the turban shell of *Turbo cornutus*), whereas phloroglucinol, eckol and the unidentified tetramer were only weakly active against the enzymes [65]. An additional way in which phlorotannins from *Ascophyllum nodosum* (400 μg/mL extract) have been demonstrated to reduce hyperglycaemia is to increase basal glucose uptake in 3 T3-L1 adipocytes; during a 20-min incubation period, glucose uptake increased by approximately threefold [68]. This action is possibly due to an increase in GLUT transporters; however this detail was not reported. Additionally, marine polyphenols from *Ecklonia cava* have activated AMPK/ACC signal transduction pathways in C_2C_{12} myoblasts [58], which resulted in increased glucose uptake into the cells and is another potential mechanism for a reduction in blood glucose levels [72] (Fig. 3.2).

3.7.1.2 Animal Studies

A diet supplemented with 0.5% w/w of a polyphenol-rich extract from *Ishige okamurae* for 6 weeks reduced hepatic G6Pase, PEPCK activity and increased hepatic glycogen production in C57BL/KsJ-db/db mice, which resulted in reduced fasting blood glucose levels [61]. The treated mice also presented with reduced hyperinsulinemia and HbA1c, compared with control [61]. Similarly, diabetic KK-A^y mice were administered 16.42 or 81.20 mg/day (0.2% or 1% of diet, respectively) of a phlorotannin extract from *Ecklonia stolonifera* for 4 weeks. The treated mice maintained blood glucose and insulin levels at a close-to-normal level in a dose-dependent manner, compared with untreated diabetic mice whose blood glucose and insulin levels increased over time [58]. Kang et al. [59] also identified that supplementation with dieckol (20 mg/kg body weight/day for 14 days), from *Ecklonia cava*, reduced blood glucose and insulin levels in C57BL/KsJ-*db/db* diabetic mice. Interestingly, a dose-dependent treatment effect on insulin levels but not blood glucose levels was observed [59]. Park et al. [64] reported that supplementation with 200 mg/kg body weight/day of an *Ecklonia cava* polyphenol extract for 7 weeks reduced fasting blood glucose in obese C57BL/6 mice compared with placebo. They also found that an extract taken from the geographical area of Gijang, Korea (79.70 mg polyphenols/g extract), improved glucose tolerance compared with an extract from the area of Jeju, Korea (68.78 mg/g polyphenols), and placebo [64].

Zhang et al. [68] reported the effects of an *Ascophyllum nodosum* extract on diabetic markers in streptozotocin-induced diabetic mice. The mice were administered 200 mg/kg body weight of a crude extract or enriched extract (purified polyphenolic fraction) orally by daily gavage for up to 4 weeks. Both extracts reduced fasting blood glucose in the diabetic mice compared with placebo. However, only the enriched extract dampened the postprandial rise in blood glucose following an oral sucrose tolerance test [68]. Similarly, a dieckol-rich extract from *Ecklonia cava* (0.5 g/100 g diet) reduced fasting blood glucose levels, HbA1c levels and plasma insulin levels in C57Bl/6/KsJ-*db/db(db/db)* mice after 6 weeks, compared with a control diet [60]. The effect observed from the dieckol-rich extract was comparable to that of rosiglitazone (0.005 g/100 g diet), a known antidiabetic drug. Glucose tolerance also improved in the mice as a result of phlorotannin supplementation; the blood glucose area under the curve (AUC) was significantly reduced following phlorotannin treatment compared with control [60]. An *Ecklonia cava* polyphenol extract (28.2 ± 0.58% polyphenols) containing dieckol, 2,7″-phloroglucinol-6,6′-bieckol, pyrygallol-phloroglucinol-6,6′-bieckol and phlorofucofuroeckol A was administered to C57BL/6 obese male mice five times a week for 12 weeks [53]. Mice received either 100 mg/kg body weight/day or 500 mg/kg body weight/day or water. Postprandial blood glucose AUC reduced following both the low-dose and high-dose treatments compared to placebo. Additionally, the high-dose group had significantly lower plasma insulin and HOMA-IR after 12 weeks compared to placebo. However polyphenol supplementation had no effect on fasting blood glucose [53]. Furthermore, when fed to streptozotocin-induced diabetic mice, a 100 mg/kg body weight single dose of DPHC diminished postprandial blood glucose AUC to 2022 (113.0) mmol/min, compared with 2210 (125.2) mmol/min in the control mice [54].

3.7.1.3 Human Studies

In a randomised controlled trial, Shin et al. [17] gave 97 overweight adults a daily dose of either 72 mg or 144 mg of a polyphenol-rich extract (polyphenol content 98.5%) from *Ecklonia cava*, or a placebo, for 12 weeks. A reduction in fasting blood glucose was observed, but only in the high-dose group [17]. Conversely, a similar length randomised controlled trial showed that 3 months of an oral supplement (500 mg/day) containing 5% marine polyphenols increased plasma insulin levels, HOMA β-cell and HOMA-IR, compared with placebo, in overweight and obese adults. However, no change was observed in fasting blood glucose levels or blood glucose levels following an oral glucose tolerance test (OGTT) [55]. Lee and Jeon [16] administered 690 mg polyphenols or a placebo to 73 adults with high fasting blood glucose (100–180 mg/dL or 5.6–10.0 mmol/L) for 12 weeks. While an improvement in postprandial blood glucose control and a significant reduction in fasting blood insulin levels were observed following supplementation, there was no change in fasting blood glucose level. Paradis et al. [63] examined the anti-hyperglycaemic effects of a phlorotannin-rich, commercially available blend of the brown algae *Ascophyllum nodosum* and *Fucus vesiculosus* in a randomised, double-blind, placebo-controlled, cross-over postprandial study in 23 adults. Participants consumed either 500 mg seaweed capsules (containing at least 10% polyphenols) or placebo capsules 30 min prior to consumption of 50 g of available carbohydrates from bread. Plasma glucose and insulin levels were measured over a 3 h period following carbohydrate ingestion. Consumption of the capsule was associated with a 12.1% reduction in insulin incremental area under the curve (iAUC) and a 7.9% increase in the Cederholm index of insulin sensitivity, compared with placebo. There was no significant effect on postprandial blood glucose iAUC [63].

3.7.1.4 Summary

Marine polyphenols inhibit the action of α-amylase and α-glucosidase in vitro and reduce the postprandial rise in blood glucose and insulin levels in animals. However, reductions in postprandial blood glucose and insulin have not been consistently demonstrated in humans (Table 3.3). This may be due to differences in the way the molecules behave, or interact with other substances, in the human gut compared with the rodent gut; however this is as yet unknown. Marine polyphenols also reduce fasting blood glucose in animals, as also shown by one study in humans [17]. Some evidence suggests that a dose-dependent relationship exists between phlorotannin intake and the anti-hyperglycaemic effects, yet the variation in dosages, timeframes and species examined between studies make interpretation difficult.

3.7.2 Anti-hyperlipidaemic Effects

Dyslipidaemia occurs during diabetes and is a well-known risk factor for CVDs [43, 73–75]. Dyslipidaemia refers to elevated levels of low-density lipoprotein cholesterol (LDL-C) and triglycerides (TG) in the blood and reduced high-density lipoprotein cholesterol (HDL-C) [76, 77]. A key preventative measure for reducing the

Table 3.3 Anti-hyperglycaemic effects of marine polyphenols [105]

Seaweed species	Polyphenol	Dosage and duration	Subject/medium	Antidiabetic effect	References
Alaria, Palmaria, Ascophyllum	Polyphenolic-rich extracts	NA	Chemical assay	α-Amylase inhibition	[62]
				α-Glucosidase inhibition	
Ascophyllum nodosum	Does not name specific polyphenols	NA	Chemical assay	Dose dependent	[52]
				Strong α-glucosidase inhibition	
				Dose-dependent α-amylase inhibition	
Eisenia bicyclis	Phloroglucinol	NA	Chemical assay	Glycosidase enzyme inhibition	[65]
	Eckol				
	Phlorofucofuroeckol A				
	Dieckol				
	8,8′-bieckol				
	An unidentified tetramer				
Ascophyllum nodosum	Does not name specific polyphenols	NA	Chemical assay	Dose-dependent α-glucosidase inhibition	[68]
		400 μg/mL extract, 20 min incubation	3 T3-L1 adipocytes	Glucose uptake stimulated	
				Fasting blood glucose reduced	
		200 mg/kg body weight for 4 weeks	Streptozotocin-diabetic mice	Postprandial blood glucose rise blunted	
Ecklonia stolonifera	Does not name specific polyphenols	NA	Chemical assay	Strong α-glucosidase inhibition in vitro	[56]
		16.42 or 81.20 mg for 4 weeks	Genetically non-insulin-dependent diabetic KK-A^y male mice	Suppressed postprandial blood glucose and insulin	

(continued)

Table 3.3 (continued)

Seaweed species	Polyphenol	Dosage and duration	Subject/medium	Antidiabetic effect	References
Ishige okamurae	Diphlorethohydroxycarmalol (DPHC)	NA	Chemical assay	Strong α-glucosidase and α-amylase inhibition	[54]
		100 mg/kg body weight (single dose)	Streptozotocin-induced diabetic mice	Supressed postprandial blood glucose	
Ecklonia cava	Does not name specific polyphenols	50–300 μg/ml, 1 h incubation	C_2C_{12} myoblasts	Activated AMPK/ACC and P13/Akt signal transduction pathways	[58]
		300 mg/kg body weight for 3 weeks	Streptozotocin-induced type 1 diabetes mellitus rats	Fasting blood glucose reduced	
				Insulin concentration increased	
Ecklonia cava	Polyphenol extract (28.2 ± 0.58% polyphenols):	100 or 500 mg/kg body weight for 12 weeks	C57BL/6 male mice	Reduced blood glucose 1 h after injection	[53]
	Dieckol			Reduced postprandial glucose AUC	
	2,7″-phloroglucinol-6,6′-bieckol,			High-dose reduced plasma insulin and HOMA-IR	
	Pyrygallol-phloroglucinol-6,6′-bieckol				
	Phlorofucofuroeckol A				
Ecklonia cava	CA extract – 68.78 mg/g polyphenols	200 mg/kg body weight for 7 weeks	C57BL/6 mice	Reduced fasting blood glucose	[64]
	G-CA extract – 79.70 mg/g polyphenols			Improved glucose tolerance	
Ecklonia cava	Dieckol-rich extract	0.5 g/100 g diet for 6 weeks	C57BL/KsJ-*db/db (db/db)* male mice	Reduced fasting blood glucose	[60]
				Reduced plasma insulin	
				Reduced HbA1c	

Ecklonia cava	Dieckol	20 mg/kg body weight for 14 days	C57BL/KsJ-*db/db*, a type II diabetes mice	Reduced blood glucose	[59]
				Dose-dependently reduced plasma insulin	
Ishige okamurae	Does not mention specific polyphenols	0.5% w/w for 6 weeks	C57BL/KsJ-*db/db* mice	Reduced G6Pase and PEPCK activity	[61]
				Increased hepatic glycogen production	
				Reduced fasting blood glucose	
				Reduced HbA1c	
Ascophyllum nodosum, Fucus vesiculosus	Commercially available blend of brown seaweeds containing a minimum of 10% polyphenols	50 mg polyphenols, single dose prior to postprandial testing	23 non-diabetic adults	Reduced postprandial insulin iAUC	[63]
				Increased insulin sensitivity	
Ecklonia cava	Polyphenol extract	72 or 144 mg/day for 12 weeks	97 non-diabetic overweight adults	Reduced fasting blood glucose (high dose only)	[17]
Ecklonia cava	Dieckol-rich extract	690 mg/day for 12 weeks	73 adults with high fasting blood glucose	Improved postprandial blood glucose control	[16]
				Reduced fasting blood insulin level	
				No change in fasting blood glucose level	
Not specified	Polyphenol containing (5%) oral supplement	25 mg polyphenols for 3 months	25 non-diabetic overweight or obese volunteers	Increased plasma insulin, HOMA β-cell and HOMA -IR	[55]

Abbreviations: *AMPK/ACC* AMP–activated protein kinase/acetyl–CoA carboxylase, *AUC* area under the curve, *G6Pase* glucose–6–phosphatase, *HbA1c* glycated haemoglobin, *HOMA–IR* homeostatic model assessment – insulin resistance, *HOMA β–cell* homeostatic model assessment – beta-cell function, *iAUC* incremental area under the curve, *NA* not available/not applicable, *P13/Akt* phosphatidylinositol 3-kinase/Akt (a serine/threonine protein kinase), *PEPCK* phosphoenolpyruvate carboxykinase risk of CVDs is to lower total cholesterol (TC), LDL-C and TG levels in the blood [78]. Polyphenols from terrestrial sources have been shown to reduce digestion and absorption of dietary lipids, decrease synthesis and secretion of apolipoprotein B, inhibit cholesterol esterification and intestinal lipoprotein production and inhibit key enzymes in lipid biosynthesis pathways [79–81] resulting in an improved lipid profile and lowered cardiovascular risk. Recent studies indicate that macroalgal polyphenols may have similar lipid-lowering actions. One of the proposed lipid-lowering actions of macroalgal polyphenols is the inhibition of the enzyme HMGCoA (3-hydroxy-3-methylglutaryl-coenzyme A) reductase in the liver and adipose tissue [66]. This is a key enzyme involved in the cholesterol production pathway and also the target of successful TC- and LDL-C-lowering drugs [82].

3.7.2.1 In Vitro Studies

Seapolynol™, a polyphenol extract containing 98.5% unspecified polyphenols, and dieckol, an isolated phlorotannin from *Ecklonia cava*, both dose-dependently (0–200 μg/mL) inhibited the activity of HMGCoA reductase in vitro [66].

Macroalgal phlorotannins have also been shown to inhibit lipid accumulation and adipocyte differentiation in 3 T3-L1 adipocytes through reducing the expression levels of adipocyte marker genes PPARγ and C/EBPα [57, 66]. This mechanism has been suggested as a way of managing obesity via reduction of the formation of mature adipocytes and adipose tissue [57, 83–85] and has been associated with reduced body weight gain in high-fat diet-fed mice [83]. This mechanism has also been associated with a reduction in the activity of TG synthetic hormones, such as diacylglycerol acyltransferase 1 (DGAT1) and glycerol-3-phosphate acyltransferase 3 (GPAT3), resulting in reduced triglyceride accumulation within adipocytes [86].

Interestingly, a reduction in serum TG levels in high-fat diet-fed mice was also observed [66]; however it was not clear how inhibition of intracellular lipid/TG accumulation and adipocyte differentiation would result in changes to serum TG levels. While the inhibition of lipid accumulation and adipocyte differentiation may contribute to a reduction in body fat, it is unlikely that this mechanism would directly reduce blood lipid levels. A combination of reduced body weight and the impact on liver enzymes may have effect on blood lipids; however this is yet to be shown.

3.7.2.2 Animal Studies

Yeo et al. [66] investigated the reduction of hyperlipidaemia in imprinting control region (ICR) mice using a generic polyphenol extract (Seapolynol – containing 98.5% polyphenols) and dieckol, from *Ecklonia cava*. Mice were fed a high-fat diet (20% fat) for 5 weeks and supplemented with Seapolynol™ (1.25, 2.5 or 5 mg/day) or dieckol (0.5, 1 or 2 mg/day) for 4 weeks, starting 1 week after high-fat diet feeding. Both treatments reduced serum TC, TG and LDL-C levels compared with mice fed a high-fat diet only [66]. Similarly, in C57BL/KsJ-*db/db (db/db)* male mice, a dieckol-rich extract from *Ecklonia cava* (0.5 g dieckol/100 g diet) reduced TC levels and free fatty acids (FFAs) after 6 weeks, compared with placebo. The same treatment reduced TG levels to an extent similar to treatment with rosiglitazone (0.005 g/100 g diet) and increased HDL-C to a greater extent than rosiglitazone treatment [60]. Eo et al. [53] examined a polyphenol extract from *Ecklonia cava* (28.2 ± 0.58% polyphenols) containing dieckol, 2,7″-phloroglucinol-6,6′-bieckol, pyrygallol-phloroglucinol-6,6′-bieckol and phlorofucofuroeckol A in high-fat diet-fed obese C57BL/6 male mice. Mice were administered 100 mg/kg body weight/day or 500 mg/kg body weight/day or water, five times per week for 12 weeks. Both doses of polyphenol extract reduced TG and TC levels compared with mice fed the high-fat diet alone. No changes were reported in HDL-C [53]. A phlorotannin-rich extract (100–250 mg/kg body weight/day) and isolations of the phlorotannins eckol and dieckol (10 or 20 mg/kg body weight/day) from *Ecklonia stolonifera* were investigated for their anti-hyperlipidaemic effects in high-cholesterol diet- or poloxamer 407-induced hyperlipidaemic rats [67]. The phlorotannin-rich ethyl acetate and *n*-butanol fractions, extracted using ethanol, reduced TG, TC and LDL-C levels and increased HDL-C levels in a dose-dependent manner after 3 days of treatment. Both eckol and dieckol isolations reduced TG, TC and LDL-C levels after 3 days. Dieckol treatment alone produced a greater hypolipidaemic effect than lovastatin (50 mg/kg) and increased HDL-C levels in the hyperlipidaemic rats after 3 days [67]. Conversely, Park et al. [64] administered polyphenol-rich extracts from *Ecklonia cava* (grown in two different geographical areas in Korea: Jeju and Gijang) to high-fat diet-induced obese C57BL/6 mice at 200 mg/kg body weight/day for 8 weeks. Treatment with the extract from Jeju had no effect on TC, TG, LDL-C or HDL-C levels; however, the extract from Gijang reduced TC level compared with placebo [64], highlighting the potential for differences in polyphenol content based on geographical location of cultivation, even within the same species of algae [5, 30].

3.7.2.3 Human Studies

In a randomised, double-blind, placebo-controlled parallel-design trial in 97 overweight adults, consumption of a phlorotannin-rich extract from *Ecklonia cava*, at doses of 72 or 144 mg polyphenols per day, reduced TC levels, LDL-C levels and TC to HDL-C ratio, in a dose-dependent manner following 12 weeks of treatment, compared with placebo. An increase in HDL-C levels was only observed following the highest dose [17]. A comparable trial in 80 adults with raised cholesterol (>200 mg/dL TC or > 110 mg/dL LDL-C) demonstrated that consumption of a dieckol-rich *Ecklonia cava* extract (400 mg/day, 8.2% dieckol) for 12 weeks resulted in reduced TC and LDL-C, compared with placebo, without change in TG or HDL-C levels [15]. Conversely, in a randomised, double-blind, placebo-controlled clinical trial of 25 overweight or obese volunteers, no changes were reported in TC, TG or HDL-C levels following 500 mg of a polyphenol-containing oral supplement (5% polyphenols) daily for 3 months. However, LDL-C levels were reduced following the supplementation treatment compared with no change in placebo group [55].

3.7.2.4 Summary

Similar to the anti-hyperglycaemic evidence, phlorotannins improved dyslipidaemia in animal models and in vitro via a number of mechanisms, although results in humans are few and inconsistent (Table 3.4). Phlorotannins have potential as an

Table 3.4 Anti-hyperlipidaemic effects of marine polyphenols

Seaweed species	Polyphenol	Dosage and duration	Subjects/medium	Anti-hyperlipidaemic effect	References
Ecklonia cava	Generic polyphenol extract (Seapolynol™ – 98.5% polyphenols) and dieckol	50 μg/mL concentration	Chemical assay	Inhibited HMGCoA reductase	[66]
		1.25, 2.5 or 5.0 mg Seapolynol™, or 0.5, 1.0 or 2.0 mg dieckol for 4 weeks	ICR mice	Reduced TC, TG and LDL-C	
Ecklonia cava	Dieckol-rich extract	0.5 g/100 g diet for 6 weeks	C57BL/KsJ-*db/db* (*db/db*) male mice	Reduced TC, TG and FFA	[60]
				Increased HDL-C	
Ecklonia cava	CA extract – 68.78 mg/g polyphenols	200 mg/kg body weight for 8 weeks	Obese C57BL/6 mice	G-CA extract reduced TC	[64]
	G-CA extract – 79.70 mg/g polyphenols				
Ecklonia cava	Polyphenol extract (28.2 ± 0.58% polyphenols):	100 or 500 mg/kg body weight for 12 weeks	Obese C57BL/6 male mice	Reduced TG and TC levels	[53]
	Dieckol			No change in HDL-C	
	2,7″-phloroglucinol-6,6′-bieckol,				
	Pyrygallol-phloroglucinol-6,6′-bieckol,				
	Phlorofucofuroeckol A				
Ecklonia stolonifera	Polyphenol extract	100–250 mg/kg body weight (polyphenol extract) or 10 or 20 mg/kg body weight (eckol and dieckol) for 3 days	Hyperlipidaemic rats	Both treatments dose-dependently reduced TC, TG and LDL-C and increased HDL-C.	[67]
	Eckol			Different extraction techniques yielded different actions	
	Dieckol				

Ecklonia cava	Polyphenol extract	72 or 144 mg for 12 weeks	97 overweight adults	Dose-dependently reduced TC, LDL-C and TC/HDL-C ratio	[17]
				High-dose increased HDL-C	
Ecklonia cava	Dieckol-rich extract (8.2% dieckol)	400 mg/day (32.8 mg dieckol) for 12 weeks	80 adults with raised cholesterol	Reduced TC and LDL-C levels	[15]
				Intervention had no effect on HDL-C or TG levels	
Not specified	Polyphenol containing (5%) oral supplement	25 mg polyphenols for 3 months	25 overweight or obese volunteers	Reduced LDL-C only	[55]

Adapted from Murray et al. [105]
Abbreviations: *TC* total cholesterol, *LDL-C* low density lipoprotein cholesterol, *HDL-C* high density lipoprotein cholesterol, *TG* triglyceride, *HMGCoA* 3-hydroxy-3-methylglutaryl-coenzyme A, *CA* Jeju geographical area, Korea, *G-CA* Gijang geographical area, Korea

Table 3.5 Health-promoting effects of marine polyphenols according to algal species

Species	Effect	Subject/medium	References
Ascophyllum nodosum	Anti-hyperglycaemic	Chemical assay	[52, 62, 68]
		3 T3-L1 adipocytes	[68]
		Diabetic mice	[68]
		Non-diabetic adults	[63]
Ecklonia cava	Anti-hyperglycaemic	C_2C_{12} myoblasts	[58]
		Diabetic rats	[58]
		Mice	[53, 64]
		Diabetic mice	[59, 60]
		Non-diabetic overweight adults	[17]
		Pre-diabetic adults	[16]
	Anti-hyperlipidaemic	Chemical assay	[66]
		3 T3-L1 preadipocytes	[66]
		Mice	[66]
		Diabetic mice	[60]
		Obese mice	[53, 64]
		Overweight adults	[17]
		Adults with raised cholesterol	[15]
Ecklonia stolonifera	Anti-hyperglycaemic	Chemical assay	[56]
		Diabetic mice	[56]
	Anti-hyperlipidaemic	3 T3-L1 preadipocytes	[57]
		Hyperlipidaemic rats	[67]
Fucus vesiculosus	Anti-hyperglycaemic	Non-diabetic adults	[63]
Ishige okamurae	Anti-hyperglycaemic	Chemical assay	[54]
		Diabetic mice	[54, 61]

Adapted from [105]

anti-hyperlipidaemic agent in humans, but due to factors such as bioavailability and dosing, which differ considerably between humans and animals, further research is required to determine a consistent effect and appropriate dosage and treatment schedule in humans.

3.7.3 Health Effects According to Algal Species

Phlorotannin-rich extracts from the *Ascophyllum nodosum*, *Ecklonia stolonifera*, *Ishige okamurae* and *Ecklonia cava* macroalgae varieties are the most predominantly tested with relation to their potential health-promoting effects (Table 3.5).

Macroalgae from the *Ascophyllum* species have been investigated in cell culture, animal models and humans and have been shown to improve diabetic risk factors. Phlorotannin-rich extracts from this species have inhibited α-amylase and α-glucosidase activity [52, 62, 68], stimulated glucose uptake [68], reduced fasting and postprandial blood glucose in diabetic mice [68] and reduced postprandial insulin and improved insulin sensitivity in non-diabetic adults [63].

Phlorotannins from the *Ecklonia stolonifera* species have exhibited both anti-hyperglycaemic and anti-hyperlipidaemic effects. These include the inhibition of α-glucosidase activity in vitro [56] and reductions in postprandial blood glucose and insulin in diabetic mice [56] and dose-dependent reductions in TC, TG and LDL-C and increasing HDL-C levels in hyperlipidaemic rats [67].

Ishige okamurae phlorotannins have demonstrated anti-hyperglycaemic effects; α-amylase and α-glucosidase inhibition has been demonstrated in vitro [54], along with reductions in postprandial blood glucose, fasting blood glucose and HbA1c levels in diabetic mice [54, 61].

Phlorotannin-rich extracts from *Ecklonia cava* are by far the most extensively tested and have demonstrated both anti-hyperglycaemic and anti-hyperlipidaemic effects. Anti-hyperglycaemic effects, including reduced fasting blood glucose and insulin levels, reduced postprandial blood glucose and improved insulin sensitivity, have been shown in animal models [53, 58–60, 64] and non-diabetic [17] and pre-diabetic adults [16]. The anti-hyperlipidaemic effects of *Ecklonia cava* phlorotannins include the reduction of TC, LDL-C, TG and FFA levels and increase of HDL-C, shown in animal models [53, 60, 64, 66] and humans [15, 17]. The inhibition of HMGCoA reductase by *Ecklonia cava* phlorotannins has also been shown in 3 T3-L1 preadipocytes [66].

While there are numerous different algal species, the research to date has tended to focus on the aforementioned species with a particular emphasis on *Ecklonia cava*, despite the fact that it is not a commonly consumed alga. Phlorotannins from *Ecklonia cava* have demonstrated all of the health effects examined in this chapter and are beginning to be tested in different human populations [15–17]. Future research is warranted to investigate the health effects of *Ecklonia cava* phlorotannins particularly in human populations, as they show potential to be used as a functional food ingredient. However, the potential of other species of macroalgae that are not yet as well investigated should not be neglected.

3.8 Macroalgae As a Functional Food Ingredient

A functional food is defined as a "natural or processed food that contains known or unknown biologically-active compounds, which in defined, effective, non-toxic amounts, provide a clinically proven and documented health benefit for the prevention, management, or treatment of chronic disease" [87]. The potential of macroalgae derived polyphenols as a functional food component and its limitation will be discussed below.

3.8.1 Safe Levels of Consumption

Healthy diet and exercise are the best methods for prevention of chronic lifestyle diseases. However in the absence of a healthy diet and exercise, drugs are the current accepted treatment for blood sugar and cholesterol control. Long-term use of oral antidiabetic and anti-hyperlipidaemic drugs can cause unpleasant side effects, including muscle cramping, fatigue, muscle breakdown, vomiting and diarrhoea [3, 43, 88, 89]. Macroalgal polyphenols are thought to be relatively safe for consumption [54, 66, 90–93] as they lack unpleasant side effects [3, 43, 63], and therefore they may be safer alternatives as functional food ingredients if efficacy can be proven. The safety of a polyphenol-rich supplement from *Fucus vesiculosus* has been demonstrated at up to 750 mg/kg/day in rats over 4 weeks [93]. Extrapolating these data to an equivalent dose in humans, based on an average body weight of 65 kg, suggests up to 48.75 g/day would be safe for consumption. This amount is far greater than what would realistically be consumed. Macroalgae-derived DPHC has also shown no cytotoxicity in human umbilical vein epithelial cells (HUVECs) at concentrations up to 3.91 mM after 20 h incubation [54]. It should be noted, however, that green tea polyphenols have been shown to cause hepatotoxicity and other adverse effects at higher doses. A dose of 500 mg/kg/day pure epigallocatechin gallate (EGCG) for 13 weeks increased bilirubin and decreased fibrinogen in rats [94]. This equates to a dose of 32.5 g/day for a 65 kg human. The risk of toxicity increased when ingested in the fasting state or over long periods of time, or when the polyphenols were administered intraperitoneally [94]. If the safety of marine polyphenols and efficacy for blood glucose or cholesterol control can be shown in a human population, then marine polyphenols would have great potential for commercialisation as a functional food ingredient. It should be a research priority to assess the safety and efficacy of marine polyphenol consumption for health promotion and disease prevention in humans.

3.8.2 Additional Bioactive Compounds in Macroalgae

While polyphenols are a key contributor to the biological activity and health benefits of brown macroalgae [4, 5], macroalgae also contain a variety of other biologically active compounds that may contribute to their value as functional food ingredients. They are a rich source of dietary fibre (33–50 g/100 g DW), in particular soluble fibre. Consumption of macroalgae alone, or its incorporation into other foods, can increase fibre intake in the diet which may help to reduce the risk of some chronic diseases that are associated with low-fibre intakes, such as diabetes and heart disease [51]. In addition to contributing to fibre content, macroalgal polysaccharides have also been shown to possess antioxidant properties and anti-tumour and anticoagulant bioactivity and to lower LDL-C in cholesterol-rich diet-fed rats [51, 95]. Fucoidans are sulfated polysaccharides found in the cell walls of brown macroalgae, but not in other algae or terrestrial plants. They are reported to have antioxidant, antiviral, anticoagulant and antiobesity effects and are being investigated as a functional food ingredient for disease prevention and health promotion [51, 95]. Other biologically active compounds found in macroalgae include omega-3 and omega-6 essential fatty acids, which help in the prevention of atherosclerosis, reduction of blood pressure, cancer prevention, promotion of bone health and improvement in brain function in children [51]. Macroalgae are a rich source of alginate, or alginic acid, which is reported to reduce cholesterol concentration, have antihypertensive effects, reduce cancer risk and contribute to dietary fibre intake [95]. Macroalgae also contain chlorophyll, carotenoids, hydroquinones, flavonoids, sterols and phospholipids, which contribute to their antioxidant capacity [51, 95]. While these health benefits may sound promising, there is as yet little evidence that these effects occur in humans, since the majority of research uses only cell culture or animal models. This is an important gap that needs to be filled before macroalgae can be used as functional food ingredients for disease prevention and health promotion in humans.

3.8.3 Availability and Sustainability of Macroalgae As a Functional Food Source

There are many environmental and economic benefits associated with marine sources, as opposed to land-based sources, to obtain biologically active compounds. With the ocean making up more than 70% of the Earth's surface [26], it provides an abundant source of marine products, and algal species are easy to harvest from the wild as well as to culture in the sea and in pools on land [33]. The cultivation of marine algae has a number of advantages over terrestrial plant cultivation such as it requires less freshwater, produces a higher biomass, can be grown in lower quality agricultural environments and can be grown in seawater avoiding the need for herbicides and pesticides [96]. Furthermore, recent advances in biotechnological tools for the extraction and identification of phlorotannins and polyphenols from macroalgae have led to an upward trend in the use of these products as functional food ingredients [3]. Therefore, there is likely to be a large market for marine polyphenols as a functional food ingredient if efficacy can be demonstrated.

3.8.4 Processing of Macroalgae for Consumption

In order to be incorporated into food products, macroalgae must be processed to ensure they are palatable for human consumption. Drying methods and other processing techniques aim to ensure the acceptability of macroalgae as a food product by altering its sensory properties, e.g. minimising textural issues, while maintaining maximum levels of biologically active compounds [51]. Higher drying temperatures (40 °C) minimise the losses of phenolic content due to drying, compared with lower temperatures (25 °C). This may be because lower drying temperatures do not completely inactivate oxidative enzymes, and some oxidation of the phenolic substances takes place [51]. This is reflected in higher losses of antioxidant activity when macroalgae were dehydrated at lower temperatures (25 °C dehydration resulted in 17.3% loss in radical scavenging ability), compared with higher temperatures (40 °C dehydration resulted in 4.5% loss in radical scavenging ability) [51].

Hydrothermal processing (treatment of macroalgae in water at 80–100 °C) is another technique used to minimise the tough texture of macroalgae to improve their sensory acceptability as a food [51]. Hydrothermal processing alone can achieve an acceptable texture within 40 min of treatment; however this results in up to 85% losses in phenolic content at both 80 and 100 °C. A combination of drying followed by hydrothermal processing can dramatically reduce these losses. When *Himanthalia elongata* was dried for 12 h at 25 °C first, an acceptable texture could be achieved with only 25 min of hydrothermal processing at 100 °C resulting in only ~9% loss in total phenolic content [51]. The pretreatment of macroalgae is essential for their successful utilisation in functional foods and to ensure the final product has acceptable sensory and textural properties. It is also vital that these treatments allow the retention of maximum levels of phenolics and other biologically active compounds [51].

3.8.5 Incorporation of Macroalgae into Foods

While maintaining high levels of bioactivity is vital in functional food development, it is also important that the resulting products are appealing to consumers and have an acceptable sensory profile [51]. Macroalgae have been successfully incorporated into a number of food products to add functional or structural properties. Macroalgae, or seaweed, as a whole component has been added to pork sausages to replace animal fat [97], beef patties to reduce salt and fat levels [98], pork patties to decrease fat content and increase fibre content [99], pasta to increase antioxidant levels [100] and noodles to increase the cooking yield [51, 101]. Products with added macroalgae have also been tested by sensory panels for their palatability and found to be acceptable, in some cases with better sensory scores than the original product [99, 102].

Meat and bread products appear to be the two main candidates for fortification using macroalgae. Macroalgae, fully dried and ground into a fine powder, has been incorporated into a wholemeal and white flour bread base mix to be made into breadsticks. It was determined that up to 10% macroalgae in the mix produced a highly acceptable product for sensory and consumer appeal [51]. When 15% or more of the mix was dried macroalgae, this resulted in a harder texture and level of aroma that were not acceptable to consumers. It is important to determine the appropriate amount of macroalgae that can be added to any particular food to get a balance between functional capacity and sensory appeal. This amount will vary depending on the acceptable sensory and texture parameters associated with individual food items [51]. The right balance will ensure a food product that is acceptable for consumption and has benefits to human health and disease prevention.

3.8.6 Commercial Potential of Functional Foods

Recent trends have shown an increase in consumer preferences for natural and sustainable health products and functional foods [33, 103]; thus there is interest in marine-based food products [3]. The global market for functional foods has experienced rapid growth in recent years, and this growth is forecast to continue [51, 104]. In the United States of America, the expected compound annual growth rate from 2016 to 2020 for functional foods is 8.8% compared with 6.0% for the nutraceutical market as a whole [51, 104]. The functional food market is often categorised as a subsection under the umbrella of nutraceuticals and pooled with the dietary supplement (tablets, capsules and liquids) market. The following data refer to the nutraceutical market as a whole, as individual data for the functional food market was not available. In 2014 the global nutraceutical market was valued at approximately US$250 billion and is forecast to be worth around US$385 billion by 2020 [104]. While this figure is an overestimation of the actual value of the functional foods market, it indicates the commercial ability and saleability of functional food products. Key factors driving this growth of the functional food industry have been the increase in population, particularly the ageing population, an increase in diet-related diseases and our understanding of how diet affects health, in addition to consumer's demand for health and wellness products [51].

3.9 Conclusion

Macroalgae show great potential as functional food ingredients. They can be grown and harvested in a sustainable and environmentally friendly way and have already been successfully incorporated into foods. Ongoing improvements in scientific technologies will allow more thorough characterisation of the polyphenol and phlorotannin content of macroalgae and improved understanding of the health-promoting effects of individual phlorotannins and phlorotannins working in synchrony. A better understanding of what happens to macroalgal polyphenols when they enter the body and how dietary intake reflects health outcomes is required. This will get clearer as more studies investigating the health effects of marine polyphenols are conducted in humans. Under experimental conditions phlorotannins from marine macroalgae have many positive health-related effects. *Ecklonia cava* has shown great potential as a source of bioactive marine polyphenols, with evidence for both anti-hyperglycaemic and anti-hyperlipidaemic effects, in the trials already completed in human populations. However, other macroalgal species should not be ignored as potential functional food ingredients. Health benefits of macroalgal polyphenols are a relatively novel area of research, with much yet to be discovered. As research in this area continues, it will be exciting to see the true potential of macroalgal polyphenols revealed.

Acknowledgements None to declare

Conflicts of Interest None to declare

References

1. Manach C, Scalbert A, Morand C et al (2004) Polyphenols: food sources and bioavailability. Am J Clin Nutr 79:727–747
2. Naczk M, Shahidi F (2004) Extraction and analysis of phenolics in food. J Chromatogr A 1054:95–111
3. Murugan AC, Karim MR, Yusoff MBM et al (2015) New insights into seaweed polyphenols on glucose homeostasis. Pharm Biol 53(8):1087–1097
4. Heffernan N, Brunton NP, FitzGerald RJ et al (2015) Profiling of the molecular weight and structural isomer abundance of macroalgae-derived phlorotannins. Mar Drugs 13:509–528
5. Bocanegra A, Bastida S, Benedi J et al (2009) Characteristics and nutritional and cardiovascular-health properties of seaweeds. J Med Food 12(2):236–258
6. Creis E, Delage L, Charton S et al (2015) Constitutive or inducible protective mechanisms against UV-B radiation in the brown alga Fucus vesiculosus? A study of gene expression and phlorotannin content responses. PLoS One 10(6):e0128003
7. Koivikko R, Loponen J, Pihlaja K et al (2007) High-performance liquid chromatographic analysis of phlorotannins from the brown alga Fucus Vesiculosus. Phytochem Anal 18:326–332
8. Crozier A, Jaganath IB, Clifford MN (2009) Dietary phenolics: chemistry, bioavailability and effects on health. Nat Prod Rep 26:1001–1043
9. D'Archivio M, Filesi C, Di Benedetto R et al (2007) Polyphenols, dietary sources and bioavailability. Ann Ist Super Sanita 43(4):348–361
10. Higdon JV, Frei B (2003) Tea catechins and polyphenols; health effects, metabolism, and antioxidant functions. Crit Rev Food Sci Nutr 43(1):89–143
11. Kris-Etherton PM, Hecker KD, Bonanome A et al (2002) Bioactive compounds in foods: their role in the prevention of cardiovascular disease and cancer. Am J Med 113(9B):71S–88S
12. Manach C, Williamson G, Morand C et al (2005) Bioavailability and bioefficacy of polyphenols in humans. I. Review of 97 bioavailability studies. Am J Clin Nutr 81(suppl):230S–242S
13. Scalbert A, Williamson G (2000) Dietary intake and bioavailability of polyphenols. J Nutr 130(8S):2073S–2085S
14. Williamson G, Manach C (2005) Bioavailability and bioefficacy of polyphenols in humans. II. Review of 93 intervention studies. Am J Clin Nutr 81(suppl):243S–255S
15. Choi E-K, Park S-H, Ha K-C et al (2015) Clinical trial of the hypolipidemic effects of a brown alga Ecklonia cava extract in patients with hypercholesterolemia. Int J Pharm 11(7):798–805
16. Lee S-H, Jeon Y-J (2015) Efficacy and safety of a dieckol-rich extract (AG-dieckol) of brown algae, Ecklonia cava, in pre-diabetic individuals: a double-blind, randomized, placebo-controlled clinical trial. Food Funct 6:853–858
17. Shin H-C, Kim SH, Park Y et al (2012) Effects of 12-week oral supplementation of Ecklonia cava polyphenols on anthropometric and blood lipid parameters in overweight Korean individuals: a double-blind randomized clinical trial. Phytother Res 26:363–368
18. Chowdhury MTH, Bangoura I, Kang J-Y et al (2014) Comparison of Ecklonia cava, Ecklonia stolonifera and Eisenia bicyclis for phlorotannin extraction. J Environ Biol 35:713–719
19. Guiry M (2017) The seaweed site: information on marine algae, 9th January 2017 [Cited 2016 10th January 2017]. Available from: http://www.seaweed.ie/index.php
20. Merchant R, Andre C (2001) A review of recent clinical trials of the nutritional supplement chlorella pyrenoidosa in the treatment of fibromyalgia, hypertension, and ulcerative colitis. Altern Ther Health Med 7(3):79–91
21. Besada V, Andrade JM, Schultze F et al (2009) Heavy metals in edible seaweeds commercialised for human consumption. J Mar Syst 75:305–313
22. Zava TT, Zava DT (2011) Assessment of Japanese iodine intake based on seaweed consumption in Japan: a literature-based analysis. Thyroid Res 4(14):1–7

23. Smithsonian Institution Department of Botany (2017) Secondary products of red algae. [Cited 2017 10th January]. Available from: https://botany.si.edu/projects/algae/economic-uses/redalgae.htm
24. Chakraborty K, Joseph D, Praveen NK (2015) Antioxidant activities and phenolic contents of three red seaweeds (Division: Rhodophyta) harvested from the Gulf of Mannar of Peninsular India. J Food Sci Technol 52(4):1924–1935
25. Farasat M, Khavari-Nejad R-A, Bagher Nabavi SM et al (2014) Antioxidant activity, total phenolics and flavonoid contents of some edible green seaweeds from northern coasts of the Persian Gulf. Iran J Pharm Res 13(1):163–170
26. Hamed I, Ozogul F, Ozogul Y et al (2015) Marine bioactive compounds and their health benefits: a review. Compr Rev Food Sci Food Saf 14:446–465
27. Shibata T, Kawaguchi S, Hama Y et al (2004) Local and chemical distribution of phlorotannins in brown algae. J Appl Phycol 16:291–296
28. Bravo L (1998) Polyphenols: chemistry, dietary sources, metabolism, and nutritional significance. Nutr Rev 56(11):317–333
29. Faller A, Fialho E (2010) Polyphenol content and antioxidant capacity in organic and conventional plant foods. J Food Compos Anal 23(6):561–568
30. Connan S, Goulard F, Stiger V et al (2004) Interspecific and temporal variation in phlorotannin levels in an assemblage of brown algae. Bot Mar 47:410–416
31. Steinberg PD (1995) Seasonal variation in the relationship between growth rate and phlorotannin production in the kelp Ecklonia radiata. Oecologia 102:169–173
32. Cuong DX, Boi VN, Van TTT et al (2015) Effect of storage time on phlorotannin content and antioxidant activity of six Sargassum species from Nhatrang Bay, Vietnam. J Appl Phycol. Published online. https://doi.org/10.1007/s10811-015-0600-y
33. Mouritsen OG, Dawezynski C, Duelund L et al (2013) On the human consumption of the red seaweed dulse (Palmaria palmata (L.) Weber & Mohr). J Appl Phycol 25:1777–1791
34. Carroll WF (2009) SUSHI: globalization through food culture: towards a study of global food networks. J East Asian Cult Int Stud 2:451–456
35. Kozu JC, Creyer EH, Burton S (2003) Making healthful food choices: the influence of health claims and nutrition information on consumers' evaluations of packaged food products and restaurant menu items. J Mark 67:19–34
36. Shim J-S, Oh K, Kim HC (2014) Dietary assessment methods in epidemiologic studies. Epidemiol Health 36:e2014009
37. Spencer JP, Abd El Mohsen MM, Minihane A-M et al (2008) Biomarkers of the intake of dietary polyphenols: strengths, limitations and application in nutrition research. Br J Nutr 99:12–22
38. Neveu V, Perez-Jimenez J, Vos F et al (2010) Phenol-explorer: an online comprehensive database on polyphenol contents in foods. Database 2010:bap024
39. Wang J, Tang L, Wang J-S (2015) Biomarkers of dietary polyphenols in cancer studies: current evidence and beyond. Oxidative Med Cell Longev 2015:1–14. https://doi.org/10.1155/2015/732302
40. Clifford MN (2004) Diet-derived phenols in plasma and tissues and their implications for health. Planta Med 70:1103–1114
41. Lee CY (2013) Challenges in providing credible scientific evidence of health benefits of dietary polyphenols. J Funct Foods 5:524–526
42. Barditch-Crovo PA, Petty BG, Gambertoglio J et al (1998) The effect of increasing gastric pH upon the bioavailability of orally-administered foscarnet. Antivir Res 38(3):209–212
43. Bahadoran Z, Mirmiran P, Azizi F (2013) Dietary polyphenols as potential nutraceuticals in management of diabetes: a review. J Diabetes Metab Disord 12(43):1–9
44. Hanhineva K, Torronen R, Bondia-Pons I et al (2010) Impact of dietary polyphenols on carbohydrate metabolism. Int J Mol Sci 11:1365–1402
45. Hursel R, Viechtbauer W, Dulloo A et al (2011) The effects of catechin rich teas and caffeine on energy expenditure and fat oxidation: a meta-analysis. Obes Rev 12:e573–e581

46. Hursel R, Viechtbauer W, Westerterp-Plantenga M (2009) The effects of green tea on weight loss and weight maintenance: a meta-analysis. Int J Obes 33:956–961
47. Rains TM, Agarwal S, Maki KC (2011) Antiobesity effects of green tea catechins: a mechanistic review. J Nutr Biochem 22:1–7
48. Willcox DC, Willcox BJ, Todoriki H et al (2009) The Okinawan diet: health implications of a low- calorie, nutrient-dense, antioxidant-rich dietary pattern low in glycemic load. J Am Coll Nutr 28(sup4):500S–516S
49. Gavrilova NS, Gavrilov LA (2012) Comments on dietary restriction, Okinawa diet and longevity. Gerontology 58(3):221–223
50. Miyagi S, Iwama N, Kawabata T et al (2003) Longevity and diet in Okinawa, Japan: the past, present and future. Asia Pac J Public Health 15(1 suppl):S3–S9
51. Abu-Ghannam N, Cox S (2014) Seaweed-based functional foods. In: Hernandez-Ledesma B, Herrero M (eds) Bioactive compounds from marine foods: plant and animal sources. Wiley-Blackwell, Madrid
52. Apostolidis E, Lee C (2010) In vitro potential of Ascophyllum nodosum phenolic antioxidant-mediated α-glucosidase and α-amylase inhibition. J Food Sci 75(3):H97–H102
53. Eo H, Jeon Y-J, Lee M et al (2015) Brown alga Ecklonia cava polyphenol extract ameliorates hepatic lipogenesis, oxidative stress, and inflammation by activation of AMPK and SIRT1 in high-fat diet-induced obese mice. J Agric Food Chem 63:349–359
54. Heo S-J, Hwang J-Y, Choi J-I et al (2009) Diphlorethohydroxycarmalol isolated from Ishige okamurae, a brown algae, a potent α-glucosidase and α-amylase inhibitor, alleviates postprandial hyperglycemia in diabetic mice. Eur J Pharmacol 615:252–256
55. Hernandez-Corona DM, Martinez-Abundis E, Gonzalez-Ortiz M (2014) Effect of Fucoidan administration on insulin secretion and insulin resistance in overweight or obese adults. J Med Food 17(7):830–832
56. Iwai K (2008) Antidiabetic and antioxidant effects of polyphenols in brown alga Ecklonia stolonifera in genetically diabetic KK-Ay mice. Plant Foods Hum Nutr 63:163–169
57. Jung HA, Jung HJ, Jeong HY et al (2014) Phlorotannins isolated from the edible brown alga Ecklonia stolonifera exert anti-adipogenic activity on 3T3-L1 adipocytes by downregulating C/EBPα and PPARγ. Fitoterapia 92:260–269
58. Kang C, Jin YB, Lee H et al (2010) Brown alga Ecklonia cava attenuates type 1 diabetes by activating AMPK and Akt signaling pathways. Food Chem Toxicol 48:509–516
59. Kang M-C, Wijesinghe W, Lee S-H et al (2013) Dieckol isolated from brown seaweed Ecklonia cava attenuates type II diabetes in db/db mouse model. Food Chem Toxicol 53:294–298
60. Lee S-H, Min K-H, Han J-S et al (2012) Effects of brown alga, Ecklonia cava on glucose and lipid metabolism in C57BL/KsJ-db/db mice, a model of type 2 diabetes mellitus. Food Chem Toxicol 50:575–582
61. Min K-H, Kim H-J, Jeon Y-J et al (2011) Ishige okamurae ameliorates hyperglycemia and insulin resistance in C57BL/KsJ-db/db mice. Diabetes Res Clin Pract 93:70–76
62. Nwosu F, Morris J, Lund VA et al (2011) Anti-proliferative and potential anti-diabetic effects of phenolic-rich extracts from edible marine algae. Food Chem 126:1006–1012
63. Paradis M-E, Couture P, Lamarche B (2011) A randomised crossover placebo-controlled trial investigating the effect of brown seaweed (Ascophyllum nodosum and Fucus vesiculosus) on postchallenge plasma glucose and insulin levels in men and women. Appl Physiol Nutr Metab 36:913–919
64. Park EY, LKim EH, Kim MH et al (2012) Polyphenol-rich fraction of brown alga Ecklonia cava collected from Gijang, Korea, reduces obesity and glucose levels in high-fat diet-induced obese mice. Evid Based Complement Alternat Med 2012:11
65. Shibata T, Yamaguchi K, Nagayama K et al (2002) Inhibitory activity of brown algal phlorotannins against glycosidases from the viscera of the turban shell Turbo cornutus. Eur J Phycol 37(4):493–500

66. Yeo A-R, Lee J, Tae IH et al (2012) Anti-hyperlipidemic effect of polyphenol extract (SeapolynolTM) and Dieckol isolated from Ecklonia cava in in vivo and in vitro models. Prev Nutr Food Sci 17:1–7
67. Yoon NY, Kim HR, Chung HY et al (2008) Anti-hyperlipidemic effect of an edible brown algae, Ecklonia stolonifera, and its constituents on Poloxamer 407-induced hyperlipidemic and cholesterol-fed rats. Arch Pharm Res 31(12):1564–1571
68. Zhang J, Tiller C, Shen J et al (2007) In vitro potential of Ascophyllum nodosum phenolic antioxidant-mediated α-glucosidase and α-amylase inhibition. Can J Physiol Pharmacol 85:1116–1123
69. Barde SR, Sakhare RS, Kanthale SB et al (2015) Marine bioactive agents: a short review on new marine antidiabetic compounds. Asian Pac J Trop Biomed 5(Suppl 1):S209–S213
70. Benalla W, Bellahcen S, Bnouham M (2010) Antidiabetic medicinal plants as a source of alpha glucosidase inhibitors. Curr Diabetes Rev 6(4):247–254
71. Kim J-S, Kwon C-S, Son KH (2000) Inhibition of alpha-glucosidase and amylase by luteolin, a flavonoid. Biosci Biotechnol Biochem 64(11):2458–2461
72. Park SH, Gammon SR, Knippers JD et al (2002) Phosphorylation-activity relationships of AMPK and acetyl-CoA carboxylase in muscle. J Appl Physiol 92(6):2475–2482
73. Grundy SM, Benjamin IJ, Burke GL et al (1999) Diabetes and cardiovascular disease: a statement for healthcare professionals from the American Heart Association. Circulation 100(10):1134–1146
74. Musunuru K (2010) Atherogenic dyslipidemia: cardiovascular risk and dietary intervention. Lipids 45(10):907–914
75. Rader DJ (2007) Effect of insulin resistance, dyslipidemia, and intra-abdominal adiposity on the development of cardiovascular disease and diabetes mellitus. Am J Med 120(3, Suppl 1):S12–S18
76. National Institutes of Health (1998) Clinical guidelines on the identification, evaluation, and treatment of overweight and obesity in adults: the evidence report. National Institutes of Health
77. National Institutes of Health (2013) Managing overweight and obesity in adults: systematic evidence review from the obesity expert panel, 2013. U.S. Department of Health and Human Services
78. World Health Organization (2015) Cardiovascular diseases (CVDs) Fact sheet No317. September 2016 [Cited 2016 September]. Available from: http://www.who.int/mediacentre/factsheets/fs317/en/
79. Koo SI, Noh SK (2007) Green tea as inhibitor of the intestinal absorption of lipids: potential mechanism for its lipid-lowering effect. J Nutr Biochem 18(3):179–183
80. Sugiyama H, Akazome Y, Shoji T et al (2007) Oligomeric procyanidins in apple polyphenol are main active components for inhibition of pancreatic lipase and triglyceride absorption. J Agric Food Chem 55(11):4604–4609
81. Vidal R, Hernandez-Vallejo S, Pauquai T et al (2005) Apple procyanidins decrease cholesterol esterification and lipoprotein secretion in Caco-2/TC7 enterocytes. J Lipid Res 46(2):258–268
82. Huntinghake DB (1992) HMG CoA reductase inhibitors. Curr Opin Lipidol 3:22–28
83. Ejaz A, Wu D, Kwan P et al (2009) Curcumin inhibits adipogenesis in 3T3-L1 adipocytes and angiogenesis and obesity in C57/BL mice. J Nutr 139(5):919–925
84. Furuyashiki T, Nagayasu H, Aoki Y et al (2004) Tea Catechin suppresses adipocyte differentiation accompanied by down-regulation of PPARγ2 and C/EBPα in 3T3-L1 cells. Biosci Biotechnol Biochem 68(11):2353–2359
85. Huang C, Zhang Y, Gong Z et al (2006) Berberine inhibits 3T3-L1 adipocyte differentiation through the PPARγ pathway. Biochem Biophys Res Commun 348(2):571–578
86. Jeon H-J, Choi H-S, Lee Y-J et al (2015) Seapolynol extracted from Ecklonia cava inhibits adipocyte differentiation in vitro and decreases fat accumulation in vivo. Molecules 20:21715–21731

87. Martirosyan DM, Singh J (2015) A new definition of functional food by FFC: what makes a new definition unique? Funct Foods Health Dis 5(6):209–223
88. Di Stasi SL, MacLeod TD, Winters JD et al (2010) Effects of statins on skeletal muscle: a perspective for physical therapists. Phys Ther 90(10):1530–1542
89. Golomb BA, Evans MA (2008) Statin adverse effects: a review of the literature and evidence for a mitochondrial mechanism. Am J Cardiovasc Drugs: Drugs Devices Intervent 8(6):373–418
90. Kang N-J, Hun S-C, Kang G-J et al (2015) Diphlorethohydroxycarmalol inhibits Interleukin-6 production by regulating NF-κB, STAT5 and SOCS1 in lipopolysaccharide-stimulated RAW264.7 cells. Mar Drugs 13:2141–2157
91. Kellogg J, Esposito D, Grace MH et al (2015) Alaskan seaweeds lower inflammation in RAW 264.7 macrophages and decrease lipid accumulation in 3T3-L1 adipocytes. J Funct Foods 15:396–407
92. Yang E-J, Moon J-Y, Kim SS et al (2014) Jeju seaweeds suppress lipopolysaccharide-stimulated proinflammatory response in RAW 264.7 murine macrophages. Asian Pac J Trop Biomed 4(7):529–537
93. Zaragoza M, Lopez D, Saiz M et al (2008) Toxicity and antioxidant activity in vitro and in vivo of two Fucus vesiculosus extracts. J Agric Food Chem 56:7773–7780
94. Mazzanti G, Menniti-Ippolito F, Moro PA et al (2009) Hepatotoxicity from green tea: a review of the literature and two unpublished cases. Eur J Clin Pharmacol 65(4):331–341
95. Holdt SL, Kraan S (2011) Bioactive compounds in seaweed: functional food applications and legislation. J Appl Phycol 23:543–597
96. Buono S, Langellotti AL, Martello A et al (2014) Functional ingredients from microalgae. Food Funct 5:1669–1685
97. Fern'andez-Mart'ın F, L'opez- L'opez I, Cofrades S et al (2009) Influence of adding sea spaghetti seaweed and replacing the animal fat with olive oil or konjac gel on pork meat batter gelation. Meat Sci 83:209–217
98. L'opez-L'opez I, Cofrades S, Yakan A et al (2009) Composition and antioxidant capacity of low-salt meat emulsion model systems containing edible seaweeds. Meat Sci 83:492–498
99. Choi Y-S, Choi J-H, Han D-J et al (2012) Effects of Laminaria japonica on the physico-chemical and sensory characteristics of reduced-fat pork patties. Meat Sci 91:1–7
100. Prabhasankar P, Ganesan P, Bhaskar N (2009) Influence of Indian brown seaweed (Sargassum marginatum) as an ingredient on quality, biofunctional and microstructure characteristics of pasta. Food Sci Technol Int 15:471–479
101. Chang H, Wu L (2008) Texture and quality properties of Chinese fresh egg noodle formulated with green seaweed (Monostroma nitidum). J Food Sci 73(8):398–404
102. Cofrades S, López-López I, Ruiz-Capillas C et al (2011) Quality characteristics of low-salt restructured poultry with microbial transglutaminase and seaweed. Meat Sci 87:373–380
103. Blandon J, Cranfield J, Henson S (2007) Functional food and natural health product issues: the Canadian and international context department of food agricultural and resource economics. Editor, International Food Economy Research Group
104. Global Nutraceuticals Market – Growth (2016) Trends and forecasts (2016–2021), Research and markets. Editor, United States of America
105. Murray M, Dordevic AL, Ryan L, Bonham MP (2016) An emerging trend in functional foods for the prevention of cardiovascular disease and diabetes: marine algal polyphenols. Crit Rev Food Sci Nutr 58(8):1342–1358
106. Kim SM, Kang K, Jeon J-S, et al (2011) Isolation of phlorotannins from Eisenia bicyclis and their hepatoprotective effect against oxidative stress induced by tert-Butyl hyperoxide. Appl Biochem Biotechnol 165:1296–1307
107. Kim J, Yoon M, Yang H, et al (2014) Enrichment and purification of marine polyphenol phlorotannins using macroporous adsorption resins. Food Chem 162:135–142

Probiotics as Functional Foods in Enhancing Gut Immunity

4

Darshika Nigam

Abstract

Probiotics as microbes when administered in sufficient amounts as functional food impose beneficial effects on gut microbiota and thus enhance health of host. The indigenous microflora of gastrointestinal tract acts as an anatomic barrier against antigens present in food, invading microorganisms which regulates the immunophysiologic mechanism. Many factors may lower the resistance of the body which may lead to inflammatory, infectious, neoplastic, and degenerative conditions. There are other means of treatment like using antibiotics, irradiation, and immunosuppressive therapy which may change normal composition of gut flora. A variety of functional properties of probiotics bound their consideration as conventional, medicinal foods, and dietary supplements.

The most commonly used probiotics are of two genera, *Lactobacillus* and *Bifidobacterium*. Healthy microflora is the chief basis of probiotic therapy in literature. Probiotic bacteria demonstrate various immunomodulatory effects and therefore may be treated as novel tool to reduce inflammation in the intestine and dysfunction of gut mucosa, including acute gastroenteritis, inflammatory bowel disease, and food allergy, and downregulate hypersensitivity reactions. A large number of probiotic effects are explained by regulating immunity, especially the balance between anti-inflammatory and proinflammatory cytokines. Probiotics stabilize microbial environment of the gut and the intestinal permeability barrier. This leads to enhanced mucosal IgA responses which promote further the immunological barrier and responses of gut mucosa. In addition, providing immunomodulatory effect on gut mucosa, probiotic therapy is now also being used to cure infections in other organs such as respiratory tract, urogenital tract, and others. This chapter focuses on roles of probiotics as functional foods.

D. Nigam (✉)
Department of Biochemistry, School of Life Sciences, Dr. Bhimrao Ambedkar University, Agra, Uttar Pradesh, India

V. Rani, U. C. S. Yadav (eds.), *Functional Food and Human Health*,
https://doi.org/10.1007/978-981-13-1123-9_4

Keywords
Probiotics · Functional food · Immunomodulation · Gastrointestinal tract · Respiratory tract · Urogenital tract · Oral infection

4.1 Introduction

In recent years, the beneficiary health reasons of functional food have been felt worldwide. Probiotic is preferred as a new and healthy component in functional food market because of its functional properties and consumer's preference. Today, probiotics as functional food ingredients are being consumed by humans as fermented milks, yogurts, baby foods, energy drinks, confectionery, and chewing gum [1, 2]. Several efforts have been attempted to affect the intestinal microbiota by functional food which is beneficiary for host's health.

Since infectious illness, malnutrition, old age, and stress decline immune system, many natural and chemical products exist with immunomodulatory properties that help in modulating the functioning of immune system under such conditions. Unfortunately, many of such immunostimulatory products leave behind deleterious side effects [3]. The manufacturing of natural food products which bear both immunoenhancing properties and free of side effects would therefore be beneficial to individuals with declined immunity. Consumption of probiotics as functional food is thus one of the most significant benefits to enhance host's immunity [4, 5]. Probiotic organisms are capable of improving human health by modifying the intestinal flora which affects the physiology, metabolism, and pathological process of the host. Some beneficial health effects of probiotics include anticarcinogenic effect, hypocholesterolemic effect, and alleviation of lactose malabsorption and allergy (Fig. 4.1). These effects are because of maintaining the balance between indigenous microbiota and inhibition of pathogenic microbial growth, thereby enhancing the innate and acquired immunity of the host.

The human body is a macrocosm of microorganisms that reside at different body sites. These sites provide an environment where specific microorganisms are more favored than others. These resident microorganisms participate in commensal, mutualistic, and parasitic relationships with the host [6]. The diverse group of commensal (nonpathogenic) bacteria may be differentiated into normal flora (native inhabitants) and transient flora. Native microorganisms colonize specific sites in the human body. Transient flora colonizes the body from the external environment and can persist until some sites are filled with native flora [7]. The normal flora may prevent the colonization of pathogenic microorganisms and thus gives health benefits to host. This phenomenon is called microbial antagonism (Fig. 4.1). When the balance between the normal microbiota and pathogenic microbes is disturbed, it may lead to diseases [8].These microorganisms colonize mainly in the mucosal surfaces of gastrointestinal (GI) tract, the upper respiratory tract, and the urogenital tract, on the skin and oral cavity [9–13]. The present chapter enlightens the relationship between probiotics and immunity at the gastrointestinal tract because probiotics primarily affect the gut and improve its

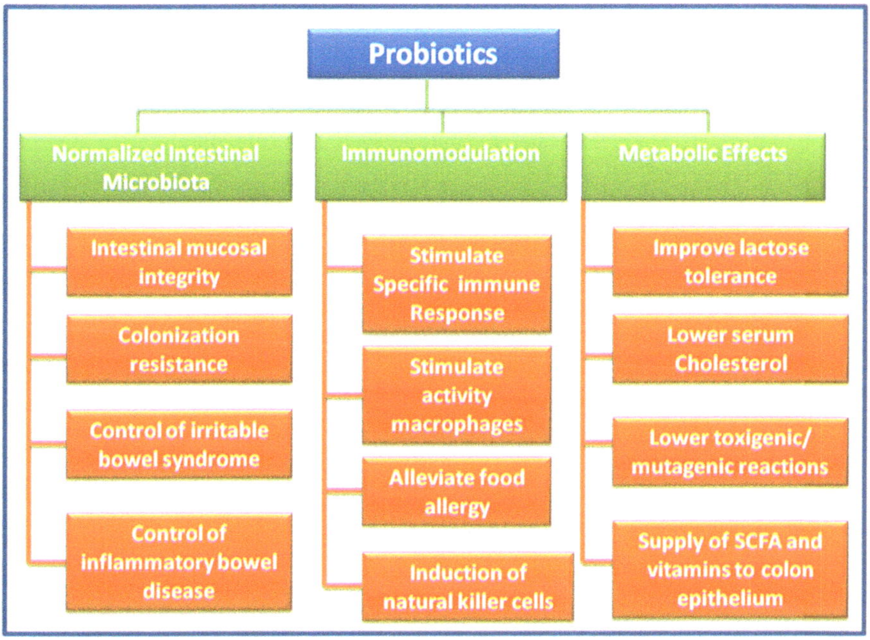

Fig. 4.1 Various roles of probiotics

morphology and functions. However, probiotics also aid in improving immune components at other mentioned mucosal surfaces. The next section of the chapter focuses the same.

4.2 Microflora of Gastrointestinal Tract and Other Organs

4.2.1 The Gastrointestinal Tract

4.2.1.1 Role of Gut Mucosa and Gut-Associated Lymphoid Tissue in Host Immune System

The main role of the GI tract is to digest and absorb the nutrients to fulfill the metabolic and physiological needs for normal growth of human beings. GI tract is constantly exposed to large number of microorganisms. GI tract provides anatomical and physiological barriers against pathogens such as mucus, saliva, stomach acid, digestive enzymes, and intestinal flora [3]. In the tract (GI), many antigens from enteric route are present in the small intestine. Moreover, antigen load composition in the small intestine constantly and rapidly changes. Mucus secreted by goblet cell of the GI tract acts as anatomical or physical barrier and is the first line of defense. It excludes most of the antigens in nonspecific manner [14]. Besides the gut defense, the villous epithelium has a special antigen transport mechanisms. Antigens are moved by transcytosis across the epithelial layer. Lysosomal processing of the antigen occurs in the main degradative pathway. As second line of defense, immune system removes antigens which penetrate the intestinal mucosa and secrete immune products to combat these antigens from microorganisms in the intestinal lumen [3].

In a newborn baby, there is a rapid change in function of gut barrier which occurs at time of birth. This happens when the gut received digesting milk in place of amniotic fluid during early days. Milk consumption starts the releasing of trophic hormones and its motility and absorption. During this postnatal period, the mucosal proteins are formed, digestive enzymes are released, and the presence of intestinal flora [15] strengthened the gut defense. Gastric acidity as first line of defense also starts secreting during the first month after birth [14].

In healthy, adult, and normal human body, the presence of good bacterial populations is able to control overgrowth of opportunistic bacteria in GI tract [8]. Maturation of small intestinal brush border further influences the epithelial cell membranes which is considered as major interface between the intracellular environment and the luminal contents. The ability of antigens to attach with the epithelial cells is dependent on rate and route of antigen transfer. This ability indicates the potency of mucosal immune responses. The intestinal mucosal layer is treated as major immune part of the body that contains various immunocompetent cells [16].

Local specific or adaptive immune system, known as gut-associated lymphoid tissue (GALT), protects the surface of mucosal membrane. GALT is the largest secondary or peripheral lymphoid tissue of the human body. It is divided into the lamina propria, which lies just below the epithelial layer and the organized lymphoid tissues, including Peyer's patches and mesenteric lymph nodes (MLNs) where lymphocytes are scattered throughout the epithelium. The outer mucosal epithelial layer contains intraepithelial lymphocytes (IELs). Many of these lymphocytes are T cells bearing gamma-delta T-cell receptors ($\gamma\partial$ TCRs) which interact with epithelial cells and protect the mucosa by killing infected cells and invite other immune cells to combat pathogens [17]. Further, the IELs mainly exhibit a suppressor and cytotoxic phenotype (CD8+ T cells), whereas the lamina propria cells exhibit a helper and inducer phenotype (CD4+ T cells). The lamina propria is enriched with lymphocytes belonging to the B-cell lineage, T_H cells, and macrophages present in loose clusters or lymphoid follicles [14, 17, 18]. Moreover, lymphocyte activation involves intestinal antigen transport to dendritic cells (DCs) and macrophages in Peyer's patches and present antigens to adjacent T helper and inducer lymphocytes. These cells distinguished into various effector cells which mediate immune suppression and also promote the differentiation of immunoglobulin A (IgA)-secreting B cells [19]. Dietary antigens, such as food proteins and probiotics, are changed into a tolerogenic form when absorb across the intestinal mucosa. This immunologic unresponsiveness toward such antigens is called oral tolerance [20]. The mechanisms of oral tolerance are not fully known. However, the major mechanisms are supposed to carry out through clonal anergy of antigen-specific T cells and B cells. MHC class I-restricted $CD4^+$ T cells and cytokines (interleukin; IL-10 and transforming growth factor; TGF-β) may intervene this mechanism with hyporesponsive functions [21]. In contrast, stimulation of antigen-specific $CD8^+$ T cells that produce inhibitory cytokines helps in gaining the tolerance. In addition of immune tolerance, protective immune responses are also the part of GI tract. The production of secretory IgA (sIgA) is the local immunoglobulin response of the mucosa of the intestine which is induced by multiple cytokines including IL-4, IL-5, IL-6, IL-10, and TGF-β [22]. These cytokines are also essential for maintaining tolerance against food antigen. Probiotics boost IgA antibody responses followed by intestinal immune exclusion and triggering subsequent elimination. Furthermore, probiotic bacteria downregulate hypersensitivity reactions to harmless antigens by modulating the immune responses. The modulation of immune response mechanisms for sIgA activation or tolerance by probiotics is highly dependent on the strains [23, 24]. T helper (T_H) lymphocytes of lamina propria have potential of cytokine production from both subsets, T_{H1} (IL-12, IFN-γ) and T_{H2} (IL-10, IL-4, IL-5), that express different immune responses [25].

The certain B lymphocytes appear in the blood 2–4 days after antigenic exposure. It reaches the highest concentration after 6–8 days and persists in the blood for 2–3 weeks. Mucosal surfaces have IgA antibody production in abundance [14, 23]. However, upper part of the human gut-associated immune system (small intestine) secretes IgA1-producing cells predominantly, while IgA2-producing cells are present in lower-part gut-associated immune system (colon). IgA2-producing cells are more resistant to bacterial proteases [23].

4.2.1.2 Intestinal Microflora and Development of the Host Immune System

To maintain good health, it is desired to have a normal and functional GI tract. The GI tract has the second largest surface area after surface area of respiratory tract, which is rich in flora containing more than 1500 various species of bacteria [4, 26]. Microflora of the GI tract is critical in the development of the host's anatomy, immunology, and physiology. It aggravates the immune system to react rapidly to infection caused by pathogens and through bacterial antagonism. It competitively inhibits the colonization of the gut by pathogenic bacteria and helping digestion [27, 28].

The gut flora develops after birth and remains almost stable for rest of the life which is for human homeostasis. The major challenge with the mucosal immune system is to differentiate between pathogens and non-pathogenic microorganisms to develop protective immunity and to maintain the integrity of the GI mucosa [5, 29].

The GI tract of a newborn is almost sterile. As infant exposes with environment, indigenous microflora colonizes the mucosal surfaces. This differs from the adult microflora [30]. The microflora expands quickly after birth and is markedly dependent on mode of delivery, hygiene level, mode of feeding, the mother's flora, genetic factors, and medication use. Infants born through vaginal delivery develop microbial colony in gut from the birth canal of the mother and the environment. These microorganisms include streptococci, coliforms, and gram-positive, nonspore-forming anaerobic bacilli [31]. An intestinal tract of infant is quickly populated with enterobacteria during the first 2 days after birth. In breastfed infants, 60%–90% of the total flora is of bifidobacteria counts. The most common species in healthy breastfed infants are bifidobacterial species which includes *Bifidobacterium bifidum*, *Bifidobacterium infantis*, *Bifidobacterium breve*, and *Bifidobacterium longum* which predominate *Lactobacillus* and *Bacteroides* increase to a smaller degree. Enterobacteria reduce in these infants [29, 32].

However, large numbers of *Lactobacillus*, *Clostridium*, and *Bacteroides* and relatively less *Bifidobacterium* species are found in formula-fed infants. In both breastfed and formula-fed infants, the microflora becomes similar when food supplements are started. *Bacteroides* and anaerobic gram-positive cocci predominate in this state [33]. Populations of *Bacteroides* and anaerobic cocci increase and sometimes exceed to *Bifidobacterium* population after the infant reaches 2 years of age, which is same as normal adult's flora. The gram-negative anaerobic counts increase, while coliforms, clostridia, and streptococci reduced to the levels as found in healthy adults. Permanent colonizing of bacteria is attained in children by 4 years of age. Infection, illness, stress, climate, changes in diet, and medication may lead to changes in these levels and physiological progression process [26, 34].

Though the composition of the microflora varies among individuals, microfloral composition in individual remains steady over a long period. The antigenic stimuli provided by colonization of commensal bacteria in gut are crucial for the development of functionally matured and balanced immune system [27]. This includes homing of B and T lymphocytes to the lamina propria, development of IgA plasmocytes, IgA secretion, and induction of tolerance toward food and microfloral antigens [10]. The microflora constitutes a large number of diverse species; many of them are recognized as pathogens. Approximately 80% of all immunologically active cells in body are present in the GALT. This is because of the development of interaction between microbe and gut immune system. Thus, microbial flora is vital for the development of immunocompetent GALT in infants [8, 27]. Intestinal bacteria are the main cause of epithelium cell functions, signaling through toll-like receptors (TLRs), which aid in T-cell activation and differentiation and B-cell responses to T-cell-dependent antigens and thus regulate the gut immune response [25]. sIgA is one of the most important components of B-cell response to gut lumen protein and pathogen antigens. Colonization also reduces the ratio of T_{H2} (proallergic) to T_{H1}(suppressive) responses that could decrease the possibilities of immune hyperreactivity, such as in allergic diseases [28, 35].

Resident bacteria, like lactobacilli and bifidobacteria, can activate antimicrobial activities by inhibiting colonization of potentially pathogenic microbes and thereby influencing both local and systemic immunities [32]. They have also been associated with the release of substances which have antimicrobial properties and the release of mucins (considered as the intestinal physiological barrier). Mucins inhibit the attachment of pathogens to mucosal lining [36]. Some bifidobacteria and lactobacilli are given orally to develop immunologic tolerance to antigens and can decrease allergic-type immune responses by enhancing the production of a balanced T_H cell response and stimulating the production of IL-10 and TGF-β [32, 37]. Another major effect of gut bacteria is the boosting of secretion of sIgA at mucosal surface and thus helps in providing protection against antigens, virulence factors, and toxins [22]. The development of IgA plasmocytes in the mucosa is most affected by the microflora. Breast milk contains sufficient amount of sIgA which is fed to the infant. Additionally, bifidobacteria stimulate the biosynthesis and secretion of sIgA [38].

Macrophage infections are effective inducers of IL-12 and IFN-γ production, whereas extracellular infections, for example, by intestinal parasites, are strong inducers of IL-4 and IL-10 production which effectively enhance cellular immune responses. The mucosal flora has the capacity to restrict or kill certain transient microbial pathogens in their habitat by competing for nutrition and releasing of suppressive factors like bacteriocins in a process called microbial interference whose mechanism needs more research [28]. Bacteriocins are proteins produced by endogenous flora which is a suitable example of intraspecies antagonistic effects [10].

4.2.1.3 Colorectal Cancer

Colorectal cancer (CRC) is one of the most common causes of cancer mortality worldwide. The development of colorectal cancer is a multifactorial process influenced by physiological, environmental, and genetic factors and to a large extent by lifestyle including diet [39]. Due to genetic mutations, thousands of abnormal cells are generated daily in our body. Our immune system clears these mutated cells to suppress the carcinogenesis. Gut flora may influence the defense against CRC by modulating immune components of the host [40–42]. Studies on fecal bacterial composition showed increased population of *Bacteroides* and *Prevotella* in CRC patients, while *Dorea* spp. and *Faecalibacterium* spp. in colonic microbiota in patients with colorectal adenoma. These species may generate carcinogens and tumor-promoting substances including heterocyclic amines and secondary bile acids [43, 44]. Other studies suggested the less diverse or altered bacterial community or high colonization of oral anaerobic bacteria *Fusobacterium nucleatum* in CRC [45]. Other gut microbiota produces beneficial metabolites such as short-chain fatty acids and is equal to prevent cancer [46, 47]. Studies investigate that the supplementation with symbiotic composition of *Lactobacillus rhamnosus* GG, *Bifidobacterium lactis* Bb12, and oligofructose-enriched inulin for 12 weeks resulted in favorable changes in the gut microbiome with high levels of lactobacilli and bifidobacteria and low levels of *Clostridium perfringens* in colorectal cancer patients [48, 49]. Intake of synbiotics also helps in reducing proliferation and DNA damage in colonic mucosal cells and fecal water-induced necrosis in colonic cells of the patients [50]. Several studies have demonstrated that *Lactobacilli casei* enhances the immunoresponsive activities of T cells, macrophages, natural killer cells, and T cells against cancer [51–53]. Probiotics have also been able to reduce side effects of radiotherapy and chemotherapy to treat CRC. In animal studies, *L. rhamnosus* GG is found to ameliorate intestinal damage from radiation in a TLR2- and COX2-dependent and MyD88-independent manner [54, 55].

4.2.1.4 Adhesion of Probiotics to Target Sites

A major selection criterion for probiotics is considered to be exclusion of pathogens competitively. Probiotics compete directly or obstruct the adhesion sites of pathogens present on gastrointestinal surface. They also affect the development of intestinal microbiota in infants. Adhesion on GI surface increases the retention time of a probiotic which is important because of the short residence time of intestinal material in it [56]. Probiotics may have a high turnover rate on the mucosal surface because of continuous displacement [28]. On the other hand, a probiotic also penetrates the mucosal layer to adhere with the epithelial cells. Once consumed, probiotics travel along the whole length of GI tract. An effective probiotic is that which must stay on desired target sites for sufficient time with sufficient concentration to obtain probiotic effects [57]. The adhesion and temporary multiplication of probiotics at the target sites result in high counts of probiotics at site of action. This achieves the desirable response even at a lower dosage [58].

As probiotic supplements, lactobacilli and bifidobacteria primarily populated at the small and the large intestine [10]. Both are capable to change immunologic responses related to allergic inflammation. Lactobacilli are investigated as ineffective against allergy generated by cow's milk [24]. Binding of probiotics preferentially on the specific antigen-processing cells including dendritic cells, macrophages, and epithelial cells may be proved as more vital than the site of adhesion [59]. *L. rhamnosus* GG has been reported effective in treating diarrhea in infants caused by rotavirus which adhere and colonize to the small intestine [57, 60]. Likewise, *Bifidobacterium lactis* Bb12 is effective in preventing and treating acute diarrhea in infants as it has strong adherent property, and *Lactobacillus bulgaricus* adhere to and colonize the intestine and help in treating antibiotic-associated diarrhea [58, 61].

4.2.2 Respiratory Tract

The most common illness among humans is viral respiratory tract infections. Probiotics may have therapeutic efficacy for treating viral respiratory tract infections as they are known to improve the immune system [62, 63]. It has been shown that *Bacillus coagulans* significantly induced TNF-α production by T cells when healthy adult is exposed to adenovirus exposure and influenza A (H3N2 Texas strain); however, *B. coagulans* did not exert much effect on other influenza strains [64]. Probiotic strains of *Lactobacillus plantarum* and *Lactobacillus paracasei* reduce the risk of acquiring common cold infections when orally administered [65].

Patients who received external ventilation are at risk to have ventilator-associated pneumonia (VAP). VAP has complex pathogenesis. This develops colonies of pathogenic bacteria in the aerodigestive tract which lead to formation of biofilms and microaspiration of contaminated secretions. Since VAP is associated with high risk of mortality and morbidity [66], therefore, new efficient VAP prevention strategies work on combating colonization and effective aspiration. It includes elevation of the head of patient's bed to drain subglottic secretion. To reduce the risk of VAP, intensive oral care and duration of mechanical ventilation should be minimized [67]. Probiotics help in reducing the occurrence of VAP through local and systemic immune defenses [68, 69]. These effects are reduction in colonization of potential pathogens, improvement in gut mucosal barrier function, reduction in bacterial translocation, and TLR-mediated enhancement of immune response. However, evidences for such prophylactic treatment through probiotics are limited but promising. *Lactobacillus rhamnosus* GG administration appears safe and effective in a patient who is at high risk for VAP [69]. The therapy may be used to avoid ICU complications like *Clostridium difficile* and ICU-associated diarrhea [70].

4.2.3 Urogenital Tract

Urinary tract infection (UTI) is the most common health problem especially in women. Members of *Enterobacteriaceae* family including *E. coli*, normal inhabitants of human intestines, are found to be the most common causative agent of UTI in several countries [71]. There is a close relation between reduction of the normal genital microflora, mainly *Lactobacillus* species and higher incidence of genital and bladder infections [72]. Various bacterial species responsible for UTI are *E. coli*, *Staphylococcus saprophyticus*, *Proteus*, and *Klebsiella*. Viruses, fungi, and parasites can also cause UTI but rarely. Although antimicrobial drugs are generally effective in eradicating the infections, chances of recurrence is high [71]. Modern concept of treating UTI is based on the use of *Lactobacillus* species which produces hydrogen peroxide, lactic acid, and bacteriocins and thereby maintains low pH of the genital area and retards growth of *E. coli*. Lactobacilli also activate TLR-2 which produces IL-10 and myeloid differentiation factor 88 [73].

The vaginal microbiota protects the area from invading pathogens that cause urinary tract infections, sexually transmitted diseases, and other diseases. *Lactobacillus acidophilus* group and *L. fermentum* are dominant in this habitat in healthy premenopausal women. *L. plantarum*, *L. jensenii*, *L. brevis*, *L. casei*, *L. salivarius*, and *L. delbrueckii* are also present [74]. Lactobacilli are supposed to interfere with pathogens by different mechanisms. The primary mechanism is competitive elimination of genitourinary pathogens from the epithelial surface receptors, whereas coaggregation of lactobacilli with some uropathogenic bacteria would result in the growth retardation of the pathogen which is the second mechanism [13, 75]. The process of coaggregation is linked to the production of antimicrobial compounds, such as lactic acid, hydrogen peroxide, and bacteriocin-like substances. In this respect, previous studies showed that *Candida albicans* and *Gardnerella vaginalis* adhere to vaginal epithelial cell surfaces and produce pathology primarily at the vaginal level. *L. acidophilus* and these pathogens compete for the same binding sites on receptors present on vaginal epithelial cells and therefore prevent pathogenesis by these microorganisms [76, 77]. The affinity of *L. acidophilus* for these receptors is stronger than the pathogens. On the other hand, *E. coli* and *Streptococcus agalactiae* do not adhere to the vaginal epithelial cells and thus are just opportunistic pathogens [13]. Except *S. agalactiae*, vaginal lactobacilli coaggregate with all of the pathogens. This suggests that coaggregation is specific because this process also hinders the access of pathogens to surface receptors present on vaginal epithelia which is another reason for the reduced adherence of *C. albicans* and *G. vaginalis* to vaginal epithelia in the presence of lactobacilli [13].

4.2.4 Oral Infection

In oral cavity, many resident bacteria include lactobacilli, streptococci, staphylococci, corynebacteria, and various anaerobes particularly bacteroides and spirochetes. The oral cavity of newborn infant rapidly becomes colonized with bacteria such as *Streptococcus salivarius* [11]. After eruption of teeth during the first year, *Streptococcus mutans*, *Streptococcus sanguinis*, and *Streptococcus gordonii* are major species which colonize the tooth surfaces, coronal plaque, in gingival crevices (supporting structures of the teeth), and on buccal and pharyngeal mucosal surfaces since the childhood [78]. They may also be found in abscesses and are etiological agents in bacterial endocarditis. During puberty bacteroides and spirochetes such as *Treponema denticola* colonize the oral cavity [79]. At all sites of colonization or infection, adherence is a prerequisite for streptococcal proliferation. In vitro adherence assays have demonstrated that *S. gordonii* has diverse adherence abilities since these bacteria adhere to surfaces coated with salivary components, such as mucins, proline-rich proteins, agglutinin, and fibronectin and bind to amylase, immunoglobulin A, and serum agglutinin [80].

Many studies suggest that probiotic *Lactobacillus reuteri Prodentis* could be effective in the treatment of both gingivitis and plaque because of its anti-inflammatory and antimicrobial effects [81, 82]. Oral administration of *L. reuteri Prodentis* improves initial-to-moderate chronic periodontitis [83]. *L. reuteri* also found high salivary *Streptococcus mutans* counts in young women [84].

4.3 Immunomodulation by Probiotics

Probiotic bacteria are shown to elevate the endogenous host defense mechanisms. In addition to stabilization of the gut microflora (nonimmunologic gut defense), probiotic bacteria enhance humoral immune responses and thus promote the immunologic barrier of the intestine [26].

Probiotics modulate non-specific or innate immunity in various ways including decreasing gut permeability, promoting production of mucin, competing with and inhibiting growth of opportunistic pathogens and increasing cytotoxic activity of natural killer cell, and activating macrophage and its phagocytic activity [36, 59]. Probiotics enhance specific or adaptive immune response by modulating inflammatory gut immune responses and raise IgA-, IgG-, and IgM-secreting plasma cells. Probiotics also increase total and specific sIgA in serum and gut lumen [38].

Oral supplements of lactobacilli can increase innate immunity of host against microbial pathogens and thereby facilitate the elimination of pathogens in the gut. A number of strains of live lactic acid bacteria (LAB) are found to stimulate the release of the proinflammatory cytokines such as TNF-α, IL-6, and IL-10 which reflect stimulation of nonspecific immunity [85]. Oral intake of *Lactobacillus casei* and *L. bulgaricus* activates the biosynthesis of macrophages, while phagocytic activity is stimulated by *L. helveticus* and *L. acidophilus*. Phagocytosis triggers the inflammatory response before antibody production by releasing lethal agents including reactive oxygen and nitrogen intermediates and degradative enzymes [37, 52, 86]. Probiotics are capable of modulating phagocytosis differently in healthy and allergic persons: in healthy people immunostimulatory effects have been detected, whereas in allergic persons, downregulation of inflammatory response was shown [87]. In rotavirus diarrhea, proinflammatory mediator fecal urease activity gets increased that leads to ammonia-induced destruction of gut mucosa and thus promotes overgrowth of urease-producing bacteria [88]. In patients with rheumatoid arthritis, bacterial composition is altered in gut as compared to healthy persons. This suggests that the intestinal microflora is responsible for inflammation beyond the gut [89]. Therapeutic use of probiotics thus may help to stabilize the gut microflora and thereby prevent the generation of inflammatory mediators by GALT, which are triggered in response to exposure of intestinal lumen by potentially harmful antigens [26, 87]. Proinflammatory cytokines such as IL-6, TNF-α, and IFN- γ may play a key role in inflammation and also stimulate production of adaptive immune cells. Intake of lactobacilli in fermented milk products or as live-attenuated bacteria potentiates production of IFN- γ by peripheral blood mononuclear cells. IFN promotes the uptake of antigens in Peyer's patches, where IgA-committed cells are generated in follicles [90, 91]. Oral administration of *Lactobacillus rhamnosus* GG (LGG) was shown to reduce high fecal concentrations of TNF-α in patients suffering from atopic dermatitis and cow milk allergy [35, 92]. In this way, probiotic bacteria may stabilize the immunologic barrier of the gut mucosa by reducing the generation of local proinflammatory molecule TNF-α and by reinforcing the systemic production of IFN. However, abnormal production of IFN interferes with the oral tolerance and disturbs gut epithelial cell integrity. Thus, immunomodulating effects shown by probiotics may depend on the immunological status of the host [20, 21].

It has been postulated that partial lactose digestion and stimulation of the intestinal lactase activity act as a mechanism against some types of diarrhea. In case of acute gastroenteritis or recurrent abdominal distress, disaccharidase activity in the small intestine produces osmotic diarrhea as transport of monosaccharides is affected [93]. Lactobacilli have active β-galactosidase which is used to decrease the lactose concentration in dairy products that may improve osmotic diarrhea due to pathogenic organisms [88, 94]. Many metabolites produced by LAB act as antimicrobial substances, such as organic acids, hydrogen peroxide, ammonia, free fatty acids, and bacteriocins. These substances are used in dairy to extend the shelf life of food and to suppress food spoilage [95]. *L. casei* strain GG (LGG) which has unique colonial morphology that makes it easy to identify in a mixed culture of other lactobacilli and *Streptococci* has the ability to produce a low-molecular-weight antibacterial substance that inhibits both gram-positive and gram-negative enteric bacteria in mice [96]. Similarly, it has been observed that IgA-specific antibody-secreting cell counts are increased in most patients suffering with rotavirus diarrhea when they received viable LGG at the convalescent stage [60].

Another mechanism used by probiotics is to block toxin-mediated diseases by modifying toxin receptors and. For example, *S. boulardii* disintegrates *Clostridium difficile* toxin receptors in the human colon and blocks cholera-induced secretion in rat jejunum by the production of polyamines [97, 98]. It has been shown that LGG and *L. plantarum* competitively inhibit the adhesion site of enteropathogenic *E. coli* 0157H7 to human colonic cancer cell HT-29 [99]. Similarly, *S. boulardii* also decreases attachment of *Entamoeba histolytica* trophozoites to erythrocytes in vitro [100]. Different strains of LAB stimulate production of

IFN-γ, IL-12, and IL-18 by human blood lymphocytes [85]. Mucosal-associated lactobacilli (mainly *L. paracasei*) can translocate over the gut barrier and enable to influence gut mucosal immune by strongly stimulating IL-12 secretion which in turn stimulates cytotoxic activity of T cells and natural killer cells. However, IL-12 may also downregulate the T_H-2 response, decreasing IL-4 and IgE production [59].

Whole bacterial cells are able to induce proinflammatory cytokines, such as TNF-α and IL-6 as well as accelerate proliferation of immune cells [101]. Contrary to this, probiotic bacteria mediate suppress cytokine production and lymphocyte proliferation which indicates that probiotic bacteria possess considerable anti-inflammatory properties as good as therapeutic efficacy. For example, *Feacalibacterium prausnitzii* suppresses release of proinflammatory cytokines IL-12 and IL-17 and T_H-17 cells which play role in inflammatory bowel disease [102, 103]. Oral administration of LGG has been shown to elevate serum IL-10 levels in atopic children. This implies that specific probiotics may exhibit anti-inflammatory effects which may be mediated through modification in intestinal microfloral niche [32, 57].

Bifidobacterium longum was shown to increase antibody response when orally incorporated in patients of allergen ovalbumin-induced lung inflammation and also stimulates IgA response to cholera toxin [104, 105]. Likewise, humoral immune response (rotavirus-specific IgA) was increased by LGG administration in children who were suffering with acute rotavirus diarrhea [106, 107]. Cow milk proteins cause type 1 hypersensitivity and lead to defective generation of local IgA responses in human infants [34]. In addition to certain gastrointestinal microfloral species which release low-molecular-weight peptides, incorporation of lactobacilli and bifidobacteria in diet stimulates immune responses through several mechanisms. However, in healthy individuals, milk protein degradative enzymes are released from intestinal bacteria into tolerogenic peptides and thereby exert suppressive effects on the lymphocyte proliferation and downregulate cow's milk allergy [108].

4.4 Sources of Probiotics as Functional Food

Yogurt, cheese, cultured buttermilk, fermented milk, and ice cream are the main sources of probiotics. Although many other nondairy probiotic food products like Japanese miso, kimchi, sauerkraut, pickles, tempeh, bread, sour dough, chocolate, olives, beer, and soy-based drinks are produced by bacterial fermentation, yogurts and fermented milks are still dominant medium as probiotics. It is because they provide a relatively low pH environment suitable for probiotic bacteria to survive [109]. Probiotic strains are also found in nondairy fermented foods like soy-based products, cabbage, legumes, sorghum, pearl millet, cereals, maize, etc. [109, 111]. Table 4.1 demonstrates various probiotic-based functional foods. Probiotic soy products, such as soya yogurts, beverages, and fruit juices, are lactose-intolerant.

Table 4.1 Various probiotic-based dairy and nondairy products

Cultured dairy/nondairy product	Beneficial bacteria
Yogurt	*L. bulgaricus, L. acidophilus, S. thermophilus*
Cheese	*L. casei, L. brevis, L. lactis, L. acidophilus, L. plantarum, S. cremoris, S. faecium*
Buttermilk	*S. cremoris*
Fermented probiotic milk	*L. casei, L. acidophilus, L. rhamnosus, L. johnsonii, B. lactis, B. bifidum, B. breve*
Fermented sausages	*P. acidilactici, P. pentosaceus, L. sakei, L. curvatus*
Fermented pickles	*Leuconostoc mesenteroides, Pediococcus cerevisiae, L. brevis, L. plantarum*
Fermented olives	*Leuconostoc mesenteroids, L. plantarum, L. pentosus*
Sauerkraut	*P. acidilactici, L. plantarum*
Fermented soy	*L. casei, L. acidophilus*
Brown rice	*L. acidophilus, L. plantarum, L. salivarius, B. lactis*
Fermented beverage	*L. delbrueckii, L. acidophilus*

Although the wide variety of probiotics is being used as functional food, the following criteria must be considered while selecting probiotic strains:

- Selection of nontoxic and nonpathogenic organism [111]
- Isolation from the same species as its targeted host [112]
- Ability to be tolerant of acid and bile at time of transit through the GI tract and not conjugating with bile acids [112, 113]
- Not carrying transferable antibiotic resistance genes and susceptible to antibiotics.
- Ability to adhere and colonize the intestinal epithelium or epithelium of other organs [110, 112]
- Ability to stabilize the normal intestinal microbiota [114]
- Ability to produce antimicrobial substances to antagonize pathogens [115]
- Potential to show beneficial effect on the host [116]
- Durability to endure the complexity of commercial handling [113]
- Possessing pleasant odor, flavors, and smooth textures [117, 118]
- Abovementioned criterion is useful for selecting probiotic strains, especially at the time for preparation of function food processing and commercial packaging. It has been identified in researches that viability of probiotics during food processing on commercial scale is an important thought for health benefits which is mentioned in the next section of the chapter.

4.5 Viability of Probiotic in Probiotic Foods

Several reports show that many manufactured health products available in the market worldwide have poor viability of probiotic than mentioned and claimed on label of packaging. As the shelf life is less, it is important to make the product sustainable in market. Therefore, it becomes important to ensure higher viability of probiotic and their ability to show probiotic effect for their long-term existence as functional foods.

Study and market survey reveal the functional foods, and other health-care products had either undetectable level or very low microbial contents. Sometimes, the identified strains in the packages do not match in quantity as declared on the label. Many surveys reveal that a number of products for which companies claim to be "probiotic" often do not fulfill the standard criteria for viable count at end of shelf life. Seeing the situation, International Scientific Association for Probiotics and Prebiotics decided that the term "probiotic" should be referred only to the products which have live and well-defined strains of microorganisms with sufficient counts [119].

Probiotic-based milk products must maintain suitable population of viable microorganisms during whole shelf life to exhibit prophylactic properties [120]. The biotherapeutic effect of probiotic is dose-dependent with a daily recommended dosage (106–109 cfu/ml) [121] to balance viability loss due to heat, pressure, and high acidity [122] and to sustain dosage up to 21 days/5 °C [30]. Fecal lactobacilli and bifidobacteria are elevated by ingesting various cultured milk products which contain 106–109 cfu/g that declines coliforms and consecutively alleviates certain human diseases. Viability of LAB is influenced by many physiological factors including gastric acidity, bile salts, and digestive enzymes [123]. It is observed that only 20–40% of the cultures survive in gastric transit in gut environment. High tolerance to low pH and bile salts is exhibited by *L. plantarum* G1, and *L. casei* G3 strains in the GI tract of rats suggest the use of these strains as functional food [124]. Thus, it is significant to retain viability of probiotic strains in all phases, i.e., during processing, storage, and transit, through GIT to manifest clinical effects, but researches indicate that even in nonviable or heat-killed probiotic strains, components of dead probiotic cells from culture can confer anti-inflammatory responses [124–127]. It is also noted that resting cells and dead cells of lactobacilli and *Bifidobacterium* strains can be able to remove cholesterol from a medium [128].

There exist many factors affecting the viability of probiotic strains in yogurt during all phases of manufacturing, storage, and GIT transit.

- Amount of acids and other substances like hydrogen peroxide produced by cultures in yogurt.
- Amount of dissolved oxygen contents in the product.
- Amount of acids such as lactic acid and acetic acid in the product.
- Quantity of fat in milk.
- Heat treatment of milk during process.
- Temperature of incubation.
- Concentration of buffers, for instance, whey protein concentrate.
- Physiological condition of probiotic cultures.
- Oxygen permeability through the packaging of the final probiotic product.
- Physical status of product storage.

4.6 Existing Regulations for Probiotics

Growing globalization of probiotics commercially resulted in globally accepted standards to ensure quality- and viability-based probiotic products to consumers as functional food and drugs. Table 4.2 demonstrates the complete summary of various classifications and standards in major country worldwide. In India, probiotics are standardized as functional food, not as pharmaceutical drugs, and are regulated by the Food Safety and Standards Act (FSSA) of 2005. According to FSSA, functional foods are defined legally, but in the categorization of food, such as nutraceuticals, biotherapeutic agent is not clear. The Prevention of Food Adulteration Act (PFA)

Table 4.2 Summary of standards adopted by major countries worldwide

Country	Mode of intake	Regulatory body	Definition key points
USA	Dietary supplements	DSHEA	Intended to supplement the diet, taken as any form [132, 139]
	Drugs	FDA	Intended for the prevention, alleviation, cure, diagnosis of disease [140]
	Biological products	BLA	Containing a virus, serum, toxin for prevention and treatment [141]
	Medical food	FDA	Dietary management of a disease, medical evaluation, supervision under physician [142]
	Live biotherapeutic	FDA	Live organisms, such as bacteria; for prevention and treatment of a disease and not vaccine [142]
India	Functional food and drugs	FSSA, PFA, FDA	Physiological functions, regulation of biorhythms, nervous system, the immune system, and defense beyond nutrient facts [129–131, 143]
China	Functional food	SFDA	Health beneficial and able to regulate health body functions [144]
Japan	Probiotic	FAO/WHO	The live microorganisms, administered in sufficient amounts for health benefit [145]
	Functional food and nutraceuticals	MHLW/ FOSHU	Products with different category as food for certain foods, with a regulatory process boosted with vitamins, minerals, and other supplements not carrying FOSHU claims, herbal supplements [136, 146]
Europe	Functional food	FUFOSE	Beneficially affects one or more functions and consumed as improved state of health and reduction of risk of disease not as medicine [147]
New Zealand and Australia	Functional food	FSANZA	Physiological roles beyond simple nutrient requirement [148]
Brazil	Functional food	ANVISA	Healthy food, physiological function, enhanced with added ingredients than normal food, beyond nutritional value [138]

Rules which decides minimum standards related to quality and content for food products regulates labeling and packaging of food products including the ingredients, date of expiry, nutritional information, manufacturing details, etc. [129, 130]. A task force was constituted by the Indian Council of Medical Research (ICMR) and Department of Biotechnology (DBT) to frame regulatory guidelines and to evaluate and set parameters to define a product/strain of probiotics. These guidelines are dealing with the use of probiotics as functional food and safety requirements associated with [131].

The USA regulates probiotics as intended usage and dealing bodies such as the Dietary Supplement Health and Education Act (DSHEA) and Food and Drug Administration (FDA). In the USA, FDA enlists microorganisms and their secreting substances, used as probiotics and in foods. DSHEA regulates a food in which probiotic is used in the form of dietary supplement. The probiotics as drugs for therapeutic purposes are regulated by FDA for ensuring safe and effective usage [132]. In case of biological products, an approval from Biologic License Application (BLA) is applied in the USA. In the USA, National Yogurt Association (NYA) forces to use "live and active culture seal." It is mandatory to declare the bacterial genera and species on the labels of probiotic-containing food products including amount of live and active cultures; however differentiation between LAB and other probiotic bacteria or assurance of viability of cultures at end of shelf life is not provided.

In the European Union, microbial cultures present in food need qualified presumption of safety (QPS) assessment in foods and food supplements [133]. European regulatory framework is still not considered as probiotics as food supplement or as are regulated by the Food Products Directive and Regulation (Regulation 178/2002/EC; Directive 2000/13/EU) and under as traditional herbal products by Herbal Medicinal Products Directive (2004/24/EC). Finally, probiotics as herbal medicinal product and registered drugs are covered under the Drug Law (65/65/EC, amended) [134, 135].

As per Japanese regulations, these probiotic products are listed in separate category of foods by the Foods for Specified Health Uses (FOSHU). The Ministry of Health and Welfare (MHLW) allows mentioning the efficacy claims of probiotic food product labels only through special permission [136]. For adopting the FOSHU label, the product must contain dietary ingredients with beneficial effects on the physiological functions and enhance and improve health issue. However, FOSHU does not allow claims of disease-risk reduction on the labels.

In China, the State Food and Drug Administration (SFDA) is the body for regulating functional foods and nutraceuticals in China. The functional food is considered as food which has special health functions or is able to supply mineral and vitamin health benefits [137].

Brazil becomes the first country to issue legislation for functional food. Probiotics are considered as functional foods. It considered being different from food and legislation address for safety and efficacy labels on food products by registering and approving health authority called National Health Surveillance Agency Brazil (ANVISA) [138].

4.7 Conclusion

It is concluded that probiotic bacteria have various immunomodulatory effects and therefore may be treated as effective means to examine intestinal inflammation and gut mucosal dysfunction including acute gastroenteritis, inflammatory bowel disease and food allergy and downregulate hypersensitivity reactions. Several probiotic effects are exhibited through keeping a balance between secretion of proinflammatory and anti-inflammatory cytokines which play indispensable role in immune regulation. Probiotics may help to balance the gut microbial milieu and mucosal permeability barrier and increase systemic and mucosal IgA immune reactions, therefore mediating the immunological barrier of the gut mucosa. In addition, providing immunomodulating effect on the gut mucosa, probiotic therapy is now also being used to cure infections in other organs such as respiratory tract, urogenital tract, and others.

It is further noted that probiotics as functional food do not exist not only as human health beneficial agents as there are evidences in literature that exist for their side effects. Although the present products in the market are excellent for human health, still there is scope for researchers to study the safety aspects involving the use of probiotics as functional food. Many epidemiological and clinical studies are done to ensure and monitor on consumer safety and their nutritional aspects too. Most of the available probiotics are "generally recognized as safe" (GRAS), but their selection and scrutiny are crucial, especially in the case of patients of immune disorders, GI disorders, or any other critical illnesses. Hence, the focus is needed to be given on identification of responders and non-responders, determining effective size, identifying strain-specific effects, and determining mechanisms to recommend future dietary usage.

The whole process of probiotic manufacturing and its trading also plays an important role in health beneficiary efficacy. Therefore, almost all countries have adopted standards for probiotic products in global markets. The viability and success of probiotic in the future as functional foods for consumers depend on many factors. Consumer's acceptance and choice of such products are the main issues. It is important to establish the effectiveness of probiotic products and the claims of manufacturers through clear, truthful, and unambiguous information on the labels of the products.

References

1. Figueroa-Gonzalez I, Quijano G, Ramirez G et al (2011) Probiotics and prebiotics-perspectives and challenges. J Sci Food Agric 91:1341–1348
2. Al-Sheraji SH, Ismail A, Manap MY et al (2013) Prebiotics as functional foods: a review. J Funct Foods 5:1542–1553
3. Ernst PB, Scicchitano R, Underdown BJ et al (1988) Immunology of the gastrointestinal tract and liver. In: Heyworth MF, Jones AL (eds) Raven Press, New York
4. Lee YK, Salminen S (2009) Incorporating probiotics into food. In: Crittenden R (ed) Handbook of probiotics and prebiotics, 2nd edn. Wiley, Hoboken, pp 60–67

5. Isolauri E, Sütas Y, Kankaanpää P et al (2001) Probiotics: effects on immunity. Am J Clin Nutr 73(2):444S–450S
6. Simon SL, Gorbach SL (1986) The human intestinal microflora. Dig Dis Sci 31(9):147S–162S
7. Ouwehand AC, Salminen S, Isolauri E (2002) Probiotics: an overview of beneficial effects. Antonie Van Leeuwenhoek 82(1–4):279–289
8. Neish AS (2009) Microbes in gastrointestinal health and disease. Gastroenterology 136(1):65–80
9. Grice EA, Kong HH, Conlan S (2009) Topographical and temporal diversity of the human skin microbiome. Science 324:1190–1192
10. Pickard KM, Bremner AR, Gordon JN et al (2004) Microbial-gut interactions in health and disease: immune responses. Best Pract Res Clin Gastroenterol 18:271–285
11. Rogers AH (ed) (2008) Molecular oral microbiology. Caister Academic Press. 1. ISBN 978-1-904455-24-0
12. Proctor DM, Relman DA (2017) The landscape ecology and microbiota of the human nose, mouth, and throat. Cell Host Microbe 21(4):421–432
13. Soledad B, Juan E, Fernando V'ZSR et al (1998) Adherence of human vaginal lactobacilli to vaginal epithelial cells and interaction with uropathogens. Infect Immun 66(5):1985–1989
14. Goldsby RA, Kindt TJ, Osborne BA et al (2003) Immunology, 5th edn. W.H. Freeman and Company, New York
15. Round JL, Mazmanian SK (2009) The gut microbiota shapes intestinal immune responses during health and disease. Nat Rev Immunol 9:313–323
16. Dongarrà M, Rizzello V, Muccio L et al (2013) Mucosal immunology and probiotics. Curr Allergy Asthma Rep 13:19–26
17. Lefrancois L (1994) Basic aspects of intraepithelial lymphocyte immunobiology. In: Ogra PL, Mestecky J, Lamm ME, Strober W, JR MG, Bienenstock J (eds) Handbook of mucosal immunology. Academic Press, San Diego
18. Suzuki K, Kawamoto S, Maruya M et al (2010) GALT: organization and dynamics leading to IgA synthesis. Adv Immunol 107:153–185
19. Kernéis S, Bogdanova A, Kraehenbuhl JP et al (1997) Conversion by Peyer's patch lymphocytes of human enterocytes into M cells that transport bacteria. Science 277:949–952
20. Pabst O, Mowat AM (2012) Oral tolerance to food protein. Mucosal Immunol 5:232–239
21. Worbs T, Bode U, Yan S et al (2006) Oral tolerance originates in the intestinal immune system and relies on antigen carriage by dendritic cells. J Exp Med 203(3):519–527
22. Mantis NJ, Rol N, Corthésy B (2011) Secretory IgA's complex roles in immunity and mucosal homeostasis in the gut. Mucosal Immunol 4:603–611
23. Brandtzaeg P (1995) Molecular and cellular aspects of the secretory immunoglobulin system. Acta Microbiol Immunol Scand (APMIS) 103(1–6):1–19
24. Vanderhoof JA (2008) Probiotics in allergy management. J Pediatr Gastroenterol Nutr 47(2):S38–S40
25. Lilic D, Cant AJ, Abinun M et al (1997) Cytokine production differs in children and adults. Pediatr Res 42(2):237–240
26. Bäckhed F, Fraser C, Ringel Y et al (2012) Defining a healthy human gut microbiome: current concepts, future directions, and clinical applications. Cell Host Microbe 12(5):611–622
27. Moreau MC, Gaboriau-Routhiau V (2001) Influence of resident intestinal microflora on the development and functions of the gut associated lymphoid tissue. Microb Ecol Health Dis 13(2):65–86
28. Bron PA, van Baarlen P, Kleerebezem M (2012) Emerging molecular insights into the interaction between probiotics and the host intestinal mucosa. Nat Rev Microbiol 10(1):66–78
29. Gronlund MM, Arvilommi H, Kero P et al (2000) Importance of intestinal colonization in the maturation of humoral immunity in early infancy: a prospective follow-up study of healthy infants aged 0–6 months. Arch Dis Child 83(3):F186–F192
30. Fanaro S, Chierici R, Guerrini P et al (2003) Intestinal microflora in early infancy: composition and development. Acta Paediatr 91(441):48–55

31. Dominguez-Bello MG, Costello EK, Contreras M et al (2010) Delivery mode shapes the acquisition and structure of the initial microbiota across multiple body habitats in newborns. Proceed Nat Acad Sci USA 107(26):11971–11975
32. Sonomoto K, Yokota A (2011) Lactic acid bacteria and bifidobacteria: current progress in advanced research. Academic Press, Caister ISBN 978-1-904455-82-0
33. Harmsen HJ, Wildeboer-Veloo AC, Raangs GC et al (2000) Analysis of intestinal flora development in breast-fed and formula-fed infants by using molecular identification and detection methods. J Pediatr Gastroenterol Nutr 30(1):61–67
34. Prakash S, Tomaro-Duchesneau C, Saha S et al (2014) Probiotics for the prevention and treatment of allergies, with an emphasis on mode of delivery and mechanism of action. Curr Pharm Des 20(6):1025–1037
35. Bjorksten B, Sepp E, Julge K et al (2001) Allergy development and the intestinal microflora during the first year of life. J Allergy Clin Immunol 108(4):516–520
36. Mack DR, Ahrne S, Hyde L et al (2003) Extracellular MUC3 mucin secretion follows adherence of Lactobacillus strains to intestinal epithelial cells in vitro. Gut 52:827–833
37. Kemgang TS, Kapila S, Shanmugam VP et al (2014) Cross-talk between probiotic lactobacilli and host immune system. J Appl Microbiol 117(2):303–319
38. Sanders ME, Tompkins T, Heimbach JT et al (2004) Weight of evidence needed to substantiate a health effect for probiotics and prebiotics, regulatory considerations in Canada, E.U. and U.S. Eur J Nutr 44:303–310
39. Boyle P, Leon ME (2002) Epidemiology of colorectal cancer. Br J Med Bull 64:1–25
40. Sobhani I, Tap J, Roudot-Thoraval F et al (2011) Microbial dysbiosis in colorectal cancer (CRC) patients. PLoS One 6(1):e16393
41. Shida K, Nomota K (2013) Probiotics as efficient immunopotentiators: Translational role in cancer prevention. Indian J Med Res 138(5):808–814
42. Khan AA, Khurshid M, Khan S et al (2013) Gut microbiota and probiotics: current status and their role in cancer therapeutics. Drug Dev Res 74(6):365–375
43. Nakamura J, Kubota Y, Miyaoka M et al (2002) Comparison of four microbial enzymes in clostridia and bacteroides isolated from human feces. Microbiol Immunol 46:487–490
44. Shen XJ, Rawls JF, Randall T et al (2010) Molecular characterization of mucosal adherent bacteria and associations with colorectal adenomas. Gut Microbes 1(3):138–147
45. Rubinstein MR, Wang X, Liu W et al (2013) Fusobacterium nucleatum promotes colorectal carcinogenesis by modulating E-cadherin/β-catenin signaling via its FadA adhesin. Cell Host Microbe 14:195–206
46. LV MF, Elmer GW (2006) Properties of Evidence-Based Probiotics for Human Health. In: Goktepe I, Juneja VK, Ahmedna M (eds) Probiotics in food safety and human health. CRC, New York, pp 109–137
47. Lan A, Lagadic-Gossmann D, Lemaire C et al (2007) Acidic extracellular pH shifts colorectal cancer cell death from apoptosis to necrosis upon exposure to propionate and acetate, major end-products of the human probiotic propionibacteria. Apoptosis 12:573–591
48. Reddy BS, Hamid R, Rao CV (1997) Effect of dietary oligofructose and inulin on colonic preneoplastic aberrant crypt foci inhibition. Carcinogenesis 18:1371–1374
49. Rafter J, Bennett M, Caderni G et al (2007) Dietary synbiotics reduce cancer risk factors in polypectomized and colon cancer patients. Am J Clin Nutr 85(2):488–496
50. Geier M, Butler R, Howarth G (2006) Probiotics, prebiotics and synbiotics: a role in chemoprevention for colorectal cancer? Cancer Biol Ther 5:1265–1269
51. Ishikawa H, Akedo I, Otani T et al (2005) Randomized trial of dietary fiber and Lactobacillus casei administration for prevention of colorectal tumors. Int J Cancer 116:762–767
52. Galdeano M, Perdigon G (2006) The probiotic bacterium Lactobacillus casei induces activation of the gut mucosal immune system through innate immunity. Clin Vaccine Immunol 13:219–226
53. Liong MT (2008) Roles of probiotics and prebiotics in colon cancer prevention: postulated mechanisms and in-vivo evidence. Int J Mol Sci 9:854–863

54. Ciorba MA, Riehl T, Hyun YJ et al (2008) M1198 Lactobacillus Rhamnosus GG Prevents Radiation Induced-Small Intestinal Injury in a MyD88 Independent, But COX2 Dependent Manner. Gastroenterology 34(1):A-359
55. Ciorba MA, Riehl TE, Rao MS et al (2012) Lactobacillus probiotic protects intestinal epithelium from radiation injury in a TLR-2/cyclo-oxygenase-2-dependent manner. Gut 61(6):829–838
56. Lebeer S, Vanderleyden J, De Keersmaecker SC (2010) Host interactions of probiotic bacterial surface molecules: comparison with commensals and pathogens. Nat Rev Microbiol 8(3):171–184
57. Yan F, Polk D (2012) Lactobacillus rhamnosus GG-an updated strategy to use microbial products to promote health. Funct Food Rev 4(2):41–48
58. Vrese M, Marteau PR (2007) Probiotics and prebiotics: effects on diarrhea. J Nutr 137 (3):803S–811S
59. Rask C, Adlerberth I, Berggren A et al (2013) Differential effect on cell-mediated immunity in human volunteers after intake of different lactobacilli. Clin Exp Immunol 172:321–332
60. Mastretta E, Longo P, Laccisaglia A et al (2002) Effect of Lactobacillus GG and breast-feeding in the prevention of rotavirus nosocomial infection. J Pediatr Gastroenterol Nutr 35(4):527–531
61. Banerjee P, Merkel GJ, Bhunia AK (2009) Lactobacillus delbrueckii ssp. bulgaricus B-30892 can inhibit cytotoxic effects and adhesion of pathogenic Clostridium difficile to Caco-2 cells. Gut Pathogens 1, 8 (open access). https://doi.org/10.1186/1757-4749-1-8
62. Lehtoranta L, Pitkäranta A, Korpela R (2014) Probiotics in respiratory virus infections. Eur J Clin Microbiol Infect Dis 33(8):1289–1302
63. Garaiova I, Muchová J, Nagyová Z et al (2014) Probiotics and vitamin C for the prevention of respiratory tract infections in children attending preschool: a randomised controlled pilot study Open. Eur J Clin Nutr. https://doi.org/10.1038/ejcn.2014.174 (online)
64. Baron MA (2009) Patented strain of Bacillus coagulans increased immune response to viral challenge. Postgrad Med 121(2):114–118
65. Berggren A, Lazou AI, Larsson N et al (2011) Randomised, double-blind and placebo-controlled study using new probiotic lactobacilli for strengthening the body immune defence against viral infections. Eur J Nutr 50(3):203–210
66. Koenig SM, Truwit JD (2006) Ventilator-associated pneumonia: diagnosis, treatment, and prevention. Clin Microbiol Rev 19(4):637–657
67. Lorente L, Lecuona M, Jiménez A et al (2007) Influence of an endotracheal tube with polyurethane cuff and subglottic secretion drainage on pneumonia. Am J Respir Crit Care Med 176(11):1079–1083
68. Bo L, Li J, Tao T et al (2014) Probiotics for preventing ventilator-associated pneumonia. The Cochrane Database Syst Rev 10. https://doi.org/10.1002/14651858.CD009066.pub2
69. Morrow LE, Kollef MH, Casale TB (2010) Probiotic prophylaxis of ventilator-associated pneumonia: a blinded, randomized, controlled trial [Randomized Controlled Trial Research Support, N.I.H., Extramural Research Support, Non-U.S. Gov't]. Am J Respir Crit Care Med 182(8):1058–1064
70. Wan-Jie G, Chun-Yin W, Rui-Xing Y (2012) Lack of efficacy of probiotics in preventing ventilator-associated pneumonia: a systematic review and meta-analysis of randomized controlled trials. Chest 142:859–868
71. Franz M, Hörl WH (1999) Common errors in diagnosis and management of urinary tract infection. I: Pathophysiology and diagnostic techniques. Nephrol Dial Transplant 14(11):2746–2753
72. Grin PM, KowalewskaPM AW et al (2013) Lactobacillus for preventing recurrent urinary tract infections in women: meta-analysis. Can J Urol 20(1):6607–6614
73. Amdekar S, Singh V, Singh DD (2011) Probiotic therapy: immunomodulating approach towards urinary tract infection. Curr Microbiol 63(5):484–490
74. Reid G (2008) Probiotic Lactobacilli for urogenital health in women. J Clin Gastroenterol 42(3):S234–S236

75. Anukam K, Osazuwa E, Ahonkhai I et al (2006) Augmentation of antimicrobial metronidazole therapy of bacterial vaginosis with oral probiotic Lactobacillus rhamnosus GR-1 and Lactobacillus reuteri RC-14: randomized, double-blind, placebo controlled trial. Microbiol Infect 8(6):1450–1454
76. Schönherr FA, Sparber F, Kirchner FR et al (2017) The intraspecies diversity of C. albicans triggers qualitatively and temporally distinct host responses that determine the balance between commensalism and pathogenicity. Mucosal Immunol. https://doi.org/10.1038/mi.2017.2
77. VodstrcilL A, Twin J, Garland SM et al (2017) The influence of sexual activity on the vaginal microbiota and Gardnerella vaginalis clade diversity in young women. PLoS One 12:e0171856. https://doi.org/10.1371/journal.pone.0171856
78. Diane ML, Howard FJ, Gerald WT (1994) Colonization of the murine oral cavity by Streptococcus gordonii. Infect Immun 62(5):2129–2131
79. Jobin MC, Amin M, Ellen RP (2008) The molecular biology of the survival and virulence of treponema denticola. Molecular oral microbiology. Caister Academic Press
80. Ichinosawa T, Ito T, Yonezawa H (2017) Molecular interaction of the analogous peptide SspB (390- T400K- 402) derived from Streptococcus gordonii surface protein peptide with periodontal bacteria. Int J Oral Med Sci 15(3–4):160–167
81. Krasse P, Carlsson B, Dhal C et al (2006) Decreased gum bleeding and reduced gingivitis by the probiotic Lactobacillus reuteri. Swed Dental J 30(2):55–60
82. Vivekananda MR, Vandana KL, Bhat KG (2010) Effect of the probiotic Lactobacilli reuteri (Prodentis) in the management of periodontal disease: a preliminary randomized clinical trial. J Oral Microbiol 2. https://doi.org/10.3402/jom.v2i0.5344
83. Vicario M, Santos A, Violant D et al (2012) Clinical changes in periodontal subjects with the probiotic Lactobacillus reuteri prodentis; a preliminary randomized clinical trial. Acta Odontol Scand 70(3):246–250
84. Romani VN, Hasslöf P, Keller MK et al (2013) Lactobacillus reuteri influences regrowth of mutans streptococci after full-mouth disinfection: a double-blind, randomized controlled trial. Caries Res 47(4):338–345
85. Miettinen M, Vuopio-Varkila J, Varkila K (1996) Production of human tumor necrosis factor alpha, interleukin-6, and interleukin-10 is induced by lactic acid bacteria. Infect Immun 64(12):5403–5405
86. Wang Y, Xie J, Wang N et al (2013) Lactobacillus casei Zhang modulate cytokine and Toll-like receptor expression and beneficially regulate poly I:C-induced immune responses in RAW264.7 macrophages. Microbiol Immunol 57:54–62
87. Ouwehand AC (1998) Antiallergic effects of probiotics. J Nutr 64:3854–3859
88. Isolauri E, Kaila M, Mykkänen H et al (1994) Oral bacteriotherapy for viral gastroenteritis. Dig Dis Sci 39(12):2595–2600
89. Feleszko W, Jaworska J (2013) Probiotics and prebiotics in immune modulation. In: Watson R, Preedy V (eds) Bioactive food as dietary interventions for arthritis and related inflammatory diseases. Academic Press, San Diego, pp 357–370
90. Chiba Y, Shida K, Nagata S et al (2010) Well-controlled proinflammatory cytokine responses of Peyer's patch cells to probiotic Lactobacillus casei. Immunology 130(3):352–362
91. Kotani Y et al (2014) Role of Lactobacillus pentosus strain b240 and the toll-like receptor 2 axis in Peyer's patch dendritic cell-mediated immunoglobulin A enhancement. PLoS One. https://doi.org/10.1371/journal.pone.0091857
92. Isolauri E (1997) Intestinal involvement in atopic disease. J R Soc Med 90:15–22
93. Rautanena T, Isolaurib E, Saloc E et al (1998) Management of acute diarrhoea with low osmolarity oral rehydration solutions and Lactobacillus strain GG. Arch Dis Child 79:157–160
94. Muñoz JA, Chenoll E, Casinos B et al (2011) Novel probiotic Bifidobacterium longum subsp. infantis CECT 7210 strain active against rotavirus infections. Appl Environ Microbiol 77:8775–8783
95. Tomás MSJ, Otero MC, Ocaña V et al (2004) Production of antimicrobial substances by lactic acid bacteria I: determination of hydrogen peroxide public health microbiology: methods and protocols series. Methods Mol Biol 268:337–346

96. Hudault S, Liévin V, Bernet-Camard MF et al (1997) Antagonistic activity exerted in vitro and in vivo by Lactobacillus casei (strain GG) against Salmonella typhimurium C5 infection. J Appl Environ Microbiol 63(2):513–518
97. Castagliuolo I, Riegler MF, Valenick L et al (1999) Saccharomyces boulardii protease inhibits the effects of clostridium difficile toxins A and B in human colonic mucosa. Infect Immun 67(1):302–307
98. Dias RS, Bambirra EA, Silva ME et al (1995) Protective effect of Saccharomyces boulardii against the cholera toxin in rats. Braz J Med Biol Res 28(3):323–325
99. Dhanani AS, Bagchi T (2013) Lactobacillus plantarum CS24.2 prevents E. coli adhesion to HT-29 cells and also down-regulates enteropathogen induced TNF-alpha and IL-8 expression. Microbiol Immunol 57:309–315
100. Rigothier MC, Maccario J, Vuong PN et al (1990) Effects of Saccharomyces boulardii yeast on trophozoites of Entamoeba histolytica in vitro and in cecal amebiasis in young rats. Ann Parasitol Hum Comp 65(2):51–60
101. Dong H, Rowland I, Yaqoob P (2012) Comparative effects of six probiotic strains on immune function in vitro. Br J Nutr 108(3):459–470
102. Tanabe S (2013) The effect of probiotics and gut microbiota on Th17 cells. Int Rev Immunol 32(5–6):511–525
103. Van Immerseel F, Ducatelle R, De Vos M et al (2010) Butyric acid-producing anaerobic bacteria as a novel probiotic treatment approach for inflammatory bowel disease. J Med Microbiol 59:141–143
104. MacSharry J, O'Mahony C, Shalaby KH et al (2012) Immunomodulatory effects of feeding with Bifidobacterium longum on allergen-induced lung inflammation in the mouse. Pulm Pharmacol Ther 25(4):325–334
105. Heikkilä JE, Nybom SMK, Salminen SJ et al (2012) Removal of cholera toxin from aqueous solution by probiotic bacteria. Pharmaceuticals 5(6):665–673
106. Kaila M, Isolauri E, Soppi E et al (1992) Enhancement of the circulating antibody secreting cell response in human diarrhea by a human lactobacillus strain. Pediatr Res 32:141–144
107. Isolauri E, Kaila M, Arvola T et al (1993) Diet during rotavirus enteritis affects jejunal permeability to macromolecules in suckling rats. Pediatr Res 33(6):548–553
108. Sütas Y, Soppi E, Korhonen H et al (1996) Suppression of lymphocyte proliferation in vitro by bovine caseins hydrolyzed with Lactobacillus casei GG-derived enzymes. J Allergy Clin Immunol 98(91):216–224
109. Anandharaj M, Sivasankari B, Rani RP (2014) Effects of probiotics, prebiotics, and synbiotics on hypercholesterolemia: a review. Chin J Biol. Article ID 572754
110. Kechagia M, Basoulis D, Konstantopoulou S et al (2013) Health benefits of probiotics: a review. ISRN Nutr. https://doi.org/10.5402/2013/481651
111. Singh K, Kallali B, Kumar A et al (2011) Probiotics: a review. Asian Pac J Trop Biomed:S287–S290
112. Maurya P, Mogra R, Bajpai P (2014) Probiotics: an approach towards health and disease. Trends Biosci 7(20):3107–3113
113. Khan RU, Naz S (2013) The applications of probiotics in poultry production. World's Poult Sci J 69:621–632
114. Parvez S, Malik KA, Ah Kang S, Kim HY (2006) Probiotics and their fermented food products are beneficial for health. J Appl Microbiol 100:1171–1185
115. Kral M, AngelovicovaM ML (2012) Application of probiotics in poultry production. Anim Sci Biotechnol 45(1):55–57
116. Fontana L, Bermudez-Brito M, Plaza-Diaz J et al (2013) Sources, isolation, characterization and evaluation of probiotics. Br J Nutr 109:S35–S50
117. Parracho H, Mc Cartney AL, Gibson GR (2007) Probiotics and prebiotics in infant nutrition. Proc Nutr Soc 66:405–411
118. Mitropoulou G, Nedovic V, Goyal A, et al (2013) Immobilization technologies in probiotic food production. J Nutr Metab Article ID 716861

119. Hill C, Guarner F, Reid G et al (2014) Expert consensus document. The International Scientific Association for Probiotics and Prebiotics consensus statement on the scope and appropriate use of the term probiotic. Nat Rev Gastroenterol Hepatol 11:506–514
120. Kurmann JA, Rasic JL (1991) The health potential of products containing bifidobacteria. In: Robinson RK (ed) Therapeutic properties of fermented milks. Elsevier Science Publishers Ltd., London, pp 117–158
121. Payne JF, Morris AEJ, Beers PJ (1998) Viability of bifidobacteria in fermented milk products. In: Sadler MJ, Saltmarsh M (eds) Functional foods: the consumer, the products and the evidence. Royal Society of Chemistry, Cambridge, pp 143–148
122. Mattila-Sandholm T, Myllarinen P, Crittenden R et al (2002) Technological challenges for future probiotic foods. Int Dairy J 12:173–182
123. Belguesmia Y, Domenger D, Caron J et al (2016) Novel probiotic evidence of lactobacilli on immunomodulation and regulation of satiety hormones release in intestinal cells. J Funct Foods 24:276–286
124. Zavisic Z, Petricevic S, Radulovic Z et al (2012) Probiotic features of two oral lactobacillus isolates. Braz J Microbiol 43:418–428
125. Adams CA (2010) The probiotic paradox: live and dead cells are biological response modifiers. Nutr Res Rev 23(1):37–46
126. Awad H, Mokhtar H, Imam SS et al (2010) Comparison between killed and living probiotic usage versus placebo for the prevention of necrotizing enterocolitis and sepsis in neonates. Pak J Biol Sci 13(6):253–262
127. Kataria J, Li N, Wynn JL et al (2009) Probiotic microbes: do they need to be alive to be beneficial? Nutr Rev 67:546–550
128. Liong MT, Shah NP (2005) Roles of probiotics and prebiotics on cholesterol: the hypothesized mechanisms. Forum Nutr 4:45–57
129. Aggarwal A, Nagori BP, Mathur V (2012) Regulatory approval of probiotics in US, Japan and India: a comparative study. Int J Sci Pharm Edu Res 2:78–86
130. Sharma S, Arora M, Baldi A (2013) Probiotics in India: current status and future prospects. Pharmaspire 1:1–11
131. Guidelines for evaluation of probiotics in food. Report of a Joint ICMR/DBT 2011. Available from.: http://icmr.nic.in/guide/PROBIOTICS_GUIDELINES.pdf
132. Food and Drug Administration. Development and approval process (drugs). Available from.: http://www.fda.gov/Drugs/DevelopmentApprovalProcess/default.htm
133. Leushner R, Robinson T, Hugas M et al (2010) Qualified Presumption of Safety (QPS): a generic risk assessment approach for biological agents notified to the European Food Safety Authority (EFSA). Trends Food Sci Technol 21:425–435
134. Coppens P, da Silva MF, Pettman S (2006) European regulations on nutraceuticals, dietary supplements and functional foods: a framework based on safety. Toxicology 221:59–74
135. Hilliam M (1998) Functional foods in Europe. World Ingred 7:40–45
136. Amagase H (2008) Current marketplace for probiotics: a Japanese perspective. Clin Infect Dis 46:S73–S75
137. Yang Y (2008) Scientific Substantiation of functional food health claims in China. J Nutr 138(6):1199S–1205S
138. Nutraceuticals in Latin America: Brazil and Beyond at Latin America: Nutraceutical Boom or Bust? Available from.: http://www.nutraceuticalsworld.com/articles/2008/11/latin-america
139. Food and Drug Administration Overview of dietary supplements. Available from: http://www.fda.gov/%20Food/
140. Food and Drug Administration Federal Food, Drug, and Cosmetic Act (FD and C Act). 21 U.S.C.321Sect201 (2004) http://www.fda.gov/regulatoryinformation/legislationfederalfooddrugandcosmeticactfdcact/default.htm
141. Dengan FH (2008) The US Food and Drug Administration and probiotics: regulatory categorization. Clin Infect Dis 46:S133–S136
142. Dengan FH (2012) Clinical studies involving probiotics: when FDA's investigational new drug rubric applies-and when it may not. Gut Microbes 3:485–489

143. Kaushik D, Kaushik N (2009) Functional food/Nutraceuticals regulation in India. Pharm Rev 7
144. State Food and Drug Administration The guideline of registration for functional food 2005. Available from: http://www.sfda.gov.cn/syjz0461/syjz0461.htm
145. FAO/WHO (2002) Guidelines for the evaluation of probiotics in food. Available from: http://www.who.int/foodsafety/fs_management/en/probiotic_guidelines.pdf
146. MHLW. Ministry of Health, Labour and Welfare: Food for Specified Health Uses (FOSHU) http://www.mhlw.go.jp/english/topics/foodsafety/fhc/02.html
147. Saxelin M (2008) Probiotic formulations and applications, the current probiotics market, and changes in the marketplace: a European perspective. Clin Infect Dis 46:S76–S79
148. Food Standards Australia New Zealand Health and related claims. http://www.foodstandards.gov.au

Flavonoids as Functional Food

5

Krunal Ramanbhai Patel, Fenisha Dilipkumar Chahwala, and Umesh C. S. Yadav

Abstract

Plants are rich source of biologically active compounds including alkaloids, flavonoids, triterpenes, phenylpropanoids, benzoic acid derivatives, stilbenes, tannins, and lignans. Importance of such plant-derived compounds is well established as nutrients and health-promoting agents. Flavonoids are one of the several important phytochemicals that possess diverse biochemical properties having positive effects on human health and thus labeled as functional food. Flavonoids are biologically active compounds known for their activities which include antioxidant, anticancer, anti-inflammatory, and cardio-protective characteristics. Technological advancement has helped in extraction, isolation, quantification, and identification of flavonoids from various natural food sources including the plants. Fruits and vegetables are major source of the flavonoids. Traditional medicinal systems as well as modern medicinal systems relay on the plants as source for the bioactive compound for disease prevention and treatment. According to their chemical nature, at least five different types of flavonoids are known, which include flavone, flavanones, anthocyanins, procyanidins, and flavonols. In the present chapter, the potential of these flavonoids as functional foods has been discussed based on the available scientific literature and epidemiological data.

Keywords

Favonoids · Functional foods · Antioxidants · Inflammation · Cardiovasclar diseases · Neuroprotection

K. R. Patel · F. D. Chahwala · U. C. S. Yadav (✉)
School of Life Sciences, Central University of Gujarat, Gandhinagar, Gujarat, India
e-mail: umeshyadav@cug.ac.in

V. Rani, U. C. S. Yadav (eds.), *Functional Food and Human Health*,
https://doi.org/10.1007/978-981-13-1123-9_5

5.1 Introduction

Food provides the energy requirement in addition to essential nutrients and satisfies the hunger which is necessary for the survival of organism. The tenet "Let food be thy medicine and medicine be thy food" was espoused by Hippocrates 2500 years ago which meant that food could be one's medicine and medicine could act as food for better health. This notion has received renewed interest recently [1]. In the ancient medicine systems including Chinese medicine system, Ayurveda, and Unani, the same principle is followed to prevent and cure various diseases and promote vigor and health. In these traditional systems, crude extract of plants or their parts, which contain vivid phytochemicals with ability to prevent and/or cure diseases, are prepared and given to patients and even used by healthy individuals to promote vitality and strength. Majority of the population in these areas express concern regarding the side effects of allopathic drugs, medical procedures, and vaccines and affordability related to the commercial drugs. For many such countries and societies, traditional medicine is the main source of cure, but for the modern healthcare system, safety and efficacy of such drugs remain a key question. Many governments in the modern times have started promoting research in the field of traditional medicine to validate as well as identify the active components in the traditional concoction which has helped in the acceptance of these alternative medicinal approach worldwide [2].

It is well established that food with balanced nutrients and caloric intake can serve as a stimulant for good health; however if taken in excess and uncontrolled manner, it could turn out as a risk factor for disease development. A balanced food intake fulfills the energy as well as nutrient requirement of individual, but changing lifestyle and food preference from balanced diet and active life to high-carbohydrate-, high-fat-containing processed diet and sedentary life has led to many lifestyle diseases including obesity, diabetes, and cardiovascular diseases (CVDs). Developed countries like Japan, the United States, and European countries have made efforts to encourage and support research on health benefits and physiologic effects of functional foods and their components. Indeed, Japanese have accepted that food can be more than just energy, nutrient supply, and gastronomic and satiety pleasure. These characteristics of foods are now covered under the term "functional food" which is widely accepted. Consumption of the functional food is expected to decrease the risk of diseases and promote health benefits. Alternatively, a food source that may contain bioactive components (nutrient or non-nutrient) that could be helpful to body in targeted manner and could improve disease symptoms or cure an individual qualified to be called a functional food. European experts at a consensus meeting in Madrid (1998) coordinated by International Life Science Institute, Europe, accepted a standard definition of functional food as "a food that satisfactorily demonstrate to affect biological pathways beneficial in a way which is relevant to either the state of well-being and health or the reduction of the risk of a disease" [1]. Japan has now established the law for the "functional food" which is known as "Food Of Specified Health Use (FOSHU)" law, and food products fulfilling these criteria can be sold with the tag of functional food [3].

5.1.1 FOSHU Criteria for Functional Food

1. Functional food should be food, not a capsule, pill, or powder. They are based on naturally accessible food components.
2. It should be consumed as normal diet.
3. It should have demarcated role on the human organism.

Based upon the source, food is broadly divided in two types, i.e., plant-derived and animal-derived food. Plants provide essential nutrient components and other chemicals as well. Fruits, vegetables, pulses, and grains contain nutrients essential for human health. Besides nutrient components such as carbohydrates, fats, proteins, and vitamins, many functional components are present in plant-derived foods such as beta-carotene, lutein, zeaxanthin, flavonoids, isothiocyanates, phenols, sulfides/thiols, etc. Some of these compounds are known to act as antioxidant, anti-inflammatory, anticancer, and antidiabetic components. These components of plants have been used as food, spices, and medicine from the ancient time. Chemicals derived from plants are known as phytochemicals which contain condense oligomeric or polymeric building blocks with phenolic features, called polyphenols. Polyphenolic compounds are aromatic secondary metabolites of plants and classified into various groups based on the aromatic rings that are present in their structure. Accordingly, polyphenols consist of wide categories of phytochemicals, e.g., phenolic acids, stilbenes, flavonoids, phenols, phenylpropanoids, benzoic acid derivatives, tannins, lignans, and lignins. Plants are rich sources of polyphenolic compounds which are potent antioxidants.

Plant-derived chemical compounds have always attracted the attention of medicine men, clinicians, researchers, and scientists. A number of phytochemicals have been identified, classified, and used for different application like coloring agents, additives, and drugs. Likewise, a new compound was isolated from oranges in 1930 which was believed to be a new class of vitamin. However, subsequent detailed investigations led to its characterization as flavonoid [4]. Till date more than 4000 flavonoids have been discovered. Flavonoids generally contain benzo-γ-pyrone structure, mostly present in plants. General chemical structure of flavonoid is C3–C6–C3. The basic chemical nature of flavonoid is aglycone. Flavonoids have a backbone of 15-carbon which contain two benzene rings linked to a heterocyclic carbon ring. Chemical composition of naturally occurring flavonoids includes aglycones, glycosides, and methylated derivatives [5].

Flavonoids, representative of plant's secondary yield, are bioactive polyphenolic compounds found in a wide variety of plant-based foods, including fruits and vegetables, tea, wine, nuts, herbs, and spices. In the model plant *Arabidopsis thaliana*, approximately 54 flavonoid molecules (11 anthocyanins, 35 flavonols, and 8 proanthocyanidins) have been isolated, which are identified as major secondary products [6]. Metabolically, flavonoids exist at the boundary of primary and secondary metabolism. According to the position of six-member ring on to benzene ring, flavonoids are divided into five different classes: (1) anthocyanins, (2) flavanols, (3) flavanones,

Table 5.1 Types and sources of flavonoids and their health-promoting effects

Sr. no.	Type and examples of flavonoids	Source	Function/health benefits	References
1.	**Anthocyanins** – cyanidin, pelargonidin, delphinidin, malvidin	Cherries, red grapes, acai, berries	Strengthen cellular antioxidant defenses; antioxidant, anti-inflammatory, antidiabetic properties	[7–10]
2.	**Flavanols** – catechins, epicatechins, epigallocatechin	Tea, cocoa, chocolate, apples, grapes	Supports maintenance of heart health, prevents endothelial dysfunctions	[8, 11, 12]
3.	**Flavanones** – hesperetin, naringenin	Citrus fruits, rosemary	Scavenges free radicals; bolster cellular antioxidant defenses, immunological functions, anticancer properties	[7, 8, 13]
4.	**Flavonols** – quercetin, kaempferol, isorhamnetin, myricetin	Onions, oregano, saffron, apples, tea, broccoli	Neutralizes free radicals which may damage cells; bolster cellular antioxidant defenses, anticancer properties	[7, 8, 14, 15]
5.	**Procyanidins**	Cranberries, cocoa, apples, strawberries, grapes, red wine, peanuts, cinnamon, tea, chocolate	Anticancer properties, anti-inflammatory	[7, 8, 16]

(4) flavonols, and (5) procyanidins. These groups have been summarized in Table 5.1 below along with their natural sources and health benefits/effects.

5.1.2 Biosynthesis of Flavonoids

In plants, flavonoids are formed by condensation of a phenylpropane (C6–C3) compound via participation of three molecules of malonyl coenzyme A. This leads to the formation of chalcones, which is cyclized under acidic condition. This leads to the formation of flavonoids that contain the basic skeleton of diphenylpropanes (C6–C3–C6), which possess different oxidation levels in the central pyran ring. Flavonoids classes generally depend on substitution and unsaturation pattern. Flavones and flavonols contain double bond between C-2 and C-3 and are present as aglycones in food. Flavones contain hydroxyl group at 3-position referred as 3-hydroxyflavones and flavonols referred as 3-deoxyflavonols as they lack it. Flavonol and flavone glycosylation depend on the action of light. They are mainly and abundantly present in leaves and fruits' skin than the other plant parts.

5.2 Effect of Various Flavonoids as Functional Food

Based on the available scientific literature such as monographs, review article, research papers, and clinical studies, different classes of flavonoids and their potential as functional food will be discussed in subsequent sections.

5.2.1 Anthocyanins

Anthocyanins are water-soluble pigments that give bright red, blue, and violet color to different parts of plants including leaves, flowers, and fruits. When glycosidic bond is present in anthocyanin, it is known as anthocyanidin. Bright red, blue, and violet skins of radish, potato, and eggplant, respectively, are due to the presence of anthocyanidin. The anthocyanin word is derived from two Greek words "anthos" which means flower and "kyanos" which means dark blue. Berries, cherries, purple grapes, red wine, plum, eggplant, purple asparagus, red cabbage, black rice, kidney beans, black beans, pomegranates, oranges, olives, red onion, fig, sweet potato, radish, mango, beets, blueberry, blackberries, strawberries, raspberries, elderberries, cranberries, bilberries, and many other blue-, purple-, or red-skinned berries are all rich sources of anthocyanins. More than 600 different types of anthocyanins have been identified from plants, of which six major aglycons are pelargonidin, peonidin, cyanidin, delphinidin, petunidin, and malvidin. Anthocyanins are markers of ripening, because most fruits accumulate these compounds only in their ripening phase [17, 18]. The chemical structure of anthocyanins is comprised of anthocyanidin aglycon and a variety of sugar moiety attached to it. The presence of oxygen atom in anthocyanin structure is responsible for its antioxidant activity. Different structures of the anthocyanins are present in cells at any given time which are temperature or pH-dependent. Variation in anthocyanin structure, their interaction with other metabolites in the intestine, affects their bioavailability, metabolism, and pharmacological response in human gut, for example, acylation in anthocyanin is known to have shorter half-life for gastrointestinal absorption as compared to non-acylated anthocyanins [19]. Anthocyanins are known for their pharmacological effects including anticancerous activity, prevention of cardiovascular disease (CVD), and control of obesity [20, 21]. These reported effects are mainly due to presence of polyphenol structure which makes them highly potent antioxidants [20]. Anthocyanins exert these effects by their ability to modulate the signal transduction pathways in cells, which induce antioxidant potential of cells, cell cycle arrest, induction of apoptosis in cancerous cells, prevention of invasion, and metastasis of cancerous cells [22].

5.2.1.1 Anthocyanins as Antioxidant

Metabolic reactions generate free radicals which lead to oxidative stress in cells impacting normal physiology. Free radicals such as reactive oxygen species (ROS)

Flavonoids as Functional Food and reactive nitrogen species (RNS) are highly reactive and unstable and tend to react with biomolecules such as lipids, nucleic acids and proteins, and cause lipid peroxidation, DNA damage and oxidation of other biomolecules. The cells have inbuilt mechanisms such as vitamins A & E, super oxide dismutase (SOD) and catalase (CAT) by which the levels of ROS and RNS are controlled and maintained at the safer levels [23]. Due to their reactive nature, these radicals cause DNA damage, lipid peroxidation, and oxidation of biomolecules. ROS and RNS are important in many physiological process including phagocytosis, but overproduction of such reactive radical harms cells and tissues [24]. As described earlier, basic structure of flavonoids and presence of various groups on that structure make them highly antioxidant in nature. For example, anthocyanins contain higher number of -OH (hydroxyl) groups present throughout the structure, which make them potent antioxidants. A number of studies have shown antioxidant potential of anthocyanins [19]. Fruits and vegetable are rich source of anthocyanin. Consumption of fruits rich in anthocyanin has been reported to bestow multiple health benefits against obesity, cancer, and diabetes.

Antioxidant potential of anthocyanins such as oenin and callistephin was tested using ARPE-19 cells. In this study, ARPE-19 (retinal pigment epithelial) cells were irradiated with blue light which caused ROS generation and increased mitochondrial redox potential. Oenin and callistephin treatment effectively reverted ROS and redox changes [25]. Anthocyanin treatment caused significant increase in SOD and catalase in streptozotocin-induced diabetes in rats [26]. Anthocyanin encounters free radical formation by inhibiting the enzyme involved in free radical formation or sequestration of cofactors involved in such reactions. Anthocyanins show antioxidant effect by capturing free radicals and/or anions, inhibition of xanthine oxidase (XO), chelation of metal ions, targeting arachidonic acid metabolism, and molecules of cell adhesion [27, 28].

Anthocyanin's phenolic ring can donate protons to free radicals and prevent oxidation by free radicals. Anthocyanin along with the other flavonoids can remove species of active oxygen, singlet oxygen, superoxide and peroxyl radicals. Antioxidant potential of anthocyanin has been studied widely using various assays and protocols, which includes measurement of free radical scavenging activity using electron spin resonance (ESR) and thiobarbituric acid reactive substance (TBARS) method [29,–30]. In 1,1-diphenyl-2-picrylhydrazy (DPPH), radical assay activity of anthocyanin was found similar to that of ascorbic acid [21].

Xanthine oxidase (XO) is an enzyme involved in metabolism of hypoxanthine to xanthine which further produces uric acid. Additionally, it generates reactive oxygen species; studies indicate anthocyanin's XO inhibitory activity in endothelial cells. Cells treated with micromolar concertation of malvidin and its glycoside inhibited XO-1 protein [27]. Further, studies by Reis et al. showed that XO inhibitory activity is attributed to the -OH in C-5 and C-7 positions in flavonoids. Oxidative stress causes lipid peroxidation that results in formation of oxidized products such as oxidized LDL [24]. Oxidation of LDL is associated with various diseases such as atherosclerosis and CVDs. Inhibition of LDL-oxidation and that of adverse effect mediated by ox-LDL were prevented using anthocyanin-rich extract [9, 31].

5.2.1.2 Anthocyanin in CVDs

Anthocyanin consumption has been shown to offer cardio-protective effect via mechanisms such as scavenging free radical which prevents LDL-oxidation or by increasing HDL cholesterol levels and suppressing LDL cholesterol levels in the circulation. A study by Zhu et al. (2014) has shown that anthocyanin-consuming individuals tend to have higher HDL as compared to placebo group [32]. Further, lower HDL-associated esterase/lactonase paraoxonase 1 (PON1) levels are associated with increased oxidative damage and diseases associated with it; however anthocyanin consumption was shown to enhance activity of PON1 [32].

Inflammation is one of the risk factors associated with CVDs. In CVDs, antioxidant enzyme (SOD and CAT) activities are decreased which were enhanced by consumption of anthocyanin-rich preparations. In vivo study using anthocyanin fractions from purple potato flakes showed antioxidant potential and inhibited linoleic acid oxidation in vitro. In addition, anthocyanin indirectly reduced hepatic lipid peroxidation in vivo by modulating the SOD expression [28]. Published data show that consumption of 375 mg anthocyanin for 6 weeks could reduce the risk of CVD by reducing oxidative stress-induced DNA damage [21]. Thus, anthocyanins offer anti-oxidative potential to the cardiovascular cells and protect them from the risks of CVDs.

5.2.1.3 Anthocyanins in Inflammation

Anti-inflammatory activity of anthocyanins including that of cyanidin, malvidin, peonidin, petunidin, and delphinidin has been widely reported [33]. Arachidonic acid metabolism, an integral part of immune system, and many intermediates of this pathway are known to play role in modulation of inflammation. Anthocyanins have been shown to modulate activities of arachidonic acid pathway enzymes such as phospholipase A2 (PLA2), cyclooxygenase (COX-2), and lipoxygenases (LOX) and thus regulate major inflammation mediators [24]. Studies by Li et al. (2014) showed COX-2 inhibitory potential of anthocyanidins in lipopolysaccharide (LPS)-induced RAW267.4 cells [34]. Similarly in murine J774 cells, inhibition of inducible nitric oxide synthase (iNOS) protein and mRNA expression was observed upon treatment with anthocyanin [17]. Anthocyanins also showed neuroprotection by inhibiting inflammation in neurons via enhancing the cellular glutathione levels and activation of nuclear factor (erythroid-derived 2)-like-2 (Nrf-2) signaling pathway [35]. Shah et al. (2016) showed that anti-inflammatory effect was also dependent on inhibition of p-NF-κB, COX-2, and caspase-3 protein expression [36]. In another study application of anthocyanin-rich fractions leads to inhibition of LPS and IFN-gamma-induced expression of iNOS, COX-2, and interleukins like IL-1β and IL-6 in macrophages [34]. These effects were mediated by the inhibition of NF-κB and AP-1 pathways which are important transcription regulators of inflammatory cascade. Anthocyanins prevent the nuclear translocation of p65, a NF-κB subunit [34]. In HAPI microglial cells, anthocyanin-rich fraction of tart cherry reduced TNF-α release [37]. Further, a clinical trial, aimed at investigating the role of anthocyanins against nonalcoholic fatty liver disease, revealed that consumption of anthocyanin (320 mg/d) for 12 weeks significantly modulated levels of plasma alanine aminotransferase, myeloperoxidase, fasting blood glucose levels, and insulin resistance [38]. In another study, subjects with hypercholesterolemia were given anthocyanin mixture containing delphinidin-3-O-b-glucoside (Dp-3 g) and cyanidin-3-O-b-glucoside (Cy-3 g) at the dose of (320 mg/d) twice a day for 24 weeks. They showed significantly reduced level of plasma IL-1β, serum high-sensitivity C-reactive protein (hsCRP), and soluble vascular cell adhesion molecule-1 (sVCAM-1) compared to placebo group [39]. These studies clearly indicate important protective roles of anthocyanins in various diseases exhibiting their antioxidant, anti-inflammatory, and antidiabetic properties.

5.2.2 Flavanols

Flavanols (3-hydroxy-2-phenylchromen-4-one) are low molecular weight polyphenols. The basic structure of flavanols consists of 3-hydroxyflavone backbone. Action of flavanols is altered by change in groups attached to its backbone. Flavanols are found in abundance from many fruits such as strawberries, lychee, grapes, and cocoa [12]. Catechins, epicatechins, and epigallocatechin are a few examples of flavanols that possess significant antioxidant and anti-inflammatory activities and are parts of functional food. Tea, one of the rich sources of flavanols, is one of the highly consumed beverages in almost every society and culture. Brewed green tea is a good source of epigallocatechin 3-gallate (EGCG) (70 mg per 100 grams), while black tea provides comparatively lesser amount of catechins (25 mg per 100 grams) compared to green tea. Blackberries are also a good source of flavanols with about 40 mg per 100 grams. Similarly, raspberries and cherries contain ~6 and 8 mg of flavanols per 100 grams, respectively [7]. Among vegetables, beans are good sources of flavanols. Cocoa and its products also contain good amount of flavanols. Quercetin is the most abundant flavanol found in fruits and possesses greatest activity against vascular complications [40]. Based upon the health-promoting effects of flavanols present in green tea, it has been proposed that regular consumption of green tea lowers the risk of developing high blood pressure and cardiovascular diseases [1]. Flavanols commonly found as monomeric, oligomeric, and polymeric forms. Monomeric form of flavanols such as catechin and epicatechin is absorbed predominantly in the intestine.

5.2.2.1 Flavanols in CVD

Cocoa flavanols have been studied in great details for their effect on cardiovascular health in various epidemiological and cohort studies of patients with respect to flavanol intake and status of cardiovascular health [41, 42]. In a randomized controlled clinical trials of 1131 participants, flavanol from cocoa products significantly improved biomarkers of lipid metabolism, systemic inflammation, and insulin resistance [42]. Daily consumption of cocoa flavanol for 8 weeks improved cognitive performance in a group of cognitively impacted older adults, without major adverse effects. Further, it also improved blood pressure, glucose levels, and insulin resistance [41]. Supplementation of dietary oligonol helped in reducing glucose, insulin levels, and inflammatory markers and improved oral glucose tolerance in high-fat diet (HFD)-induced obesity in mice. At the cellular level, flavanols decreased fat area, adipocyte, and adipokines like leptin which are known to be regulated by PPAR gene expression [43], suggesting the role of flavanols in regulating obesity, diabetes, and CVD-related gene expressions.

5.2.2.2 Flavanols in Inflammation

Leucocyte adhesion is one of the risk factors for developing CVDs and cell surface receptors such as CD40, CD36, and VLA-4 that help in this process. Flavanols are effective inhibitors of these molecules that lower their concentration in vivo. Flavanol metabolites have potential to counter the inflammation via regulating NF-κB signaling, regulation of the molecules involved in cell adhesion, and MAPK cell signaling pathways [11, 44]. In another study, flavanol-rich lychee fruit extract has shown anti-inflammatory activity mediated by decreasing inhibitory NF-κB phosphorylation, thus suppressing the NF-κB activation and its nuclear translocation [45]. Flavanols and their related oligomers isolated from cocoa have shown potential to alter immune system-related signaling molecules and inflammation in vitro in phytohemagluttinin (PHA)-stimulated human peripheral blood mononuclear cells (PBMC) and caused inhibition of TNF-α secretion and cytokine release [46].

5.2.2.3 Flavanols in Endothelial Dysfunction

Endothelial dysfunction, measured as flow-mediated dilation (FMD), is associated with the development of arterial hypertension and coronary vascular abnormalities in the long run. Generation of ROS, diminished activity of antioxidant enzyme system, inhibition of angiotensin-converting enzyme, and endothelial nitric oxide synthetase (eNOS) activation are responsible for FMD. In a study by Jumar and Schmieder (2016), consumption of cocoa flavanol could inhibit the risk factors associated with endothelial dysfunctions [11]. Further, another study showed that administration of flavanols could also improve ischemic heart disease via counteracting FMD and endothelial dysfunctions [47]. Flavanols have also shown vasodilatory activity which is majorly endothelium dependent. However, flavanol's endothelial-independent mechanism of action remains unclear. Several mechanisms have been proposed for explaining the mechanism of vasodilatory activity of flavanols. One of the mechanisms is concentration-dependent inhibition of Ca^{2+} uptake in aorta via altering the mobilization of calcium from the sarcoplasmic reticulum by flavanols. Flavanols are known to prevent phosphorylation of L-type calcium channels through inhibition of protein kinase C (PKC) [48].

5.2.3 Flavanones

Flavanones are predominantly found in majority of citrus fruits such as oranges and thus are referred as citrus flavonoids. Among the citrus flavonoids, the flavanone, hesperidin is the major flavonoid present in oranges, tangerines, grapefruits, and lemons. The flavanones are present predominantly in the skins of the citrus fruits.

Hesperidin is structurally a flavanone β-glycoside at C-7 of hesperetin aglycone, which is 3′,5,7-trihydroxy-4-methoxy flavanone. Its sugar moiety is a disaccharide rutinose composed of rhamnose and glucose, that is, O-α-L-rhamnosyl-(1 → 6)-glucose. Hesperidin has shown a wide variety of physiological activities including apoptosis induction, antiproliferative effects against cancer cells, and antibacterial, antiviral, and antifungal activities [13]. Orange juice has been found to significantly inhibit atherosclerosis and lower the cholesterol levels in patients [49]. In an in vivo study using mandarin juice enriched with hesperidin inverse, relation of juice consumption and colon cancer incidence in rats was observed indicating its anticancer properties [50].

5.2.3.1 Flavanone in Neuroprotection

Accumulation of metals in the body is associated with various biological complications such as Alzheimer's disease, reproductive toxicity, microcytic anemia, osteomalacia, dialysis encephalopathy, cytoskeletal pathologies, and hepato-renotoxicity in humans. Study by Jangra et al. (2015) showed that chronic treatment of hesperidin displayed inhibitory effect in $AlCl_3$-induced neurotoxicity by inhibiting oxidative-nitrosative stress and inflammation in the hippocampus due to toxic effect of $AlCl_3$ [51]. In an experimental model of streptozotocin-induced diabetes, oxidative stress and its effect in central nervous system (CNS) were significantly altered in hesperidin-treated animals as measured by downregulation of markers of neurotoxicity and oxidative stress [52]. In another study, Hesperetin, a metabolized form of hesperidin, showed the effectiveness on oxidative stress-induced neural damage using PC12 cells. Further, hesperidin was also found effective against oxidative stress, amyloid-beta associated neurotoxicity, and glutamate-induced excitotoxicity [53]. Hesperidin could also prevent neurotoxicity induced by cisplatin, which is known to cause significant peroxidation and inhibits antioxidant defense system in rats [54]. Thus, flavanones could impart significant neuroprotection in the experimental models showing their importance as functional food.

5.2.3.2 Flavanone in Inflammation

Inflammatory responses are in most cases regulated at transcriptional levels by transcription factors such as AP-1, NF-κB, and others. Hesperidin has been well documented for NF-κB inhibitory properties in the literatures [55]. It has been shown that such compounds can effectively inhibit nuclear translocation of redox-sensitive transcription factors, thereby inhibiting the transcription of genes involved in inflammatory process. In RAW 264.7 cells, treatment with hesperetin showed the strong inhibition of NF-κB and other mediators of inflammation including COX-2, iNOS, IL-6, and TNF-α. On the other hand, hesperetin also activates Nrf2/HO-1 pathway which plays important role in countering inflammation. Nrf2 is responsible for maintenance of antioxidant status of cells, and hesperetin treatment caused upregulation of Nrf2 expressions [56].

5.2.3.3 Flavanone in Cancer

Citrus flavonoids are also reported to possess anticancerous effects. For example, hesperetin has been shown to inhibit cancer cell proliferation, apoptosis, and release of angiogenic growth factors and various inflammatory markers in colon cancer [57]. In this study cell proliferation marker PCNA was decreased in the hesperetin treatment group as compared to the 1,2-dimethylhydrazine-treated group. Hesperetin also decreased the angiogenic factors such as basic fibroblast growth factor (bFGF), vascular endothelial growth factor (VEGF), and epidermal growth factor (EGF). Further, the expression of COX-2 mRNA was also affected in hesperetin-supplemented rats [57]. In another animal study, supplementation of hesperidin significantly improved chemical-induced oxidative stress and subsequent damage in mice lung [14]. Supplementation of hesperidin in benzo(a)pyrene (B(a)P)-treated mice leads to decreased lipid peroxides (LPO), carcinoembryonic antigen (CEA), and serum marker enzymes aryl hydrocarbon hydroxylase (AHH) as compared to the mice treated with (B(a)P) alone. Targeted drug therapy using hesperidin in combination with the established anticancer drug doxorubicin has shown higher cytotoxicity on MCF-7 breast cancer cells and HeLa cells [58]. These evidences clearly depict anticancer roles of flavanones.

5.2.3.4 Flavanone in Nephrotoxicity

Flavanones have also shown potential to prevent trichloroethylene (TCE)-induced nephrotoxicity in Wistar rats. TCE treatment in rats caused significant reduction in antioxidant enzymes and enhanced lipid peroxidation, while antioxidant system and lipid peroxidation improved in rats treated with hesperidin. Further, creatinine, kidney injury molecule (KIM-1), caspases and bax expression decreased in hesperidin-treated rats as compared with only TCE-treated rats [59]. Similar study addressing nephrotoxicity and preventive effects of hesperidin were studied using cisplatin as a nephrotoxicity inducer. Cisplatin increased serum creatinine, blood urea nitrogen, and serum sodium, and decreased serum total protein, SOD, and creatinine clearance, whereas administration of hesperidin prevented cisplatin-induced changes and improved kidney function [60].

5.2.4 Flavonols

Flavonols are present in variety of diets as diverse glycosides. All the vegetables and fruits provide flavonols; however onions, grapes, cider, wine, and tea are among the richest sources of flavonols. Flavonols are derived from 3-hydroxyflavone. The common flavonols are quercetin, isorhamnetin, fisetin, myricetin, morin, kaempferol, and tamarixetin. Quercetin is the most abundant flavonol found in plants. Daily intake of flavonols has been estimated as 20–35 mg per day, of which 65% is quercetin and its glycosides [61]. The presence of multiple hydroxyl group in structure of quercetin is responsible for its vivid physiological effects including antioxidant, anti-inflammatory, anticancer, and antiviral effects [15]. Quercetin has lower water solubility; hence its physiological effects are limited, but its formulation with other proteins or lipids may improve its bioavailability. Chemical modification and its association with proteins or different formulation like soy protein isolates improves the ability of quercetin as an antioxidant as well as its solubility and potency in the system [62]. Flavonols have been studied widely for their beneficiary effects on human health which includes antioxidant, antimicrobial, antiproliferative, anticancer, cardio-protective, and prevention of neurodegeneration [63].

5.2.4.1 Flavonols in Inflammation

Inflammation is an important physiological process which is regulated tightly, or it could lead to diseased conditions. Inflammation is associated with many diseases including cancer, obesity, atherosclerosis, colitis, and inflammatory bowel disease. Colitis is one such inflammatory disorder in which myeloperoxidase activity is increased. In acetic acid-induced colitis model of rats, quercetin administration reduced myeloperoxidase activity and improved the antioxidant capacity [64]. In study of healthy volunteers, flavonol administration inhibited LPS-induced TNF-α in blood and improved the total plasma antioxidants in vitro [65]. In viral infection study, pretreatment of airway epithelial cells reduced AKT phosphorylation, viral uptake by cells, and IL-8 response, whereas postinfection administration of flavonols has their effects on viral load, IL-8, and IFN response in airway epithelial cells. In rats, similar infection and flavonol administration reduced replication of viral genetic material [65], which suggests that these compounds act differently at various stages of disease.

Fisetin (3,3′,4′,7-tetrahydroxyflavone) is flavonol found in fruits and vegetables including strawberry, apple, cucumber, onion, persimmon, and grapes. Fisetin and its anti-inflammatory activity have been well studied. Fisetin showed inhibitory effect on NF-κB activation and its translocation to the nucleus. Further, fisetin in dose-dependent manner modulated interleukin levels in bronchoalveolar lavage fluid [66]. Kaempferol, a flavonol, also showed NF-κB inhibitory activity in LPS-induced RAW264.7 cells. Kaempferol via modulation of inflammation by NF-κB, p38, and ERK1/2 pathway found effective in preventing or curing diseases like Japanese encephalomyelitis and rheumatoid arthritis [67].

Quercetin treatment is effective in mice exposed to elastase/LPS; quercetin treatment prevented oxidative stress, lung inflammation, and goblet cell metaplasia. It also effectively inhibited MMP9 and MMP12 expression, which are markers of epithelial to mesenchymal transformation [68]. Quercetin has anti-inflammatory activity which depends on the inhibition of NF-κB pathway. Quercetin inhibits nuclear translocation and DNA binding of NF-κB. Further, quercetin and its other derivatives have been shown to possess anti-inflammatory activity via PKA and PKC signaling pathway [69]. In rat model of adjuvant arthritis, quercetin administration resulted in amelioration of oxidative stress and inflammation. Inhibition of inflammation in quercetin-treated animal was mainly dependent on the NF-κB inhibition and resulted in lowered level of C-reactive protein, monocyte chemotactic protein-1 (MCP-1), and improved antioxidant capacity [70].

Flavonol fraction isolated from date fruit was shown to have antiatherogenic activity, measured by their ability to scavenge free radicals, removal of fat from macrophages, and reduced ox-LDL formation [71]. Differentiation of osteoblasts is inhibited in the osteoporosis which is contributed by the high oxidative stress. Flavonol treatment in rat's osteoblasts showed upregulation of antioxidant gene response where ERK1/2 and NF-κB were suppressed, while Nrf2 was not affected in these cells [72].

5.2.4.2 Flavonol in Cancer

Epidemiological data show that consumption of flavonols reduces the risk of developing cancer, mostly due to their antioxidant activity. Flavonol consumption decreased the cell proliferation markers in cancer cells and had direct effect on several protein kinases. Diverse mechanisms of action make flavonols suitable molecule against cancer [73]. Quercetin also found effective agent against herbicide-induced oxidative stress and other modulations in lung cancer (A549) cell line. Quercetin inhibited herbicide paraquat dichloride-induced cytotoxicity and activation of transcription factor Nrf2 and its target gene HO-1 expression [74]. Fisetin treatment inhibited growth and survival of gastric cancer cells and generated ROS which led to DNA damage and ultimately apoptosis of the cancer cells [75]. These studies clearly indicate that flavonols could be useful as functional food for their anticancer properties.

5.2.4.3 Flavonol in Endothelial Dysfunction

Accumulation of lipids and fibrous materials in large arteries are major contributors of atherosclerosis and CVDs in the long run. At such sites, inflammation in endothelial cells leads to endothelial dysfunction and modulation of sVCAM-1, sICAM-1, and sE-selectin which are risk factors of CVDs. In pre-hypertensive men and women, quercetin administration blocked inflammation and endothelial dysfunction-related markers which suggested that quercetin might be useful in countering CVDs [76]. Similarly, in a cohort study of elderly women, flavonol intake showed reduced risk of atherosclerosis, which suggests that flavonol-rich diet may play preventive role against atherosclerosis and CVDs [77]. In a randomized clinical trial, supplementation of mono- and oligomeric flavonols showed significant decrease in serum total cholesterol and LDL in volunteers with elevated baseline levels. Further, supplementation of flavonoids showed anti-inflammatory effect in leucocytes treated with bacterial endotoxin ex vivo [78].

5.2.5 Procyanidins

Procyanidins (PCs) are the major flavonoid found mainly in fruits, legumes, cereals, and beverages such as tea, wine, and cocoa. Fruits such as apple, curry, barley, currant, strawberry, red grapes, and walnut provide 60 to 100 mg/100 g procyanidins. Cacao beans, chocolate, cinnamon, and sorghum are richer sources of procyanidins and provide more than 1000 mg/100 g of food. Many epidemiological studies have described health beneficiary effects of flavonoids isolated from the wine, chocolate, and cocoa [79]. Procyanidins have been found to show protection against CVDs due to its antioxidant activity, modification of blood lipid profile, and anti-inflammatory activity [80].

5.2.5.1 Procyanidins in Neurological Disorders

Alzheimer's disease is caused by accumulation of plaques made up of amyloid-β (Aβ) proteins. Consumption of apple-derived procyanidins was shown to attenuate formation of insoluble plaque of Aβ proteins in a dose-dependent manner. Further, it stimulated proliferation of neuronal cells in vitro in PC-12 cells which suggests that foods rich in procyanidins contents may be neuroprotective [81]. Yokukansan, a traditional Japanese medicine containing procyanidin B1, is used for the treatment of neurosis, insomnia, and night crying and irritability in children. Procyanidin B1 derived from the plant used to make Yokukansan showed effectiveness against Aβ oligomer-induced apoptosis in primary neuron culture by suppressing caspase-3 [82].

5.2.5.2 Procyanidins in Cancer

Procyanidins have also been shown to play role in cancer prevention by inhibiting matrix metalloproteinase (MMP)-mediated proteolysis and enhancing cross-linking of vascular extracellular matrix in vitro and inhibition of angiogenesis in vivo [83]. Procyanidin fraction of grapes has stronger antioxidant activity which was able to induce protein expression of Nrf2 in human hepatocarcinoma (HepG2) cells and antioxidant enzymes including NAD(P)H:quinone oxidoreductase1 and hemeoxygenase1 via p38 and PI3K/Akt pathway [84]. Further, procyanidins were shown to affect the activity of phosphatases of regenerating liver (PRLs). These are important therapeutic targets against cancer which are inhibited using micromolar concentration of procyanidins in PRL-1 and PRL-3 overexpressing stable HEK293 cell line. Inhibition of such enzymes resulted in prevention of migration of cancer cells [85].

5.2.5.3 Procyanidins in Inflammation

Oligomeric procyanidins isolated from non-ripe apple peel have shown antiviral activities in human peripheral blood mononuclear cells (PBMCs). Further, dengue virus-infected PBMC showed decreased viral load when treated with procyanidins. Procyanidins increased expression of *STAT1* transcripts and MHC I and TNF-α protein production in PBMCs [16]. Procyanidins isolated from wild grapes showed anti-inflammatory activity in LPS-treated RAW 264.7 cells and lowered NO, prostaglandin E_2, and ROS levels. Further, procyanidins also affected the expression of COX-2, iNOS, TNF-α, and IL-1β by regulating p38 MAPK/NF-κB pathway [86]. These studies thus propound that procyanidins could have anti-inflammatory activities and may be effective in inflammatory diseases.

5.2.5.4 Procyanidins in Metabolic Disorders

Procyanidins extracted from the grapes have been shown to regulate triglyceridemia in normolipidemic animals by causing significant downregulation of transcription factor, namely, steroid response element-binding protein 1 (SREBP1) and its target genes involved in lipogenesis [87]. This study suggests that consumption of functional food containing procyanidin-rich diet could have implication on body weight management and can also prevent from cytotoxic effects of oxidative stress and inhibit the proliferation of preadipocytes [88]. In an in vitro study using cocoa extract and procyanidins, procyanidins were found potent inhibitors of digestive enzymes including pancreatic α-amylase, secreted phospholipase, and pancreatic lipase, which suggest that procyanidins can function as anti-obesity agent along with providing the low caloric diet [89] and thus could be excellent functional food.

In cholesterol-fed rabbit model of atherosclerosis, proanthocyanidin-rich extract isolated from the grape seeds was effective in prevention of severe atherosclerosis in the aorta, although it had no effect on LDL level. Further, ox-LDL-positive macrophages in atherosclerotic lesion were decreased in proanthocyanidin-fed rabbits, and it also prevented oxidation of LDL in human plasma in vitro [90]. Additionally, grape seed-derived proanthocyanidin extract demonstrated dose-dependent decrease in 12-O-tetradecanoylphorbol-13-acetate (TPA)-induced ROS production and lipid peroxidation in peritoneal macrophages.

Insulin resistance is one of the major contributing factors in the onset of diabetes. In vivo model of diabetes and obesity showed modulation of glucose uptake after treating the animals with procyanidins. Procyanidin treatment caused significantly modulated lipogenesis, oxidative stress, and inflammatory state in various tissues. Further, procyanidins showed positive effect on insulin secretion and β-cell mass in pancreas [91].

5.3 Present Scenario of Flavonoids as Functional Food

Chocolate, tea, wine, and cocoa are rich in the flavonoid content, and these food items and beverages are part of many traditional as well as contemporary diets. These food items provide the required amount of flavonoids which have been proved in many clinical studies [92]. The studies have focused on the identification of health beneficiary effects of flavonoids in different diseases and in different populations.

Tea (*Camellia sinensis*) is the second most consumed beverage after water. Tea is rich source of flavonoids and provides the sufficient amount of flavonoids to tea-consuming populations. Both green and black tea contain high amount of flavonoids [93]. It is recommended by the dieticians that two to three servings of tea could contribute enough flavonoids on daily basis [94]. Tea-consuming population could provide vital information about the effect of flavonoids on the health and disease conditions. Indeed, many in vitro and in vivo studies have indicated the effect of tea-derived flavonoids [95]. Several clinical studies show the correlation between beneficial effects of tea intake against various diseases including cardiovascular diseases [93, 96]. Regular tea intake has been associated with comparatively lower cholesterol level; however it has failed to show consistent results in clinical trials [97]. Hodgson and Croft (2010) reviewed the effect of tea flavonoids on cardiovascular health in which they listed results of various clinical trials and epidemiological data. Ten cohort studies and seven case control studies have revealed that consumption of three cups of tea could decrease the incidence rate of myocardial infarction by 11% [98]. Meta-analysis by Huxley and Neil (2003) showed significant relation between the dietary intake from tea, wine, fruits, and vegetable and reduction in the CHD-related mortalities in the seven cohort study of men and women [96].

A study using C3H mice showed enhancement in the effects of tamoxifen on mammary tumor when used in combination with green tea-derived catechin and showed synergistic effect of anticancer drug and green tea extracts [99]. In another in vivo study using 7,12-dimethylbenz[a]anthracene (DMBA) as an inducer of cancer in Sprague-Dawley rat, green tea-derived catechin treatment improved survival rate as compared to rats on the basal diet [99]. Although the effects of green tea-derived catechins are promising in in vivo study using animal model; however, more detailed studies in patients involving tea flavonoids and their preventive effects on diseases are required to be investigated.

Cocoa-derived flavonoids are vastly studied in clinical setup for their effect on cardiovascular health and related diseases. Healthy individuals were incorporated in the clinical study in which the effect of cocoa flavonoids was observed [100]. The study revealed that supplementation with epicatechin did not alter the flow-mediated dilation, and plasma glucose level significantly improved along with fasting plasma insulin level and significantly decreased insulin resistance. Epicatechin supplementation did not changed other risk factors associated with CVDs including nitric oxide level, endothelin 1 or blood lipid profile, and arterial stiffness [100]. Heiss et al. (2015) studied the impact of cocoa flavanol intake on cardiovascular health of healthy individuals where volunteers with no previous sign and history of CVD were incorporated [101]. Two-week intervention of cocoa flavonol caused significant improvement in FMD in the young as well as elderly volunteers. In elderly individuals, intake of cocoa flavonoids reverted the age-related burden of CVDs [101]. Small amount of flavonoid-rich chocolate consumption in healthy adults reduced the risk factors associated with the cardiovascular diseases. Similar results were found in Zutphen Elderly study where chocolate consumption resulted in the lower cardiovascular mortalities. Authors concluded that chocolate could be beneficial for the health if taken in the moderate amount with minimum fats and sugars in diet [102]. In the study by Flammer et al. (2007), consumption of dark chocolate caused significant reduction in the platelet adhesion in less than 3 hours. Since chocolate contains high amount of flavonoids, it could possibly be responsible for these outcomes. Dark chocolate contains high amount of polyphenols per gram than other polyphenol-rich food items such as wine, berries, or tea. Therefore, even a small amount of dark chocolate could be sufficient for the significant effect on endothelia dysfunction and platelet activation [103].

In another study, volunteers were given 1.4 g cocoa extract in calorie-restricted diet. Researchers in this study found that the plasma monoamines were altered among which homovanillic acids were significantly upregulated and overall symptoms of depression among patients decreased [104]. Similarly, effect of flavonoid-rich chocolate and risk factors of CVDs were studied in postmenopausal women with type 2 diabetes undergoing hypoglycemic therapy and treatment with statin. In this study, researchers found that, compared to the placebo group, flavonoid-rich chocolate consumption resulted in a significant reduction in estimated peripheral insulin resistance and thus insulin sensitivity was enhanced. These changes led to decrease in the levels of insulin. Total cholesterol to HDL cholesterol ratio and LDL levels were also decreased. Further, 1 year intervention with flavan-3-ols and isoflavones caused improvement in insulin resistance and the lipid profile which could cause delay in the onset of CVD in patients with type 2 diabetes [105]. Consumption of flavanone-rich grape fruit juice caused decrease in the arterial stiffness in the healthy postmenopausal women. However, other factors of vascular health did not show significant change [106]. Cocoa consumption with hypocaloric diet caused significant improvement in ox-LDL levels in middle-aged male volunteers [107].

Wine, derived from grapes (*Vitis vinifera*), is another rich source of polyphenols and is widely consumed in many parts of the world. In a randomized trial performed for 2 years with 108 patients having carotid atherosclerosis, patients were divided in two groups, i.e., lifestyle change group and no lifestyle change group. Lifestyle changes included modified diet, sessions of physical exercise, and daily intake of red wine. Intake of red wine along with the lifestyle changes showed improvement of LDL/HDL ratio in patients [108]. In another randomized crossover trials with healthy individuals, consumption of red wine along with Mediterranean meal showed improved status of oxidized LDL and antioxidant gene expression [109].

However, many of these human clinical trials remained inconclusive due to inadequate sample size, limitation of study design, and other factors such as baseline risk factors in study population. For example, study with broader outcome is less likely to show significance between tea consumption and disease prevention, whereas in case of restricted outcome, results are more significant [110, 111].

5.4 Summary

Flavonoids are undoubtedly very useful bioactive compounds with the many beneficial effects. Some of these flavonoids have been investigated in detail, and some of them have returned promising results in clinical studies. Cocoa, chocolate, tea, and coffee are flavonoid-rich food items and consumed by the vast population around the world. Vegetables and fruits also provide good amount of flavonoids. Flavonoids have been shown to affect multiple targets in the cells; hence they have impact on many cellular pathways and thus are beneficial for maintaining the homeostatic condition in cells. This could be the reason behind the antioxidant, anti-inflammatory, anticancer, antidiabetic, and cardio-protective effects displayed by flavonoids containing functional foods. Studies have been performed for the effects of flavonoid on human health promotion and disease prevention. Many clinical studies remained inconclusive about the effects of the single flavonoid and its role in disease prevention. Effects of flavonoids may change according to genetic makeup and environmental factors as well as human lifestyle and food habits. Further, consumption of flavonoids alone or in combination with the drug can be helpful in treating or controlling the disease. In conclusion, addition of flavonoid-rich diet may give promising health-promoting effects and may act as functional food.

Acknowledgment Dr. Umesh C S Yadav acknowledges the award of Ramanujan Fellowship by DST, Govt. of India. KRP and FDC would like to acknowledge University Grant Commission (UGC) for providing NON-NET fellowship and NFOBC fellowship.

References

1. Hasler CM (2002) Functional foods: benefits, concerns and challenges-a position paper from the american council on science and health. J Nutr 132:3772–3781. https://doi.org/10.1002/mus.20330
2. Pandey MM, Rastogi S, Rawat AKS (2013) Indian traditional ayurvedic system of medicine and nutritional supplementation. Evid Based Complement Alternat Med 2013:376327. https://doi.org/10.1155/2013/376327
3. Hasler C (1998) Functional foods: their role in disease prevention and health promotion. Food Technol 52:61–72
4. Kumar S, Pandey AK (2013) Chemistry and biological activities of flavonoids: an overview. Sci World J 2013:1–16. https://doi.org/10.1155/2013/162750
5. Cirmi S, Ferlazzo N, Lombardo GE et al (2016) Chemopreventive agents and inhibitors of cancer hallmarks: may citrus offer new perspectives? Nutrients 8:1–38. https://doi.org/10.3390/nu8110698
6. Saito K, Yonekura-Sakakibara K, Nakabayashi R et al (2013) The flavonoid biosynthetic pathway in Arabidopsis: structural and genetic diversity. Plant Physiol Biochem 72:21–34. https://doi.org/10.1016/j.plaphy.2013.02.001
7. Bhagwat S, Haytowitz DB, Holden JM (2011) USDA Database for the Flavonoid Content of Selected Foods Release 3.2
8. Grotewold E (2006) The science of flavonoids. Springer. https://doi.org/10.1007/978-0-387-28822-2
9. Davinelli S, Bertoglio JC, Zarrelli A et al (2015) A randomized clinical trial evaluating the efficacy of an anthocyanin–Maqui berry extract (delphinol®) on oxidative stress biomarkers. J Am Coll Nutr 34:28–33. https://doi.org/10.1080/07315724.2015.1080108
10. Song Y, Huang L, Yu J (2016) Effects of blueberry anthocyanins on retinal oxidative stress and inflammation in diabetes through Nrf2/HO-1 signaling. J Neuroimmunol 301:1–6. https://doi.org/10.1016/j.jneuroim.2016.11.001
11. Jumar A, Schmieder RE (2016) Cocoa flavanol cardiovascular effects beyond blood pressure reduction. J Clin Hypertens 18:352–358. https://doi.org/10.1111/jch.12715
12. Hagiwara K, Okura M, Sumikawa Y et al (2016) Biochemical effects of the flavanol-rich lychee fruit extract on the melanin biosynthesis and reactive oxygen species. J Dermatol 43:1174–1183. https://doi.org/10.1111/1346-8138.13326
13. Suzuki H, Asakawa A, Kawamura N et al (2014) Hesperidin potentiates ghrelin signaling. Recent Pat Food Nutr Agric 6:60–63. https://doi.org/10.2174/2212798406666140825120623
14. Kamaraj S, Ramakrishnan G, Anandakumar P et al (2009) Antioxidant and anticancer efficacy of hesperidin in benzo(a)pyrene induced lung carcinogenesis in mice. Investig New Drugs 27:214–222. https://doi.org/10.1007/s10637-008-9159-7

15. Wang Y, Wang X (2015) Binding, stability, and antioxidant activity of quercetin with soy protein isolate particles. Food Chem 188:24–29. https://doi.org/10.1016/j.foodchem.2015.04.127
16. Kimmel EM, Jerome M, Holderness J et al (2011) Oligomeric procyanidins stimulate innate antiviral immunity in dengue virus infected human PBMCs. Antivir Res 90:80–86. https://doi.org/10.1016/j.antiviral.2011.02.011
17. Miguel MG (2011) Anthocyanins: antioxidant and/or anti-inflammatory activities. J Appl Pharm Sci 1:7–15
18. Jaakola L (2013) New insights into the regulation of anthocyanin biosynthesis in fruits. Trends Plant Sci 18:477–483. https://doi.org/10.1016/j.tplants.2013.06.003
19. J a N, B a C, Kurilich AC (2012) Anthocyanin kinetics are dependent on anthocyanin structure. Br J Nutr 107:504–509. https://doi.org/10.1017/S000711451100314X
20. Lin BW, Gong CC, Song HF, Cui YY (2016) Effects of anthocyanins on the prevention and treatment of cancer. Br J Pharmacol. https://doi.org/10.1111/bph.13627
21. Kuntz S, Kunz C, Herrmann J et al (2014) Anthocyanins from fruit juices improve the antioxidant status of healthy young female volunteers without affecting anti-inflammatory parameters: results from the randomised, double-blind, placebo-controlled, cross-over ANTHONIA (ANTHOcyanins in Nutrition Investigation Alliance) study. Br J Nutr 112:925–936. https://doi.org/10.1017/S0007114514001482
22. Lim S, Xu J, Kim J et al (2013) Role of anthocyanin-enriched purple-fleshed sweet potato p40 in colorectal cancer prevention. Mol Nutr Food Res 57:1908–1917. https://doi.org/10.1002/mnfr.201300040
23. Goetz ME, Judd SE, Safford MM et al (2016) Dietary flavonoid intake and incident coronary heart disease: The REasons for Geographic and Racial Differences in Stroke (REGARDS) study. Am J Clin Nutr 104:1236–1244. https://doi.org/10.3945/ajcn.115.129452
24. Ferreira Reis J, Vinicius V, Monteiro S et al (2016) Action mechanism and cardiovascular effect of anthocyanins: a systematic review of animal and human studies. J Transl Med 14:315. https://doi.org/10.1186/s12967-016-1076-5
25. Yacout SM, Gaillard ER (2016) The anthocyanins, oenin and callistephin, protect RPE cells against oxidative stress. Photochem Photobiol 38:42–49. https://doi.org/10.1111/php.12683
26. Stull A (2016) Blueberries' impact on insulin resistance and glucose intolerance. Antioxidants 5:44. https://doi.org/10.3390/antiox5040044
27. Huang W, Zhu Y, Li C et al (2016) Effect of blueberry anthocyanins malvidin and glycosides on the antioxidant properties in endothelial cells. Oxidative Med Cell Longev. https://doi.org/10.1155/2016/1591803
28. Han KH, Sekikawa M, Shimada K et al (2006) Anthocyanin-rich purple potato flake extract has antioxidant capacity and improves antioxidant potential in rats. Br J Nutr 96:1125–1133. https://doi.org/10.1017/Bjn20061928
29. Hwang JW, Kim EK, Lee SJ et al (2012) Anthocyanin effectively scavenges free radicals and protects retinal cells from H2O2-triggered G2/M arrest. Eur Food Res Technol 234:431–439. https://doi.org/10.1007/s00217-011-1648-9
30. Muselík J, García-Alonso M, Martín-López MP et al (2007) Measurement of antioxidant activity of wine catechins, procyanidins, anthocyanins and pyranoanthocyanins. Int J Mol Sci 8:797–809. https://doi.org/10.3390/i8080797
31. Chang YC, Huang KX, Huang AC et al (2006) Hibiscus anthocyanins-rich extract inhibited LDL oxidation and oxLDL-mediated macrophages apoptosis. Food Chem Toxicol 44:1015–1023. https://doi.org/10.1016/j.fct.2005.12.006
32. Zhu Y, Huang X, Zhang Y et al (2014) Anthocyanin supplementation improves HDL-Associated paraoxonase 1 activity and enhances cholesterol efflux capacity in subjects with hypercholesterolemia. J Clin Endocrinol Metab 99:561–569. https://doi.org/10.1210/jc.2013-2845
33. Kim B, Lee SG, Park YK et al (2016) Blueberry, blackberry, and blackcurrant differentially affect plasma lipids and pro-inflammatory markers in diet-induced obesity mice. Nutr Res Pract 10:494–500. https://doi.org/10.4162/nrp.2016.10.5.494

34. Li L, Wang L, Wu Z et al (2014) Anthocyanin-rich fractions from red raspberries attenuate inflammation in both RAW264.7 macrophages and a mouse model of colitis. Sci Rep 4:6234. https://doi.org/10.1038/srep06234
35. Ferrari D, Speciale A, Cristani M et al (2016) Cyanidin-3-O-glucoside inhibits NF-kB signalling in intestinal epithelial cells exposed to TNF- α and exerts protective effects via Nrf2 pathway activation. Toxicol Lett 264:51–58. https://doi.org/10.1016/j.toxlet.2016.10.014
36. Shah SA, Amin FU, Khan M et al (2016) Anthocyanins abrogate glutamate-induced AMPK activation, oxidative stress, neuroinflammation, and neurodegeneration in postnatal rat brain. J Neuroinflammation 13:286. https://doi.org/10.1186/s12974-016-0752-y
37. Shukitt-Hale B, Kelly ME, Bielinski DF, Fisher DR (2016) Tart cherry extracts reduce inflammatory and oxidative stress signaling in microglial cells. Antioxidants. https://doi.org/10.3390/antiox5040033
38. Zhang PW, Chen FX, Li D et al (2015) A CONSORT-compliant, randomized, double-blind, placebo-controlled pilot trial of purified anthocyanin in patients with nonalcoholic fatty liver disease. Medicine (Baltimore) 94:1–8. https://doi.org/10.1097/MD.0000000000000758
39. Zhu Y, Ling W, Guo H et al (2013) Anti-inflammatory effect of purified dietary anthocyanin in adults with hypercholesterolemia: a randomized controlled trial. Nutr Metab Cardiovasc Dis 23:843–849. https://doi.org/10.1016/j.numecd.2012.06.005
40. Leo C-H, Woodman OL (2015) Flavonols in the prevention of diabetes-induced vascular dysfunction. J Cardiovasc Pharmacol 65:532–544. https://doi.org/10.1097/FJC.0000000000000180
41. Mastroiacovo D, Kwik-uribe C, Grassi D et al (2015) Cocoa flavanol consumption improves cognitive function, blood pressure control, and metabolic profile in elderly subjects: the Cocoa, Cognition, and Aging (CoCoA) Study – a randomized controlled trial. Am J Clin Nutr 101:538–548. https://doi.org/10.3945/ajcn.114.092189.1
42. Lin X, Zhang I, Li A et al (2016) Cocoa flavanol intake and biomarkers for cardiometabolic health: a systematic review and meta-analysis of randomized controlled trials. J Nutr 146:2325–2333. https://doi.org/10.3945/jn.116.237644
43. Liu HW, Wei CC, Chen YJ et al (2016) Flavanol-rich lychee fruit extract alleviates diet-induced insulin resistance via suppressing mTOR/SREBP-1 mediated lipogenesis in liver and restoring insulin signaling in skeletal muscle. Mol Nutr Food Res 60:2288–2296. https://doi.org/10.1002/mnfr.201501064
44. Claude S, Boby C, Rodriguez-Mateos A et al (2014) Flavanol metabolites reduce monocyte adhesion to endothelial cells through modulation of expression of genes via p38-MAPK and p65-Nf-kB pathways. Mol Nutr Food Res 58:1016–1027. https://doi.org/10.1002/mnfr.201300658
45. Yamanishi R, Yoshigai E, Okuyama T et al (2014) The anti-inflammatory effects of flavanol-rich lychee fruit extract in rat hepatocytes. PLoS One. https://doi.org/10.1371/journal.pone.0093818
46. Mao TK, van de Water J, Keen CL et al (2002) Modulation of TNF-α secretion in peripheral blood mononuclear cells by cocoa flavanols and procyanidins. Dev Immunol 9:135–141. https://doi.org/10.1080/1044667031000137601
47. De Palma R, Sotto I, Wood EG et al (2016) Cocoa flavanols reduce N-terminal pro-B-type natriuretic peptide in patients with chronic heart failure. ESC Hear Fail 3:97–106. https://doi.org/10.1002/ehf2.12077
48. Duarte J, Vizcaíno FP, Utrilla P et al (1993) Vasodilatory effects of flavonoids in rat aortic smooth muscle. Structure-activity relationships. Gen Pharmacol Vasc Syst 24:857–862. https://doi.org/10.1016/0306-3623(93)90159-U
49. Assini JM, Mulvihill EE, Huff MW (2013) Citrus flavonoids and lipid metabolism. Curr Opin Lipidol 24:34–40. https://doi.org/10.1097/MOL.0b013e32835c07fd
50. Wang YQ, Zhang HM, Zhang GC et al (2007) Interaction of the flavonoid hesperidin with bovine serum albumin: a fluorescence quenching study. J Lumin 126:211–218. https://doi.org/10.1016/j.jlumin.2006.06.013

51. Jangra A, Kasbe P, Pandey SN et al (2015) Hesperidin and silibinin ameliorate aluminum-induced neurotoxicity: modulation of antioxidants and inflammatory cytokines level in mice hippocampus. Biol Trace Elem Res 168:462–471. https://doi.org/10.1007/s12011-015-0375-7
52. Ashafaq M, Varshney L, Khan MHA et al (2014) Neuromodulatory effects of hesperidin in mitigating oxidative stress in streptozotocin induced diabetes. Biomed Res Int. https://doi.org/10.1155/2014/249031
53. Hwang SL, Shih PH, Yen GC (2012) Neuroprotective effects of citrus flavonoids. J Agric Food Chem 60:877–885. https://doi.org/10.1021/jf204452y
54. Kamisli S, Ciftci O, Kaya K et al (2013) Hesperidin protects brain and sciatic nerve tissues against cisplatin-induced oxidative, histological and electromyographical side effects in rats. Toxicol Ind Health 31:841–851. https://doi.org/10.1177/0748233713483192
55. Parhiz H, Roohbakhsh A, Soltani F et al (2015) Antioxidant and anti-inflammatory properties of the citrus flavonoids hesperidin and hesperetin: an updated review of their molecular mechanisms and experimental models. Phytother Res 29:323–331. https://doi.org/10.1002/ptr.5256
56. Ren H, Hao J, Liu T et al (2016) Hesperetin suppresses inflammatory responses in lipopolysaccharide-induced RAW 264.7 cells via the inhibition of NF-κB and activation of Nrf2/HO-1 pathways. Inflammation 39:964–973. https://doi.org/10.1007/s10753-016-0311-9
57. Nalini N, Aranganathan S, Kabalimurthy J (2012) Chemopreventive efficacy of hesperetin (citrus flavonone) against 1,2-dimethylhydrazine-induced rat colon carcinogenesis. Toxicol Mech Methods 22:397–408. https://doi.org/10.3109/15376516.2012.673092
58. Meiyanto E, Hermawan A, Anindyajati (2012) Natural products for cancer-targeted therapy: citrus flavonoids as potent chemopreventive agents. Asian Pacific J Cancer Prev 13:427–436. https://doi.org/10.7314/APJCP.2012.13.2.427
59. Siddiqi A, Nafees S, Rashid S et al (2015) Hesperidin ameliorates trichloroethylene-induced nephrotoxicity by abrogation of oxidative stress and apoptosis in wistar rats. Mol Cell Biochem 406:9–20. https://doi.org/10.1007/s11010-015-2400-8
60. Kamel KM, Abd El-Raouf OM, Metwally SA et al (2014) Hesperidin and rutin, antioxidant citrus flavonoids, attenuate cisplatin-induced nephrotoxicity in rats. J Biochem Mol Toxicol 28:312–319. https://doi.org/10.1002/jbt.21567
61. Perez-Vizcaino F, Duarte J (2010) Flavonols and cardiovascular disease. Mol Asp Med 31:478–494. https://doi.org/10.1016/j.mam.2010.09.002
62. Lee YH, Kim HJ, Yoo H et al (2015) Synthesis of (2-amino)ethyl derivatives of quercetin 3-O-methyl ether and their antioxidant and neuroprotective effects. Bioorganic Med Chem 23:4970–4979. https://doi.org/10.1016/j.bmc.2015.05.023
63. Jeganathan B, Punyasiri PAN, Kottawa-Arachchi JD et al (2016) Genetic Variation of Flavonols Quercetin, Myricetin, and Kaempferol in the Sri Lankan Tea (Camellia sinensis L.) and Their Health-Promoting Aspects. Int. J Food Sci 2016:6057434. https://doi.org/10.1155/2016/6057434
64. Guazelli CFS, Fattori V, Colombo BB et al (2013) Quercetin-loaded microcapsules ameliorate experimental colitis in mice by anti-inflammatory and antioxidant mechanisms. J Nat Prod 76:200–208. https://doi.org/10.1021/np300670w
65. Boots AW, Wilms LC, Swennen ELR et al (2008) In vitro and ex vivo anti-inflammatory activity of quercetin in healthy volunteers. Nutrition 24:703–710. https://doi.org/10.1016/j.nut.2008.03.023
66. Goh FY, Upton N, Guan S et al (2012) Fisetin, a bioactive flavonol, attenuates allergic airway inflammation through negative regulation of NF-κB. Eur J Pharmacol 679:109–116. https://doi.org/10.1016/j.ejphar.2012.01.002
67. Rajendran P, Rengarajan T, Nandakumar N et al (2014) Kaempferol, a potential cytostatic and cure for inflammatory disorders. Eur J Med Chem 86:103–112. https://doi.org/10.1016/j.ejmech.2014.08.011
68. Ganesan S, Faris AN, Comstock AT et al (2010) Quercetin prevents progression of disease in elastase/LPS-exposed mice by negatively regulating MMP expression. Respir Res 11:131. https://doi.org/10.1186/1465-9921-11-131

69. Kim BH, Choi JS, Yi EH et al (2013) Relative antioxidant activities of quercetin and its structurally related substances and their effects on NF-κB/CRE/AP-1 signaling in murine macrophages. Mol Cells 35:410–420. https://doi.org/10.1007/s10059-013-0031-z
70. Gardi C, Bauerova K, Stringa B et al (2015) Quercetin reduced inflammation and increased antioxidant defense in rat adjuvant arthritis. Arch Biochem Biophys 583:150–157. https://doi.org/10.1016/j.abb.2015.08.008
71. Borochov-Neori H, Judeinstein S, Greenberg A et al (2015) Antioxidant and antiatherogenic properties of phenolic acid and flavonol fractions of fruits of "Amari" and "Hallawi" date (Phoenix dactylifera L.) varieties. J Agric Food Chem 63:3189–3195. https://doi.org/10.1021/jf506094r
72. Messer JG, Hopkins RG, Kipp DE (2015) Quercetin metabolites up-regulate the antioxidant response in osteoblasts isolated from fetal rat calvaria. J Cell Biochem 116:1857–1866. https://doi.org/10.1002/jcb.25141
73. Lea MA (2015) Flavonol regulation in tumor cells. J Cell Biochem 116:1190–1194. https://doi.org/10.1002/jcb.25098
74. Zerin T, Kim YS, Hong SY, Song HY (2013) Quercetin reduces oxidative damage induced by paraquat via modulating expression of antioxidant genes in A549 cells. J Appl Toxicol 33:1460–1467. https://doi.org/10.1002/jat.2812
75. Sabarwal A, Agarwal R, Singh RP (2017) Fisetin inhibits cellular proliferation and induces mitochondria-dependent apoptosis in human gastric cancer cells. Mol Carcinog 56:499–514. https://doi.org/10.1002/mc.22512
76. Dower JI, Geleijnse JM, Gijsbers L et al (2015) Supplementation of the pure flavonoids epicatechin and quercetin affects some biomarkers of endothelial dysfunction and inflammation in (pre)hypertensive adults: a randomized double-blind, placebo-controlled, crossover trial. J Nutr 145:1459–1463. https://doi.org/10.3945/jn.115.211888
77. Ivey KL, Lewis JR, Prince RL, Hodgson JM (2013) Tea and non-tea flavonol intakes in relation to atherosclerotic vascular disease mortality in older women. Br J Nutr 110:1648–1655. https://doi.org/10.1017/S0007114513000780
78. Weseler AR, Ruijters EJB, Drittij-Reijnders MJ et al (2011) Pleiotropic benefit of monomeric and oligomeric flavanols on vascular health – a randomized controlled clinical pilot study. PLoS One. https://doi.org/10.1371/journal.pone.0028460
79. Martinez-Micaelo N, González-Abuín N, Ardèvol A et al (2012) Procyanidins and inflammation: molecular targets and health implications. Biofactors 38:257–265. https://doi.org/10.1002/biof.1019
80. Quesada IM, Del Bas JM, Bladé C et al (2007) Grape seed procyanidins inhibit the expression of metallothione in genes in human HepG2 cells. Genes Nutr 2:105–109. https://doi.org/10.1007/s12263-007-0027-4
81. Shimizu T, Toda T, Sunagawa T et al (2011) Apple procyanidins suppress amyloid β-protein aggregation. Biochem Res Int. https://doi.org/10.1155/2011/784698
82. Kanno H, Kawakami Z, Tabuchi M et al (2015) Protective effects of glycycoumarin and procyanidin B1, active components of traditional Japanese medicine yokukansan, on amyloid β oligomer-induced neuronal death. J Ethnopharmacol 159:122–128. https://doi.org/10.1016/j.jep.2014.10.058
83. Zhai WY, Jia CP, Zhao H, Sen XY (2011) Procyanidins inhibit tumor angiogenesis by crosslinking extracellular matrix. Chin J Cancer Res 23:99–106. https://doi.org/10.1007/s11670-011-0099-y
84. Bak MJ, Jun M, Jeong WS (2012) Procyanidins from wild grape (Vitis amurensis) seeds regulate ARE-mediated enzyme expression via Nrf2 coupled with p38 and PI3K/Akt pathway in HepG2 cells. Int J Mol Sci 13:801–818. https://doi.org/10.3390/ijms13010801
85. Stadlbauer S, Rios P, Ohmori K et al (2015) Procyanidins negatively affect the activity of the phosphatases of regenerating liver. PLoS One 10:1–18. https://doi.org/10.1371/journal.pone.0134336
86. Bak M-J, Truong VL, Kang H-S, et al (2013) Anti-inflammatory effect of procyanidins from wild grape (Vitis amurensis) seeds in LPS-induced RAW 264.7 cells. Oxid Med Cell Longev. https://doi.org/10.1155/2013/409321

87. Bas JM, Del RML, Vaqué M et al (2009) Dietary procyanidins enhance transcriptional activity of bile acid-activated FXR in vitro and reduce triglyceridemia in vivo in a FXR-dependent manner. Mol Nutr Food Res 53:805–814. https://doi.org/10.1002/mnfr.200800364
88. Marimoutou M, Le Sage F, Smadja J et al (2015) Antioxidant polyphenol-rich extracts from the medicinal plants Antirhea borbonica, Doratoxylon apetalum and Gouania mauritiana protect 3T3-L1 preadipocytes against H2O2, TNFα and LPS inflammatory mediators by regulating the expression of superoxide dismut. J Inflamm 12:10. https://doi.org/10.1186/s12950-015-0055-6
89. Gu Y, Hurst WJ, Stuart DA, Lambert JD (2011) Inhibition of key digestive enzymes by cocoa extracts and procyanidins. J Agric Food Chem 59:5305–5311. https://doi.org/10.1021/jf200180n
90. Yamakoshi J, Kataoka S, Koga T, Ariga T (1999) Proanthocyanidin-rich extract from grape seeds attenuates the development of aortic atherosclerosis in cholesterol-fed rabbits. Atherosclerosis 142:139–149. https://doi.org/10.1016/S0021-9150(98)00230-5
91. Gonzalez-Abuin N, Pinent M, Casanova-Marti A et al (2015) Procyanidins and their healthy protective effects against type 2 diabetes. Curr Med Chem 22:39–50
92. Grassi D, Desideri G, Di GP et al (2013) Tea, flavonoids, and cardiovascular health: endothelial protection. Am J Clin Nutr 98:1660–1666. https://doi.org/10.3945/ajcn.113.058313
93. Hodgson JM (2006) Effects of tea and tea flavonoids on endothelial function and blood pressure: A brief review. Clin Exp Pharmacol Physiol 33:838–841
94. Chun OK, Chung SJ, Song WO (2007) Estimated dietary flavonoid intake and major food sources of U.S. adults. J Nutr 137:1244–1252
95. Hodgson JM (2008) Review article tea flavonoids and cardiovascular disease. Asia Pac J Clin Nutr 17:288–290
96. Huxley RR, Neil HAW (2003) The relation between dietary flavonol intake and coronary heart disease mortality: a meta-analysis of prospective cohort studies. Eur J Clin Nutr 57:904–908. https://doi.org/10.1038/sj.ejcn.1601624
97. Troup R, Hayes JH, Raatz SK et al (2015) Effect of black tea intake on blood cholesterol concentrations in individuals with mild hypercholesterolemia: a diet-controlled randomized trial. J Acad Nutr Diet 115:264–271.e2. https://doi.org/10.1016/j.jand.2014.07.021
98. Hodgson JM, Croft KD (2010) Tea flavonoids and cardiovascular health. Mol Asp Med 31:495–502. https://doi.org/10.1016/j.mam.2010.09.004
99. Yiannakopoulou EC (2014) Effect of green tea catechins on breast carcinogenesis: a systematic review of in-vitro and in-vivo experimental studies. Eur J Cancer Prev 23:84–89. https://doi.org/10.1097/CEJ.0b013e328364f23e
100. Dower JI, Geleijnse JM, Gijsbers L et al (2015) Effects of the pure flavonoids epicatechin and quercetin on vascular function and cardiometabolic health: a randomized, double-blind, placebo-controlled, crossover trial. Am J Clin Nutr 101:914–921. https://doi.org/10.3945/ajcn.114.098590
101. Heiss C, Sansone R, Karimi H et al (2015) Impact of cocoa flavanol intake on age-dependent vascular stiffness in healthy men: a randomized, controlled, double-masked trial. Age (Dordr) 37:9794. https://doi.org/10.1007/s11357-015-9794-9
102. Bordeaux B, Yanek LR, Moy TF et al (2007) Casual chocolate consumption and inhibition of platelet function. Prev Cardiol 10:175–180. https://doi.org/10.1111/j.1520-037X.2007.06693.x
103. Flammer AJ, Hermann F, Sudano I et al (2007) Dark chocolate improves coronary vasomotion and reduces platelet reactivity. Circulation 116:2376–2382. https://doi.org/10.1161/CIRCULATIONAHA.107.713867
104. Ibero-Baraibar I, Perez-Cornago A, Ramirez MJ et al (2016) An increase in plasma homovanillic acid with Cocoa extract consumption is associated with the alleviation of depressive symptoms in overweight or obese adults on an energy restricted diet in a randomized controlled trial. J Nutr 146:897S–904S. https://doi.org/10.3945/jn.115.222828
105. Curtis PJ, Sampson M, Potter J et al (2012) Chronic ingestion of flavan-3-ols and and lipoprotein status and attenuates with type 2 diabetes. Diabetes Care 35:226–232. https://doi.org/10.2337/dc11-1443.

106. Habauzit V, Verny MA, Milenkovic D et al (2015) Flavanones protect from arterial stiffness in postmenopausal women consuming grapefruit juice for 6 mo: A randomized, controlled, crossover trial. Am J Clin Nutr 102:66–74. https://doi.org/10.3945/ajcn.114.104646
107. Ibero-Baraibar I, Abete I, Navas-Carretero S et al (2014) Oxidised LDL levels decreases after the consumption of ready-to-eat meals supplemented with cocoa extract within a hypocaloric diet. Nutr Metab Cardiovasc Dis 24:416–422. https://doi.org/10.1016/j.numecd.2013.09.017
108. Droste DW, Iliescu C, Vaillant M et al (2013) A daily glass of red wine associated with lifestyle changes independently improves blood lipids in patients with carotid arteriosclerosis: results from a randomized controlled trial. Nutr J 12:147. https://doi.org/10.1186/1475-2891-12-147
109. Di Renzo L, Carraro A, Valente R et al (2014) Intake of red wine in different meals modulates oxidized LDL level, oxidative and inflammatory gene expression in healthy people: a randomized crossover trial. Oxidative Med Cell Longev. https://doi.org/10.1155/2014/681318
110. Arab L, Liebeskind DS (2010) Tea, flavonoids and stroke in man and mouse. Arch Biochem Biophys 501:31–36. https://doi.org/10.1016/j.abb.2010.03.015
111. Arab L, Khan F, Lam H (2013) Tea consumption and cardiovascular disease risk 1 – 3. Am J Clin Nutr 98:1651S–1659S. https://doi.org/10.3945/ajcn.113.059345

6 Comprehensive Assessment of Curcumin as a Functional Food

Aditi Jain, Sharad Saxena, and Vibha Rani

Abstract

The importance of bioactive compounds present in natural sources has withdrawn ample attention in human nutrition and established as "functional food" in the field of food chemistry and as "preventive medicine" in the field of pharmacology and healthcare. Curcumin is one such promising and well-studied natural bioactive plant compound that is present in *Curcuma longa* and known for providing various protective effects in different diseased states. This chapter highlights the present understanding of various protective effects of curcumin in wide range of diseases including cancer, cardiovascular diseases, diabetes, obesity, Alzheimer's disease, etc. The major emphasis is on the molecular pathways associated with curcumin-mediated effects. The significance of its unique structure attributing to its function and present advances in curcumin applications to overcome its limitations has also been discussed in detail.

Keywords

Curcumin · Functional food · Anticancer · Antioxidant · Diabetes · Cardiomyopathy

6.1 Introduction

Medicinal science has grown exponentially, but one thing that is common between the prehistoric times as well as present days are the importance of natural products for defense against numerous diseases. In fact, many present-day drugs have the base of natural products known for various medicinal values [5, 22, 48]. In the

A. Jain · S. Saxena · V. Rani (✉)
Department of Biotechnology, Jaypee Institute of Information Technology, Noida, Uttar Pradesh, India
e-mail: Vibha.rani@jiit.ac.in

V. Rani, U. C. S. Yadav (eds.), *Functional Food and Human Health*,
https://doi.org/10.1007/978-981-13-1123-9_6

present chapter, curcumin, a natural product with wide history of its application in various health issues is discussed. It is a major constituent of the perennial herb/spice named as turmeric (*Curcuma longa*). Roughley and Whiting in 1973 reported that curcumin was first isolated in the year 1815 and structurally defined it as bis-α, β-unsaturated β- diketone, which exists in equilibrium with its enol tautomer in turmeric [20]. Curcumin is known to be used as a remedy for different wounds, burns, infections, and skin diseases and is known to be associated with various biological activities including antioxidant, anti-inflammatory, antitumor, antiangiogenesis, antimicrobial, antiviral, antidiabetic, etc. These properties make curcumin a very potential and promising functional food and nutraceutical ingredient. As mentioned in previous chapters, functional food products are broadly defined as the food components that deliver additional health benefits away from sustaining hunger and providing the basic nutritional values. Functional food displays beneficial effects on various body functions by improving the state of health and reducing the disease risk apart from its nutritional importance. Recent trends indicate substantial awareness in elderly population for consuming functional food product rich diet worldwide [67]. Curcumin is an established chemopreventive therapeutic agent with diverse pharmacological effects against chronic diseases. Curcumin belongs to the group of compounds called "curcuminoids" along with demethoxycurcumin, bis-demethoxycurcumin, and cyclic curcumin. Turmeric contains approximately 80% curcumin, 18% demethoxycurcumin, and 2% bis-demethoxycurcumin [2]. Scientific studies have established that oral consumption of up to the dose of 12,000 mg curcumin per day by humans is nontoxic and safe [35]. Also, a number of phase I and II clinical trials have been conducted to see the effects of curcumin for the treatment of cancer and other chronic diseases [21, 25, 58].

This chapter conducts a systemic review of various beneficial effects of curcumin consumption as a functional food in the prevention and cure of chronic diseases. It consists an elaborated comprehension of the significance of curcumin's unique structure, current understanding of the multi-linked molecular pathways, and various therapeutic effects in the treatment and prevention of a wide range of diseases. Present advances in terms of delivering techniques of curcumin for improved outcomes have also been reviewed in detail in the present chapter.

6.2 Structural Characteristics of Curcumin

More than 30,000 publications and extensive research in relation with curcumin exhibit its significance in the fields of medicinal chemistry, healthcare, food chemistry, pharmacology, and analytical chemistry. Curcumin is a very promising natural compound and has been studied for broad range of biological and chemical properties. From the wide history of turmeric usage in the ancient times to the reported and evidenced medicinal properties of curcumin, its importance has grown widely.

Curcumin is also known as diferuloyl methane, and its IUPAC name is (1E,6E)-1,7-bis(4-hydroxy-3-methoxyphenyl)-1,6-heptadiene-3,5-dione. It is soluble in organic solvents, alkaline solvent, and solvents with extreme acidic pH. It is sparingly soluble in hydrocarbon solvents and insoluble in water [66]. Curcumin is a crystalline compound with a characteristic yellow color. It has a symmetric structure with two phenolic groups and one diketone moiety. These functional groups are presented as two aromatic rings containing o-methoxy phenolic groups with a seven carbon linker having α,β-unsaturated β-diketone moiety. The β-diketo group present in curcumin exhibits keto-enol tautomerism as per the acidity of the solution, where it is present in the keto and enol forms in acidic/neutral and alkaline solutions, respectively (Fig. 6.1). The heptadienone bonds of methoxyphenol rings in the keto form of curcumin make it majorly available in the cell membranes as a result of the presence of highly activated carbon atom [56]. The keto-enol tautomers of curcumin also display metal-chelating ability, and their interactions with biomolecules are contributed to its very sensitive spectroscopic properties.

The higher stability of curcumin at low pH is because of its conjugated diene structure that plays the role of a potent H-atom donor. Also, at higher pH, the enol forms act as an electron donor similar to that of phenolic antioxidants [69]. The chemical reactivity of curcumin with reactive oxygen species (ROS) and hydrogen donation reactions leads to its oxidation and contributes to the well-established ROS-scavenging potential in biological system [55]. The presence of ortho-methoxy group and the structure–activity relationship of curcumin play a crucial role in displaying the characteristic antioxidant activity. The interactions between the hydrogen bonds, phenolic OH, and ortho-methoxy groups of curcumin lead to the abstraction of H-atom by free radicals and majorly contribute to its antioxidant characteristics. Methoxy substitutions on the aromatic rings of curcumin differentiate the interactions of different curcumin forms with the nucleophiles via Michael reaction inside the cellular microenvironment because of different physicochemical and physiological activities. As a result of different methoxy substitutions, when curcumin is administered orally, it undergoes glucuronidation and sulfation; and when administered intravenously or intraperitoneally, it undergoes reduction [62]. Also, different curcumin analogs display distinct biological activities that are based on different cell/tissue/organism type. Curcumin is reported as a robust antioxidative and anti-inflammatory compound and possesses multidimensional therapeutic actions for the treatment of various chronic diseases.

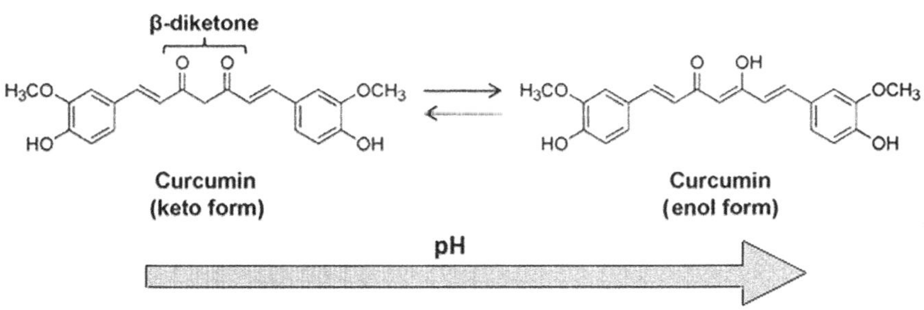

Fig. 6.1 Curcumin tautomers present at different pH of solvents. Keto form predominates in acidic and neutral solvents, whereas enol form is present majorly in alkaline solutions

6.3 Preventive Role of Curcumin in Chronic Diseases

Curcumin is known to be a pleiotropic compound that modulates molecular signaling pathways involved in cell survival, inflammation, apoptosis, etc. Various studies have confirmed the beneficial pharmacological effects and therapeutic properties of curcumin. These properties are associated with the well-established anti-oxidative and anti-inflammatory characteristics. Being a natural compound, curcumin is safe and nontoxic compound and hence holds exceptional pharmacological safety profile. Antibacterial, antiviral, and anticancer activities of curcumin makes it a potential contrivance to be used against various malignant diseases including diabetes, allergies, arthritis, Alzheimer's disease, and other chronic illnesses [64].

6.3.1 Anticancer Effects of Curcumin

Cancer is characterized by an atypical increase in cell proliferation as a result of epigenetic and genetic mutations in cells. These mutations interrupt with the cell cycle and regulatory proteins balance within a cell. Curcumin is known to modulate regulatory proteins of various signaling pathways through various molecular mechanisms. Literature provides in-depth exploration of multi-furious targets of curcumin responsible for mediating protective effects against broad range of cancers including gastrointestinal, breast, genitourinary, bone, gynecological, hematological, pulmonary, thymic, brain, etc. The molecular mechanisms of action of curcumin in treating different types of cancers still remain under investigation. Curcumin mediates its effects by constraining cell signaling pathways at several levels. Curcumin inhibit cancerous growth mainly by altering cell cycle and binding with molecular targets including transcription factors like NF-kB, STAT3, b-catenin, and AP-1; growth factors like EGF, PDGF, and VEGF; enzymes like COX-2, iNOS, and MMPs; signaling kinases like cyclin D1, CDKs, Akt, PKC, and AMPK; and inflammatory cytokines like TNF, MCP, IL-1, and IL-6. It is also responsible for the upregulation of proapoptotic proteins and downregulation of antiapoptotic proteins. Safe effects of curcumin administration and consumption have been proven by numerous studies on animal models and humans. Its high doses are tolerable and display no significant toxic effects [59]. Curcumin affects with several vital pathways in cancer cells and results in displaying carcinogenic properties [26]. In a clinical trial with patients having high-risk or premalignant lesions, the effect of curcumin was studied, and 1000–8000 mg of curcumin was administered daily. Histological images displayed improvement in the lesions in the time period of 3 months [10]. In another study, downregulation of COX-2 was reported in anticancer activity of curcumin in colon tumors [25]. Curcumin also upregulates Akt protein kinase, thereby promoting cell survival and preventing apoptosis. Genes mediating apoptosis in stress conditions like P53 gene and their downstream targets are found to be overexpressed in tumors treated with curcumin [12]. As cancerous mass is dependent on the blood supply for continuous division and survival of cells, they enter the circulation for metastasis; and these properties are inhibited by curcumin by interfering the pathways involved in angiogenesis like inhibition of FGF-induced neovascularization, ligands of vascular endothelial growth factor, and angiopoietin 1 and 2. Curcumin is also shown to regulate cell adhesion molecules and indirectly protecting the process of metastasis by cancer cells [4]. Effects of curcumin on prostate cancer cell invasion were studied by Hong et al. in in vitro and in vivo models where it significantly reduced the tumor size and MMP activity. The metastatic nodules were found to be significantly reduced in curcumin-treated groups as compared to the untreated group, in vivo [28]. In another study by Chendil et al., the radiosensitizing outcomes of curcumin in p53 mutant prostate cancer cell line PC-3 were studied, and curcumin was hypothesized as a potential radiosensitizer as it overcame the effects of radiation-induced pro survival gene expression in prostate cancer [11]. Curcumin also showed the direct anti-inflammatory activity by regulating MKP5 in prostate cells that may contribute to its chemopreventive actions [49]. These onco-pharmacological characteristics of curcumin make it an efficacious cancer therapeutic agent.

6.3.2 Curcumin-Mediated Effects in Diabetes

Antidiabetic effects of curcumin are reported where anti-inflammatory and anti-oxidative activities contribute majorly [45, 76]. Hyperglycemic stress induces the NF-kB transcription and pro-inflammatory cytokines production that promotes insulin resistance in cells [3]. The α- and β-unsaturated diketone moieties of curcumin inhibit the NF-κB activation and transcription by blocking TNF, phorbol ester, and hydrogen peroxide production, thereby reducing the hyperglycemia-induced effects in cells [32, 36, 52]. Curcumin interventions in prediabetic population were studied, and curcumin extracts were given for 9 months, and significant delay in the development of type 2 diabetes was observed [13]. In another randomized clinical trial by Chuengsamarn et al., curcuminoid extracts were shown to reduce the atherogenic risk among type 2 diabetic patients [14]. Depression is considered as an important risk factor for developing diabetes, and therapeutic effects of curcumin were studied in vivo in rats induced with chronic mild stress. Curcumin was found to upregulate insulin receptor substrate phosphorylation and protein kinase B, and it is reported to downregulate glycogen synthase kinase and phosphoenolpyruvate carboxykinase. These modulations result in improved hepatic glycogen content and reversing the insulin resistance and chronic mild stress-induced metabolic abnormalities suggesting the possible application of curcumin for treating depression and/or associated metabolic disorders associated with diabetic conditions [60].

6.3.3 Protective Properties of Curcumin in Diabetic Cardiomyopathy and Other Cardiac Complications

Literature and case reports strongly support the pathogenesis of cardiovascular disorders, such as cardiomyopathy induced by diabetes results in inducible iNOS and endothelial eNOS generation which play a central role in developing cardiac complications and heart failure [16]. Farhangkhoee et al. suggested that curcumin treatment in the diabetic rats results in the higher levels of eNOS mRNA and iNOS mRNA in the myocardial tissues compared to the control rats. Curcumin also halted eNOS mRNA and iNOS mRNA upregulation thereby decreasing the oxidative DNA impairment [17]. In a study, the effect of bioactive complex of curcumin along with other polyphenols like quercetin, selenium, and catechins was studied on cardiovascular risk markers in a healthy population. These polyphenols were included in the diet of healthy individuals for 2 months, and the levels of total cholesterol, HDL-cholesterol, C-reactive protein, homocysteine, vitamin B12, folic acid, cysteine, vitamin B6, and asymmetric dimethylarginine were closely monitored over the time period of study. The significant reduction in total cholesterol and LDL cholesterol along with other biomarkers was recorded resulting in the better lipid profile, thereby reducing the risk of cardiovascular abnormalities [44]. Role of curcumin for preventing vascular aging was supported by reducing arterial stiffening and endothelial dysfunction in a recent in vivo study [18]. Protective effects of curcumin have been reported in drug-induced cardiotoxicity caused by different classes of drugs by enhancing cellular glutathione S transferase and decreasing lipid peroxidation as a result of ROS-scavenging properties of curcumin. In case of patients having comorbidity due to diabetes and cardiomyopathy, curcumin has shown to downregulate NOS and NO production and reduced abnormal accumulation of various connective tissue constituents in the endothelia, thereby stabilizing lysosomal membranes. It also controls hypertrophy in the aging heart by hindering the expression of adenoviral transcription coactivator, p300 [72].

6.3.4 Role of Curcumin in Other Diseases

Various other physiological and pharmacological effects of curcumin have been reported in other diseased states apart from cancer, cardiac diseases, and diabetes. Obesity is another major factor of developing diabetes, and curcumin is shown to modulate adipokines regulation in diabetes by affecting the levels of adiponectin, leptin, resistin, and visfatin, in various in vitro, in vivo, and clinical studies [23, 29, 30, 68]. In addition to target obesity-induced diabetes, curcumin is reported to act as an anti-obesity molecule where it targets NF-κB activation, reducing adipokines and plasminogen activator inhibitor type-1 and inducing adiponectin expression, the principal anti-inflammatory agent secreted by adipocytes [8].

Curcumin has also been extensively studied for its therapeutic effects on nervous system. Therapeutic role of curcumin has been recently reported in Alzheimer's disease, and it has been found to improve the impaired insulin signaling and increase glucose metabolism by modulating insulin-like growth factor expression in the brain of mice [70]. Cholesterol plays a critical role in the growth of the brain, and high levels of cholesterol are associated with neurodegenerative diseases. Tian et al. showed that curcumin activates the ABCA1 promoter and increases the apoA1-dependent cellular cholesterol efflux from the brain, thereby facilitating proper cholesterol transportation and reducing the detrimental effects of cholesterol in the brain [65]. Its role is as an LXR agonist and has been reported by other researchers as well in inflammatory diseases. Curcumin is shown to bind to aggregated beta molecules and phosphorylate tau protein and improve senescent brain alterations and neurodegeneration. Brain plaques and cerebral amyloid angiopathy of Alzheimer's patients have been showing improved results upon curcumin treatment [46]. ROS-induced stress leads to neuronal alterations and cognitive dysfunction leading to chronic cerebral ischemia. Curcumin upregulated the expression of uncoupling protein 2, thereby reducing oxidative stress [40]. Hence, role of curcumin as a potential neuroprotective agent is well-established.

Curcumin has also shown protective effects in facultative upper respiratory tract infections by mediating antibacterial effects and inhibiting bacterial growth, adherence, invasion, and pro-inflammatory activation and suppressing the release of IL-8 in vitro [43]. Curcumin has also displayed protective effects in skin diseases caused by fractionated irradiation and carbon dioxide lasering [31, 42].

6.4 Recent Advancement in Curcumin Delivery in Biological Systems

Curcumin has gone through detailed research in past three decades, and its limitations have also been elaborated in addition to its various preventive activities. Its poor bioavailability, low aqueous solubility, susceptibility toward degradation under alkaline conditions, speedy metabolism, and faster body elimination rates account for the major and very critical impediments to its applications for the betterment of diseased states [54, 64]. Several studies have depicted the variations in the bioavailability of curcumin when administered orally and intravenously [27]. Yang et al. [74] observed that the bioavailability of curcumin in rats was only 1% when administered orally as compared to intravenous administration. Similar scenario was observed in humans where traces of curcumin (0.006 ± 0.005 mg/ml at 1 h) were detected when administered with oral dose of 2 g [61]. These preclinical and clinical results depict that curcumin goes through extensive metabolism throughout digestive tract; therefore, some modifications/formulations are required to be implemented in the crude basic form of curcumin [9, 10, 74].

The limitations tagged with crude form of curcumin promoted the researchers to develop novel nanoformulations aimed for improved water solubility, bioavailability, pharmacokinetics, and additional mechanisms for cellular uptake [71]. The term "nanocurcumin" was first used to describe nanoparticle-encapsulated formulation of curcumin which shows comparable in vitro therapeutic efficacy to free curcumin [6, 47]. Since then, liposomes, polymer conjugates, hydrogels, polyethylene glycols, nanogels, and nanoparticles based on curcumin nanoformulations have been extensively explored (Table 6.1). Liposomes protect the encapsulated curcumin form hydrolysis, thereby maintaining the therapeutic drug levels for longer duration. Polymer conjugates (polycurcumins) and pegylation enhance

Table 6.1 Curcumin formulations used for various diseases

Formulation	Disease	Target Model	Outcome	References
Nanosuspension	Cervical, breast cancer	Cell line (MCF7), animal (rabbit)	Increased solubility, dissolution, stability, anticancer ability	Gao et. al. [19]
	Liver cancer	Cell line (HepG2), animal (rat)	Increased solubility, half time, stability	Li et al. [39]
Liposome	Status epilepticus	Animal (mice)	Anticonvulsant activity	Agarwal et al. [1]
	Breast cancer	Cell line (MCF7)	Increased cancer cell cytotoxicity	Hasan et al. [24]
Nanoemulsion	Cancer	Animal (mice)	Increased bioavailability	Zhongfa et al. [77]
	Control	Human intestine mimic	Increased bioavailability	Sari et al. [57]
Nanoparticles	Osteosarcoma	Cell line (U2OS)	Apoptotic cell death	Peng et al. [53]
	Bacterial	*E. coli, S. aureus*	Efficient antibacterial activity	Pandit et al. [50]
Micelle	Lung, breast cancer	Animal (mice)	Inhibit tumor growth	Li et al. [37]
	Colon cancer	Cell line (CT26 cells), animal (mice)	Induced apoptosis of cancer cells, increased uptake of curcumin	Yang et al. [73]
Sonocrystallized curcumin	Gastric cancer	Cell line (HeLa)	Increased solubility and dissolution	Khan et al. [33]
Carbon nanotubes	Prostate cancer	Cell line (PC-3)	Increased solubility, stability, inhibit cancer cell growth	Li et al. [38]
Lipid-based oral formulations	Parkinson's disease	Animal (rat)	Increased bioavailability	Borker et al. [7]; Kundu et al. [34]
Amphiphilic peptide	Acute lung injury	Cell line (HEK293), Animal (mice)	Higher transfection efficiency	Park et al. [51]
	Cervical cancer	Cell line (HeLa)	Improved cellular uptake	Song et al. [63]
PEGylated curcumin	Breast cancer	Cell line (T47D cells)	Enhanced aqueous and organic solubility	Farajzadeh et al. [15]
	Hepatic steatosis	Animal (mice)	Reverse hepatic steatosis	Liu et al. [41]

drug-loading capacity, stabilize the curcumin, and improve water solubility and cellular retention of curcumin. Few of nanoformulations have been approved by the Food and Drug Administration (FDA) and are available commercially among which albumin-bound paclitaxel (PTX) poly (lactide-co-glycolide) (PLGA) nanoformulation is commonly used in cancer treatment. A very recent report has demonstrated that poly(lactic-co-glycolic acid)-based curcumin nanoformulation inhibits cell growth, induces apoptosis, and arrests the cell cycle in cervical cancer cell lines by modulating the level of miR-21 (an onco-miRNA), miR-214 (a tumor suppressor), and transcription factors [75]. Overall, nanoformulations provide a beneficial approach for a rational strategy to use curcumin as a broad spectrum drug.

6.5 Conclusion

Since curcumin influences multiple molecular targets including transcription factors, their receptors, cytokines, growth factors, enzymes, and small noncoding RNA such as microRNA, it has exhibited excellent therapeutic benefits in medicinal applications. However, due to several limitations of crude curcumin mentioned above, it is yet to be completely recognized for clinical applications. In recent reports, nanoformulations of curcumin have depicted excellent pharmacological effects; the toxicity of nanoformulations administered in high dose is a concern to resolve. Further preclinical and clinical studies will help in gaining the in-depth knowledge of curcumin nanoformulations which can be translated into potential therapeutic tools.

Acknowledgment We acknowledge the Jaypee Institute of Information Technology for providing the infrastructure and literature support for conducting the detailed study presented in the chapter.

References

1. Agarwal NB, Jain S, Nagpal D et al (2013) Liposomal formulation of curcumin attenuates seizures in different experimental models of epilepsy in mice. Fundam Clin Pharmacol 27(2):169–172
2. Anand P, Thomas SG, Kunnumakkara AB et al (2008) Biological activities of curcumin and its analogues (congeners) made by man and mother nature. Biochem Pharmacol 76:1590–1611
3. Baker RG, Hayden MS, Ghosh S (2011) NF-κB, inflammation and metabolic disease. Cell Metab 13(1):11–22
4. Bhandarkar SS, Arbiser JL (2007) Curcumin as an inhibitor of angiogenesis. Adv Exp Med Biol 595:185–195
5. Balunas MJ, Kinghorn AD (2005) Drug discovery from medicinal plants. Life Sci 78:431–441
6. Bisht S, Feldmann G, Soni S et al (2007) Polymeric nanoparticle encapsulated curcumin ("nanocurcumin"): a novel strategy for human cancer therapy. J Nanobiotechnol 5:3
7. Borkar N, Andersson DR, Yang M et al (2017) Efficacy of oral lipid-based formulations of apomorphine and its diester in a Parkinson's disease rat model. J Pharm Pharmacol 69(9):1110–1115
8. Bradford PG (2013) Curcumin and obesity. Biofactors 39(1):78–87

9. Chang MT, Tsai TR, Lee CY et al (2013) Elevating bioavailability of curcumin via encapsulation with a novel formulation of artificial oil bodies. J Agric Food Chem 61:9666–9671
10. Cheng AL, Hsu CH, Lin JK et al (2001) Phase I clinical trial of curcumin, a chemopreventive agent, in patients with high-risk or premalignant lesions. Anticancer Res 21:2895–2900
11. Chendil D, Ranga RS, Meigooni D et al (2004) Curcumin confers radiosensitizing effect in prostate cancer cell line PC-3. Oncogene 23:1599–1607
12. Choudhuri T, Pal S, Das T et al (2005) Curcumin selectively induces apoptosis in deregulated cyclin D1-expressed cells at G2 phase of cell cycle in a p53-dependent manner. J Biol Chem 280:20059–20068
13. Chuengsamarn S, Rattanamongkolgul S, Luechapudiporn R et al (2012) Curcumin extract for prevention of type 2 diabetes. Diabetes Care 35(11):2121–2127
14. Chuengsamarn S, Rattanamongkolgul S, Phonrat B et al (2014) Reduction of atherogenic risk in patients with type 2 diabetes by curcuminoid extract: a randomized controlled trial. J Nutr Biochem 25(2):144–150
15. Farajzadeh R, Pilehvar-Soltanahmadi Y, Dadashpour M et al (2017) Nano-encapsulated metformin-curcumin in PLGA/PEG inhibits synergistically growth and hTERT gene expression in human breast cancer cells. Artif Cells Nanomed Biotechnol 5:1–9
16. Farhangkhoee H, Khan ZA, Mukherjee S et al (2003) Heme oxygenase in diabetes-induced oxidative stress in the heart. J Mol Cell Cardiol 35(12):1439–1448
17. Farhangkhoee H, Khan ZA, Chen S et al (2006) Differential effects of curcumin on vasoactive factors in the diabetic rat heart. Nutr Metab (Lond) 3:27
18. Fleenor BS, Sindler AL, Marvi NK et al (2013) Curcumin ameliorates arterial dysfunction and oxidative stress with aging. Exp Gerontol 48:269–276
19. Gao Y, Li Z, Sun M (2011) Preparation and characterization of intravenously injectable curcumin nanosuspension. Drug Deliv 18(2):131–142
20. Goel A, Kunnumakkara AB, Aggarwal BB (2008) Curcumin as "Curecumin": from kitchen to clinic. Biochem Pharmacol 75:787–809
21. Gupta SC, Patchva S, Aggarwal BB (2013) Therapeutic roles of curcumin: lessons learned from clinical trials. AAPS J 15:195–218
22. Gurib-Fakim A (2006) Medicinal plants: traditions of yesterday and drugs of tomorrow. Mol Asp Med 27:1–93
23. Hajavi J, Abbas-Momtazi A, Johnston TP et al (2017) Curcumin: a naturally occurring modulator of adipokines in diabetes. J Cell Biochem. https://doi.org/10.1002/jcb.26121
24. Hasan M, Belhaj N, Benachour H et al (2014) Liposome encapsulation of curcumin: physico-chemical characterizations and effects on MCF7 cancer cell proliferation. Int J Pharm 461(1–2):519–528
25. Hatcher H, Planalp R, Cho J et al (2008) Curcumin: from ancient medicine to current clinical trials. Cell Mol Life Sci 65:1631–1652
26. Heger M, van Golen RF, Broekgaarden M et al (2014) The molecular basis for the pharmacokinetics and pharmacodynamics of curcumin and its metabolites in relation to cancer. Pharmacol Rev 66:222–307
27. Helson L (2013) Curcumin (diferuloylmethane) delivery methods: a review. Biofactors 39:21–26
28. Hong JH, Ahn KS, Bae E et al (2006) The effects of curcumin on the invasiveness of prostate cancer in vitro and in vivo. Prostate Cancer Prostatic Dis 9:147–152
29. Ismail NA, Ragab S, Abd El Baky ANE et al (2014) Effect of oral curcumin administration on insulin resistance, serum resistin and fetuin-A in obese children: randomized placebo-controlled study. RJPBCS 5:887–896
30. Ismail NA, Abd El Dayem SM, Salama E et al (2016) Impact of curcumin intake on gluco-insulin homeostasis, leptin and adiponectin in obese subjects. RJPBCS 7:1891–1897
31. Jagetia GC, Rajanikant GK (2012) Acceleration of wound repair by curcumin in the excision wound of mice exposed to different doses of fractionated γ radiation. Int Wound J 9(1):76–92
32. Jiménez-Flores LM, Lopez-Briones S, Macías-Cervantes MH et al (2014) A PPARγ, NF-κB and AMPK-dependent mechanism may be involved in the beneficial effects of curcumin in the diabetic db/db mice liver. Molecules 19(6):8289–8302

33. Khan MA, Akhtar N, Sharma V et al (2015) Product development studies on sonocrystallized curcumin for the treatment of gastric cancer. Pharmaceutics 7(2):43–63
34. Kundu P, Mohanty C, Sahoo S (2012) Antiglioma activity of curcumin loaded lipid nanoparticles and its enhanced bioavailability in brain tissue for effective glioblastoma therapy. Acta Biomater 8(7):2670–2687
35. Lao CD, Ruffin MT 4th, Normolle D et al (2006) Dose escalation of a curcuminoid formulation. BMC Complement Altern Med 6:10
36. Leclercq IA, Farrell GC, Sempoux C et al (2004) Curcumin inhibits NF-kappa B activation and reduces the severity of experimental steatohepatitis in mice. J Hepatol 41(6):926–934
37. Li H, Zhang N, Hao Y et al (2014a) Formulation of curcumin delivery with functionalized single-walled carbon nanotubes: characteristics and anticancer effects in vitro. Drug Deliv 21(5):379–387
38. Li X, Chen T, Xu L et al (2014b) Preparation of curcumin micelles and the in vitro and in vivo evaluation for cancer therapy. J Biomed Nanotechnol 10(8):1458–1468
39. Li X, Yuan H, Zhang C et al (2016) Preparation and in-vitro/in-vivo evaluation of curcumin nanosuspension with solubility enhancement. J Pharm Pharmacol 68(8):980–988
40. Liu L, Zhang P, Li Y et al (2012) Curcumin protects brain from oxidative stress through inducing expression of UCP2 in chronic cerebral hypo perfusion aging-rats. Mol Neurodegener 7:S10
41. Liu Y, Cheng F, Luo Y (2017) PEGylated curcumin derivative attenuates hepatic steatosis via CREB/PPAR-γ/CD36 pathway. Biomed Res Int 2017:8234507
42. Lopez-Jornet P, Camacho-Alonso F, Jimenez-Torres MJ et al (2011) Topical curcumin for the healing of carbon dioxide laser skin wounds in mice. Photomed Laser Surg 29(12):809–814
43. Luer S, Troller R, Jetter M et al (2011) Topical curcumin can inhibit deleterious effects of upper respiratory tract bacteria on human oropharyngeal cells *in vitro*: potential role for patients with cancer therapy induced mucositis? Support Care Cancer 19(6):799–806
44. Madaric A, Kadrabova J, Krajcovicova-Kudlackova M et al (2013) The effect of bioactive complex of quercetin, selenium, catechins and curcumin on cardiovascular risk markers in healthy population after a two month consumption. Bratisl Lek Listy 114(2):84–87
45. Meng B, Li J, Cao H (2013) Antioxidant and anti-inflammatory activities of curcumin on diabetes mellitus and its complications. Curr Pharm Des 19(11):2101–2113
46. Mutsuga M, Chambers JK, Uchida K et al (2012) Binding of curcumin to senile plaques and cerebral amyloid angiopathy in the aged brain of various animals and to neurofibrillary tangles in Alzheimer's brain. J Vet Med Sci 74(1):51–57
47. Naksuriya O, Okonogi S, Schiffelers RM et al (2014) Curcumin nanoformulations: a review of pharmaceutical properties and preclinical studies and clinical data related to cancer treatment. Biomaterials 35(10):3365–3383
48. Newman DJ, Cragg GM (2007) Natural products as sources of new drugs over the last 25 years. J Nat Prod 70(3):461–477
49. Nonn L, Duong D, Peehl DM (2007) Chemopreventive anti-inflammatory activities of curcumin and other phytochemicals mediated by MAP kinase phosphatase-5 in prostate cells. Carcinogenesis 28(6):1188–1196
50. Pandit RS, Gaikwad SC, Agarkar GA (2015) Curcumin nanoparticles: physico-chemical fabrication and its in vitro efficacy against human pathogens. 3. Biotech 5(6):991–997
51. Park JH, Kim HA, Park JH et al (2012) Amphiphilic peptide carrier for the combined delivery of curcumin and plasmid DNA into the lungs. Biomaterials 33(27):6542–6550
52. Peng YM, Zheng JB, Zhou YB, Li J (2013) Characterization of a novel curcumin analog P1 as potent inhibitor of the NF-κB signaling pathway with distinct mechanisms. Acta pharmacol Sin 34(7):939–950
53. Peng SF, Lee CY, Hour MJ (2014) Curcumin-loaded nanoparticles enhance apoptotic cell death of U2OS human osteosarcoma cells through the Akt-Bad signaling pathway. Int J Oncol 44(1):238–246
54. Prasad S, Tyagi AK, Aggarwal BB (2014) Recent developments in delivery, bioavailability, absorption and metabolism of curcumin: the golden pigment from golden spice. Cancer Res Treat 46(1):2–18

55. Priyadarsini KI (2014) The chemistry of curcumin: from extraction to therapeutic agent. Molecules 19(12):20091–20112
56. Parimita SP, Ramshankar YV, Suresh S et al (2007) Redetermination of curcumin: (1E,4Z,6E)-5-hydroxy-1,7-bis(4-hydroxy-3-methoxy- phenyl)hepta-1,4,6- trien-3-one. Acta Cryst E63:860–862
57. Sari TP, Mann B, Kumar R et al (2015) Preparation and characterization of nanoemulsion encapsulating curcumin. Food Hydrocoll 43:540–546
58. Shehzad A, Wahid F, Lee YS (2010) Curcumin in cancer chemoprevention: molecular targets, pharmacokinetics, bioavailability, and clinical trials. Arch Pharm (Weinheim) 343(9):489–499
59. Shehzad A, Lee J, Lee YS (2013) Curcumin in various cancers. Biofactors 39(1):56–68
60. Shen JD, Wei Y, Li YJ et al (2017) Curcumin reverses the depressive-like behavior and insulin resistance induced by chronic mild stress. Metab Brain Dis 32(4):1163–1172
61. Shoba G, Joy D, Joseph T et al (2007) Influence of piperine on the pharmacokinetics of curcumin in animals and human volunteers. Planta Med 64(4):353–366
62. Somparn P, Phisalaphong C, Nakornchai S (2007) Comparative antioxidant activities of curcumin and its demethoxy and hydrogenated derivatives. Biol Pharm Bull 30(1):74–78
63. Song SJ, Lee S, Ryu KS et al (2017) Amphiphilic peptide nanorods based on oligophenylalanine as a biocompatible drug carrier. Bioconjug Chem. https://doi.org/10.1021/acs.bioconjchem.7b00247
64. Stanic Z (2017) Curcumin, a compound from natural sources, a true scientific challenge – a review. Plant Foods Hum Nutr 72(1):1–12
65. Tian M, Wang L, Yu G et al (2012) Curcumin promotes cholesterol efflux from brain through LXR/RXR-ABCA1-apoA1 pathway in chronic cerebral hypoperfusion aging-rats. Mol Neurodegener 7(1):S7
66. Tonnesen HH, Masson M, Loftsson T (2002) Studies of curcumin and curcuminoids. XXVII. Cyclodextrin complexation: solubility, chemical and photochemical stability. Int J Pharm 244:127–135
67. Van der Zanden LD, van Kleef E, de Wijk RA et al (2014) Knowledge, perceptions and preferences of elderly regarding protein-enriched functional food. Appetite 80:16–22
68. Varemo L, Henriksen TI, Scheele C et al (2017) Type 2 diabetes and obesity induce similar transcriptional reprogramming in human myocytes. Genome Med 9(1):47
69. Wang YJ, Pan MH, Cheng AL et al (1997) Stability of curcumin in buffer solutions and characterization of its degradation products. J Pharm Biomed Anal 15:1867–1876
70. Wang P, Su C, Feng H et al (2017) Curcumin regulates insulin pathways and glucose metabolism in the brains of APPswe/PS1dE9 mice. Int J Immunopathol Pharmacol 30(1):25–43
71. WHO technical report series-922 (2003) Evaluation of certain food additives and contaminants: sixty-first report of the Joint FAO/WHO Expert Committee on Food Additives, Rome, Italy
72. Wongcharoen W, Phrommintikul A (2009) The protective role of curcumin in cardiovascular diseases. Int J Cardiol 133:145–151
73. Yang X, Li Z, Wang N (2015) Curcumin-encapsulated polymeric micelles suppress the development of colon cancer in vitro and in vivo. Sci Rep 5:10322
74. Yang KY, Lin LC, Tseng TY et al (2007) Oral bioavailability of curcumin in rat and the herbal analysis from Curcuma longa by LC-MS/MS. J Chromatogr B Analyt Technol Biomed Life Sci 853:183–189
75. Zaman MS, Chauhan N, Yallapu MM et al (2016) Curcumin nanoformulation for cervical cancer treatment. Sci Rep 6:20051
76. Zhang DW, Fu M, Gao SH et al (2013) Curcumin and diabetes: a systematic review. Evid Based Complement Alternat Med 2013:636053
77. Zhongfa L, Chiu M, Wang J et al (2012) Enhancement of curcumin oral absorption and pharmacokinetics of curcuminoids and curcumin metabolites in mice. Cancer Chemother Pharmacol 69(3):679–689

Resveratrol: A Miracle Drug for Vascular Pathologies

7

Shishir Upadhyay, Kunj Bihari Gupta, Sukhchain Kaur, Rubal, Sandeep Kumar, Anil K. Mantha, and Monisha Dhiman

Abstract

Cardiovascular diseases (CVDs) are multifactorial noncommunicable diseases that are responsible for most prominent health problems worldwide in the twenty-first century. The genetic factors, environmental factors, change in diet, lifestyle, lack of physical activities, stress, and high blood pressure are the key risk factors for CVDs, and diseases like diabetes also contribute to the progression of CVDs. Platelet aggregation, vascular endothelial dysfunction, and imbalance in nitric oxide (NO) levels are the key events in cardiovascular pathologies that results in inflammation and oxidative stress that ultimately leads to death. To counteract the pathogenicity of CVDs, the use of phytochemicals is advancing as the conventional drugs have multiple side effects. Experimental demonstrations have showed that phytochemicals exhibit numerous cardioprotective properties with limited side effects. This chapter is focused on the use of resveratrol (3,5,4′-trihydroxystilbene), a phytochemical well known for its cardioprotective, antioxidant, anti-inflammatory, anti-atherosclerotic properties *in vitro* and *in vivo*. Existing systemic studies revealed that resveratrol could target various signaling pathways associated with cell growth and proliferation, inflammation, and mitochondrial functioning by modulating PGC-1α and SIRT-1 activity and also improves remodeling in the heart by activating adenosine monophosphate kinase (AMPK). Resveratrol can act as an inhibitor of migration and proliferation of aortic vascular smooth muscle cell by decreasing the cross talk between an inducer of matrix metalloproteinases (MMPs) and IL-18. Resveratrol improves

S. Upadhyay · A. K. Mantha
Department of Animal Sciences, School of Basic and Applied Sciences, Central University of Punjab, Bathinda, Punjab, India

K. B. Gupta · S. Kaur · Rubal · S. Kumar · M. Dhiman (✉)
Department of Biochemistry and Microbial Sciences, School of Basic and Applied Sciences, Central University of Punjab, Bathinda, Punjab, India
e-mail: monisha.dhiman@cup.edu.in

V. Rani, U. C. S. Yadav (eds.), *Functional Food and Human Health*, https://doi.org/10.1007/978-981-13-1123-9_7

the systolic performance of heart by regulating diastolic function and thus prevents heart failure risk. Scientific literature shows that the use of resveratrol as miracle drug for vascular pathogenesis can revamp cardiac health which will shed light on the path to make treatment strategies for medication of vascular-related disorders.

Keywords
CVDs · Inflammation · Oxidative stress · Phytochemicals · Resveratrol

7.1 Introduction

CVDs are related to heart and blood vessels and include coronary heart disease (heart attack), cerebrovascular disease (stroke), high blood pressure (hypertension), peripheral artery disease, rheumatic heart disease, congenital heart disease, and heart failure. CVDs are not only developing challenges for health but also affecting the socioeconomic pillar in the twenty-first century, particularly low and middle-income countries. According to a report of the American College of Cardiology (published in 2012), worldwide CVD-related deaths are expected to surpass 23.6 million by 2030 based on a higher prevalence of atherosclerosis, hypertension, stroke, ischemic heart diseases, and heart failure [1]. According to WHO factsheet, CVD is engulfing annually ~17.5 million lives globally (68% of all deaths) and 3.7 million lives in South-East Asia Region (SEAR). India is the second leading country among SEAR for premature (between 30 and 70 years of age) noncommunicable disease (NCD) mortality [2]. The main risk factors for premature deaths in CVDs include smoking, higher intake of carbohydrates and low-density lipoprotein (LDL)-rich diet, family history, lifestyle, and diabetes. In the pathogenesis of CVDs, oxidative stress and inflammation are important events that further leads to a variety of complications. During oxidative stress, oxidant enzymes such as NADPH oxidase, xanthine oxidase, and electron transport chain (ETC) continuously produce the reactive oxygen species/reactive nitrogen species (ROS/RNS) that reduces the antioxidant capacity. The increased oxidative stress causes the lowering of eNOS that ultimately affects the synthesis and availability of NO in the endothelial cells finally lead to endothelial dysfunction. In March 2002, the Science Advisory and Coordinating Committee of American Heart Association (AHA) and the Centers for Disease Control and Prevention (CDC) established a relationship between inflammatory markers [C-reactive protein (hs-CRP), serum amyloid A (SAA), white blood cell count, fibrinogen] with CVDs [3].

Plant secondary metabolite- or phytochemical-based therapies in the treatment of CVDs are being practiced widely. These phytochemicals possess potential to increase the levels of antioxidants, quench the inflammatory responses, lower the ROS/RNS levels, and act as antiplatelet agents, ischemic preconditioning, and angiogenic properties. Carotenoids, flavonoids, and polyphenols are main dietary phytochemicals which are easily found in various fruits and vegetables.

Resveratrol (3,5,4′-trihydroxystilbene) is a natural polyphenol and mainly present in *Polygonum cuspidatum*, grapes, peanuts, and berries, as well as in the products manufactured from these mentioned sources such as red wine. Resveratrol has often been linked to the phenomenon of "French paradox" that French people have a low risk of CVDs due to their moderate consumption of red wine [4]. Resveratrol has pleiotropic effects on the inflammatory responses in CVDs. It increases the bioavailability of NO, causing vasodilation and, thereby, reducing the risk of atherosclerosis [5]. It is demonstrated that the expression of antioxidant enzymes such as superoxide dismutase (SOD), glutathione peroxidase (GSH-Px) catalase (CAT), and silent mating type information regulation 2 homolog sirtuin (SIRT) 1, 3, and 4 gets upregulated [6, 7] and levels of thiobarbituric acid-reactive substances (TBARS) significantly decreased, suggesting an antioxidant potential of resveratrol. The antioxidative, anti-inflammatory, antithrombotic, and antihyperlipidemic properties and a key role in maintaining glucose homeostasis have drawn the attention for the current antihypertensive pharmacotherapeutic approaches based on resveratrol [8, 9]. Resveratrol regulates gene expression, translation, cellular signaling (mTOR and AMPK), enzymatic pathways, apoptosis, inflammation-related miRNA, and cytokine expression needed for cardiovascular remodeling [10]. Besides cardiovascular remodeling, resveratrol can also be used in potential deadly inflammation in appendicitis, cancer, and diabetes [11, 12].

7.2 Vascular Pathologies

Vascular system is a stupendous entangled network of arteries and veins, collectively referred as blood vessels, is also called as circulatory system which is a closed and dynamic system in vertebrates. Around sixteenth century BC, Ebers Papyrus first described the partial relation of heart and vessels following which the contribution from other scholars like Sushruta, Hippocratean, Herophilus, Erasistratus, and Galen led to the concept of a vascular system with arteries (carrying brighter and thinner blood) and veins (carrying darker red blood). Greek physician, Avicenna, postulated the complete cardiac cycle. The role of other organs such as lungs and liver was established, and thereafter, William Harvey narrated the complete systemic circulation.

Human heart is an obligate aerobic, most active, energy-consuming organ with minimal antioxidative defense. A number of defects, namely, atherosclerosis peripheral artery/venous disease, aneurysm, Raynaud's phenomenon, Buerger's disease, peripheral venous disease, varicose veins, blood clots in veins, lymphedema, etc., can be the consequence of a disturbed state of vascular mechanism. Most of the issues arise due to an imbalance in the homeostasis of endothelial cells in the large and medium muscular arteries. Factors like pathogens, lifestyle, and genetic consequences lead to ox LDL (oxidized low-density lipoprotein) particles or their glycation which further gets triggered by inflammatory environment in endothelial cells. Invading macrophages or lymphocytes as well as the dividing smooth muscle cells stimulate the thickening of the vessel wall and plaque formation which narrows the lumen, causing an obstruction in the blood flow and creates a hypoxic starved condition for organs and several associated pathologies.

7.3 Cardio Vascular Diseases (CVDs)

(a) *Hypertension*

Hypertension or high blood pressure, a common vascular disorder, is characterized by blood pressure > 140–160(diastolic)/90–100 (systolic) mm of Hg. Though itself is not risky but a prevalent condition in CVDs, ischemic heart disease [13], strokes [14], and other peripheral vascular diseases including heart failure, aneurysms, atherosclerosis, chronic kidney disease, and pulmonary embolism makes it one of the most critical disorder. Hypertension is also a risk factor for cognitive impairment and dementia [15]. Aortic wall stiffness due to hypertension in the vessels network of perfused organs may pulsatile the hemodynamic forces which increases the pulse pressure and pulse wave velocity resulting in hypertensive retinopathy and nephropathy [16].

(b) *Atherosclerosis*

Atherosclerotic lesions (atheroma) are slow, silent, chronic inflammatory ailment of arteries. A cascade is initiated by disturbances in endothelial cell lining which is characterized by arterial stiffness due to ROS imbalance (enhanced angiotensin II and NADPH oxidase activity), and a ceased NO (vasodilator or endothelium derived relaxing factor) bioavailability [17]. Oxidative stress in endothelial cells lesions is responsible for LDL deposition underneath and signaling for other white blood cell (WBC). Monocytes or macrophages engulf the LDLs, form "foam cells," and accumulate to form a "fatty streak" that secretes inflammatory cytokines and C-reactive proteins (CRP). The contact of platelets leads to secretion of platelet-derived growth factors (PDGFs) which causes the smooth muscle cells to migrate from *tunica interna* to *tunica media*. These further promote the thickening of arterial walls due to secretion of collagen and calcification or crystallization to form a fibrous cap around the fatty streak. The entire process leads to development of an atherosclerotic plaque which when comes in contact with other blood cells, makes the clot and occludes the blood vessels (Fig. 7.1). The clot in the network of blood vessels is fatal due to (1) angina and myocardial infarction, (2) stroke and cerebral atrophy, and (3) peripheral ischemia. Sometimes the plaques split and get stuck into small vessels to block them, which is referred as embolism [18].

Atherosclerosis is affected by age, lifestyle, family background, diet, or other health conditions such as hyperglycemia and hypertension. Atherosclerosis is reported to be predominant in males than females, in postmenopause women than premenopause, and in aged than young people. The age dependent atherosclerosis shows an eccentric distribution with associated inflammation in young population. Young generation shows thrombosis, whereas old people shows predominant plaque

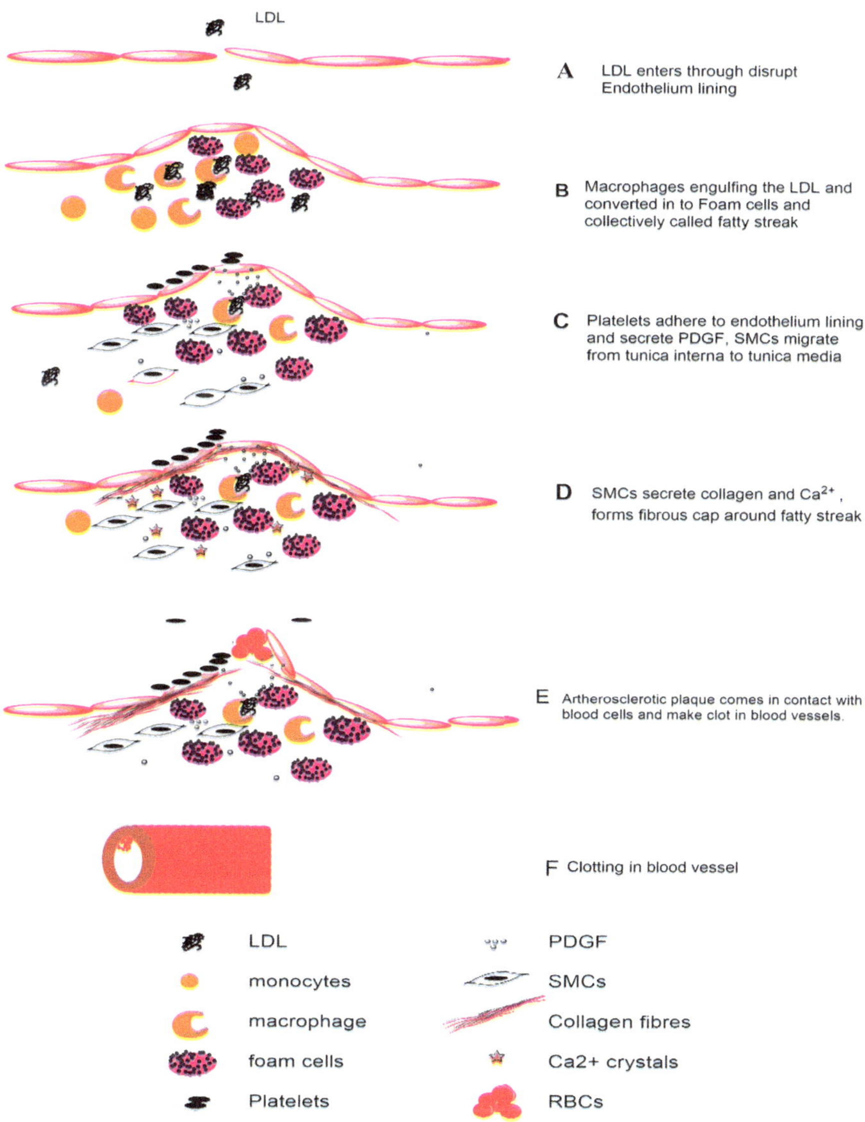

Fig. 7.1 Diagrammatic Representation of Mechanism of Atherosclerosis Formation

hemorrhage [19]. Hypertension, hyperlipidemia, hyperglycemia, hyperhomocysteinemia, infections, and smoking and drug abuse are known as atherogenic factors causing subtle endothelial injury and subsequent inflammation [19]. The severity increases due to its association with increased risk of vascular dementia [20], myocardial infarction [21], atherothrombotic occlusion [22], retinopathy [23], and Alzheimer's disease [24].

(c) *Aneurysm*

Aneurysm, another vascular inflammatory disease, occurs commonly in arteries and is categorized by abnormal balloon-like dilatation in the wall of a blood vessel. Aneurysm appears with the advent of atherosclerosis and has a fatal effect on thoracic and abdominal aorta and on the brain through circle of Willis. The arteries rupture, and clots are formed in those area that look like fusiform hence named fusiform aneurism and also called pseudoaneurysms [25]. Aneurysm and atherosclerosis are closely related, but it is not clear whether both need the same inflammatory conditions or not. Some of the inflammatory stages such as monocyte activation, macrophage and T-cell infiltration, vascular oxidative stress, and matrix proteolysis play a major role in abdominal aortic aneurysm (AAA), coronary aortic aneurysm (CAA), and thoracic aortic aneurysm (TAA) development [26]. On the basis of histopathological studies, 5 grades of "inflammatory" aneurysm in patients have been defined: grade A or mixed acute/chronic inflammation, grade 0 or no inflammation, grade 1 or mild chronic, grade 2 or moderate chronic, and grade 3 or severe chronic inflammation or symptomatic aneurysm [27].

CAA is a defect in one or more of coronary artery of heart which is either congenital or environmental [28]. Kawasaki Disease (KD) is the most common cause of CAA worldwide, particularly in children, but its mechanism is still unclear. It is an example of collective manifestation of inflammation and connective tissues where infiltrated inflammatory cells (macrophages and $CD8^+$T-lymphocytes) actively participate in pathogenesis, and degrade the elastic tissue in *vascular media* and extracellular matrix (ECM) [29].

In AAA and TAA, the perforin-positive $CD3^+$ T cells migrate from adventitia to media, localize to *vasa vasorum*, and make the SMCs liable for apoptosis [30]. The elevated levels of IL-1β, IL-8, IL-6, TNF-α, IFN-γ, MCP-1, and inflammatory cells in AAA patients indicate the central role of inflammation. IL-8, IL-1β, and IL-6 may act as chemotactic factors as well as activators of various immune cells such as neutrophils, macrophage, T and B cells, and endothelial cells [31, 32]. IFN-γ (a pro-inflammatory) and IL-4 (anti-inflammatory) cytokines have complex effects on the development of allograft aorta and aorta aneurysm, IL-4 act as an important stimulus to AAA formation, and IFN-γ attenuate collagenolytic and elastolytic activity [33]. The levels of C-reactive protein and D-dimer levels are higher during the development of AAA and may act as prognostic markers [34].

Marfans' syndrome is a genetic inflammatory aortic aneurysm, which is caused by mutations in *FBN-1*, a gene that encodes for an essential component of elastic fibers called fibrillin-1. A malfunctioned congenital bicuspid aortic valve ascends to aortic aneurysm and might cause premature death. Many other factors such as distorted components of ECM and cytokines specifically TGF-β have also been recognized as pathogenic factors [35–37].

(d) *Other Associated Pathologies*

The mechanisms by which various factors cause CVDs are thought to be the same as inflammation and oxidative stress. Platelets, also called thrombocytes, are blood component which mediate (along with the coagulation factors) clumping and clotting in blood vessel during injuries. The inflammation in platelets causes acute myocardial infarction, stroke, and other vascular traumas. Under normal conditions, platelets are engaged to the site of vascular injury for forming a hemostatic plug. Prothrombotic factors such as thromboxane A2 and adenosine diphosphate are secreted by the platelets to further recruit more and more of these platelets to the injured spot [38, 39]. During CVD, vascular trauma promotes P-selectin to the outer membrane in activated platelets, adheres with the monocyte which shows conversion to macrophage, and forms a fatty streak. PDGF, a mitogenic chemokine, secreted by activated platelets, helps in proliferation and migration of vascular SMCs. Mean platelet volume (MPV) is an important biological variable, which determines the platelet reactivity. MPV is increased in patients after myocardial infarction and is a predictor of a further ischemic event. Changes not only in platelets but also in the parental megakaryocyte (MK) are associated with chronic and acute vascular events [40].

The renin-angiotensin system (RAS) is also crucial in the pathophysiology of a CVDs. Angiotensin II, a vasoactive hormone, is an important component of RAS that controls the migration of immune cells and endothelial wall functioning during vascular inflammatory conditions. *In vivo* and *in vitro* studies with RAS inhibitors demonstrated a ceased vascular inflammation-mediated oxidative stress and endothelial wall dysfunctions along with markers ICAM-1, VCAM-1, TNF, IL-6, and CRP [41, 42].

Prostacyclin is a metabolic product of arachidonic acid formed by the action of cyclooxygenase-2 (COX-2) in vascular endothelial cells. It provides protection against atherothrombosis and vasodilation and has antiplatelet-aggregator activity. Any mutation or deficiency can lead to atherosclerosis, pulmonary artery disease, platelet aggregation, and enhanced proliferative response to carotid vascular injury with *intima media* thickness [43, 44].

Endothelin (ET), a vasoconstrictor protein, is a prognostic marker of vascular dysfunction [45]. It is a short peptide (21-aa long) that maintains ion and water homeostasis by regulating the osmoregulator components such as RAS, vasopressin, and natriuretic peptide, thus influencing blood pressure [46]. The role of endothelin in pathophysiology is unclear, but it has been proposed to stimulate inflammation, oxidative stress, platelet activation, endothelial cell proliferation, and artery calcification [45, 47, 48]. ET 1 displays an association with hypertension, cardiac arrest, atherosclerosis, stroke, and idiopathic cardiomyopathy. In a recent study, AM-36, a successful stroke treatment drug was utilized to directly target ET-1 [49]. Vascular endothelial dysfunctions may also arise due to an imbalance between NO and ET-1 [50].

NO is a vital endothelial signal-transducing molecule. In endothelial cells, NO is synthesized from L-arginine-NO pathway by eNOS (endothelial NO synthase enzyme). NO is described as antihypertensive, antithrombotic, and anti-atherosclerotic molecule. Its imbalance or inadequate availability leads to atherosclerosis and other vascular inflammatory disorders. During atherosclerosis, a distressed endothelial lining perturbs the NO homeostasis through "eNOS uncoupling" mechanism. The resultant oxidative stress promotes the production of oxLDL, superoxide radicals, and peroxynitrite generation followed by a cascade of changes leading to uncoupling of eNOS factor. Conclusively, component of RAS, many metabolic pathways and bioavailability of NO act as major episodes during atherosclerosis [51].

7.4 Factors Responsible for CVDs

CVDs depend on multiple factors encompassing complex mechanisms. Some common factors playing central role in vascular complications will be discussed in the following section.

(a) *Genetic Predisposition*: Genome-wide linkage analyses and gene association studies can provide evidence for the familial basis of CVDs. The inherited genes attributed to the modification in lipid metabolism, response to risk factor for CVD development and progression [52]. Some common examples that are known to pass from generation to generation include high blood pressure or coronary artery disease, hypertrophic cardiomyopathy, dilated cardiomyopathy, arrhythmogenic right ventricular cardiomyopathy, long QT syndrome, and Brugada syndrome [53]. A prior knowledge of family history can estimates the risk susceptibility that can further aid for the development of preventive strategies.
(b) *Environmental Factors*: Clinical and epidemiological studies have realized that environmental factors such as tobacco smoking [54], polluted drinking water [55], residential air pollution [56], and aircraft noise [57] can have adverse effects on cardiovascular system. Air pollutants are reported to cause serious, harmful damage to the cardiovascular system. Artery blockage leading to congestive heart failure (arterial occlusion) and myocardial infarction are some manifestations of poor quality of breathing air. Several community-intervention studies on water as well as water with calcium (Ca^{++}) and magnesium (Mg^{++}) deposits showed protective role against CVDs [58]. While long exposure to low-level inorganic arsenic (average 95 μg/L) in drinking water is associated with increased risk for CVD in the population [59, 60].
(c) *Lifestyle-Based Factors*: Lifestyle choices can affect the cardiac health and associated problems. Physicians always recommend change in lifestyle to the CVD patients. Lifestyle includes diet, physical activities, and stress. French paradox exemplifies the relation of CVD with diet. It is due to the lack of saturated and trans fats such as butter, cheese, and cream in the French diet; the American diet includes greater amounts of saturated fats with trans fats that may have association with cardiac health risks. Alcohol and tobacco consumption leads to increment in blood pressure by increasing the cholesterol level and hardening or disrupting the endothelium lining, while moderate alcohol con-

sumption significantly reduced the risk of CVDs [19, 26]. Along with diet, physical activity also supports the healthy cardiac life. Daily physical exercise and sports minimize the incidence of cardiac issue by decreasing the level of oxLDLs which accumulate during obesity. Exercise also significantly lowers the stress level by increasing the endorphin (peptide hormone) level in blood makes cardiac and brain tissues fit.

7.5 Phytochemicals and CVDs

Many phytoextracts and their bioactive compounds are often administered along with conventional drugs to treat CVDs. Carotenoids, flavonoids, isoflavones, isothiocyanates, and polyphenols have been isolated from different plant sources. Generally, these are ROS quenchers that enhance nitric oxide (NO) bioavailability and have presented promising results in inflammatory vascular trauma such as hyperlipidemia, hypertension, and atherosclerosis [61, 62]. Carotenoids (astaxanthin, lycopene) [61], flavonoids and isoflavones (Genistein, Quercetin) [63–65], polyphenols (Curcumin, Resveratrol) [66], and isothiocyanate (sulforaphane) [67] are the modulators of various enzymes, e.g., lipoxygenase, NADPH oxidase, xanthine oxidase hemeoxygenase-1 (HO-1), and SOD. These phytochemicals maintain vascular homeostasis through platelet aggregation and leukocyte-endothelium interaction and improve the endothelial barrier during inflammatory CVDs [68]. Besides these, *in vivo* studies have also supported their potential role in controlling the signaling proteins associated with inflammation involving, phosphatidylinositol 3-kinase, PTEN, NF-κB-inducing kinase, Akt phosphorylation, IL-8 production, NF-kB/p65 nuclear translocation, Nrf2, and redox signaling in CVDs [69, 70]. On the basis of dietary classification, the phytochemicals are grouped into seven major categories as described in Fig. 7.2.

Studies have verified the significance of naturally occurring polyphenols in encouraging cardiovascular health especially in aged patients. Polyphenols such as curcumin, epigallocatechin gallate (EGCG), and resveratrol are known for their

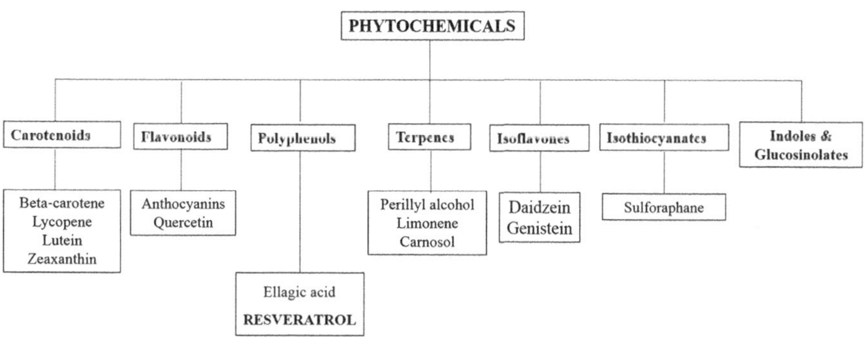

Fig. 7.2 Classification of dietary phytochemicals

beneficial effects on cardiovascular fitness and their important role in declining the effects of aging [71]. *Curcumin* is active against acute coronary syndrome, diabetic cardiomyopathy, and ischemia-induced vascular dementia [72] and enhances the levels of HDL and HO-1 [73, 74]. *Catechins* (epigallocatechin-3-*O*-gallate) are primarily found in green tea. It is found to directly boost the endothelial cell function by promoting NO signaling, lowering the concentration of endothelin-1, and preventing the cardiovascular abnormalities [75, 76]. In a study involving salt-induced hypertensive rats, it is established that a high dose of epicatechin lowers the systolic blood pressure, plasma endothelin-1, and malondialdehyde (MDA) and prevents the proteinuria condition and urinary iso-prostaglandin F2α excretion in animals [77]. Stilbenes (resveratrol) is a natural ingredient of grapevine, mulberries, and peanuts. It is assumed to have antiatherogenic activities, i.e., it can inhibit the oxidation of LDL and aggregation of platelets and regulate the proliferation of vascular smooth muscle cells [74, 78]. TNF-α induces endothelial activation and vascular inflammation to promote vascular aging and atherogenesis. Resveratrol reduces the TNF-α-induced signal transduction in human coronary arterial endothelial cells (HCAECs) by controlling the NF-κB activity [79]. It further upregulates the levels of endogenous antioxidants and phase 2 enzymes such as GST, cytochrome P450 in cultured aortic smooth muscle cells (ASMCs) and also provides protection against electrophile-mediated oxidative attack [80].Consequent demonstrations confirmed that resveratrol upsurges the transcriptional activity of Nrf2 in endothelial cells which is associated with the activation of NOX and HO-1 gene targets [81]. An upregulation of these targets is directly linked with a healthy endothelial cell environment whenever there is a rise of intracellular oxidative stress [82]. Several accounts with bioactivity of resveratrol in cardiovascular remodeling are available. Keeping in mind the paybacks of phytocompounds, our review is an attempt to cover the chemistry, bioavailability, and the mechanism behind the action of a popular antioxidant, resveratrol, in response to CVD and the associated therapeutic strategies.

7.6 Chemistry of Resveratrol and Derived Natural Products

Resveratrol (5,3,4′-trihydroxystilbene), a non-flavonoid polyphenol, mainly responsible for "French Paradox" (French people have a relatively low incidence of heart disease while having a diet which has relatively higher content of saturated fats) [83]. It is a secondary metabolite (natural product) found in food-based beverages including red wine, grapes, a variety of berries, and peanuts. The chemical structure of resveratrol was characterized from the root extract of *Veratrum grandiflorum* in the year 1940 by Takaoka [84]. Its basic structure has two phenolic rings joined together by a double styrene bond, having a molecular weight of 228.25 g/mol. This double bond is responsible for the isometric *cis* and *trans* forms of resveratrol. In red wine it is found in both *cis* and *trans* isomeric forms, but the *trans* form is more biologically active and is more investigated. *Trans* form can change into *cis* form by the process of *photoisomerization* after exposure of sun light (Fig. 7.3) [85].

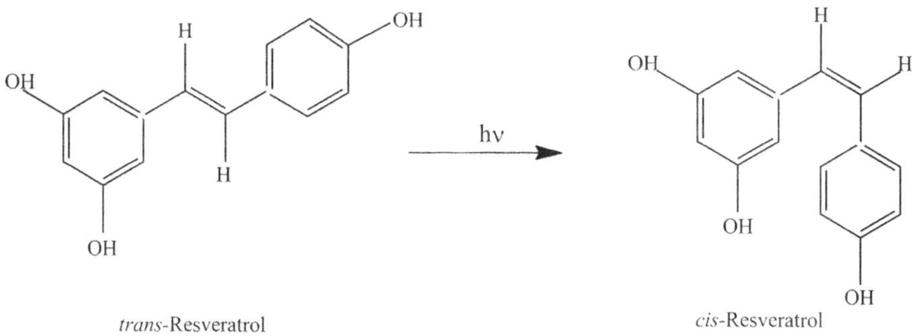

Fig. 7.3 Isomerization of *trans* to *cis* form of resveratrol after the exposure of sunlight (hν)

The biological synthesis of resveratrol and its different derivatives are either constitutive or inducible which is controlled by a multigene key enzyme, stilbene synthase (STS). STS catalyzes the condensation of three molecules of coumaroyl-CoA to form resveratrol [86]. The synthesis of resveratrol is strongly enhanced by both biotic stress (i.e., fungal attacks) and abiotic stress (i.e., UV irradiation, physical strain, and other environmental stress conditions). In nature the richest source of resveratrol is the root of Japanese knotweed (*Polygonum cuspidatum*), which is a traditional medicine used in the treatment of inflammation and several CVDs.

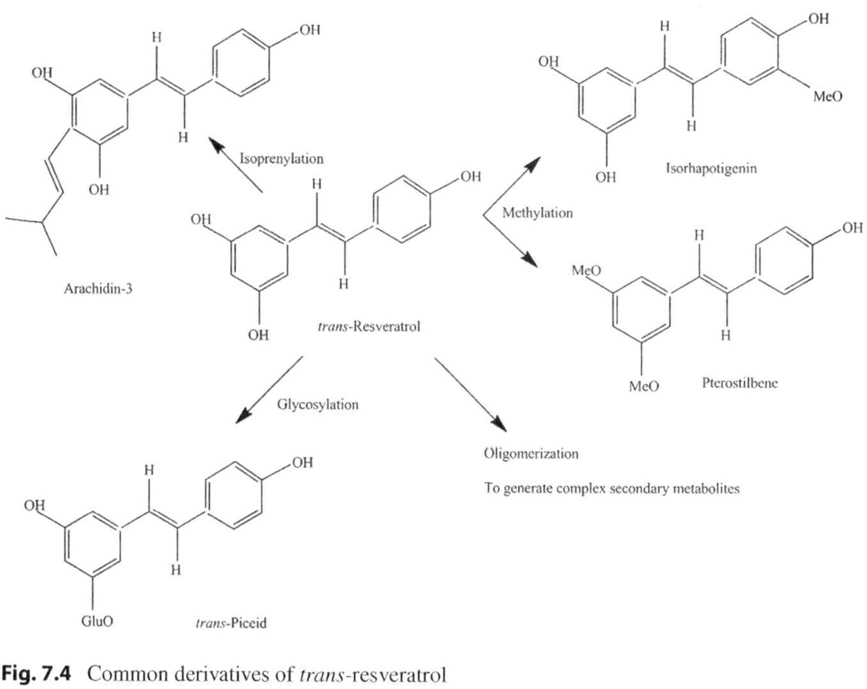

Fig. 7.4 Common derivatives of *trans*-resveratrol

Probably grapes are the main source of resveratrol which is widely consumed by human beings.

After the biosynthesis of resveratrol, it is modified into various structural derivatives. The monomer residue can form pterostilbene, isorhapotigenin, and piceid, and different monomeric forms can be oligomerized to form complex polyphenolic compounds (Fig. 7.4). These complex secondary metabolites may be composed of 2–8 monomeric resveratrol units which are collectively known as "resveratrol oligomers" [87, 88].

(a) *Bioavailability of Resveratrol*

Resveratrol is poorly soluble in water (<0.05 mg/mL) but highly soluble in organic solvents like ethanol (>50 mg/mL). Due to its chemical nature (lipophilic), resveratrol easily interacts with fatty acids, and it was found that majority of free resveratrol binds to human plasma lipoproteins and limits the bioavailability [89]. The main reason for poor bioavailability of resveratrol is that it gets rapidly absorbed in the plasma and then gets metabolized very quickly, mainly to its sulfo- and glucuro-conjugates which are easily excreted out from the body via urine [90]. From *in vitro* studies, it is clear that minimum 5 μM/L concentration of resveratrol is required for the chemopreventive effects in cells [91]. The oral bioavailability of resveratrol is low (<1%) due to rapid and extensive metabolism in the intestine and liver. Low bioavailability and rapid clearance of resveratrol from circulation are factors that may reduce its usefulness. In human studies, it is found that when a single oral dose of resveratrol is given, about 75% of its metabolites are excreted out via urine, indicating its low bioavailability [84, 92]. There is no evidence to show that resveratrol undergoes enzymatic oxidation in the human body. Three different metabolites, namely, sulfate conjugate, glucuronic acid conjugate, and hydrogenation of the aliphatic double bond of resveratrol, have been identified in liquid

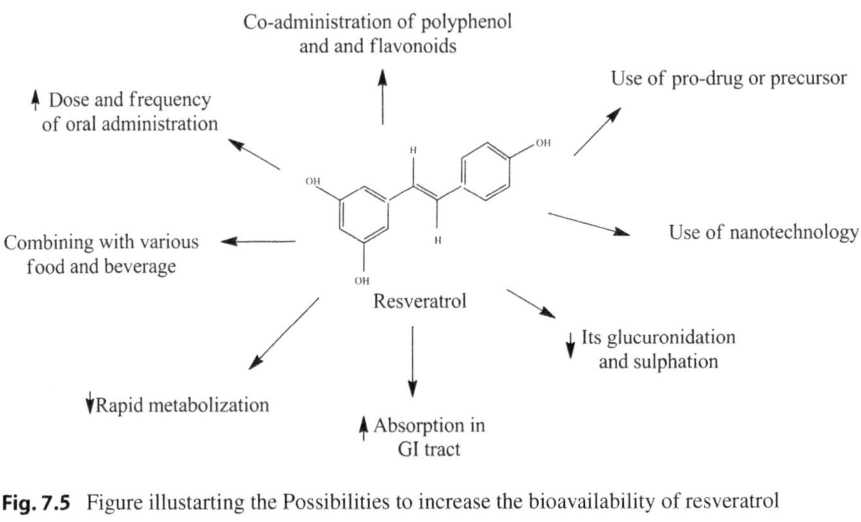

Fig. 7.5 Figure illustarting the Possibilities to increase the bioavailability of resveratrol

chromatography/mass spectrometry (LC/MS)-based analysis of urine samples of resveratrol consuming humans [93, 94].

(b) *How to Increase Its Bioavailability?*

The bioavailability of resveratrol in the human body can be increased by increase in oral dose concentration and frequency, the combination with various food and beverages, and increase in the absorption of resveratrol in the gastrointestinal tract, protecting its rapid metabolization (Fig. 7.5). Some consequent research in cellular and animal models showed that co-administration of polyphenol (e.g., piperine) along with resveratrol is used to increase its bioavailability via inhibiting its glucuronidation and increasing its pharmacokinetic parameters [95]. Similarly, co-administration of different flavonoids (e.g., quercetin, myricetin, catechin, kaempferol, fisetin, apigenin, etc.) along with resveratrol increases its bioavailability in the model organism by inhibiting the glucuronidation and sulfation [96, 97].

Another approach to increase the bioavailability of resveratrol is the use of pro-drug or precursor of resveratrol [98]. Many recent studies are focusing on the use of nanotechnology to increase the bioavailability of resveratrol with slight or no side effect [99]. Indeed, nanoparticles can increase solubility and stability of resveratrol by enhancing its absorption, protecting it from premature degradation in the body, and prolonging its circulation time [100].

(c) *Biological Effects of Resveratrol*

Resveratrol has pleiotropic effect in the biological system. Resveratrol is involved in many signaling pathways and other cellular processes. It shows anticancerous property by blocking the cell proliferation of transformed cells, and it can also

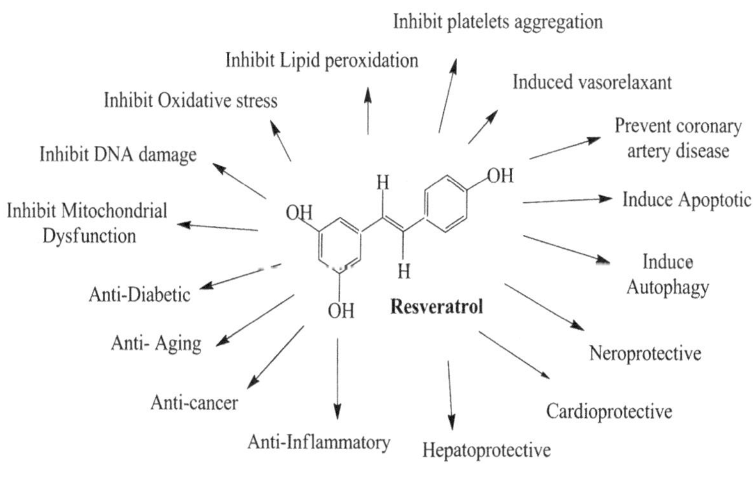

Fig. 7.6 Biological effects of *trans*-resveratrol

induce cell death by different pathways like autophagy, apoptosis, and senescence [101]. It is reported to arrest the cell cycle growth at the G1/S phase and inhibit the cellular proliferation in tumor cells [102]. It interacts with many receptor kinases and influences TGF-β super family and exhibits antiproliferative activity [103]. Resveratrol plays a significant role in ROS regulation and control oxidative stress by scavenging the superoxide radicals and hydrogen peroxide (H_2O_2) and by inducing the production of major cellular antioxidants. It has metal chelating property by which it prevents LDL oxidation and shows hepatoprotective effect [104]. Numerous studies have demonstrated that resveratrol exhibits anti-inflammatory activity by inhibiting iNOS, cyclooxygenase-1 and 2, (COX-1 and 2) and activates IL-1β, IL-6, and TNF-α cytokines [105, 106]. Studies have also showed that resveratrol plays an important role in neuroprotection via preventing the mitochondrial dysfunction and maintaining the genome integrity, act as anti-aging, anti-diabetic, and has beneficial effects on many biological processes (Fig. 7.6) [107].

In addition, other biological significances of resveratrol are well studied and mainly being the cardioprotective effect. Platelets stick to the vascular endothelial surface and activate the process of thrombus formation, and their aggregation creates vascular occlusion. Platelets have also been linked to the synthesis of eicosanoids from arachidonic acid that contributes to platelet adhesion [108]. Few studies have showed that resveratrol decreases the platelet aggregation and ultimately prevent coronary artery disease by inhibiting the eicosanoid synthesis [109].

7.7 Resveratrol and CVDs

(a) *Cardioprotective Effects of Resveratrol*

CVDs such as hypertension, stroke, atherosclerosis, heart failure, and ischemic heart diseases are the major cause of mortality in the worldwide. It is well documented that resveratrol decreases atherosclerosis by vascular relaxation and by inhibiting formation of platelets, thus displays cardioprotective role. Many epidemiological studies suggested that red wine consumption may reduce the risk of CVDs, because of the presence of this polyphenol resveratrol [110].

(b) *Resveratrol as a Vascular Relaxant*

Resveratrol relaxes human internal thoracic artery, mesenteric and uterine artery, and coronary and retinal artery [111, 112]. It potentially improves vascular response in diabetic rat model study [113]. The complete mechanism of vascular relaxant induced by resveratrol is still not fully elucidated. It is suggested that it synthesizes NO by upregulating the different NOSs by activating SIRT-1-induced deacetylation process and maintains critical concentration of NO by inhibiting enzyme-mediated neutralization process which relaxes the blood vessels [114]. The other suggested mechanism of vascular relaxation is via different ion channels present in the vascular smooth muscle cells. Resveratrol maintains Ca^{2+} and K^+ channel and helps in vascular relaxation [115]. Apart from these roles, resveratrol inhibit the growth of fibroblast in cardiac cells by stimulating the angiotensin II, epidermal growth factor-β (EGF- β), and TGF-β and ultimately prevent the heart failure.

(c) *Antioxidant Properties of Resveratrol*

Resveratrol acts as an antioxidant by exhibiting protective effects by upregulating the levels of endogenous cellular antioxidant system and indirectly involves in the scavenging activity of the harmful ROS. The role of ROS is well illustrated in CVDs, and are well known for impairment of cardiac contraction, cardiac hypertrophy, and heart failure [116]. The antioxidant mechanisms rendered by resveratrol play an important role in enzymatic pathways, gene expression, cellular signaling, and apoptosis which additionally can increase the current antihypertensive pharmacotherapeutic approaches. Furthermore, resveratrol regulates the activity and expression of antioxidant enzymes, namely, SOD, CAT, glutathione peroxidase (GPx), and SIRT, by the way of transcriptional regulation through forkhead box protein O (FOXO), Nrf2, and activator protein (AP) 1 or via other enzymatic modulations [117].

Resveratrol restores the dimethyl arginine dimethyl amino hydrolase (DDAH) activity which is an enzyme that degrades asymmetric dimethyl arginine (ADMA) an endogenous inhibitor of eNOS thereby resveratrol could improves the bioavailability of NO and decreases endothelial dysfunctioning reported in diabetes. It is observed that cardiac stem cells when treated with resveratrol *in vitro* begin to manifest substantial proliferation, differentiation, and survival toward cardiomyocytes. However, *in vivo* resveratrol shows antithrombotic property in carotid artery. Resveratrol accounts for pharmacological approach for the metabolic syndrome treatment that is associated with CVDs and its increased risk, by taking into the account that the resveratrol beneficial effects on inflammation, dyslipidemia, diabetes, hypertension, and obesity [9].

(d) *Resveratrol in Platelet Aggregation*

Resveratrol helps in reducing Ca^{2+} and in preventing platelet activation. Resveratrol also blocks COX thus inhibiting TxA2 formation [118]. Resveratrol stimulates human BH3-interacting domain death agonist cleavage, mitochondrial membrane potential indulgence, caspase-3, caspase-9, and caspase-8 and cytochrome C release in washed human platelets that further helps to prevent pathological clotting and endorse platelet apoptosis [119]. In addition, resveratrol also found to inhibit protein kinase C and phospholipase C activities, decrease the load of platelet H_2O_2, increase the NO level, and decrease superoxide anion in the platelets, thus inhibiting the platelet aggregation [120].

(e) *Resveratrol in Lipid Metabolism*

Resveratrol can alter the low-density lipoproteins (LDL) lipid profile by decreasing plasma LDL and triglycerides level and increasing high-density lipoproteins (HDL). Resveratrol can potentiate the pravastatin hypocholesterol eminent action, by downregulating the 3-hydroxy-3-methylglutaryl-Co-A (HMG-CoA) reductase enzyme responsible for the intervention in the first step of cholesterol biosynthesis [9]. In fat diet-induced obesity mouse model study, the resveratrol was found to play a significant role to promote fat mobilization in white adipocyte tissues and hence protects mice from obesity by activating SIRT-1 and prevents proliferation and differentiation of preadipocytes [121]. Resveratrol assists in lowering of serum LDL, very-low-density lipoprotein (VLDL), and triglycerides (TG). Consequently, reduction in LDL oxidation in combination with these effects thereby helps in prevention of proliferation of smooth muscle cells. On the contrary, resveratrol shows statin-like effect in a hyperlipidemia model by HMG-CoA reductase downregulation thereby increasing the ratio of apolipoprotein (Apo) A1 to Apo B, that is, negatively related to risk factors such as metabolic syndrome and CVDs. Alternatively, HDL helps in protecting from cardiovascular risk factors through the process of reverse transport. Therefore, resveratrol helps to reduce cholesteryl esters transport(RCT) from HDL to LDL/VLDL and helps in enhancement of RCT [122].

(f) *Resveratrol in Endothelial Function*

Platelet activation, leukocyte adherence to vascular endothelial cells, impaired coagulation, thrombosis, atherosclerosis, and abnormal vasoconstriction are the characteristics of endothelial dysfunction. Abnormalities in endothelium are generally related with cardiovascular risk factors such as hypertension, insulin resistance, and hypercholesterolemia [123]. Phosphorylation of endothelial NO synthase is found to be stimulated by resveratrol via AKt and phsophatidylinositol-3-hydroxy kinase. The mechanism of action of resveratrol in endothelial function involves with the activation of SIRT-1. Activation of SIRT-1 further activates AMP-dependent protein kinase and decreases the p53 activity, a regulator of the cell cycle and apoptosis and also activates a regulator of cellular energy status AMP-dependent protein kinase [124]. Thus, resveratrol play a pivotal role in regulation of endothelial functioning during CVDs.

(g) *Resveratrol in Cardiovascular Remodeling and Heart Failure*

Resveratrol activates PGC-1α and SIRT-1 thus bestows to mitochondrial function that further can influence metabolism of whole body. Resveratrol activates adenosine monophosphate-activated protein kinase (AMPK) which prevent heart, liver, aorta, and pancreas damage in mice fed with high-calorie-rich diet [125]. Resveratrol helps to improve remodeling of sympathetic neural system and leads to ventricular arrhythmias [126]. Resveratrol not only helps in delaying the cardiac remodeling process but also plays significant role in directly improving the cardiac function of failing the heart [127]. Resveratrol increases the systolic performance of

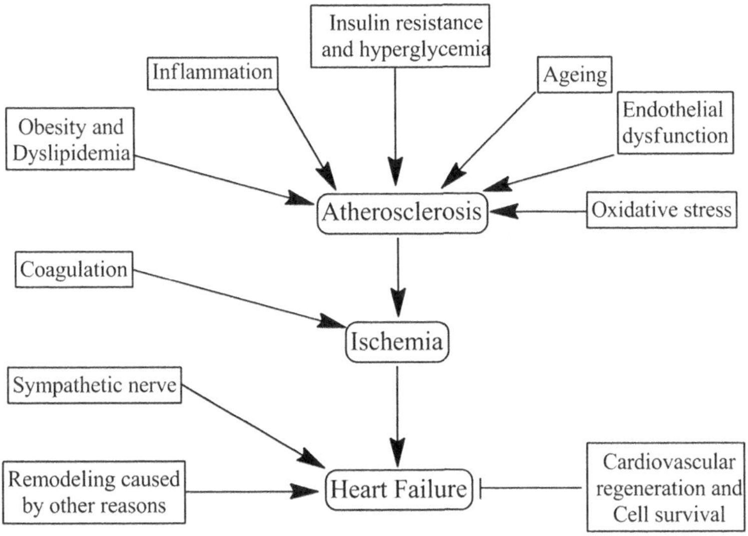

Fig. 7.7 Multiple cellular targets of resveratrol in cardiovascular diseases

heart by improving diastolic function thus prevents heart failure risk [128]. Resveratrol exhibits an inhibitory effect of oxidative stress on liver kinase B1(LKB1) and also favors AMPK activation which ultimately prevents cardiac tissue remodeling [129]. Resveratrol also plays significant role in inhibiting migration and proliferation of aortic vascular smooth muscle cells by decreasing the cross talk between an inducer of MMPs and IL-18 [130]. Antiproliferative properties of resveratrol are associated with the induction of p53 in SMCs and inhibition of DNA synthesis [131].

Further resveratrol manifests significant role in CVDs by maintaining cardiovascular system stationarity. The innumerable risk factors are obesity and dyslipidemia, insulin resistance and hyperglycemia, oxidative stress, endothelial dysfunction, aging, and inflammation that contribute in atherosclerosis, ischemia, and finally toward heart failure (Fig. 7.7). To recapitulate it can be said that resveratrol can interfere in the functioning of these factors thus either it slows down or completely prevent the occurrence/incidence of CVDs [132].

7.8 Resveratrol as an Emerging Drug for CVDs

Resveratrol regulates gene expression, translation, cellular signaling, and some other cardiovascular remodeling processes indirectly targeting the miRNA machinery. Resveratrol modulates about 50 differentially expressed inflammation-related miRNA such as miR-21, miR-20b, miR-27a, miR-9 miR-181b miR-30c2, miR-155, and miR-34a which deals with different metabolic functions which may play key role in reducing I/R injury [133–135]. The upregulation of Dicer1 functions with resveratrol treatment in endothelial cells of aged mice brain also supports this fact

[136]. However, in cerebral ischemic tolerance with resveratrol, preconditioning does not alter miRNA profiles [137].

Resveratrol also alters the serum lipid profile by inhibiting lipogenesis, fat accumulation by lipolysis. Resveratrol also accustoms lipid metabolism and increases the apo-AI/apo-B ratio, and HDL-to-LDL ratio, which stimulate the mobilization of excess cholesterol into the body system [138–140]. Long-term resveratrol intervention downregulates the expression of key pro-inflammatory cytokines (TNF-α, IL-1β, IL-6, and platelet-derived growth factor-α/β) [141].

7.9 Summary and Conclusion

CVDs have become one of the prevalent life-threatening episodes globally. Daily routine, diet, environmental factors, and genetic predisposition are often held accountable, further rendering the CVDs tough to deal with. It is evident that the usage of natural supplements is capable of refurbishing the cardiac functions. Study on resveratrol has been encouraging in curbing various physiological disorders together with cancer, diabetes, and AD and, in this manner, extending the longevity. Being anti-inflammatory and antioxidant, resveratrol supports the vascular homeostasis during pathology; still, the biological intervention is complicated and is less explored. After the famous "French paradox," a comprehensive study confirmed its direct/indirect role in regulating the gene expression, translation, cellular signaling metabolic functions, and other cardiovascular remodeling processes. This bioactive compound is now considered as a promising therapeutic tool not only due to its limiting toxicity and side effects but also due to its bioavailability and targeting at different molecular levels associated with the betterment of human health.

Acknowledgment S.U., K.B.G. acknowledges the financial support from ICMR (New Delhi). Rubal and S. Kaur acknowledges UGC (New Delhi) for RGNF-JRF. S. Kumar is supported by the CUPB institutional fellowship for PhD. AKM acknowledges the Alzheimer's Association, USA (NIRG-11-203527), and M.D. acknowledges the DST (Fast-Track: (SB/YS/LS-107/2013). Because of the limited focus of the manuscript, many appropriate references could not be included, for which the authors apologize.

References

1. Laslett LJ et al (2012) The worldwide environment of cardiovascular disease: prevalence, diagnosis, therapy, and policy issues: a report from the American College of Cardiology. J Am Coll Cardiol 60(25):S1–S49
2. World Health Organization (2014) Global status report on alcohol and health 2014. World Health Organization, Geneva
3. Pearson TA et al (2003) Markers of inflammation and cardiovascular disease. Circulation 107(3):499–511
4. Lippi G et al (2010) Moderate red wine consumption and cardiovascular disease risk: beyond the "French paradox". In: Seminars in thrombosis and hemostasis. Thieme Medical Publishers, New York

5. Ramesh Vidavalur M et al (2006) Significance of wine and resveratrol in cardiovascular disease: French paradox revisited. Exp Clin Cardiol 11(3):217–225
6. Ballini A et al (2017) Resveratrol in vascular diseases and therapeutics. Vasc Dis Ther 2:1–2
7. Zhu L, Luo X, Jin Z (2008) Effect of resveratrol on serum and liver lipid profile and antioxidant activity in hyperlipidemia rats. Asian Aust J Animal Sci 21(6):890
8. Borriello A et al (2010) Dietary polyphenols: focus on resveratrol, a promising agent in the prevention of cardiovascular diseases and control of glucose homeostasis. Nutr Metab Cardiovasc Dis 20(8):618–625
9. Bonnefont-Rousselot D (2016) Resveratrol and cardiovascular diseases. Nutrients 8(5):250
10. Tomé-Carneiro J et al (2013) One-year supplementation with a grape extract containing resveratrol modulates inflammatory-related microRNAs and cytokines expression in peripheral blood mononuclear cells of type 2 diabetes and hypertensive patients with coronary artery disease. Pharmacol Res 72:69–82
11. Szkudelski T, Szkudelska K (2011) Anti-diabetic effects of resveratrol. Ann NY Acad Sci 1215(1):34–39
12. Vallianou NG et al (2015) Resveratrol and cancer. Hosp Chron 10(3):137
13. Collaboration, P.S. (2002) Age-specific relevance of usual blood pressure to vascular mortality: a meta-analysis of individual data for one million adults in 61 prospective studies. Lancet 360(9349):1903–1913
14. Gareth B, Lip GY, O'Brien E (2001) The ABC of hypertension: the pathophysiology of hypertension. BMJ 322:912–916
15. Tzourio C (2007) Hypertension, cognitive decline, and dementia: an epidemiological perspective. Dialogues Clin Neurosci 9(1):61–70
16. Franklin SS (2005) Arterial stiffness and hypertension. Hypertension 45(3):349–351
17. Park S, Lakatta EG (2012) Role of inflammation in the pathogenesis of arterial stiffness. Yonsei Med J 53(2):258–261
18. Rudijanto A (2007) The role of vascular smooth muscle cells on the pathogenesis of atherosclerosis. Acta Med Indones 39(2):86–93
19. Kitulwatte ID, Pollanen MS (2015) A comparative study of coronary atherosclerosis in young and old. Am J Forensic Med Pathol 36(4):323–326
20. van Oijen M et al (2007) Atherosclerosis and risk for dementia. Ann Neurol 61(5):403–410
21. Heeringa J et al (2007) Subclinical atherosclerosis and risk of atrial fibrillation: the Rotterdam study. Arch Intern Med 167(4):382–387
22. Yasaka M, Yamaguchi T, Shichiri M (1993) Distribution of atherosclerosis and risk factors in atherothrombotic occlusion. Stroke 24(2):206–211
23. Klein R et al (2002) The association of atherosclerosis, vascular risk factors, and retinopathy in adults with diabetes: the atherosclerosis risk in communities study. Ophthalmology 109(7):1225–1234
24. Casserly I, Topol EJ (2004) Convergence of atherosclerosis and Alzheimer's disease: inflammation, cholesterol, and misfolded proteins. Lancet 363(9415):1139–1146
25. Befeler B et al (1977) Coronary artery aneurysms: study of their etiology, clinical course and effect on left ventricular function and prognosis. Am J Med 62(4):597–607
26. Kaneko H et al (2011) Resveratrol prevents the development of abdominal aortic aneurysm through attenuation of inflammation, oxidative stress, and neovascularization. Atherosclerosis 217(2):350–357
27. Rijbroek A et al (1994) Inflammation of the abdominal aortic aneurysm wall. Eur J Vasc Surg 8(1):41–46
28. Cohen P, O'Gara PT (2008) Coronary artery aneurysms: a review of the natural history, pathophysiology, and management. Cardiol Rev 16(6):301–304
29. Brown TJ et al (2001) CD8 T lymphocytes and macrophages infiltrate coronary artery aneurysms in acute Kawasaki disease. J Infect Dis 184(7):940–943
30. He R et al (2006) Characterization of the inflammatory and apoptotic cells in the aortas of patients with ascending thoracic aortic aneurysms and dissections. J Thorac Cardiovasc Surg 131(3):671–678 e2

31. Szekanecz Z et al (1994) Human atherosclerotic abdominal aortic aneurysms produce interleukin (IL)-6 and interferon-gamma but not IL-2 and IL-4: the possible role for IL-6 and interferon-gamma in vascular inflammation. Inflamm Res 42(3):159–162
32. Koch A et al (1993) Enhanced production of the chemotactic cytokines interleukin-8 and monocyte chemoattractant protein-1 in human abdominal aortic aneurysms. Am J Pathol 142(5):1423
33. Shimizu K et al (2004) Th2-predominant inflammation and blockade of IFN-γ signaling induce aneurysms in allografted aortas. J Clin Invest 114(2):300–308
34. Parry DJ et al (2010) Markers of inflammation in men with small abdominal aortic aneurysm. J Vasc Surg 52(1):145–151
35. Radonic T et al (2012) Inflammation aggravates disease severity in Marfan syndrome patients. PLoS One 7(3):e32963
36. Nataatmadja M et al (2003) Abnormal extracellular matrix protein transport associated with increased apoptosis of vascular smooth muscle cells in Marfan syndrome and bicuspid aortic valve thoracic aortic aneurysm. Circulation 108(10 suppl 1):II-329–II-334
37. Wang Y et al (2010) TGF-β activity protects against inflammatory aortic aneurysm progression and complications in angiotensin II–infused mice. J Clin Invest 120(2):422–432
38. Anahita Dua M et al (2013) A review of the role of platelets in vascular trauma patients compared to patients with chronic vascular disease. Vasc Dis Manag 10(11):E240–E243
39. Reinhart W (2013) Platelets in vascular disease. Clin Hemorheol Microcirc 53(1–2):71–79
40. van der Loo B, Martin JF (1997) 6 megakaryocytes and platelets in vascular disease. Baillieres Clin Haematol 10(1):109–123
41. Schmidt-Ott KM, Kagiyama S, Phillips MI (2000) The multiple actions of angiotensin II in atherosclerosis. Regul Pept 93(1):65–77
42. Pacurari M et al (2014) The renin-angiotensin-aldosterone system in vascular inflammation and remodeling. Int J Inflamm 2014:1
43. Arehart E et al (2007) Prostacyclin, atherothrombosis, and cardiovascular disease. Curr Med Chem 14(20):2161–2169
44. Nagaya N (2010) Orally active prostacyclin analogue for cardiovascular disease. Int Angiol: J Int Union Angiol 29(2 Suppl):14–18
45. Böhm F, Pernow J (2007) The importance of endothelin-1 for vascular dysfunction in cardiovascular disease. Cardiovasc Res 76(1):8–18
46. Agapitov AV, Haynes WG (2002) Role of endothelin in cardiovascular disease. J Renin-Angiotensin-Aldosterone Syst 3(1):1–15
47. Qing P et al (2015) Association of big endothelin-1 with coronary artery calcification. PLoS One 10(11):e0142458
48. Chester AH, Yacoub MH (2014) The role of endothelin-1 in pulmonary arterial hypertension. Glob Cardiol Sci Pract 2014:29
49. Kuhn H et al (2013) The role of endothelin in stroke. Issues 7(1)
50. Schaller BJ (2006) The role of endothelin in stroke: experimental data and underlying pathophysiology. Arch Med Sci 2(3):146
51. Li H, Förstermann U (2013) Uncoupling of endothelial NO synthase in atherosclerosis and vascular disease. Curr Opin Pharmacol 13(2):161–167
52. Scheuner MT (2001) Genetic predisposition to coronary artery disease. Curr Opin Cardiol 16(4):251–260
53. Wilde AA, Behr ER (2013) Genetic testing for inherited cardiac disease. Nat Rev Cardiol 10(10):571–583
54. Enstrom JE, Kabat GC (2006) Environmental tobacco smoke and coronary heart disease mortality in the United States—a meta-analysis and critique. Inhal Toxicol 18(3):199–210
55. Sauvant M-P, Pepin D (2002) Drinking water and cardiovascular disease. Food Chem Toxicol 40(10):1311–1325
56. Lee B-J, Kim B, Lee K (2014) Air pollution exposure and cardiovascular disease. Cardiovasc Dis 3:24

57. Eriksson C et al (2007) Aircraft noise and incidence of hypertension. Epidemiology 18(6):716–721
58. Momeni M et al (2014) Does water hardness have preventive effect on cardiovascular disease? Int J Prev Med 5(2):159
59. Tsuji JS et al (2014) Association of low-level arsenic exposure in drinking water with cardiovascular disease: a systematic review and risk assessment. Toxicology 323:78–94
60. James KA et al (2015) Association between lifetime exposure to inorganic arsenic in drinking water and coronary heart disease in Colorado residents. Environ Health Perspect (Online) 123(2):128
61. Maria AG, Graziano R, Nicolantonio DO (2015) Carotenoids: potential allies of cardiovascular health? Food Nutr Res 59(1):26762c
62. Dhiman M et al (2015) Oxidative stress and inflammation in cardiovascular diseases: two sides of the same coin. In: Free radicals in human health and disease. Springer, New Delhi, pp 259–278
63. Si H, Liu D (2009) Isoflavone genistein protects human vascular endothelial cells against tumor necrosis factor-α-induced apoptosis through the p38β mitogen-activated protein kinase. Apoptosis 14(1):66
64. Babu PVA et al (2012) Genistein prevents hyperglycemia-induced monocyte adhesion to human aortic endothelial cells through preservation of the cAMP signaling pathway and ameliorates vascular inflammation in obese diabetic mice. J Nutr 142(4):724–730
65. Shen Y et al (2013) Dietary quercetin attenuates oxidant-induced endothelial dysfunction and atherosclerosis in apolipoprotein E knockout mice fed a high-fat diet: a critical role for heme oxygenase-1. Free Radic Biol Med 65:908–915
66. Tangney CC, Rasmussen HE (2013) Polyphenols, inflammation, and cardiovascular disease. Curr Atheroscler Rep 15(5):324
67. Evans PC (2011) The influence of sulforaphane on vascular health and its relevance to nutritional approaches to prevent cardiovascular disease. EPMA J 2(1):9
68. Si H, Liu D (2007) Phytochemical genistein in the regulation of vascular function: new insights. Curr Med Chem 14(24):2581–2589
69. Agarwal M et al (2012) Dynamic action of carotenoids in cardioprotection and maintenance of cardiac health. Molecules 17(4):4755–4769
70. Thimmulappa RK et al (2002) Identification of Nrf2-regulated genes induced by the chemopreventive agent sulforaphane by oligonucleotide microarray. Cancer Res 62(18):5196–5203
71. Khurana S et al (2013) Polyphenols: benefits to the cardiovascular system in health and in aging. Nutrients 5(10):3779–3827
72. Alwi I et al (2008) The effect of curcumin on lipid level in patients with acute coronary syndrome. Acta Med Indones 40(4):201–210
73. Li Y (2011) Protective effects of curcumin on brain vascular dementia by chronic cerebral ischemia in rats and study of the molecular mechanism. Alzheimers Dement 7((4):e47–e48
74. Zhirongwang YH et al (2002) Effects of red wine and wine polyphenol resveratrol on platelet aggregation in vivo and in vitro. Int J Mol Med 9:77–79
75. Shenouda SM, Vita JA (2007) Effects of flavonoid-containing beverages and EGCG on endothelial function. J Am Coll Nutr 26(4):366S–372S
76. Loke WM et al (2008) Pure dietary flavonoids quercetin and (−)-epicatechin augment nitric oxide products and reduce endothelin-1 acutely in healthy men. Am J Clin Nutr 88(4):1018–1025
77. Gómez-Guzmán M et al (2012) Epicatechin lowers blood pressure, restores endothelial function, and decreases oxidative stress and endothelin-1 and NADPH oxidase activity in DOCA-salt hypertension. Free Radic Biol Med 52(1):70–79
78. Zou J et al (2000) Effects of resveratrol on oxidative modification of human low density lipoprotein. Chin Med J 113(2):99–102
79. Csiszar A et al (2006) Resveratrol attenuates TNF-α-induced activation of coronary arterial endothelial cells: role of NF-κB inhibition. Am J Phys Heart Circ Phys 291(4):H1694–H1699

80. Li Y, Cao Z, Zhu H (2006) Upregulation of endogenous antioxidants and phase 2 enzymes by the red wine polyphenol, resveratrol in cultured aortic smooth muscle cells leads to cytoprotection against oxidative and electrophilic stress. Pharmacol Res 53(1):6–15
81. Ungvari Z et al (2010) Resveratrol confers endothelial protection via activation of the antioxidant transcription factor Nrf2. Am J Phys Heart Circ Phys 299(1):H18–H24
82. Ungvari Z et al (2007) Resveratrol increases vascular oxidative stress resistance. Am J Phys Heart Circ Phys 292(5):H2417–H2424
83. Renaud Sd, de Lorgeril M (1992) Wine, alcohol, platelets, and the French paradox for coronary heart disease. Lancet 339(8808):1523–1526
84. Gambini J et al (2015) Properties of resveratrol: in vitro and in vivo studies about metabolism, bioavailability, and biological effects in animal models and humans. Oxidative Med Cell Longev 2015:1
85. Bernard E, Britz-McKibbin P, Gernigon N (2007) Resveratrol photoisomerization: an integrative guided-inquiry experiment. J Chem Educ 84(7):1159
86. Giovinazzo G et al (2012) Resveratrol biosynthesis: plant metabolic engineering for nutritional improvement of food. Plant Foods Hum Nutr 67(3):191–199
87. Keylor MH, Matsuura BS, Stephenson CR (2015) Chemistry and biology of resveratrol-derived natural products. Chem Rev 115(17):8976–9027
88. He S, Yan X (2013) From resveratrol to its derivatives: new sources of natural antioxidant. Curr Med Chem 20(8):1005–1017
89. Urpí-Sardà M et al (2005) Uptake of diet resveratrol into the human low-density lipoprotein. Identification and quantification of resveratrol metabolites by liquid chromatography coupled with tandem mass spectrometry. Anal Chem 77(10):3149–3155
90. Cottart CH et al (2010) Resveratrol bioavailability and toxicity in humans. Mol Nutr Food Res 54(1):7–16
91. Boocock DJ et al (2007) Phase I dose escalation pharmacokinetic study in healthy volunteers of resveratrol, a potential cancer chemopreventive agent. Cancer Epidemiol Prev Biomark 16(6):1246–1252
92. Walle T (2011) Bioavailability of resveratrol. Ann N Y Acad Sci 1215(1):9–15
93. Walle T et al (2004) High absorption but very low bioavailability of oral resveratrol in humans. Drug Metab Dispos 32(12):1377–1382
94. Yu C et al (2002) Human, rat, and mouse metabolism of resveratrol. Pharm Res 19(12):1907–1914
95. Johnson JJ et al (2011) Enhancing the bioavailability of resveratrol by combining it with piperine. Mol Nutr Food Res 55(8):1169–1176
96. De Santi C et al (2000) Glucuronidation of resveratrol, a natural product present in grape and wine, in the human liver. Xenobiotica 30(11):1047–1054
97. De Santi C et al (2000) Sulphation of resveratrol, a natural compound present in wine, and its inhibition by natural flavonoids. Xenobiotica 30(9):857–866
98. Biasutto L, Zoratti M (2014) Prodrugs of quercetin and resveratrol: a strategy under development. Curr Drug Metab 15(1):77–95
99. Smoliga JM, Blanchard O (2014) Enhancing the delivery of resveratrol in humans: if low bioavailability is the problem, what is the solution? Molecules 19(11):17154–17172
100. Wang S et al (2014) Application of nanotechnology in improving bioavailability and bioactivity of diet-derived phytochemicals. J Nutr Biochem 25(4):363–376
101. Delmas D, Solary E, Latruffe N (2011) Resveratrol, a phytochemical inducer of multiple cell death pathways: apoptosis, autophagy and mitotic catastrophe. Curr Med Chem 18(8):1100–1121
102. Casanova F et al (2012) Resveratrol chemosensitizes breast cancer cells to melphalan by cell cycle arrest. J Cell Biochem 113(8):2586–2596
103. Varoni EM et al (2016) Anticancer molecular mechanisms of resveratrol. Front Nutr 3:8
104. Gülçin İ (2010) Antioxidant properties of resveratrol: a structure–activity insight. Innovative Food Sci Emerg Technol 11(1):210–218

105. Donnelly LE et al (2004) Anti-inflammatory effects of resveratrol in lung epithelial cells: molecular mechanisms. Am J Phys Lung Cell Mol Phys 287(4):L774–L783
106. Das S, Das DK (2007) Anti-inflammatory responses of resveratrol. Inflamm Allergy Drug Targets (Form Curr Drug Targets Inflamm Allergy) 6(3):168–173
107. Ungvari Z et al (2011) Mitochondrial protection by resveratrol. Exerc Sport Sci Rev 39(3):128
108. Bhat KP, Kosmeder JW, Pezzuto JM (2001) Biological effects of resveratrol. Antioxid Redox Signal 3(6):1041–1064
109. Yang T et al (2015) Properties and molecular mechanisms of resveratrol: a review. Die Pharmazie-Int J Pharm Sci Res 70(8):501–506
110. De Lorgeril M et al (2002) Mediterranean diet and the French paradox. Cardiovasc Res 54(3):503–515
111. Novakovic A et al (2006) The mechanism of endothelium-independent relaxation induced by the wine polyphenol resveratrol in human internal mammary artery. J Pharmacol Sci 101(1):85–90
112. Jäger U, Nguyen-Duong H (1999) Relaxant effect of trans-resveratrol on isolated porcine coronary arteries. Arzneimittelforschung 49(03):207–211
113. Silan C (2008) The effects of chronic resveratrol treatment on vascular responsiveness of streptozotocin-induced diabetic rats. Biol Pharm Bull 31(5):897–902
114. Gojković-Bukarica L et al (2013) Cardiovascular effects of resveratrol. Vojnosanit Pregl 70(12):1145–1150
115. Reuter S et al (2010) Oxidative stress, inflammation, and cancer: how are they linked? Free Radic Biol Med 49(11):1603–1616
116. Spanier G et al (2009) Resveratrol reduces endothelial oxidative stress by modulating the gene expression of superoxide dismutase 1 (SOD1), glutathione peroxidase 1 (GPx1) and NADPH oxidase subunit (Nox4). J Physiol Pharmacol 60(Suppl 4):111–116
117. Wong R et al (2011) Acute resveratrol supplementation improves flow-mediated dilatation in overweight/obese individuals with mildly elevated blood pressure. Nutr Metab Cardiovasc Dis 21(11):851–856
118. Shen MY et al (2007) Inhibitory mechanisms of resveratrol in platelet activation: pivotal roles of p38 MAPK and NO/cyclic GMP. Br J Haematol 139(3):475–485
119. Wang H et al (2012) Resveratrol in cardiovascular disease: what is known from current research? Heart Fail Rev 17(3):437–448
120. Gresele P et al (2008) Resveratrol, at concentrations attainable with moderate wine consumption, stimulates human platelet nitric oxide production. J Nutr 138(9):1602–1608
121. Mader I et al (2010) Identification of a novel proapoptotic function of resveratrol in fat cells: SIRT1-independent sensitization to TRAIL-induced apoptosis. FASEB J 24(6):1997–2009
122. Weber O et al (2010) Cholesteryl ester transfer protein and its inhibition. Cell Mol Life Sci 67(18):3139–3149
123. Grover-Páez F, Zavalza-Gómez AB (2009) Endothelial dysfunction and cardiovascular risk factors. Diabetes Res Clin Pract 84(1):1–10
124. Dohadwala MM, Vita JA (2009) Grapes and cardiovascular disease. J Nutr 139(9):1788S–1793S
125. Das DK, Mukherjee S, Ray D (2010) Resveratrol and red wine, healthy heart and longevity. Heart Fail Rev 15(5):467–477
126. Dolinsky VW et al (2009) Resveratrol prevents the prohypertrophic effects of oxidative stress on LKB1. Circulation 119(12):1643–1652
127. Xin P et al (2010) Favorable effects of resveratrol on sympathetic neural remodeling in rats following myocardial infarction. Eur J Pharmacol 649(1):293–300
128. Pagliaro B et al (2015) Phytochemical compounds and protection from cardiovascular diseases: a state of the art. Biomed Res Int 2015:1
129. Noga AA et al (2007) Expression of an active LKB1 complex in cardiac myocytes results in decreased protein synthesis associated with phenylephrine-induced hypertrophy. Am J Phys Heart Circ Phys 292(3):H1460–H1469

130. Venkatesan B et al (2009) Resveratrol blocks interleukin-18-EMMPRIN cross-regulation and smooth muscle cell migration. Am J Physiol Heart Circ Physiol 297(2):H874–H886
131. Wang Z et al (2006) Regulation of proliferation and gene expression in cultured human aortic smooth muscle cells by resveratrol and standardized grape extracts. Biochem Biophys Res Commun 346(1):367–376
132. Gurusamy N et al (2010) Red wine antioxidant resveratrol-modified cardiac stem cells regenerate infarcted myocardium. J Cell Mol Med 14(9):2235–2239
133. Raj P, Zieroth S, Netticadan T (2015) An overview of the efficacy of resveratrol in the management of ischemic heart disease. Ann N Y Acad Sci 1348(1):55–67
134. Mukhopadhyay P et al (2010) Restoration of altered microRNA expression in the ischemic heart with resveratrol. PLoS One 5(12):e15705
135. Lancon A et al (2012) Control of MicroRNA expression as a new way for resveratrol to deliver its beneficial effects. J Agric Food Chem 60(36):8783–8789
136. Ungvari Z et al (2013) Aging-induced dysregulation of dicer1-dependent microRNA expression impairs angiogenic capacity of rat cerebromicrovascular endothelial cells. J Gerontol Ser A Biol Med Sci 68(8):877–891
137. Lopez MS, Dempsey RJ, Vemuganti R (2016) Resveratrol preconditioning induces cerebral ischemic tolerance but has minimal effect on cerebral microRNA profiles. J Cereb Blood Flow Metab 36(9):1644–1650
138. Tanko Y et al (2016) Resveratrol protects rabbits against cholesterol diet-induced hyperlipidaemia. Niger J Physiol Sci 31(1):71–75
139. Mendes KL et al (2016) Distinct metabolic effects of resveratrol on lipogenesis markers in mice adipose tissue treated with high-polyunsaturated fat and high-protein diets. Life Sci 153:66–73
140. Lima LM, Carvalho MDG, Sousa MO (2007) Apo B/apo AI ratio and cardiovascular risk prediction. Arq Bras Cardiol 88(6):e187–e190
141. Csiszar A et al (2009) Resveratrol prevents monocrotaline-induced pulmonary hypertension in rats. Hypertension 54(3):668–675

Plant-Derived Drug Molecules as Antibacterial Agents

8

Gauri Gaur, Utkrishta L. Raj, Shweta Dang, Sanjay Gupta, and Reema Gabrani

Abstract

With the advent of time, many microbial strains have developed resistance against various antibiotics. The treatment of infectious diseases gets more difficult when the bacteria form multidrug-resistant biofilms. Bacteria like *Escherichia coli*, *Staphylococcus epidermidis*, *Staphylococcus aureus*, and *Listeria monocytogenes* are capable of forming biofilm in vitro and cause several chronic infections like inflammatory bowel disease, colorectal cancer, catheter infection, and listeriosis. As these microorganisms are difficult to eliminate with antibiotics, there is an insistent need for alternative sources of antimicrobial therapies. Plants are one of the major sources of chemical compounds with higher therapeutic potential since many years. A large spectrum of plant-derived compounds and their secondary metabolites have antimicrobial activity and are being widely studied for their antimicrobial, anti-inflammatory, and antiviral properties. This chapter focuses on the various antibacterial properties of these plant-derived compounds and their activity alone or in combination with other compounds or antibiotics as possible antibacterial agents. It also focuses on the nanoencapsulation of these compounds to improve their bioavailability.

Keywords

Antimicrobial resistance · Biofilm · Nanoencapsulation · Synergy

---

Authors Gauri Gaur and Utkrishta L. Raj have equally contributed to this chapter.

G. Gaur · U. L. Raj · S. Dang · S. Gupta · R. Gabrani (✉)
Department of Biotechnology, Jaypee Institute of Information Technology, Noida, Uttar Pradesh, India
e-mail: reema.gabrani@jiit.ac.in

V. Rani, U. C. S. Yadav (eds.), *Functional Food and Human Health*,
https://doi.org/10.1007/978-981-13-1123-9_8

8.1 Introduction

Post the discovery of penicillin in 1928, the living style has changed due to the production and commercialization of many more antibiotics. The production of antibiotics is estimated to be 100,000 tons annually worldwide, and this has affected the life of bacteria on earth [1]. With the advent of time, the microbial strains have developed resistance against these antibiotics. Subsequently, infections caused by bacteria got complicated due to multidrug resistance and biofilm formation, which makes it difficult for the antibiotics to penetrate deep into the formed biofilms. Biofilm is wherein the microorganisms make up community by the mechanism called quorum sensing in which the microbes communicate with each other by producing autoinducer molecules. Therefore, treatment of patients with these microbes becomes more complicated and leads to longer hospital stay and therefore affects the health as well as wealth of the patient [2]. One such case is where the *Staphylococcus aureus* is not only resistant to methicillin but also to other antimicrobial agents like aminoglycosides, macrolides, tetracycline, chloramphenicol, and lincosamides and also to disinfectants which makes them an important source of hospital-acquired infections [3]. The data indicates in 15 European countries, more than 10% infection in the bloodstream is caused due to methicillin-resistant strains of *S. aureus*, and in many countries the resistance rates are nearly 50%. According to the data sets, this developed resistance claims many lives in Europe and the United States [4]. The problem is grave in certain Asian countries where treating some of the severe diseases like TB has become difficult, and the data indicates 480,000 new cases in 2013, out of which more than half went untreated [5].

The widespread unnecessary and excessive use of antibiotics and the prolonged duration of disease can contribute toward antimicrobial resistance which further increases the severity of disease with severe complications and increased mortality rate and healthcare costs [6, 7]. To overcome the side effects of conventional medicines and other related issues, researchers are interested in medicinal plants for natural drug development. The selection of natural drugs from medicinal plants can be done with the help of ethnobotany, which is a scientific study of folk medicines in different cultures. An alternative way for plant selection can be done based on pharmacological activity of plant or perform a randomized search to consider the bioactive species for further study [8]. The plant extracts and phytochemicals have been utilized for microbiological applications, metabolic disorders, and as anticancer agents.

A large spectrum of plant-derived resin and their secondary metabolites have antimicrobial activity. The group of compound like quinones, tannins, coumarins, alkaloids, and essential oils are studied for their antimicrobial, anti-inflammatory, and antiviral properties. These compounds alone or in combination with other phytochemicals or antibiotics are widely studied these days. Nonetheless, there are certain limitations to these natural compounds to be used directly like insolubility in water, less stability, etc. [9, 10]. Encapsulating the phytochemicals in the form of nanoemulsions formulation plays an important role to address these limitations and improve bioavailability. The nanoemulsions may also help in improving the antimicrobial property of the compound [11].

8.2 Antimicrobial Resistance

Microorganisms have been part of the natural environment since 3.8 billions of years ago and have a distinct genetic and metabolic diversity. They undergo various adaptations in order to cope up with the changing habitat condition, and most importantly, the pathogenic bacteria develop resistance to antibiotics/drugs that might hinder their growth. There have been reports indicating the increase in the prevalence of antimicrobial resistance among the pathogens with time in many areas around the world including the developing countries [11]. The Centers for Disease Control and Prevention (CDC) estimates that there are approximately 2 million ailments and 23,000 deaths each year in the United States due to antibiotic resistance [12]. The changes in the microbe's environment, unlimited use of antimicrobials, and non-compliant of antibiotic regime are responsible for the increase in the drug resistance among the microbes [13]. The augment in the resistance can be due to many reasons like incessant use of antimicrobials for prevention and restricting infection, long-term hospitalization, increased use of the invasive medical devices, and antibiotics used for agriculture. According to the Center for Disease Dynamics, Economics & Policy, 2015 report, increased access to antibiotics due to rise in income and surge in demand of animal proteins leading to incremental use of antibiotics in agriculture has contributed significantly to global scale-up in consumption of antibiotics and development of resistance [14]. The mechanism behind the resistance generally involves genetic mutations, which avoid the entry of antibiotic or inactivate the antimicrobial.

8.2.1 Consequences of Antimicrobial Resistance

Antimicrobial resistance is acknowledged as one of the major worldwide issues. The possible mechanism for the development of resistance can be related to the natural phenomenon of evolution. The microorganisms evolve with time, and this might lead to the overexpression of antimicrobial resistance traits and, therefore, will not be affected by the drugs. The overuse of antimicrobial drugs can also be the cause of antimicrobial drug resistance [15].

The National Strategy for Combating Antibiotic-Resistant Bacteria (CARB) emphasizes the need to address antimicrobial resistance [14]. Antimicrobial resistance may lead to the following consequences:

1. Resistant microbes become more dominant: These microorganisms do not respond to the medication; their growth results in serious health issues like patients infected with MRSA (methicillin-resistant *S. aureus*) face higher morbidity as compared to persons infected with nonresistant strains.

2. Impedes the control mechanism of communicable diseases: The decrease in the efficiency of the treatment marks the infection for a longer period and may risk its spread, e.g., the emergence of *Mycobacterium tuberculosis* multidrug resistance.
3. Increase in the healthcare costs: Prolonged infection increases the cost of treatment, which adds financial burden on people and government.
4. Antimicrobial resistance is a risk to modern medicine: Any new development in the field of medicine is at risk of developing resistance.

Thus, it can be concluded that the diseases caused due to resistant microorganisms do not respond to antimicrobials, which may lead to persistent infection that can even cause death of the patients. New resistance mechanisms have come up which make the current generation's antibiotics virtually ineffective [16, 17].

8.3 Biofilms

Biofilms are the assembly of microorganisms that irreversibly attaches to the surface by producing a polymeric substance known as exopolysaccharide (EPS). EPS is composed of extracellular DNA, proteins, and polysaccharide produced by the microbial cell. It is well known that bacteria in order to survive in nature follow a mechanism known as quorum sensing through which these bacteria interact with each other in concentration-dependent manner and results in various phenotypes like bioluminescence in the case of *Vibrio fischeri* and virulence in the case of *Pseudomonas aeruginosa* and *Agrobacterium tumefaciens* [18, 19]. Biofilms are also one of the phenotypes resulting from quorum sensing and are shown by various pathogenic bacteria including *P. aeruginosa*, *S. aureus*, *V cholera*, and *E. coli*. The EPS produced by the microbes in the biofilm is a polymeric conglomeration, which consists of various components like extracellular DNA, proteins, and polysaccharides. The biofilms when compared to the planktonic species have distinct physiology as they are formed in response to various factors like surface recognition by the cells, the nutritional factors, and, in some cases, treating planktonic species [20]. Biofilms are formed in a series of steps involving (1) deposition of the conditioning films, (2) attachment of the microbes to this conditioning film, (3) growth and colonization of the bacteria, and (4) dispersion (Fig. 8.1).

Biofilms have adherent cells which attach to the EPS, and its formation is a complex mechanism, but it generally consists of four main steps [21, 22]:

- Habituation: Herein, the substratum is altered which permits the microbes to affix to the surface. The characteristics of the substratum may significantly affect the rate and extent of microorganism's attachment. The coarse and hydrophobic surfaces are more prone to develop biofilms within short duration [23, 24]. The surface characteristics of microbial cells also play a major role in the attachment to the surfaces like the presence of pili, flagella, fimbriae, or glycocalyx. Korber et al. compared the wild-type and mutant strains of biofilm forming microorganism and

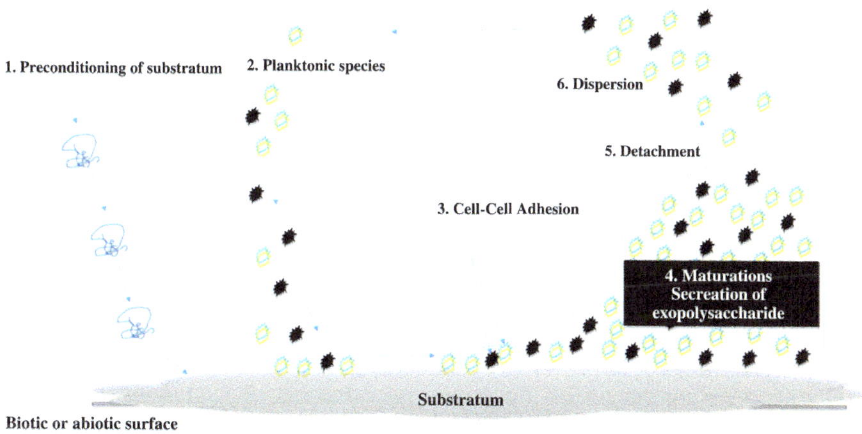

Fig. 8.1 Steps in biofilm formation

showed that the presence of flagella on the cell surface aided the Gram-negative bacteria to stick to the surfaces [25]. Another study displayed the role of fimbriae, peripheral protein structure of bacteria, in the process of attachment [26].

- Growth: The production of polysaccharides facilitates bacteria to attach to the surface to allow colonies to grow [27].
- Biofilm formation: In this stage, the microorganisms start cooperating with each other by producing autoinducers that indeed lead to the expression of certain genes that are specific to biofilm formation. The microorganisms secrete an exopolysaccharide (EPS) matrix, which helps in stabilizing the network of microbes. Biofilm generally consists of heterogeneous structure and has the capability of internal mass transport.
- Biofilm dispersion: In the final stage of biofilm, the cells from the mature biofilm will be dispersed, and the attached cells will move on to a new surface and will colonize [28].

8.3.1 Factors Influencing the Biofilm Formation

The surface properties of microbial cell and the medium characteristics are important for attachment of microbes. The type of substratum is also crucial like the rough surface which is prone to microbial attachment due to less shear force and also because of the larger surface area. Moreover, the physical characters of the outer surface also influence the attachment. This may include the hydrophobic, nonpolar surfaces like Teflon and plastic to which the microbes attach rapidly [29, 30]. The surface material when exposed to the aqueous medium results in the formation of conditioning film that is organic in nature which consists of the polymers from the medium that will ultimately result in chemical modification and will cause the microbes to grow rapidly. The surface energy of the medium to some extend affects the hydrodynamic interactions of microbes with the surface by varying the characteristics of the substratum [31].

Finally, the pH, nutrient level, ionic strength, and temperature of the medium also affect the microbial attachment to substratum [32, 33]. Fletcher conducted a study and discovered that rise in certain ions such as sodium, calcium, etc. of the medium favor the adherence of *Pseudomonas fluorescens* to the top surface [34]. Cowan et al. elicited that rise in nutrient concentration is linked to enhanced attachment of bacterial cells [35]. A study has shown that mutated *P. aeruginosa* deprived of producing acyl-homoserine-lactone (AHL) formed a less differentiated, flat biofilm in a continuous flow system thus implying that the cell communication is necessary for biofilm maturation.

Hydrophobicity of the cell surface is an important aspect with respect to adhesion as it gets enhanced with the nonpolar nature of both the cell surface and substratum surface [36].

Korber et al. concluded that the motile microbes like *P. fluorescens* are much more capable of binding to the surface as compared to the non-motile microbes. Non-motile bacteria do attach to the surface but at a slower rate. Flagella on the other hand helps in the early bonding of the microbes to the surface by overpowering the repellent forces linked to the substructure [25].

Apart from all these factors, the attachment of the microbes to the surface is genetically regulated at the initial interaction phase with the substratum. One such study conducted by Davies and Geesey stated that *algC* is upregulated in *P. aeruginosa* by distinct microbial cells within minutes of attachment [37]. Another suggested that about 22% of the genes are being unregulated and 16% of the genes are downregulated at the time of biofilm formation [38]. Pulcini showed that certain genes like *algD*, *algU*, and *rpoS* influence the formation of biofilm by regulating polyphosphokinase (PPK) synthesis in *P. aeruginosa* [39].

8.3.2 Consequences of Biofilm

The biofilm formed develops resistance to antibiotics among the bacteria and therefore aggravates the infection and hinders the cure. Biofilms cause prolonged infections, which are characterized by persistent inflammation and tissue damage in spite of being treated with antibiotics. The biofilm formation of uropathogenic *E. coli* strains is one of the main causes of urinary tract infection (UTI) that is considered as one of the major public health issues. Biofilm formation by various pathogens may lead to diseases like cystic fibrosis (CF), bacterial vaginosis, infection of the middle ear, and medical device-related infections, and biofilms may occur on contact lenses [40]. Biofilms can also grow on surfaces like membranes that are used in medical devices like central venous catheters, endotracheal tubes, mechanical heart valves, pacemakers, etc. therefore, posing a serious issue for both the environment and human health [41]. Certain diseases aggravated by bacterial biofilm are enlisted in Table 8.1.

Table 8.1 Diseases caused by bacterial biofilm

Disease	Pathogenic bacteria	Affected body part or symptoms	References
Tuberculosis	*M. tuberculosis*	Lungs	[42]
Typhoid fever	*Salmonella enterica* serotype *Typhi*	Abdominal pain, constipation, headaches, and skin rash with rose-colored spots	[43]
Salmonellosis	*Salmonella* spp.	Stomach infection, fever 12–72 h after infection	[44]
Diphtheria	*Corynebacterium diphtheriae*	Sore throat, fever, patches in the throat. It may mask the human air channel, which results in barking cough. Complications can cause myocarditis and affect other organs owed to low platelet levels	[45, 46]
Cholera	*Vibrio cholerae*	Small intestine	[47]
Bacteremia	*S. aureus*, *Enterococci*, *E. coli*	High fever	[48]
Necrotizing fasciitis	Group A *Streptococcus* (group A strep), *Klebsiella*, *Clostridium*, *E. coli*, *S. aureus*, and *Aeromonas hydrophila*	Skin infection that spreads quickly and kills the body's soft tissue	[49]
Bacterial pneumonia	*Streptococcus pneumoniae*, *Bacillus anthracis*	Pneumonia, fever, rigors, cough, runny nose	[50]
Rheumatic fever	*Streptococcus pyogenes*	Heart, joints, skin, and brain	[51]
Listeriosis	*L. monocytogenes*	Central nervous system and bacteremia in those who are immunocompromised, pregnant women, and those at the extremes of age, as well as gastroenteritis in healthy persons who have been severely infected	[52]
Melioidosis	*Burkholderia pseudomallei*	Chest pain, cough; skin infection, lung infection	[53]
Tooth decay	*Streptococcus mutans* and *Streptococcus sobrinus* and *Lactobacilli*	Breakdown of teeth	[54]
Pneumonic plague	*Yersinia pestis*	Lung infection	[55]

8.4 Major Groups of Antibacterial Compounds Derived from Plants

Plants are being extensively explored as the possible therapeutic method as they can kill the microorganisms with diverse mechanisms of action with minimal chance for bacteria to develop resistance to it. The phytochemicals have broadly been classified

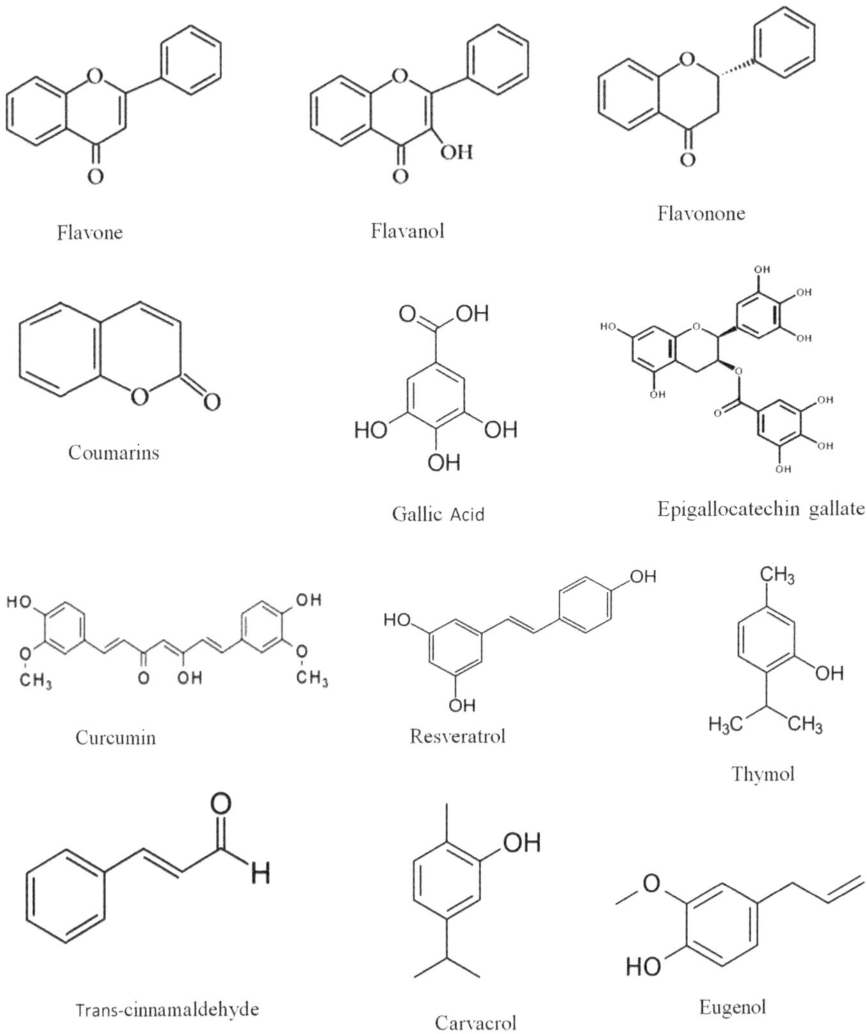

Fig. 8.2 The chemical structures of various plant-derived compounds

based on their chemical structures. Some of the chemical structures are illustrated in Fig. 8.2. The mechanism of action of different classes of plant compounds and their therapeutic role are presented in Table 8.2.

8.4.1 Phenolic and Polyphenols

Polyphenols, secondary metabolites, are pervasively present in all higher plants, consisting of a single-substituted phenolic ring, predominantly classified as flavonoids and non-flavonoids. Flavonoids have 15-carbon chemical structure that

Table 8.2 Antibacterial activity of plant-associated compounds

Plant-derived antibacterial compounds	Groups	Examples	Antibacterial activity	Mode of mechanism	References
Flavonoid	Flavones	Luteolin, apigenin, tangeretin, chrysin, rutin	*S. aureus*, *E. coli*	Inhibition of efflux pumps	[56, 57]
				Interference with DNA synthesis by inhibiting DNA gyrase activity	
	Flavanones	Fisetin, hesperetin, naringenin, naringin, taxifolin, eriodictyol, homoeriodictyol	Methicillin-resistant *S. aureus*	Alteration of membrane fluidity	[58]
	Flavan-3-ols	Catechin, epicatechin, epigallocatechin, epigallocatechin gallate	*V. cholerae*, *S. mutans*, *Shigella*	Damage the bacterial cell wall by forming a complex with the phospholipids, inhibit DNA synthesis	[57, 58]
	Anthocyanins	Cyanidin, delphinidin, malvidin, pelargonidin, peonidin, petunidin	*Aspergillus flavus*	Reducing membrane fluidity of bacterial cells	[57]
			Penicillium sp.		
			Cladosporium sp.		
Quinones		Anthraquinone	*S. aureus*	Associates with cell membrane inactivate enzymes	[59]
		Benzoquinone	*P. aeruginosa*		
		Naphthoquinone	*Bacillus subtilis*		
		Plastoquinone	*E. coli*		
		Pyrroloquinoline quinone			
Tannins	Hydrolyzable and condensed tannins	Tannic acid	*Bacteroides fragilis*, *S. aureus*, *Clostridium perfringens*, *E. coli*, and *Enterobacter cloacae*	Inactivate microbial adhesins, enzymes, cell electron transport system, or complexation of metal ions	[60]
		Gallic acid			
		Proanthocyanidins			

(continued)

Table 8.2 (continued)

Plant-derived antibacterial compounds	Groups	Examples	Antibacterial activity	Mode of mechanism	References
Coumarins		Ammoresinol, curcumin, Anthogenol, warfarin, Agasyllin	*S. aureus*, *S. typhimurium*, *S. enteritidis*, *A. hydrophila*, *E. coli*, *Shigella* sp., *Yersinia* sp., and *Vibrio parahaemolyticus*	Damage the bacterial cell wall, interaction with eukaryotic DNA	[61]
Essential oils		Carvacrol, thymol, eugenol, linalol, perillaldehyde, trans-cinnamaldehyde, cinnamic acid, ϒ-terpinene, 1,8-cineole	*S. aureus*, *P. aeruginosa*, *E. coli*, *L. monocytogenes*, *S. epidermidis*	Permeabilization of the membranes, reduced membrane potential, damage to the proton pump, and loss of ATP pool	[62]
Alkaloids		Solamargine, berberine, allocryptopine, calafatine, jatrorrhizine, palmatine, protopine, reticuline, thalifendine	*P. aeruginosa*, *H. pylori*	Intercalate with DNA leading to impaired cell division and subsequent cell death, loss of cell integrity by depolarization of the cytoplasmic membrane	[63]
Lectins and polypeptides		Fabatin, mannose-specific agglutinin, concanavalin A	*S. aureus* *B. subtilis* *E. coli* *P. aeruginosa*	Affect the ion channels in the bacterial membrane or restrict binding of bacterial proteins to host receptors	[59]

contains a three-ring core, two phenyl rings, and heterocyclic ring [64]. Over 4000 different flavonoids have been recognized in various fruits, vegetables, and drinks and are widely known for their antioxidant properties. Based on the oxidation state of the central pyran ring, flavonoids are further grouped into several subclasses flavonols, flavones, flavanones, isoflavones, catechins, anthocyanidins, and chalcones [57]. Tannins and flavonoids are high-molecular-weight complex polymers [65].

Flavones contain one carbonyl group, and flavonols have additional hydroxyl group. They are effective antibacterial agents as they are able to interact with the extracellular proteins as well as bacterial cell wall components [66, 67]. Catechins, the reduced form of flavonoids, are widely studied due to their wide distribution in green tea. Tea due to the presence of catechins exhibits antimicrobial activity and inhibited *Vibrio cholerae*, *Streptococcus mutans*, and other microorganisms [66, 68]. Anthocyanidins are positively charged aromatic compounds, and chalcones are aromatic ketone compounds [67].

On the other hand, non-flavonoids are divided into phenolic acids, stilbenes, and lignans. Phenolic acids are subdivided into gallic acid and protocatechuic acid (derived from benzoic acid) and coumaric, caffeic, and ferulic acid (obtained from cinnamic acid). Stilbenes primarily consist of resveratrol and lignans synthesized from two phenylpropane units linked by β-β' bond between the central atoms of the side chains [59, 65, 69].

The phenolic toxicity to microorganism can be due to various reasons like their antioxidative nature, able to scavenge free radicals and chelate metals, interact with various pathways, and inhibit some cellular components [59]. Flavonols and tannins have been widely reported for their wide spectrum as well as increased antimicrobial activity. They have also elicited inhibition of biofilm formation, reduced adhesion, and are able to neutralize bacterial toxins and also have the ability to exhibit synergistic activity with antibiotics [65, 68].

8.4.2 Quinones

Quinones are a class of compounds with two phenyl rings and ketone groups. They are ubiquitously present and are known for their antibacterial activity. Free radicals which are generated by quinones complex permanently with nucleophilic groups in proteins, inhibit their function and thus lead to their inactivation. Quinones most likely bind the exposed molecules on bacterial cell [59]. The bactericidal nature is attributed to target the protein synthesis and has led to the development of many successful clinical drugs. Anthraquinone isolated from *Cassia italica* is an example of quinones that showed both bacteriostatic activity against *Bacillus anthracis*, *Corynebacterium pseudodiphthericum*, and *P. aeruginosa* and also bactericidal activity against *Burkholderia pseudomallei* [64]. Another compound, 7-epi-clusianone, isolated form *Rheedia brasiliensis*, can be used to control *S. mutans* biofilms as it showed antibacterial activity at low dosage against *S. mutans* [70].

8.4.3 Tannins

Tannins are water-soluble polymeric polyphenols, formed either by the condensation of the flavonoid derivatives present in the woody part of the plant or by polymerization of quinone units. They have high molecular weight ranging from 500 to 3000 Da and are present in most components of a plant [59]. They exhibit the property of astringency, by virtue of which they bind by non-specific bonds and precipitate proteins such as gelatin from the solution due to the formation of cross-linkages with various functional groups such as the hydroxyl group, leading to enzyme inhibition and substrate deprivation [71]. It is thus used for tanning animal hides into leather. Tannins are categorized according to their structure into two groups – proanthocyanidins or condensed tannins and hydrolyzable tannins. Condensed tannins are derived from the condensation of flavon units such as flavan-3-ols and flavan-3,4-diols. Hydrolyzable tannins are esters of phenolic acids, which are either gallic acid or other phenolic acids derived from the oxidation of gallic acid and a polyol, generally D-glucose [59, 64].

The antibacterial action of tannins may be attributed to their property to ameliorate microbial adhesins, enzymes, cell electron transport system, or complexation of metal ions such as iron. Tannins such as tannic acid are known to inhibit bacteria such as *Bacteroides fragilis*, *S. aureus*, *Clostridium perfringens*, *E. coli*, and *Enterobacter cloacae*, among others [72]. Condensed tannins have been shown to bind to the cell walls of certain bacteria, restrict the growth and enzymatic activity, and are thus used for treating diarrhea and gastric ulcers. It is also known to regulate physiological effects like accelerate blood clotting to prevent hemorrhages, reduce blood pressure, and also moderate immune responses [73]. Tannins, found in various beverages such as tea and wine, are known to be antimutagenic and anticarcinogenic due to their antioxidative property [74].

8.4.4 Coumarins

Coumarins are aromatic benzopyrone compounds made of a merged phenyl and α-pyrone ring. Among over 1300 coumarin compounds being identified, they are known for their anticoagulating, anti-inflammatory, anticarcinogenic, and vasodilatory activities [61]. Several coumarins are known to possess antibacterial activity. 7-amino-4-methylcoumarin, an amino coumarin, is known to show antibacterial activity in vitro against various microorganisms like *S. aureus*, *E. coli*, *Salmonella typhimurium*, *Aeromonas hydrophila*, *Yersinia sp.*, *Shigella sp.*, and *Vibrio parahaemolyticus* [61]. The antibacterial activity of coumarins changed with the position and type of the functional group. Studies have shown that the presence of a methoxy group at C-7 and an OH group at either C-6 or C-8 position is consistently effective against tested bacteria [59]. An aromatic dimethoxy group is more effective against those microorganisms that necessitate the special nutrient requirements (beta-hemolytic *Streptococcus*, *Streptococcus pneumoniae*, and *Haemophilus influenzae*). Thus, coumarins can show broad-spectrum antibacterial activity attributed to its structural arrangements, two functional methyl groups and one additional aromatic ring [75].

8.4.5 Lectins and Polypeptides

Polypeptides show inhibitory properties against bacteria by preventing the formation of ion channels in the membrane or by active inhibition of binding of the bacterial proteins to target polysaccharide receptors [59]. Thionins are proteinaceous in nature and are known to be toxic to bacteria by damaging their cell wall. Fabatin, recently identified peptide, inhibits *E. coli*, *P. aeruginosa*, and *Enterococcus hirae* [64].

8.4.6 Alkaloids

Alkaloids, a large group of heterocyclic nitrogen compounds, are therapeutically substantial plant compounds. Pure alkaloids isolated from plants and their synthetic derivatives are known for their pain relieving, antispasmodic, and antibacterial effects. The antibacterial activity dates back to the 1940s. They are known to effectively inhibit the growth of pathogenic bacteria like *P. aeruginosa*, *Pneumoniae*, and *Helicobacter pylori* [64]. Alkaloids were found to intercalate with bacterial DNA leading to reduced replication and subsequently cell death. Polyamine alkaloid squalamine's mechanism of action is via destruction of the membrane integrity by penetrating the lipopolysaccharide monolayers, leading to depolarization of the cytoplasmic membrane and causing leakage of cytoplasmic content [63]. Alkaloids are medicinal compounds used across the world as they are constituents of antibacterial drugs, morphine being the first medically useful drug. They have also stimulated the synthesis of other significant drugs like quinolones, metronidazole, and bedaquiline. The drugs like linezolid and trimethoprim consist of scaffold substructures of alkaloids [63].

8.4.7 Essential Oils

Essential oils are naturally volatile aromatic compounds obtained from plants as secondary metabolites. Although they are generally extracted by steam distillation, other methods comprise extraction by solvent, oil extraction, resin tapping, and cold pressing. Currently, out of the 3000 known essential oils, 300 are currently being used commercially in pharmaceutical, plant-obtained products, food, cosmetic, and other industries. Furthermore, they are used in spa and most commonly in aromatherapy [76].

Some essential oils are also known for their antimicrobial, anti-inflammatory, pain relieving, and anti-spasm properties. Studies have revealed that many essential oils can thwart the spread of drug-resistant pathogenic strains like *Staphylococcus* and *Streptococcus* [77].

Essential oils are composite mixtures containing about 20–60 compounds in various concentrations. They distinctively have some major components at higher level (20–70%), and the other compounds are present in very less quantity. For example, the essential oil obtained from *Origanum compactum* has carvacrol (30%) and thymol (27%) as its main constituents, and ϒ-terpinene and p-cymene are present in trace amounts; linalol, the essential oil isolated from *Coriandrum sativum*, is the major component, and E-2-decanal is found in traces; and trans-cinnamaldehyde (65%) is the main constituent of the *Cinnamomum zeylandicum* essential oil [62]. The biological role and the functional activity are governed by the main constituents of essential oils.

The essential oil is composed of two groups of diverse biosynthetic origin. The major class consists of terpenes and terpenoids, whereas the other class is of aromatic and aliphatic compounds. Terpenes are made by combining numerous 5-carbon-base (C5) units commonly known as isoprene. Although they are mainly present as monoterpenes (C10) and sesquiterpenes (C15), other variants hemiterpenes (C5), diterpenes (C20), triterpenes (C30), and tetraterpenes (C40) also exist [78]. Constituting 90% of the essential oil, monoterpenes are formed by linking two isoprene units, whereas sesquiterpenes are formed by coupling three isoprene units. A terpene having an oxygen group is known as terpenoid. Terpenoids are synthesized from acetate units and are extensively branched and are cyclized. Some of the common sources for terpenes are coriander, eucalyptus, citrus fruits, lavender, lemongrass, resin, mandarin, mint, petitgrain, pine, lilacs, rosemary, sage, liver oil, and thyme. Certain known terpenoids are menthol, citral, camphor, curcuminoids, cannabinoids, farnesol, and artemisinin [59, 76, 78].

On the other hand, aromatic compounds are derived from phenylpropane and less common than terpenes. They are derived from tyrosine and phenylalanine amino acids and generally have aromatic ring and cinnamic acid. Examples include cinnamaldehyde, coumaric acid, eugenol, resveratrol, and caffeic acid obtained from common sources of cinnamon, clove, parsley, fennel, nutmeg, etc. [76].

Essential oils seem to target the microorganisms by diverse mechanism of action due to the large number of constituents. They are able to pass through membranes due to their lipophilic nature causing irreversible damage to it. This leads to the leakage of the intracellular content and hence cellular lysis [76]. The permeabilization of the membranes in bacteria specifically causes the loss of ions, alteration in membrane potential, disruption of the proton pump, and also reduced ATP pool. They can also change mitochondrial membranes and DNA which affect the cellular respiration [59, 79]. It has been argued that targeting the organisms at various sites reduces its chances of developing drug resistance, and attacking the organism at various sites can cooperate in synergistic manner. Different components of essential oils often interact in a complex way which leads to varied pharmacological effect and therapeutic outcome. These component interactions could have synergistic, additive, or antagonistic effect on the growth of microorganism [79].

8.5 Synergy and Nanoencapsulation as Advanced Strategies

The plant compounds used as antimicrobial agents at times have limitations in terms of dosage, efficacy, patient compliance, and bioavailability. The concept of synergism can help to combat the resistance against the antimicrobial agents. Synergism is when the positive interaction between two agents results in inhibition of the target organism. These interactions can be between two different phytochemicals, phytochemical with an antibiotic, or antibiotics with a chemical agent [80]. This kind of synergistic activity has advantages like improved efficiency of the antimicrobial agent, reduced toxicity, enhanced stability as well as the bioavailability of the agent, and finally the reduced effective dosage [80].

Other than this nanotechnology approaches have also been proved to be effective against microorganisms. Nanoencapsulation of the antimicrobial agents controls their release into a system, and this strategy also protects the natural compounds from thermal or photodegradation and therefore enhances the stability and function of the compound [81]. Further, nanoencapsulation strategy also enhances the antimicrobial activity by increasing the bioavailability of these phytochemicals through diffusion across biological membranes attributed to their size [82].

Nanoencapsulation strategy includes:

- *Polymeric nanoparticles*: It covers two classes, namely, nanocapsules and nanospheres. Nanocapsules mainly consist of a wall which is polymeric in nature and a central core which is primarily the oil; on the other hand, nanosphere consists of a matrix system where the essential oils are either linked to the matrix or to the oily inner core [83]. The association with the polymeric nanoparticles carries along various benefits that include precise release, improved water solubility, reduction in the cytotoxicity, and enhanced antimicrobial activity [84, 85]. Further, nanoencapsulating the antimicrobial also improves its water solubility [86].
- *Lipid nanoparticles*: Lipid nanoparticles may include liposomes, noisome, and solid lipid nanoparticles (SLN). Liposomes and noisome are generally the colloidal link between the amphiphilic lipids which organize themselves into a bilayer vesicles and therefore can take both the hydrophilic and hydrophobic compounds. SLN are solid in nature at room temperature with a lipid core, which makes them a perfect carrier of lipophilic compounds like essential oil [87].
- *Nanoemulsion*: The nanoemulsion is the thermodynamically stable homogeneous mixture of two insoluble liquids, which are generally stabilized using suitable emulsifiers with a particle size less than 100 nm. These generally require less energy to be prepared [88, 89]. Surfactant is used to reduce the interfacial tension. The emulsions are usually prepared by mixing the surfactant into the oil and water by mechanical methods which result in the dispersion of these two fluids. These emulsions appear to be transparent or milky and usually have a size

around 0.1 to 1 micron in diameter. Emulsions are usually kinetically stable and thermodynamically unstable [90]. There are three types of nanoemulsions.

- *Oil-in-water*: This mainly consists of oil that is surrounded by surfactant molecule and is dispersed in water, which acts as a continuous phase. These have higher interaction volume as compared to water-in-oil nanoemulsion [91]. This system is mostly preferred for the hydrophobic drugs, as oil-in-water nanoemulsion will enable them to be dissolved in aqueous fluids [92].
- *Water-in-oil*: They generally consist of water, surrounded by surfactant molecules with oil as the continuous phase. This type of system is usually used for oral and parental administration and can be destabilized by the biological systems that are usually aqueous.
- *Bicontinuous*: This system consists of equal amount of oil and water where both exist as continuous phase [93]. The transition from oil-in-water phase to water-in-oil phase may pass through the continuous system [94]. They usually show non-Newtonian flow and plasticity, and therefore this property allows them to be used as tropical or intravenous route of administration [95].

8.6 Selected Phytochemicals as Antimicrobial Agents

8.6.1 Epigallocatechin Gallate

Epigallocatechin gallate (EGCG), an ester of epigallocatechin and gallic acid, is a catechin that is present in tea. It is being extensively investigated due to its antitumorigenic, antimicrobial, and anti-inflammatory properties. EGCG shows antibacterial properties as it interacts with the phospholipids of the bacterial cell wall, which prevents the adherence of the bacteria to the cell wall, the production of toxic metabolites by downregulating certain enzymes, and influences the cell signaling and cell cycle [96]. EGCG has been effective against many multidrug-resistant Gram-positive and Gam-negative bacteria like *S. aureus*, *Helicobacter pylori*, *E. coli*, *P. aeruginosa*, and *Stenotrophomonas maltophilia* among various others [97, 98]. It also shows anti-biofilm activities against *S. mutans*, *S. aureus* at sub-inhibitory levels, and *E. coli* [99, 100].

Synergy and Nanoencapsulation EGCG has been reported to act synergistically with β-lactams against MRSA by binding to the peptidoglycan of the cell wall. EGCG synergistically in combination with antibiotics like penicillin and tetracycline and also with ampicillin and sulbactam has shown to inhibit *S. aureus* [101, 102].

The use of cationic lipid nanoparticles has been tested and recognized as a favorable strategy for release over longer duration and site-specific binding of EGCG for the treatment of ocular diseases such as age-related macular edema, diabetic retinopathy, and inflammatory disorders [103]. These disorders are mainly associated with antioxidant and anti-inflammatory pathways. Also, for localizing the nanoencapsulated EGCG to the infection site for its specific anti-urease property, a fucose–chitosan/gelatin nanoparticle has been prepared to encapsulate EGCG and make direct contact with *H. pylori* on the gastric epithelium. Analysis has shown that the nanoencapsulated EGCG is specific in binding to *H. pylori* at the target site and ameliorating the infection. Meanwhile, in vivo clearance assays results were very encouraging when performed on gastric-infected mouse model. The data demonstrated significant reduction of *H. pylori* induced infection and inflammation of gastric lining using EGCG nanoparticles as compared to EGCG alone [104].

8.6.2 Curcumin

Curcumin is a curcuminoid constitute of *Curcuma longa Linn*, commonly known as turmeric. It consists of two phenolic rings attached by two α, β-unsaturated carbonyl groups. Curcumin has been known to show anti-inflammatory, antibacterial, antiviral, antitumor, and anticarcinogenic activities. Curcumin shows antibacterial activity by causing loss of integrity of the bacterial membrane [105]. Its effectiveness has been shown against bacteria like *Bacillus subtilis, B. coagulans, Enterococcus faecalis, S. aureus, E. coli, H. pylori, and P. aeruginosa.* In the case of *B. subtilis,* curcumin strongly inhibited the formation of the FtsZ in the cytokinetic Z-ring which caused the suppression of bacterial cell proliferation by affecting cell division [106]. It has also shown anti-biofilm activity against *P. aeruginosa* by inhibiting some of quorum-sensing genes and also by downregulating virulence factors like acyl homoserine lactone (AHL) production, elastase/protease activity, and pyocyanin biosynthesis [107]. Furthermore, it inhibits the biofilms formed by uropathogens such as *E. coli, P. aeruginosa, Proteus mirabilis,* and *Serratia marcescens.* Curcumin has been shown to alter quorum sensing by interrupting biofilm main component exopolysaccharide production and extracellular transport and also affect the motility of uropathogens [108].

In an in vivo study, curcumin has shown to eradicate 65 isolates of *H. pylori* in C57BL/6 mice by inhibiting NF-B activation and release of IL-8 and cell scattering, which consequently led to a reduction in gastric inflammation [109].

Synergy and Nanoencapsulation Curcumin in combination with oxacillin, ampicillin, ciprofloxacin, and norfloxacin synergistically potentiates antibacterial activity against MRSA by reducing the MIC of these antibiotics [110].

Complexes of curcumin with cobalt nanoparticles showed enhanced antibacterial activity against *E. coli* by the ability of curcumin being able to penetrate the cell wall and cobalt showing chelating properties [111]. Curcumin encapsulated into a silane-hydrogel nanoparticle has shown to inhibit MRSA and *P. aeruginosa* [112]. Combinational therapy of silver nanoparticles and curcumin nanoparticles has also shown to eradicate mature biofilms or restrict the biofilm formation of both Gram-negative and Gram-positive bacteria [113].

8.6.3 Resveratrol

Resveratrol (trans 3, 4′, 5-trihydroxystilbene) is a phytoalexin found in food and beverages such as grapes, red wine, blueberries, raspberries, peanuts, and mulberries. It has shown biological activities such as antioxidant, anticancer, antimicrobial, and anti-inflammatory. The antitumor activity of resveratrol is linked to its ability to arrest cell cycle and induce cell death. The antibacterial mechanism is not very clear, but recent reports have shown it to be caused due to DNA fragmentation. Resveratrol has shown to inhibit Z-ring formation leading to the arrest of cell growth. Additionally, it has shown antibacterial effect against *H. pylori*, *S. aureus*, *E. faecalis*, *P. aeruginosa*, *Haemophilus ducreyi*, *Arcobacter butzleri*, and *Arcobacter cryaerophilusa* [114].

Synergy and Nanoencapsulation Oxyresveratrol reduced bacterial counts of MRSA below the detectable limit after overnight incubation when combined with ciprofloxacin or gentamicin [115].

To improve the solubility, stability, and cellular uptake of resveratrol, nanoencapsulation using chitosan (CS) and γ-poly (glutamic acid) (γ-PGA) was done [116].

8.6.4 Trans–Cinnamaldehyde

Trans-cinnamaldehyde is obtained from the bark of cinnamon tree and plants that come under the genus *Cinnamomum* which belongs to Lauraceae family. It is used extensively as a flavorant in beverages, ice creams, meat products, and candies. As an antibacterial compound, trans-cinnamaldehyde has been found to effectively prevent the growth of uropathogenic *E. coli* biofilm, responsible for urinary tract infection, on urinary catheters without any cytotoxic effects on human bladder epithelial cells [117]. Cinnamon oil has shown antibacterial activity against planktonic bacteria and biofilms of clinical *S. epidermidis* strains, and decrease in biofilm formation was directly correlated to reduction in *ica* expression [118]. Cinnamaldehyde interferes with the quorum sensing of the bacteria, which might also inhibit their biofilm formation [119].

Synergy and Nanoencapsulation Polyethylene-co-vinylacetate films with 3.5 wt% of carvacrol and 7 wt% of cinnamaldehyde were prepared and were found to have a significant bactericidal against *S. aureus* and *E. coli*. The formulation had bacteriostatic effect against *S. epidermidis* and *L. monocytogenes*. Nanoencapsulated carvacrol and cinnamaldehyde diminish the intermolecular forces of polymer chains, thus considerably enhances the extensibility and elasticity of the film. Cinnamaldehyde was also reported to reduce the surface tension and interfere with the adhesion process. The films allowed slow and steady diffusion of the bioactive compounds which could damage the bacterial cell wall and hence had detrimental effect on bacterial growth [120].

Spherical poly (DL-lactide-co-glycolide) (PLGA) nanoparticles with entrapped eugenol and trans-cinnamaldehyde were prepared by emulsion evaporation method for future antimicrobial delivery applications and were found to effectively control the growth of *Salmonella* spp. and *Listeria* spp. [121].

8.6.5 Carvacrol

Carvacrol, main component in the essential oil of *Origanum vulgare*, is one of the most common bioactive components of essential oils. It is a monoterpenoid phenol (5-isopropyl-2-methylphenol) and has been widely reported for its antimicrobial properties. The antibacterial activity of carvacrol is associated with changes in the structural and functional properties of the cytoplasmic membrane. This interaction causes reduced integrity, non-functionality of the proton motive force, and hence leakage of intracellular content. It also results in alteration in membrane permeability and ion transport and denaturation of membrane proteins and decreases respiration that eventually causes cell death. Carvacrol was shown to inhibit both planktonic and biofilm of *S. aureus* and *S. epidermidis* strains [122]. It reduces motility and invasiveness in bacteria and hence biofilm formation. It was also able to inhibit the formation of biofilms of *Chromobacterium violaceum*, *Salmonella enterica* subsp. enterica serovar Typhimurium, and *S. aureus* at sublethal concentrations by disrupting quorum sensing [123].

Synergy and Nanoencapsulation The encapsulated carvacrol in Ca-alginate microcapsules was shown to be rather effective against *E. coli* K88 [124]. In another study, the encapsulated carvacrol and eugenol were found to be effective against biofilm formed by *L. monocytogenes* and *E. coli*. The bactericidal effects were observed within initial growth period and resulted in increase in active component concentrations [125]. Carvacrol, p-cymene, thymol, and γ-terpinene were successfully encapsulated in phosphatidyl choline-based liposomes and it resulted in improved antioxidant and antimicrobial activities as observed against selected microbes [126].

In a recent study, carvacrol–encapsulated PLGA nanoparticles were notably effective in inhibiting biofilm formation of *S. epidermidis* by reducing the elasticity and mechanical stability of the biofilms. This could possibly allow the active agents to penetrate into the core of the bacterial biofilms [127].

In another study, cinnamaldehyde and carvacrol film made with PLGA were shown to inhibit biofilms formed by *E. coli*, *S. aureus*, and *P. aeruginosa*. PLGA films incorporated with cinnamaldehyde and carvacrol alter its surface properties and render them less hydrophobic and thus more resistant to bacterial adherence. The presence of 0.1% carvacrol in PLGA film interfered the biofilm development by both *E. coli* and *S. aureus*, whereas the more resistant *P. aeruginosa* required higher concentration of 1% carvacrol in PLGA film to inhibit its biofilm formation [128].

8.6.6 Thymol

Thymol is a natural monoterpene phenol, present in oil of thyme, and is extracted from *Thymus vulgaris*. It shows the properties of reducing antibiotic resistance in drug-resistant pathogens and is also a powerful antioxidant. It has been shown to inhibit the growth of both Gram-positive and Gram-negative bacteria. It shows antibacterial activity by decreasing enzyme activity and by disrupting the membrane integrity by altering protein reactions. It also inhibits the glucose uptake of the bacteria [129].

Synergy and Nanoencapsulation Thymol and carvacrol in combination showed complete inhibition of *P. aeruginosa* and *S. aureus* due to loss of membrane integrity, which in turn affects the pH homeostasis and the equilibrium of inorganic ions [130].

Chlorhexidine digluconate (CHG) is widely used as disinfectant/antiseptic; antimicrobial activity in combination with eucalyptol oil, tea tree oil, and thymol against *S. epidermidis* was evaluated against both planktonic culture and biofilm formed by the organism. It was observed that CHG in combination with thymol and tea tree oil was not synergistic against planktonic and biofilm formed by *S. epidermidis* [131]. Treatment of *S. typhimurium* with cinnamaldehyde/thymol, cinnamaldehyde/carvacrol, and thymol/carvacrol revealed that in paired combinations, MIC dose of cinnamaldehyde, thymol, and carvacrol could be decreased from 200 mg/L, 400 mg/L, and 400 mg/L to 100 mg/L each, respectively [132]. A study has shown that liposomal incorporation of carvacrol and thymol increases antimicrobial activity. Thymol exhibited stronger activity than carvacrol, against most bacteria [133].

8.6.7 Eugenol

Eugenol, the essential component of clove oil, is a phenolic compound known for its various pharmacological actions, including anti-inflammatory, antimicrobial and antitumor effects. It is a phenyl propene (2-methoxy-4-(prop-2-en-1-yl) phenol) and is known to reduce the concentration of *L. monocytogenes* and *E. coli* O157:H7 [134]. At sub-MIC concentrations, it inhibits the production of virulence factors and biofilm formation and was thus identified as a novel quorum-sensing inhibitor [135]. Chemical properties such as low solubility, strong odor, and prone to sublimation make the approval of carvacrol difficult despite its antibacterial properties [136].

Synergy and Nanoencapsulation Eugenol has been shown to exhibit synergy with not only other phytochemicals but also with antimicrobial peptides and antibiotics. Polyhydroxybutyrate-based antimicrobial films containing eugenol and pediocin restricted pathogens like *S. aureus*, *E. coli*, *S. typhimurium*, *B. cereus*, and fungi [137].

Eugenol-loaded oil-in-water nanoemulsion was synthesized using sesame as oil phase, Tween20/Tween80 as nonionic surfactant, and water by ultrasound cavitation method. It demonstrated antibacterial activity against *S. aureus* and decreased the bacterial population in orange juice [138]. In another study, carvacrol and eugenol encapsulated in micellar nonionic surfactant solutions were tested against two strains of *L. monocytogenes* and two strains of *E. coli* O157:H7 grown as biofilm and were found to inhibit *E. coli* more efficiently than *L. monocytogenes* [125].

8.7 Recent Patents

Due to the various antimicrobial properties of plant-derived products, the interest in studying them has increased in various fields. US 9,636,361 assigned to LiveLeaf, Inc. in 2017 claims a method of killing bacteria with a plant-based biocidal solution. The composition of the plant-based biocidal solution is a bioactive material and a plant-based substance from the cellular material of a plant. This solution can form a cluster, a complex, or an aggregate for providing sufficient bioactive material to overcome the defense mechanism of the target [139]. Another patent assigned to Slippery Rock University Foundation, Inc. in US 9,545,386 claims methods of treating infectious diseases caused by a parasitic microorganism like trypanosomes, bacteria, fungi, and nematodes. The patent assigned in 2017 claims that by administering an effective amount in unit dosage form of a C.sub.8-C.sub.16-alpha, beta-unsaturated aliphatic aldehyde, such as trans-2-dodecenal, the diseases can be treated [140].

US 9,192,635 assigned to LiveLeaf, Inc. in 2015, claims method of treating damaged mucosal or gastrointestinal tissue by administering a mixture of pomegranate and green tea extracts and releasably bound hydrogen peroxide. The claim is related to site-activated binding systems that selectively increase the bioactivity of phenolic compounds bound to a reactive oxygen species at the target sites [141]. Another patent assigned to the Hong Kong Polytechnic University in US 8,980,848 claims the synthesis and use of bis-flavonoid compounds. This patent assigned in 2015 claims that these compounds inhibit multidrug resistance in chemotherapy and protozoan infection [142]. Another patent assigned to LiveLeaf Inc. (2014) in US 20140072660 claims the method and composition of mixture of polyphenol and hydrogen peroxide to treat or prevent a bacterial infection [143].

In US 7,557,139, derivatives of triptolide (a diterpenoid) alone or in combination or alteration with other anti-autoimmune or anti-inflammatory compounds were claimed. The patent assigned to Emory University in 2009 claimed the synthetic methods and compositions for treatment of autoimmune and anti-inflammatory disorders [144].

Further in US 6,149,912, the treatment with plant-derived alkaloids to enhance innate immunity defense mechanism is claimed. The patent assigned to Michael J. Gubarev and Elena Y. Enioutina in 2000 claims that the administration of these compounds is known to enhance the antimicrobial activities of the innate system [145].

8.8 Conclusion

The spread of drug-resistant microorganisms has gained attention and is a global health issue. It limits the efficacy of the currently available drugs and hence causes the treatment failure of various diseases. According to the WHO, antimicrobial resistance has reached distressing levels in many parts of the world, and there is an urgent need to reinforce and manage collaboration to deal with this issue, else it threatens the achievements of modern medicine. As a result, there is an urgent need for more effective agents to limit the emergence and spread of these organisms. Different plant compounds are being considered as a source of unexplored chemical compounds with higher therapeutic potential. Traditionally, plants are in used to treat and cure diseases since the ancient times, long before the discovery of modern medicine. This untapped source for therapy is now being extensively explored and studied as these compounds can kill the microorganism with diverse mechanisms of action with minimal chance for development of resistance. The main mechanisms of antibacterial action of these compounds are disrupting the integrity of the cell membrane, inhibition of protein and nucleic acid synthesis, and suppression of metabolic pathways. As a result, interest in natural product-based drug discovery, new approaches for the identification, characterization, and resupply of natural products has developed. The large spectrum of activity of these bioactive compounds along with their low toxicity proves that they have potential application as antimicrobial agents. But there are major problems related to these compounds in terms of stability, solubility, efficacy, and low bioavailability for effectual and secure use in humans and animals.

An alternative is to use these compounds in combination with various antibiotics and other natural compounds. With this concept of combination, the word synergism has emerged wherein these antimicrobial agent when used together results in a positive interaction which ultimately inhibits the growth of microbes. The synergistic combination can expand the antimicrobial spectrum, prevent the appearance of resistant mutants, and curtail toxicity. This also gives an alternative to monotherapy for patients with invasive infections that are difficult to eradicate. Moreover, the effective treatment for a given disease mitigates the side effects of the synthetic drug.

The emergence of nanotechnology has opened a vast opportunity for the researchers to explore new ways by which the growth of pathogenic microbes is stopped. Encapsulating the active compound or its synergistic combination with other plant compound or antibiotic in nanoparticles/nanoemulsions has shown great potential as an effective delivery system.

Nanoemulsions encompassing the active ingredient are thermodynamically driven to fuse with lipid-containing organisms. When sufficient nanoparticles combine with the pathogens, they discharge part of the energy trapped within the emulsion. Both the active constituent and the energy released are required to target the pathogen, resulting in cell lysis and cell death. They exhibit enhanced shelf stability and have unique property of being selectively toxic to bacterial cells. They confer a number of advantages for herbal drugs, including enhancement of solubility and bioavailability; masking of color, taste, and smell; and improved pharmacological activity and shelf life. Thus nanoemulsions considerably minimize the side effects upon systemic administration of traditional antimicrobial agents. They can help to overcome the resistance developed by pathogenic bacteria to various antibiotics.

Even though a huge number of natural antimicrobial are known till date, only a limited number of natural antimicrobial compounds are currently in use for commercial applications. Further studies with different compounds in unexplored/diverse plants on different species of bacteria are required to be carried out to obtain additional clues to design or discover successful novel drugs to combat antibiotic resistance of bacteria and control infectious diseases. The current scientific developments and trends show that plant products will be an important source of novel drugs in the future.

References

1. Nikaido H (2009) Multidrug resistance in bacteria. Annu Rev Biochem 78:119–146
2. Davey ME, O'toole GA (2000) Microbial biofilms: from ecology to molecular genetics. Mol Biol Rev 64(4):847–867
3. de Lencastre H, Oliveira D, Tomasz A (2007) Antibiotic resistant *Staphylococcus aureus*: a paradigm of adaptive power. Curr Opin Microbiol 10(5):428–435
4. European Centre for Disease Prevention and Control Antimicrobial Resistance Interactive (EARS-Net) Database (2013) http://ecdc.europa.eu/en/activities/surveillance/EARS-Net/Pages/Database.aspx. Accessed 15 Feb 2017
5. World Health Organization (2014) Global tuberculosis report 2014. WHO, Geneva. http://www.who.int/tb/publications/global_report/en/. Accessed 12 Jan 2017
6. Van Boeckel TP, Gandra S, Ashok A et al (2014) Global antibiotic consumption 2000 to 2010: an analysis of national pharmaceutical sales data. Lancet Infect Dis 14(8):742–750
7. Llor C, Bjerrum L (2014) Antimicrobial resistance: risk associated with antibiotic overuse and initiatives to reduce the problem. Ther Adv Drug Saf 5(6):229–241
8. Tiwari BK, Valdramidis VP, O'Donnell CP et al (2009) Application of natural antimicrobials for food preservation. J Agric Food Chem 57(14):5987–6000
9. Dorman HJ, Deans SG (2000) Antimicrobial agents from plants: antibacterial activity of plant volatile oils. J Appl Microbiol 88(2):308–316
10. Cocchiara J, Letizia CS, Lalko J et al (2005) Fragrance material review on cinnamaldehyde. Food Chem Toxicol 45(6):867–923
11. Tadros T, Izquierdo P, Esquena J, Solans C (2004) Formation and stability of nano-emulsions. Adv Colloid Interf Sci 108–109:303–318
12. Center for Disease Control and Prevention, Office of Infectious Disease (2013) Antibiotic resistance threats in the United States. http://www.cdc.gov/drugresistance/threat-report-2013. Accessed 28 Jan 2017
13. Byarugaba DK (2005) Antimicrobial resistance and its containment in developing countries. In: Gould IM, van der Meer JWM (eds) Antibiotic policies: theory and practice. Springer, New York
14. Center for Disease Dynamics, Economics & Policy (2015) State of the world's antibiotics, 2015. CDDEP: Washington, DC. http://cddep.org/publications/state_worlds_antibiotics_2015#sthash.5RFGA8Oh.dpb. Accessed 17 Feb 2017
15. Walsh C (2000) Molecular mechanisms that confer antibacterial drug resistance. Nature 406(6797):775–781

16. World Health Organization (2016) Antibiotic resistance fact sheet 2016. WHO, New York. http://www.who.int/mediacentre/factsheets/fs194/en/. Accessed 12 Jan 2017
17. Ventola CL (2015) The antibiotic resistance crisis: part 1: causes and threats. Pharm Ther 40(4):277–283
18. Lade H, Paul D, Kweon JH (2014) Quorum quenching mediated approaches for control of membrane biofouling. Int J Biol Sci 10(5):550–565
19. Tannières M, Lang J, Barnier C et al (2017) Quorum-quenching limits quorum-sensing exploitation by signal-negative invaders. Sci Rep 7:40126
20. Petrova OE, Sauer K (2016) Escaping the biofilm in more than one way: desorption, detachment or dispersion. Curr Opin Microbiol 30:67–78
21. Costerton JW, Lewandowski Z, Caldwell DE et al (1995) Microbial Biofilms. Annu Rev Microbiol 49:711–745
22. Donlan RM (2002) Biofilms: microbial life on surfaces. Emerg Infect Dis 8(9):881–890
23. Fletcher M, Loeb GI (1979) Influence of substratum characteristics on the attachment of a marine pseudomonad to solid surfaces. Appl Environ Microbiol 37(1):67–72
24. Hall-Stoodley L, Costerton JW, Stoodley P (2004) Bacterial biofilms: from the natural environment to infectious diseases. Nat Rev Microbiol 2(2):95–108
25. Korber DR, Lawrence JR, Sutton B, Caldwell DE (1989) Effect of laminar flow velocity on the kinetics of surface recolonization by Mot+ and Mot− *Pseudomonas fluorescens*. Microb Ecol 18(1):1–19
26. Prüß BM (2017) Involvement of two-component signaling on bacterial motility and biofilm development. J Bacteriol 199(18):e00259–e00217
27. Kokare CR, Chakraborty S, Khopade AN, Mahadik KR (2009) Biofilm: importance and application. Indian J Biotechnol 8(2):159–169
28. Stoodley P, Purevdorj-Gage B, Costerton JW (2005) Clinical significance of seeding dispersal in biofilms: a response. Microbiology 151(11):3453–3453
29. Myllymaa K, Levon J, Tiainen VM et al (2013) Formation and retention of staphylococcal biofilms on DLC and its hybrids compared to metals used as biomaterials. Colloids Surf B Biointerfaces 101:290–297
30. Karimi A, Karig D, Kumar A, Ardekani AM (2015) Interplay of physical mechanisms and biofilm processes: review of microfluidic methods. Lab Chip 15(1):23–42
31. Bendinger B, Rijnaarts HH, Altendorf K, Zehnder AJ (1993) Physicochemical cell surface and adhesive properties of coryneform bacteria related to the presence and chain length of mycolic acids. Appl Environ Microbiol 59(11):3973–3977
32. Ramli NS, Eng Guan C, Nathan S, Vadivelu J (2012) The effect of environmental conditions on biofilm formation of Burkholderia pseudomallei clinical isolates. PLoS One 7(9):e44104
33. Liu R, Zhu J, Yu Z et al (2014) Molecular analysis of long-term biofilm formation on PVC and cast iron surfaces in drinking water distribution system. J Environ Sci 26(4):865–874
34. Fletcher M (1988) Attachment of *Pseudomonas fluorescens* to glass and influence of electrolytes on bacterium-substratum separation distance. J Bacteriol 170(5):2027–2030
35. Cowan MM, Warren TM, Fletcher M (1991) Mixed species colonization of solid surfaces in laboratory biofilms. Biofouling 3(1):23–34
36. Hanlon GW, Denyer SP, Hodges NA et al (2004) Biofilm formation and changes in bacterial cell surface hydrophobicity during growth in a CAPD model system. J Pharm Pharmacol 56(7):847–854
37. Davies DG, Geesey GG (1995) Regulation of the alginate biosynthesis gene algC in *Pseudomonas aeruginosa* during biofilm development in continuous culture. Appl Environ Microbiol 61:860–867
38. Prigent-Combaret C, Vidal O, Dorel C, Lejeune P (1999) Abiotic surface sensing and biofilm-dependent regulation of gene expression in *Escherichia coli*. J Bacteriol 181:5993–6002
39. Pulcini E (2001) The effects of initial adhesion events on the physiology of *Pseudomonas aeruginosa*. Dissertation, Montana State University

40. Imamura Y, Chandra J, Mukherjee PK et al (2008) *Fusarium* and *Candida albicans* biofilms on soft contact lenses: model development, influence of lens type, and susceptibility to lens care solutions. Antimicrob Agents Chemother 52(1):171–182
41. Donlan RM (2001) Biofilm formation: a clinically relevant microbiological process. Clin Infect Dis 33(8):1387–1392
42. Harries AD, Dye C (2006) Tuberculosis. Ann Trop Med Parasitol 100(5–6):415–431
43. Bhan MK, Bahl R, Bhatnagar S (2005) Typhoid and paratyphoid fever. Lancet 366(9487):749–762
44. Adcox HE, Vasicek EM, Dwivedi V et al (2016) Salmonella extracellular matrix components influence biofilm formation and gallbladder colonization. Infect Immun 84(11):3243–3251
45. Atkinson W (2012) Epidemiology and prevention of vaccine- preventable diseases, 12th edn. Public Health Foundation, Washington, DC
46. World Health Organization (2006) Diphtheria vaccine. Wkly Epidemiol Rec 81(3):24–32
47. Finkelstein RA (1996) Cholera, Vibrio cholerae O1 and O139, and other pathogenic vibrios. In: Baron S (ed) Medical microbiology, 4th edn. University of Texas, Galveston. Available online http://gsbs.utmb.edu/microbook/ch024.htm
48. Myers AL, Hall M, Williams DJ et al (2013) Prevalence of bacteremia in hospitalized pediatric patients with community-acquired pneumonia. Pediatr Infect Dis J 32(7):736–740
49. Center for Disease Control and Prevention (2016) Necrotizing fasciitis: a rare disease, especially for the healthy. https://www.cdc.gov/features/necrotizingfasciitis/. Accessed 20 Feb 2017
50. Mosier DA (1997) Bacterial pneumonia. Vet Clin North Am Food Anim Pract 13(3):483–493
51. Marijon E, Mirabel M, Celermajer DS, Jouven X (2012) Rheumatic heart disease. Lancet 379(9819):953–964
52. Baron S (1996) Medical microbiology, 4th edn. University of Texas Medical Branch, Galveston
53. Cheng AC, Currie BJ (2005) Melioidosis: epidemiology, pathophysiology, and management. Clin Microbiol Rev 18(2):383–416
54. Conrads G, de Soet JJ, Song L et al (2014) Comparing the cariogenic species *Streptococcus sobrinus* and *S. mutans* on whole genome level. J Oral Microbiol 6:26189
55. Smiley ST (2008) Immune defense against pneumonic plague. Immunol Rev 225(1):256–271
56. Singh M, Kaur M, Silakari O (2014) Flavones: an important scaffold for medicinal chemistry. Eur J Med Chem 84:206–239
57. Nijveldt RJ, van Nood E, van Hoorn DE et al (2001) Flavonoids: a review of probable mechanisms of action and potential applications. Am J Clin Nutr 74(4):418–425
58. Kumar S, Pandey AK (2013) Chemistry and biological activities of flavonoids: an overview. Sci World J 2013:162750
59. Cowan MM (1999) Plant products as antimicrobial agents. Clin Microbiol Rev 12(4):564–582
60. Bag A, Chattopadhyay RR (2014) Efflux-pump inhibitory activity of a gallotannin from *Terminalia chebula* fruit against multidrug-resistant uropathogenic *Escherichia coli*. Nat Prod Res 28(16):1280–1283
61. Jain PK, Joshi H (2012) Coumarin: chemical and pharmacological profile. J Appl Pharm Sci 2(6):236–240
62. Burt S (2004) Essential oils: their antibacterial properties and potential applications in foods—a review. Int J Food Microbiol 94(3):223–253
63. Cushnie TP, Cushnie B, Lamb AJ (2014) Alkaloids: an overview of their antibacterial, antibiotic-enhancing and antivirulence activities. Int J Antimicrob Agents 44(5):377–386
64. Upadhyay A, Upadhyaya I, Kollanoor-Johny A, Venkitanarayanan K (2014) Combating pathogenic microorganisms using plant-derived antimicrobials: a mini-review of the mechanistic basis. Biomed Res Int 2014:761741
65. Daglia M (2012) Polyphenols as antimicrobial agents. Curr Opin Biotechnol 23(2):174–181
66. Pistelli L, Giorgi I (2012) Antimicrobial properties of flavonoids. In: Patra A (ed) Dietary phytochemicals and microbes. Springer, Dordrecht

67. Borges A, Abreu AC, Dias C et al (2016) New perspectives on the use of phytochemicals as an emergent strategy to control bacterial infections including biofilms. Molecules 21(7):E877
68. Cushnie TT, Lamb AJ (2005) Antimicrobial activity of flavonoids. Int J Antimicrob Agents 26(5):343–356
69. Rentzsch M, Wilkens A, Winterhalter P (2009) Non-flavonoid phenolic compounds. In: Moreno-Arribas MV, Polo C (eds) Wine chemistry and biochemistry. Springer, New York
70. Saleem M, Nazir M, Ali MS et al (2010) Antimicrobial natural products: an update on future antibiotic drug candidates. Nat Prod Rep 27(2):238–254
71. Scalbert A (1991) Antimicrobial properties of tannins. Phytochemistry 30(12):3875–3883
72. Akiyama H, Fujii K, Yamasaki O et al (2001) Antibacterial action of several tannins against *Staphylococcus aureus*. J Antimicrob Chemother 48(4):487–491
73. Khanbabaee K, van Ree T (2001) Tannins: classification and definition. Nat Prod Rep 18(6):641–649
74. Majed F, Rashid S, Khan AQ et al (2015) Tannic acid mitigates the DMBA/croton oil-induced skin cancer progression in mice. Mol Cell Biochem 399(1–2):217–228
75. Kayser O, Kolodziej H (1999) Antibacterial activity of simple coumarins: structural requirements for biological activity. Z Naturforsch 54(3–4):169–174
76. Bakkali F, Averbeck S, Averbeck D, Idaomar M et al (2008) Biological effects of essential oils–a review. Food Chem Toxicol 46(2):446–475
77. Kavanaugh NL, Ribbeck K (2012) Selected antimicrobial essential oils eradicate *Pseudomonas* spp. and *Staphylococcus aureus* biofilms. Appl Environ Microbiol 78(11):4057–4061
78. Swamy MK, Akhtar MS, Sinniah UR (2016) Antimicrobial properties of plant essential oils against human pathogens and their mode of action: an updated review. Evid Based Complement Alternat Med 2016:3012462
79. Nazzaro F, Fratianni F, De Martino L et al (2013) Effect of essential oils on pathogenic bacteria. Pharmaceuticals 6(12):1451–1474
80. Levinson W, Jawetz E (2002) Medical microbiology and immunology: examination and board review, 7th edn. Lange Medical Books/McGraw-Hill, New York
81. Adwan G, Abu-Shanab B, Adwan K (2010) Antibacterial activities of some plant extracts alone and in combination with different antimicrobials against multidrug-resistant *Pseudomonas aeruginosa* strains. Asian Pac J Trop Med 3(4):266–269
82. Liang R, Xu S, Shoemaker CF et al (2012) Physical and antimicrobial properties of peppermint oil nanoemulsions. J Agric Food Chem 60(30):7548–7555
83. Chifiriuc C, Grumezescu V, Grumezescu AM et al (2012) Hybrid magnetite nanoparticles/*Rosmarinus officinalis* essential oil nanobiosystem with antibiofilm activity. Nanoscale Res Lett 7(1):209
84. Elumalai EK, Ramachandran M, Thirumalai T, Vinothkumar P (2011) Antibacterial activity of various leaf extracts of *Merremia emarginata*. Asian Pac J Trop Biomed 1(5):406–408
85. Zakaria Z, Sreenivasan S, Mohamad M (2007) Antimicrobial activity of *Piper ribesoides* root extract against *Staphylococcus aureus*. J Appl Biol Sci 1(3):87–90
86. Wu Y, Luo Y, Wang Q (2012) Antioxidant and antimicrobial properties of essential oils encapsulated in zein nanoparticles prepared by liquid–liquid dispersion method. LWT Food Sci Technol 48(2):283–290
87. Zhang L, Pornpattananangku D, Hu CM, Huang CM et al (2010) Development of nanoparticles for antimicrobial drug delivery. Curr Med Chem 17(6):585–594
88. Sahu P, Das D, Mishra VK et al (2017) Nanoemulsion: a novel eon in cancer chemotherapy. Mini Rev Med Chem 17(18):1778–1792
89. Nirmal C, Puvanakrishnana R (1996) Effect of curcumin on certain lysosomal hydrolases in isoproterenol-induced myocardial infarction in rats. Biochem Pharmacol 51(1):47–51
90. Capek I (2004) Degradation of kinetically-stable o/w emulsions. Adv Colloid Interf Sci 107(2–3):125–155
91. Ghaderi L, Moghimi R, Aliahmadi A et al (2017) Development of antimicrobial nanoemulsion-based delivery systems against selected pathogenic bacteria using a thymol-rich thymus daenensis essential oil. J Appl Microbiol 123(4):832–840

92. Gupta A, Badruddoza AZM, Doyle PS (2017) A general route for nanoemulsion synthesis using low-energy methods at constant temperature. Langmuir 33(28):7118–7123
93. Lawrence MJ, Rees GD (2000) Micro emulsion-based media as novel drug delivery system. Adv Drug Deliv Rev 45(1):89–121
94. Wang L, Mutch KJ, Eastoe J et al (2008) Nanoemulsions prepared by a two-step low-energy process. Langmuir 24(12):6092–6099
95. von Corswant C, Thorén P, Engström S (1998) Triglyceride-based micro emulsion from intervenes administration of sparingly soluble substances. J Pharm Sci 87(2):200–208
96. Taylor PW, Hamilton-Miller JMT, Paul D, Stapleton PD (2005) Antimicrobial properties of green tea catechins. Food Sci Technol Bull 2:71–81
97. Vidigal PG, Müsken M, Becker KA et al (2014) Effects of green tea compound epigallocatechin-3-gallate against *Stenotrophomonas maltophilia* infection and biofilm. PLoS One 9(4):e92876
98. Jeon J, Kim JH, Lee CK et al (2014) The antimicrobial activity of (−)-epigallocatehin-3-gallate and green tea extracts against *Pseudomonas aeruginosa* and *Escherichia coli* isolated from skin wounds. Ann Dermatol 26(5):564–569
99. Chakrawarti L, Agrawal R, Dang S et al (2016) Therapeutic effects of EGCG: a patent review. Expert Opin Ther Pat 26(8):907–916
100. Serra DO, Mika F, Richter AM, Hengge R et al (2016) The green tea polyphenol EGCG inhibits *E. coli* biofilm formation by impairing amyloid curlifibre assembly and down-regulating the biofilm regulator CsgD via the σE -dependent sRNA RybB. Mol Microbiol 101(1):136–151
101. Zhao WH, Hu ZQ, Okubo S et al (2001) Mechanism of synergy between epigallocatechin gallate and beta-lactams against methicillin-resistant *Staphylococcus aureus*. Antimicrob Agents Chemother 45(6):1737–1742
102. Hemaiswarya S, Kruthiventi AK, Doble M et al (2008) Synergism between natural products and antibiotics against infectious diseases. Phytomedicine 15(8):639–652
103. Fangueiro JF, Calpena AC, Clares B et al (2016) Biopharmaceutical evaluation of epigallocatechingallate-loaded cationic lipid nanoparticles (EGCG-LNs): *In vivo*, *in vitro* and *ex vivo* studies. Int J Pharm 502(1):161–169
104. Lin YH, Feng CL, Lai CH et al (2014) Preparation of epigallocatechin gallate-loaded nanoparticles and characterization of their inhibitory effects on *Helicobacter pylori* growth in vitro and in vivo. Sci Technol Adv Mater 15(4):045006
105. Tyagi P, Singh M, Kumari H et al (2015) Bactericidal activity of curcumin I is associated with damaging of bacterial membrane. PLoS One 10(3):e0121313
106. Rai D, Singh JK, Roy N, Panda D (2008) Curcumin inhibits FtsZ assembly: an attractive mechanism for its antibacterial activity. Biochem J 410(1):147–155
107. Chang C-Y, Krishnan T, Wang H et al (2014) Non-antibiotic quorum sensing inhibitors acting against *N*-acyl homoserine lactone synthase as druggable target. Sci Rep 4:7245
108. Packiavathy IA, Priya S, Pandian SK, Ravi AV et al (2014) Inhibition of biofilm development of uropathogens by curcumin–an anti-quorum sensing agent from *Curcuma longa*. Food Chem 148:453–460
109. Moghadamtousi SZ, Kadir HA, Hassandarvish P et al (2014) A review on antibacterial, antiviral, and antifungal activity of curcumin. Biomed Res Int 2014:186864
110. Mun SH, Joung DK, Kim YS et al (2013) Synergistic antibacterial effect of curcumin against methicillin-resistant *Staphylococcus aureus*. Phytomedicine 20(8–9):714–718
111. Hatamie S, Nouri M, Karandikar SK et al (2012) Complexes of cobalt nanoparticles and polyfunctional curcumin as antimicrobial agents. Mater Sci Eng C 32(2):92–97
112. Krausz AE, Adler BL, Cabral V et al (2015) Curcumin-encapsulated nanoparticles as innovative antimicrobial and wound healing agent. Nanomedicine 11(1):195–206
113. Loo CY, Rohanizadeh R, Young PM et al (2015) Combination of silver nanoparticles and curcumin nanoparticles for enhanced anti-biofilm activities. J Agric Food Chem 64(12):2513–2522

114. Hwang D, Lim YH (2015) Resveratrol antibacterial activity against Escherichia coli is mediated by Z-ring formation inhibition via suppression of FtsZ expression. Sci Rep 5:10029
115. Joung DK, Choi SH, Kang OH et al (2015) Synergistic effects of oxyresveratrol in conjunction with antibiotics against methicillin-resistant *Staphylococcus aureus*. Mol Med Rep 12(1):663–667
116. Jeon YO, Lee JS, Lee HG (2016) Improving solubility, stability, and cellular uptake of resveratrol by nanoencapsulation with chitosan and γ-poly (glutamic acid). Colloids Surf B Biointerfaces 147:224–233
117. Amalaradjou MA, Narayanan A, Baskaran SA, Venkitanarayanan K et al (2010) Antibiofilm effect of trans-cinnamaldehyde on uropathogenic *Escherichia coli*. J Urol 184(1):358–363
118. Nuryastuti T, van der Mei HC, Busscher HJ et al (2009) Effect of cinnamon oil on icaA expression and biofilm formation by *Staphylococcus epidermidis*. Appl Environ Microbiol 75(21):6850–6855
119. Niu C, Afre S, Gilbert ES (2006) Sub-inhibitory concentrations of cinnamaldehyde interfere with quorum sensing. Lett Appl Microbiol 43(5):489–494
120. Nostro A, Scaffaro R, D'Arrigo M et al (2012) Study on carvacrol and cinnamaldehyde polymeric films: mechanical properties, release kinetics and antibacterial and antibiofilm activities. Appl Microbiol Biotechnol 96(4):1029–1038
121. Gomes C, Moreira RG, Castell-Perez E (2011) Poly (DL-lactide-co-glycolide) (PLGA) nanoparticles with entrapped trans-cinnamaldehyde and eugenol for antimicrobial delivery applications. J Food Sci 76(2):16–24
122. Nostro A, Sudano Roccaro A, Bisignano G et al (2007) Effects of oregano, carvacrol and thymol on *Staphylococcus aureus* and *Staphylococcus epidermidis* biofilms. J Med Microbiol 56(4):519–523
123. Burt SA, Ojo-Fakunle VT, Woertman J, Veldhuizen EJ (2014) The natural antimicrobial carvacrol inhibits quorum sensing in *Chromobacterium violaceum* and reduces bacterial biofilm formation at sub-lethal concentrations. PLoS One 9(4):e93414
124. Wang Q, Gong J, Huang X et al (2009) *In vitro* evaluation of the activity of microencapsulated carvacrol against *Escherichia coli* with K88 pili. J Appl Microbiol 107(6):1781–1788
125. Pérez-Conesa D, Cao J, Chen L et al (2011) Inactivation of *Listeria monocytogenes* and *Escherichia coli* O157:H7 biofilms by micelle-encapsulated eugenol and carvacrol. J Food Prot 74(1):55–62
126. Miladi H, Zmantar T, Kouidhi B et al (2017) Synergistic effect of eugenol, carvacrol, thymol, p-cymene and γ-terpinene on inhibition of drug resistance and biofilm formation of oral bacteria. Microb Pathog 112:156–163
127. Iannitelli A, Grande R, Di Stefano A et al (2011) Potential antibacterial activity of carvacrol-loaded poly(DL-lactide-co-glycolide) (PLGA) nanoparticles against microbial biofilm. Int J Mol Sci 12(8):5039–5051
128. Zodrow KR, Schiffman JD, Elimelech M (2012) Biodegradable polymer (PLGA) coatings featuring cinnamaldehyde and carvacrol mitigate biofilm formation. Langmuir 28(39):13993–13999
129. Evans JD, Martin SA (2000) Effects of thymol on ruminal microorganisms. Curr Microbiol 41(5):336–340
130. Lambert RJ, Skandamis PN, Coote PJ, Nychas GJ et al (2001) A study of the minimum inhibitory concentration and mode of action of oregano essential oil, thymol and carvacrol. J Appl Microbiol 91(3):453–462
131. Karpanen TJ, Worthington T, Hendry ER et al (2008) Antimicrobial efficacy of chlorhexidine digluconate alone and in combination with eucalyptus oil, tea tree oil and thymol against planktonic and biofilm cultures of *Staphylococcus epidermidis*. J Antimicrob Chemother 62(5):1031–1036
132. Zhou F, Ji B, Zhang H et al (2007) The antibacterial effect of cinnamaldehyde, thymol, carvacrol and their combinations against the foodborne pathogen *Salmonella typhimurium*. J Food Saf 27(2):124–133

133. Liolios CC, Gortzi O, Lalas S et al (2009) Liposomal incorporation of carvacrol and thymol isolated from the essential oil of *Origanum dictamnus* L. and *in vitro* antimicrobial activity. Food Chem 112(1):77–83
134. Gill AO, Holley RA (2004) Mechanisms of bactericidal action of cinnamaldehyde against *Listeria monocytogenes* and of eugenol against *L. monocytogenes* and *Lactobacillus sakei*. Appl Environ Microbiol 70(10):5750–5755
135. Zhou L, Zheng H, Tang Y et al (2013) Eugenol inhibits quorum sensing at sub-inhibitory concentrations. Biotechnol Lett 35(4):631–637
136. Zhang P, Zhang E, Xiao M et al (2013) Enhanced chemical and biological activities of a newly biosynthesized eugenol glycoconjugate, eugenol α-D-glucopyranoside. Appl Microbiol Biotechnol 97(3):1043–1050
137. Narayanan A, Neera M, Ramana KV (2013) Synergized antimicrobial activity of eugenol incorporated polyhydroxy butyrate films against food spoilage microorganisms in conjunction with pediocin. Appl Biochem Biotechnol 170(6):1379–1388
138. Ghosh V, Mukherjee A, Chandrasekaran N (2014) Eugenol-loaded antimicrobial nanoemulsion preserves fruit juice against, microbial spoilage. Colloids Surf B Biointerfaces 114:392–397
139. LiveLeaf, Inc (2017) Method of killing a bacteria with a plant-based biocidal solution. USPatent 9, 636, 361, 2 May 2017
140. Slippery Rock University Foundation, Inc (2017) Methods of treating infectious diseases. US Patent 9,545,386, 17 Jan 2017
141. LiveLeaf, Inc (2015) Method of treating damaged mucosal or gastrointestinal tissue by administering a composition comprising a mixture of pomegranate and green tea extracts and releasably bound hydrogen peroxide. US Patent 9,192,635, 24 Nov 2015
142. The Hong Kong Polytechnic University (2015) Flavonoid dimers and their use. US Patent 8,980,848, 17 March 2015
143. Liveleaf, Inc (2014) Combining a polyphenol with hydrogen peroxide to treat or prevent a bacterial infection. US Patent 20140072660, 13 March 2014
144. Emory University (2009) Triptolide analogs for the treatment of autoimmune and inflammatory disorders. US Patent 7,557,139, 7 July 2009
145. Gubarev MJ, Enioutina EY (2000) Method to enhance innate immunity defense mechanisms by treatment with plant-derived alkaloids. US Patent 6,149,912, 21 Nov 2000

Omega-3 Fatty Acids and Its Role in Human Health

9

Darshika Nigam, Renu Yadav, and Udita Tiwari

Abstract

Fats (triglyceride) are the main energy storage macromolecules in case of animal and have served not only as a structural component of cell membrane, in which they are anchored by phospholipid molecules. Fats are esters of fatty acids condensed with glycerol molecules. Fatty acids have been categorized into essential and nonessential fatty acids on the basis of the ability of an organism to biosynthesize them. Essential fatty acids (linolenic, linoleic, and arachidonic acids) cannot be biosynthesized by an organism so must be provided through diet. One of the essential families of fatty acids for humans is the omega-3 fatty acids which include α-linolenic acid (ALA), eicosapentaenoic acid (EPA), and docosahexaenoic acid (DHA). These essential nutrients enhance quality of life by supporting hormones, inflammation, regulating blood coagulation, mental health of an individual. The present chapter explores in detail the biological functions and dietary benefits of the omega-3 fatty acids in regulating metabolism of the body and its protective role in prevention many diseases.

Keywords

Omega-3 fatty acids, α-linolenic acid · Eicosapentaenoic acid · Docosahexaenoic acid · Human health · Functional food

9.1 Introduction

Omega-3 fatty acids are classified as polyunsaturated fatty acids (PUFA) consisting of double bond with cis configuration at the third carbon atom from the end of the carbon chain [1]. Omega-3 is treated as an essential fatty acid (EFA)

D. Nigam (✉) · R. Yadav · U. Tiwari
Department of Biochemistry, School of Life Sciences, Dr. Bhimrao Ambedkar University, Agra, Uttar Pradesh, India

V. Rani, U. C. S. Yadav (eds.), *Functional Food and Human Health*,
https://doi.org/10.1007/978-981-13-1123-9_9

because human body is unable to form them and therefore required to be added in diet [2]. It consists of alpha-linolenic acid (ALA) (food source, plant oils), docosahexaenoic acid (DHA), and eicosapentaenoic acid (EPA) (food source, marine oils like fatty fish oils) [1]. Naturally, human body is unable to form fatty acids but able to form 20- and 22-C omega-3 fatty acids from its primitive 18-carbon ALA. Conversion of ALA into EPA and then to DHA is low yield process, where only 5–10% and 2–10% are formed, respectively [2]. Little amounts of omega-3 fatty acids in diet are able to contribute to healthy growth and to prevent many disease conditions. The primary sources of omega-3 fatty acids are flaxseed, chia seeds, walnuts, soybeans, microalgae oil, cold water fatty fish like salmon, sardine, and krill, etc. In the food chain, microalgae, such as zooplankton and phytoplankton, are consumed by fish which in turn becomes chief sources of EPA and DHA [1, 3].

During 1970s, researchers observed that Greenland Eskimos and Alaskans had less cardiovascular diseases as they consumed large amount of seal and whale blubber which are rich in omega-3 fatty acids as compared to other American and Caucasian counterparts [4].

9.2 Classification of Fatty Acids and Forms of Omega-3 Fatty Acids

Fatty acids are classified as saturated and unsaturated fatty acids based on the degree of saturation. Saturated fatty acids (SFAs) are further classified on the basis of length of chain into short, medium, and long chain as shown in Fig. 9.1. Unsaturated fatty acids are monounsaturated (MUFA) and polyunsaturated (PUFA) on the basis of their configuration of double bonds. MUFAs are normally called as omega-9 fatty acids. PUFAs are further classified into omega-3 and omega-6 fatty acids on the basis of location of double bond. Linoleic acid (LA) and arachidonic acid (AA) are common omega-6 fatty acids. Their major dietary sources are vegetable oils including sunflower, corn, safflower, peanut, and olive oil [4].

Omega-3 fatty acids, which are the prime concern of the chapter, have beneficial effects for many biological systems, including smooth muscle cells, immune reactions, endothelial cells, blood platelets, liver cells, neuron cells, heart cells, osteoblasts, and adipocytes. They are involved in promoting apoptosis and regulation of growth of various cell types in human body [5, 6]. Although the biological effects of omega-3 fatty acids are not apparently known yet, it is possible that eicosanoids, derived from the omega-3 fatty acid, are vital for cell membrane formation and act as ligands for transcription factors [13]. ALA, EPA, and DHA are the most important bioactive forms of omega-3 fatty acids. ALA is an essential fatty acid as mammalian cells lack enzyme to desaturate fatty acids after ninth carbon from the carboxyl end of fatty acid molecule. ALA is converted into EPA and then DHA by metabolic pathway which is readily utilized in the body. It is important to discuss these three forms of omega-3 fatty acids here.

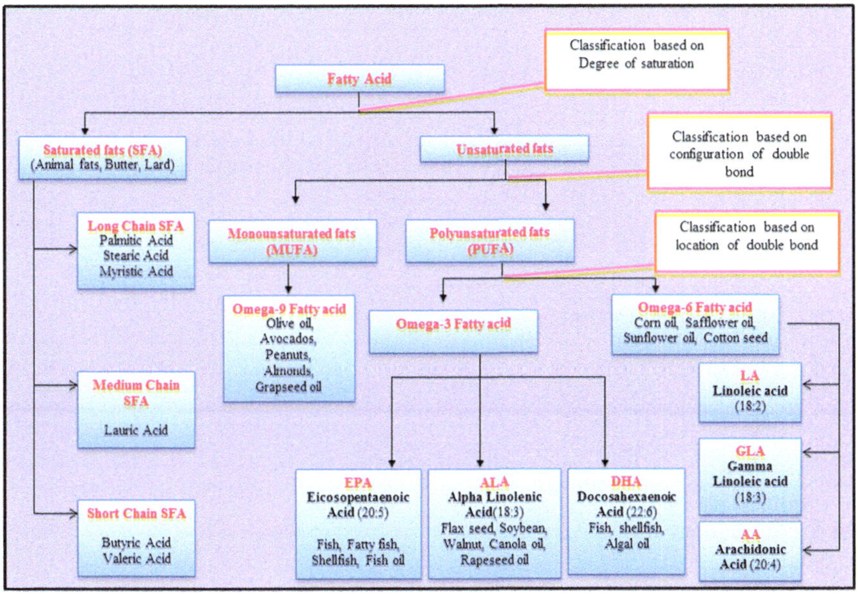

Fig. 9.1 Classification of fatty acids

9.2.1 Alpha-Linolenic Acid (ALA)

Alpha-linolenic acid (ALA) is one of the two essential fatty acids (omega 3) as it cannot be produced by human body. ALA is a carboxylic acid with 18-carbon chain and 3 cis double bonds, named as *all-cis*-9,12,15-octadecatrienoic acid with lipid number, 18:3. It is an isomer gamma-linolenic acid (GLA), a polyunsaturated $n-6$ (omega-6) fatty acid (as shown in Fig. 9.2). Though, absence of 12- and 15-desaturase enzymes which are required for de novo synthesis of ALA from stearic acid, therefore ALA can only be obtained by humans through their diets. ALA is found in vegetable oils (rapeseed or canola oil), seeds (chia seed, flaxseed, soybean), and nuts (walnuts) and many common vegetable oils [7].

9.2.2 Eicosapentaenoic Acid (EPA)

Eicosapentaenoic acid (EPA) is a 20-carbon chain carboxylic acid with 5 cis double bonds, named as *all-cis*-5,8,11,14,17-icosapentaenoic acid with lipid number, 20:5. EPA acts as precursor for prostaglandin-3, thromboxane-3, and leukotriene-5. EPA can be synthesized by humans from dietary ALA but with efficiency much lower than the absorption of EPA from food containing it. It is readily available from fish, fish oil (cod liver oil, herring, salmon, sardine, etc.), algal oil, various types of edible seaweed, phytoplankton, and human breast milk [8].

EPA plays a vital role in many metabolic processes. EPA competes with the overproduction of AA (omega-6 fatty acid) which is an inflammatory molecule and,

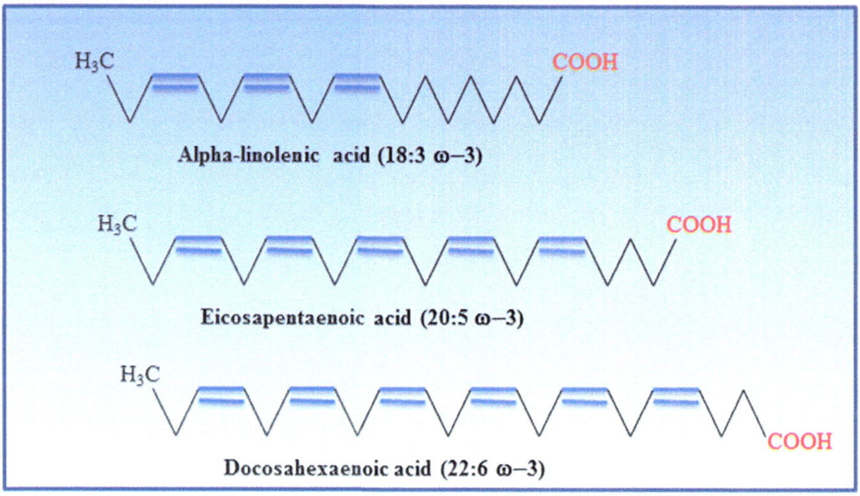

Fig. 9.2 Structure of ALA, EPA, and DHA

therefore, counters excessive inflammatory reactions throughout the body. Thus, it is recommended that the amount of EPA should be higher in a person's diet to subsidize AA-related inflammation [9].

9.2.3 Docosahexaenoic Acid (DHA)

Docosahexaenoic acid (DHA) is a key structural component of the human brain, skin, and retina. It is a 22-carbon chain carboxylic acid compound with 6 cis double bonds, named as *all-cis*-4,7,10,13,16,19-docosahexaenoic acid with lipid number, 22:6. Although the role of DHA is not clear as anti-inflammatory compound, it plays important role in health of various tissues especially in brain health. DHA is found to be abundant in seafood (fatty fish, shellfish, etc.). DHA contribute around 60% of PUFA in the retina and 40% of PUFAs in the brain. DHA comprises half of the weight of a brain cell membrane [10].

EPA and DHA are the most important bioactive forms of omega-3 fatty acids. DHA is very important to carry out physiological functions of the brain, while EPA plays major role in anti-inflammation; thus these two fatty acids can play role in regulating whole-body functions. Despite playing different roles in contributing to better health, both EPA and DHA share many similar beneficial effects. Thus, EPA and DHA, together or separately, are beneficial in various functions of human body. EPA is also important for brain health as it reduces neuro-inflammation. However, the level of EPA in the brain is lower than DHA because it can be depleted fast from the brain. In contrast, DHA is incorporated in membranes of all brain cells. Therefore, many researchers believe that EPA is needed more than DHA for normal brain functions [10].

9.2.4 Omega-6 to Omega-3 Ratio

Another class of polyunsaturated fatty acids is the omega-6 fatty acids. They are essential for normal growth in young children and beneficial for skin integrity and kidney function. It also helps in reducing nerve pain and symptoms of ADHD (attention deficit hyperactivity disorder), fighting inflammation, curing arthritis, controlling high blood pressure, supporting bone health, and lowering risk of heart attack. Some of the major dietary sources of omega-6 fatty acids are grapeseed, safflower, poppyseed oil, sunflower oil, corn oil, cottonseed oil, walnut oil, sesame oil, and soybean oil. Omega-3 and omega-6 fatty acids both are, therefore, necessary for good health. The ideal dietary ratio of omega-6 to omega-3 fatty acids is 1:1, but consuming too much omega-6 fatty acids in comparison to omega-3 fatty acids can disturb fatty acid balance, which may lead to several negative effects [11]. Excessive dietary consumption of omega-6 fatty acids is detrimental and promotes many diseases including cardiovascular disease, diabetes, inflammatory and autoimmune diseases, cancer, etc. Effects of certain medicines are also increased or decreased by some omega-6 fatty acids, such as GLA [5, 6, 12, 13].

Both the omega-3 and omega-6 fatty acids and their intermediate components compete for the common enzymes like delta-6 desaturase, elongase, and 5-desaturase in sequential metabolic pathway as depicted in Fig. 9.3 [14]. By the cascade of reactions catalyzed by common enzymes, ALA is converted into EPA and subsequently into DHA, whereas linoleic acid (LA) is converted into arachidonic acid (AA) by same enzyme cascade. DHA and EPA produce anti-inflammatory molecules such as

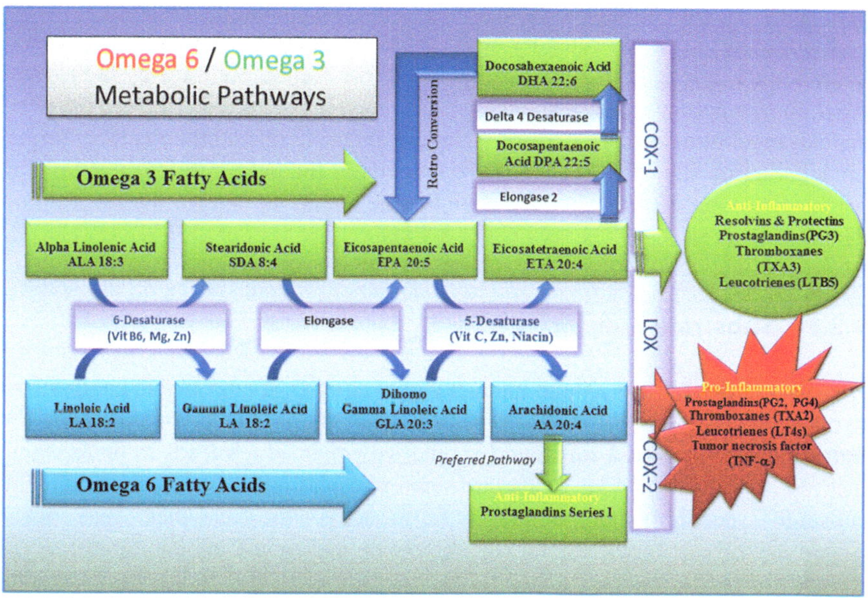

Fig. 9.3 Metabolic pathways of omega-3 and omega-6 fatty acids

3 series prostaglandins (PG3s), 5 series leukotrienes (LT5s), resolvins (RvDs and RvEs), and protectins, while AA has very potent pro-inflammatory effect due to the formation of 2 and 4 series prostaglandins and 4 series leukotrienes (LTB4, LTC4, LTD4, LTE4). Since the ALA has higher affinity for the enzyme delta-6 desaturase, the formation of anti-inflammatory molecules is higher than pro-inflammatory molecules by LA metabolic pathway. Therefore, higher intake of omega-3 than omega-6 acids is recommended for maximum health benefits [14–16].

9.3 Mechanism of Action of Omega-3 Fatty Acids

There are several mechanisms through which omega-3 fatty acids can affect the physiology of biological systems followed by their health.

9.3.1 Eicosanoids

Eicosanoids are 20-carbon atoms made by the enzymatic or non-enzymatic oxidation of arachidonic acid or other PUFAs. There are many subfamilies of eicosanoids which include prostaglandins, prostacyclins, leukotrienes, thromboxanes, resolvins, eoxins, and lipoxins. Eicosanoids are important signaling molecules which play role in many physiological and pathological processes like platelet aggregation; chemotaxis of blood cells; regulating cell growth and differentiation including cancer; controlling blood pressure; modulating flow of blood to tissues; triggering and inhibiting inflammation, fever, allergy, and other immune responses; regulating the abortion of pregnancy and child birth; and contributing to the perception of pain [16, 17].

Eicosanoids are a sub-category of oxylipins, i.e., oxidized fatty acids of diverse carbon units in length, and are distinguished from other oxylipins by their overwhelming importance as cell signaling molecules. In performing these roles, eicosanoids most often act as autocrine signaling agents to impact their cells of origin or as paracrine signaling agents to impact cells in the proximity of their cells of origin. Eicosanoids may also act as endocrine agents to control the function of distant cells [18].

9.3.2 Substrate Specificity

Omega-3 fatty acids are able to interact with a number of enzymes in comparison to other fatty acids to perform their mechanism of action. In other words, omega-3 fatty acids are preferred substrate for other enzymes which leads to preferential incorporation of these fatty acids into some phospholipids. For example, altered substrate specificity of omega-3 fatty acids for enzymes like acyl-CoA:diacylglycerol acyltransferase (ADGAT) and acyl-CoA:cholesterol acyltransferase (ACAT) shows that little triacylglycerol (TAG) is incorporated into very-low-density lipoproteins (VLDL), and therefore, EPA and DHA are preferentially found in phospholipids [19].

9.3.3 Acylation of Proteins

Acylation of proteins is vital for anchoring specific proteins in cellular membranes with viral or other proteins by different covalent bonds. It is necessary for the signaling function of the proteins. Saturated fatty acids are often linked to proteins though omega-3 fatty acids may also acylate proteins. PUFAs (AA and EPA) inhibit palmitoylation and change membrane localization of a protein kinase (Fyn), and C14:2 may be acylated to a protein kinase (Fyn). This alters its raft localization and promotes reduction of T cell signal inflammation and transduction [20].

9.3.4 Membrane Fluidity

High dietary intake of very long-chain omega-3 fatty acids result in more incorporation of EPA and DHA into membrane phospholipids. This incorporation changes some physical characteristics of cell membranes which thereby causes changes in membrane protein functions and affects the fluidity and permeability of the mitochondrial and cellular membranes [21]. Higher DHA level in the body changes the composition of the cell membrane of red blood cell and makes it more pliable and adaptable. This reduces the risk of sudden cardiac arrest and ischemic heart disease and provides protection against heart arrhythmias [22, 23].

Fish oil reduces blood viscosity and high incorporation of long-chain omega-3 fatty acids into low-density lipoprotein (LDL) and changes its physical properties. This also promotes reduction in melting point of core cholesteryl esters. High levels of DHA in phosphatidylethanolamine and phosphatidylserine present in specific areas of outer segments of retinal rods that help in increasing light transduction [24].

9.3.5 Lipid Peroxidation

Products of lipid peroxidation and signaling molecules in certain cells are promoted by the degree of unsaturation of omega-3 fatty acids. The degree of lipid peroxidation is determined in vivo by appearance of thiobarbituric acid reactive substances (TBARS), lipid peroxides, degradation and diene conjugation of PUFA, and antibodies against oxidized LDL. Omega-3 fatty acids affect the modification of LDL of mononuclear cells which in turn may release enzymes or H_2O_2 that may modify LDL and thereby enhanced uptake of LDL by macrophages in endothelial cells. This may develop atherosclerosis [25]. However some study suggests that dietary PUFA stimulates proteolysis of apoB100 degradation and, thus, may cause hypolipidemic effects in hepatic lipid peroxidation [26].

DHA is highly susceptible to free radical-mediated peroxidation, which gives rise to a wide array of bioactive lipid peroxidation products. One of these species is cyclopentenone neuroprostanes reactive and structurally similar to anti-inflammatory cyclopentenone prostaglandins. Some of these neuroprostanes are potent inhibitors of nuclear factor kB (NF-kB) signaling in a variety of cell types and may contribute to the anti-inflammatory actions of DHA [27, 28]. NF-kB is a key mediator of the inflammatory response and also regulates a variety of pro-inflammatory proteins at transcriptional level which include cyclooxygenase-2 (COX-2) and inducible nitric oxide synthase (iNOS) and in response to antigenic stimuli such as lipopolysaccharide [29].

In contrast to this, reduced formation of lipid peroxidation products and superoxide scavenging activity by omega-3 fatty acids DHA suggest antioxidative ability of these fatty acids which prevents onset of many diseases [30].

9.3.6 Gene Interactions

It is evident that inflammation is key cause of many chronic diseases, such as coronary heart disease, diabetes, and many other diseases. Dietary intake of omega-3 fatty acids plays an important role in the prevention of such diseases especially in persons with genetic variations [31]. Transcription of certain genes is altered when fatty acids or their derivatives interact with nuclear receptor proteins. One such example is the peroxisome proliferator-activated receptors (PPARs) that alter expression levels of genes involved in lipid metabolism. Ligands for PPAR are those fatty acids that are blocked from being beta-oxidized and eicosanoids. Prostaglandin J2 (PGJ2), an eicosanoid derived from AA, EPA, and DHA, down-regulates lipopolysaccharide-induced activation of NF-kB and monocytes via PPAR-gamma and PPAR-alpha activation. This could play an important role in preventing many diseases caused by inflammation such as brain damage, ischemia, and nephropathy [32–34].

In addition EPA, DHA, and other PUFA may also suppress several glycolytic and lipogenic genes independent of PPAR which encode protein L-pyruvate kinase and fatty acid synthase, respectively [35].

PUFAs have marked effects on gene expression by regulating the activity of other hepatic transcription factors, including liver X receptors (LXRs) (alpha and beta), retinoid X receptor (RXR), hepatic nuclear factor (HNF)-4 alpha, and sterol regulatory element-binding proteins (SREBPs) 1 and 2. These transcription factors play key roles in hepatic carbohydrate, fatty acid, triglyceride, cholesterol, and bile acid metabolism. Omega-3 fatty acids are feed-forward activators of PPARs, whereas are feed-back inhibitors of LXRs and SREBPs [36].

9.4 Physiological Roles and Health Benefits of Omega-3 Fatty Acids

Omega-3 fatty acids are probably known best for their heart health benefits. EPA and DHA help in lowering triglyceride levels which are linked to increased risk of cardiovascular disease. Further, recommended doses of DHA increase the level LDL, VLDL, in blood. Higher levels of LDL are debated for a known risk factor

Table 9.1 Sources and health benefits of omega-3 fatty acids

Fatty acid	Food sources	Application in health and disease
Docosahexaenoic acid (DHA)	Cold water fatty fish and fish oil	Helps in visual, cognitive, and brain development and cardiovascular functions
		Reducing risk of neurological disorders and some forms of cancer
Eicosapentaenoic acid (EPA)	Cold water fatty fish and fish oil, algae, and breast milk	Helps in brain development, anti-inflammatory, potential use in arthritis, asthma, and other inflammatory conditions
Alpha-linolenic acid (ALA)	Canola oil, walnuts, chia seed, flaxseed, flax oil	Reduces risk of degenerative disease including heart disease and cancer arthritis, skin conditions, diabetic neuropathy, immune function, and premenstrual syndrome

for atherosclerosis; hence increasing the amount of EPA in their dietary supplements is recommended in comparison to DHA [5, 6, 11, 12]. However, a growing body of research suggests that omega-3 fatty acids may also provide other benefits such as lowering the risk of developing certain cancers and neurological disorders, retarding cognitive decline in aged persons, improving memory, fighting metabolic syndrome and diabetes, and improving healthy pregnancy outcomes such as better motor skills, higher mental processing scores, and visual acuity among infants [10]. Table 9.1 shows the sources and health benefits of omega-3 fatty acids. Recommended amounts of DHA help in resisting cancer and also resist inflammatory molecules to trigger the inflammatory response [6]. In the present section of the chapter, the role of omega-3 fatty acid is enlightened.

9.4.1 Anti-Inflammatory Activity

Inflammation is basically a source of many diseases such as atherosclerosis, rheumatoid arthritis, asthma, inflammatory bowel disease, ulcerative colitis, Crohn's disease, diabetes, psoriasis, lupus erythematosus, depression, multiple sclerosis, migraine headaches, and many more [9, 11, 37]. These diseases occur due to incorrect activation of T cells which results in destruction of healthy host tissues. The inflammatory responses are generated mainly by the subsequent conversion of arachidonic acid into eicosanoids such as prostaglandin PGE2, leukotriene B4, and thromboxane A2 [38]. Supplements with omega-3 fatty acids such as fish oil in the diet will reduce the inflammatory responses by decreasing the conversion of arachidonic acid to prostaglandins and leukotrienes [39]. Omega-3 fatty acids also reduce serum levels of pro-inflammatory cytokines, tumor necrosis factor-alpha (TNF-alpha), C-reactive protein (CRP), and interleukins IL-6 and IL-1 which are the major cause of rheumatoid arthritis and other diseases [40–43]. There are evidences supported by the National Center for Complementary and Integrative Health that incorporation of dietary supplementation of omega-3 fatty acids especially fish oil may be beneficial for rheumatoid arthritis [31] (as shown in Table 9.2).

Table 9.2 Omega-3 fatty acids help in curing the listed diseases

Brain disorder
Alzheimer's and Parkinson's disease, behavioral disorder, dementia, hyperactivity and learning disorders, multiple sclerosis, stroke, schizophrenia, autism
Cardiovascular disorders
Congenital heart disease, angina, atherosclerosis, high blood pressure
Immune disorders
AIDS, allergies, immune deficiencies, systemic lupus erythematosus, inflammatory condition, infection
Disorders in females
Breast cancer, breast cysts, endometriosis, menopause
Gastrointestinal disorders
Inflammatory bowel syndrome, malnutrition, diabetes, Crohn's diseases
Skin disorders
Eczema, acne, psoriasis, dermatitis
Eye disorders
Vision acuity disorders, glaucoma, macular degeneration, retinitis pigmentosa
Cancers and other diseases
Carcinoma and leukemia, cystic fibrosis, arthritis, obesity, post-viral fatigue, Reye's syndrome, renal disease

9.4.2 Cardiovascular Disease

It has been observed that trans-fatty acids, cholesterol, and saturated fats are mainly responsible for atherosclerosis. Supplementation of omega-3 fatty acids (when taken more than 1 gram everyday at least for 1 year) will be protective against CVD including myocardial infarction and sudden cardiac death [44–47]. This supplementation also lowers systolic and diastolic blood pressure in hypertension patients [48]. Many researchers support that when diet is supplemented with EPA and DHA, it increases fibrin breakdown, blood circulation, vascular permeability, and HDL levels, improves plaque stability, and decreases the production of thromboxane A2, PGE2, IL-1, IL-2, IL-6, endothelial activation, and cholesterol levels [10, 49, 50].

Omega-3 fatty acids reduce triglyceride levels by increasing the expression of hydrolytic enzyme lipoprotein lipase (LPL) and, hence, reducing lipid levels. This reduction in triglyceride levels also contributes to decrease in the production of VLDL [51, 52]. It has been observed that patients suffering from metabolic syndrome when supplemented with EPA and DHA daily reduce body weight and systolic blood pressure and show decrease in LDL cholesterol, triglycerides, and C-reactive protein [53].

9.4.3 Diabetes

There is significant increase in the plasma level of glucose, fructosamine, triglycerides, and LDL, catalase activity, nitric oxide levels, and release of IL-1 and TNF-alpha cytokines in diabetes and decrease in HDL level [54, 55]. It has been observed that the diabetes can be treated with supplementation of omega-3 fatty acids in diet which lowers the blood sugar levels and also reduces the hazardous effects of pro-inflammatory molecules which are the main factor in the pathogenesis of type 1 and type 2 diabetes [54–57]. There are evidences that patients with type 2 diabetes when supplemented with fish oil of 6 g daily have improved insulin resistance and also lowered triglyceride and TNF-alpha levels [58]. Those people who suffer from type 2 diabetes have high risk of developing cardiovascular disease (CVD). It can be minimized by consumption of fish oil [59–62].

9.4.4 Obesity

Supplementation of omega-3 fatty acid (EPA and DHA) in the diet drastically reduces the uptake of lipid into adipocytes and synthesis of fatty acids and increases the lipid oxidation which will be beneficial for the people suffering from obesity and metabolic syndrome [63–65]. Daily intake of omega-3 fatty acids increases the lipoprotein lipase gene expression which thereby reduces triglyceride, cholesterol, and VLDL levels [51]. Fish oil in the diet enhances protein incorporation of metabolism and increases muscle mass as well as insulin sensitivity in young and aged persons [66, 67].

9.4.5 Brain Development and Neurological Disorders

Total lipid composition of adult brain is approximately 50–60% lipid, and DHA constitutes 10–20% of this lipid content [68]. DHA helps in increased formation of synaptic membranes and levels of pre- and postsynaptic membrane proteins, myelination, number of dendritic spines, and synapses in hippocampus, a memory region of the brain, parietal cortex, and hypothalamus [69–72]. Supplementation of omega-3 fatty acids especially DHA in diet is capable of increasing dopamine levels and acetylcholine levels by inhibiting pro-inflammatory molecule TNF-alpha which in turn increases the level of nitric oxide synthesis in the brain [73] and, therefore, prevents neuronal apoptosis and enhances memory [74, 75]. DHA is very essential for the brain development in fetus and later boosts hand-eye coordination, fine motor skills, communication, and IQ level. Therefore, pregnant women are recommended to have DHA in their diet throughout pregnancy as well as lactation period [76–79].

9.4.5.1 Stroke

Stroke occurs either due to blood supply to the brain is blocked as result of a blood clot, called an ischemic stroke, or blood vessel ruptures within the brain, called a hemorrhagic stroke. Increased consumption of omega-3 fatty acids reduces the risk of ischemic stroke. Omega-3 fatty acids stop platelet aggregation and also reduce whole blood viscosity [80].

9.4.5.2 Alzheimer's Disease

Among elderly persons the most common neurodegenerative disease is Alzheimer's disease (AD). In AD, progressive impairment of cognitive functions occur which is the commonest cause of dementia and is accompanied by behavioral dysfunction. DHA and other omega-3 fatty acids could reduce the secretion of amyloid-beta plaques which are the causative agents of the disease and, thus, keep the brain cells intact and reduce cognitive decline. This induces expression of anti-apoptotic and neuroprotective genes [81–83].

9.4.5.3 Parkinson's Disease

Parkinson's disease (PD) is a well-known movement disorder in elderly persons. It is characterized by the loss of dopaminergic neurons in substantia nigra and other regions of the brain [83]. PD can be treated by dopamine replacement therapy using L-dopa (levodopa). The major side effect of L-dopa is the development of dyskinesia (involuntary movement). Major evidences support that supplementation of DHA (100 mg/kg) in the diet along with levodopa reduces the chances of dyskinesia [84, 85].

9.4.5.4 Schizophrenia

Schizophrenia is a severe mental disorder that usually appears in late adolescence or early adulthood. It is characterized by positive or negative symptoms. Positive symptoms include hallucinations and delusions, whereas negative symptoms are apathy, lack of communication, and depression [86]. These patients show reduction in dopaminergic neurons of the prefrontal cortex having negative symptoms of schizophrenia, while the level of dopaminergic neurons of limbic systems increases showing positive symptoms of schizophrenia. Supplementation of omega-3 fatty acids in the diet helps to increase the level of dopamine in the prefrontal cortex and decrease the level in limbic system. In omega-3 fatty acid, EPA (10 g) was found to be more effective than DHA in the treatment of schizophrenia [87].

9.4.5.5 Depression

Depression or major depressive disorder is a common but serious mood disorder which can happen at any age but often begins in adolescence or in adulthood. It is caused by a combination of genetic, biological, psychological, and environmental factors [88].

There are evidences supporting a strong negative correlation between fish consumption and prevalence of depression [88, 89]. Dietary intake of 2 g fish oil for 6 months helps in mood stabilization [90]. Alteration of membrane fluidity to optimal level by omega-3 fatty acids influences signaling within and between neurons of depressed patients. Pro-inflammatory cytokines including IL-2,-6, TNF-alpha, and IFN-gamma activate hypothalamic-pituitary axis and alter metabolism of neurotransmitters which in turn cause depression. Dietary intake of all the three omega-3 fatty acids improves symptoms of depression by modulating cytokines [91].

9.4.5.6 Attention Deficit Hyperactivity Disorder

Approximately 8–12% of children suffer from attention deficit hyperactivity disorder (ADHD) worldwide. The symptoms of ADHD include difficulty in paying attention, listening, reading, writing, and finishing tasks [92, 93]. ADHD patients have deficiency of DHA. Moreover, lack of omega-3 fatty acids in diet further worsens symptoms of ADHD. Researches show that supplementation of omega-3 fatty acids for 3 months improves the.

It has been stated that the lack of omega-3 fatty acids influences the brain in such a way that may cause and/or worsen symptoms of ADHD (both children and adults) having lower plasma concentrations of DHA omega-3 compared to age- and sex-matched controls [94]. Additional studies have demonstrated the efficacy of both EPA and DHA omega-3 fatty acids in reducing ADHD-related symptoms [94, 95]. Also children (5–12 years of age) with developmental coordination disorder, which is also associated with difficulties in learning and behavior, showed significant improvements in reading, spelling, and behavior after 3 months treatment with omega-3 fatty acid supplements [96].

9.4.5.7 Autism

Autism is a neuro-developmental disorder whose cardinal symptoms are verbal and nonverbal communication, impaired social interaction, and repetitive behavior, commonly noticed in child before or during age of 3 years. Autism is observed in both children and adults having significantly lower levels of omega-3 fatty acids [97, 98]. Supplementation of fish oil in the diet will be more beneficial for the treatment of autistic spectra disorders such as improvement in the behavioral benefits including aggression, tantrums, and self-injurious behavior [99, 100].

9.4.6 Eye Health and Disease

DHA and EPA in diet are essential for proper visual development and retinal function [101]. Both omega-3 fatty acids help in protecting from age-related macular degeneration, retinitis pigmentosa, dry eye syndrome, and diabetic nephropathy and decreasing intraocular fluid pressure which results in glaucoma. DHA prevents the degeneration photoreceptor cells blinding eye diseases, macular degeneration, and retinitis pigmentosa [101, 102]. Omega-3 fatty acid supplementation in infants improves visual acuity (clarity of vision) [103, 104].

9.4.7 Endometriosis

Endometriosis is one of the most common diseases of women associated with pelvic pain due to endometrium lining which is normally present inside the uterus and grows outside the uterine cavity. High production of the pro-inflammatory cytokine, IL-1 beta, results in induction of COX 2 which thereby synthesizes high amount of prostaglandin, a causative agent of pelvic pain [105]. Intake of EPA and DHA reduces thickening of interstitium of endometriotic tissues and cysts by downregulating genes of immunomodulator and pro-inflammatory cytokine IL-1 beta and matrix metalloproteinases [106, 107].

9.4.8 Cancer

Cancer (multistage process) can be treated if detected in early and premalignant stages. Many evidences support the supplementation of omega-3 fatty acids in diet suppresses the development of cancer and tumor size at different sites due to anti-inflammatory, antiproliferative, and anti-metastatic properties of DHA. DHA also reduces harmful side effects of chemotherapeutic drugs [108–111].

9.4.9 Skin Health and Diseases

Fish intake is beneficial for skin health which contains omega-3 fatty acids along with vitamin D and selenium. Omega-3 fatty acids reduce inflammation-related skin problems such as acne, psoriasis, eczema, and other types of dermatitis [112]. These fatty acids also protect skin from UV radiation and reduce the risk of skin cancer [113].

9.5 Sources of Omega-3 Fatty Acids as Functional Food

Omega-3 fatty acids are found in both plants and animals. Plant sources of omega-3 fatty acids include flaxseed, butternuts, walnuts, almonds, chia seeds, rapeseed, vegetables like soybeans and garden peas, and grains like oats, and little amount is found in some berries and algae (Table 9.3) [8, 114]. Fishes like salmon, menhaden, sardine, cod, herring, mackerel, tuna (both blue and yellow fin) trout, and halibut; crustaceans like krill, crab, shrimp, and lobsters; and mollusks like mussel, oyster, octopus, and conch are prime animal sources of omega-3 fatty acids (Table 9.4) [8, 115].

9.5.1 Alpha-Linolenic Acid (ALA)

The most prominent omega-3 fatty acid in the human diet is ALA. It is found in plant products mainly in vegetable oils, seeds, nuts, legumes, grains, fruits, and some wild plants like purslane (*Claytonia perfoliata*) [8, 114].

ALA is obtained from the thylakoid membranes peas and other legumes. Soybeans and its processed products like tofu and soy–based milk can be consumed to increase linolenic acid intake the diet. Flax seeds are the richest source of ALA content (Table 9.3 showing selected food sources of ALA). Walnuts, butternuts, and chia (chia sage) are also rich source of ALA. Margarines, mayonnaise, soybean, and rapeseed oils are rich sources with high caloric content. Spinach, broccoli, beans, peas, wheat, rice, oat, berries, and some food like avocado also contain sources of ALA with less caloric contents [114].

Table 9.3 ALA content of various foods and oils [116]

Source (100 g raw edible portion)	ALA (g)
Nuts and seeds	
Flaxseed	22.8
Butternuts	8.7
Walnuts, English and Persian	6.8
Chia seeds	3.9
Beechnuts	1.7
Almonds	0.4
Vegetables	
Soybeans, green	3.2
Soybeans, mature seeds, sprouted	2.1
Seaweed, spirulina	0.8
Radish seeds, sprouted	0.7
Purslane	0.4
Broccoli	0.1
Legumes	
Soybeans	1.6
Beans, common	0.6
Cowpeas	0.3
Peas, garden	0.2
Grains	
Oats, germ	1.4
Wheat, germ	0.7
Corn, germ	0.3
Fruit	
Kiwi fruit seed	0.6
Raspberries	0.1
Strawberries	0.1

9.5.2 Eicosapentaenoic Acid (EPA) and Docosahexaenoic Acid (DHA)

DHA is the chief component of the human brain and retina. EPA and DHA are present in fatty cold water fish and marine products and algae. Wide range of fish and fish oil such as salmon, herring, mackerel, anchovies, menhaden, and sardines are rich source of EPA and DHA as listed in Table 9.4. These fish contain two- to five-fold greater amount of DHA than EPA [8, 117].

Another source of omega-3 fatty acids are seafood products like crustaceans and mollusks. Lobsters, crab, shrimp, and krill are the major crustaceans containing sufficient amount of EPA and DHA [117]. Similarly, major mollusks like mussels, oyster, clam, and octopus are also good source of EPA and DHA. These fish and other marine organisms gain most of the DHA from phytoplankton and algae (green, red, and brown species) as they move up the food chain [118].

Table 9.4 DHA and EPA contents in various food sources [116]

Source (100 g portion)	EPA + DHA(g)
Fish	
Salmon, Atlantic, farmed, cooked, dry heat	2.147
Herring, Atlantic, cooked, dry heat	2.014
Mackerel, Pacific, cooked, dry heat	1.848
Anchovy, European	1.449
Sardine, Atlantic	0.982
Trout, cooked, dry heat	0.936
Tuna, cooked, dry heat	0.328
Crustaceans	
Spiny lobster, mixed species, cooked, moist heat	0.480
Crab, Alaska king, cooked, moist heat	0.413
Shrimp, mixed species, cooked, moist heat	0.315
Mollusks	
Mussel, blue, cooked, moist heat	0.782
Oyster, eastern, farmed, cooked, dry heat	0.440
Octopus, common, cooked, moist heat	0.314
Clam, mixed species, cooked, moist heat	0.284
Conch, baked or broiled	0.120
Microalgae	
Thraustochytrium aureum	6–7 (DHA)
Phaeodactylum tricornutum	3 (EPA)
Crypthecodinium cohnii	5 (DHA)
Ulkenia	5 (DHA)
Schizochytrium limacinum	15 (DHA)
Nannochloropsis sp.	4 (EPA)
Porphyridium cruentum	3 (EPA)

Higher amounts of PUFAs with predominant omega-3 fatty acids are present in algae relative to the total lipid content [118, 119]. Many auto- and heterotrophic marine microalgal species of different classes like Chrysophyceae and Bacillariophyceae (diatoms) produce both EPA and DHA. Prasinophyceae, Cryptophyceae, Xanthophyceae, Rhodophyceae, Eustigmatophyceae, and Glaucophyceae classes are EPA-rich sources, whereas Prymnesiophyceae, Dinophyceae, and Euglenophyceae classes have significant amount of DHA. Microalgal species *Crypthecodinium*, *Schizochytrium*, *Ulkenia*, and *Thraustochytrium* contain high concentration of DHA, while alga *Nannochloropsis*, *Chlorella*, *Phaeodactylum tricornutum*, and *Monodus* are rich source of EPA [120–123]. *Crypthecodinium cohnii*, heterotrophic algae, is used in many infant formulas to enrich with DHA. *Thraustochytrium aureum* and *Phaeodactylum tricornutum* contain approximately 50% DHA and 30–40% EPA of the total fatty acids, respectively, under optimal conditions [124, 125]. Thus, these algae are being produced commercially to manufacture functional foods. *Pavlova lutheri* microalga contains both EPA and DHA in high concentration and, thus, is a valuable for commercial purpose [126].

Algal oil containing DHA is produced from *Schizochytrium limacinum* and other algal species by fermentation process. Dry cell weight of *S. limacinum* can comprise approximately 30% DHA of the total fatty acids and can produce more 11 g of DHA for every liter of media [127].

9.6 Omega-3 Fatty Acids as Functional Food

Seeing the overall health benefits of omega-3 fatty acids, modern food and beverage products are available in market with added omega-3 fatty acids [128]. The US Food and Drug Administration (FDA) recommends omega-3-enriched fish and algal oils to be used for food fortification and declared as "generally recognized as safe" (GRAS) [129]. Food technologists around the world are developing omega-3 fatty acids rich functional food in form of fat-free mayonnaise, low fat cottage cheese with flax seed oil, and salad dressing using macadamia nut oil vinaigrette, milk-based products, breakfast cereals, table spreads, juices, sauces, infant formulas, and baby foods [129–131]. Omega-3 fatty acid-rich functional food is now being prepared by adding fatty seafood and flaxseeds in form of granola, muffins, and bean dip. The Department of Defense of various countries have taken initiative to provide fortified foods like omega-3-enriched chicken, eggs, salad dressing with soybean and flaxseed oil, and fruit beverages for war fighter which improve cognitive functions and provide protection from diseases and speedy wound healing [129, 132].

Omega-3 fatty acids are added to animal feed by biotechnology- and biodelivery-enabled processes to make animal's tissues enriched with such fatty acids especially in eggs, hen, pork, and beef products [132, 133]. Biotechnologists are developing cultivation of high ALA, EPA, or DHA yielding crops, yet their market version is very little [134].

Dietary recommendations of food agencies in various countries stress upon the balance of omega-3 and omega-6 in food items. The prime aim of such recommendation is lowering dietary omega-6 fatty acids in contrast with increasing level of omega-3 fatty acids. Nowadays, the researchers have developed new variants of foods and oils like peanuts, peanut butter and peanut, safflower, and sunflower oils with less omega-6 and more monounsaturated oleic acid to increase the omega balance to make more positive scores [135]. In the animal product section, the foods with positive balance scores are added in recipes. Scallops (balance score: +15), shrimp (+29), crab (+30), flounder (+30), mussels (+32), white tuna (+46), sea bass (+49), sardines (+53), striped bass (+55), Coho salmon (+57), Eastern oysters (+57), kippered herring (+70), and canned pink salmon (+83) are examples of such foods. Marine plants are becoming available to consumers as they are good source of omega-3 fatty acids which imbibe with traditional food to increase omega balance score [135, 136].

9.7 Dietary Recommendations for Omega-3 Fatty Acids

In the United States, Recommended Dietary Allowances (RDAs) suggest amount of omega-3 fatty acids for different gender, age, and groups in their official publication of "Acceptable Macronutrient Distribution Ranges" (AMDRs). The FDA recommends total of 3 g per day of combined DHA and EPA can be consumed by adults with not more than 2 g from dietary supplements. These recommendations are mentioned further with the help of Table 9.5 [138].

Similarly, the American Heart Association (AHA) recommends (as mentioned in Table 9.6) dietary consumption of two servings (about 8 ounces) of oily fish or seafood per week with a warning of contamination of fish with traces of mercury which has its toxic effect especially for pregnant and lactating women (up to 12 ounces) [139]. Shrimp, canned light tuna, salmon, pollock, and catfish are the major source of intakes which are less contaminated by mercury [140].

The National Cholesterol Education Program Adult Treatment Panel III (NCEP ATP-III) recommends supplement dose of 2–4 g of EPA and DHA per day for persons with high blood triglycerides [141]. It is noted that omega-3 fatty acids from seafoods or fish oil supplements help in reducing risk of coronary events. The present statement also embeds a warning that relatively high intakes (1–2 g/day) of

Table 9.5 ALA, EPA, and DHA dietary intake for different age and gender [137]

Life age	Stage	Omega-3 fatty acids	Males (g/day)	Females (g/day)
Infants	0–6 months	ALA, EPA, DHA (mother milk)	0.5	0.5
Infants	7–12 months	ALA, EPA, DHA	0.5	0.5
Children	1–3 years	ALA	0.7	0.7
Children	4–8 years	ALA	0.9	0.9
Children	9–13 years	ALA	1.2	1.0
Adolescents	14–18 years	ALA	1.6	1.1
Adults	>19 years	ALA	1.6	1.1
Pregnancy	All ages	ALA		1.4
Breast-feeding	All ages	ALA		1.3

Table 9.6 Guidelines for EPA and DHA intake by different organizations

Organization	Recommendation
American Heart Association	0.5–1.0 g/day
British Nutrition Foundation Task Force	1.0–1.5 g/day
UK Department of Health	0.2 g/day
World Health Organization	0.7 g/day
Institutes of Medicine Dietary Reference Intakes	0.11–0.16 g/day

omega-3 fatty acids for either primary or secondary prevention of congenital heart disease (CHD) should be properly studied and clinically trialed before its intake [141]. On the other hand, scientists impress upon the balance of omega-6 and omega-3 fatty acids for maximizing the benefits of these fats [142–144].

Like in the United States, the European Commission and the European Food Safety Authority (EFSA) recommend "adequate intakes" (AI) for the essential fatty acids LA and ALA and omega-3 fatty acids EPA and DHA. An AI is proposed by intake of 4% of whole energy and an ALA intake of 0.5% of whole energy of dietary intake. An AI of 250 mg/day is suggested for both EPA and DHA combined [145].

The World Health Organization (WHO) recommends (Table 9.6) an "Acceptable Macronutrient Distribution Range" (AMDR) for omega-3 fatty acid dietary intake of 0.5–2% of total energy component of intake [146]. Their AMDR for EPA plus DHA is decided as 0.250–2 g/day for primary and secondary prevention of CHD, respectively [147].

9.8 Conclusion

Omega-3 fatty acids play important roles in the modulation and prevention of human diseases especially mental and coronary heart diseases. Certainly, the evidence is now strong that omega-3 fatty acids are beneficial in human development not only in infancy but throughout life. As omega-3 fatty acids are essential food, various food products are fortified with omega-3 fatty acids to improve the nutritional profile. Scientists and food technologists continued their efforts to extract benefits of omega-3 fatty acids as functional food. Various food and health organizations around the globe have recommended daily reference intake for omega-3 fatty acids for all groups of age, both genders and motherhood.

References

1. Chan EJ, Leslie C (2009) What can we expect from omega-3 fatty acids? Cleve Clin J Med 76(4):245–251
2. Anderson BM, Ma DWL (2009) Are all n-3 polyunsaturated fatty acids created equal? Lipids Health Dis 8:33–39
3. Doughman SD, Krupanidhi S, Carani BS (2007) Omega-3 fatty acids for nutrition and medicine: considering microalgae oil as a vegetarian source of EPA and DHA. Curr Diabetes Rev 3(3):198–203
4. Fodor JG, Helis E, Yazdekhasti N et al (2014) "Fishing" for the origins of the "Eskimos and Heart Disease" story: facts or wishful thinking? Can J Cardiol 30(8):864–868
5. Maillard V, Bougnoux P, Ferrari P et al (2002) N-3 and N-6 fatty acids in breast adipose tissue and relative risk of breast cancer in a case-control study in tours, France. Int J Cancer 98(1):78–83
6. Ferrucci L, Cherubini A, Bandinelli S et al (2006) Relationship of plasma polyunsaturated fatty acids to circulating inflammatory markers. J Clin Endocrinol Metab 91(2):439–446
7. FAO/WHO Fats and fatty acids in human nutrition. Report of an expert consultation. FAO Food and Nutrition Paper 91 Rome 2011. ISSN:0254-4725

8. Garg ML, Wood LG, Singh H et al (2006) Means of delivering recommended levels of long chain n-3 polyunsaturated fatty acids in human diets. J Food Sci 71(5):66–71
9. Zaciragic A, Huskic J, Hadzovic-Dzuvo A et al (2007) Serum C-reactive protein concentration and measures of adiposity in patients with type 2 diabetes mellitus. Bosn J Basic Med Sci 7(4):322–327
10. Gogus U, Smith C (2010) n-3 Omega fatty acids: a review of current knowledge. Int J Food Sci Technol 45:417–436
11. Simopoulos AP (2002) The importance of the ratio of omega-6/omega-3 essential fatty acids. Biomed Pharmacother 56(8):365–379
12. De Lorgeril M, Renaud S, Mamelle N et al (1994) Mediterranean alpha-linolenic acid-rich diet in secondary prevention of coronary heart disease. Lancet 343(8911):1454–1459
13. Raheja BS, Sadikot SM, Phatak RB et al (1993) Significance of the N-6/N-3 ratio for insulin action in diabetes. Ann N Y Acad Sci 683:258–271
14. Portolesi R, Powell BC, Gibson RA (2007) Competition between 24:5n-3 and ALA for Delta 6 desaturase may limit the accumulation of DHA in HepG2 cell membranes. J Lipid Res 48(7):1592–1598
15. Cleland LG, James MJ, Neumann MA et al (1992) Linoleate inhibits EPA incorporation from dietary fish-oil supplements in human subjects. Am J Clin Nutr 55(2):395–399
16. Calder PC (2010) Omega-3 fatty acids and inflammatory processes. Nutrients 2:355–374
17. Tapiero H, Ba GN, Couvreur P et al (2002) Polyunsaturated fatty acids (PUFA) and eicosanoids in human health and pathologies. Biomed Pharmacother 56:215–222
18. Calder PC (2009) Polyunsaturated fatty acids and inflammatory processes: new twists in an old tale. Biochimie 1(6):791–795
19. Li D, Moorman R, Vanhercke T et al (2016) Classification and substrate head-group specificity of membrane fatty acid desaturases. Comput Struct Biotechnol J 14:341–349
20. Liang X, Nazarian A, Erdjument-Bromage H (2001) Heterogeneous fatty acylation of Src family kinases with polyunsaturated fatty acids regulates raft localization and signal transduction. J Biol Chem 276(33):30987–30994
21. Carughi A, Huynh LT, Perelman D (2010) Effect of omega-3 fatty acid supplementation on indicators of membrane fluidity. FASEB J (Meeting Abstract Supplement) 939:12
22. Siscovick DS, Raghunathan TE, King I (1995) Dietary intake and cell membrane levels of long-chain n-3 polyunsaturated fatty acids and the risk of primary cardiac arrest. JAMA 274:1363–1367
23. Albert CM, Campos H, Stampfer MJ (2002) Blood levels of long-chain n-3 fatty acids and the risk of sudden death. N Engl J Med 346:1113–1118
24. Bicknell IR, Darrow R, Barsalou L et al (2002) Alterations in retinal rod outer segment fatty acids and light-damage susceptibility in P23H rats. Mol Vis 8(34):333–340
25. Das UN (1999) Essential fatty acids, lipid peroxidation and apoptosis. Prostaglandins Leukot Essent Fatty Acids 61:157–163
26. Yap SC, Choo YM, Hew NF et al (1995) Oxidative susceptibility of low density lipoprotein from rabbits fed atherogenic diets containing coconut, palm, or soybean oils. Lipids 30:1145–1150
27. Musiek ES, Brooks JD, Joo M et al (2008) Electrophilic cyclopentenone neuroprostanes are anti-inflammatory mediators formed from the peroxidation of the −3 polyunsaturated fatty acid docosahexaenoic acid. J Biol Chem 283(29):19927–19935
28. Komatsu W, Ishihara K, Murata M et al (2003) Docosahexaenoic acid suppresses nitric oxide production and inducible nitric oxide synthase expression in interferon-γ plus lipopolysaccharide-stimulated murine macrophages by inhibiting the oxidative stress. Free Radic Biol Med 34:1006–1016
29. Karin M, Ben-Neriah Y (2000) Phosphorylation meets ubiquitination: the control of NF-κB activity. Annu Rev Immunol 18:621–663
30. Pogozheva AV, Martynova EA, Samsonov MA (1994) Dietary effects of PUFA omega-3 on lipid peroxidation and antioxidant system in patients with IHD, hyperlipoproteinemia and hypertension. Vopr Pitan 4:40–42

31. Simopoulos AP (1991) Omega-3 fatty acids in health and disease and in growth and development. Am J Clin Nutr 54:438–463
32. Li H, Ruan XZ, Powis SH et al (2005) EPA and DHA reduce LPS-induced inflammation responses in HK-2 cells: evidence for a PPAR-gamma-dependent mechanism. Kidney Int 67(3):867–874
33. Bernardo A, Levi G, Minghetti L (2000) Role of the peroxisome proliferator-activated receptor-gamma (PPAR-gamma) and its natural ligand 15-deoxy-Delta12, 14-prostaglandin J2 in the regulation of microglial functions. Eur J Neurosci 12(7):2215–2223
34. Bogna G (2014) Peroxisome proliferator-activated receptors and their ligands: nutritional and clinical implications – a review. Nutr J 13:17
35. Jump DB, Tripathy S, Depner CM (2013) Fatty acid–regulated transcription factors in the liver. Annu Rev Nutr 33:249–269
36. Rudkowska I, Garenc C, Couture P et al (2009) Omega-3 fatty acids regulate gene expression levels differently in subjects carrying the PPARa L162V polymorphism. Genes Nutr 4:199–205
37. Danner M, Kasl SV, Abramson JL et al (2003) Association between depression and elevated C-reactive protein. Psychosom Med 65(3):347–356
38. Song C, Li X, Leonard BE et al (2003) Effects of dietary n-3 or n-6 fatty acids on interleukin-1beta-induced anxiety, stress, and inflammatory responses in rats. J Lipid Res 44(10):1984–1991
39. James MJ, Gibson RA, Cleland LG (2000) Dietary polyunsaturated fatty acids and inflammatory mediator production. Am J Clin Nutr 71:343S–348S
40. Caughey GE, Mantzioris E, Gibson RA et al (1996) The effect on human tumor necrosis factor alpha and interleukin 1 beta production of diets enriched in n-3 fatty acids from vegetable oil or fish oil. Am J Clin Nutr 63(1):116–122
41. Phillips T, Childs AC, Dreon DM et al (2003) A dietary supplement attenuates IL-6 and CRP after eccentric exercise in untrained males. Med Sci Sports Exerc 35(12):2032–2037
42. Venkatraman JT, Chu WC (1999) Effects of dietary omega-3 and omega-6 lipids and vitamin E on serum cytokines, lipid mediators and anti-DNA antibodies in a mouse model for rheumatoid arthritis. J Am Coll Nutr 18(6):602–613
43. Norris JM, Yin X, Lamb MM et al (2007) Omega-3 polyunsaturated fatty acid intake and islet autoimmunity in children at increased risk for type 1 diabetes. JAMA: J Am Med Assoc 298(12):1420–1428
44. Rizos EC, Ntzani EE, Bika E, Kostapanos MS et al (2012) Association between Omega-3 fatty acid supplementation and risk of major cardiovascular disease events a systematic review and meta-analysis. JAMA 308(10):1024–1033
45. Kwak SM, Myung SK, Lee YJ et al (2012) Efficacy of omega-3 fatty acid supplements (eicosapentaenoic acid and docosahexaenoic acid) in the secondary prevention of cardiovascular disease: a meta-analysis of randomized, double-blind, placebo-controlled trials. Arch Int Med 172(9):686–694
46. Billman GE (2013) The effects of omega-3 polyunsaturated fatty acids on cardiac rhythm: a critical reassessment. Pharmacol Ther 140(1):53–80
47. Miller PE, Van Elswyk M, Alexander DD (2014) Long-chain omega-3 fatty acids eicosapentaenoic acid and docosahexaenoic acid and blood pressure: a meta-analysis of randomized controlled trials. Am J Hypertens 27(7):885–896
48. Geleijnse JM, Giltay EJ, Grobbee DE et al (2002) Blood pressure response to fish oil supplementation: metaregression analysis of randomized trials. J Hypertens 20(8):1493–1499
49. Morris MC, Sacks F, Rosner B (1993) Does fish oil lower blood pressure? A meta-analysis of controlled trials. Circulation 88(2):523–353
50. Mori TA, Bao DQ, Burke V, Puddey IB, Beilin LJ (1993) Docosahexaenoic acid but not eicosapentaenoic acid lowers ambulatory blood pressure and heart rate in humans. Hypertension 34(2):253–260
51. Harris WS, Connor WE, Illingworth DR et al (1990) Effects of fish oil on VLDL triglyceride kinetics in humans. J Lipid Res 31(9):1549–1558

52. Nestel PJ, Connor WE, Reardon MF et al (1984) Suppression by diets rich in fish oil of very low density lipoprotein production in man. J Clin Invest 74(1):82–89
53. Ebrahimi M, Ghayour-Mobarhan M, Rezaiean S et al (2009) Omega-3 fatty acid supplements improve the cardiovascular risk profile of subjects with metabolic syndrome, including markers of inflammation and auto-immunity. Acta Cardiol 64(3):321–327
54. Wall R, Ross RP, Fitzgerald GF et al (2010) Fatty acids from fish: the anti-inflammatory potential of long-chain omega-3 fatty acids. Nutr Rev 68(5):280–289
55. Miles EA, Calder PC (2012) Influence of marine n-3 polyunsaturated fatty acids on immune function and a systematic review of their effects on clinical outcomes in rheumatoid arthritis. Br J Nutr 107(Suppl 2):S171–S184
56. Robinson LE, Mazurak VC (2013) n-3 Polyunsaturated fatty acids: relationship to inflammation in health adults and adults exhibiting features of metabolic syndrome. Lipids 48(4):319–332
57. Li K, Huang T, Zheng J et al (2014) Effect of marine-derived n-3 polyunsaturated fatty acids on C-reactive protein, interleukin 6 and tumor necrosis factor α: a meta-analysis. PLoS One 9(2):e88103
58. Farmer A, Montori V, Dinneen S et al (2001) Fish oil in people with type 2 diabetes mellitus. Cochrane Database Syst Rev 3:CD003205
59. MacLean CH, Mojica WA, Morton SC et al (2004) Effects of omega-3 fatty acids on lipids and glycemic control in type II diabetes and the metabolic syndrome and on inflammatory bowel disease, rheumatoid arthritis, renal disease, systemic lupus erythematosus, and osteoporosis. Evid Rep Technol Assess 89:1–4
60. Hartweg J, Farmer AJ, Perera R et al (2007) Metaanalysis of the effects of n-3 polyunsaturated fatty acids on lipoproteins and other emerging lipid cardiovascular risk markers in patients with type 2 diabetes. Diabetologia 50(8):1593–1602
61. Hu FB, Cho E, Rexrode KM et al (2003) Fish and long-chain omega-3 fatty acid intake and risk of coronary heart disease and total mortality in diabetic women. Circulation 107(14):1852–1857
62. Nettleton JA, Katz R (2005) n-3 Long-chain polyunsaturated fatty acids in type 2 diabetes: a review. J Am Diet Assoc 105(3):428–440
63. Li J, Huang CJ, Xie D (2008) Anti-obesity effects of conjugated linoleic acid, docosahexaenoic acid, and eicosapentaenoic acid. Mol Nutr Food Res 52(6):631–645
64. Weintraub MS, Zechner R, Brown A et al (1988) Dietary polyunsaturated fats of the W-6 and W-3 series reduce postprandial lipoprotein levels. Chronic and acute effects of fat saturation on postprandial lipoprotein metabolism. J Clin Invest 82(6):1884–1893
65. Woodman RJ, Mori TA, Burke V et al (2002) Effects of purified eicosapentaenoic and docosahexaenoic acids on glycemic control, blood pressure, and serum lipids in type 2 diabetic patients with treated hypertension. Am J Clin Nutr 76(5):1007–1015
66. Gingras A, White PJ, Chouinard PY et al (2007) Long-chain omega-3 fatty acids regulate bovine whole-body protein metabolism by promoting muscle insulin signalling to the AktmTOR- S6K1 pathway and insulin sensitivity. J Physiol 579:269–284
67. Metter EJ, Talbot LA, Schrager M et al (2002) Skeletal muscle strength as a predictor of all-cause mortality in healthy men. J Gerontol A Biol Sci Med Sci 57(10):B359–B365
68. Yehuda S, Rabinovitz S, Mostofsky DI (1999) Essential fatty acids are mediators of brain biochemistry and cognitive functions. J Neurosci Res 56(6):565–570
69. Wurtman RJ, Ulus IH, Cansev M et al (2006) Synaptic proteins and phospholipids are increased in gerbil brain by administering uridine plus docosahexaenoic acid orally. Brain Res 1088(1):83–92
70. Innis SM (2007) Dietary (n-3) fatty acids and brain development. J Nutr 137(4):855–859
71. Georgieff MK, Innis SM (2005) Controversial nutrients that potentially affect preterm neurodevelopment: essential fatty acids and iron. Pediatr Res 57:99R–103R
72. Ahmad A, Moriguchi T, Salem N (2002) Decrease in neuron size in docosahexaenoic acid-deficient brain. Pediatr Neurol 26(3):210–218

73. Das UN, Fams (2003) Long-chain polyunsaturated fatty acids in the growth and development of the brain and memory. Nutrition (Burbank, Los Angeles County, Calif.) 19(1):62–65
74. Lauritzen I, Blondeau N, Heurteaux C et al (2000) Polyunsaturated fatty acids are potent neuroprotectors. EMBO J 19(8):1784–1793
75. Kim HY, Akbar M, Lau A et al (2000) Inhibition of neuronal apoptosis by docosahexaenoic acid (22:6n-3). Role of phosphatidylserine in antiapoptotic effect. J Biol Chem 275(45):35215–35223
76. Giedd JN, Blumenthal J, Jeffries NO et al (1999) Brain development during childhood and adolescence: a longitudinal MRI study. Nat Neurosci 2(10):861–863
77. Innis SM (2005) Essential fatty acid transfer and fetal development. Placenta 26(Suppl A):S70–S75
78. Dunstan JA, Simmer K, Dixon G et al (2008) Cognitive assessment of children at age 2(1/2) years after maternal fish oil supplementation in pregnancy: a randomised controlled trial. Arch Dis Child Fetal Neonatal 93(1):F45–F50
79. Hibbeln JR, Davis JM, Steer C et al (2007) Maternal seafood consumption in pregnancy and neurodevelopmental outcomes in childhood (ALSPAC study): an observational cohort study. Lancet 369(9561):578–585
80. He K, Song Y, Daviglus ML et al (2004) Fish consumption and incidence of stroke: a meta-analysis of cohort studies. Stroke; J Cereb Circ 35(7):1538–1542
81. Lukiw WJ, Cui JG, Marcheselli VL et al (2005) A role for docosahexaenoic acid-derived neuroprotectin D1 in neural cell survival and Alzheimer disease. J Clin Invest 115(10):2774–2783
82. Schaefer EJ, Bongard V, Beiser AS et al (2006) Plasma phosphatidylcholine docosahexaenoic acid content and risk of dementia and Alzheimer disease: the Framingham heart study. Arch Neurol 63(11):1545–1550
83. Nigam D (2015) Chapter 11: Role of free radicals and oxidative stress in neurodegenerative disorders. In: Rani V, Yadav UCS (eds) Free radicals in human health and disease. Springer, New York, pp 143–158
84. Samadi P, Gregoire L, Rouillard C et al (2006) Docosahexaenoic acid reduces levodopa-induced dyskinesias in 1-methyl-4-phenyl-1,2,3,6-tetrahydropyridine monkeys. Ann Neurol 59(2):282–288
85. Julien C, Berthiaume L, Hadj-Tahar A et al (2006) Postmortem brain fatty acid profile of levodopa-treated Parkinson disease patients and parkinsonian monkeys. Neurochem Int 48(5):404–414
86. Tandon R, Nasrallah HA, Keshavan MS (2009) Schizophrenia, "just the facts" 4. Clinical features and conceptualization. Schizophr Res 110(1–3):1–23
87. Peet M, Brind J, Ramchand CN et al (2001) Two double blind placebo-controlled pilot studies of eicosapentaenoic acid in the treatment of schizophrenia. Schizophr Res 49(3):243–251
88. Tanskanen A, Hibbeln JR, Tuomilehto J et al (2001) Fish consumption and depressive symptoms in the general population in Finland. Psychiatr Serv (Washington, DC) 52(4):529–531
89. Noaghiul S, Hibbeln JR (2003) Cross-national comparisons of seafood consumption and rates of bipolar disorders. Am J Psychiatry 160(12):2222–2227
90. Osher Y, Bersudsky Y, Belmaker RH (2005) Omega-3 eicosapentaenoic acid in bipolar depression: report of a small open-label study. J Clin Psychiatry 66(6):726–729
91. Logan AC (2004) Omega-3 fatty acids and major depression: a primer for the mental health professional. Lipids Health Dis 3:25–33
92. Trull TJ, Verges A, Wood PK et al (2012) The structure of diagnostic and statistical manual of mental disorders (4_{th} edition, text revision) personality disorder symptoms in a large national sample. Pers Disord 3(4):355–369
93. Mitchell EA, Aman MG, Turbott SH et al (1987) Clinical characteristics and serum essential fatty acid levels in hyperactive children. Clin Pediatr 26(8):406–441
94. Richardson AJ, Puri BK (2002) A randomized double-blind, placebo-controlled study of the effects of supplementation with highly unsaturated fatty acids on ADHD-related symptoms in children with specific learning difficulties. Prog Neuro-Psychopharmacol Biol Psychiatry 26(2):233–239

95. Joshi K, Lad S, Kale M et al (2006) Supplementation with flax oil and vitamin C improves the outcome of attention deficit hyperactivity disorder (ADHD). Prostaglandins Leukot Essent Fatty Acids 74(1):17–21
96. Richardson AJ, Montgomery P (2005) The Oxford-Durham study: a randomized, controlled trial of dietary supplementation with fatty acids in children with developmental coordination disorder. Pediatric 115(5):1360–1366
97. Bell JG, MacKinlay EE, Dick JR et al (2004) Essential fatty acids and phospholipase A2 in autistic spectrum disorders. Prostaglandins Leukot Essent Fatty Acids 71(4):201–204
98. Meguid NA, Atta HM, Gouda AS et al (2008) Role of polyunsaturated fatty acids in the management of Egyptian children with autism. Clin Biochem 41(13):1044–1048
99. Amminger GP, Berger GE, Schafer MR et al (2007) Omega-3 fatty acids supplementation in children with autism: a double blind randomized, placebo-controlled pilot study. Biol Psychiatry 61(4):551–553
100. Meiri G, Bichovsky Y, Belmaker RH (2009) Omega 3 fatty acid treatment in autism. J Child Adolesc Psychopharmacol 19(4):449–451
101. Bazan NG (2006) Cell survival matters: docosahexaenoic acid signaling, neuroprotection and photoreceptors. Trends Neurosci 29(5):263–271
102. Wu J, Cho E, Edward L et al (2017) Dietary intakes of eicosapentaenoic acid and docosahexaenoic acid and risk of age-related macular degeneration. Opthalmology 124(5):634–643
103. Birch EE, Castaneda YS, Wheaton DH et al (2005) Visual maturation of term infants fed long chain polyunsaturated fatty acid-supplemented or control formula for 12 mo. Am J Clin Nutr 81(4):871–879
104. SanGiovanni JP, Parra-Cabrera S, Colditz GA et al (2000) Meta-analysis of dietary essential fatty acids and long-chain polyunsaturated fatty acids as they relate to visual resolution acuity in healthy preterm infants. Pediatric 105(6):1292–1298
105. Abu JI, Konje JC (2000) Leukotrienes in gynaecology: the hypothetical value of anti-leukotriene therapy in dysmenorrhoea and endometriosis. Hum Reprod Update 6(2):200–205
106. Netsu S, Konno R, Odagiri K et al (2008) Oral eicosapentaenoic acid supplementation as possible therapy for endometriosis. Fertil Steril 90:1496–1502
107. Fujiwara H, Konno R, Netsu S et al (2004) Localization of mast cells in endometrial cysts. Am J Reprod Immunol (New York, NY: 1989) 51(5):341–344
108. Berquin IM, Edwards IJ, Chen YQ (2008) Multi–targeted therapy of cancer by omega –3 fatty acids. Cancer Lett 269(2):363–377
109. Donat-Vargas C, Berglund M, Glynn A et al (2017) Dietary polychlorinated biphenyls, long-chain n-3 polyunsaturated fatty acids and incidence of malignant melanoma. Eur J Cancer 72:137–143
110. Bosetti C, Negri E, Franceschi S et al (2001) Diet and ovarian cancer risk: a case-control study in Italy. Int J Cancer 93(6):911–915
111. Chavarro JE, Stampfer MJ, Li H et al (2007) A prospective study of polyunsaturated fatty acid levels in blood and prostate cancer risk. cancer epidemiology, biomarkers & prevention: a publication of the American Association for Cancer Research. Cosponsored by Am Soc Prevent Oncol 16(7):1364–1370
112. Boelsma E, Hendriks HFJ, Roza L (2001) Nutritional skin care: health effects of micronutrients and fatty acids. Am J Clin Nutr 73(5):853–864
113. Pilkington SM, Rhodes LE (2010) Omega-3 fatty acids and skin. In: Krutmann J, Humbert P (eds) Nutrition for healthy skin. Springer, New York, pp 91–107
114. Zevenbergen HA, de Bree M, Zeelenberg K et al (2009) Foods with a high fat quality are essential for healthy diets. Ann Nutr Metab 54(1):15–24
115. Rubio-Rodrguez N, Beltrn S, Jaime I et al (2010) Production of omega-3 polyunsaturated fatty acid concentrates: a review. Innovative Food Sci Emerg Technol 11:1–12
116. Kris-Etherton PM, Harris WS, Appel LJ (2003) Fish consumption, fish oil, omega-3 fatty acids, and cardiovascular disease. Arterioscler Thromb Vasc Biol 23:20–30
117. Tur JA, Bibiloni MDM, Sureda A et al (2012) Dietary sources of omega 3 fatty acids: public health risks and benefits. British J Nutr 107(S2):S23–S52

118. Plaza M, Cifuentes A, Ibanez E (2008) In the search of new functional food ingredients from algae. Trends Food Sci Technol 19:31–39
119. Sanchez-Machado DI, Lopez-Hernandez J, Paseiro-Losada P et al (2004) An HPLC method for quantification of sterols in edible seaweeds. Biomed Chrom 18(3):183–190
120. Yongmanitchai W, Ward OP (1991) Growth of and omega-3 fatty acid production by Phaeodactylum tricornutum under different culture conditions. Appl Environ Microbiol 57(2):419–425
121. Lemahieua C, Bruneela C, Termote-Verhalle R et al (2013) Impact of feed supplementation with different omega-3 rich microalgae species on enrichment of eggs of laying hens. Food Chem 141(4):4051–4059
122. Burja AM, Radianingtyas H, Windust A et al (2006) Isolation and characterization of polyunsaturated fatty acid producing thraustochytrium species: screening of strains and optimization of omega-3 production. Appl Microbiol Biotechnol 72(6):1161–1169
123. Zhu L, Zhang X, Ji L (2007) Changes of lipid content and fatty acid composition of schizochytrium limacinum in response to different temperatures and salinities. Process Biochem 42(2):210–214
124. Ward OP, Singh A (2005) Omega-3/6 fatty acids: alternative sources of production. Process Biochem 40(12):3627–3652
125. Hur B, Cho D, Kim H (2002) Effect of culture conditions on growth and production of docosahexaenoic acid (DHA) using Thraustochytrium aureum ATCC 34304. Biotechnol Bioprocess Eng 7(1):10–15
126. Carvalho AP, Malcata FX (2005) Optimization of omega-3 fatty acid production by microalgae: crossover effects of CO2 and light intensity under batch and continuous cultivation modes. Mar Biotechnol (NY) 7(4):381–388
127. Zhao B, Li Y, Mbifile MD et al (2017) Improvement of docosahexaenoic acid fermentation from Schizochytrium sp. AB-610 by staged pH control based on cell morphological changes. Engg Life Sci 17(9):981–988. https://doi.org/10.1002/elsc.201600249
128. Marriott BP, Yu K, Majchrzak-Hong S et al (2014) Understanding diet and modeling changes in the omega-3 and omega-6 fatty acid composition of U.S. Garrison Foods for active duty personnel. Mil Med 179:168–175
129. US Food and Drug Administration, Center for Food Safety and Applied Nutrition. Agency response letter: GRAS Notice No. GRN 000080 (2001) http://www.fda.gov/ohrms/dockets/dockets/95s0316/95s-0316-rpt0354-061-Ref-F-FDAResponse-Ltr-g80-vol273.pdf
130. Wood KE, Mantzioris E, Gibson RA et al (2013) Incorporating macadamia oil and butter to reduce dietary omega-6 polyunsaturated fatty acid intake. Nutr Diet 70:94–100
131. Larque E, Demmelmair H, Koletzko B (2002) Perinatal supply and metabolism of long-chain polyunsaturated fatty acids: importance for the early development of the nervous system. Ann N Y Acad Sci 967:299–310
132. Lands WE (2000) Commentary on the workshop statement essentiality of and recommended dietary intakes for omega-6 and omega-3 fatty acids. Prostaglandins Leukot Essent Fatty Acids 63:125–126
133. Howe PRC, Downing JA, Grenyer BFS et al (2002) Tuna fishmeal as a source of DHA for n-3 PUFA enrichment of pork, chicken, and eggs. Lipids 37:1067–1076
134. Wu G, Truksa M, Datla N et al (2005) Stepwise engineering to produce high yields of very long-chain polyunsaturated fatty acids in plants. Nat Biotechnol 23(8):1013–1017
135. Bibus D, Lands B (2015) Balancing proportions of competing omega-3 and omega-6 highly unsaturated fatty acids (HUFA) in tissue lipids. Prostaglandins Leukot Essent Fatty Acids 99:19–23
136. Feldstein AE, Mori TA, Barden A et al (2013) Targeted alteration of dietary n-3 and n-6 fatty acids for the treatment of chronic headaches: a randomized trial. Pain 154:2441–2451
137. Food and Nutrition Board, Institute of Medicine (2002) Dietary reference intakes: energy, carbohydrate, fiber, fat, fatty Acids, cholesterol, protein, and amino acids. Part 1 and 2. National Academy Press, Washington DC

138. Bi X, Siow PC, Lim SW, Henry CJ (2017) Dietary fatty acids open analysis and its relevance to human health. SM J Nut Metab 1(1):1005–1017
139. Lichtenstein AH, Appel LJ, Brands M et al (2006) Diet and lifestyle recommendations revision 2006: a scientific statement from the American Heart Association nutrition committee. Circulation 114:82–96
140. Food and Drug Administration and Environmental Protection Agency. FDA and EPA Announce the Revised Consumer Advisory on Methylmercury in Fish
141. National Cholesterol Education Program (NCEP) Expert Panel on detection, evaluation, and treatment of high blood cholesterol in adults (adult treatment panel III) final report (2002) Circulation 106(3):143–3421
142. Harris WS, Assaad B, Poston WC (2006) Tissue omega-6/omega-3 fatty acid ratio and risk for coronary artery disease. Am J Cardiol 98(4A):19i–26i
143. Harris WS (2006) The omega-6/omega-3 ratio and cardiovascular disease risk: uses and abuses. Curr Atheroscler Rep 8(6):453–459
144. Simopoulos AP (2004) Omega-6/omega-3 essential fatty acid ratio and chronic diseases. Food Rev Intl 20(1):77–90
145. European Food Safety Authority Panel on Dietetic Products, Nutrition, and Allergies (NDA) (2010) Scientific opinion on dietary reference values for fats, including saturated fatty acids, polyunsaturated fatty acids, monounsaturated fatty acids, trans fatty acids, and cholesterol. EFSA J 8(3):107
146. FAO/WHO (2008) Interim summary of conclusions and dietary recommendations on total fat & fatty acids. Joint FAO/WHO expert consultation on fats and fatty acids in human nutrition. WHO, Geneva, pp 1–14
147. International Society for the Study of Fatty Acids and Lipids. Recommendations for intake of polyunsaturated fatty acids in healthy adults. Available at: Intake of Polyunsaturated Fatty Acids in Healthy Adults. Available at: http://www.issfal.org/statements/pufa-recommendations

Part II

Functional Food and Human Health

Phytochemicals and Human Health

10

Krishnendu Sinha, Sayantani Chowdhury, and Parames C. Sil

Abstract

The World Health Organization (WHO) defined health as "a state of complete physical, mental, and social well-being and not merely the absence of disease or infirmity." Any disturbance in this well-being leads to ill-health and a related condition called pathophysiology. Disease conditions, xenobiotics, and environmental and social stresses are the most common causes behind these pathophysiological conditions, and this can be generalized from recent studies that in most of the cases ROS plays the pivotal role as the main effector. However, fortunately in many cases, these health problems are preventable. Reasonable cost, presence in the daily consumables, and negligible side effects make the naturally occurring plant-derived compounds interesting and attractive for pharmacological study in recent years. Primarily for the defense purpose, plants yield assorted types of low-molecular-weight products. These are generally termed as phytochemicals. Among them, a group of secondary metabolites associated with a polyphenolic group have been named flavonoids and are of pronounced interest due to their implausible pharmacological properties. Flavonoids are widely accepted as potent antioxidant agents which can prevent injury caused by free radicals by scavenging of ROS, activation of antioxidant enzymes, and inhibiting oxidases. In addition, increase in antioxidant properties of low-molecular antioxidants, metal chelating activity, and reduction of α-tocopheryl radicals and mitigation of oxidative stress caused by NO also plays important role. In this chapter, we have summarized most of the findings, if not all, available till date related to five very noticeable phytochemicals, namely, morin, quercetin, rutin,

K. Sinha
Department of Zoology, Jhargram Raj College, Jhargram, West Bengal, India

Division of Molecular Medicine, Bose Institute, Kolkata, West Bengal, India

S. Chowdhury · P. C. Sil (✉)
Division of Molecular Medicine, Bose Institute, Kolkata, West Bengal, India
e-mail: parames@jcbose.ac.in

V. Rani, U. C. S. Yadav (eds.), *Functional Food and Human Health*,
https://doi.org/10.1007/978-981-13-1123-9_10

mangiferin, and myricetin. Hope this chapter will help readers in understanding the utmost importance of the phytochemicals and will motivate them to further dig into the mechanistic study to fetch more novel information.

Keywords
Phytochemicals · Morin · Quercetin · Rutin · Mangiferin · Myricetin

10.1 Introduction

Rational cost, present mostly in consumables, and negligible side effects make the naturally occurring compounds fascinating and attractive for pharmacological study in recent years. Plants yield diverse types of low-molecular-weight products mainly for the defense purpose. Among them, a group of secondary metabolites associated with a polyphenolic group have been named flavonoids and are of pronounced interest due to their implausible pharmacological properties. The term flavonoid is derived from the Latin word "flavus," meaning yellow. These are phenolic substances show various biological activities like antiallergenic, anti-inflammatory, antiviral, gastroprotective, cardioprotective, renoprotective, neuroprotective, vasodilating actions, etc.[1]. Approximately, more than 3000 varieties of flavonoids have been recognized. The flavonoids consist of six major subgroups, namely, flavone, flavonol, flavanone, chalcone, anthocyanins, and isoflavonoids. Along with carotenes, flavonoids are responsible for the coloring of fruits, herbs, and vegetables. The most significant dietary sources are fruits, soybean, and tea, where green and black tea contains as much as about 25% flavonoids. Other important sources of flavonoids are citrus fruits (rutin and hesperidin), apple (quercetin), flowers, red wine, nuts, herbs, vegetables, fruits, seeds, spices, stems, etc.

The concept of oxidative stress or the imbalance between prooxidants and antioxidants in a living system has been comprehensively associated with the biomedical sciences since last two decades. Oxidative stress plays a significant role in the pathophysiology of highly prevalent diseases such as hypertension, diabetes, acute renal failure, atherosclerosis, Alzheimer's, Parkinson's diseases, etc. In typical physiological conditions, ROS are unceasingly produced and are excellently removed by several antioxidant defense systems (e.g., antioxidant proteins, enzymes, vitamins, etc.). However, an increased ROS levels in the cell have a considerable impact which leads to defective cellular functions, disease, and aging.

Flavonoids are widely accepted as potent antioxidant agents which can prevent injury caused by free radicals by scavenging of ROS, activation of antioxidant enzymes, and inhibiting oxidases. In addition, increase in antioxidant properties of low-molecular antioxidants, metal chelating activity, and reduction of α-tocopheryl radicals plays an important role.

This chapter aims to present a brief idea of the beneficial role of naturally occurring phytochemicals (Table 10.1) in relation to human health. It is believed that this will inspire readers and researchers in the field of applied pharmacology, ethnobotany, and other related fields of research. Here we would like to discuss the beneficial efficacy of several important flavonoids in the light of numerous up-to-date reports.

10.2 Morin

Morin [morin hydrate: 2-(2,4-dihydroxyphenyl)-3,5,7-trihydroxy-4H-chromen-4-one hydrate, 2′,3,4′,5,7-Pentahydroxyflavone] belongs to the group of flavonols found regularly in the branches of the family members of Moraceae like Osage orange (*Maclura pomifera*), white mulberry (*Morus alba* L.), fig (*Chlorophora tinctoria*), almond (*Psidium guajava*), mill (*Prunus dulcis*), old fustic (*Maclura tinctoria*), etc. [2, 3]. Morin exhibits different types of pharmacologically important properties like free radical scavenging activity, anti-inflammatory property, xanthine oxidase inhibitor property, gastroprotective property, hepatoprotective property, anticancer property, etc. It also possesses several add-on health benefits. Also, an accumulative number of studies showed that morin suggestively modulates different cell signaling pathways related to chronic pathophysiological conditions. We will discuss few of them in the following section.

Excitotoxicity (i.e., excessive glutamate receptors activation) leads to acute and chronic neurological disorders including stroke. In vitro model of excitotoxic neuronal death involving NMDA receptor over activation has already showed the neuroprotective role of morin [4]. In PC12 neuronal differentiated cells, Zhang et al. showed the neuroprotective role of morin on 1-methyl-4-phenylpyridinium ion-mediated apoptotic cell death as well as in an in vivo model (1-methyl-4-phenyl-1, 2, 3, 6-tetrahydropyridine (MPTP)-induced murine model of Parkinson's disease). They found that upon addition of morin in the system, there was a significant attenuation of the MPP + –induced loss of cell viability, apoptosis, and inhibition of ROS formation in the in vitro model, whereas morin significantly attenuated the MPTP-induced dopaminergic neuronal death, nigrostriatal lesions, striatal dopamine depletion, and permanent behavioral deficits in vivo [5]. In another study, Subash and Subramaniam evaluated the chronotherapeutic effect of morin on ammonium chloride (AC)-induced hyperammonemia using rat model. Ammonia is considered as a persuasive neurotoxin. It has been strongly associated in the pathogenesis of hepatic encephalopathy. In hyperammonemic rats, the chronotherapeutic role of the molecule was suggested due to the temporal variations of antioxidants, lipid peroxidation, urea cycle enzymes, metabolic enzymes involved in morin degradation, and the temporal variation in the bioavailability of morin [6].

Table 10.1 Depicts the chemical structure, nature, as well as the biological properties of morin, quercetin, rutin, mangiferin, and myricetin

Name of the molecule and structure	Molecular formula	Molar mass	IUPAC name	Solubility	Biochem/physiol actions
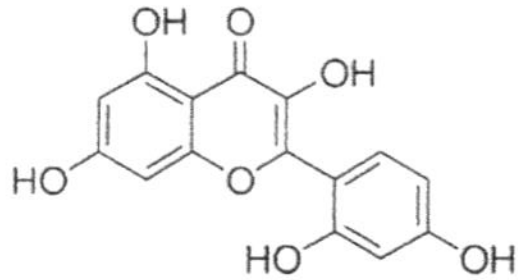	$C_{15}H_{12}O_8$. $_XH_2O$	302.24 g Mol^{-1}	2-(2,4-dihydroxyphenyl)-3,5,7-trihydroxychromen-4-one;hydrate	Methanol (50 mg/ml); water (0.25 mg/ml, 20 °C; 0.94 mg/ml, 100 °C); aqueous alkaline solutions; ether and acetic acid (sparingly soluble)	Free radical scavenging activity, anti-inflammatory property, xanthine oxidase inhibitor property, gastroprotective property, hepatoprotective property, anticancer property, etc.
Quercetin 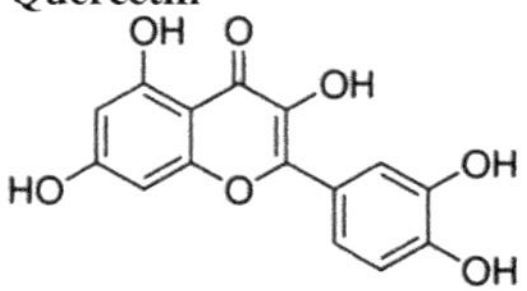	$C_{15}H_{10}O_7$	302.238 g Mol^{-1}	2-(3,4-dihydroxyphenyl)-3,5,7-trihydroxychromen-4-one	Very soluble in ether, methanol; soluble in ethanol, acetone, pyridine, acetic acid; soluble in alcohol and glacial acetic acid; in water, 60 mg/L at 16 °C	Cardiovascular protection, antiviral, anti-inflammatory activity, antitumor, anticancer, anti-ulcer, anti-allergy, antidiabetic, gastroprotective effects, antihypertensive, immunomodulatory, anti-infective, etc.

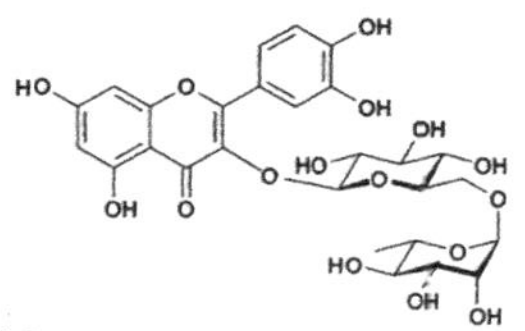	$C_{27}H_{30}O_{16}$	610.521 g Mol^{-1}	2-(3,4-dihydroxyphenyl)-5,7-dihydroxy-3-[(2S,3R,4S,5S,6R)-3,4,5-trihydroxy-6-[[(2R,3R,4R,5R,6S)-3,4,5-trihydroxy-6-methyloxan-2-yl]oxymethyl]oxan-2-yl]oxychromen-4-one	Pyridine solubility: 50 mg/mL; DMSO: Soluble; aqueous base: Soluble; water solubility 125 mg/L	Antimicrobial, antifungal, antiallergic properties, free radicals scavenging activity, antidiabetic, neuroprotective effects, etc.
Mangiferin 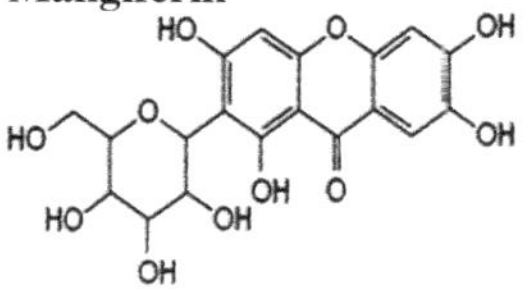	$C_{19}H_{18}O_{11}$	422.342 g Mol^{-1}	1,3,6,7-tetrahydroxy-2-[(2S,3R,4R,5S,6R)-3,4,5-trihydroxy-6-(hydroxymethyl)oxan-2-yl]xanthen-9-one	Solubility: 20 mg/mL DMSO	Antistimulatory, antimodulatory, antioxidative, antidiabetic, dyslipidemic, antiallergic, analgesic, anticancer, anti-HIV properties, etc.
Myricetin 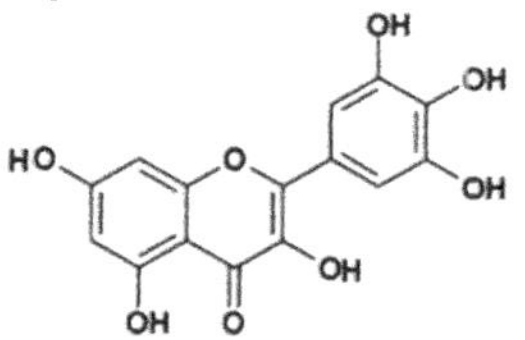	$C_{15}H_{10}O_8$	318.237 g Mol^{-1}	3,5,7-trihydroxy-2-(3,4,5-trihydroxyphenyl)-4H-1-benzopyran-4-one	Sparingly soluble in boiling water; soluble in alcohol	Anti-inflammatory, antioxidant, antidiabetic, anticancer, iron-chelating activities, etc.

A contemporary study was conducted to assess the antiarthritic effect of an ethanolic extract of *Ficus exasperata* (FEE) in an arthritis rat model in which the arthritis was induced by the Freund's adjuvant and dexamethasone and methotrexate were used as positive controls [7]. Along with these positive controls, FEE also showed significant antiarthritic properties by preventing the arthritic edema in the lateral paw of the animals and also the spread of the edema from the lateral to the contralateral paws. The principal active compound in the FEE is believed to be the morin. Zeng et al. also found morin effective against type II collagen-induced arthritis (CIA) in rats in terms of attenuating arthritic development specified by reduction of paw swelling and arthritis scores [5]. Sultana and Rasool ascertained the effectiveness of morin-NSAID combinatorial therapy in subduing the pathogenesis of rheumatoid arthritis (RA) in rats where they found that imbalances in the paw edema, levels/activities of elastase, inflammatory mediators (TNFα, IL1β, PGE2, VEGF, and MCP1), glycoproteins (hexosamine and hexose), urinary constituents (hydroxyproline and glycosaminoglycans), reactive oxygen species (LPO and NO), proinflammatory cytokines (IL1β, TNFα, IL17, MCP1, and IL6), inflammatory enzymes (iNOS and COX2), RANKL, and transcription factors AP1 and NFkB p65 were elevated in case of RA whereas regulated back effectually to the basal level by morin and indomethacin [8].

Morin also have efficient gastroprotective activity. In a study, we have shown that morin considerably ameliorates nonsteroidal anti-inflammatory drug (NSAID)-induced gastropathyin SD rats. We found that the gastroprotective action of morin is primarily accredited to its persuasive antioxidant and anti-inflammatory nature[1]. In another study, Galvez et al. reported that morin possesses intestinal anti-inflammatory activity. They found that colonic insult with trinitrobenzenesulfonic acid induces myeloperoxidase activity, colonic leukotriene B4 and IL-1β levels, oxidative stress, and colonic nitric oxide synthase activity, whereas morin effectively reduces the changes [9]. Morin hydrate also has been shown to constrain nitric oxide synthase activity and the leukotriene B4 synthesis [10].

Cardiovascular diseases (CVD) are the main cause of chief death worldwide due to its byzantine nature. Among CVD, myocardial infarction (MI) is a foremost one. Al-Numair et al. showed that morin protects cardiovascular system in isoproterenol-induced myocardial infarction by chiefly scavenging free radicals [11]. They suggest that morin supplement on a daily basis significantly decreases the activities of cardiac marker enzymes such as lactate dehydrogenase, aspartate transaminase, creatine kinase, and creatine kinase-MB in serum. They also showed that the activity of sodium-potassium-dependent adenosine triphosphatase was decreased, whereas calcium-dependent adenosine triphosphatase and magnesium-dependent adenosine triphosphatase were found to increase in the heart as well as the levels of glycoprotein containing hexose, hexosamine, fucose, and sialic acid decreased both in the heart and serum. Prahalathan et al. recently proved the protective effect of morin against deoxycorticosterone acetate (DOCA)-induced hypertension in male Wistar rats where they showed that the morin effectively lowered the increased systolic and diastolic blood pressure in association with considerably increased systolic and diastolic blood, ALT, GGT, AST, ALP, urea, uric acid, and creatinine levels in the plasma of hypertensive rats [12].

Morin also has significant effect on diabetes and related pathophysiology. Noor et al., from transmission electron microscopy (TEM) and right-angle light scattering, showed that morin hydrate inhibits amyloid formation by human islet amyloid polypeptide (IAPP, amylin), and not only that it even disaggregates preformed IAPP amyloid fibers [13]. IAPP is related to the formation of amyloid in islets in type 2 diabetes and in the transplantation of islet cell which in turn leads to graft failure. Human IAPP is extremely amyloidogenic and has fewer inhibitors, whereas specific substitution pattern on the B-ring makes morin hydrate a novel type of IAPP amyloid inhibitor [3, 13]. Vanitha et al. observed that morin administration resulted in significant reduction in blood glucose levels, an increase in the levels of serum insulin in type 1 diabetic experimental rats [14]. Morin dose-dependently ameliorated the altered levels of glucose-6-phosphate dehydrogenase, fructose-1,6-bisphosphatase, glucose-6-phosphatase, and hexokinase in the liver and significantly preserved insulin-positive cells as well as protected the overall morphology in the pancreatic islets of diabetic rats [14]. Abuohashish et al. found that both the anti-inflammatory and antioxidant properties of morins are useful against diabetic osteopenia in rats [15]. A study showed that when morin was administered in diabetic rats, there was a significant attenuation of bone loss which was evident at bone turnover parameters level which included BALP, OC, CTX, and DPD [15]. They also found that morin brings back the changes occurred in diabetic rats in respect to serum levels of glucose, TBARS, IL-1β, IL-6, TNF-α (which were significantly elevated), and that of insulin and GSH (which were decreased) [15].

Besides the abovementioned activity, morin also possesses significant immunoregulatory activity. In an independent study, Kim et al. showed that morin repressed IgE-mediated allergic responses by inhibiting production of IL-4 and TNF-α and degranulation of antigen (Ag)-stimulated mast cells in a mice model. They also found that morin inhibited the phosphorylation of spleen tyrosine kinase (Syk) (which plays a very important role in the Syk activation) and activation of linker for activation of T cells (LAT) in rat bone marrow-derived mast cells (BMMCs) and basophilic leukemia (RBL)-2H3 cells, along with the inhibition of p38, Akt, and MAPKs. Their results suggest that inhibition of Fyn kinase in mast cells by the morin is mainly responsible for the action described above [16]. The notion is also supported by the findings of Fang et al. which indicated that morin might have the ability to regulate immune response through modulating the cytokine profiles displayed in chronic immunotoxic pathophysiology where they have shown that morin and it's sulfated or glucuronidated derivatives were operative on LPS-activated RAW 264.7 cells by tumbling NO, TNF-α, and IL-12 production. They also showed the reduced phagocytic activity of peripheral blood cells in the morin-treated cells in respect to control. These activities such as reduced macrophagic phagocytic activities, lowering of NO production, etc. resembled to LPS-resistant state, and this is very important to treat various chronic autoimmune diseases [17].

Morin exlung parenchyma and airways isatocytes against chemically produced rat tongue carcinogenesis and blocked phorbol ester-mediated transformation

[18, 19]. Besides, it inhibited the lipoxygenase pathway and the peroxisome proliferator-activated receptor-mediated differentiation of keratinocyte [3, 20]. In context to the release of inflammatory cytokines from mast cells such as IL-6, IL-8, TNF, etc., the inhibitory role of morin was reported [21]. Morin hydrate exhibited anticancer activity in an in vivo study, i.e., in cancer models (viz., inhibition of COLO205 cells growth in nude mice) [22]. Morin hydrate downregulated NF-κB which in turn inhibited the inflammatory gene cascade along with the downregulation of several factors related to NF-κB which take part in cell survival such as X-chromosome-linked IAP, inhibitor of apoptosis protein 1 and 2 and BcL-xL, in cell proliferation, viz., cyclooxygenase-2 and cyclin D1, and invasion such as matrix metalloproteinase-9 [3, 23].

From the diverse studies reviewed here, morin emerged as a valuable natural flavone in the management of different chronic pathophysiological conditions. Different studies also showed that it could be used as an exceptional and novel pathological detection tool. Nevertheless, in future more studies are required on the morin in understanding the precise molecular mechanism of its action and to discover new possibilities.

10.3 Quercetin

Quercetin is one of the significant bioflavonoids present in many medicinally important plants and has numerous beneficial effects including cardiovascular protection, antiviral, anti-inflammatory activity, antitumor, anticancer, anti-ulcer, anti-allergy, antidiabetic, gastroprotective effects, antihypertensive, immunomodulatory, anti-infective, etc.[24, 25, 26]. Quercetin has also proved to be beneficial against environmental stress which causes formation of free radicals such as smoking and pollution [27, 28].

The name quercetin (3,3′,4′,5,7-pentahydroxyflavone) is derived from the Latin word "quercetum" that means oak forest. It fits in to the class of compounds called flavonols [24]. The compound is yellow in color and quite soluble in alcohol and lipids where as poorly soluble in hot water and virtually insoluble in cold water. This dietary flavonoid is mainly found in citrus fruits, seeds, green leafy vegetables, nuts, olive oil, apples, broccoli, barks, green tea, onions, red grapes, red wine, berries, dark cherries, buckwheat, cranberries, blueberries, etc. [27].

Researchers have shown the impact of quercetin as systemic anti-inflammatory agents [29]. Elevated C-reactive protein (CRP) levels are associated with numerous diseases states which involve inflammatory conditions, whereas intake of quercetin can lower the levels of the inflammatory risk factor, CRP. In preclinical in vitro studies, García-Mediavilla et al. showed that quercetin significantly reduces the levels of inflammatory mediators such as COX-2, NO synthase, and CRP in human hepatocyte-derived cell line [30]. Whereas, quercetin, in a dose-dependent manner, inhibited both acute and chronic inflammation related to arthritis and also showed significant antiarthritic activity against adjuvant-induced arthritis, in murine models [31, 32].

In a study including human as a model, Askari et al. found that 2-month quercetin supplementation (at a dose of 500 mg) in healthy male nonprofessional athletes with consistent workout showed a significant decrease in the levels of CRP, [whereas, quercetin remain ineffective (at a dose of 500 mg/day) in altering the levels of CRP in female patients, with rheumatoid arthritis (RA)[27, 33, 34]]. But in another study it has been shown that xanthine oxidase inhibition activity of quercetin make it able to prevent the accumulation of uric acid, which may help the subjects who are suffering from gout[35].

Quercetin also possesses a wide spectrum of biological activities which may have a positive influence on cardiovascular diseases. A study on 30 men suffering from coronary heart disease (CHD) showed that the consumption of red grape polyphenol extract rich in quercetin caused an increase in flow-mediated dilation of major arteries, which in fact a persuasive indicator of improved endothelial health [36]. The molecule inhibited the aggregation of platelets and improved the endothelium. It protected against CHD and reduced the mortality risk caused due to low-density lipoprotein (LDL). The molecule exhibited a vasorelaxant effect on isolated arteries which in turn helped in lowering blood pressure and prevented the cardiac hypertrophy onset [27, 37]. Quercetin prevented LDL and cholesterol damage, and in context to this, it was observed that consuming high flavonoid containing food supplements could lower cholesterol. It was observed that LDL oxidation was inhibited on consumption of quercetin along with an alcohol-free red wine extract containing quercetin [38]. In a 6-week clinical trial, quercetin reduced oxidized LDL levels in plasma and systolic blood pressure in subjects who were at high risk of heart disease [39]. Quercetin has a feature to induce apoptosis in mature fat cells while inhibits fat accumulation in maturing human fat cells [40, 41]. It also blocks the uptake of glucose from the blood and the fat cell production while enhances the fat cell necrosis [27, 42, 43].

Neuro-inflammatory processes in the central nervous system are the ultimate fate of neurodegenerative diseases such as Alzheimer's and Parkinson's disease as well as neuronal injury associated with stroke [44]. Quercetin and ascorbic acid combinatorial therapy effectively reduces the incidence of oxidative damage to neurovascular structures in the skin and prevents damage to neurons. Quercetin is also capable to protect brain cells against the oxidative stress-induced tissue damages which sequentially leads to Alzheimer's and other neurological conditions [24, 27].

The anticancer effect of quercetin includes the suppression of antiproliferative growth factor and antioxidant [45]. It possesses anticarcinogenic property and also acts as an apoptosis inductor; decreases the tumor growth of colon, liver, brain, and other tissues; and restricts the spread of malignant cells [46, 47]. Combination therapy of quercetin along with curcumin decreased both the size and number of rectal and ileal adenomas with minimal and/or no adverse effects [48].

Quercetin inhibits chemical carcinogen, hexavalent chromium (Cr[VI]), induced cell transformation, ROS generation, and MicroRNA-21 (miR-21) elevation in human colon cancer Caco-2 cells [27, 49, 50]. It also has beneficial effect on prostate cancer which was proved through in vitro and in vivo cancer studies [51].

Quercetin also has significant gastroprotective activity. It can inhibit lipid peroxidation and gastric acid secretion of gastric cells, and in doing so, it can serve as a gastroprotective agent. *Helicobacter pylori* infection, a potent ulcer forming agent, can also be inhibited by quercetin [24]. Suzuki et al. showed that quercetin dose-dependently inhibits ethanol-induced gastric mucosal injury in rats. The results suggest the anti-ulcer activity of quercetin through its free radical scavenging activity or its enhanced production in gastric mucus [52, 53].

The molecule exhibited anti-allergy effect by inhibiting histamine release from the mast cells and several allergic molecules and in the prevention of bronchitis and asthma [27, 54].

Progressive chronic inflammatory disorder of lung parenchyma and airways is collectively known as chronic obstructive pulmonary disease (COPD). However, unfortunately, therapies for COPD are said to be partially effective with opportunities of side effects. But as a hope, recent evidences are showing that the quercetin supplementation is beneficial in COPD. It has been shown previously that a fourfold increase in plasma quercetin levels ominously decreased lung inflammation and prevented the disease progression. Perhaps the anti-inflammatory property of quercetin is attributed to its beneficial role in pulmonary disorders [55]. Also a clinical trial of 12 weeks duration proved quercetin (at a dose of 1000 mg/day) beneficial against upper respiratory tract infection rates in middle- and older-age patients [56].

10.4 Rutin

Rutin (3,3′,4′,5,7-pentahydroxyflavone-3-rhamnoglucoside) is a flavonoid that is prevalent in the plant kingdom [57]. Rutin has a wide range of pharmacological properties. It has widely been exploited as human medicine and nutrition. It has proved to have antimicrobial, antifungal, and antiallergic properties and hence widely used as alternative medicine. However, current research has shown its multispectrum pharmacological benefits for the treatment of various chronic diseases, such as cancer, diabetes, hypertension, and hypercholesterolemia [58]. Recently, Patil et al. showed the potential of rutin in alleviating the radiation-induced cytogenetic damage and mortality. They have speculated that the activity might be attributed to the free radicals scavenging activity of rutin [59, 60]. Park et al. showed that rutin has an osteoblast stimulant property. The study showed that it could induce bone development via the differentiation of human MG-63 osteosarcoma cells [61]. Rutin also has antidiabetic activity. It is potent for glycemic control by enhancing the activity insulin-mediated receptor kinase and thereby triggering the insulin pathway which in turn would increase both the translocation of glucose transporter 4 as well as glucose uptake [62]. From a study Niture et al. speculate that rutin inhibits inflammatory cytokines and also improves the antioxidant and plasma lipid profiles in high-fat diet. It also did the same in streptozotocin-induced type 2 diabetic model. Hence they projects rutin as a diabetic modulator together with customary antidiabetic drugs [59, 63]. Rutin has significant beneficial effects over several neurological disorders. It has been proven to protect against the neurodegenerative effects of prion buildup by increasing the production of neurotropic factors and also by inhibiting apoptotic pathway activation in neuronal cells [64]. Nieoczym et al. showed that it also has a weak and temporary anticonvulsant effect [65]; whereas, Qu et al. showed that rutin has multi-target therapeutic potential for cognitive discrepancies related to conditions involving enduring cerebral hypoperfusion, such as Alzheimer's disease and vascular dementia [66]. Rutin was found to provide protection against trimethyltin-induced impairment of spatial memory and the involvement of dopaminergic system, as well as synaptophysin has been suggested in trimethyltin-triggered damage of neuronal cells in the hippocampus [59, 62, 61]. Rutin also can modulate NO production and thus can modulate NO-related physiological and/or pathophysiological processes. Ugusman et al. showed that rutin increases nitric oxide production in human endothelial cells and thus improved endothelial function [67].In ultraviolet B-irradiated skin of mouse, rutin exerted anti-inflammatory effects by interfering with the cyclooxygenase-2 expression as well as nitric oxide synthase production [68]. Rutin provided protection against high cholesterol diet-mediated hepatotoxicity and inflammation [69].

10.5 Mangiferin

2-*C*-β-D-glucopyranosyl-1,3,6,7-tetrahydroxyxanthone or mangiferin ($C_{19}H_{18}O_{11}$) having a molecular weight of 422.35 and a melting point of 271 °C (anhydrous) is a naturally occurring *C*-glucoside xanthone. This xanthone derivative is obtained from the fruits, leaves, bark, and roots of *Mangifera indica* and belongs to Anacardiaceae family. Mangiferin has been reported to exhibit a diversified use in therapeutics. It possesses antistimulatory [70], antimodulatory [71], antioxidative [72, 73], antidiabetic, dyslipidemic, antiallergic, analgesic, anticancer, and anti-HIV properties [74–78].

Evidences suggest that mangiferin possesses iron-complexing ability which in turn is believed to be the mechanism for protection against Fe^{2+} citrate-mediated lipid peroxidation in rat liver [79]. Mangiferin, at a concentration of 10 μM, reflected amelioration against Fe^{2+} citrate-triggered swelling of mitochondria followed by loss of mitochondrial membrane potential. Iron citrate-triggered antimycin A-insensitive consumption of oxygen in mitochondria was found to be inhibited by the xanthone derivative. On the other hand, mangiferin induced oxygen consumption by stimulating Fe^{2+} autoxidation and prevented the reduction of Fe^{3+}ascorbate. The mangiferin-Fe^{2+}/Fe^{3+} absorption spectra suggested the possibility of transient charge transfer complex formation between mangiferin and Fe^{2+}, Fe^{2+} oxidation acceleration, and formation of a stable complex, i.e., Fe^{3+}-mangiferin complex. This complex fails to take part in Fenton-type reactions and propagation of lipid peroxidation, thus suggesting that the antioxidant activity of mangiferin is contributed due to its iron-chelating activity and is not contributed due to scavenging of free radicals. In this respect, the molecule has pharmacological relevance in terms of chelation therapy in diseases related to iron overload or abnormal distribution of intracellular iron [77].

Studies have revealed that in OF1 mice, mangiferin showed protection by restoring the altered levels of antioxidant enzymes, viz., superoxide dismutase, etc., prevented protein oxidation significantly in terms of the net sulfhydryl group protein content, reduced lipid peroxidation confirmed through 4-hydroxy-alkenals and malondialdehyde assays, checked the release of cytochrome *C*, and inhibited apoptotic cell death [72]. The strong antioxidant activity of mangiferin has been confirmed through 1,1-diphenyl-2-picrylhydrazyl radical scavenging assay [80]. Even from the structural point of view, this property of the molecule is well understood as it possesses four phenolic H-atoms of which two can readily be involved in interacting with free radicals to form phenoxy radicals which in turn are resonance stabilized [77, 80].

Inflammation involves several mediator, viz., nitric oxide (NO), which is synthesized by NO synthase and prostanoids synthesis through cyclooxygenase (COX-2). Dilatation of arterioles occurs due to an increase in blood permeability and blood flow which is a vascular event of inflammation [81, 82]. Beltran et al. showed that in spontaneous hypersensitive and normotensive rats, mangiferin inhibited the IL-1β mediated induction of COX-2 and iNOS; however, no effect was exerted with the molecule in absence of IL-1β on iNOS and COX-2 of noradrenaline-mediated vasoconstriction from mesenteric arteries of vascular smooth muscle cells [83]. In activated macrophages, mangiferin significantly reduced the level of transcription of iNOS as well as the production of NO [84]. In regard to the anti-inflammatory mechanism of mangiferin, inhibition of NF-κB activation is believed to be involved which in turn regulates the activation of the promoter for iNOS and COX-2 genes [85].

NF-κB regulates the cascade of gene encoding inflammatory enzymes, proinflammatory cytokines, adhesion molecules, and chemokine which are overexpressed as a response to inflammation [86]. Evidence suggests that the activation of NF-κB is concomitantly associated with ROS production [87]. Mangiferin was observed to suppress the activation of NF-κB which in turn inhibits the inflammatory gene cascade, increased intracellular antioxidant levels, and triggered anticancer drug-mediated cellular death, indicating a plausible role as a combination therapy in treating cancer [88]. This ameliorative property of mangiferin is believed to be mediated through free radical quenching and increasing intracellular GSH which in turn interferes with TNF-mediated NF-κB activation [89]. Microarray data showed that magniferin countered NF-κB activation induced by IL-1, TNF, and LPS through TNF receptor-mediated factor 6; inhibited NF-κB triggered activation of signaling molecules, viz., RelA and RelB (genes of Rel/NF-κB/κB); and hindered the expression of toll-like receptors such as JNK1 and JNK2 in a thioglycollate-triggered murine macrophages elicited with gamma interferon and lipopolysaccharide model [71]. Mangiferin was also found to inhibit TNF-mediated protein synthesis.It also trigger DNA damage,apoptotic cell death, and inhibit an array of proinflammatory cytokines, viz., IL-1, IL-6, IL-1α, macrophage colony stimulating factors, adhesion molecules, etc. [71]. Studies hence revealed that mangiferin modulated the expression of an array of genes involved in regulating inflammation, apoptosis, tumorigenesis, viral replication, and many others and thus depicting its immunomodulatory property.

Exposure to the industrial and environmental pollutant, Hg has been reported to affect the central nervous system and varied target organs, viz., the liver, kidney, gastrointestinal tract, etc. Evidences suggest that Hg(II) generates ROS through thiol complexation, peroxidizes lipid molecules of mitochondrial membrane which leads to a drop of mitochondrial membrane potential, interferes with the sulfhydryl (-SH) group, and hence results in depletion of intracellular thiols, e.g., GSH [90]. Hg(II) interferes with the conformation and activity of proteins through side-chain modifications and compromises the activities of antioxidant enzymes, viz., CAT, SOD, GST, GR, etc. [91]. These Hg(II)-mediated alterations have been reported to result in cellular death. In this regard, mangiferin has drawn attention due to its antioxidant property. The cytoprotective effect of mangiferin has been invested on $HgCl_2$-induced human liver carcinoma HepG2 cells [92]. It has shown that a 2 h pretreatment of mangiferin to $HgCl_2$ exposure at varied concentrations inhibited apoptotic cell death significantly together with a decrease in the enhanced ROS levels and reversing the activities of the antioxidant enzymes. The study suggested that the protective role of mangiferin might be due to the $HgCl_2$-induced ROS quenching, restoration of altered mitochondrial membrane potential, and intracellular antioxidant activities [77].

The universally acclaimed toxic metal, Pb, has been extensively reported to affect the endocrine, reproductive, as well as central nervous system [93]. Lead toxicity induces oxidative stress which in turn imbalances the prooxidant-antioxidant levels, affects cell membrane, interferes with transcription, disrupts the synthesis of protein, etc. [77, 93, 94]. Pb(II) interferes with –SH groups of biomolecules, calcium homeostasis, and lipid peroxidation through ROS production [95, 96]. Pal et al. have reported the ameliorative effect of mangiferin in Pb-induced hepatotoxicity [76].

Following $Pb(NO_3)_2$-induced hepatic dysfunction, posttreatment with mangiferin reduced increased ROS production; repaired the altered antioxidant machineries, viz., levels of SOD, CAT, GSH, etc.; and restored the altered mitochondrial membrane potential. The molecule significantly restored the increased levels of serum hepatic markers, viz., alanine aminotransferase and alkaline phosphatase. Mangiferin effectively downregulated the altered expressions of MAPKs, phospho-ERK1/2, phospho-JNK, and phospho-p38, inhibited the translocation of NF-κB, and reduced apoptotic hepatic cell death. In vitro studies with primary hepatocytes also reflected the beneficial role of the xanthone derivative against Pb(II)-triggered cytotoxicity [76].

Diabetes mellitus, one of the most prevalent endocrine metabolic disorder, is fundamentally associated with hyperglycemia due to defects in the secretion of insulin or varying degree of endogenous insulin resistance and results in β-cell destruction or dysfunction. Hyperglycemia is associated with excessive ROS generation and attenuation of antioxidant machineries [97]. In experimental animals, streptozotocin (STZ) is an established inducer of diabetes and is reported to induce oxidative stress through free radical generation which in turn leads to diabetic complications [98]. In regard to this, diabetic nephropathy draws special attention which is a common consequence of both type 1 and 2 diabetes and is characterized by increased secretion of urinary albumin, mesangial thickness, thickness of basement membrane, glomerular hypertrophy and hyperfiltration, and extracellular matrix protein accumulation [77, 99]. In diabetic nephropathy, the plausible cause of ROS generation is due to the activation of advanced glycation end products, polyol pathways, increased activity of xanthine oxidase and nitric oxide synthase, glucose autoxidation, and deficiency in the mitochondrial respiratory chain [100]. In this regard, mangiferin attracts attention in the context of diabetic complications due to its antioxidant and hypoglycemic effects [101]. However, post-treatment with mangiferin on STZ-triggered diabetic rats reduced the altered antioxidant levels in both renal and cardiac tissue, reduced lipid peroxidation and glycosylation of hemoglobin, lowered the creatine phosphokinase level, and restored the STZ-induced altered levels of triglycerides, high- and low-density lipoprotein, cholesterol. Studies suggested that this antidiabetic property of mangiferin could be due to alternate mechanisms apart from insulin release and/or secretion from pancreatic β-cell [102]. Enhancement of peripheral glucose utilization, stimulation of glycogenic and glycolytic processes, and reduction of glycemia through glucose intake/uptake could be the contributing factors for the extrapancreatic actions [103]. Another study showed that treatment with mangiferin inhibits α-glucosidases [104]. Significant reduction in total cholesterol, low-density lipoprotein cholesterol, and total triglycerides with concomitant increase in high-density lipoprotein cholesterol reflected the antiatherogenic and antihyperlipidemic properties of mangiferin in diabetic animals. Mangiferin has been reported to improve oral glucose tolerance without interfering with the basal plasma glucose levels [102]. Sellamuthu et al. have reported that in diabetic rats, oral administration of mangiferin for a month reduced blood glucose levels and glycosylation of hemoglobin, whereas, the levels of hemoglobin and insulin were increased [105]. The activities of pyruvate kinase, hexokinase, glucose-6-phosphate dehydrogenase, and glycogen synthase significantly increased in diabetic rats following mangiferin administration. On the other hand, the molecule significantly reduced the altered activities of glucose-6-phosphate, lactate dehydrogenase, glycogen phosphorylase, and fructose-1,6-diphosphate in the hepatic tissue of diabetic animals, thus suggesting the antihyperglycemic effect of the molecule. Miura et al. reported that in KK-Ay mice, mangiferin diminished the increased glucose level almost by 56% and decreased the levels of triglycerides almost by 70% and blood cholesterol almost by 40% [89]. Thus findings from several studies suggested the antihyperlipidemic, antidiabetic, antihyperglycemic, and antiatherogenic properties of mangiferin in regard to diabetes and/or its associated complications.

10.6 Myricetin

Myricetin is a member of the tree families likes Myricaceae, Polygonaceae, Anacardiaceae, Primulaceae, and Pinaceae and is commonly found in vegetables, berries, wines, and teas [106, 107]. This plant-derived phenolic compound occurs both in glycosidically bound and free forms. It is used in beverages and food and exhibits diversified properties such as anti-inflammatory, antioxidant, antidiabetic, anticancer, iron-chelating activities, etc. In several diseases related to central nervous system, viz., Alzheimer's and Parkinson's, this molecule has been found to provide protection. As preservative, myricetin has been found to increase the shelf life of oil- and fat-containing food. Its interference with RNA polymerases, DNA polymerases, telomerase, kinases, transcriptase, and helicases has been reported [107].

Extensive studies suggested the antioxidant property of myricetin. In regard to the scavenging activity of the molecule, it inhibited DPPH radical significantly by 71.5% when used at a concentration of 1 mg/ml [108], whereas at a concentration of 40 g/ml, 3.2 g/ml, 32 g/ml, and 320 g/ml, the inhibition was found to be 78%, 85.6%, 92.8%, and 96.9%, respectively [109]. The inhibition of DPPH radicals by myricetin has been reported to be polyphenol oxidase mediated [110]. At 0.32 g/ml, 3.2 g/ml, and 32 g/ml, myricetin scavenged superoxide radicals by 24.6%, 79.5%, and 96.4%, respectively [111]. Significant inhibition has also been reported in respect to TEAC activity and FRAP assay [112]. The molecule, at a varied concentration significantly reduced both ascorbic acid- and ferrous sulfate-mediated peroxidation of lipid, inhibited oleic acid triggered over accumulation of triglyceride in HepG2 cells, and decreased the production of NP in the liver, brain cortex, lungs, kidney, and blood in experimental rats, exhibited NO-scavenging activity, reduced collagenase in dermal fibroblasts in human, and inhibited peroxyl radical generation [113–115, 119]. Myricetin was found to inhibit thiyl radical which acts as a catalyst during cis-trans fatty acid isomerization. In addition, the molecule was able to scavenge hydroxyl radicals which were generated through UV photolysis of hydrogen peroxide [116]. It was observed to regulate the activities and expressions of antioxidant enzymes and ROS production and hence ameliorate H_2O_2-mediated cellular death by regulating MAPK and PI3K/Akt signaling pathways [116, 117], prevented H_2O_2-triggered breakage of DNA strand in human lymphocytes and colonocytes [118], inhibited lipid peroxidation [107], and stimulated DNA repair following Fe(III)-triggered genotoxicity. Morel et al. suggested that formation of phenoxy radical could be the mode of protective action of myricetin against Fe-induced lipid peroxidation in rat hepatocytes [106], whereas another study revealed the SIN-1-mediated DNA strand breakage inhibition [108] which in turn inhibited NO and O^{2-} generation. In case of sickle cell anemia, myricetin provided protection against the red blood cells [109]. In phenazine methosulfate and xanthine oxidase system, the molecule could significantly combat against increased superoxide anions generation [110, 111]. Other probable mechanisms of myricetin'santioxidant response are mediated through altered Nrf2 activity, increased glutathione level, decreased malondialdehyde production, and decreased leakage of lactate dehydrogenase [112]. At a lower concentration, myricetin was found to be an inhibitor of Fe-mediated lipid peroxidation, whereas a higher concentration displayed peroxidant effect against the formation of hydroxyl radicals. It is worth mentioning that in presence of bleomycin (antiviral and antitumor drug), the molecule behaved as a prooxidant (at a higher concentration) whereas behaved as an antioxidant at lower concentrations [113]. In type 2 diabetes, the molecule significantly mitigated altered protein carbonylation and lipid peroxidation [114] and reduced Ca^{2+}-induced oxidative metabolism and free radical production in the rodents' brain neurons, subjected to ischemia [119]. Xie et al. reported that that the hydroxy functional group in C-41 position in myricetin is responsible for its activity against lipid peroxide radical [115]. The structure as well as activity analysis of the molecule has revealed that the free radical scavenging activity depends on the free radical variant such as the catechol moiety of B-ring was found to be responsible for scavenging of DPPH, the hydroxyl moiety present at C-41 position was observed to be associated with xanthine oxidase generated reduction, whereas the presence of double bond at C-2-C-3 position, 3-hydroxy groups, and catechol in B-ring was found to be an attributing factor for the reducing property of myricetin [111, 116, 117].

Myricetin has been reported to be cytotoxic toward skin, hepatic, colon, and pancreatic cancer cells and plays a key role in the initiation as well as progression of cancer. The hydroxyl group of B-ring, C-2-C-3 double bond, and aromatic B-ring at C-2 has been reported to be responsible for the cytotoxic effect of the molecule. The antiproliferative activity was found against human acute leukemia cells on one hand, while the cytotoxic effect was observed in chronic leukemia cells as well as in normal peripheral blood mononuclear cells [118]. Myricetin displayed a dose-dependent effect both in vitro and in vivo in producing topo-triggered chemotherapeutics and carcinogenic effects [120]. In this context, it has been suggested that the inhibition of topo I and II was due to carbonyl moiety at C-4; hydroxyl substitution at C-3, C-7, C-31, C-41, and C-2-C-3 saturation; and hydroxyl group in B-ring [131]. Through regulation of JAK1, MEK, MKK4, and Akt kinase activity, myricetin was found to provide protection against skin cancer [122] and attenuated the induction of activator protein-1 or c-Fos activation by tumor promoter [122]. In EGF-triggered mouse JB6 P+ cells, the molecule inhibited JAK1/STAT3 signaling pathway which in turn blocked the transformation of cells and also inhibited both the transcriptional activity and DNA binding as well as phosphorylation of STAT3 at Ser727 and Tyr705 [123]. Ichimatsu et al. reported that myricetin blocked the EGF-mediated mouse epidermal cell transformation which in turn suppressed activator protein-1 [124]. Xu et al. showed a strong dose-dependent inhibitory activity of myricetin against human prostate cancer PC-3 cells and synergistically decreased cell proliferation resulted in apoptosis [125]. In bladder cancer T24 cells, the molecule significantly decreased the viability and proliferation along with the migration of the cells by reducing the expression of MMP-9 [126]. Moreover, the molecule triggered cell cycle arrest at G2/M phase, induced apoptosis by decreasing cyclin-dependent kinase cdc2 and cyclin B1 expression, and inhibited Akt phosphorylation whereas increased p38 MAPK phosphorylation. In HepG2 cells, myricetin decreased the cancer cell proliferation, arrested G2/M phase, increased p21/p53 signaling cascade, and decreased cyclin B1 and cdc2 expression [127]. In esophageal adenocarcinoma cells, this molecule upregulated Tyr15/Thr14 phosphorylated p27 and cdc2, downregulated CDK7 kinase and CDK7 kinase-induced cdc2 phosphorylation, and induced apoptosis ([111],). It also arrested G2/M cell cycle via GADD upregulation and cyclin B1 downregulation[128, 129]. Myricetin is effectively protected against medulloblastoma by inhibiting HGF/Met pathway and prevented actin-rich membrane ruffles formation [130]. The molecule was found to be effective in causing metastasis of human lung carcinoma cells by interfering with the invasion, migration, and adhesion of cancer cells and inhibited MMP-2, phosphorylation of ERK1/2, NF-κB activation, c-Jun, and c-Fos. Myricetin induced cellular death via apoptosis and decreased the activity of PI3K in pancreatic cancer cells [130, 131]. An in vivo study showed that the molecule was potent enough to reduce the orthopic pancreatic tumor metastatic spread and tumor regression whereas was nontoxic toward untreated healthy cells. In regard to prostate cancer, the molecule served as a potent chemotherapeutic agent, whereas in colon cancer, it induced DNA condensation and cytotoxicity [132]. The compound stimulated apoptotic-inducing factor release from mitochondria and pro-carcinogen 2-amino-1-methyl-6-phenylimidazo basolateral uptake through MRP2-induced excretion of the former to the lumen from intestinal cells [133].

In vivo studies have revealed that deoxycorticosterone acetate-induced oxidative stress and hypertension were reduced following treatment with myricetin in the cardiac tissue of rats [134]. The molecule effectively reversed the otherwise altered vascular reactivity, systolic blood pressure, heart rate, levels and/or content of intracellular antioxidant enzymes, and thiobarbituric acid-reactive substances. Godse et al. showed that myricetin significantly decreased catecholamine-induced vascular reactivity and systolic blood pressure, whereas, in case of fructose rich diet, the molecule effectively lowered the blood pressure and reversed the sugar-induced altered metabolic pathway [135].

Research conducted both in vitro and in vivo confirmed the immunomodulatory role of myricetin by modulating the responses toward the immune system through antibody formation or inhibition of WBC activity. The molecule modulated LPS-triggered bone marrow-derived dendritic cells activation in mice on the one hand while on the other decreased TNF, IL-12, and IL-6 secretion along with the inhibition of cell proliferation, MHC class II, CD86, and CD40 and blockage of migratory and endocytic capacity of the cells [136]. Kang et al. elucidated the probable mode of this action of the molecule through inhibiting IL-12 production in macrophages by NF-κB downregulation [136]. The molecule stimulated cytosolic free calcium production in cultured endothelial cells of bovine and endothelium-mediated contractile responses in aortic rings of rats [137]. The molecule exhibited immunosuppressive effect by suppressing the secretion and expression of IL-2 as well as blocked CD69 expression of CD3+ T cell and mouse lymphocytes proliferation and thus reflecting its immunomodulatory activity [138].

In insulin-independent diabetes, myricetin showed protection by enhancing the glucose uptake which is independent of insulin receptors [139]. The molecule was also found to stimulate the uptake of D-3-O-methylglucose and D-glucose in the adipocyte tissue of rats. Intraperitoneal myricetin administration to streptozotocin-induced diabetic rat decreased hyperglycemia significantly whereas increased hepatic glucose-6-phosphate and glycogen content. The antidiabetic activity of the molecule was related to glycogen metabolism [140]. Islet amyloid polypeptide (IAPP) aggregation which leads to pancreatic β cell death in case of type II diabetes was inhibited by myricetin by preventing thioflavin T binding and formation of fiber [141]. Plasma glucose level was decreased, whereas insulin resistance was increased through β-endotrophin production in insulin-resistant and fructose-induced experimental animals following myricetin uptake. The molecule significantly restored the altered expression and hence functions of insulin receptor phosphorylation, insulin receptor substrate-1, Akt, and its substrate together with the translocation of glucose transporter subtype 4 [142]. Another study showed that myricetin reduced the sensitivity of insulin by regulating post-receptor signaling of insulin, GLUT-4 activity, and IRS-1-associated PI3K and decreased both the glucose-insulin index and plasma glucose concentration in the muscles of experimental animals [143]. It was observed that in the skeletal muscle cells, under hyperinsulinemic state, the molecule improved low-dose insulin-triggered uptake of glucose [111]. Myricetin displayed cytoprotection against cytokine-mediated cellular death in insulin-secreting cells, increased the viability of cells, whereas decreased cellular apoptosis, reduced cytokine-induced increased NF-κB expression, triggered accumulation of NO, stimulated ROS generation, and increased the release of cytochrome *c* in to cytosol from mitochondria. The molecule showed protection against diabetic nephropathy by reducing glomerulosclerosis, urine volume, protein excretion, and blood urea nitrogen. The altered levels of glutathione peroxidase and xanthine oxidase in the renal tissues were restored following the administration of myricetin in diabetic rats [144]. The molecule stimulated the biosynthesis of cholesterol in the hepatocytes of rats at a lower concentration whereas inhibited the biosynthesis of cholesterol at a higher concentration in HepG2 cells [145]. Chang et al. showed that myricetin decreased intracellular triglyceride accumulation in high-fat diet-fed rats, reduced body weight, and decreased visceral fat pad weights and the levels of plasma lipid [146].

10.7 Conclusion

Extensive studies reveal the pharmacological activities of the phytochemicals. There is no doubt that phytochemicals hold a potential in protecting against several diseases and/or pathophysiological conditions. These molecules may work either through synergistic interaction or when administered unaccompanied. However, regardless of the studies (both in vivo and in vitro) which have elucidated on the role of phytochemicals related to human health, further extensive research is required to understand the specific molecular mechanism of the phytochemicals. Detailed preclinical studies and its clinical experiments are needed to provide a basis for potential expediency of these gifts of nature in the welfare of human health.

References

1. Sinha K, Sadhukhan P, Saha S, Pal PB, Sil PC (2015) Morin protects gastric mucosa from nonsteroidal anti-inflammatory drug, indomethacin induced inflammatory damage and apoptosis by modulating NF-kappaB pathway. Biochim Biophys Acta 1850:769–783
2. Basile A, Sorbo S, Giordano S, Ricciardi L, Ferrara S, Montesano D, Castaldo Cobianchi R, Vuotto ML, Ferrara L (2000) Antibacterial and allelopathic activity of extract from Castanea sativa leaves. Fitoterapia 71(Suppl 1):S110–S116
3. Sinha K, Ghosh J, Sil PC (2016) Morin and its role in chronic diseases. Adv Exp Med Biol 928:453–471
4. Gottlieb M, Leal-Campanario R, Campos-Esparza MR, Sanchez-Gomez MV, Alberdi E, Arranz A, Delgado-Garcia JM, Gruart A, Matute C (2006) Neuroprotection by two polyphenols following excitotoxicity and experimental ischemia. Neurobiol Dis 23:374–386
5. Zeng N, Tong B, Zhang X, Dou Y, Wu X, Xia Y, Dai Y, Wei Z (2015) Antiarthritis effect of morin is associated with inhibition of synovial angiogensis. Drug Dev Res 76(8):463–473
6. Subash S, Subramanian P (2012) Chronotherapeutic effect of morin in experimental chronic hyperammonemic rats. Int J Nutr Pharmacol Neurol Dis 2:266
7. Amo-Barimah A, Woode E, Boakye-Gyasi E, Ainooson G, Abotsi W (2010) Antiarthritic and antioxidant effects of the leaf extract of *Ficus exasperata* P. Beauv (Moraceae). Pharm Res 2:89. https://doi.org/10.4103/0974-8490.62958
8. Sultana F, Rasool M (2015) A novel therapeutic approach targeting rheumatoid arthritis by combined administration of morin, a dietary flavanol and non-steroidal anti-inflammatory drug indomethacin with reference to pro-inflammatory cytokines, inflammatory enzymes, RANKL and transcription factors. Chem Biol Interact 230:58–70
9. Galvez J, Coelho G, Crespo ME, Cruz T, Rodriguez-Cabezas ME, Concha A, Gonzalez M, Zarzuelo A (2001) Intestinal anti-inflammatory activity of morin on chronic experimental colitis in the rat. Aliment Pharmacol Ther 15:2027–2039
10. Hogaboam CM, Jacobson K, Collins SM, Blennerhassett MG (1995) The selective beneficial effects of nitric oxide inhibition in experimental colitis. Am J Phys 268:G673–G684
11. Al-Numair KS, Chandramohan G, Alsaif MA (2012) Pretreatment with morin, a flavonoid, ameliorates adenosine triphosphatases and glycoproteins in isoproterenol-induced myocardial infarction in rats. J Nat Med 66:95–101
12. Prahalathan P, Kumar S, Raja B (2012) Effect of morin, a flavonoid against DOCA-salt hypertensive rats: a dose dependent study. Asian Pac J Trop Biomed 2:443–448
13. Noor H, Cao P, Raleigh DP (2012) Morin hydrate inhibits amyloid formation by islet amyloid polypeptide and disaggregates amyloid fibers. Protein Sci 21:373–382
14. Vanitha P, Uma C, Suganya N, Bhakkiyalakshmi E, Suriyanarayanan S, Gunasekaran P, Sivasubramanian S, Ramkumar K (2014) Modulatory effects of morin on hyperglycemia by attenuating the hepatic key enzymes of carbohydrate metabolism and β-cell function in streptozotocin-induced diabetic rats. Environ Toxicol Pharmacol 37:326–335
15. Abuohashish HM, Al-Rejaie SS, Al-Hosaini KA, Parmar MY, Ahmed MM (2013) Alleviating effects of morin against experimentally-induced diabetic osteopenia. Diabetol Metab Syndr 5:5
16. Kim JM, Lee EK, Park G, Kim MK, Yokozawa T, Yu BP, Chung HY (2010) Morin modulates the oxidative stress-induced NF-kappaB pathway through its anti-oxidant activity. Free Radic Res 44:454–461
17. Fang SH, Hou YC, Chang WC, Hsiu SL, Chao PD, Chiang BL (2003) Morin sulfates/glucuronides exert anti-inflammatory activity on activated macrophages and decreased the incidence of septic shock. Life Sci 74:743–756

18. Hsiang CY, Wu SL, Ho TY (2005) Morin inhibits 12-O-tetradecanoylphorbol-13-acetate-induced hepatocellular transformation via activator protein 1 signaling pathway and cell cycle progression. Biochem Pharmacol 69:1603–1611
19. Kawabata K, Tanaka T, Honjo S, Kakumoto M, Hara A, Makita H, Tatematsu N, Ushida J, Tsuda H, Mori H (1999) Chemopreventive effect of dietary flavonoid morin on chemically induced rat tongue carcinogenesis. Int J Cancer 83:381–386
20. Thuillier P, Brash AR, Kehrer JP, Stimmel JB, Leesnitzer LM, Yang P, Newman RA, Fischer SM (2002) Inhibition of peroxisome proliferator-activated receptor (PPAR)-mediated keratinocyte differentiation by lipoxygenase inhibitors. Biochem J 366:901–910
21. Kempuraj D, Madhappan B, Christodoulou S, Boucher W, Cao J, Papadopoulou N, Cetrulo CL, Theoharides TC (2005) Flavonols inhibit proinflammatory mediator release, intracellular calcium ion levels and protein kinase C theta phosphorylation in human mast cells. Br J Pharmacol 145:934–944
22. Chen YC, Shen SC, Chow JM, Ko CH, Tseng SW (2004) Flavone inhibition of tumor growth via apoptosis in vitro and in vivo. Int J Oncol 25:661–670
23. Gopal JV (2013) Morin hydrate: botanical origin, pharmacological activity and its applications: a mini-review. Pharm J 5:123–126
24. Lakhanpal P, Rai DK (2007) Quercetin: a versatile flavonoid. Internet J Med Updat 2:22–37
25. Salvamani S, Gunasekaran B, Shaharuddin NA, Ahmad SA, Shukor MY (2014) Antiartherosclerotic effects of plant flavonoids. Biomed Res Int 2014:480258
26. Sultana B, Anwar F (2008) Flavonols (kaempeferol, quercetin, myricetin) contents of selected fruits, vegetables and medicinal plants. Food Chem 108:879–884
27. Anand David AV, Arulmoli R, Parasuraman S (2016) Overviews of biological importance of quercetin: a bioactive flavonoid. Pharmacogn Rev 10:84–89
28. Begum AN, Terao J (2002) Protective effect of quercetin against cigarette tar extract-induced impairment of erythrocyte deformability. J Nutr Biochem 13:265–272
29. Chun OK, Chung SJ, Claycombe KJ, Song WO (2008) Serum C-reactive protein concentrations are inversely associated with dietary flavonoid intake in U.S. adults. J Nutr 138:753–760
30. Garcia-Mediavilla V, Crespo I, Collado PS, Esteller A, Sanchez-Campos S, Tunon MJ, Gonzalez-Gallego J (2007) The anti-inflammatory flavones quercetin and kaempferol cause inhibition of inducible nitric oxide synthase, cyclooxygenase-2 and reactive C-protein, and down-regulation of the nuclear factor kappaB pathway in Chang liver cells. Eur J Pharmacol 557:221–229
31. Guardia T, Rotelli AE, Juarez AO, Pelzer LE (2001) Anti-inflammatory properties of plant flavonoids. Effects of rutin, quercetin and hesperidin on adjuvant arthritis in rat. Farmaco 56:683–687
32. Mamani-Matsuda M, Kauss T, Al-Kharrat A, Rambert J, Fawaz F, Thiolat D, Moynet D, Coves S, Malvy D, Mossalayi MD (2006) Therapeutic and preventive properties of quercetin in experimental arthritis correlate with decreased macrophage inflammatory mediators. Biochem Pharmacol 72:1304–1310
33. Askari G, Ghiasvand R, Feizi A, Ghanadian SM, Karimian J (2012) The effect of quercetin supplementation on selected markers of inflammation and oxidative stress. J Res Med Sci 17:637–641
34. Javadi F, Eghtesadi S, Ahmadzadeh A, Aryaeian N, Zabihiyeganeh M, Foroushani AR, Jazayeri S (2014) The effect of quercetin on plasma oxidative status, C-reactive protein and blood pressure in women with rheumatoid arthritis. Int J Prev Med 5:293–301
35. Ahmad NS, Farman M, Najmi MH, Mian KB, Hasan A (2008) Pharmacological basis for use of Pistacia integerrima leaves in hyperuricemia and gout. J Ethnopharmacol 117:478–482
36. Lekakis J, Rallidis LS, Andreadou I, Vamvakou G, Kazantzoglou G, Magiatis P, Skaltsounis AL, Kremastinos DT (2005) Polyphenolic compounds from red grapes acutely improve endothelial function in patients with coronary heart disease. Eur J Cardiovasc Prev Rehabil 12:596–600
37. Edwards RL, Lyon T, Litwin SE, Rabovsky A, Symons JD, Jalili T (2007) Quercetin reduces blood pressure in hypertensive subjects. J Nutr 137:2405–2411

38. Chopra M, Fitzsimons PE, Strain JJ, Thurnham DI, Howard AN (2000) Nonalcoholic red wine extract and quercetin inhibit LDL oxidation without affecting plasma antioxidant vitamin and carotenoid concentrations. Clin Chem 46:1162–1170
39. Egert S, Bosy-Westphal A, Seiberl J, Kurbitz C, Settler U, Plachta-Danielzik S, Wagner AE, Frank J, Schrezenmeir J, Rimbach G, Wolffram S, Muller MJ (2009) Quercetin reduces systolic blood pressure and plasma oxidised low-density lipoprotein concentrations in overweight subjects with a high-cardiovascular disease risk phenotype: a double-blinded, placebo-controlled cross-over study. Br J Nutr 102:1065–1074
40. Ahn J, Lee H, Kim S, Park J, Ha T (2008) The anti-obesity effect of quercetin is mediated by the AMPK and MAPK signaling pathways. Biochem Biophys Res Commun 373:545–549
41. Park HJ, Yang JY, Ambati S, Della-Fera MA, Hausman DB, Rayalam S, Baile CA (2008) Combined effects of genistein, quercetin, and resveratrol in human and 3T3-L1 adipocytes. J Med Food 11:773–783
42. Strobel P, Allard C, Perez-Acle T, Calderon R, Aldunate R, Leighton F (2005) Myricetin, quercetin and catechin-gallate inhibit glucose uptake in isolated rat adipocytes. Biochem J 386:471–478
43. Yang JY, Della-Fera MA, Rayalam S, Ambati S, Hartzell DL, Park HJ, Baile CA (2008) Enhanced inhibition of adipogenesis and induction of apoptosis in 3T3-L1 adipocytes with combinations of resveratrol and quercetin. Life Sci 82:1032–1039
44. Choi GN, Kim JH, Kwak JH, Jeong C-H, Jeong HR, Lee U, Heo HJ (2012) Effect of quercetin on learning and memory performance in ICR mice under neurotoxic trimethyltin exposure. Food Chem 132:1019–1024
45. Lamson DW, Brignall MS (2000) Antioxidants and cancer, part 3: quercetin. Altern Med Rev J Clin Ther 5:196–208
46. Akan Z, Garip AI (2013) Antioxidants may protect cancer cells from apoptosis signals and enhance cell viability. Asian Pac J Cancer Prev 14:4611–4614
47. Vasquez-Garzon VR, Arellanes-Robledo J, Garcia-Roman R, Aparicio-Rautista DI, Villa-Trevino S (2009) Inhibition of reactive oxygen species and pre-neoplastic lesions by quercetin through an antioxidant defense mechanism. Free Radic Res 43:128–137
48. Cruz-Correa M, Shoskes DA, Sanchez P, Zhao R, Hylind LM, Wexner SD, Giardiello FM (2006) Combination treatment with curcumin and quercetin of adenomas in familial adenomatous polyposis. Clin Gastroenterol Hepatol 4:1035–1038
49. Han M, Song Y, Zhang X (2016) Quercetin suppresses the migration and invasion in human Colon Cancer Caco-2 cells through regulating toll-like receptor 4/nuclear factor-kappa B pathway. Pharmacogn Mag 12:S237–S244
50. Pratheeshkumar P, Son YO, Divya SP, Wang L, Turcios L, Roy RV, Hitron JA, Kim D, Dai J, Asha P, Zhang Z, Shi X (2016) Quercetin inhibits Cr(VI)-induced malignant cell transformation by targeting miR-21-PDCD4 signaling pathway. Oncotarget 8(32):52118
51. Yang F, Song L, Wang H, Wang J, Xu Z, Xing N (2015) Quercetin in prostate cancer: chemotherapeutic and chemopreventive effects, mechanisms and clinical application potential (Review). Oncol Rep 33:2659–2668
52. Alarcon De La Lastra C, Martin MJ, Motilva V (1994) Antiulcer and gastroprotective effects of quercetin: a gross and histologic study. Pharmacology 48:56–62
53. Suzuki Y, Ishihara M, Segami T, Ito M (1998) Anti-ulcer effects of antioxidants, quercetin, alpha-tocopherol, nifedipine and tetracycline in rats. Jpn J Pharmacol 78:435–441
54. Coles L (2000) Quercetin: a review of clinical applications. Natural medicine Online
55. Ferry DR, Smith A, Malkhandi J, Fyfe DW, Detakats PG, Anderson D, Baker J, Kerr DJ (1996) Phase I clinical trial of the flavonoid quercetin: pharmacokinetics and evidence for in vivo tyrosine kinase inhibition. Clin Cancer Res 2:659–668
56. Heinz SA, Henson DA, Austin MD, Jin F, Nieman DC (2010) Quercetin supplementation and upper respiratory tract infection: a randomized community clinical trial. Pharmacol Res 62:237–242
57. Hosseinzadeh H, Nassiri-Asl M (2014) Review of the protective effects of rutin on the metabolic function as an important dietary flavonoid. J Endocrinol Investig 37:783–788

58. Sharma S, Ali A, Ali J, Sahni JK, Baboota S (2013) Rutin : therapeutic potential and recent advances in drug delivery. Expert Opin Investig Drugs 22:1063–1079
59. Al-Dhabi NA, Arasu MV, Park CH, Park SU (2015) An up-to-date review of rutin and its biological and pharmacological activities. EXCLI J 14:59–63
60. Patil SL, Rao NB, Somashekarappa HM, Rajashekhar KP (2014) Antigenotoxic potential of rutin and quercetin in Swiss mice exposed to gamma radiation. Biom J 37:305–313
61. Park SE, Sapkota K, Choi JH, Kim MK, Kim YH, Kim KM, Kim KJ, Oh HN, Kim SJ, Kim S (2014) Rutin from Dendropanax morbifera Leveille protects human dopaminergic cells against rotenone induced cell injury through inhibiting JNK and p38 MAPK signaling. Neurochem Res 39:707–718
62. Hsu CY, Shih HY, Chia YC, Lee CH, Ashida H, Lai YK, Weng CF (2014) Rutin potentiates insulin receptor kinase to enhance insulin-dependent glucose transporter 4 translocation. Mol Nutr Food Res 58:1168–1176
63. Niture NT, Ansari AA, Naik SR (2014) Anti-hyperglycemic activity of rutin in streptozotocin-induced diabetic rats: an effect mediated through cytokines, antioxidants and lipid biomarkers. Indian J Exp Biol 52:720–727
64. Na JY, Kim S, Song K, Kwon J (2014) Rutin alleviates prion peptide-induced cell death through inhibiting apoptotic pathway activation in dopaminergic neuronal cells. Cell Mol Neurobiol 34:1071–1079
65. Nieoczym D, Socala K, Raszewski G, Wlaz P (2014) Effect of quercetin and rutin in some acute seizure models in mice. Prog Neuro-Psychopharmacol Biol Psychiatry 54:50–58
66. Qu J, Zhou Q, Du Y, Zhang W, Bai M, Zhang Z, Xi Y, Li Z, Miao J (2014) Rutin protects against cognitive deficits and brain damage in rats with chronic cerebral hypoperfusion. Br J Pharmacol 171:3702–3715
67. Ugusman A, Zakaria Z, Chua KH, Nordin NA, Abdullah Mahdy Z (2014) Role of rutin on nitric oxide synthesis in human umbilical vein endothelial cells. ScientificWorldJournal 2014:169370
68. Choi KS, Kundu JK, Chun KS, Na HK, Surh YJ (2014) Rutin inhibits UVB radiation-induced expression of COX-2 and iNOS in hairless mouse skin: p38 MAP kinase and JNK as potential targets. Arch Biochem Biophys 559:38–45
69. Sikder K, Kesh SB, Das N, Manna K, Dey S (2014) The high antioxidative power of quercetin (aglycone flavonoid) and its glycone (rutin) avert high cholesterol diet induced hepatotoxicity and inflammation in Swiss albino mice. Food Funct 5:1294–1303
70. Rajendran P, Rengarajan T, Nandakumar N, Divya H, Nishigaki I (2015) Mangiferin in cancer chemoprevention and treatment: pharmacokinetics and molecular targets. J Recept Signal Transduct Res 35:76–84
71. Yang Z, Weian C, Susu H, Hanmin W (2016) Protective effects of mangiferin on cerebral ischemia-reperfusion injury and its mechanisms. Eur J Pharmacol 771:145–151
72. Benard O, Chi Y (2015) Medicinal properties of mangiferin, structural features, derivative synthesis, pharmacokinetics and biological activities. Mini-Rev Med Chem 15:582–594
73. Ghosh S, Banerjee S, Sil PC (2015) The beneficial role of curcumin on inflammation, diabetes and neurodegenerative disease: a recent update. Food Chem Toxicol 83:111–124
74. Ghosh M, Das J, Sil PC (2012) D(+) galactosamine induced oxidative and nitrosative stress-mediated renal damage in rats via NF-kappaB and inducible nitric oxide synthase (iNOS) pathways is ameliorated by a polyphenol xanthone, mangiferin. Free Radic Res 46:116–132
75. Li AN, Li S, Zhang YJ, Xu XR, Chen YM, Li HB (2014) Resources and biological activities of natural polyphenols. Nutrients 6:6020–6047
76. Pal PB, Sinha K, Sil PC (2013) Mangiferin, a natural xanthone, protects murine liver in Pb(II) induced hepatic damage and cell death via MAP kinase, NF-kappaB and mitochondria dependent pathways. PLoS One 8:e56894
77. Saha S, Sadhukhan P, Sil PC (2016a) Mangiferin: a xanthonoid with multipotent anti-inflammatory potential. Biofactors 42:459–474
78. Selles AJ, Villa DG, Rastrelli L (2015) Mango polyphenols and its protective effects on diseases associated to oxidative stress. Curr Pharm Biotechnol 16:272–280

79. Menkovic N, Juranic Z, Stanojkovic T, Raonic-Stevanovic T, Savikin K, Zdunic G, Borojevic N (2010) Radioprotective activity of Gentiana lutea extract and mangiferin. Phytother Res 24:1693–1696
80. Sadhukhan P, Saha S, Sil P (2015) Targeting oxidative stress: a novel approach in mitigating cancer. Biochem Anal Biochem 4. https://doi.org/10.4172/2161-1009.1000236
81. Apontes P, Liu Z, Su K, Benard O, Youn DY, Li X, Li W, Mirza RH, Bastie CC, Jelicks LA, Pessin JE, Muzumdar RH, Sauve AA, Chi Y (2014) Mangiferin stimulates carbohydrate oxidation and protects against metabolic disorders induced by high-fat diets. Diabetes 63:3626–3636
82. Marquez L, Garcia-Bueno B, Madrigal JL, Leza JC (2012) Mangiferin decreases inflammation and oxidative damage in rat brain after stress. Eur J Nutr 51:729–739
83. Pal PB, Ghosh S, Sil PC (2015) Beneficial effect of naturally occurring antioxidants against oxidative stress–mediated organ dysfunctions. In: Bioactive natural products: chemistry and biology. Wiley-VCH Verlag GmbH & Co. KGaA, Weinheim
84. Lim J, Liu Z, Apontes P, Feng D, Pessin JE, Sauve AA, Angeletti RH, Chi Y (2014) Dual mode action of mangiferin in mouse liver under high fat diet. PLoS One 9:e90137
85. Guo F, Huang C, Liao X, Wang Y, He Y, Feng R, Li Y, Sun C (2011) Beneficial effects of mangiferin on hyperlipidemia in high-fat-fed hamsters. Mol Nutr Food Res 55:1809–1818
86. Xing X, Li D, Chen D, Zhou L, Chonan R, Yamahara J, Wang J, Li Y (2014) Mangiferin treatment inhibits hepatic expression of acyl-coenzyme a:diacylglycerol acyltransferase-2 in fructose-fed spontaneously hypertensive rats: a link to amelioration of fatty liver. Toxicol Appl Pharmacol 280:207–215
87. Mirza RH, Chi N, Chi Y (2013) Therapeutic potential of the natural product mangiferin in metabolic syndrome. J Nutr Ther 2:74–79
88. Asthana RK, Gupta R, Agrawal N, Srivastava A, Chaturvedi U, Kanojiya S, Khanna AK, Bhatia G, Sharma VL (2014) Evaluation of antidyslipidemic effect of mangiferin and amarogentin from Swertia chirayita extract in HFD induced Charles Foster rat model and in vitro antioxidant activity and their docking studies. Int J Pharm Sci Res 5:3734
89. Saha S, Sadhukhan P, Sinha K, Agarwal N, Sil PC (2016b) Mangiferin attenuates oxidative stress induced renal cell damage through activation of PI3K induced Akt and Nrf-2 mediated signaling pathways. Biochem Biophys Rep 5:313–327
90. Wolfender J-L, Urbain A, Hostettmann K (2015) Profiling, isolation, chemical characterisation and distribution of Gentianaceae constituents. In: The Gentianaceae-volume 2: biotechnology and applications. Springer, Berlin/Heidelberg
91. Crockett SL, Poller B, Tabanca N, Pferschy-Wenzig EM, Kunert O, Wedge DE, Bucar F (2011) Bioactive xanthones from the roots of *Hypericum perforatum* (common St John's wort). J Sci Food Agric 91:428–434
92. Abbaskhan A, Siddiqui H, Anjum S, Orhan I, Gurbuz I, Ayanoglud F (2010) New and known constituents from Iris unguicularis and their antioxidant activity. Heterocycles 82:813–824
93. Xu L, Li A, Sun A, Liu R (2010) Preparative isolation of neomangiferin and mangiferin from Rhizoma anemarrhenae by high-speed countercurrent chromatography using ionic liquids as a two-phase solvent system modifier. J Sep Sci 33:31–36
94. Viswanadh EK, Rao BN, Rao BS (2010) Antigenotoxic effect of mangiferin and changes in antioxidant enzyme levels of Swiss albino mice treated with cadmium chloride. Hum Exp Toxicol 29:409–418
95. Kammalla AK, Ramasamy MK, Inampudi J, Dubey GP, Agrawal A, Kaliappan I (2015) Comparative pharmacokinetic study of mangiferin after oral administration of pure mangiferin and US patented polyherbal formulation to rats. AAPS PharmSciTech 16:250–258
96. Wang H, Ye G, Ma CH, Tang YH, Fan MS, Li ZX, Huang CG (2007) Identification and determination of four metabolites of mangiferin in rat urine. J Pharm Biomed Anal 45:793–798
97. Rashid K, Sinha K, Sil PC (2013) An update on oxidative stress-mediated organ pathophysiology. Food Chem Toxicol 62:584–600

98. Sinha K, Pal PB, Sil PC (2013) Mangiferin, a naturally occurring xanthone C-glycoside, ameliorates lead (Pb)-induced murine cardiac injury via mitochondria-dependent apoptotic pathways. Signpost Open Access J Org Biomol Chem 1:47–63
99. Faizi S, Zikr-Ur-Rehman S, Ali M, Naz A (2006) Temperature and solvent dependent NMR studies on mangiferin and complete NMR spectral assignments of its acyl and methyl derivatives. Magn Reson Chem 44:838–844
100. Barreto JC, Trevisan MT, Hull WE, Erben G, De Brito ES, Pfundstein B, Wurtele G, Spiegelhalder B, Owen RW (2008) Characterization and quantitation of polyphenolic compounds in bark, kernel, leaves, and peel of mango (*Mangifera indica* L.). J Agric Food Chem 56:5599–5610
101. Sekar M (2015) Molecules of interest-mangiferin-a review. Ann Res Rev Biol 5:307
102. Danthu P, Lubrano C, Flavet L, Rahajanirina V, Behra O, Fromageot C, Rabevohitra R, Roger E (2010) Biological factors influencing production of xanthones in Aphloia theiformis. Chem Biodivers 7:140–150
103. Sethiya NK, Mishra S (2014) Investigation of mangiferin, as a promising natural polyphenol xanthone on multiple targets of Alzheimer's disease. J Biol Act Prod Nat 4:111–119
104. Ahmad A, Padhye S, Sarkar FH (2012) Role of novel nutraceuticals garcinol, plumbagin and mangiferin in the prevention and therapy of human malignancies: mechanisms of anticancer activity. In: Nutraceuticals and cancer. Springer, Dordrecht
105. Joubert E, Otto F, Grüner S, Weinreich B (2003) Reversed-phase HPLC determination of mangiferin, isomangiferin and hesperidin in Cyclopia and the effect of harvesting date on the phenolic composition of C. genistoides. Eur Food Res Technol 216:270–273
106. Morel I, Abalea V, Sergent O, Cillard P, Cillard J (1998) Involvement of phenoxyl radical intermediates in lipid antioxidant action of myricetin in iron-treated rat hepatocyte culture. Biochem Pharmacol 55:1399–1404
107. Sahu SC, Gray GC (1993) Interactions of flavonoids, trace metals, and oxygen: nuclear DNA damage and lipid peroxidation induced by myricetin. Cancer Lett 70:73–79
108. Chen W, Li Y, Li J, Han Q, Ye L, Li A (2011) Myricetin affords protection against peroxynitrite-mediated DNA damage and hydroxyl radical formation. Food Chem Toxicol 49:2439–2444
109. Henneberg R, Otuki MF, Furman AE, Hermann P, Do Nascimento AJ, Leonart MS (2013) Protective effect of flavonoids against reactive oxygen species production in sickle cell anemia patients treated with hydroxyurea. Rev Bras Hematol Hemoter 35:52–55
110. Robak J, Gryglewski RJ (1988) Flavonoids are scavengers of superoxide anions. Biochem Pharmacol 37:837–841
111. Semwal DK, Semwal RB, Combrinck S, Viljoen A (2016) Myricetin: a dietary molecule with diverse biological activities. Nutrients 8:90
112. Xinhuai Z, Xin Z (2009) Comparisons of cytoprotective effects of three flavonoids against human hepatocytes oxidative injury induced by hydrogen peroxide or carbon tetrachloride in vitro. J Med Plant Res 3:776–784
113. Laughton MJ, Halliwell B, Evans PJ, Hoult JR (1989) Antioxidant and pro-oxidant actions of the plant phenolics quercetin, gossypol and myricetin. Effects on lipid peroxidation, hydroxyl radical generation and bleomycin-dependent damage to DNA. Biochem Pharmacol 38:2859–2865
114. Pandey KB, Mishra N, Rizvi SI (2009) Myricetin may provide protection against oxidative stress in type 2 diabetic erythrocytes. Z Naturforsch C 64:626–630
115. Xie H-J, Mou W-S, Lin F-R, Xu J-H, Lei Q-F (2013) Radical scavenging activity of myricetin. Acta Phys -Chim Sin 29:1421–1432
116. Mira L, Fernandez MT, Santos M, Rocha R, Florencio MH, Jennings KR (2002) Interactions of flavonoids with iron and copper ions: a mechanism for their antioxidant activity. Free Radic Res 36:1199–1208
117. Justino GC, Vieira AJ (2010) Antioxidant mechanisms of Quercetin and Myricetin in the gas phase and in solution--a comparison and validation of semi-empirical methods. J Mol Model 16:863–876

118. Romanouskaya TV, Grinev VV (2009) Cytotoxic effect of flavonoids on leukemia cells and normal cells of human blood. Bull Exp Biol Med 148:57–59
119. Oyama Y, Fuchs PA, Katayama N, Noda K (1994) Myricetin and quercetin, the flavonoid constituents of Ginkgo biloba extract, greatly reduce oxidative metabolism in both resting and Ca(2+)-loaded brain neurons. Brain Res 635:125–129
120. Lopez-Lazaro M, Willmore E, Austin CA (2010) The dietary flavonoids myricetin and fisetin act as dual inhibitors of DNA topoisomerases I and II in cells. Mutat Res 696:41–47
121. Constantinou A, Mehta R, Runyan C, Rao K, Vaughan A, Moon R (1995) Flavonoids as DNA topoisomerase antagonists and poisons: structure-activity relationships. J Nat Prod 58:217–225
122. Kang NJ, Jung SK, Lee KW, Lee HJ (2011) Myricetin is a potent chemopreventive phytochemical in skin carcinogenesis. Ann N Y Acad Sci 1229:124–132
123. Kumamoto T, Fujii M, Hou DX (2009) Myricetin directly targets JAK1 to inhibit cell transformation. Cancer Lett 275:17–26
124. Ichimatsu D, Nomura M, Nakamura S, Moritani S, Yokogawa K, Kobayashi S, Nishioka T, Miyamoto K (2007) Structure-activity relationship of flavonoids for inhibition of epidermal growth factor-induced transformation of JB6 Cl 41 cells. Mol Carcinog 46:436–445
125. Xu R, Zhang Y, Ye X, Xue S, Shi J, Pan J, Chen Q (2013) Inhibition effects and induction of apoptosis of flavonoids on the prostate cancer cell line PC-3 in vitro. Food Chem 138:48–53
126. Sun F, Zheng XY, Ye J, Wu TT, Wang J, Chen W (2012) Potential anticancer activity of myricetin in human T24 bladder cancer cells both in vitro and in vivo. Nutr Cancer 64:599–606
127. Zhang ZT, Cao XB, Xiong N, Wang HC, Huang JS, Sun SG, Wang T (2010) Morin exerts neuroprotective actions in parkinson disease models in vitro and in vivo. Acta Pharmacol Sin 31(8):900–906
128. Benavente-Garcia O, Castillo J, Lorente J, Alcaraz M, Yanez I, Martinez C, Vicente V, Lozano J (2005) Antiproliferative activity of several phenolic compounds against melanoma cell lines: relationship between structure and activity. Agro Food Ind Hi Tech 16:30–34
129. Walker EH, Pacold ME, Perisic O, Stephens L, Hawkins PT, Wymann MP, Williams RL (2000) Structural determinants of phosphoinositide 3-kinase inhibition by wortmannin, LY294002, quercetin, myricetin, and staurosporine. Mol Cell 6:909–919
130. Labbe D, Provencal M, Lamy S, Boivin D, Gingras D, Beliveau R (2009) The flavonols quercetin, kaempferol, and myricetin inhibit hepatocyte growth factor-induced medulloblastoma cell migration. J Nutr 139:646–652
131. Shih YW, Wu PF, Lee YC, Shi MD, Chiang TA (2009) Myricetin suppresses invasion and migration of human lung adenocarcinoma A549 cells: possible mediation by blocking the ERK signaling pathway. J Agric Food Chem 57:3490–3499
132. Kim ME, Ha TK, Yoon JH, Lee JS (2014) Myricetin induces cell death of human colon cancer cells via BAX/BCL2-dependent pathway. Anticancer Res 34:701–706
133. Schutte ME, Van De Sandt JJ, Alink GM, Groten JP, Rietjens IM (2006) Myricetin stimulates the absorption of the pro-carcinogen PhIP. Cancer Lett 231:36–42
134. Borde P, Mohan M, Kasture S (2011) Effect of myricetin on deoxycorticosterone acetate (DOCA)-salt-hypertensive rats. Nat Prod Res 25:1549–1559
135. Godse S, Mohan M, Kasture V, Kasture S (2010) Effect of myricetin on blood pressure and metabolic alterations in fructose hypertensive rats. Pharm Biol 48:494–498
136. Kang BY, Kim SH, Cho D, Kim TS (2005) Inhibition of interleukin-12 production in mouse macrophages via decreased nuclear factor-kappaB DNA binding activity by myricetin, a naturally occurring flavonoid. Arch Pharm Res 28:274–279
137. Jimenez R, Andriambeloson E, Duarte J, Andriantsitohaina R, Jimenez J, Perez-Vizcaino F, Zarzuelo A, Tamargo J (1999) Involvement of thromboxane A2 in the endothelium-dependent contractions induced by myricetin in rat isolated aorta. Br J Pharmacol 127:1539–1544
138. Cho YC, Yoon G, Lee KY, Choi HJ, Kang BY (2007) Inhibition of interleukin-2 production by myricetin in mouse EL-4 T cells. Arch Pharm Res 30:1075–1079

139. Ong KC, Khoo HE (1996) Insulinomimetic effects of myricetin on lipogenesis and glucose transport in rat adipocytes but not glucose transport translocation. Biochem Pharmacol 51:423–429
140. Ong KC, Khoo HE (2000) Effects of myricetin on glycemia and glycogen metabolism in diabetic rats. Life Sci 67:1695–1705
141. Zelus C, Fox A, Calciano A, Faridian BS, Nogaj LA, Moffet DA (2012) Myricetin inhibits islet amyloid polypeptide (IAPP) aggregation and rescues living mammalian cells from IAPP toxicity. Open Biochem J 6:66–70
142. Tzeng TF, Liou SS, Liu IM (2011) Myricetin ameliorates defective post-receptor insulin signaling via beta-endorphin signaling in the skeletal muscles of fructose-fed rats. Evid Based Complement Alternat Med 2011:150752
143. Liu IM, Tzeng TF, Liou SS, Lan TW (2007) Improvement of insulin sensitivity in obese Zucker rats by myricetin extracted from Abelmoschus moschatus. Planta Med 73:1054–1060
144. Ozcan F, Ozmen A, Akkaya B, Aliciguzel Y, Aslan M (2012) Beneficial effect of myricetin on renal functions in streptozotocin-induced diabetes. Clin Exp Med 12:265–272
145. Gebhardt R (2003) Variable influence of kaempferol and myricetin on in vitro hepatocellular cholesterol biosynthesis. Planta Med 69:1071–1074
146. Chang CJ, Tzeng TF, Liou SS, Chang YS, Liu IM (2012) Myricetin increases hepatic peroxisome proliferator-activated receptor alpha protein expression and decreases plasma lipids and adiposity in rats. Evid Based Complement Alternat Med 2012:787152

Natural Therapeutics for Alzheimer's Disease

11

Shweta Dang, Deeksha Mehtani, Atinderpal Kaur, and Reema Gabrani

Abstract

Alzheimer's disease (AD) is a neurodegenerative disease which is serious, persistent and progressive and is linked with deterioration of memory and cognition. Commonly, Alzheimer's is the reason to cause dementia in aged people. The pathogenesis of this disease is linked with the buildup of amyloid beta (Aβ) plaques and neurofibrillary tangles (NFTs) in brain tissues, and also the tau protein gets hyper-phosphorylated in neurons. The generation of reactive oxygen species (ROS) as a result of oxidative stress is regarded as the main cause of AD. The present treatment offers only symptomatic relief which turns down the rate of cognitive destruction related with AD. Inhibition of the enzyme acetylcholinesterase (AChE) is believed as one of the key therapeutic approach contributing only symptomatic relief and modest disease modifying result. None of the drugs currently available could delay or halt the progression of AD. Several compounds showed positive results in preclinical studies but failed in clinical trials as they had limited targeting because of their inability to cross blood-brain barrier (BBB). Several problems exist in the development of new therapeutics. Medicinal plants have been reported for promising anti-AD action in many preclinical and clinical trials. Natural compounds provide various structural characteristics and biological activities and therefore are an attractive source for developing compounds against AD. Advance in extraction and separation method leads to the generation of natural products as potential therapeutics. Various medicinal plants also in their basic structure or as secluded compounds have demonstrated to lessen the pathological characteristics related with AD. In this chapter an effort has been made to focus on natural substances having role in anti-Alzheimer's therapy with their source, mechanism of action and limitations.

S. Dang (✉) · D. Mehtani · A. Kaur · R. Gabrani
Department of Biotechnology, Jaypee Institute of Information Technology, Noida, Uttar Pradesh, India
e-mail: shweta.dang@jiit.ac.in

V. Rani, U. C. S. Yadav (eds.), *Functional Food and Human Health*,
https://doi.org/10.1007/978-981-13-1123-9_11

Keywords
Alzheimer's disease · Berberine · Herbal formulation · Huperzine · Phytochemical

11.1 Introduction

One of the common neurodegenerative disorders, Alzheimer's disease (AD), is most frequent among aged and is characterized by progressive decline in cognition and memory loss, along with behaviour changes, because of the formation of outsized senile plaques in patient's brain [1]. Many biochemical changes happening within the cell are also identified to be responsible to induce neuronal cell death such as oxidative stress, inflammation, disturbances in cellular metabolism and accumulation of amyloid-beta peptides that finally lead to the programmed cell death of neurons in AD [2]. Histopathological characteristics of AD comprise amyloid-beta (Aβ) plaque deposition and formation of neurofibrillary tangles (NFTs) [3].

Senile plaques are produced by insoluble amyloid-beta peptide aggregates from the cleavage of amyloid precursor protein (APP). These senile plaques highly harm the synapses [4]. However, neurofibrillary tangles (NFTs) are nothing, but the aggregates of hyper-phosphorylated tau proteins which are microtubule-associated proteins and NFTs lead to the synaptic toxicity [4].

11.1.1 Amyloid Cascade Hypothesis

The key event that triggers the commencement and succession of Alzheimer's disease is formation of amyloid plaques by the buildup of Aβ proteins on the outside surface of neurons which leads to degeneration and killing of neurons. The killing and degeneration of neurons ultimately lead to dementia [5].

11.1.2 Formation of Amyloid–Beta Protein

Pathogenesis of AD involves the genesis, abnormal aggregation and buildup of reactive amyloid beta (Aβ) peptide produced due to the abnormal processing of amyloid precursor protein (APP). The generation of Aβ from APP proteolysis is evidenced to be the critical process linked with the generation of plaques in the brain of AD patients [6]. APP is metabolized sequentially by a series of proteases, including α-secretase, β-secretase and the intra-membranous γ-secretase complex. The process of sequential proteolysis of APP which is a transmembrane protein includes two pathways: the amyloidogenic and the non-amyloidogenic. In the pathway which is known as amyloidogenic, here APP that is cleaved by β-secretase produces soluble peptide APPβ (APPsβ) and cleavage of left-out fragment can be carried out by γ-secretase yielding the intracellular domain of APP and sticky Aβ peptide.

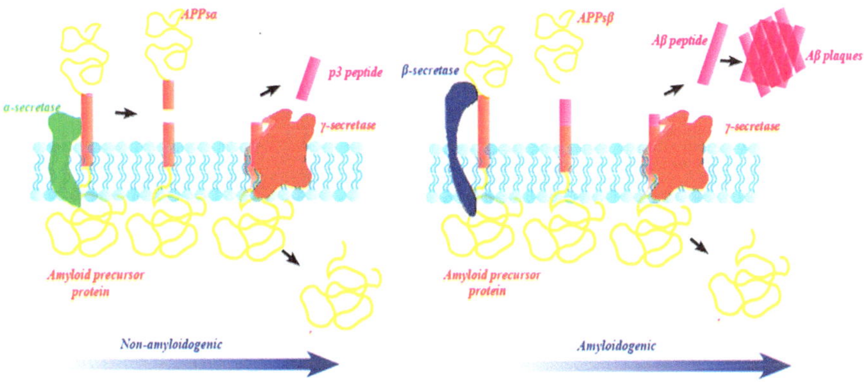

Fig. 11.1 Representation showing how the secretase enzymes lead to the chronological cleavage of amyloid precursor protein in non-amyloidogenic and amyloidogenic pathways

Many Aβ peptides stick together and form amyloid plaques. Whereas, in non-amyloidogenic pathway, APP is first cleaved by α-secretase generating a soluble fragment of APP (APPsα), the left-out part is additionally cleaved by γ-secretase and releases the peptide known as p3 peptide [7] (Fig. 11.1).

11.1.3 Formation of Neurofibrillary Tangles (NFTs)

In normal mature neuron, three major microtubule-associated proteins are present—microtubule-associated protein (MAP) tau, MAP 1 (A/B) and MAP 2. These three MAPs provide the promotion and stability of microtubules [8]. The degree of phosphorylation determines the biological activity of tau protein. Disintegration of microtubule assembly happens because the tau gets hyper-phosphorylated [9]. In Alzheimer's disease (AD), tauopathies are likely to happen in which tau protein is abnormally hyper-phosphorylated and mount up as intracellular tangles of paired helical filaments (PHF) in the brain. This characteristic lesion in the brain directly associates with dementia in these patients [3, 4].

11.1.4 Mitochondrial Dysfunction and Oxidative Stress in AD

Mitochondria are well known as the powerhouse of the cell because it generates energy and participates in many physiological functions. Mitochondrial dysfunction and oxidative stress promote intracellular accumulation of Aβ peptides considered as one of the key processes associated with the pathogenesis of AD [10]. Dysfunction in mitochondria includes abnormalities in ATP production, oxidative stress and membrane potential damage [10, 11].

Oxidative stress is originated due to disparities in production of reactive oxygen species (ROS) and anti-oxidative defence, strongly associated in AD pathogenesis.

The brain is prone to oxidative damage because of lack of antioxidant systems compared to other organs. The accumulation of these Aβ peptides in the brain promotes the oxidative stress. It is long known that oxidative stress augments the Aβ peptide formation, by decreasing the activity of α-secreatse and promoting cleavage of APP by β-secreatse and thereafter by γ-secreatse. Oxidative stress enhances BACE expression by activation of redox-sensitive activator protein (AP1) and nuclear factor (NF)-kappa B transcription factors. Therefore oxidative stress is also playing a key role in AD pathogenesis [12].

Mitochondria are pivotal for the synthesis of ATP, calcium homeostasis, cell survival and apoptosis. Mitochondrial respiratory chain majorly produces ROS inside the cell, and therefore the environment inside mitochondria is vulnerable to oxidative stress. Deficiency in mitochondrial cytochrome oxidase (complex IV) which is an electron transport enzyme enhances the ROS production and reduces stored energy. In due course, this contributes to the neurodegenerative process [12]. It has been evidenced that intracellular uptake of Aβ peptides by neurons may directly disrupt the functioning of mitochondria which leads to the scarcity of energy metabolism and ultimately death of neurons. The existence of Aβ in mitochondria was linked to lack of ATP production and augmented mitochondrial ROS production [10].

11.1.5 Cholinesterase Enzyme Activity in AD

The two chief types of cholinesterases (ChEs) are acetylcholinesterase (AChE) and butyrylcholinesterase (BuChE) which are genetically and structurally diverse. BuChE is apparently found in neurons and glial cells. In Alzheimer's patients, BuChE activity increases, whereas AChE activity decreases [13]. AChE is there in the CNS (central nervous system) and catalyses the hydrolysis of the acetylcholine (ACh) to choline and therefore returns a stimulated cholinergic neuron to its dull and resting state. In AD, there's a shortage of acetylcholine in the brain. Thus, AChE and BuChE are imperative features to be studied in the pathogenesis of AD [13, 14].

11.2 Limitation of Current Therapy for AD

Till date there is no therapy that can cure AD completely. There are five drugs approved by FDA which provide the modest symptomatic benefit but does not delay disease progression [15]. Acetylcholinesterase (AChE) inhibitors are the main class to which drug molecules like tacrine, donepezil, rivastigmine and galantamine belong to and are available in market for AD. Acetylcholinesterase (AChE) inhibitors reduce the rate of acetylcholine degradation and lead to increase in its concentration in the brain. Shortage of acetylcholine (ACh) leads to decline in cognition [15, 16]. Donepezil is commending for the treatment of advanced stages of AD [17]. NMDA receptor antagonist is another class of drug molecule, and memantine is approved by the US Food and Drug Administration (FDA), is able to work significantly in patients with modest to severe AD and perks up language function and cognitive ability [18, 19].

None of the drugs mentioned above could delay or halt the progression of AD [20]. Several problems exist in development of new therapeutics. Several compounds showed positive results in preclinical studies but failed in clinical trials as they had limited targeting because of their inability to pass blood-brain barrier (BBB). Although many compounds are reported for their neuroprotective effects in cellular and animal models of AD [21, 22], many failed in clinical trials, and also some of them are promising candidates for AD treatment [23, 24].

11.3 Natural Products for the Treatment of AD

Phytochemicals are important source of protective compounds for the generation of anti-AD agents. Plant-produced natural compounds provide various structural characteristics and biological activities and therefore are an attractive source for developing compounds against AD. Advancement in isolation procedures leads to the growth of natural products as budding therapeutics [25]. In the past few years, many researchers have tried to estimate the consequences of total plant extract on AD as well as isolated the bioactive substances having the protective role [26, 27]. The chapter focuses on natural substances having role in anti-Alzheimer's therapy with their source, mechanism of action and limitations (Table 11.1).

11.3.1 Role of Alkaloids

11.3.1.1 Huperzine A

Huperzine A (HupA) (Fig. 11.2) is a lycopodium alkaloid secluded for the first time in year 1986 from a Chinese herb named *Huperzia serrata* [56]. It is a chemically exceptional molecule in contrast to other substances under research for AD. It is a reversible, compelling and selective acetylcholinesterase (AChE) inhibitor [57].

HupA exhibits cholinergic and non-cholinergic effects. Moreover, it acts as both competitive and reversible acetylcholinesterase inhibitor [58] and showed maximum activity in both in vitro and in vivo studies in contrast to donepezil, rivastigmine, tacrine as well as galantamine [57]. Via inhibiting acetylcholinesterase, HupA enhances the synaptic acetylcholine release and cholinergic neurotransmission. Cholinergic system is an imperative part of the neuronal circuitry that amends cognition. ACh is regarded to be a neurotransmitter and can also be purposed as a cytokine in a range of neuroprotective pathways [59]. Cholinergic system is also involved in the direction of nerve growth factor synthesis and discharge. Nerve growth factor (NGF) promotes the continued existence and outgrowth of vital cholinergic neurons [65], while some AChE inhibitors also exert NGF-like activity [66]. HupA is a valuable AChE inhibitor and holds NGF-inducing commotion and a capacity to bring on certain NGF-like effects. In an in vitro study, it has been seen that HupA induced neurite outgrowth and stimulated expression of NGF and

Table 11.1 Natural substances having role in anti-Alzheimer's therapy, their mechanism of action and limitations

Type	Name	Structure	Source	Mechanism of action	Limitation	Reference
Flavonoid	Resveratrol	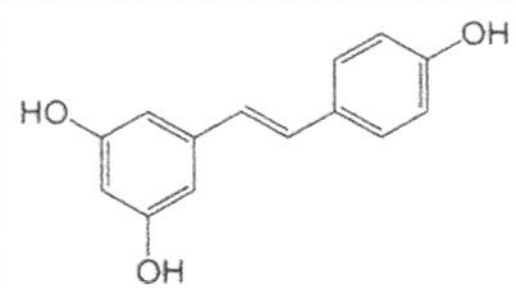	Red wine	Clearance of amyloid-beta peptide, reduce oxidative damage, enhance heme oxygenase HO-1 expression	Poor bioavailability and fast absorption and clearance	[28–30]
	Kaempferol	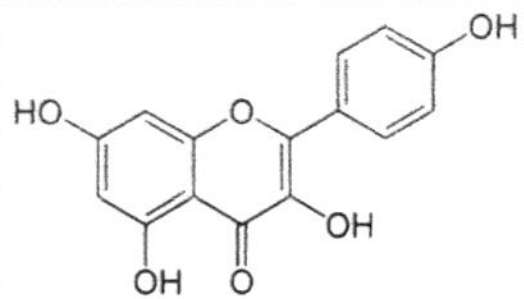	*Ginkgo biloba*	Reduces oxidative stress, reduces neuroinflammation	Bioavailability is poor at ~ 2%, low to moderate absorption, high first-pass metabolism by glucuronidation and additional pathways in the gut as well as liver	[31–33]
	Isorhamnetin	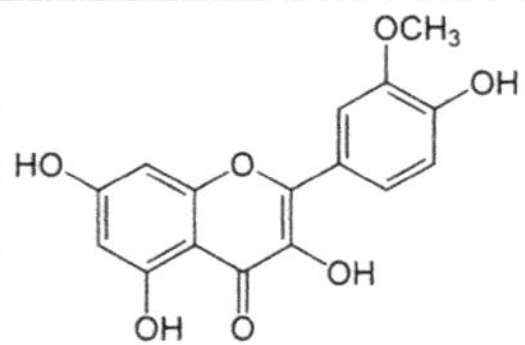	*Ginkgo biloba*	Reduces oxidative stress, up-regulates synthesis of heme oxygenase 1 enzyme	Low bioavailability and insoluble in water	[31]
	Quercetin	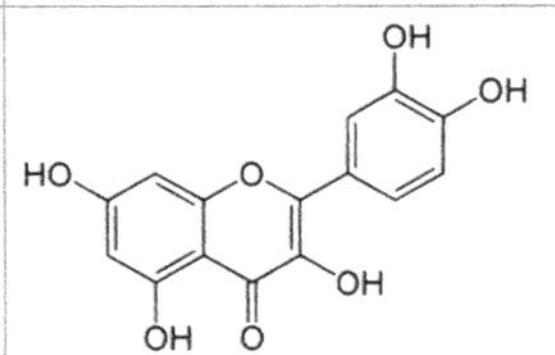	*Ginkgo biloba*, apple, cherries	Reduces oxidative stress, enhances hippocampal neurogenesis and synaptogenesis, decreases amyloidosis, decreases tauopathy, decreases astrogliosis and microgliosis in the hippocampus and amygdala	Poor solubility and bioavailability around 2%	[31, 34–36]

	EGCG	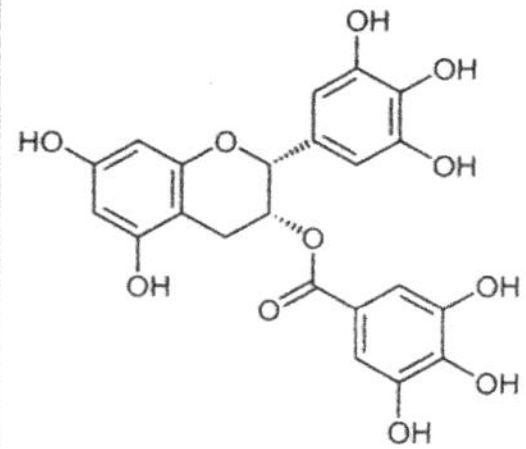	Green tea	Protects against ab (amyloid beta)-induced cytotoxicity, acetylcholinesterase inhibitors, scavenging of reactive oxygen species induced by Aβ	Bioavailability around 1.6%, absorbed but extensively glucuronidated	[37–40]
	Naringenin	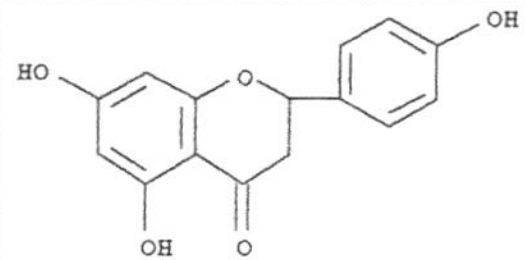	Citrus fruits	Protects against oxidative stress, protects brain from cognitive deficits and neuronal injury	Low solubility, minimal bioavailability and largely hydrophobic ring structure	[41, 42]
	Catechin	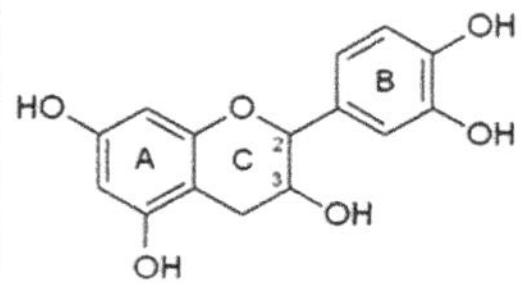	Green tea	Protects against ab-induced cytotoxicity, acetylcholinesterase inhibitors	Absolute bioavailability of 0.39%, absorption is slow, high first-pass effect and wide tissue distribution	[38, 43]
	Curcumin	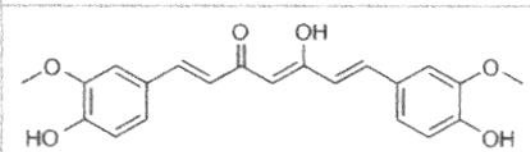	Turmeric	Decreases Aβ plaques, delays neuronal degradation, anti-inflammatory, antioxidant, decreases formation of microglia, improves patient's memory	Poor bioavailability, reduced absorption, quick metabolism and speedy systemic elimination	[44, 45]

(continued)

Table 11.1 (continued)

Type	Name	Structure	Source	Mechanism of action	Limitation	Reference
Alkaloid	Berberine	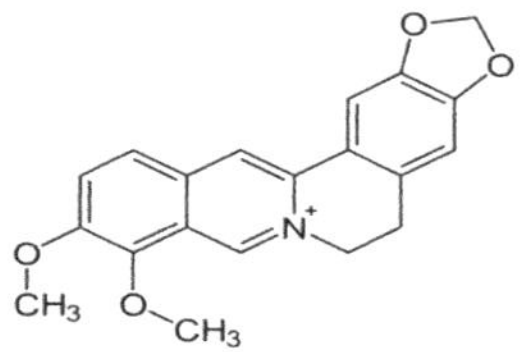	Chinese herb *Rhizoma coptidis*	Antioxidant, AChE and BChE inhibitory, MAO inhibitory, reduces Aβ level and also lowers cholesterol levels	Substrate of p-gp, intestinal absorption is very low, absolute bioavailability of 0.68%	[46–55]
	Huperzine A	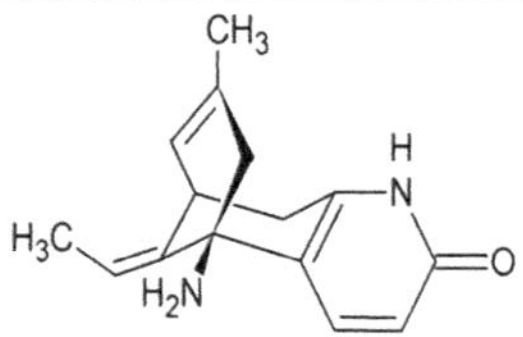	*Huperzia serrata*	Cholinesterase inhibitor, non-cholinergic effects, attenuates mitochondrial dysfunction	Nausea, vomiting and diarrhoea. In extremely huge doses, it can cause indistinct talking, muscle twitching, drooling, prominent blood pressure and a sluggish heart rate	[56–61]
Terpenoids	Ginsenoids	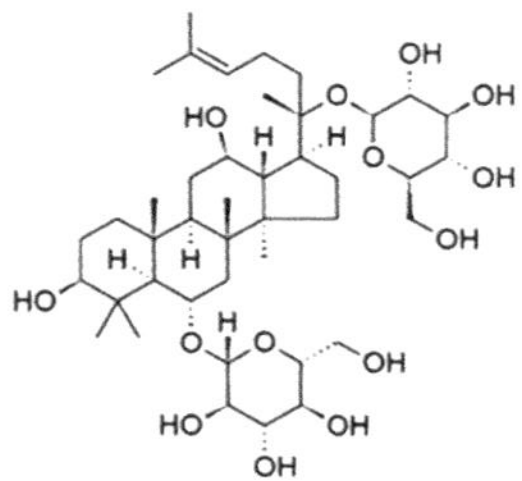	*Panax* ginseng	Reduces Aβ levels by enhancing neprilysin gene expression, which is a rate-limiting enzyme in Aβ degradation	Poor membrane permeability and active biliary excretion	[35, 62]

	Ginkgolide A	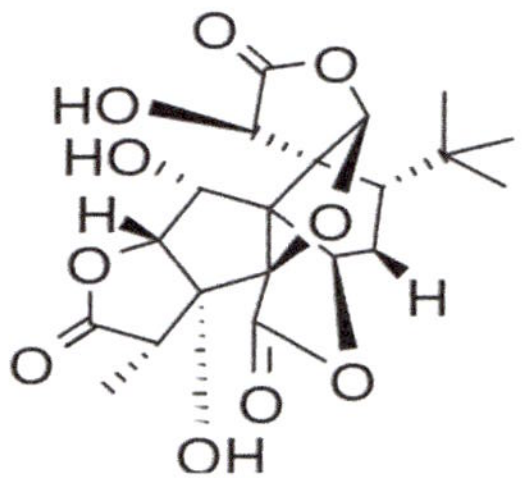	*Gingko biloba*	Platelet activating factor (PAF) antagonist. Also has antioxidant capacities, anxiolytic effects	Absolute bioavailability (61.1 ± 10.4)%	[31, 63]
	Ginkgolide B	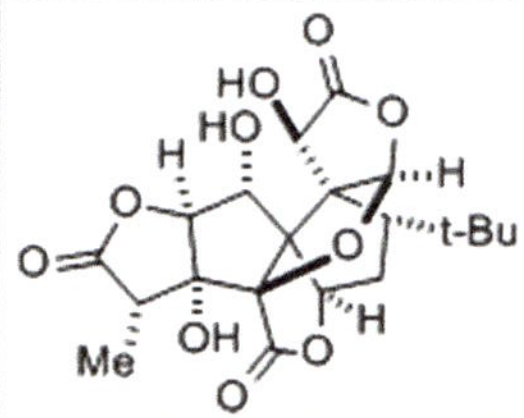	*Gingko biloba*	PAF antagonist, attenuates glutamate-induced neuronal damage, increases the production of brain-derived neurotrophic factor, reverses Aβ-induced decrease of ACh release	Absolute bioavailability (27.2 ± 7.7)%	[35, 63, 64]
	Bilobalide	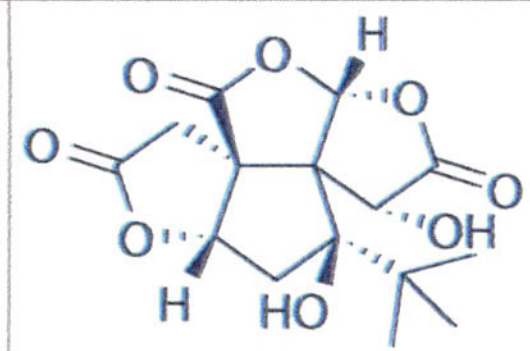	*Gingko biloba*	Reduces synaptic loss induced by Aβ, enhances hippocampal neurogenesis and synaptogenesis, reduces neuronal inflammation, anti-apoptotic properties	Absolute bioavailability (56.2 ± 4.4)%	[31, 35, 63]

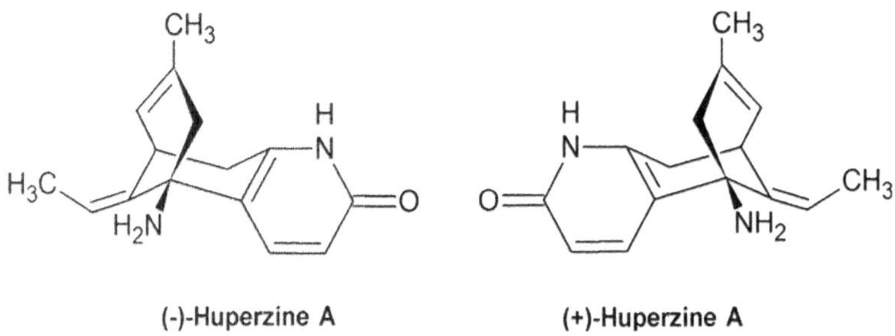

Fig. 11.2 Stereoisomer of Huperzine A [56]

secretion of NGF in cultured rat cortical astrocytes [60]. Potential cholinergic effects of HupA might be useful in maintaining and restoring the normal functioning of neural cells in neurodegenerative disease.

However, HupA was found to exhibit some added benefits that are independent of AChE inhibition. This differentiates HupA treatment from other acetylcholine inhibitory drugs [59]. HupA is able to reduce mitochondrial dysfunction. In a study, it was observed that administering HupA to isolated mitochondria blocked swelling caused by normal osmosis and increased the rate of ATP production, indicating that HupA protects mitochondria against Aβ by preserving the integrity of membrane and improving energy metabolism [61]. In vitro studies have provided important clues which suggest that HupA is able to protect the neuronal cells from mitochondrial dysfunction [67] and could be an alternate way to slow down the progression of intracellular Aβ neurotoxicity during decline in the buildup of intracellular or mitochondrial Aβ [68].

In an in vivo study on healthy human volunteers, HupA was orally administered in the form of tablet at a single dose of 400 mcg. HupA is swiftly immersed and extensively circulated in the body. After 5 to 10 min of administration, HupA started to appear in plasma. Peak concentrations of HupA reached to C_{max} of 2.59 ± 0.37 ng/ml in 58.33 ± 3.89 min. The area below plasma vs time curve $AUC_{(0\text{-}t)}$ was found to be 1986.96 ± 164.57 μg/l min, and the area below plasma from zero to infinity $AUC_{(0\text{-infinity})}$ for HupA was 2450.34 ± 233.32 μg/l.min. HupA was metabolized in the liver and excreted in the urine [69].

11.3.1.2 Berberine

Berberine is a bitter-tasting, plant isoquinoline alkaloid having a molecular weight of 353.36. It can be isolated from the roots, rhizomes and stem bark of the plants, such as *Hydrastis canadensis* (goldenseal), *Berberis aquifolium* (Oregon grape), *Coptis chinensis* (coptis or golden thread), *Berberis aristata* (tree turmeric) and *Berberis vulgaris* (barberry) [70]. Berberine possesses many therapeutic properties, like antioxidant, anti-cancerous [46], antiobesity, helpful in congestive heart failure and anti-inflammatory [71], and plays a role in neurodegenerative diseases and cardiovascular diseases. This alkaloid has been found to possess multiple neuroprotective effects and is involved in the improvement of survival, function of neurons.

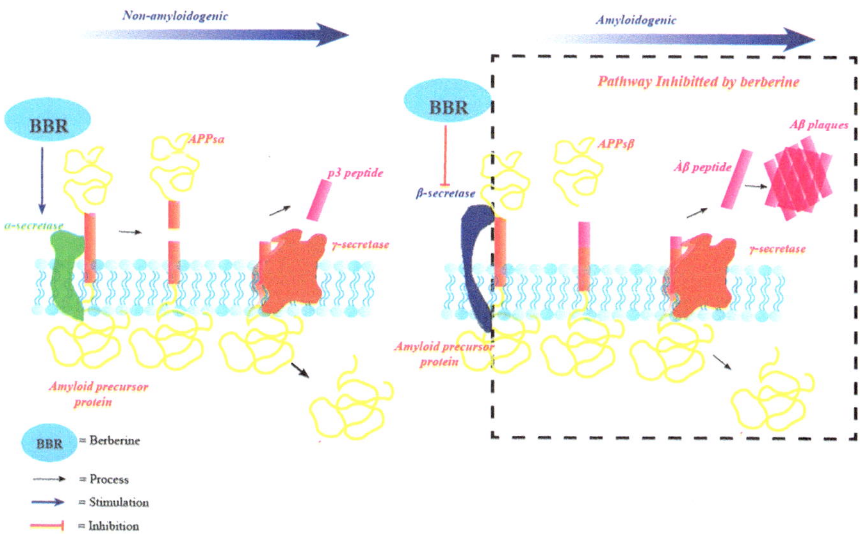

Fig. 11.3 Possible mechanism of action by which berberine reduces Aβ levels. Here, BBR refers to berberine, and APP refers to amyloid precursor protein

Also it has been evidenced that berberine has a protective role in the process of neurodegeneration [72].

Pathogenesis of AD entails the genesis, buildup and abnormal aggregation of reactive amyloid beta (Aβ) peptide formed due to the abnormal cleavage and processing of amyloid precursor protein (APP). It has been evidenced that the production of Aβ by APP proteolysis is the critical process in the development of plaques in the brain of AD patients [73].

Researchers suggested that berberine is involved in decreasing the formation of Aβ by modifying the processing of APP. An in vitro study using HEK293 cells demonstrated that berberine inhibits the expression of beta-secretase enzyme by activating ERK1/2 pathway and leads to decrement in the production of amyloid beta [47]. An in vivo study also demonstrated that berberine chloride treatment protected aged rabbit hippocampus from degeneration and altered the behaviour. It was suggested that berberine chloride decreased the activity of β-site amyloid precursor protein-cleaving enzyme-1 [48]. The schematic presentation shows the probable mechanism of alteration of APP processing by berberine (Fig. 11.3).

AChE and BuChE are two important factors in the pathogenesis of AD [50]. Several studies are reported to provide evidence where berberine acts as an inhibitor of cholinesterase enzymes [51–54]. In an in vitro study, it has been proved that berberine which is one of an alkaloid present in *Rhizoma coptidis* is able to inhibit AChE as well as BuChE with $IC_{50\ \mathrm{values}}$ of 0.44 ± 0.04 μM and 3.44 ± 0.26 μM, respectively [51]. A study using physical chemistry investigated that during the interaction of AChE with berberine, there are events which lead to structural changes in AChE which leads to its inhibition and IC_{50} of berberine against AChE as

0.67 μmol/L(0.25 μg/mL) was reported [52]. Also a docking study found the molecular basis of the inhibitory effects of berberine against the enzymes (AChE, BuChE, MAO-A, MAO-B) implicated in the pathogenesis of AD [54]. Thus, berberine is acting as a two fold inhibitor of AChE and BChE and may enrich lives of AD patients. Therefore, berberine possesses numerous actions which possibly will be drawn in anti-AD potential, including antioxidant action, AChE and BChE inhibitory activity, MAO inhibitory activity and its abilities to diminish Aβ level.

Moreover, berberine is not cytotoxic or mutagenic at doses used in clinical situations, while higher doses can cause such side-effects which include gastrointestinal discomfort, low blood pressure, flu-like symptoms and cardiac damage. Usage of berberine should be avoided in pregnancy, because of its potential to cause uterine contractions. The therapeutic dose reported for most of the clinical situations is 200 mg orally for two to four times daily [70].

11.3.2 Role of Flavonoids and Terpenoids

11.3.2.1 Resveratrol

Resveratrol (3,5,4′-trihydroxystilbene) is a phytoalexin, naturally existing plant polyphenol, belonging to stilbene family and formed in retort to injury in grapevines, legumes plus pines [74]. It has a molecular weight equivalent to 228.24 g/mole. It can be isolated from the seeds of grapes and peanuts and also a valuable constituent of red wine [75, 76]. It has been studied for its therapeutic properties like antioxidant, anti-cancerous, anti-inflammatory, cardio-protective roles and so on [77]. Resveratrol has neuroprotective function against AD. It promotes the clearance of intracellular A*β* by activating AMPK independently of SIRT1 without any influence on the generation of A*β*. Resveratrol lessens the expression of Nox4, a ROS-producing enzyme, but increases the expression of SOD1 and GPx1, ROS-inactivating enzymes. It influences Bcl-2 expression, inhibits Bax expression, blocks JNK activation and thereby suppresses the apoptosis occurring because of oxidative stress [78]. Neuroprotective functions of resveratrol in AD pathogenesis are shown in Fig. 11.4.

An in vitro study proved that resveratrol upholds removal of Alzheimer's disease amyloid peptides. It was shown in the study that resveratrol sturdily trims down Aβ produced by diverse cell lines which expressed wild-type or Swedish mutant APP_{69}. It was suggested that resveratrol advances the intracellular deprivation of the amyloid peptide by a mechanism that associates the proteasome [28]. A study using a rat model of AD demonstrates that resveratrol shields rats from Aβ-induced neurotoxicity by the decline of iNOS expression and lipid peroxidation. It was also found that resveratrol is beneficial for the revival of learning and recollection, inhibits the assembly of iNOS, prevents lipid peroxidation result accumulation and enhances the expression of HO-1 and is also secluded from Aβ-induced neuronal loss. Resveratrol could also undo the Aβ-induced MDA overproduction [29].

Moreover, after oral administration of resveratrol, ~75% of the dose is absorbed mainly by trans-epithelial diffusion [30]. Resveratrol shows high absorption but low

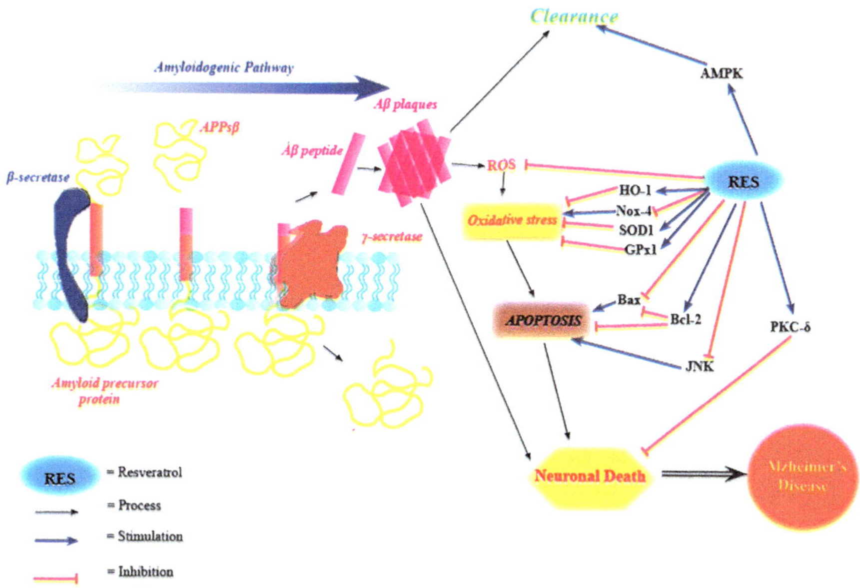

Fig. 11.4 Representation of neuroprotective functions of resveratrol in AD pathogenesis

bioavailability [79]. In subsequent absorption, resveratrol faces fast and wide metabolism which leads to its stumpy bioavailability. Oral bioavailability is reported to be less than 1%. It is chiefly metabolized in the intestine and liver. Chief metabolites of resveratrol are glucuronides and sulphates. Colonic bacteria also metabolize resveratrol. The reported plasma levels of the conjugated metabolites are higher as compared to the parent compound in animal and human studies [30].

An everyday dosage of 200–400 mg of trans-resveratrol is successful. Elevated doses may be required for cancer therapy [80]. Taking resveratrol all along with curcumin, quercetin and piperine is helpful in lowering the dose and increasing bioavailability [81]. Taking resveratrol with quercetin lowers the metabolism [80].

11.3.2.2 *Ginkgo biloba* Extract

Ginkgo biloba is the aged existing plant dating back 200 million years ago also known as living fossil and possesses diverse organic and pharmacological activities [82]. *Ginkgo biloba* leaves extract named as EGb761 is standardized to contain multiple compounds. It has therapeutic role in neurodegenerative, sensory and vascular diseases. Its Extract contains 24% flavonoid glycosides (containing quercetin, kaempferol, isorhamnetin), 6% terpenoids (3.1% ginkgolides A, B, C and J and 2.9% bilobalide) and 5–10% organic acids, out of which flavonoids and terpenoids are suggested to be pharmacologically active ingredients. EGb 761 does not have any activating or inhibitory actions; rather it has regulatory mechanisms which help the organism to adapt to the circumstances [83].

Oxidative stress plays a chief role in the pathogenesis of AD, EGb761 has free radical scavenging activity. There are numerous studies evidencing the anti-oxidative role of the extract [84–86]. When the hippocampus of rats was treated with EGb761 extract, the activity of catalase (CAT) and superoxide dismutase (SOD) were increased, and the decrease in lipid peroxidation in the hippocampus was observed. These indicated a potential role meant for the extract in the cure of diseases concerning free radicals and oxidative damage [84]. Also it has been shown that when cerebellar granule cells were pre-treated with the antioxidant EGb76, it effectively attenuates the oxidative damage induced by $H_2O_2/FeSO_4$. Also the extract disallowed cells from apoptotic cell death. These results suggested that EGb761 might be used as a possible drug for neuronal diseases in which too much production of reactive oxygen species occurs [85]. A study used PC12 cells to examine the shielding features of EGb 761 on H_2O_2 and antimycin stressed mitochondria and also inspected the effects of extract on ROS. They reported that EGb 761 protected mitochondria from the hit of hydrogen peroxide, antimycin and Aβ. Also the extract was able to reduce ROS levels. Their study emphasizes that *Ginkgo biloba* exerts neuroprotective properties like protection against Aβ toxicity and anti-apoptotic properties; these are most likely because of the preventive effects of extract on mitochondria [86]. Thus *Ginkgo biloba* extract may be helpful in treating oxidative stress caused due to plaque formation in Alzheimer's disease. The flavonoid fraction of the extract is chiefly liable for the anti-oxidative effects. Furthermore, EGb761 restrain the development of amyloid-beta fibrils, which are the contributory attribute of AD. A study on N2a cell line which was stably expressing Swedish mutant APP695 and the exon-9 deletion mutant PS1 found decreased amyloid-beta fibrillogenesis in the existence of EGb761. It was established that EGb761 was able to attenuate mitochondrion-initiated apoptosis and reduced the activity of caspase 3 in the cells. The results suggested that neuronal damage in AD might be due to the amyloid-beta toxicity and initiated apoptosis by the mitochondria. EGb761 attenuated apoptosis and directly inhibited aggregation of amyloid-beta fibrils, proving the neuroprotective effects of EGb761 [87].

Quercetin, a flavonoid present in *Ginkgo biloba* extract, is shown to reduce extracellular β-amyloidosis, tauopathy, astrogliosis and microgliosis in the hippocampus and the amygdale by reducing paired helical filament (PHF), reducing Aβ levels by decreasing BACE1-mediated cleavage of APP. These results imply that quercetin undoes histological characteristics of AD and protects cognitive function in elderly AD mice [34].

Ginkgolides A and B protect neuronal cells from synaptic damage and increase endurance against Aβ-induced toxicity. Ginkgolide B saves hippocampal neurons from Aβ-induced apoptosis by raising the production of brain-derived neurotrophic factors. Furthermore, ginkgolide J is the main compelling inhibitor of Aβ-induced hippocampal neuronal cell death amid the ginkgolides in EGb761. Moreover, bilobalide trims down Aβ-induced synaptic loss and later improves hippocampal neurogenesis and synaptogenesis [88]. Loss of synapses, an early incident throughout the succession of disease, is the main reason to lead. Therefore for the avoidance and treatment of AD, synaptogenesis and neurogenesis may perhaps provide as a beneficial target. A study tested that bilobalide and quercetin, the constituents of *Ginkgo biloba* extract, augmented cell proliferation in the hippocampal neurons, improved phosphorylation of cyclic-AMP response element-binding protein (CREB) and raised the levels of brain-derived neurotrophic factor in mice brain. Furthermore, both the components restored Aβ-stimulated synaptic loss and phosphorylation of CREB. The study advised that improved neurogenesis and synaptogenesis by bilobalide and quercetin is because of the signalling pathway interceded by phosphorylation of CREB [35]. A study also reported that ginkgolides also act as PAF receptor antagonists and may guard in opposition to the synapse harm and the cognitive loss seen all through the premature stages of AD [64].

Furthermore, *Ginkgo biloba* extract is well absorbed by administering via oral route. The bioavailability of the components in extract is as follows: flavonol-glycosides is 60%, ginkgolides is 98% and bilobalide is 70% [89]. Absorption of flavonol-glycosides and terpene-lactones together is mostly in the small intestine. Also there is a likelihood of a gastric absorption for ginkgolide B [89]. Daily dosage is 40–80 mg of unvarying extract for two to three times daily. For the patients with Alzheimer's disease, recommended dosage is around 240 mg every day. In unceasing situation the extract must be given for no less than 6–8 weeks before assessment of effectiveness [90].

11.4 Conclusion

Natural products are an attractive source for developing anti-Alzheimer's therapeutics. FDA has only approved five drugs till date to treat patients with AD. This chapter has discussed about some phytochemicals and their targets, particularly alkaloids, flavonoids and terpenoids, which can be generated as budding anti-AD agents.

Alkaloids, flavonoids and terpenoids are widely available in natural foods. Therefore, using these natural compounds as therapeutics for treatment of AD is considered as an attractive alternative. These phytochemicals have been successfully demonstrated to have preventive and beneficial properties against the general mechanisms of AD as discussed above. Means to deliver these organic bioactive compounds to the brain successfully must be measured. Also, a combination of two or more phytochemicals with the aim of acting synergistically might be considered to be utilized as an alternative instead of a solitary compound. Advancement in understanding the mechanisms exhibited by these natural products in treatment of AD represents a promising goal in developing novel approach for management and treatment of neurodegenerative diseases.

References

1. LaFerla FM, Oddo S (2005) Alzheimer's disease: Aβ, tau and synaptic dysfunction. Trends Mol Med 11(4):170–176
2. Sadigh-Eteghad S et al (2015) Amyloid-beta: a crucial factor in Alzheimer's disease. Med Princ Pract 24:1–10

3. Serrano-Pozo A et al (2011) Neuropathological alterations in Alzheimer disease. Cold Spring Harb Perspect Med 1(1):a006189
4. Spires-Jones TL, Hyman BT (2014) The intersection of amyloid beta and tau at synapses in Alzheimer's disease. Neuron 82(4):756–771
5. Reitz C (2012) Alzheimer's disease and the amyloid cascade hypothesis: a critical review. Int J Alzheimer's Dis 2012:Article ID 369808, 11 pages
6. Zhang H et al (2012) Proteolytic processing of Alzheimer's B-amyloid precursor protein. J Neurochem 120(Suppl 1):9–21 PMC
7. O'Brien RJ, Wong PC (2011) Amyloid precursor protein processing and Alzheimer's disease. Annu Rev Neurosci 34:185–204
8. Iqbal K et al (2005) Tau pathology in Alzheimer disease and other tauopathies. Biochim Biophys Acta (BBA) – Mol Basis Dis 1739(2–3):198–210
9. Gong C-X, Iqbal K (2008) Hyperphosphorylation of microtubule-associated protein tau: a promising therapeutic target for Alzheimer disease. Curr Med Chem 15(23):2321–2328
10. Misonou H, Kawashima MM, Ihara Y (2000) Oxidative stress induces intracellular accumulation of amyloid β-protein (Aβ) in human neuroblastoma cells. Biochemistry 39(23):6951–6959
11. Federico A et al (2012, 9 pages) Mitochondria, oxidative stress and neurodegeneration. J Neurol Sci 322:254. https://doi.org/10.1016/j.jns.2012.05.030
12. Zhao Y, Zhao B (2013) Oxidative stress and the pathogenesis of Alzheimer's disease. Oxid Med Cell Longev 2013:Article ID 316523, 10 pages. https://doi.org/10.1155/2013/316523
13. Giacobini E (2004) Cholinesterase inhibitors: new roles and therapeutic alternatives. Pharmacol Res 50:433–440
14. Tabet N (2006) Acetylcholinesterase inhibitors for Alzheimer's disease: anti-inflammatories in acetylcholine clothing! Age Ageing 35(4):336–338. https://doi.org/10.1093/ageing/afl027
15. Mielke MM et al (2012) Effects of FDA approved medications for Alzheimer's disease on clinical progression. Alzheimer's Dement: J Alzheimer's Assoc 8(3):180–187
16. Čolović MB et al (2013) Acetylcholinesterase inhibitors: pharmacology and toxicology. Curr Neuropharmacol 11(3):315–335
17. Winblad B (2009) Donepezil in severe Alzheimer's disease. Am J Alzheimer's Dis Other Demen® 24(3):185–192
18. Van Marum RJ (2009) Update on the use of memantine in Alzheimer's disease. Neuropsychiatr Dis Treat 5:237–247
19. Puangthong U, Hsiung G-YR (2009) Critical appraisal of the long-term impact of memantine in treatment of moderate to severe Alzheimer's disease. Neuropsychiatr Dis Treat 5:553–561
20. Bond M (2012) The effectiveness and cost-effectiveness of donepezil, galantamine, rivastigmine and memantine for the treatment of Alzheimer's disease (review of technology appraisal no. 111): a systematic review and economic model. Health Technol Assess 16(21):1–470
21. Irannejad H et al (2010) Synthesis and in vitro evaluation of novel 1,2,4-triazine derivatives as neuroprotective agents. Bioorg Med Chem 18(12):4224–4230
22. Zhao B (2009) Natural antioxidants protect neurons in Alzheimer's disease and Parkinson's disease. Neurochem Res 34(4):630–638
23. Winblad B et al (2012) Safety, tolerability, and antibody response of active Aß immunotherapy with CAD106 in patients with Alzheimer's disease randomised, double-blind, placebo-controlled, first-in-human study. Lancet Neurol 11(7):597–604
24. Frakey LL, Salloway S, Buelow M, Malloy P (2012) A randomized, double-blind, placebo-controlled trial of modafinil for the treatment of apathy in individuals with mild-to-moderate Alzheimer's disease. J Clin Psychiatry 73(6):796–801
25. Sasidharan S et al (2011) Extraction, isolation and characterization of bioactive compounds from plants' extracts. Afr J Tradit Complement Altern Med 8(1):1–10
26. Asadi S et al (2010) In vitro antioxidant activities and an investigation of neuroprotection by six Salvia species from Iran: a comparative study. Food Chem Toxicol 48(5):1341–1349
27. Kuruuzum-Uz A et al (2012) Investigation on anti-inflammatory and antiulcer activities of Anchusa azurea extracts and their major constituent rosmarinic acid. Z Naturforsch C 67(7–8):360–366

28. Marambaud P, Zhao H, Davies P (2005) Resveratrol promotes clearance of Alzheimer's disease amyloid-peptides. J Biol Chem 280(45):37377–37382
29. Huang T-C, Kwok-Tung L, Wo Y-YP, Yao-Ju W, Yang Y-L (2011) Resveratrol protects rats from ab-induced neurotoxicity by the reduction of iNOS expression and lipid peroxidation. PLoS One 6(12):e29102. https://doi.org/10.1371/journal.pone.0029102
30. Walle T (2011) Bioavailability of resveratrol. Ann N Y Acad Sci 1215:9–15
31. Nash and Shah (2015) Current perspectives on the beneficial role of Ginkgo biloba in neurological and cerebrovascular disorders. Integr Med Insight 10:1–9. https://doi.org/10.4137/IMI.S25054
32. Park S, Sapkota K, Kim S, Kim H, Kim S (2011) Kaempferol acts through mitogen-activated protein kinases and protein kinase B/AKT to elicit protection in a model of neuroinflammation in BV2 microglial cells. Br J Pharmacol 164(3):1008–1025. https://doi.org/10.1111/j.1476-5381.2011.01389.x
33. Barve A, Chen C, Hebbar V, Desiderio J, Saw CL-L, Kong A-N (2009) Metabolism, oral bioavailability and pharmacokinetics of chemopreventive kaempferol in rats. Biopharm Drug Dispos 30(7):356–365. https://doi.org/10.1002/bdd.677
34. Maria S-GA, Ignacio M-MJ, Ramírez-Pineda Jose R, Marisol L-R, Edison O, Patricia C-GG (2015) The flavonoid quercetin ameliorates Alzheimer's disease pathology and protects cognitive and emotional function in aged triple transgenic Alzheimer's disease model mice. Neuropharmacology 93:134–145. https://doi.org/10.1016/j.neuropharm.2015.01.027
35. Tchantchou F, Lacor PN, Cao Z, Lao L, Hou Y, Cui C, Klein WL, Luo Y (2009) Stimulation of neurogenesis and synaptogenesis by bilobalide and quercetin via common final pathway in hippocampal neurons. J Alzheimers Dis 18(4):787–798
36. Monograph quercetin (2011) Altern Med Rev 16(2):172–194
37. Choi YT, Jung CH, Lee SR, Bae JH, Baek WK, Suh MH, Park J, Park CW, Suh SI (2001) The green tea polyphenol (-)-epigallocatechin gallate attenuates beta-amyloid-induced neurotoxicity in cultured hippocampal neurons. Life Sci 70(5):603–614
38. Okello EJ, Leylabi R, McDougall GJ (2012) Inhibition of acetylcholinesterase by green and white tea and their simulated intestinal metabolites. Food Funct 3(6):651–661. https://doi.org/10.1039/c2fo10174b
39. Lambert JD, Hong J, Kim DH, Mishin VM, Yang CS (2004) Piperine enhances the bioavailability of the tea polyphenol (-)-epigallocatechin-3-gallate in mice. J Nutr 134(8):1948–1952
40. Lambert JD, Lee MJ, Lu H, Meng X, Hong JJ, Seril DN, Sturgill MG, Yang CS (2003) Epigallocatechin-3-gallate is absorbed but extensively glucuronidated following oral administration to mice. J Nutr 133(12):4172–4177
41. Khan MB, Khan MM, Khan A, Ahmed ME, Ishrat T, Tabassum R, Vaibhav K, Ahmad A, Islam F (2012) Naringenin ameliorates Alzheimer's disease (AD)-type neurodegeneration with cognitive impairment (AD-TNDCI) caused by the intracerebroventricular-streptozotocin in rat model. Neurochem Int 61(7):1081–1093. https://doi.org/10.1016/j.neuint.2012.07.025
42. Shulman M, Cohen M, Soto-Gutierrez A et al (2011) Enhancement of naringenin bioavailability by complexation with hydroxypropoyl-β-cyclodextrin. PLoS One 6(4):e18033. https://doi.org/10.1371/journal.pone.0018033
43. Zhu M, Chen Y, Li RC (2000) Oral absorption and bioavailability of tea catechins. Planta Med 66(5):444–447
44. Mishra S, Palanivelu K (2008) The effect of curcumin (turmeric) on Alzheimer's disease: an overview. Ann Indian Acad Neurol 11(1):13–19. https://doi.org/10.4103/0972-2327.40220
45. Anand P, Kunnumakkara AB, Newman RA, Aggarwal BB (2007) Bioavailability of curcumin: problems and promises. Mol Pharm 4(6):807–818. https://doi.org/10.1021/mp700113r
46. Imanshahidi M, Hosseinzadeh H (2008) Pharmacological and therapeutic effects of Berberis vulgaris and its active constituent, berberine. Phytother Res 22:999–1012
47. Zhu F et al (2011) Decrease in the production of beta-amyloid by berberine inhibition of the expression of beta secretase in HEK293 cells. BMC Neurosci 12:125
48. Panahi N, Mahmoudian M, Mortazavi P, Hashjin GS (2013) Effects of berberine on β-secretase activity in a rabbit model of Alzheimer's disease. Arch Med Sci 1:146–150

49. Cai Z, Wang C, Yang W (2016) Role of berberine in Alzheimer's disease. Neuropsychiatr Dis Treat 12:2509–2520
50. Rees TM et al (2003) The role of acetylcholinesterase in the pathogenesis of Alzheimer's disease. Drugs Today (Barc) 39(1):75–83
51. Jung HA et al (2009) Anti-Alzheimer and antioxidant activities of Coptidis Rhizoma alkaloids. Biol Pharm Bull 32(8):1433–1438
52. Xiang J, Yu C, Yang F, Yang L, Ding H (2009) Conformation-activity studies on the interaction of berberine with acetylcholinesterase: physical chemistry approach. Prog Nat Sci 19(12):1721–1725
53. Huang L, Luo Z, He F, Lu J, Li X (2010) Synthesis and biological evaluation of a new series of berberine derivatives as dual inhibitors of acetylcholinesterase and butyrylcholinesterase. Bioorg Med Chem 18(12):4475–4484. https://doi.org/10.1016/j.bmc.2010.04.063
54. Ji H-F, Shen L (2012) Molecular basis of inhibitory activities of berberine against pathogenic enzymes in Alzheimer's disease. Sci World J 2012:Article ID 823201, 4 pages. https://doi.org/10.1100/2012/823201
55. Chen W, Miao Y-Q, Fan D-J, Yang S-S, Lin X, Meng L-K, Tang X (2011) Bioavailability study of berberine and the enhancing effects of TPGS on intestinal absorption in rats. AAPS PharmSciTech 12(2):705–711. https://doi.org/10.1208/s12249-011-9632-z
56. Ferreira A, Rodrigues M'r, Fortuna A, Falcao A'l, Alves G (2016) Huperzine A from Huperzia serrata: a review of its sources, chemistry, pharmacology and toxicology. Phytochem Rev 15(1):51–85. https://doi.org/10.1007/s11101-014-9384-y
57. Tang XC, Han YF (1999) Pharmacological profile of Huperzine A, a novel acetylcholinesterase inhibitor from Chinese herb. CNS Drug Rev 5(3):281–300
58. Wang R, Yan H, Tang X-c (2006) Progress in studies of huperzine A, a natural cholinesterase inhibitor from Chinese herbal medicine. Acta Pharmacol Sin 27(1):1–26
59. Zhang H-y (2012) New insights into huperzine A for the treatment of Alzheimer's disease. Acta Pharmacol Sin 33:1170–1175
60. Tang L-l, Wang R, Xi-can T (2005) Effects of huperzine A on secretion of nerve growth factor in cultured rat cortical astrocytes and neurite outgrowth in rat PC12 cells. Acta Pharmacol Sin 26(6):673–678
61. Gao X, Zheng CY, Yang L, Tang XC, Zhang HY (2009) Huperzine A protects isolated rat brain mitochondria against β-amyloid peptide. Free Radic Biol Med 46(11):1454–1462
62. Liu H, Yang J, Du F, Gao X, Ma X, Huang Y, Xu F, Niu W, Wang F, Mao Y, Sun Y, Lu T, Liu C, Zhang B, Li C (2009) Absorption and disposition of ginsenosides after oral administration of Panax notoginseng extract to rats. Drug Metab Dispos 37(12):2290–2298. https://doi.org/10.1124/dmd.109.029819 Epub 2009 Sep 28
63. Si HH, Geng T, Sun XP, Zhao J, Xue J (2015) Absolute bioavailability of ginkgolide compounds in rats. Zhongguo Zhong Yao Za Zhi 40(14):2882–2886
64. Bate C, Tayebi M, Williams A (2008) Ginkgolides protect against amyloid-beta1-42-mediated synapse damage in vitro. Mol Neurodegener 3(1):1. https://doi.org/10.1186/1750-1326-3-1
65. Triaca V, Calissano P (2016) Impairment of the nerve growth factor pathway driving amyloid accumulation in cholinergic neurons: the incipit of the Alzheimer's disease story? Neural Regen Res 11(10):1553–1556
66. Shigeta K, Ootaki K, Tatemoto H, Nakanishi T, Inada A, Muto N (2002) Potentiation of nerve growth factor-induced neurite outgrowth in PC12 cells by a Coptidis Rhizoma extract and protoberberine alkaloids. Biosci Biotechnol Biochem 66:2491–2494
67. Gao X, Tang XC (2006) Huperzine A attenuates mitochondrial dysfunction in beta-amyloid-treated PC12 cells by reducing oxygen free radicals accumulation and improving mitochondrial energy metabolism. J Neurosci Res 83(6):1048–1057
68. Lei Y et al (2015) Involvement of intracellular and mitochondrial Aβ in the ameliorative effects of Huperzine A against oligomeric Aβ42-induced injury in primary rat neurons. PLoS One 10(5):e0128366. https://doi.org/10.1371/journal.pone.0128366
69. Kumar P, Jhanjee A, Bhatia MS, Verma D (2011) Huperzisne. Delhi Psychiatry J 14(1):177–179
70. Berberine monograph (2000) Altern Med Rev 5(2):175–177

71. Kuo CL, Chi CW, Liu TY (2004) The anti-inflammatory potential of berberine in vitro and in vivo. Cancer Lett 203:127–137
72. Ahmed T, Gilani AU, Abdollahi M, Daglia M, Nabavi SF, Nabavi SM (2015) Berberine and neurodegeneration: a review of literature. Pharmacol Rep 67(5):970–979
73. Selkoe DJ, Hardy J (2016) The amyloid hypothesis of Alzheimer's disease at 25 years. EMBO Mol Med 8:595–608
74. Burns J, Yokota T, Ashihara H, Lean ME, Crozier A (2002) Plant foods and herbal sources of resveratrol. J Agric Food Chem 50(11):3337–3340
75. Chen RS, Wu PL, Chiou RY (2002) Peanut roots as a source of resveratrol. J Agric Food Chem 50:1665–1667
76. Soleas GJ, Diamandis EP, Goldberg DM (1997) Wine as a biological fluid: history, production, and role in disease prevention. J Clin Lab Anal 11:287–313
77. Baur JA, Sinclair DA (2006) Therapeutic potential of resveratrol: the in vivo evidence. Nat RevlDrug Discov 5:493–456
78. Ma T, Tattn M-S, Yu J-T, Tan L (2014) Resveratrol as a therapeutic agent for Alzheimer's disease. BioMed Res Int 2014:Article ID 350516, 13 pages. https://doi.org/10.1155/2014/350516
79. Walle T, Hsieh F, DeLegge MH, Oatis JE Jr, Kristina Walle U (2004) High absorption but very low bioavailability of oral resveratrol in humans. Drug Metab Dispos 32(12):1377–1382
80. Resveratrol monograph (2010) Altern Med Rev 15(2)
81. Johnson JJ, Nihal M, Siddiqui IA, Scarlett CO, Bailey HH, Mukhtar H, Ahmad N (2011) Enhancing the bioavailability of resveratrol by combining it with piperine. Mol Nutr Food Res 55(8):1169–1176
82. Jacobs BP, Browner WS (2000) Ginkgo biloba: a living fossil. Am J Med 108:341–342
83. Christen Y, Maixent JM (2002) What is Ginkgo biloba extract EGb 761? An overview–from molecular biology to clinical medicine. Cell Mol Biol 48(6):601–611
84. Bridi R, Crossetti FP, Steffen VM, Henriques AT (2001) The antioxidant activity of standardized extract of Ginkgo biloba (EGb 761) in rats. Phytother Res 15(5):449–451
85. Wei T, Ni Y, Hou J, Chen C, Zhao B, Xin W (2000) Hydrogen peroxide-induced oxidative damage and apoptosis in cerebellar granule cells: protection by Ginkgo biloba extract. Pharmacol Res 41:427–433
86. Eckert A, Keil U, Kressmann S, Schindowski K, Leutner S, Leutz S, Müller WE (2003) Effects of EGb 761® Ginkgo biloba extract on mitochondrial function and oxidative stress. Pharmacopsychiatry 36:15–23
87. Luo Y, Smith JV, Paramasivam V, Burdick A, Curry KJ, Buford JP, Khan I, Netzer WJ, Xu H, Butko P (2002) Inhibition of amyloid-aggregation and caspase-3 activation by the Ginkgo biloba extract EGb761. PNAS 99(19):12197–12202
88. Yoo K-Y, Park S-Y (2012) Terpenoids as potential anti-Alzheimer's disease therapeutics. Molecules 17:3524–3538
89. Kleijnen J, Knipschild P (1992) Ginkgo biloba. Lancet 340:1136–1139
90. Monograph Ginkgo biloba (1998) Altern Med Rev 3(1):54–57

Fiber in Our Diet and Its Role in Health and Disease

12

Dipeeka Mandaliya, Sweta Patel, and Sriram Seshadri

Abstract

Lifestyle and dietary habits directly influence the health of an individual constraining to understand the role of different types of diet components. Recently, dietary fibers are being studied comprehensively to understand its role in prevention of heart diseases, obesity, diabetes, cancer, etc. Dietary fibers provide nutrition to the gut microbiota directly and to intestinal epithelial cells indirectly via its fermentation products mainly short-chain fatty acids (SCFAs). Dietary fiber consumption alters gut microbiota, and its fermentation product mainly short-chain fatty acids modulates disease condition by maintaining energy homeostasis and immune response. Dietary fibers reduce serum LDL cholesterol and blood pressure preventing heart diseases. Dietary fiber consumption also helps in reducing BMI via inducing satiety signals and provides bulking effect. SCFAs act on different tissues via GPCR receptors and modulate disease condition. In case of obesity and diabetes, SCFA improves insulin secretion and glucose homeostasis via gut-brain axis.

Keywords

Dietary fibers · SCFA · Type 2 diabetes · GPCR · Gut microbiota · Metabolism.

12.1 Introduction

Lifestyle and dietary habits directly influence the health of an individual constraining to understand the role of different types of diet components. Recently, dietary fibers are being studied comprehensively to understand its role in prevention of

D. Mandaliya · S. Patel · S. Seshadri (✉)
Institute of Science, Nirma University, Ahmedabad, Gujarat, India
e-mail: sriram.seshadri@nirmauni.ac.in

V. Rani, U. C. S. Yadav (eds.), *Functional Food and Human Health*,
https://doi.org/10.1007/978-981-13-1123-9_12

heart diseases, obesity, diabetes, cancer, etc. Plant non-starch polysaccharides like cellulose, hemicellulose, pactin, β-glucan, and fibers contained in wheat bran, oats, etc. are the examples of dietary fibers. Dietary fibers can be defined as plant carbohydrates and lignins which cannot be digested and absorbed in small intestine by the conventional digestive enzymes [31]. Dietary fibers provide nutrition to the gut microbiota directly and to intestinal epithelial cells indirectly via its fermentation products mainly short-chain fatty acids (SCFAs). The beneficial effects of dietary fibers can be explained as they protect mucus membrane of host from being degraded by the microbiota. In case of fiber deficiency, the microbiota degrades the colonic mucus layer and thus providing access to the pathogenic bacteria for infecting the host [9]. The SCFAs regulate metabolic as well as immune response of an individual by maintaining energy and immune homeostasis, respectively.

12.2 Types of Dietary Fibers

Dietary fibers can be defined as plant polysaccharides, lignin, or oligosaccharides such as resistant starch and inulin that cannot be digested by the human digestive enzymes [31]. Undigested or partially digested dietary fibers are fermented by the gut microbiota in small intestine or colon. On the basis of solubility, dietary fibers can be classified as soluble and insoluble dietary fibers. Fruits and vegetables are rich source of soluble dietary fibers (pectin, inulin, etc.), while wheat bran, oats, and barley have more amount of insoluble dietary fibers (cellulose, hemicellulose, etc.). Almost all high-fiber-containing foods provide rich source of soluble and insoluble fibers to some extent [21].

Soluble dietary fibers such as viscous or fermentable fibers like β-glucan and pectin have very important role in insulin-mediated glucose homeostasis in diabetic individuals. While soluble fibers are fermented in colon and their fermentation products have the beneficial effects, insoluble fibers like wheat bran provide bulking effect and partially fermented in colon. Daily consumption of guar gum, pectin, β-glucan, and hydroxypropyl methylcellulose reduced serum LDL cholesterol and thus prevents atherosclerosis and cardiovascular diseases. Psyllium husk is the most important component of fiber-supplemented products that lowers blood pressure, cholesterol, sugar, and risk of heart diseases [2]. Guar gum and other soluble fibers reduce gastric acid production and prevent duodenal ulcer disease [32].

12.3 Health Benefits of Dietary Fibers

The dietary fiber offers health benefits related to metabolic regulation, energy homeostasis, and immune regulation. Dietary fibers aid in maintaining good health and combat various diseases like obesity, diabetes, dyslipidemia, hypertension, colon cancer, etc. The beneficial effects of dietary fibers are due to its viscous nature improving satiety and control body weight gain. Dietary fibers inhibit the absorption of cholesterol and fat and prevent bile acid recycling in the liver. The bile acid formation in the liver utilizes cholesterol and thus lowers blood cholesterol [2]. Dietary fiber intake has inverse correlation with risk of cardiovascular disease and coronary heart disease [18]. Reduced levels of C-reactive protein (CRP), the inflammatory marker, have been observed in high-fat diet consuming rats following fiber ingestion. Dietary fibers maintain glucose and energy homeostasis via hypothalamic regulation in the brain by modulating neuropeptides affecting gluconeogenesis in the brain and intestine [6].

Soluble fibers pectin and guar but not bran have been shown to exert hypocholesterolemic effects in human. It has been reported in animal studies that diet-induced rise in cholesterol and atherosclerosis is inhibited by dietary fibers. Animal experiments suggest that some components of the complex mixture of substances called fibers could reduce cholesterol levels to a modest extent and inhibit atherosclerosis induced by diet. In man the data center on the effects of fiber on plasma cholesterol levels and some fibers such as pectin or guar exert significant hypocholesterolemic effects, whereas others, such as bran, do not. Oats and barley mediated reduction in total, and LDL cholesterol is due to the presence of soluble fiber, β-glucan, in it. Daily intake of 3 g oat β-glucan for 13 years has been shown to significantly reduce total and LDL cholesterol in hypercholesterolemic or normocholesterolemic individuals [24]. It has been shown that oats and buckwheat consumption is associated with decrease in body mass index (BMI), serum triglyceride, LDL cholesterol, total cholesterol, HDL cholesterol, and blood pressure. Daily intake of 9–30 g guar gum, 12–24 g pectin, 5 g β-glucan, and 5 g hydroxypropyl methylcellulose in nondiabetic individuals has been reported to reduce LDL cholesterol by 10.6%, 13%, 11.1%, and 8.5%, respectively. Blood LDL cholesterol-lowering efficiency of dietary fibers has been shown to reduce risk of cardiovascular diseases and coronary heart disease. Dietary fibers such as β-glucan from processed oats and barley foods also lower post-prandial blood glucose in healthy as well as diabetic subjects [34].

The dietary fibers delay gastric emptying and increase satiating hormones that create sense of fullness. High-fiber consumption is associated with incretin gut hormone secretion from intestinal L cells regulating insulin secretion and glucose homeostasis [1]. A high intake of dietary fiber was associated with a reduced risk of pancreatic cancer [19]. High-fiber diet containing low levels of red meat and alcohol has been reported to minimize risk of colorectal cancer. The beneficial effects of these three components might be because of gut microbiota alteration and their metabolites affecting balance between health and disease like colorectal cancer [35].

12.4 Dietary Fibers and Gut Microbiota

Gut microbiota is well known for its pivotal role in maintaining gut immune homeostasis via regulation of inflammation. Gut microbiota alteration has also been shown to induce obesity and insulin resistance via modulation of gut permeability, systemic inflammation, and energy metabolism. Dietary fiber has been shown to increase the abundance of *Bacteroidetes* while decrease the abundance of *Firmicutes* with elevated levels of propionate in plasma [6]. The gut microbiota-mediated alterations are thought to be due to its fermentation products, i.e., short-chain fatty acids (SCFAs), derived from dietary fibers.

One of the mechanisms by which dietary fibers prevent disease risk has been reported by Mahesh et al. In gnotobiotic mice colonized with fully sequenced human gut microbiota, the dietary fiber-deprived diet consumption leads to degradation of host colonic mucus membrane. In such condition, the gut microbiota utilizes mucus glycoproteins as nutrient source and degrades mucus barrier providing access to mucosal pathogen *Citrobacter rodentium* leading to colitis. Thus, dietary fibers have prominent role in diet-mediated gut microbiota shift and preventing intestinal permeability-mediated pathogenic invasion [9]. Gut microbiota alteration using spectrum-specific antibiotics lowers LPS levels in circulation and reduces GPCR expression as well as pro-inflammatory cytokines [28, 29].

12.5 Dietary Fibers and Short-Chain Fatty Acids

Our gut microbiota ferments dietary fibers that are incompletely hydrolyzed due to lack of the appropriate enzymes, which results in the production of SCFAs [7]. Although common SCFAs include formic, acetic, propionic, butyric, isobutyric, valeric, isovaleric, and caproic acids, 90–95% of the SCFAs present in the colon are constituted by acetate, propionate, and butyrate. Most SCFAs production occurs in the caecum and proximal colon and utilized as energy sources by intestinal epithelial cells and liver [23]. SCFAs with concentrations of about 60% acetate, 25% propionate, and 15% butyrate are found in the colon. Butyrate oxidation is the major source of energy for colonic epithelial cells. Propionate, entering the portal circle, is primarily utilized in gluconeogenesis in the liver, whereas acetate enters the blood circulation and gets access to different tissues [27]. SCFAs are rapidly transferred from gut to the bloodstream, and the usual concentration in peripheral blood is around 100–150 μM for acetate, 4–5 μM for propionate, and 1–3 μM for butyrate.

SCFAs are recognized by free fatty acid receptors FFAR2 and FFAR3. Acetate and propionate activate Ffar2 more specifically than butyrate, while butyrate and propionate are more specific for Ffar3 than acetate [3]. Both receptors are expressed in a variety of cells, including colonic enteroendocrine L cells, mucosal mast cells, adipose tissue, neutrophils, and monocytes. The intracellular signaling cascade triggers inositol 1,4,5-trisphosphate formation, intracellular [Ca^{+2}] mobilization, activation of extracellular signal-regulated kinase 1/2, and inhibition of intracellular cAMP accumulation. FFAR3 is involved in regulation of leptin production by adipocytes as well as expression and secretion of peptide yy (PYY). FFAR2 seems to be more involved in obesity and insulin resistance via regulation of glucagon like peptide-1 (GLP-1) secretion and modulation of inflammation. FFAR2 activation suppresses insulin signaling and fat accumulation in adipocytes and positively influences metabolism of unincorporated lipids and glucose in other tissues [4].

12.5.1 SCFAs and Metabolism

The SCFAs play vital roles in the pathophysiology of obesity and type 2 diabetes. Butyrate and propionate control blood glucose and body weight by activating gluconeogenesis in intestine via gut-brain neural circuit [6]. Acetate modulates the expression of neuropeptides regulating appetite in hypothalamus via TCA cycle activation [12]. GPR43 has a key role in fat accumulation regulation as indicated by Ffar2-deficient mice being obese on normal diet while Ffar2-overexpressing mice being protected from diet-induced obesity. Ffar2 inhibits Akt phosphorylation and insulin signaling in adipose tissues while promotes fatty acis oxidation in muscles and lowers triglyceride in the liver [16].

The gut microbiota-produced SCFAs regulate PYY secretion from intestinal L cells via Ffar3 as indicated by increased PYY secretion from wild-type germ-free mice after colonization with specific microbes but not from Ffar3 knockout mice [13]. Propionate-induced FFAR3 activation in sympathetic ganglia resulted in increased heart rate and energy expenditure via sympathetic outflow [15]. The SCFAs abrogate insulin resistance by FFAR3 activation in peripheral nerve via gut-brain neural circuit [6].

Propionate regulates body weight and lipid metabolism in adipose tissue and liver in overweight adult individuals [5]. Butyrate upregulates the expression of peroxisome proliferator-activated receptor-gamma coactivator 1α (PGC-1α) and the phosphorylation of adenosine monophosphate-activated kinase (AMPK) in muscle and liver tissues as well as PGC-1α and mitochondrial uncoupling protein-1 (UCP-1) in brown adipose tissues and thus improves fat oxidation and thermogenesis [10]. SCFA maintains the balance between lipid oxidation and lipogenesis via PPARγ regulation in high-fat diet-induced obesity [8]. Butyrate treatment has been shown to increase AMPK activity and accelerated the assembly of tight junctions. Butyrate enhances the intestinal barrier function, at least partly, by increasing expression of some major tight junction proteins and facilitating the assembly of tight junctions [25].

Pancreatic beta cell mass and function are damaged in case of inflammation, insulin resistance, and stress which can be protected by SCFA-mediated secretion of GLP-1 [27]. SCFAs have been reported to increase incretin levels when orally administered in mice. Butyrate induces GLP-1, GIP, and PYY secretion along with increased concentrations of insulin and amylin via stimulating pancreatic beta cells. Propionate administration also showed elevated levels of GIP, insulin, and amylin without affecting GLP-1 and PYY. Butyrate and propionate exert its beneficial effect on body weight and GIP stimulation via FFAR3 [17]. GLP-1 improves insulin resistance via enhanced secretion of insulin and suppression of glucagon secretion by activation of GLP-1 receptor which also plays role in beta cell proliferation via activation of PKB and PDX1 and apoptosis inhibition [11, 26]. Thus, SCFAs improve pancreatic insulin secretion and maintain lowered blood glucose levels.

12.5.2 SCFAs and Immune Regulation

Among SCFAs, butyrate has been shown to prove the best anti-inflammatory molecule as it can affect immune cell adhesion, migration, and cytokine expression and induces T cell energy to prevent antigen-mediated obesity-associated inflammation [22]. In obese mice, butyrate also enhances the immune homeostasis maintaining T regulatory cells (T reg cells) which inhibit effector T cells [22]. Saemann et al. showed that, in *Staphylococcus aureus*-infected PBMC-derived monocytes, butyrate inhibited IL-2 and IFN-γ while upregulated IL-4- and IL-10-mediated T regulatory cells [30]. SCFAs inhibit inflammatory TNFα release by LPS induction in human blood-derived neutrophils, and they also inhibit NFkB activation in human colon adenocarcinoma cell line [33].

SCFAs, particularly propionic acid and butyric acid, have anti-inflammatory effects via downregulation of pro-inflammatory cytokines TNF-α, IL-1β, and IL-6 [20]. SCFA receptor FFAR2 influences the differentiation and activation of monocyte- and neutrophil-mediated inflammation. SCFA-mediated activation of FFAR2 was shown to trigger recruitment of circulating leukocytes to the inflammatory site via activation of intracellular signaling pathways including MAPK, protein kinase C (PKC), and phospholipase C (PLC) [36]. Butyrate has been shown to downregulate TNF-α-mediated expression of adhesion molecule VCAM-1 which prevents leucocyte migration [22]. Propionate and butyrate but not acetate inhibit inflammatory cytokine release, NFkB pathway, and immune-related gene expression in vitro [33]. SCFA, mainly butyrate, has been shown to induce IL-10-mediated T regulatory cell function and suppression of inflammation. SCFA-mediated HDAC inhibition leads to mTOR pathway activation and IL-10 production [14].

12.6 Conclusion

High-fiber consumption improves health, prevents disease development, and helps cure various diseases. The dietary fiber offers health benefits related to metabolic regulation, energy homeostasis, and immune regulation. Dietary fibers aid in maintaining good health and combat various diseases like obesity, diabetes, dyslipidemia, hypertension, colon cancer, etc. (Fig. 12.1). Increasing the intake of high-fiber foods or fiber supplements lowers blood pressure, improves blood glucose homeostasis for diabetic individuals, improves serum lipoprotein, and reduces weight gain. Inulin and certain soluble fibers have been reported to enhance immune function in humans. Dietary fibers prevent mucus membrane digestion as well as maintain intestinal membrane integrity preventing pathogen invasion and disease development.

The fermentation product of dietary fibers, SCFAs, is thought to be the major factor providing beneficial effects of dietary fibers. Butyrate and propionate regulate insulin secretion and glucose homeostasis via incretin production. SCFAs further increase pancreatic β cell mass and insulin secretion as well as reduce glucagon

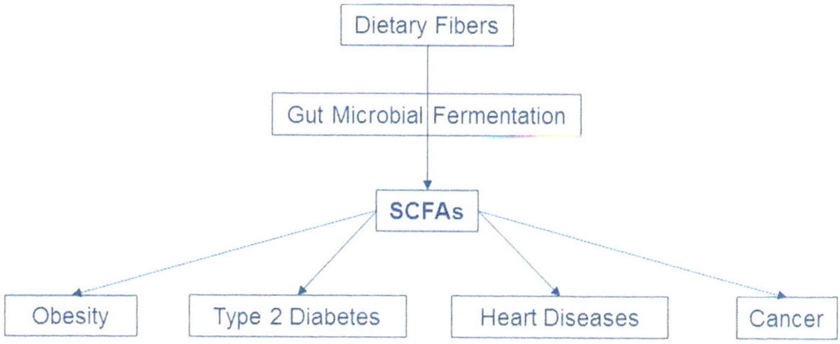

Fig. 12.1 Role of dietary fiber in health. Dietary fiber fermentation products, i.e., SCFAs, have a role in prevention of obesity, diabetes, heart diseases, cancer, etc.

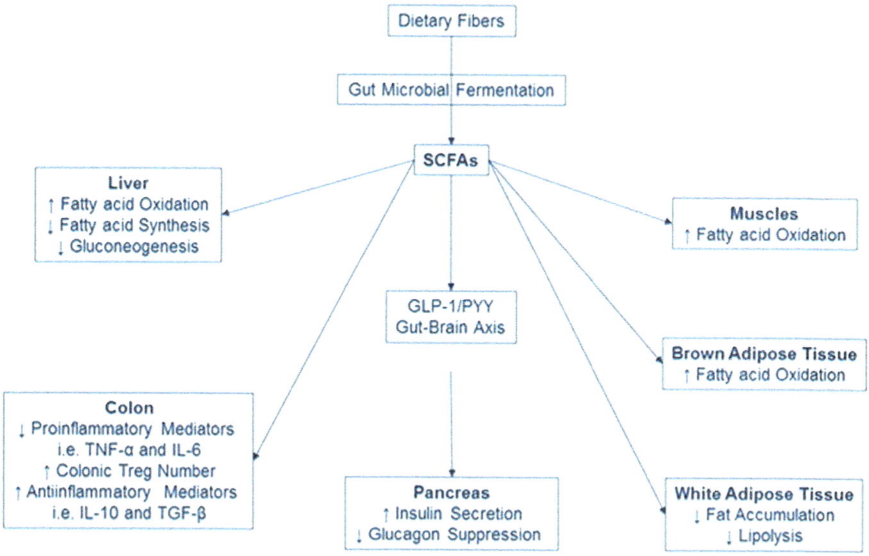

Fig. 12.2 Metabolism of action of dietary fiber. SCFAs modulate fat metabolism leading to reduced serum free fatty acid by inhibiting adipose tissue lipolysis as well as reduced fat accumulation by improving fat oxidation in the liver, muscles, and brown adipose tissue. SCFA-mediated HDAC inhibition leads to suppression of inflammation (TNF-α, IL-6), induction of colonic T regulatory cells and secretion of anti-inflammatory IL-10 and TGF-β. SCFAs also improve insulin secretion and inhibit glucagon suppression via gut-brain axis

secretion and thus regulate glucose metabolism. Butyrate enhances fatty acid oxidation in the liver, muscles, and adipose tissues. SCFAs, especially butyrate, seem to exert broad anti-inflammatory activities via downregulation of pro-inflammatory cytokines TNF-α, IL-1β, and IL-6. SCFA-mediated HDAC inhibition leads to mTOR pathway activation and IL-10 production that induce T regulatory cells and suppress inflammation (Fig. 12.2).

Thus, high-fiber-containing foods and supplements are recommended to be consumed daily for health benefits and prevention of metabolic as well as inflammatory diseases.

References

1. Anderson JW (2008) Dietary fiber and associated phytochemicals in prevention and reversal of diabetes. In: Pasupuleti VK, Anderson JW (eds) Nutraceuticals, glycemic health and type 2 diabetes. Blackwell Publishing Professional, Ames, pp 111–142
2. Anderson JW, Baird P, Davis RH Jr, Ferreri S, Knudtson M, Koraym A et al (2009) Health benefits of dietary fiber. Nutr Rev 67(4):188–205 Epub 2009/04/02
3. Brown AJ, Goldsworthy SM, Barnes AA, Eilert MM, Tcheang L, Daniels D et al (2003) The orphan G protein-coupled receptors GPR41 and GPR43 are activated by propionate and other short chain carboxylic acids. J Biol Chem 278(13):11312–11319 Epub 2002/12/24
4. Byrne CS, Chambers ES, Morrison DJ, Frost G (2005) The role of short chain fatty acids in appetite regulation and energy homeostasis. Int J Obes 39(9):1331–1338 Epub 2015/05/15
5. Chambers ES, Viardot A, Psichas A, Morrison DJ, Murphy KG, Zac-Varghese SE et al (2015) Effects of targeted delivery of propionate to the human colon on appetite regulation, body weight maintenance and adiposity in overweight adults. Gut 64(11):1744–1754 Epub 2014/12/17
6. De Vadder F, Kovatcheva-Datchary P, Goncalves D, Vinera J, Zitoun C, Duchampt A et al (2014) Microbiota-generated metabolites promote metabolic benefits via gut-brain neural circuits. Cell 156(1–2):84–96 Epub 2014/01/15
7. den Besten G, van Eunen K, Groen AK, Venema K, Reijngoud DJ, Bakker BM (2013) The role of short-chain fatty acids in the interplay between diet, gut microbiota, and host energy metabolism. J Lipid Res 54(9):2325–2340 Epub 2013/07/04
8. den Besten G, Bleeker A, Gerding A, van Eunen K, Havinga R, van Dijk TH et al (2015) Short-chain fatty acids protect against high-fat diet-induced obesity via a PPARgamma-dependent switch from lipogenesis to fat oxidation. Diabetes 64(7):2398–2408 Epub 2015/02/20
9. Desai MS, Seekatz AM, Koropatkin NM, Kamada N, Hickey CA, Wolter M et al (2016) A dietary fiber-deprived gut microbiota degrades the colonic mucus barrier and enhances pathogen susceptibility. Cell 167(5):1339–53 e21 Epub 2016/11/20
10. Donohoe DR, Garge N, Zhang X, Sun W, O'Connell TM, Bunger MK et al (2011) The microbiome and butyrate regulate energy metabolism and autophagy in the mammalian colon. Cell Metab 13(5):517–526 Epub 2011/05/03
11. Farilla L, Bulotta A, Hirshberg B, Li Calzi S, Khoury N, Noushmehr H et al (2003) Glucagon-like peptide 1 inhibits cell apoptosis and improves glucose responsiveness of freshly isolated human islets. Endocrinology 144(12):5149–5158 Epub 2003/09/10
12. Frost G, Sleeth ML, Sahuri-Arisoylu M, Lizarbe B, Cerdan S, Brody L et al (2014) The short-chain fatty acid acetate reduces appetite via a central homeostatic mechanism. Nat Commun 5:3611 Epub 2014/05/02
13. Ichimura A, Hasegawa S, Kasubuchi M, Kimura I (2014) Free fatty acid receptors as therapeutic targets for the treatment of diabetes. Front Pharmacol 5:236
14. Kim CH, Park J, Kim M (2014) Gut microbiota-derived short-chain fatty acids, T cells, and inflammation. Immune Netw 14(6):277–288 Epub 2015/01/01
15. Kimura I, Inoue D, Maeda T, Hara T, Ichimura A, Miyauchi S et al (2011) Short-chain fatty acids and ketones directly regulate sympathetic nervous system via G protein-coupled receptor 41 (GPR41). Proc Natl Acad Sci U S A 108(19):8030–8035 Epub 2011/04/27
16. Kimura I, Ozawa K, Inoue D, Imamura T, Kimura K, Maeda T, Terasawa K, Kashihara D, Hirano K, Tani T et al (2013) The gut microbiota suppresses insulin-mediated fat accumulation via the short-chain fatty acid receptor GPR43. Nat Commun 4:1829

17. Lin HV, Frassetto A, Kowalik EJ Jr, Nawrocki AR, Lu MM, Kosinski JR et al (2012) Butyrate and propionate protect against diet-induced obesity and regulate gut hormones via free fatty acid receptor 3-independent mechanisms. PloS One 7(4):e35240 Epub 2012/04/17
18. Ma Y, Griffith JA, Chasan-Taber L, Olendzki BC, Jackson E, Stanek EJ et al (2006) Association between dietary fiber and serum C-reactive protein. Am J Clin Nutr 83(4):760–766
19. Mao QQ, Lin YW, Chen H, Qin J, Zheng XY, Xu X et al (2017) Dietary fiber intake is inversely associated with risk of pancreatic cancer: a meta-analysis. Asia Pac J Clin Nutr 26(1):89–96 Epub 2017/01/05
20. Mattace Raso G, Simeoli R, Russo R, Iacono A, Santoro A, Paciello O et al (2013) Effects of sodium butyrate and its synthetic amide derivative on liver inflammation and glucose tolerance in an animal model of steatosis induced by high fat diet. PLoS One 8(7):e68626 Epub 2013/07/19
21. McKee LH, Latner TA (2000) Underutilized sources of dietary fiber: a review. Plant Foods Hum Nutr (Dordrecht, Netherlands) 55(4):285–304 Epub 2000/11/22
22. Meijer K, de Vos P, Priebe MG (2010) Butyrate and other short-chain fatty acids as modulators of immunity: what relevance for health? Curr Opin Clin Nutr Metab Care 13(6):715–721 Epub 2010/09/09
23. Mortensen PB, Clausen MR (1996) Short-chain fatty acids in the human colon: relation to gastrointestinal health and disease. Scand J Gastroenterol Suppl 216:132–148 Epub 1996/01/01
24. Othman RA, Moghadasian MH, Jones PJ (2011) Cholesterol-lowering effects of oat beta-glucan. Nutr Rev 69(6):299–309 Epub 2011/06/03
25. Peng L, Li ZR, Green RS, Holzman IR, Lin J (2009) Butyrate enhances the intestinal barrier by facilitating tight junction assembly via activation of AMP-activated protein kinase in Caco-2 cell monolayers. J Nutr 139(9):1619–1625 Epub 2009/07/25
26. Perfetti R, Hui H (2004) The role of GLP-1 in the life and death of pancreatic beta cells. Horm Metab Res = Hormon- und Stoffwechselforschung = Hormones et metabolisme 36(11–12):804–810 Epub 2005/01/19
27. Puddu A, Sanguineti R, Montecucco F, Viviani GL (2014) Evidence for the gut microbiota short-chain fatty acids as key pathophysiological molecules improving diabetes. Mediat Inflamm 2014:162021 Epub 2014/09/13
28. Rajput P, Prajapati B, Jena P, Seshadri S (2015a) Targeting type 2 diabetes and obesity: the role of gut microflora produced short chain fatty acids (SCFAs) in adiposity and inflammation. J Diabetes Metab 6(8):78
29. Rajput P, Prajapati B, Jena P, Seshadri S (2015b) The role of gut microflora produced short chain fatty acids (SCFAs) in adiposity and inflammation in obesity and type 2 diabetes. J Diabetes Metab 5(6):300
30. Saemann MD, Bohmig GA, Osterreicher CH, Burtscher H, Parolini O, Diakos C et al (2000) Anti-inflammatory effects of sodium butyrate on human monocytes: potent inhibition of IL-12 and up-regulation of IL-10 production. FASEB J: Off Publ Fed Am Soc Exp Biol 14(15):2380–2382 Epub 2000/10/12
31. Smith CE, Tucker KL (2011) Health benefits of cereal fibre: a review of clinical trials. Nutr Res Rev 24(1):118–131 Epub 2011/02/16
32. Smits BJ (1974) The irritable bowel syndrome. Practitioner 213(1273):37–46 Epub 1974/07/01
33. Tedelind S, Westberg F, Kjerrulf M, Vidal A (2007) Anti-inflammatory properties of the short-chain fatty acids acetate and propionate: a study with relevance to inflammatory bowel disease. World J Gastroenterol 13(20):2826–2832 Epub 2007/06/15
34. Tosh SM (2013) Review of human studies investigating the post-prandial blood-glucose lowering ability of oat and barley food products. Eur J Clin Nutr 67(4):310–317 Epub 2013/02/21
35. Tuan J, Chen YX (2016) Dietary and lifestyle factors associated with colorectal Cancer risk and interactions with microbiota: Fiber, red or processed meat and alcoholic drinks. Gastrointest Tumor 3(1):17–24 Epub 2016/10/11
36. Vinolo MAR, Rodrigues HG, Nachbar RT, Curi R (2011) Regulation of inflammation by short chain fatty acids. Nutrients 3(10):858–876

13 Metabolic Syndrome and Nutritional Interventions

Bhawna Kumari, Akanksha Sharma, and Umesh C. S. Yadav

Abstract

Metabolic syndrome is a cluster of pathophysiological conditions that are associated with many other diseases including dyslipidaemia, cardiovascular diseases, insulin resistance and type 2 diabetes. Obesity is one of the many factors implicated in the progression of metabolic syndrome. Although some genetic components have been shown to be responsible for obesity, recent increase in the rate of obesity globally is mainly associated with the altered life-style (mostly sedentary) and food habit which includes consuming many and frequent meals during the day and energy-rich Westernized diet which are usually deficient of fibres and supply excessive calorie in shorter duration. Therefore, the weight loss by hypocaloric diet and life-style modifications are recommended to manage obesity and associated complications that come under metabolic syndrome. Recent focus on the functional food-derived nutrient components including polyphenols such as alkaloids, flavonoids, terpenes, saponins, etc. has advocated nutritional intervention in patients as a preventive and therapeutic approach for metabolic syndrome and life-style-associated diseases. In the present chapter, recent insight in the field of nutraceuticals and metabolic syndrome has been discussed.

Keywords

Metabolic disorder · Nutraceuticals · Dyslipidaemia · Phytochemicals · Nutritional intervention

B. Kumari · A. Sharma
Metabolic Disorders and Inflammatory Pathologies Laboratory, School of Life Sciences, Central University of Gujarat, Gandhinagar, Gujarat, India

U. C. S. Yadav (✉)
School of Life Sciences, Central University of Gujarat, Gandhinagar, Gujarat, India
e-mail: umeshyadav@cug.ac.in

V. Rani, U. C. S. Yadav (eds.), *Functional Food and Human Health*,
https://doi.org/10.1007/978-981-13-1123-9_13

13.1 Introduction

Metabolic syndrome (MS) is a multifaceted entity which involves a cluster of disorders. Reaven (1996) for the first time recognized a collection of risk factors of associated cardiovascular diseases, namely, dyslipidaemia (low HDL or elevated level of plasma cholesterol and/or triglycerides), hypertension (high blood pressure) and hyperglycemia in certain individuals (Fig. 13.1) [1]. He coined the term 'syndrome X' for this group of risk factors [2]. In 2001, the National Cholesterol Education Program Adult Treatment Panel Third (NCEP-ATP-III) defined MS for the purpose of diagnosis [3]. NCEP recommended that an individual to be diagnosed with MS should have at least three of the following five factors above a threshold level: (a) waist circumference (≥94 cm for men and ≥ 80 cm for women), (b) serum triglyceride concentration (≥2.5 mmol/L), (c) serum high-density lipoprotein (HDL) concentration (<1.3 mmol/L for women<1.0 mmol/L for men), (d) blood pressure (≥135/85 mmHg) and (e) fasting serum glucose level (6.1 mmol/L). Moreover, the International Diabetes Federation (IDF) in 2005 redefined the qualification for a person to be affected by MS and included central obesity (waist circumference ≥ 94 cm for men and ≥ 80 cm for women) plus any two other risk factors mentioned in the NCEP recommendation.

The changing global life-style and Westernized food habits have unfolded MS as a highly heterogeneous and very complex disease condition. The rapidly changing food habits and life-style of the people have led to increased prevalence of obesity and MS; these include lack of exercise, intake of calorie-rich diet and overeating along with the interplay of other environmental and genetic factors that every individual is endowed with. Although MS is a well-studied phenomenon now, it is difficult to identify the prevalence of MS worldwide because of the difference in the criteria used to diagnose it by different organizations and countries. Early and accurate identification of the individuals with MS is important. Early diagnosis may help reduce the probability of developing other associated diseases in these individuals. If diagnosed at an early stage, the patients can be treated through nutritional intervention and pertinent modification in their life-style, food habits and diet chart.

Diet and nutrition are important components and determinants of the human health. Since obesity is the pivotal component in the MS, nutritional interventions

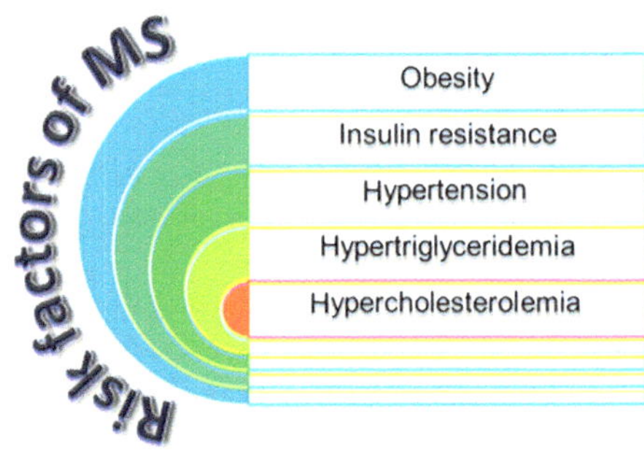

Fig. 13.1 List of risk factors of metabolic syndrome

along with other therapeutic strategies to control obesity and enhance weight loss in affected individuals may help reduce the incidence of obesity-related complications.

13.2 Therapeutic Strategies for Metabolic Syndromes

Metabolic syndrome is a chronic condition which develops over a period of time due to negligence of the health and intake of high-calorie diet along with sedentary life-style. Although prevention always should be the primary measure to decrease the disease incidence, however, in case of acute and severe diseases associated with MS, this approach may not be very effective. Thus, in the serious pathophysiological conditions associated with MS, therapeutic approaches such as pharmacological approaches including drug and surgical treatment and non-pharmacological approaches such as life-style and food habit changes and diet and nutritional intervention may be used.

13.2.1 Pharmacological Therapeutic Strategies

The awareness about MS and its associated diseases have increased the interest in pharmacological therapy. The pharmacological treatment through drug therapy deals with individual components of metabolic syndrome. Various drugs are known which are used separately or in combinations for treatment of MS. They have specific metabolic targets. Table 13.1 explicates various known drugs that are classified based on their respective targets in metabolic syndrome.

As discussed above, drug therapy is the primary therapeutic strategy used to treat MS; however, recurrence of medical complication in MS is common. In the cases of recurrence, surgery is opted as the effective and durable treatment in MS, particularly in complications where obesity is the underlying cause behind individual MS-associated diseases. A number of clinical studies showed that pathophysiological conditions related to MS were resolved after bariatric surgery in many patients such as type 2 diabetes [15, 16]. Studies also demonstrated decreased death rate in patients with obesity after weight loss through surgery [17, 18]. The National Institute of Health (NIH), USA, in 1991 recommended surgery for patients with BMI $\geq$ 35 with obesity and associated diseases and also for those with BMI $\geq$ 40 with central obesity; however no comorbid conditions could be treated with surgical approach [19]. Nonetheless, these evidences suggest efficiency of surgical intervention in treatment of obesity and MS. Many bariatric surgery approaches including laparoscopic adjustable gastric banding (LAGB), laparoscopic Roux-en-Y gastric bypass (RYGBP), and biliopancreatic diversion with duodenal switch and sleeve gastrectomy have been used to treat obesity-related complications (Table 13.2).

Despite various surgical procedures being used to treat obesity-linked diabetes and metabolic syndrome globally, there is no single usual procedure for such treatments. In addition, the surgical process is associated with risks and complications. The types of complication depend on specific procedure being used for surgery. For

Table 13.1 List of different drugs and their targets and effect

	Drugs	Target	Effect	References
Antiobesity drugs	Sibutramine	Control reuptake of norepinephrine and serotonin	Reduced appetite Increased energy expenditure	[4]
	Orlistat	Enzyme lipase inhibitor	Reduces fat absorption	[5]
	Metformin	Antidiabetic	Reduces body weight	[6]
	Rimonabant	Cannabinoid (CB1-Type) receptor antagonist	Counteract hyperphagia	[7]
Antidiabetic drugs	Thiazolidinediones (glitazones)	Peroxisome Proliferator-Activated Receptor subtype g (PPAR-g) is activated	Reduces insulin resistance	[8]
	Metformin	Hepatic glucose production reduces		[6]
	Doxazosin	a1-Adrenoreceptor antagonists		[9]
	Rosiglitazone	Inhibitor of hepatic lipoprotein overproduction		[10]
	Exenatide	GLP-1 receptor agonists	Inhibits beta-cell death	[11]
	Sitagliptin	DPP-4 inhibitors	Antihyperglycemic agents	[12]
Atherogenic dyslipidemia	Statins (e.g. atorvastatin)	Specific inhibitors of HMG-CoA reductase	Ameliorates low-density lipoprotein (LDL) levels and lipoprotein production	[13]
	Ezetimibe	Cholesterol absorption inhibitors	Inhibits cholesterol absorption in intestine	[14]

example, the restrictive procedures used for surgery generally results into malabsorption of many nutrients that causes nutritional deficiency. Early complications include problems in wound healing, incisional hernias, etc., whereas late complications involve gastrointestinal disorder and vitamin deficiency [20]. Due to these shortcomings of surgical therapies, alternative approaches are required to be identified.

13.2.2 Non-pharmacological Therapeutic Strategies

The pharmacological and surgical treatments, in addition to having their own side effects, are also expensive, invasive and extremely painful. Therefore, non-pharmacological treatments are recommended over the drug and surgical

Table 13.2 Different bariatric surgical procedures

S. no.	Types of bariatric surgeries	Techniques	References
1.	Laparoscopic adjustable gastric banding (LAGB)	Includes encompassing the upper portion of the stomach with a liquid-filled tube	[20]
2.	Laparoscopic Roux-en-Y gastric bypass (RYGBP)	Includes the use of a surgical stapler to create a small and vertical gastric sack	[21]
3.	Biliopancreatic diversion (BPD)	Includes horizontal gastrectomy leaving only a functional upper stomach	[20]
4.	BPD with duodenal switch	Includes vertical gastrectomy leaving a gastric 'sleeve' of 150 to 200 ml	[22]
5.	Sleeve gastrectomy	Includes sleeve gastrectomy (SG), followed by duodenoileostomy/ileoileostomy performed a few months apart	[23]

interventions. The non-pharmacological treatments include life-style and food habit modifications, diet and nutrition selection based upon their calorie values and glucose index, increased physical activity, quitting smoking and alcohol habits and use of nutraceuticals.

Life-style modification therapy plays the central role in the MS treatment. A clinical study with 41 obese patients, with high (81%) prevalence of MS, was performed to observe the effect of diet and life-style modification therapy. The two-third of the patients who achieved 10% weight loss did not meet the MS diagnostic criteria compared to those patients who lost less than 10% [24]. Nowadays, life-style modification therapy is used in combination with pharmacotherapy and shows additive effect [25]. Various neurotransmitters including amino acids and neuropeptides play important role in food intake behaviour and digestion [26]. Thus, the drugs modulating this behaviour could be used in combination therapy [27]. The clinical data show that long-term weight reduction and life-style modification decreased the prevalence of type 2 diabetes and MS [28, 29].

Life-style and diet modification regimen endorses a low-calorie diet to induce calorie deficiency which enhances the weight loss. These programmes generally require regular medical monitoring and nutritional supplementation along with medical counselling. The principle of life-style and diet modifications involves behavioural therapy, which is based on learning theory or behaviourism [30]. This theory states that the behaviour for obesity is always learnt, albeit unconsciously, and hence can be modified [31]. The aim of the therapy is to modify the eating and exercise habits of the patient. This kind of therapy involves various strategies to deliver the treatment through following procedures:

(a) *Behavioural procedure* involves self-monitoring of eating habit and exercises. The patients involved in life-style modification are encouraged for 'goal setting' [30]. Apart from this the patients with MS are instructed to develop a control over the triggers for food consumption or physical inactivity [32]. In addition, alternative behavioural strategies are used to divert the patient's mind which may replace eating drives with some other activities such as exercising, jogging, dancing or other physically engaging hobbies [30].

(b) *Cognitive procedure* includes proactive problem-solving and cognitive restructuring sessions performed by clinicians during treatment regimen [33]. The proactive problem-solving session is used to address the incident which hampers the weight loss adherence of the patients. These sessions are necessarily done with clinicians [34]. Cognitive restructuring is used to recognize the dysfunctional thought which reduces the efforts by patients aimed at weight loss [33].

The procedures discussed above are strategic therapy regimens to deliver the planned treatment benefits to patient in guided manner with the aim to achieve significant weight loss. However, critical shortage of trained clinicians in this area has created a resource gap which needs to be bridged with more trained staff.

13.2.3 Phytochemicals and Nutrigenomics

Naturally derived chemicals from plants constitute the diet and nutrition, and a large and diverse group of such compounds was used in therapy of various complications. These phytochemicals usually determine the colour, flavour, odour of fruits, for example, blueberries' dark skin and garlic's strong odour, etc. are provided by the phytochemicals [35]. Different classes of phytochemicals have been identified having positive effects on human health (Fig. 13.2).

Using plants for medicinal purposes is as old as the humanity itself. Human beings learned to use plant parts such as barks, leaves, seeds, fruits, etc. to find cure for several illnesses in addition to their dietary needs [36]. The ancient Chinese, Indians, Egyptians, Native Americans and Babylonians were known to practice herbal medicine. Modern drugs are being developed from phytochemicals that are derived from various herbs; e.g. phytochemicals from *Stephania glabra* and alkaloid forskolin from the plant *Coleus forskohlii* have been identified as adenylate cyclase and nitric oxide activators, which may help in preventing obesity and atherosclerosis. With the advent of industrialization and establishment of pharmacology as a leading and effective branch of medicinal therapeutics, the practice of herbal medicine had declined, even though the native population in the various countries continued using respective traditional methods of treatments as they could not afford or access the modern medicines. However, with time the synthetic pharmaceutical drug revealed their unwanted and serious side effects with short-term benefits, and sometimes the drug resistance became rampant. In this background, the herbal medicines and phytochemicals have again gained importance and new impetus, and more and more studies are being planned and funded to explore these compounds.

For a long time, nutritionists and traditional medicine practitioners have focused on preventing nutritional deficiency but did not focus on other beneficial effects of nutrients. When most of the known vitamins (cofactors) were already discovered

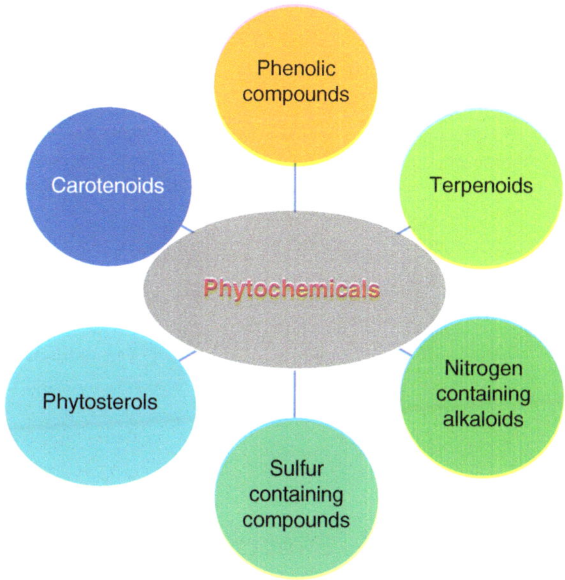

Fig. 13.2 Different classes of phytochemicals

and identified by biochemists, search for other novel dietary components declined [37]. However, emergence of phytochemicals and their regained popularity have provided evidence for their role in prevention and treatment of various metabolic syndromes. Numerous studies have shown that natural diets such as berries, leafy vegetables, legumes, plant-derived flavouring materials, grains, etc. contain phytochemicals which could decrease the risk of heart diseases, neurodegenerative diseases, cancer, type 2 diabetes, etc. Some of these studies have been summarized in Table 13.3 below.

In case of metabolic syndrome, polyphenols affect different risk factors of MS by different mechanisms. Many studies have indicated the effects of polyphenols in MS-associated diseases such as diabetes. They have been shown to possess hypoglycemic effect by reducing the absorption of carbohydrates in the intestine; modulating glucose-metabolizing enzymes, β-cell homeostasis and insulin action; and increasing insulin release and anti-oxidative and anti-inflammatory functions [43–46]. Some other plant-derived polyphenols including green tea-derived catechins and epicatechins, chlorogenic acid, ferulic acid, caffeic acid, tannic acid, quercetin and naringenin could inhibit the Na^+-dependent glucose transporters and interfere with glucose absorption in the intestine [46, 47].

Another MS-associated complication, dyslipidemia, has been defined as incidence of increased lipid levels in the circulation. It is characterized by enhanced levels of triglycerides, LDL and lower concentration of HDL. The reduction in the level of these lipids is the aim of all therapies aimed at weight loss or obesity. Further, control of triglycerides and/or total cholesterol in circulation in humans as well as animal models of MS was exhibited by flavonoid-containing food or pure flavonoid without showing any significant side effects as has been usually associated with synthetic anticholesterol drugs [48–55].

Table 13.3 Different classes of phytochemicals and their respective sources and effects in human health and diseases

S. no.	Polyphenols	Source	Effect	References
1.	Flavonoids/ flavones	Black tea, green tea, apple, red wine, cocoa, orange, thyme, tomato, onion	Anti-oxidative, anti-apoptotic, anticancer, anti-inflammatory, anti-angiogenesis	[38]
2.	Curcuminoids	Turmeric, mustard	Anti-oxidative, anti-apoptotic, anticancer, anti-inflammatory, epigenetic modulation	[39, 40]
3.	Phenolic acids	Berries, pineapple, walnuts, lemon, banana, tea		[41]
4.	Stilbenes	Grapes, wine, berries	Anti-oxidative, anti-proliferative, anticancer, anti-inflammatory, modulation of lipid metabolism, anti-cyclooxygenase	[42]

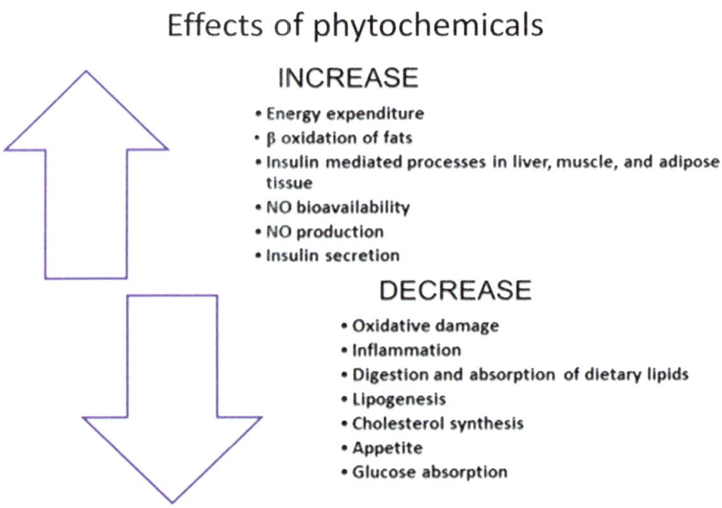

Fig. 13.3 Effects of phytochemicals on various metabolic parameters

The phytochemicals are known to decrease disease promoting indicators in MS such as glucose absorption, lipid and cholesterol absorption and synthesis, oxidative stress and inflammation on one hand and increase health promoting indicators such as insulin secretion and sensitization, energy expenditure, nitric oxide production and bioavailability on the other hand (Fig. 13.3).

Several factors affecting metabolism of phytochemicals which play crucial role in their effectiveness are discussed briefly below:

Bioavailability the percentage of phytochemicals consumed and actually used by an individual varies significantly [56]. Not much evidence is available regarding storage of phytochemicals in the body; studies indicate that plant-derived polyphenols are not stored for extended period in the tissues [57]. Food processing and methods of consumption determine the bioavailability and metabolism of the phytochemicals present in the food items. The methods of cooking and food processing can decrease the concentration of phytochemicals or may change the chemical nature of the phytochemicals present in the final food product which affects their absorption and bioavailability.

Genetic Makeup of the individual and the gut microflora are important factors that determine the extent of phytochemical metabolism. Various studies have demonstrated that alterations in the composition of our microbiomes (microorganisms present inside our intestine) correlate well with various disease states, raising the possibility that manipulation of these microbial communities could be useful to treat diseases [58]. For example, gut microbiota could be improved by consuming probiotics and prebiotics. Microorganisms in the gut, mainly *Bacteroidetes* and *Firmicutes*, play an important role in absorption of nutrients, xenobiotic metabolism, mucosal barricading and neovascularisation. Further, the diet consumed by an individual regulates the type of microorganisms present in the gut [59, 60].

It is now known that certain micronutrients and macronutrients influence expression of various genes through activation or inhibition of transcription factors that are involved in inflammatory and metabolic stress. Such correlations are studied in a new branch of science called *nutrigenomics*, which deals with the effects of dietary biomolecular components affecting the genome, proteome and metabolome [61]. For example, polyunsaturated fatty acids (PUFAs) regulate the transcription factors including peroxisome proliferator-activated receptors (PPARs). PPAR, which acts as nutrient sensors in healthy persons, mediates the negative feedback system that deals with variations in blood free fatty acid levels. In diabetics of obese individuals, this feedback system is strained because of metabolic and inflammatory stress. Interestingly, increased free fatty acids in circulation, cytokine-induced downregulation of PPARα and other nuclear receptors have been shown to cause inflammatory stress condition [61].

Thus, understanding how nutrients influence certain specific set of genes and their mechanisms at the molecular level may allow development of dietary intervention strategies for treating metabolic syndromes and maintaining homeostasis. The metabolic dysregulation leading to diet-induced diseases may be understood by identification of the interaction between pathways of nutrient-induced signalling and pro-inflammatory cascade. Further, assessing the genotypes associated with risk factors responsible for diet-induced diseases including obesity, diabetes and hypertension may help in defining their impact on human health and in their use in nutrition systems biology. These studies may help identify new biomarkers of early diagnosis of diet-induced metabolic dysregulation and disease susceptibility [61].

13.2.4 Personalized Medicine and Nutrigenomics

The genetic makeup of every individual is different and unique, and so is the effect of nutrients on their genes. The interaction of the genes and nutrients is studied in the super specialized area called 'nutrigenetics', which deals with the effects of genetic variations on influence of nutrients on human health and diseases [62]. It allows us to identify genetic variation patterns and suitable diet accordingly. Genotypes in diverse environmental conditions are affected differently; the study of these interactions in metabolic syndromes may help increase the understanding of biological pathways and how environmental factors affect them. Such understanding may help devise personalized medicine based on individual patient's genetic makeup.

Personalized medicine or precision medicine is an emerging practice in health sciences. Medicines and diets tailored to an individual's need suitable for its genotype seem promising in the near future. Dietary recommendations may not be universal after all. It depends on how genes and nutrients interact in different individuals. Cytochrome P4502D6, a well-known polymorphic metabolic enzyme, responsible for 'overdose' or amplification of the effects of debrisoquine (a hypertension drug) was discovered by Gaedigk et al. in 1991 [63]; however, genetic reason could be elucidated much later. This polymorphism was found to be a clinically significant genetic change in the drug metabolism characterized by extensive metabolizer and the poor metabolizer phenotypes [64]. Subsequently, human CYP2D6 cDNA was cloned from a human liver lgt 11 cDNA library, which when expressed showed the expected bufuralol hydroxylase activity [65]. Further, it was confirmed by altered RFLP patterns in poor metabolizers. In 1990, CYP2D6*3 and CYP2D6*4 alleles were fully characterized [66] which clearly showed the link between the gene variation and the drug molecule metabolism. The effects of genetic changes have been studied widely and implemented in the drug development, drug metabolism and gene-drug interactions [67]. In the past few decades, the effects of genetic variation and gene-nutrient interactions have received much attention to determine nutritional requirement and chronic disease management [68].

Similarly, angiotensin-converting enzyme (ACE) that converts inactive angiotensin I to active angiotensin II contains an insertion/deletion polymorphism on chromosome 17q23 in intronic region 16 and has been reported in different diseases [69]. Further, ACE insertion/deletion polymorphism has been found to be associated with diabetes [70], hypertension [71], coronary heart disease [72, 73] and diabetic nephropathy [74]. For example, diabetes patients with kidney complications have shown increased ACE activity as compared to diabetic patients without any kidney complications [75]. In a recent study, many RAS components were found in adipocytes, suggesting that local RAS could be responsible for adipocyte physiology and likely for the obesity and obesity-related hypertension [76]. Thus, a close relationship between ACE genotype polymorphism and metabolic syndrome is apparent [77]. Higher than 160 single nucleotide ACE gene polymorphisms are known now. Similarly, studies suggest that difference in plasma lipid concentration in different individuals in response to change in dietary fat may indicate the involvement of genotype [78].

Linoleic acid and alpha-linolenic acid are crucial fatty acids that are required for proper growth and development of humans [79]. Omega-6, a type of linoleic acid, and omega-3, alpha linoleic acids, are metabolically as well as physiologically distinct fatty acids, which humans cannot synthesize, and therefore they must receive them from foods. Desaturase enzymes such as deltas 5 and 6 desaturases are encoded by FADS1 and FADS2 gene, respectively, that are required for the metabolism of omega-6 and omega-3 fatty acids [80]. These genes are located on the chromosome 11 (11q12–13.1) in the desaturase cluster [81]. Schaeffer et al. (2006) unambiguously showed that gene cluster of FADS1 and FADS2 genetically controls fatty acid composition of serum phospholipids [82]. Approximately 26 genetic polymorphisms in this gene cluster metabolize omega-6 and omega-3 fatty acids differently, which indicates that different amounts of dietary PUFAs or LC-PUFAs are required to attain comparable effects. Therefore, it is likely that at the same dietary intake of linoleic acid and alpha-linolenic acid, the health effects could vary due to metabolic differences [68]. These evidences indicate that personalized nutritional intervention may resolve MS-related health derangement in accordance with treatment response of different individuals according to their genotype.

13.3 Combined Nutritional and Exercise Interventions Reverse Metabolic Syndrome

The importance of diet modifications and exercise interventions has been appreciated well as a therapy regime. Intake of different types of nutrients has different effects on metabolic syndrome. Various nutrient groups have been identified to play the interventional role in reversal of metabolic syndrome. For the last two decades, Mediterranean diet has received attention because of balance in nutrient patterns. However, consumption of diet containing one or the other dietary component in excess leads to deranged metabolism and health issues. For example, consumption of carbohydrates is implicated as a perilous factor for developing obesity and other associated pathologies; however, the excess consumption of wrong sugar instead of complex sugar has been shown to cause the adverse health consequences. Consumption of complex sugar is always recommended in large proportion to decrease calorie value of the diet [83]. Fats, generally referring to oils, lipids and waxes, are another major component of human diet. Among unsaturated and saturated fats, saturated fats are involved in dyslipidaemia and atherosclerosis. Protein is usually required in moderate quantity, and increased rate of protein intake in case of obesity and kidney disease can induce detrimental effect [84]. Intake of edible greens, fruits and olive oils provides good quantity of beta-carotene, vitamins, antioxidants and minerals which provide beneficial effect on health [85].

In addition to a balanced diet, exercise plays an important role in reversing metabolic syndrome. Regular exercise increases the energy expenditure and helps in weight reduction and fat distribution. A study by Abramson et al. (2002)

observed relationship between physical activity and inflammation in the body during obesity. The study included 3638 people of US population (both men and women). The data indicated anti-inflammatory effect of physical exercises and helped in reduction of risks for heart disease evaluated by measuring C-reactive protein, white blood cell count and fibrinogen level [86]. Another study indicated an inverse relationship between physical exercise and fibrinogen level in the body [87] suggesting the impact of exercise in decreasing the inflammation and associated complications. Recently, Yamaoka and Tango (2012) showed in a clinical study that life-style and diet modifications are more effective in regulating the adverse symptoms in MS patients compared to control [88]. These studies thus indicate that nutritional and life-style modifications in the MS patients may be beneficial in resolving MS.

13.4 Nutrition Guidelines for Metabolic Disorder

Nutritional recommendations do exist in the traditional medicine system; however, the first modern dietary guidelines to combat were given by Detection, Evaluation, and Treatment of High Blood Cholesterol in Adults (Adult Treatment Panel [ATP]) mandated by the National Heart, Lung, and Blood Institute (NHLBI), USA, and American Heart Association (AHA) [89]. The diet includes a balanced intake of fats, carbohydrate, protein, fibres and vitamins. Each component of diet is important and plays critical role in metabolism. The diet recommended by ATP panel includes fat intake of 25–35% of total energy which must contain less than 7% of saturated fat. The polyunsaturated and monounsaturated fat must constitute 10% and 20%, respectively. It should include very limited quantities of cholesterol, refined sugar and trans fats. The recommendation also suggests large intake of fruits, vegetables and cereals which serve as a reservoir of vitamins, minerals and antioxidants. Further, the quantity of carbohydrates should not be more than 60% of the diet. Moreover, fibre is an important part of diet which should be consumed between 20 and 30 g/day. Protein should be 15% of total energy. The patients with MS therefore must ensure balanced energy intake through diet and energy expenditure of the body by involving physical exercise in the daily routine.

13.5 Role of Metabolic Diet in Pathophysiology of Diseases

The advancement of nutrition science in the past few decades has increased the understanding about specific nutrients and dietary patterns on health outcomes. The dietary components have been used as therapeutic agents for many MS-associated diseases. It is clearly understood that dietary components do not affect the body weight alone but also modulate adipose tissue metabolism, glucose homeostasis, liver and kidney function and gut microbiota. The most famous early clinical trial recorded in 1747 A.D. was performed by Captain James Lind of England for scurvy disease, where soldiers receiving citrus fruits treatment were shown to recover from the disease. It was later confirmed in 1932 that the component responsible for cure of scurvy was vitamin C [90]. Hence, the role of nutrients in the therapeutics intervention is not new. Indeed, nutrition intervention has been studied for treatment for many diseases.

The cardiovascular diseases (CVDs) are complex and involve all the complications of the heart or blood vessels [91]. CVD is associated with sedentary life-style and unregulated nutritional intake. Increased physical activity and changed life-style and occupational modification decrease the CVD-related mortality rate [92]. Regular increase in physical activity has favourable effect on weight reduction and also reduces blood pressure and LDL level [93]. The diets which increase the risk of CVD include intake of excessive trans-fat, dietary cholesterol, fats rich in lauric acids, high sodium intake and high alcohol intake. On the contrary, diet components that reduce the risk of CVD include fruits, berries, vegetables, fish and fish oils, potassium, fibres, nuts, folate, plant sterols, flavonoids and soy products [94]. These dietary components reduce the blood pressure and liver fat synthesis and improve endothelial function. They also maintain adipocyte and cardiac function and hamper blood coagulation in blood vessels and vascular adhesion [90].

Diabetes is an ancient disease and has been described in Egyptian manuscripts [95]. The Indian physicians and surgeons, *Sushruta* and *Charaka*, identified two types of diabetes in 400–500 A.D [96]. In diabetes, ability to produce or response to insulin is impaired which causes imbalance in the glucose homeostasis in the body. The resultant hyperglycaemia leads to various diabetic complications such as insulin resistance, neuropathy and nephropathy. The connection between diabetes and obesity is well established. The prolonged high-glucose circulation in the body and high free fatty acids during obesity increase the chronic low-grade inflammation. The pancreas secretes increased amount of insulin that decreases the circulating high glucose, but due to insulin resistance, glucose level does not recede leading to further increase in insulin. As a result, a condition called hyperinsulinemia develops in the body. The hyperinsulinemia takes its toll on pancreas and decreases insulin production with time. Hyperglycaemia and low-grade inflammation during obesity with hyperinsulinemia produce metabolic stress in pancreatic beta cells, which eventually results into beta-cell apoptosis [97] that leads to type 2 diabetes.

Life-style modification has been proposed as a crucial aspect of diabetes therapy [33]. The overweight/obese patients with diabetes are prescribed with hypocaloric diet to reduce weight, which was demonstrated to enhance insulin sensitivity and reduce insulin resistance [98]. Low calorie and energy restriction has been shown to activate insulin receptor by tyrosine kinase which reduces glucose level in circulation [99]. Additionally, fruits and vegetables are recommended for diabetics as the source of vitamins, minerals and important phytochemicals.

Cancer is one of the prominent diseases which has been appreciably linked with obesity and type 2 diabetes [100]. Cancer is the second major cause of mortality worldwide [101]. Although cancer has not been considered as

MS-related disease, its association with MS has been lately corroborated. Several pathophysiological processes have been proposed for obesity-linked cancers, including pancreatic, breast and colon cancer [102]. Diabetes, which emanates from MS, has been associated with different cancers as well [103]. Despite available drugs and treatment options, the incidence and deaths due to cancer have not reduced. Instead, the available treatments cause severe side effects which ruins patient's quality of life, socially and economically. Some of the dietary components, which are part of daily food habit and used regularly, have shown positive results concerning anticancer properties. Among many available dietary components, curcumin is one of the most studied phenolic compounds which is known to have anti-cancerous and anti-inflammatory effects. Many molecular targets have been shown to be controlled by curcumin. One such target is activator protein-1 (AP-1) which is known for its mitogenic effects. Curcumin inhibits AP-1 factor which down regulates cyclin D1 and inhibits tumorigenesis [104].

13.6 Economic Impact of Nutritional Therapy

The world has achieved tremendous economic growth in last few decades, yet a few countries are struggling with poverty and malnutrition. Additionally, in the developing countries, there is vast disparity between rural and urban areas in terms of availability of food items and nutrition. Still, due to concomitant changes in dietary habits and life-style, obesity and MS have become widespread, and pharmacological treatment suffers from insufficiency due to uneven distribution of economic wealth. Since modern pharmacotherapy and surgery are expensive options and cannot be afforded by millions of people across the continents, the nutritional intervention and life-style modifications could be a cost-effective approach to treat MS. The cost utility of different dietary patterns used as a therapy or prevention has been calculated, and the Mediterranean diet and intensive life-style change are found highly cost-effective interventions [105]. Nutritional therapy could be more cost-effective where the crops and plants are grown and processed locally for the population.

13.7 Limitations and Shortcomings

Nutritional intervention in preventing and treating metabolic syndromes is promising as has been highlighted in many studies as discussed above. However, dietary recommendations cannot be universal, and various factors such as genetics, environment and gender play an important role in its effectiveness and therefore must be considered while prescribing such therapies. The current knowledge on nutrigenomics is limited, and thus its application is not sufficient in clinical practices. Further, the role of phytochemicals present in foods is not clear, whether they show their effects independently orin combination with other compounds, and thus needs to be investigated [106]. Though personalized medicines and diets have been used in the traditional medicine systems and are gaining popularity lately, to implement them on a global scale has not been possible due to involvement of expensive techniques and exasperating clinical trials. The recent trends suggest that although some of the plant-based medicines have shown positive effects against the disease, the same are not affordable by the common people. Therefore, the argument that herbal-based drug would be economically affordable becomes a fallacy. Further, awareness about life-style modification of the industrialized societies will help a lot in reversing metabolic syndromes. Many countries have started promoting the practice of yoga in curriculum or otherwise, especially after the declaration of June 21 of every month as 'International Yoga Day' by United Nations General Assembly (UNGA). However, yoga should be practiced in the care of trained and expert healthcare professionals, food and nutrition professionals and genetic counsellors as well as teachers in the education sector, to achieve maximum benefit, especially to the MS patients.

13.8 Conclusion and Future Perspectives

Metabolic syndrome has emerged as a global epidemic of twenty-first century. The growing problems of public health indicate that nutritional intervention and life-style modification should become a part of global healthcare system. Many bioactive components present in the regular diets have been identified to serve as therapeutics in several diseases, including MS. Further studies, clinical trials and support from the governmental and non-government bodies are required to promote the practice of nutrition-based therapeutics. Elucidation of the gene-diet interactions, altered genetic background of patients with the MS and type of nutritional intervention and life-style modification required will improve the effectiveness of the nutritional intervention in the amelioration of MS-related diseases. This could be understood better through studying nutrient-gene interaction in detail along with the affected signalling pathways. Merely, data collection and epidemiology may not be sufficient; rather functional analysis of biomarkers, diet sensors and risk factors should also be addressed. Thorough clinical trials are required to improve the understanding and to realize full potential of nutritional intervention as therapeutic strategy to treat MS.

Acknowledgement The Ramanujan Fellowship to UCSY from DST, Government of India, is acknowledged thankfully.

References

1. Grundy SM et al (2004) Implications of recent clinical trials for the National Cholesterol Education Program Adult Treatment Panel III guidelines. Circulation 110:227
2. Reaven GM (1988) Role of insulin resistance in human disease. Diabetes 37:1595
3. Evaluation and Treatment of High Blood Cholesterol in Adults Expert Panel on Detection (2001) Executive Summary of The Third Report of The National Cholesterol Education

Program (NCEP) Expert Panel on Detection, Evaluation, And Treatment of High Blood Cholesterol In Adults (Adult Treatment Panel III). JAMA 285(19):2486–2497
4. Luque CA, Rey JA (2002) The discovery and status of sibutramine as an anti-obesity drug. Eur J Pharmacol 440(2–3):119–128
5. Xiao D et al (2013) Carboxylesterase-2 is a highly sensitive target of the antiobesity agent orlistat with profound implications in the activation of anticancer prodrugs. Biochem Pharmacol 85(3):439–447
6. Poolsup N, Suksomboon N, Setwiwattanakul W (2012) Efficacy of various antidiabetic agents as add-on treatments to metformin in type 2 diabetes mellitus: systematic review and meta-analysis. ISRN Endocrinol 2012:798146
7. Van Gaal LF et al (2005) Effects of the cannabinoid-1 receptor blocker rimonabant on weight reduction and cardiovascular risk factors in overweight patients: 1-year experience from the RIO-Europe study. Lancet 365(9468):1389–1397
8. Yki-Järvinen H (2004) Thiazolidinediones. N Engl J Med 351(11):1106–1118
9. Giorda C et al (1995) Alpha 1-blocker doxazosin improves peripheral insulin sensitivity in diabetic hypertensive patients. Metabolism 44(5):673–676
10. Carpentier A et al (2002) Ameliorated hepatic insulin resistance is associated with normalization of microsomal triglyceride transfer protein expression and reduction in very low density lipoprotein assembly and secretion in the fructose-fed hamster. J Biol Chem 277(32):28795–28802
11. Knop FK, Bronden A, Vilsboll T (2017) Exenatide: pharmacokinetics, clinical use, and future directions. Expert Opin Pharmacother 18:555
12. Barnett A (2006) DPP-4 inhibitors and their potential role in the management of type 2 diabetes. Int J Clin Pract 60(11):1454–1470
13. Mangaloglu L et al (2002) Treatment with atorvastatin ameliorates hepatic very-low-density lipoprotein overproduction in an animal model of insulin resistance, the fructose-fed Syrian golden hamster: evidence that reduced hypertriglyceridemia is accompanied by improved hepatic insulin sensitivity. Metabolism 51(4):409–418
14. Bass A, Hinderliter AL, Lee CR (2009) The impact of ezetimibe on endothelial function and other markers of cardiovascular risk. Ann Pharmacother 43(12):2021–2030
15. Buchwald H et al (2009) Weight and type 2 diabetes after bariatric surgery: systematic review and meta-analysis. Am J Med 122(3):248–256 e5
16. Sjostrom CD et al (1999) Reduction in incidence of diabetes, hypertension and lipid disturbances after intentional weight loss induced by bariatric surgery: the SOS Intervention Study. Obes Res 7(5):477–484
17. Flum DR, Dellinger EP (2004) Impact of gastric bypass operation on survival: a population-based analysis. J Am Coll Surg 199(4):543–551
18. Sowemimo OA et al (2007) Natural history of morbid obesity without surgical intervention. Surg Obes Relat Dis 3(1):73–77
19. Williams S, Cunningham E, Pories WJ (2012) Surgical treatment of metabolic syndrome. Med Princ Pract 21(4):301–309
20. Keidar A (2011) Bariatric surgery for type 2 diabetes reversal: the risks. Diabetes Care 34(Suppl 2):S361–S266
21. Mason EE, Ito C (1967) Gastric bypass in obesity. Surg Clin North Am 47(6):1345–1351
22. Marceau P et al (1998) Biliopancreatic diversion with duodenal switch. World J Surg 22(9):947–954
23. Cătoi AF et al (2016) Effects of sleeve gastrectomy on insulin resistance. Clujul Med 89(2):267–272
24. Muzio F et al (2005) Long-term effects of low-calorie diet on the metabolic syndrome in obese nondiabetic patients. Diabetes Care 28(6):1485–1486
25. Wadden TA et al (2005) Randomized trial of lifestyle modification and pharmacotherapy for obesity. N Engl J Med 353(20):2111–2120
26. Berthoud HR (2002) Multiple neural systems controlling food intake and body weight. Neurosci Biobehav Rev 26(4):393–428

27. Bray GA (2008) Lifestyle and pharmacological approaches to weight loss: efficacy and safety. J Clin Endocrinol Metab 93(11 Suppl 1):S81–S88
28. Knowler WC et al (2002) Reduction in the incidence of type 2 diabetes with lifestyle intervention or metformin. N Engl J Med 346(6):393–403
29. Orchard TJ et al (2005) The effect of metformin and intensive lifestyle intervention on the metabolic syndrome: the diabetes prevention program randomized trial. Ann Intern Med 142(8):611–619
30. Fabricatore AN (2007) Behavior therapy and cognitive-behavioral therapy of obesity: is there a difference? J Am Diet Assoc 107(1):92–99
31. Ferster CB, Nurnberger JI, Levitt EB (1996) The control of eating. 1962. Obes Res 4(4):401–410
32. Gruber KJ, Haldeman LA (2009) Using the family to combat childhood and adult obesity. Prev Chronic Dis 6(3):A106
33. Grave RD et al (2010) Lifestyle modification in the management of the metabolic syndrome: achievements and challenges. Diabetes Metab Syndr Obes 3:373–385
34. D'Zurilla TJ, Goldfried MR (1971) Problem solving and behavior modification. J Abnorm Psychol 78(1):107–126
35. Shahidi F, Ambigaipalan P (2015) Phenolics and polyphenolics in foods, beverages and spices: antioxidant activity and health effects – a review. J Func Foods 18(Part B):820–897
36. Messina M, Messina V (1996) Nutritional implications of dietary phytochemicals. Adv Exp Med Biol 401:207–212
37. Hegsted DM (1994) A look back at lessons learned and not learned. J Nutr 124(9 Suppl):1867S–1870S
38. Sauer S, Plauth A (2017) Health-beneficial nutraceuticals—myth or reality? Appl Microbiol Biotechnol 101(3):951–961
39. Agrawal DK, Mishra PK (2010) Curcumin and its analogues: potential anticancer agents. Med Res Rev 30(5):818–860
40. Leu TH, Maa MC (2002) The molecular mechanisms for the antitumorigenic effect of curcumin. Curr Med Chem Anticancer Agents 2(3):357–370
41. Dai J, Mumper RJ (2010) Plant Phenolics: extraction, analysis and their antioxidant and anticancer properties. Molecules 15(10):7313
42. Park EJ, Pezzuto JM (2015) The pharmacology of resveratrol in animals and humans. Biochim Biophys Acta 1852(6):1071–1113
43. Iwai K et al (2006) α-Glucosidase inhibitory and Antihyperglycemic effects of polyphenols in the fruit of Viburnum dilatatum Thunb. J Agric Food Chem 54(13):4588–4592
44. Iwai K (2008) Antidiabetic and antioxidant effects of polyphenols in brown alga Ecklonia stolonifera in genetically diabetic KK-A(y) mice. Plant Foods Hum Nutr 63(4):163–169
45. Cabrera C, Artacho R, Gimenez R (2006) Beneficial effects of green tea--a review. J Am Coll Nutr 25(2):79–99
46. Kobayashi Y et al (2000) Green tea polyphenols inhibit the sodium-dependent glucose transporter of intestinal epithelial cells by a competitive mechanism. J Agric Food Chem 48(11):5618–5623
47. Johnston K et al (2005) Dietary polyphenols decrease glucose uptake by human intestinal Caco-2 cells. FEBS Lett 579(7):1653–1657
48. Rivera L et al (2008) Quercetin ameliorates metabolic syndrome and improves the inflammatory status in obese Zucker rats. Obesity (Silver Spring) 16(9):2081–2087
49. Bose M et al (2008) The major green tea polyphenol, (−)-Epigallocatechin-3-Gallate, inhibits obesity, metabolic syndrome, and fatty liver disease in high-fat–fed mice. J Nutr 138(9):1677–1683
50. Egert S et al (2010) Serum lipid and blood pressure responses to quercetin vary in overweight patients by apolipoprotein E genotype. J Nutr 140(2):278–284
51. Mulvihill EE et al (2009) Naringenin prevents dyslipidemia, apolipoprotein B overproduction, and hyperinsulinemia in LDL receptor-null mice with diet-induced insulin resistance. Diabetes 58(10):2198–2210

52. Yokozawa T, Kim HJ, Cho EJ (2008) Gravinol ameliorates high-fructose-induced metabolic syndrome through regulation of lipid metabolism and proinflammatory state in rats. J Agric Food Chem 56(13):5026–5032
53. Shabrova EV et al (2011) Insights into the molecular mechanisms of the anti-atherogenic actions of flavonoids in normal and obese mice. PLoS One 6(10):e24634
54. Mulvihill EE et al (2010) Naringenin decreases progression of atherosclerosis by improving dyslipidemia in high-fat–fed low-density lipoprotein receptor–null mice. Arterioscler Thromb Vasc Biol **30**(4):742
55. Huang HC, Lin JK (2012) Pu-erh tea, green tea, and black tea suppresses hyperlipidemia, hyperleptinemia and fatty acid synthase through activating AMPK in rats fed a high-fructose diet. Food Funct 3(2):170–177
56. Gordon MH (2012) Significance of dietary antioxidants for health. Int J Mol Sci 13(1):173–179
57. Erdman JW et al (2007) Flavonoids and heart health: proceedings of the ILSI North America flavonoids workshop, May 31–June 1, 2005, Washington, DC. J Nutr 137(3):718S–737S
58. Kasubuchi M et al (2015) Dietary gut microbial metabolites, short-chain fatty acids, and host metabolic regulation. Nutrients 7(4):2839–2849
59. Ley RE et al (2006) Microbial ecology: human gut microbes associated with obesity. Nature 444(7122):1022–1023
60. Yoo J, Kim S (2016) Probiotics and prebiotics: present status and future perspectives on metabolic disorders. Nutrients 8(3):173
61. Muller M, Kersten S (2003) Nutrigenomics: goals and strategies. Nat Rev Genet 4(4):315–322
62. Fenech M et al (2011) Nutrigenetics and nutrigenomics: viewpoints on the current status and applications in nutrition research and practice. J Nutrigenet Nutrigenomics 4(2):69–89
63. Gaedigk A et al (1991) Deletion of the entire cytochrome P450 CYP2D6 gene as a cause of impaired drug metabolism in poor metabolizers of the debrisoquine/sparteine polymorphism. Am J Hum Genet 48(5):943–950
64. Ingelman-Sundberg M (2005) Genetic polymorphisms of cytochrome P450 2D6 (CYP2D6): clinical consequences, evolutionary aspects and functional diversity. Pharmacogenomics J 5(1):6–13
65. Gonzalez FJ et al (1988) Characterization of the common genetic defect in humans deficient in debrisoquine metabolism. Nature 331(6155):442–446
66. Heim M, Meyer UA (1990) Genotyping of poor metabolisers of debrisoquine by allele-specific PCR amplification. Lancet 336(8714):529–532
67. Kalow W (1986) Ethnic differences in reactions to drugs and xenobiotics. Caffeine and other drugs. Prog Clin Biol Res 214:331–341
68. Simopoulos AP (2010) Genetic variants in the metabolism of omega-6 and omega-3 fatty acids: their role in the determination of nutritional requirements and chronic disease risk. Exp Biol Med (Maywood) 235(7):785–795
69. Sameer AS et al (2011) ACE I/D polymorphism in hypertensive patients of Kashmiri population. Cardiol Res 1(1):1–7
70. Rahimi Z (2012) ACE insertion/deletion (I/D) polymorphism and diabetic nephropathy. J Nephropathol 1(3):143–151
71. Perticone F et al (1998) Angiotensin-converting enzyme gene polymorphism is associated with endothelium-dependent vasodilation in never treated hypertensive patients. Hypertension 31(4):900
72. Tiret L et al (1993) Deletion polymorphism in angiotensin-converting enzyme gene associated with parental history of myocardial infarction. Lancet 341(8851):991–992
73. Ruiz J et al (1994) Insertion/deletion polymorphism of the angiotensin-converting enzyme gene is strongly associated with coronary heart disease in non-insulin-dependent diabetes mellitus. Proc Natl Acad Sci U S A 91(9):3662–3665
74. Navis G et al (1999) Angiotensin-converting enzyme gene I/D polymorphism and renal disease. J Mol Med 77(11):781–791
75. Rahimi Z et al (2011) The frequency of factor V Leiden mutation, ACE gene polymorphism, serum ACE activity and response to ACE inhibitor and angiotensin II receptor

antagonist drugs in Iranians type II diabetic patients with microalbuminuria. Mol Biol Rep 38(3):2117–2123
76. Engeli S, Negrel R, Sharma AM (2000) Physiology and pathophysiology of the adipose tissue renin-angiotensin system. Hypertension 35(6):1270–1277
77. Lee YJ, Tsai JC (2002) ACE gene insertion/deletion polymorphism associated with 1998 World Health Organization definition of metabolic syndrome in Chinese type 2 diabetic patients. Diabetes Care 25(6):1002–1008
78. Connor WE, Connor SL (1997) Should a low-fat, high-carbohydrate diet be recommended for everyone? The case for a low-fat, high-carbohydrate diet. N Engl J Med 337(8):562–563; discussion 566-7
79. Heird WC, Lapillonne A (2005) The role of essential fatty acids in development. Annu Rev Nutr 25:549–571
80. Cho HP, Nakamura M, Clarke SD (1999) Cloning, expression, and fatty acid regulation of the human delta-5 desaturase. J Biol Chem 274(52):37335–37339
81. Marquardt A et al (2000) cDNA cloning, genomic structure, and chromosomal localization of three members of the human fatty acid desaturase family. Genomics 66(2):175–183
82. Schaeffer L et al (2006) Common genetic variants of the FADS1 FADS2 gene cluster and their reconstructed haplotypes are associated with the fatty acid composition in phospholipids. Hum Mol Genet 15(11):1745–1756
83. Pitsavos C et al (2006) Diet, exercise and the metabolic syndrome. Rev Diabet Stud 3(3):118–126
84. Anderson CA, Miller ER 3rd (2006) Dietary recommendations for obese patients with chronic kidney disease. Adv Chronic Kidney Dis 13(4):394–402
85. Trichopoulou A, Lagiou P (1997) Healthy traditional Mediterranean diet: an expression of culture, history, and lifestyle. Nutr Rev 55(11 Pt 1):383–389
86. Abramson JL, Vaccarino V (2002) Relationship between physical activity and inflammation among apparently healthy middle-aged and older US adults. Arch Intern Med 162(11):1286–1292
87. MacAuley D et al (1996) Physical activity, physical fitness, blood pressure, and fibrinogen in the Northern Ireland health and activity survey. J Epidemiol Community Health 50(3):258–263
88. Yamaoka K, Tango T (2012) Effects of lifestyle modification on metabolic syndrome: a systematic review and meta-analysis. BMC Med 10(1):138
89. Third Report of the National Cholesterol Education Program (NCEP) Expert Panel on Detection, Evaluation, and Treatment of High Blood Cholesterol in Adults (Adult Treatment Panel III) (2002) Third Report of the National Cholesterol Education Program (NCEP) Expert Panel on Detection, Evaluation, and treatment of high blood cholesterol in adults (Adult Treatment Panel III) final report. Circulation 106(25):3143–3421
90. Mozaffarian D, Ludwig DS (2010) Dietary guidelines in the 21st century–a time for food. JAMA 304(6):681–682
91. Myers J (2003) Exercise and cardiovascular health. Circulation 107(1):e2, 2e
92. Paffenbarger RSJ et al (1993) The Association of Changes in physical-activity level and other lifestyle characteristics with mortality among men. N Engl J Med 328(8):538–545
93. Mann S, Beedie C, Jimenez A (2014) Differential effects of aerobic exercise, resistance training and combined exercise modalities on cholesterol and the lipid profile: review, synthesis and recommendations. Sports med 44(2):211–221
94. Srinath Reddy K, Katan MB (2004) Diet, nutrition and the prevention of hypertension and cardiovascular diseases. Public Health Nutr 7(1A):167–186
95. MacCracken J, Hoel D (1997) From ants to analogues. Puzzles and promises in diabetes management. Postgrad Med 101(4):138–140 143–5, 149–50
96. Tipton CM (2008) Susruta of India, an unrecognized contributor to the history of exercise physiology. J Appl Physiol (1985) 104(6):1553–1556

97. Westley RL, May FE (2013) A twenty-first century cancer epidemic caused by obesity: the involvement of insulin, diabetes, and insulin-like growth factors. Int J Endocrinol 2013:632461
98. Riccardi G, Rivellese AA (2000) Dietary treatment of the metabolic syndrome – the optimal diet. Br J Nutr 83(Suppl 1):S143–S148
99. Wirth A (1995) Non-pharmacological therapy of metabolic syndrome. Herz 20(1):56–69
100. Garg SK et al (2014) Diabetes and cancer: two diseases with obesity as a common risk factor. Diabetes Obes Metab 16(2):97–110
101. Global Burden of Disease Cancer Collaboration (2015) The global burden of cancer 2013. JAMA Oncol 1(4):505–527
102. Renehan AG et al (2008) Body-mass index and incidence of cancer: a systematic review and meta-analysis of prospective observational studies. Lancet 371(9612):569–578
103. Vigneri P et al (2009) Diabetes and cancer. Endocr Relat Cancer 16(4):1103–1123
104. Shanmugam MK et al (2015) The multifaceted role of curcumin in cancer prevention and treatment. Molecules 20(2):2728–2769
105. Dalziel K, Segal L (2007) Time to give nutrition interventions a higher profile: cost-effectiveness of 10 nutrition interventions. Health Promot Int 22(4):271–283
106. Hoffman R, Gerber M (2015) Food processing and the Mediterranean diet. Nutrients 7(9):7925–7964

Herbal Intervention in Cardiovascular Diseases

14

Johnna Francis Varghese, Rohit Patel, Mohit Singh, and Umesh C. S. Yadav

Abstract

Cardiovascular diseases (CVDs) are the most prevalent cause of morbidity and mortality worldwide. Oxidative stress and chronic inflammation are the major causes of CVDs. In the vasculature, inflammation occurs due to injury, oxidation of lipids, infection, etc. that result in damage to cells of the blood vessels and progression towards plaque formation. Inflammation portrays an important role in all stages of plaque formation which leads to cardiac arrest or heart failure, peripheral vascular diseases and stroke and could be fatal. Unlike plaque formation, the birth defect in cardiovascular system also leads to short and poor quality of life. Among all preventive and therapeutic strategies for CVDs, herbal interventions are known to be effective with little or no side effects. Herbal interventions have been in use for medical treatments from the beginning of civilisation. WHO report states that approximately 80% of the global population depends on the herbal resources, of which many have moved to clinical use in the recent times. Phytochemicals have a wide array of properties especially anti-inflammatory, anti-oxidative and anti-lipidaemic which are effective in preventing and treating CVDs. They increase the expression of antioxidant and anti-inflammatory molecules mainly via nuclear factor erythroid 2-related factor 2 (Nrf2) activation and reduce the levels of cholesterol, low-density lipoprotein (LDL) and inflammatory cytokines, hence effective in treatment of CVDs. In this chapter, herbal derivative-induced alteration in the mechanisms pertaining to pathophysiology of CVDs has been discussed.

J. F. Varghese · R. Patel · M. Singh
Metabolic Disorders and Inflammatory Pathologies Laboratory, School of Life Sciences, Central University of Gujarat, Gandhinagar, Gujarat, India

U. C. S. Yadav (✉)
School of Life Sciences, Central University of Gujarat, Gandhinagar, Gujarat, India
e-mail: umeshyadav@cug.ac.in

V. Rani, U. C. S. Yadav (eds.), *Functional Food and Human Health*,
https://doi.org/10.1007/978-981-13-1123-9_14

Keywords
Cardiovascular diseases (CVDs) · Phytochemicals · Oxidative stress · Inflammation · ROS · Nuclear factor erythroid 2-related factor 2 (Nrf2)

14.1 Introduction

Cardiovascular diseases (CVDs) have become one of the most prominent global public health concerns leading to maximum number of morbidity and mortality. Cardiovascular system is highly complex where several signalling molecules and mechanisms operate in unison to maintain its physiology. These mechanisms and molecules get altered in unfavourable conditions and trigger the establishment of cardiovascular pathophysiology. CVDs are usually associated with the partial or complete blockage of the blood vessels that affect the structure and function of the heart as well as other organs. The reason for the blockage is mainly plaque formation or atherosclerosis which accounts for approximately 42.5% of total CVD deaths globally [1].

Broadly, there are six types of CVDs classified as ischemic heart disease (IHD), cerebrovascular diseases, peripheral vascular disease, heart failure, rheumatic heart disease (RHD) and congenital heart disease (CHD), among which the first four are solely attributed to plaque formation (Fig. 14.1) [2]. RHD is an autoimmune disease which is caused by certain bacterial infection that results in irreversible damage of valves of the heart [3]. In contrast, CHD is the improper development of the heart and nearby blood vessels, present at birth of the individual [2].

Multiple factors have been identified that contribute towards the pathophysiology of CVDs. Initially, CVDs were tagged as lipid-related diseases with increased presence of cholesterol, triglyceride and low-density lipoprotein (LDL) in blood circulation [4]. However, studies have recently established significant role played by metabolic syndrome (MS) and inflammatory agents in developing CVDs. MS is an array of disease conditions such as diabetes, obesity, hypercholesterolaemia and hyperinsulinaemia [5] that exacerbates the risk of CVDs by provoking oxidative stress through production of reactive oxygen species (ROS). ROS has been identified as a potent stimulator of vascular dysfunction and damage. ROS can stimulate the oxidation of LDL which itself acts as a potent atherogenic molecule to induce dysfunction in endothelial cells (ECs) and also in conversion of macrophage into foam cell (FC), both are critical features in the initiation of atherosclerosis [6]. Additionally, ROS also reacts with nitric oxide (NO) and forms peroxynitrite, turning off the cGMP production in smooth muscle cells (SMCs) and ECs that adversely affect the vasculature [7]. Apart from inducing oxidative stress, MS partakes in establishment of inflammatory condition via increased inflammatory cytokines such as tumour necrosis factor (TNF)-α, interleukin (IL)-6, IL-1 and many more in the blood circulation [8]. These inflammatory agents have now been recognised as critical components in the occurrence and development of the pathophysiology of CVDs.

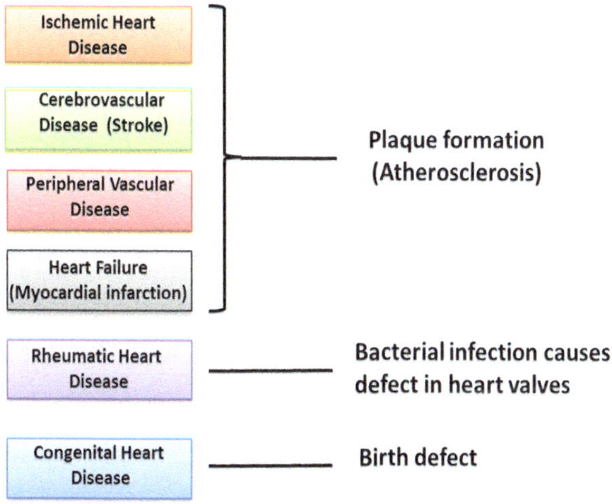

Fig. 14.1 Diagram showing types of CVDs and their causes

Despite advances in understanding the pathophysiology of CVDs, their prevention and treatment continues to be a major concern in the pharmaceutical fraternity. During the past decades, vast extent of improvement has been made in treatment regimen of CVDs. Several therapies are well established in treating CVDs including β-blockers, angiotensin-converting enzyme (ACE) inhibitors, pre-protein convertase subtilisin kexin (PCSK-9) inhibitors and statins [9–12]. Additionally, surgical procedures and insertion of stents are recommended in severe cases that reduce the death rates of patients suffering from CVDs. However, due to the side effects caused by existing synthetic drugs and the development of in-stent restenosis, treating CVDs still faces many challenges. In the face of these challenges, utilisation of medicines derived from natural sources such as herbal products offers an alternative therapeutic approach for treating CVDs.

Since the beginning of civilisation, herbal products have been an integral part of the human consumption, especially for their medicinal values. The traditional medicine practices such as Ayurveda, Siddha, Unani, Chinese and Homoeopathy have long used herbal products. In the present scenario, a new concept has been developed in the western world that 'natural' is a better remedy than 'synthetic or chemical'. Many conventional drugs still in use are plant based such as quinine, aspirin, morphine, etc. Phytochemicals derived from herbal products have been identified as pharmacologically active components with diverse benefits. Some of the well-known properties of phytochemicals are antioxidant, anti-inflammatory and antiplatelet aggregation, indicating multi-faceted role of the phytochemicals for CVDs. They can alter the lipid levels as well as inflammatory response which are major components responsible for the development of CVDs. Epidemiological data indicate that fruit- and vegetable-rich diets are critical components for maintaining cardiovascular health, e.g. numerous plant-derived phytochemicals such as anthocyanins, carotenoids, sterols and polyphenols have been shown to reduce the risk of CVDs and aid

Table 14.1 Natural sources of different phytochemicals and their effectiveness in the treatment of CVDs

Sr. No.	Types of drug	Phytochemicals	Sources	Preventive mechanism	Reference
1	Anticoagulants and antiplatelet	Acetylsalicylic acid (aspirin)	Willow	Prevent clot formation in arteries	[14–19]
		Warfarin	Sweet clover		
		Ellagic acid	Walnuts, pecans, berries and grapes		
2	Angiotensin-converting enzyme (ACE) inhibitors and angiotensin II receptor blockers	Cyanidin	Bamboo shoot	Lower blood pressure and mediate smooth blood flow	[20, 21]
		Ursolic acid	Mirabilis jalapa, apple, bilberries		
3	Beta blockers (antiarrhythmic)	Berberine	Oregon grape, barberry and tree turmeric	Regulate heart beat and lower blood pressure	[22]
4	Aldosterone antagonists/diuretics	Apigenin	Apples, oranges, celery, onions, oregano, parsley, tarragon, basil, cilantro and wheat sprouts	Remove excess water through urine and thereby lower blood pressure	[23–25]
		Furosemide	Digitalis		
5	Anti-lipidaemic and cholesterol-lowering drugs	Allinin	Green garlic	Lower free plasma cholesterol levels	[26–32]
		Gugulipid	Indian bdellium-tree		
		Berberine	Barberry, Oregon grape		
		Curcumin	Turmeric		
		Resveratrol	Skins of grapes and berry		
		Ursolic acid	Mirabilis Jalapa, apple, bilberries		
		Zerumbone	Ginger		

in cardioprotection [13]. In a number of studies, the phytochemicals have been shown to possess similar mode of actions in treating CVDs as those of the well-known synthetic drugs (Table 14.1). Indeed the phytochemicals have become a major area of exploration and research for amelioration of CVDs.

14.2 Herbal Interventions in Plaque Formation

A number of phytochemicals with beneficial effects have shown antioxidant, anti-inflammatory, antithrombotic and anti-lipidemic activities, which hold them at par with the synthetic drugs available for CVDs therapy. Below a brief discussion has been included on some of the promising phytochemicals, e.g. ellagic acid, isothiocyanate, anthocyanins, zerumbone, curcumin, apigenin and hesperidin, which primarily focuses on their beneficial effects against atherosclerotic plaque formation.

14.2.1 Ellagic Acid

Ellagic acid (EA), a naturally occurring phenolic compound, is widely available in fruits and nuts, exclusively in raspberries [33]. It has shown a protective role in various chronic inflammatory diseases. EA portrays an atheroprotective effect through various mechanisms, for example, in a study in mice with high-fat diet (HFD), EA displays atheroprotective effect by upregulating Nrf2 [34]. Nrf2 is a redox-sensitive transcription factor and regulates the endothelial atheroprotective nature against oxidative stress [35]. In ECs, Nrf2 activates the expression of defence enzyme heme oxygenase-1 (HO-1) which functions as an anti-oxidative, anti-inflammatory and cytoprotective molecule [36–38] (Fig. 14.2). Therefore, upregulation of Nrf2 markedly offers protection to ECs during oxidative stress and inflammation.

Presence of atherogenic molecules induces ROS production and downstream activation of extracellular signal-regulated kinase (ERK) and NF-κB. This is followed by overexpression of adhesion markers such as intercellular adhesion molecule-1 (ICAM-1) and vascular cell adhesion molecule-1 (VCAM-1), responsible for directing monocytes to the lesion site and their subsequent recruitment into the sub-endothelial space [39–42]. The presence of modified lipids such as oxLDL induces ERK/NF-κB pathway that further increases the expression of matrix metalloproteinases (MMP)1 and MMP9, which are responsible for the destabilisation of atherosclerotic plaque [43]. These phenomena are abrogated by EA via inhibition of ROS by increased Nrf2/HO-1 signalling pathway [43] (Fig. 14.2).

Uncontrolled uptake of oxLDL by cells present in the arterial lining is a common sight. This phenomenon is facilitated by specific receptors expressed on the cell surface such as lectin-like oxidised low-density lipoprotein receptor-1 (LOX-1). Overexpression of LOX-1 in ECs and subsequent activation of NADPH oxidase (NOX) lead to ROS production. This occurs exclusively in the presence of oxLDL, where oxLDL-induced ROS mediates an inflammatory response via phosphorylating p38 mitogen-activated protein kinase (MAPK) and subsequently activating NF-κB. It then triggers an inflammatory response by expressing downstream molecules such as IL-6, IL-8, ICAM-1 and VCAM-1. Vasodilatory property of ECs is hampered by oxLDL-induced ROS via decreased expression of eNOS and increased formation of peroxynitrite which is toxic to the cells. Vasoconstriction is also induced by endothelin-1 expressed in the presence of oxLDL. These adverse effects are ameliorated by EA by reducing the ROS production [42]. Thus, EA portrays cardioprotective response exclusively via targeting Nrf2- and ROS-mediated signalling.

14.2.2 Isothiocyanate

Isothiocyanate is a group of cardioprotective phytochemicals belonging to the organosulfur family, majorly present in cruciferous vegetables, such as cauliflower, broccoli, cabbage and turnip, and widely accepted as a beneficial dietary component. Similar to EA, in the presence of oxLDL, isothiocyanate also augments the expression of Nrf2 which drives the activation of HO-1 and works as a cardioprotective agent in ECs [44]. In addition, Nrf2 activates glutamate–cysteine ligase (GCL)-catalytic subunit (GCLC) and GCL-modifier subunit (GCLM), which in turn catalyse de novo synthesis of glutathione that decreases ROS production and pro-inflammatory activation of NF-κB in ECs [44] (Fig. 14.2). GCLC is a vital subunit, and loss of its activity leads to excessive oxidative damage to ECs [45, 46].

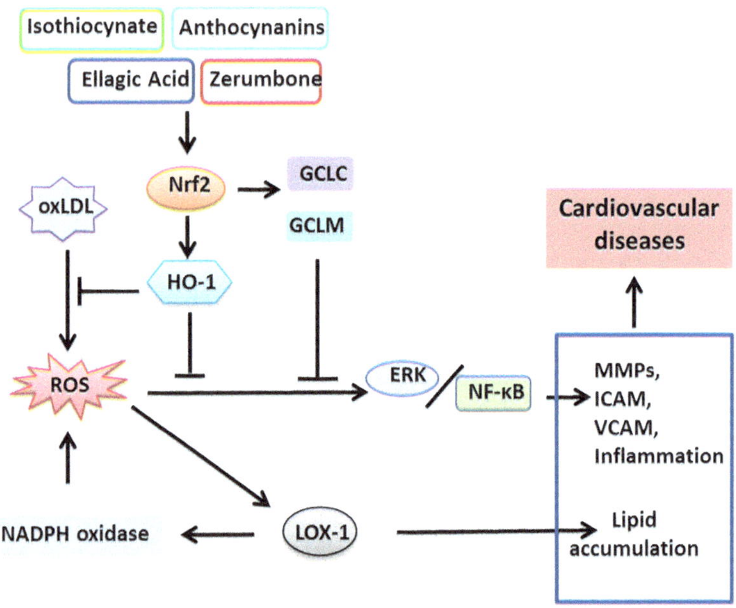

Fig. 14.2 Schematic representation of inhibition of ROS/NF-κB signalling pathway in CVDs by ellagic acid, zerumbone, isothiocyanate and anthocyanins, which inhibit ROS production and ROS-mediated activation of ERK/NF-κB and expression of LOX-1 through Nrf2/HO-1 signalling

Apart from the activating Nrf2 signalling, isothiocyanate inhibits the pro-inflammatory response in ECs. They impede the TLR4/MyD88-mediated activation of the effector proteins, viz. p38, JNK and their downstream transcription factor NF-κB. Activation and nuclear translocation of NF-κB mediate expression of ICAM-1 and VCAM-1 on the surface of ECs which is hampered by isothiocyanate [47]. In ECs, p38 MAPK activation in the presence of atherogenic stress agents overexpresses inflammatory markers such as COX-2 and iNOS. COX-2 enhances monocyte adhesion to the endothelium, and iNOS acts as a pro-inflammatory molecule. This inflammatory pathway is also inhibited by isothiocyanates [48]. Therefore, isothiocyanates could be effective agents to prevent alteration in signalling pathways during CVDs and thus could be potential therapeutic agents.

14.2.3 Anthocyanins

Anthocyanins, the water-soluble vacuolar pigments belonging to the family of flavonoids, are widely available in berries, grapes, plums, prunes, cereals and vegetables. Anthocyanins are efficient in maintaining the healthy lipid profile. Clinical trials in hypercholesterolaemic patients showed that consumption of anthocyanins significantly increased the HDL levels and decreased the LDL levels. Anthocyanins also subsided the serum levels of important inflammatory markers such as the C-reactive protein (CRP), soluble VCAM-1 and IL-1β. In support to the above-mentioned clinical trials, in vitro studies in liver cells also suggested anti-inflammatory action of anthocyanins by inhibiting IL-6- and IL-1β-induced CRP production in the presence of excessive cholesterol [49]. Atherogenic molecules such as oxLDL and saturated fatty acids trigger oxidative and inflammatory responses in ECs and macrophages. One such response is increased NADPH oxidase activity via LOX-1 leading to ROS production (Fig. 14.2). This is followed by ERK1/2 phosphorylation, nuclear translocation of NF-κB and enhanced expression of pro-inflammatory genes such as TNF-α, IL-1β, ICAM-1, VCAM-1 and E-selectins that are commonly observed in inflamed macrophages.

Macrophages are an integral part of the innate immune system and also implicated in plaque development. Anthocyanins have been shown to inhibit these pro-inflammatory pathways in macrophages [50]. Studies using RAW264.7 cells and bone marrow-derived macrophages from $Nrf2^{(-/-)}$ mice in the presence of atherogenic stress have indicated that anthocyanins also curb oxidative stress-mediated ROS production in macrophages through the activation of Nrf2 [50] (Fig. 14.2). Similarly, TNF-α-induced oxidative stress in ECs is prevented by anthocyanins via activation of Nrf2 which binds to the antioxidant responsive elements (ARE) and expresses cytoprotective enzymes such as HO-1 and NADPH dehydrogenase quinone (NQO)-1 [51]. Apart from anti-oxidative effects, anthocyanins also maintain the endothelial vasodilatory properties by activating the NO-cGMP signalling pathway via NO production [52] and regulate the blood pressure in hypercholesterolaemic condition.

14.2.4 Zerumbone

Zerumbone, a naturally available cyclic sesquiterpene, is widely present in edible plant *Zingiber zerumbet* (bitter ginger) and a well-known anti-inflammatory agent. Zerumbone also possesses anti-lipidemic property. Oral administration of zerumbone effectively regulates the dyslipidaemic condition in hyperlipidaemic hamsters [53]. It elevates the serum levels of HDL whereas decreases the levels of LDL and triglycerides. Apart from this, zerumbone also decreased the expression of certain important lipogenic proteins such as fatty acid synthase, malic enzyme, sterol-regulatory element-binding protein and 3-hydroxy-3-methyl-glutaryl-CoA reductase at transcriptional levels [53].

One of the important phenomena during plaque formation is monocyte transvasation in to the sub-endothelial space, followed by differentiation into macrophages which take up and accumulate modified lipids and convert into FC. A plethora of scavenger receptors are expressed in macrophages during hypercholestrolaemic condition which facilitate the uptake of modified lipids converting them into FC [54]. The FCs further secrete inflammatory molecules and increase the plaque burden [55]. Zerumbone has been shown to attenuate the transcriptional activity of activator protein-1 (AP-1) and NF-κB in monocyte-derived macrophages that results in decreased expression of the scavenger receptors thereby reducing the excessive lipid uptake [56]. Along with the inhibition of modified lipid uptake, zerumbone plays an appreciable role in inducing cholesterol efflux through increased expression of efflux transporter proteins such as ATP-binding cassette transporter-1 (ABCA-1) thereby reduces the overburden of lipids [57]. Further, by inhibiting the conversion of macrophages into FCs, zerumbone inhibits the cascade of events that lead to atherosclerosis. Apart from inhibiting lipid accumulation, zerumbone also acts as an antioxidant and anti-inflammatory agent which regulates the Nrf2/HO-1 signalling and ROS production as discussed earlier [58] (Fig. 14.2).

14.2.5 Curumin

Curcumin, a polyphenolic derivative, is extensively present in *Curcuma longa* (turmeric) which is an important part of South Asian daily diet. Comprising of a wide range of pharmacological effects, curcumin possesses anti-lipidaemic property, one of the functions through which it displays atheroprotection. Curcumin, when administered to hypercholesterolaemic rabbits, maintains the serum lipid profile by decreasing the levels of total cholesterol, LDL and triglycerides [59]. In hypercholesterolaemic rabbits, curcumin inhibits nuclear translocation of p65 subunit of NF-κB that inhibits the expression of pro-inflammatory cytokines such as TNF-α, IL-1β and IL-6 circulating in the blood. Curcumin also acts by inhibiting the expression of adhesion molecules such as ICAM-1, VCAM-1, P-selectins and monocyte chemottractant protein (MCP)-1 in the aorta [59] (Fig. 14.3). Apart from the regulation of pro-inflammatory response, curcumin also prevents high-fat-induced fibrosis and hypertrophy of cardiac cells by decreasing the expression of tissue growth

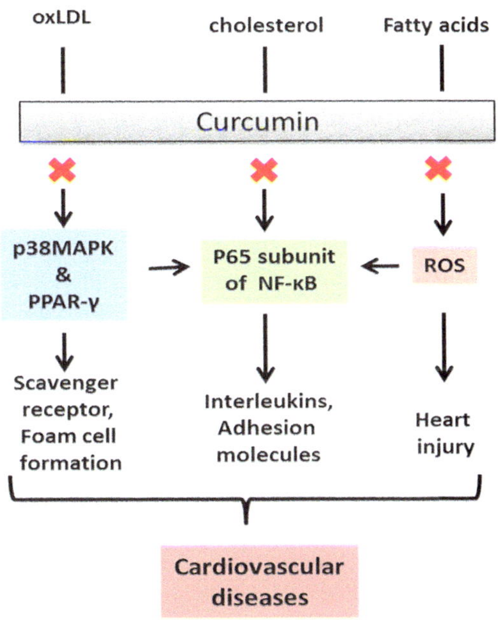

Fig. 14.3 Curcumin inhibits pathological signalling in CVDs. Activation of p38MAPK, NF-κB and ROS in the presence of oxLDL, cholesterol and fatty acids, respectively, is inhibited by curcumin that leads to overall reduction in the factors that contribute to CVDs

factor (TGF)-β, collagen-1 and MMP9 and thereby prevents heart failure [60]. Additionally, curcumin displays its anti-apoptotic activity in cardiac cells in the presence of excessive free fatty acids by increasing the expression of Bcl-2 and decreasing the expression of pro-apoptotic proteins such as Bax and casapse-3. This results in decreased amount of apoptotic cell debris, thereby decreasing atherosclerotic plaque burden [60]. Further, curcumin prevents free fatty acid-mediated ROS production and brings about its antioxidant activity through the activation of Nrf2 and its downstream defence enzymes HO-1, GCLC and NQO-1 in cardiac cells [60]. Curcumin also inhibits macrophage foam cell formation by decreasing the expression of scavenger receptors. Curcumin impedes p38 MAPK phosphorylation and peroxisome proliferator-activated receptors (PPAR)-ϒ expression which are responsible for the abrogation of CD-36 scavenger receptor expression [61]. Therefore, curcumin is an effective agent for the survival of cardiac cells in the presence of atherogenic stimuli. It also inhibits inflammation-associated events responsible for plaque formation (Fig. 14.3).

14.2.6 Apigenin

Apigenin, a naturally occurring flavone, is widely available in various edible sources such as parsley, celery, cloves, peppermint and thyme. It has received extensive attention for its varied pharmacological effects. In CVDs apigenin portrays its antioxidant activity by inducing expression of superoxide dismutase (SOD) in hyperglycaemic conditions in rats with type 2 diabetes. Additionally, it also decreases the expression of malondialdehyde (MDA), a reactive electrophilic species which propagates oxidative stress [62]. Apigenin depicts its anti-inflammatory response by decreasing the activation of NF-κB and expression of ICAM-1 on ECs when subjected to free fatty acids [62]. Under stressful conditions dysregulation of mitochondrial respiratory chain in ECs leads to ROS production and membrane depolymerisation. This leads to mitochondrial dysregulation and activation of the intrinsic apoptotic pathway via increased caspase-3 activity. Apigenin acts by reversing the dysregulation of mitochondrial respiratory chain thereby inhibiting apoptosis [63].

Survival of macrophages residing in the sub-endothelial space plays a critical role in establishing the inflammatory environment responsible for plaque formation. Apigenin reverses the oxLDL-induced survival in peritoneal macrophages derived from ApoE$^{(-/-)}$ mice. It induces apoptosis in macrophages by supressing the phosphorylation of Akt at Ser473, which thereby downregulates plasminogen activator inhibitor (PAI-2), a well-known anti-apoptotic protein [64]. In vitro studies suggest that apigenin also decreases the production of inflammatory cytokine such as IL-1β, IL-6 and TNF-α in macrophages [65]. Apigenin has also been shown to disintegrate NLRP3 which indeed lead to the inhibition of caspase-1 activation and IL-1β secretion. It also prevents IL-6 and IL-1β through inhibiting Erk1/2 activation. Apart from these beneficial effects, apigenin significantly impedes the NF-κB activation in macrophages in the presence of various cytokines [65]. Based on the above-mentioned evidences, apigenin could prove to be an efficient anti-inflammatory phytochemical with potential in treating CVDs.

14.2.7 Hesperidin

Hesperidin is a flavanone glucoside exclusively found in citrus plants such as lemons and oranges. Hesperidin has been found to show protective effect by stimulating PI3K and activating its downstream signalling molecules Akt and AMPK in patients with MS [66]. These molecules phosphorylate and activate eNOS to produce NO in ECs, hence maintaining the proper vasoconstriction and relaxation in the vasculature [67] (Fig. 14.4). During CVDs, enhanced expression of inflammatory cytokines, such as TNF-α, leads to an increased expression of adhesion molecules including VCAM-1 in ECs, which are responsible for monocyte adhesion to the endothelial lining. This pro-inflammatory action of TNF-α is ameliorated by hesperidin through PI3K/Akt-dependent pathway [66]. Further, in cardiac tissues of senescent rat heart under oxidative stress, hesperidin upregulates Nrf2 expression, which then binds to ARE and upregulates the expression of antioxidant enzymes such as catalase, SOD and glutathione s-transferase (GST) [68]. Oral administration of hesperidin to diabetic rats with cardiac ischemia and reperfusion injury exhibited anti-apoptotic effects on the myocardium tissue by upregulating Bcl-2 and downregulating Bax. Anti-apoptotic activity of hesperidin is also displayed by the overexpression of PPAR-γ [69]. In addition to these cardioprotective actions, hesperidin renders its antioxidant activity through curbing the ROS-mediated upregulation of NF-κB and subsequent generation of excessive pro-inflammatory molecules.

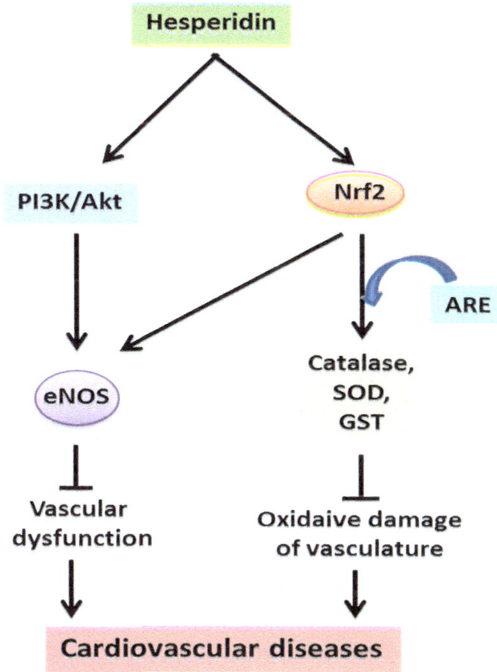

Fig. 14.4 Hesperidin-mediated regulation of vascular dysfunction. Hesperidin increases PI3K/Akt pathway that results in elevated expression of eNOS. Nrf2 induced by hesperidin treatment binds with ARE and induces expression of antioxidant enzymes that are catalase, SOD and GST. These enzymes inhibit the oxidative damage and maintain vascular function

14.3 Herbal Interventions in Rheumatic Heart Disease

Rheumatic heart disease (RHD) is an autoimmune disease caused by bacterial infection of the cardiac tissue. A well-known causative agent of RHD is *Streptococcus pyogenes*, which is a gram-positive bacteria [70]. Infection of *Streptococcus* results in inflammation and rheumatic fever in non-treated throat infection and susceptible children. Although rheumatic fever can be cured using antibiotics, however, chronic rheumatic fever may result in endocarditis and damage of the heart valve [71, 72]. It can lead to increased chest pain, excessive fatigue, shortness of breath, heart failure and fatality. For clear understanding, based on its occurrence, RHD can be divided into two parts, that is, infection-induced and autoimmunity-induced.

14.3.1 Herbal Interventions in Infection-Induced RHD

In infection-induced RHD, very few studies have suggested the use of phytochemicals or extract of different parts of herbal species, specifically in the RHD caused by infection of *S. pyogenes*. For example, the extracts from fruits and branch of *Terminalia* species (*Terminalia arjuna*, *Terminalia catappa* and *Terminalia chebula*) which include triterpenoids, tannins, saponins, phytosteroids, phenolic compounds and flavonoids have been shown to inhibit the growth of *S. pyogenes* [73]. A similar mode of action has been observed with alkaloids, terpenoids and other compounds from *Melodinus* which showed an inhibitory effect against this bacteria in

RHD [74, 75]. Since not many studies have been done in this field, more research required to be undertaken for a better understanding of the mechanism involved in preventing *S. pyogenes* infection-induced RHD. Additionally, phytochemicals can be screened for their potential effects in preventing or treating RHD caused by *S. pyogenes* infection.

14.3.2 Herbal Interventions in Autoimmunity-Induced RHD

Autoimmunity which results from chronic rheumatic fever leads to cardiac dysfunction as discussed above. One of the major cardiac problems is endocarditis which damages cardiac tissue structure and function and is mostly irreversible. However, some phytochemicals discussed below have shown beneficial effects in preventing the adverse consequences of autoimmunity-induced RHD.

14.3.2.1 Apigenin

Apigenin has been shown to play immunomodulatory role in cardiac protection [76] (Fig. 14.5). A study in mice indicates that apigenin regulates Th (T helper) cell proliferation and cytokines released from Th1 and Th2 cells such as TNF-α, IFN-γ, IL-2, IL-12 and IL-4, IL-5, IL-10 and IL-13, respectively [76, 77]. Th cells are central effectors in autoimmune response. The alteration in Th1 and Th2 regulation and consequent release of different cytokines as mentioned above increase the susceptibility towards RHD [78]. Apigenin regulates and controls the synthesis of these cytokines and helps maintain the cardiac function by modulating the altered immunity.

14.3.2.2 Berberine

Berberine, an alkaloid, is found in roots and stem bark of many plants including berberis and coptis. Berberine has shown its beneficial effect in immune-mediated disorders and reduction of autoimmune myocarditis [79]. In the experimental rat model of autoimmune carditis, berberine regulated the JAK-STAT pathway followed by reduction of cytokines such as IL-17 and IFN-γ secreted from Th17 and Th1, respectively. Similar to Th1 response, Th17 response has been found involved in autoimmune disease including RHD. Berberine by modulating the activation of STATs, viz. STAT1, STAT3 and STAT4, reduced the Th1 and Th17 cell inflammation and thereby prevented fatality in rats [79] (Fig. 14.5).

14.3.2.3 Oleanolic Acid

Oleanolic acid, a triterpenoid, found in olive oil, garlic and American pokeweed, displayed protective effects against endocarditis [80]. The cytokines released during autoimmune response in mice induced dystrophic calcification and fibrosis that lead to cardiac dysfunction. Oleanolic acid intervention in experimental mice reduced the expression of inflammatory molecules from Th cells [80], suggesting its role in regulating cardiac function and reducing the risk of RHD.

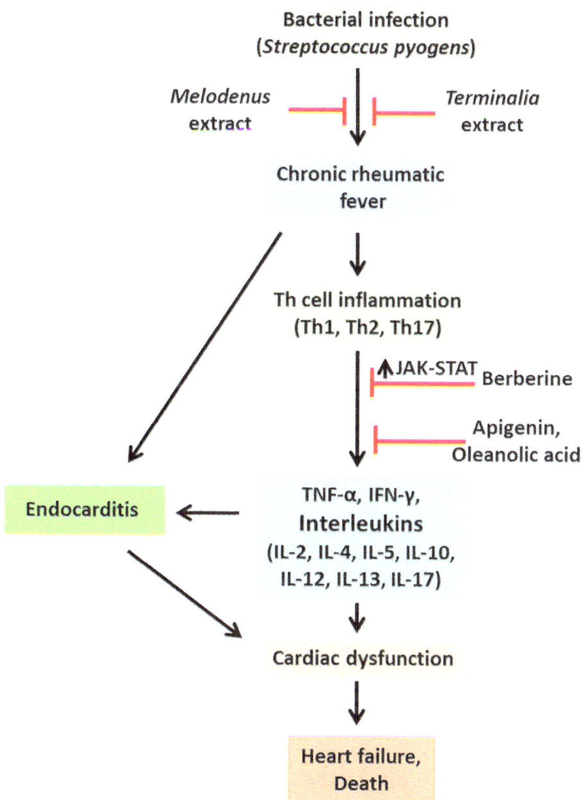

Fig. 14.5 Schematic diagram representing pathways in cardiovascular complications during RHD and the target sites of phytochemicals. Bacterial infection leads to chronic fever which is inhibted by extract from *Melodinus* and *Terminalia*. Further, Th cell inflammation, that results in production of different cytokines, is inhibited by apigenin and oleanolic acid. Berberine also inhibits the inflammation via upregulation of JAK-STAT pathway

Thus, phytochemicals and extracts of different parts of plants have shown promising results in the experimental models of autoimmune-induced RHD. However, proper mechanism of the pathophysiology of RHD is still not well understood. Further studies should reveal various therapeutic interventional strategies using herbal products for prevention and therapy of RHD.

14.4 Herbal Interventions in Congenital Heart Disease

Congenital heart disease (CHD) is the defect in structure or/and function of the heart and is present at birth [81]. It includes various abnormalities which result in pulmonary arterial hypertension (PAH), arrhythmias, infective endocarditis, anticoagulation, congestive heart failure, etc. [82]. Patients suffering from CHD take medicines or undergo surgery to repair this defect although the treatment options depend on the extent of the abnormality. For example, patients suffering from PAH associated with CHD have less chances of survival even after the correction of small defects [83]. Therefore, it is important for patients to take medication lifelong to manage the disease condition and increase the chance of survival [83–85]. Having long-term medication may result in serious side effects and development of other complications. The clinical trials and medical databases have accounted that long-term usage of statins among patients causes myalgia, hypothyroidism and myopathy. Some statins potentially turn out to be diabetogenic when administered at higher doses for longer duration of time [86].

A number of phytochemicals including those mentioned above may be used to reduce the severities associated with CHD. However, very few studies have been undertaken regarding the use of phytochemical in CHD. Some of the phytochemicals identified such as resveratrol and pomegranate juice can be used as an alternative to available synthetic drugs to treat this disease.

14.4.1 Resveratrol

Resveratrol, a polyphenolic compound mostly present in red wine and edible deep-coloured fruit sources such as blueberries, cranberries and cocoa, has been identified to render benefits exclusively in maintaining diastolic function, controlling hypertension and lessening congestive heart failure. A study in mice model shows resveratrol functions via activating Nrf2 transcription factor under hypertensive condition. It decreases ROS production and increases a number of detoxifying enzymes and substrate which help in maintaining the cardiovascular system by reducing cardiac fibrosis, improving structure and function of the heart by remodelling and enhancing diastolic function of blood vessels [87, 88].

14.4.2 Pomegranate Juice

Administration of pomegranate juice has also been shown to regulate the systolic blood pressure in CHD patients [89]. A study by Aviram and Dornfeld indicated that patients consuming about 50 ml of pomegranate juice daily for 2 weeks showed reduction in the blood pressure [90]. However, the longer duration (1 year) administration of pomegranate juice reduced that systolic blood pressure by 21% [91]. The authors have shown that pomegranate juice directly interacts with angiotensin-converting enzyme (ACE), thereby reducing the blood pressure. Further, in vitro as well as clinical study with healthy male volunteer indicated that pomegranate juice did not show harmful effect as it does not affect cytochrome P450-3A pathway. Hence, it can be used for longer duration and with other therapy to control hypertension.

14.5 Recent Advances in Herbal Interventions in CVDs

Herbal-derived medicines have a long history to cure numerous diseases including CVDs. Nonetheless, they are not as well accredited as the allopathic drugs due to their slow mode of action. This calls for directing the attention towards enhancing the capabilities of the herbal medicines by identifying their molecular targets, mode of action and appropriate dosing in CVDs.

Chemical or synthetic drugs available in the market are not appreciable enough in combating cardiovascular diseases, a primary reason being their side effects. Statins, one of the well-known atheroprotective drugs, when used at high doses for longer duration show adverse effects as discussed earlier [92]. Further, usage of statin or any phytochemicals alone has not been adequately effective as that of their combinations. For example, combinatorial treatment of statin along with a phytochemical such as cyanidin-3-glucoside (C3G, a type of anthocyanin) showed a synergistic effect on curbing both the inflammation and oxidative stress caused due to angiotensin II in smooth muscle cells in vitro [93].

A similar approach can also be introduced by combination of various other phytochemicals discussed in this chapter. In a study, the combination of hesperidin and naringenin treatment in stroke-prone hypertensive rats showed synergistic effect in ameliorating the blood pressure when compared with their individual mode of action. Together, both the drugs were more effective as antioxidants and helped in maintaining the systolic blood pressure [94]. Further, in a randomised trial, two different types of anthocyanins such as delphinidin-3-O-β-glucoside and cyanidin-3-O-β-glucoside, when administered together to the patients, showed a remarkable decrease in the expression of inflammatory cytokines in liver cells and porcine iliac ECs under hyperlipidaemic conditions in comparison to the individual anthocyanins [49]. Thus, these studies clearly indicate the better prognosis of intervention with herbal compounds in CVDs, especially when they are administered together.

Apart from medication, surgical procedures are critical options suggested for treating CVDs. Placing stents in the plaque-prone blood vessels is a common procedure used in treating atherosclerosis. A major drawback of using stents is the development of in-stent restenosis. To overcome this condition, drug-eluting stents (DES) have grabbed an enormous amount of attention in the recent times [95]. For example, an in vivo study in pigs demonstrated that curcumin-loaded nanoparticles, when administered along with the stent in the arteries, prevented in-stent restenosis by decreasing inflammation and accelerating the healing process [96]. Further, bioavailability and efficacy of phytochemicals can be increased by adding certain carriers or modern delivery vehicles such as nanoparticles. For example, epigallocatechin-3-gallate-conjugated gold nanoparticles have shown better anti-thrombotic, anti-inflammatory and antioxidant effects in ECs and VSMCs [97]. These particles are more efficiently internalised in these cells and inhibit their migration, thereby maintaining phenotypic integrity, and thus prevent the growth and progression of the plaque [97].

14.6 Conclusion

CVDs offer greater challenge due to diverse components and mechanisms involved in its pathophysiology. A major challenge in treating CVDs is to identify key target and develop therapeutic agents or a combinatorial therapeutic strategy which could include clinically effective and readily available phytochemicals. Herbal interventions stand out due to their biocompatible property and therefore minimal side effects. Phytochemicals can increase the quality of life and life span of patients suffering from CVDs, if they become the part of daily diet as functional food. Targeting the individual factors of MS such as hyperlipidaemia, hyperinsulinaemia, hyperglycaemia and obesity using effective and target-oriented phytochemicals may significantly contribute towards atheroprotection. The phytochemicals that specifically target key signalling molecules such as Nrf2, NF-κB and ROS are of prime importance and can be developed for clinical use in combatting CVDs.

Acknowledgement Dr. Umesh C S Yadav acknowledges the award of Ramanujan Fellowship by DST, Govt. of India. Junior Research Fellowship to RP from SERB and to JFV from GSBTM is also acknowledged.

References

1. Cannon B (2013) Cardiovascular disease: biochemistry to behaviour. Nature 493(7434):S2–S3
2. Six Types of Cardiovascular Disease. 2010 23.07.2010
3. Marijon E et al (2012) Rheumatic heart disease. Lancet 379(9819):953–964
4. Arsenault BJ, Boekholdt SM, Kastelein JJ (2011) Lipid parameters for measuring risk of cardiovascular disease. Nat Rev Cardiol 8(4):197–206
5. Qiao Q et al (2007) Metabolic syndrome and cardiovascular disease. Ann Clin Biochem 44(Pt 3):232–263
6. Sauer H, Shah AM, Laurindo FRM (2010) Studies on cardiovascular disorders, oxidative stress in applied basic research and clinical practice. Humana Press, New York
7. Wang S et al (2014) 6-Gingerol attenuates hydrogen peroxide-induced DNA damage in human umbilical vein endothelia cells. Food Sci Technol Res 20(5):947–954
8. Taube A et al (2012) Inflammation and metabolic dysfunction: links to cardiovascular diseases. Am J Physiol Heart Circ Physiol 302(11):H2148–H2165
9. Sipahi I et al (2007) Beta-blockers and progression of coronary atherosclerosis: pooled analysis of 4 intravascular ultrasonography trials. Ann Intern Med 147(1):10–18
10. Sheppard RJ, Schiffrin EL (2013) Inhibition of the renin-angiotensin system for lowering coronary artery disease risk. Curr Opin Pharmacol 13(2):274–279
11. Steg PG, Ducrocq G (2016) Future of the prevention and treatment of coronary artery disease. Circ J 80(5):1067–1072
12. Ferdowsian HR, Barnard ND (2009) Effects of plant-based diets on plasma lipids. Am J Cardiol 104(7):947–956
13. Vasanthi HR, ShriShriMal N, Das DK (2012) Phytochemicals from plants to combat cardiovascular disease. Curr Med Chem 19(14):2242–2251
14. Brotons C et al (2015) A systematic review of aspirin in primary prevention: is it time for a new approach? Am J Cardiovasc Drugs 15(2):113–133
15. Xie M et al (2014) Aspirin for primary prevention of cardiovascular events: meta-analysis of randomized controlled trials and subgroup analysis by sex and diabetes status. PLoS One 9(10):e90286

16. Eikelboom JW et al (2012) Antiplatelet drugs: antithrombotic therapy and prevention of thrombosis, 9th ed: American College of Chest Physicians Evidence-Based Clinical Practice Guidelines. Chest 141(2 Suppl):e89S–e119S
17. Witt DM et al (2016) Guidance for the practical management of warfarin therapy in the treatment of venous thromboembolism. J Thromb Thrombolysis 41(1):187–205
18. Neidecker M et al (2012) Use of warfarin in long-term care: a systematic review. BMC Geriatr 12:14
19. Smith SA, Travers RJ, Morrissey JH (2015) How it all starts: initiation of the clotting cascade. Crit Rev Biochem Mol Biol 50(4):326–336
20. Ojeda D et al (2010) Inhibition of angiotensin convertin enzyme (ACE) activity by the anthocyanins delphinidin- and cyanidin-3-O-sambubiosides from Hibiscus sabdariffa. J Ethnopharmacol 127(1):7–10
21. Atsumi Shimada MI (2014) Angiotensin I-Converting Enzyme (ACE) inhibitory activity of Ursolic acid isolated from Thymus vulgaris, L. Food Sci Technol Res 20(3):711–714
22. Lau CW et al (2001) Cardiovascular actions of berberine. Cardiovasc Drug Rev 19(3):234–244
23. Ohlsson A et al (2010) Mixture effects of dietary flavonoids on steroid hormone synthesis in the human adrenocortical H295R cell line. Food Chem Toxicol 48(11):3194–3200
24. Hasegawa E et al (2013) Effect of polyphenols on production of steroid hormones from human adrenocortical NCI-H295R cells. Biol Pharm Bull 36(2):228–237
25. Paterna S et al (2015) Hypertonic saline in conjunction with high-dose furosemide improves dose-response curves in worsening refractory congestive heart failure. Adv Ther 32(10):971–982
26. Yang Y et al (2011) Effects of some common food constituents on cardiovascular disease. ISRN Cardiol 141:1–16
27. Urizar NL, Moore DD (2003) GUGULIPID: a natural cholesterol-lowering agent. Annu Rev Nutr 23:303–313
28. Doggrell SA (2005) Berberine–a novel approach to cholesterol lowering. Expert Opin Investig Drugs 14(5):683–685
29. Alwi I et al (2008) The effect of curcumin on lipid level in patients with acute coronary syndrome. Acta Med Indones 40(4):201–210
30. Militaru C et al (2013) Oral resveratrol and calcium fructoborate supplementation in subjects with stable angina pectoris: effects on lipid profiles, inflammation markers, and quality of life. Nutrition 29(1):178–183
31. Kunkel SD et al (2012) Ursolic acid increases skeletal muscle and brown fat and decreases diet-induced obesity, glucose intolerance and fatty liver disease. PLoS One 7(6):e39332
32. Hemn HO et al (2015) Antihypercholesterolemic and antioxidant efficacies of zerumbone on the formation, development, and establishment of atherosclerosis in cholesterol-fed rabbits. Drug Des Devel Ther 9:4173–4208
33. Landete J (2011) Ellagitannins, ellagic acid and their derived metabolites: a review about source, metabolism, functions and health. Food Res Int 44(5):1150–1160
34. Ding Y et al (2014) Dietary ellagic acid improves oxidant-induced endothelial dysfunction and atherosclerosis: role of Nrf2 activation. Int J Cardiol 175(3):508–514
35. Dai G et al (2007) Biomechanical forces in atherosclerosis-resistant vascular regions regulate endothelial redox balance via phosphoinositol 3-kinase/Akt-dependent activation of Nrf2. Circ Res 101(7):723–733
36. Taha H et al (2010) Role of heme oxygenase-1 in human endothelial cells: lesson from the promoter allelic variants. Arterioscler Thromb Vasc Biol 30(8):1634–1641
37. Saragih H et al (2014) PECAM-1-dependent heme oxygenase-1 regulation via an Nrf2-mediated pathway in endothelial cells. Thromb Haemost 111(6):1077–1088
38. Chung HT, Pae HO, Cha YN (2008) Role of heme oxygenase-1 in vascular disease. Curr Pharm Des 14(5):422–428
39. Lawrence T (2009) The nuclear factor NF-kappaB pathway in inflammation. Cold Spring Harb Perspect Biol 1(6):a001651

40. Claude S et al (2014) Flavanol metabolites reduce monocyte adhesion to endothelial cells through modulation of expression of genes via p38-MAPK and p65-Nf-kB pathways. Mol Nutr Food Res 58(5):1016–1027
41. Zhang LP et al (2016) Glabridin attenuates lipopolysaccharide-induced acute lung injury by inhibiting p38MAPK/ERK signaling pathway. Oncotarget 8:18935–18942
42. Lee WJ et al (2010) Ellagic acid inhibits oxidized LDL-mediated LOX-1 expression, ROS generation, and inflammation in human endothelial cells. J Vasc Surg 52(5):1290–1300
43. Kuo MY et al (2011) Ellagic acid inhibits oxidized low-density lipoprotein (OxLDL)-induced metalloproteinase (MMP) expression by modulating the protein kinase C-alpha/extracellular signal-regulated kinase/peroxisome proliferator-activated receptor gamma/nuclear factor-kappaB (PKC-alpha/ERK/PPAR-gamma/NF-kappaB) signaling pathway in endothelial cells. J Agric Food Chem 59(9):5100–5108
44. Huang CS et al (2013) Isothiocyanates protect against oxidized LDL-induced endothelial dysfunction by upregulating Nrf2-dependent antioxidation and suppressing NFkappaB activation. Mol Nutr Food Res 57(11):1918–1930
45. Yet SF et al (2003) Absence of heme oxygenase-1 exacerbates atherosclerotic lesion formation and vascular remodeling. FASEB J 17(12):1759–1761
46. Yang Y et al (2002) Initial characterization of the glutamate-cysteine ligase modifier subunit Gclm(−/−) knockout mouse. Novel model system for a severely compromised oxidative stress response. J Biol Chem 277(51):49446–49452
47. Shan Y et al (2012) Sulphoraphane inhibited the expressions of intercellular adhesion molecule-1 and vascular cell adhesion molecule-1 through MyD88-dependent toll-like receptor-4 pathway in cultured endothelial cells. Nutr Metab Cardiovasc Dis 22(3):215–222
48. Shan Y et al (2010) Protective effect of sulforaphane on human vascular endothelial cells against lipopolysaccharide-induced inflammatory damage. Cardiovasc Toxicol 10(2):139–145
49. Zhu Y et al (2013) Anti-inflammatory effect of purified dietary anthocyanin in adults with hypercholesterolemia: a randomized controlled trial. Nutr Metab Cardiovasc Dis 23(9):843–849
50. Lee SG et al (2014) Berry anthocyanins suppress the expression and secretion of proinflammatory mediators in macrophages by inhibiting nuclear translocation of NF-kappaB independent of NRF2-mediated mechanism. J Nutr Biochem 25(4):404–411
51. Speciale A et al (2013) Cyanidin-3-O-glucoside counters the response to TNF-alpha of endothelial cells by activating Nrf2 pathway. Mol Nutr Food Res 57(11):1979–1987
52. Zhu Y et al (2011) Purified anthocyanin supplementation improves endothelial function via NO-cGMP activation in hypercholesterolemic individuals. Clin Chem 57(11):1524–1533
53. Tzeng TF et al (2014) Lipid-lowering effects of zerumbone, a natural cyclic sesquiterpene of Zingiber zerumbet Smith, in high-fat diet-induced hyperlipidemic hamsters. Food Chem Toxicol 69:132–139
54. Moore KJ, Tabas I (2011) Macrophages in the pathogenesis of atherosclerosis. Cell 145(3):341–355
55. Rudijanto A (2007) The role of vascular smooth muscle cells on the pathogenesis of atherosclerosis. Acta Med Indones 39(2):86–93
56. Eguchi A et al (2007) Zerumbone suppresses phorbol ester-induced expression of multiple scavenger receptor genes in THP-1 human monocytic cells. Biosci Biotechnol Biochem 71(4):935–945
57. Zhu S, Liu JH (2015) Zerumbone, a natural cyclic Sesquiterpene, promotes ABCA1-dependent cholesterol efflux from human THP-1 macrophages. Pharmacology 95(5–6):258–263
58. Shin JW et al (2011) Zerumbone induces heme oxygenase-1 expression in mouse skin and cultured murine epidermal cells through activation of Nrf2. Cancer Prev Res (Phila) 4(6):860–870
59. Um MY et al (2014) Curcumin attenuates adhesion molecules and matrix metalloproteinase expression in hypercholesterolemic rabbits. Nutr Res 34(10):886–893
60. Zeng C et al (2015) Curcumin protects hearts from FFA-induced injury by activating Nrf2 and inactivating NF-kappaB both in vitro and in vivo. J Mol Cell Cardiol 79:1–12

61. Min KJ et al (2013) Curcumin inhibits oxLDL-induced CD36 expression and foam cell formation through the inhibition of p38 MAPK phosphorylation. Food Chem Toxicol 58:77–85
62. Ren B et al (2016) Apigenin and naringenin regulate glucose and lipid metabolism, and ameliorate vascular dysfunction in type 2 diabetic rats. Eur J Pharmacol 773:13–23
63. Duarte S et al (2013) Apigenin protects endothelial cells from lipopolysaccharide (LPS)-induced inflammation by decreasing caspase-3 activation and modulating mitochondrial function. Int J Mol Sci 14(9):17664–17679
64. Zeng P et al (2015) Apigenin attenuates Atherogenesis through inducing macrophage apoptosis via inhibition of AKT Ser473 phosphorylation and downregulation of plasminogen activator Inhibitor-2. Oxidative Med Cell Longev 2015:1
65. Zhang X et al (2014) Flavonoid Apigenin inhibits lipopolysaccharide-induced inflammatory response through multiple mechanisms in macrophages. PLoS One 9(9):e107072
66. Rizza S et al (2011) Citrus polyphenol hesperidin stimulates production of nitric oxide in endothelial cells while improving endothelial function and reducing inflammatory markers in patients with metabolic syndrome. J Clin Endocrinol Metab 96(5):E782–E792
67. Cacicedo Jé M et al (2011) Acute exercise activates AMPK and eNOS in the mouse aorta. Am J Physiol Heart Circ Physiol 301(4):H1255–H1265
68. Elavarasan J et al (2012) Hesperidin-mediated expression of Nrf2 and upregulation of antioxidant status in senescent rat heart. J Pharm Pharmacol 64(10):1472–1482
69. Agrawal YO et al (2014) Hesperidin produces cardioprotective activity via PPAR-gamma pathway in ischemic heart disease model in diabetic rats. PLoS One 9(11):e111212
70. Guilherme L, Kalil J (2010) Rheumatic fever and rheumatic heart disease: cellular mechanisms leading autoimmune reactivity and disease. J Clin Immunol 30(1):17–23
71. Zuhlke L et al (2015) Characteristics, complications, and gaps in evidence-based interventions in rheumatic heart disease: the global rheumatic heart disease registry (the REMEDY study). Eur Heart J 36(18):1115–122a
72. Six Types of Cardiovascular Diseases. 2010 23.07.2010
73. Mitchell Henry Wright MSJA, Lee CJ, Courtney R, Greene AC, Cock IE (2016) Qualitative phytochemical analysis and antibacterial activity evaluation of Indian Terminalia spp. against the pharyngitis causing pathogen Streptococcus pyogenes. Pharmacogn Commun 6(2):85–92
74. Lu Y, Khoo TJ, Wiart C (2014) The genus Melodinus (Apocynaceae): chemical and pharmacological perspectives. Sci Res 5(5):540–550
75. Jiang J-H, Zhang W-D, Chen Y-G (2015) Phytochemical and pharmacological properties of the Genus Melodinus – a review. Trop J Pharm Res 14(12):2325–2344
76. Zhang S et al (2016) Apigenin attenuates experimental autoimmune myocarditis by modulating Th1/Th2 cytokine balance in mice. Inflammation 39(2):678–686
77. Barnes AG et al (2007) Bacillus subtilis spores: a novel microparticle adjuvant which can instruct a balanced Th1 and Th2 immune response to specific antigen. Eur J Immunol 37(6):1538–1547
78. Liu Y et al (2009) The effect of Gd@C82(OH)22 nanoparticles on the release of Th1/Th2 cytokines and induction of TNF-alpha mediated cellular immunity. Biomaterials 30(23–24):3934–3945
79. Liu X et al (2016) Protective mechanisms of berberine against experimental autoimmune myocarditis in a rat model. Biomed Pharmacother 79:222–230
80. Martin R et al (2014) Oleanolic acid modulates the immune-inflammatory response in mice with experimental autoimmune myocarditis and protects from cardiac injury. Therapeutic implications for the human disease. J Mol Cell Cardiol 72:250–262
81. What Are Congenital Heart Defects? 2011 1.07.2011
82. The Impact of Congenital Heart Defects. 2015 21.10.2015
83. Manes A et al (2014) Current era survival of patients with pulmonary arterial hypertension associated with congenital heart disease: a comparison between clinical subgroups. Eur Heart J 35(11):716–724
84. Murphy JG et al (1993) Long-term outcome in patients undergoing surgical repair of tetralogy of Fallot. N Engl J Med 329(9):593–599

85. Congenital heart defects in children. Diseases and conditions 04.02.2016
86. Ramkumar S, Raghunath A, Raghunath S (2016) Statin therapy: review of safety and potential side effects. Acta Cardiol Sin 32(6):631–639
87. Javkhedkar AA et al (2015) Resveratrol restored Nrf2 function, reduced renal inflammation, and mitigated hypertension in spontaneously hypertensive rats. Am J Physiol Regul Integr Comp Physiol 308(10):R840–R846
88. Sung MM et al (2015) Resveratrol treatment of mice with pressure-overload-induced heart failure improves diastolic function and cardiac energy metabolism. Circ Heart Fail 8(1):128–137
89. Stowe CB (2011) The effects of pomegranate juice consumption on blood pressure and cardiovascular health. Complement Ther Clin Pract 17(2):113–115
90. Aviram M, Dornfeld L (2001) Pomegranate juice consumption inhibits serum angiotensin converting enzyme activity and reduces systolic blood pressure. Atherosclerosis 158(1):195–198
91. Aviram M et al (2004) Pomegranate juice consumption for 3 years by patients with carotid artery stenosis reduces common carotid intima-media thickness, blood pressure and LDL oxidation. Clin Nutr 23(3):423–433
92. Laufs U et al (2015) Treatment options for statin-associated muscle symptoms. Dtsch Arztebl Int 112(44):748–755
93. Pantan R et al (2016) Synergistic effect of atorvastatin and Cyanidin-3-glucoside on angiotensin II-induced inflammation in vascular smooth muscle cells. Exp Cell Res 342(2):104–112
94. Ikemura M et al (2012) Preventive effects of hesperidin, glucosyl hesperidin and naringin on hypertension and cerebral thrombosis in stroke-prone spontaneously hypertensive rats. Phytother Res 26(9):1272–1277
95. Hao H et al (2011) Drug-eluting stent: importance of clinico-pathological correlations. Circ J 75(7):1548–1558
96. Lu Q et al (2015) Accelerated recovery of endothelium function after stent implantation with the use of a novel systemic nanoparticle curcumin. Biomed Res Int 2015:291871
97. Khoobchandani M et al (2016) Laminin receptor-avid Nanotherapeutic EGCg-AuNPs as a potential alternative therapeutic approach to prevent restenosis. Int J Mol Sci 17(3):316

Therapeutic Potential of Phytoestrogens

15

Atiya Fatima, Asrar Alam, and Ram Singh

Abstract

Phytoestrogens are naturally occurring constituents of plants present in the significant proportion of our diet. They have been extensively studied due to their potential as pharmacological targets and nutraceutical benefits. The potential pharmacological applications of these molecules include cardioprotection, antimicrobial, anticancer, anti-obesity, antiosteoporosis, antidiabetic, and neuroprotection. Phytoestrogens are polyphenolic nonsteroidal compounds of plant origin with estrogen-like biological activity. They mimic estradiol-like effects in several tissue/tissues of the mammalian body. The health benefits accredited to them are due to their ability to mimic estrogenic actions. In this chapter, we aim to provide comprehensive coverage of the pharmacological aspects of the most pronounced phytoestrogens of our daily life. We will discuss different classes of phytoestrogens under the subcategory of flavonoids and non-flavonoids. Numerous plant-derived compounds like genistein, daidzein, 8-prenylnaringenin, equol, quercetin, coumestrol, isoliquiritigenin, and resveratrol belonging to different classes of phytoestrogens will be discussed with their therapeutic values and clinical applications in a broad range of health-related problems and disorders.

Keywords

Phytoestrogens · Flavonoids · Estrogenic activity · RBA (Receptor binding affinity) · Genistein · Resveratrol

A. Fatima (✉) · R. Singh
Department of Applied Chemistry, Delhi Technological University, New Delhi, India

A. Alam
Laboratory of Vaccinology and Applied Immunology, Kanazawa University School of Pharmacy, Kanazawa, Japan

V. Rani, U. C. S. Yadav (eds.), *Functional Food and Human Health*,
https://doi.org/10.1007/978-981-13-1123-9_15

15.1 Introduction

Phytoestrogens are naturally occurring constituents of plants, having structural similarity to mammalian female sex hormones estrogens (xenoestrogens). They exhibit estradiol-like effects in one or more target tissues in mammalian body [1]. Under the classical definition of phytoestrogens fall those compounds, which can induce estrus, exert estrogenic effects on the central nervous system (CNS), and stimulate growth of the genital tract of female animals. Broadly, the phytoestrogen can be defined as the chemicals which can display effects suggestive of estrogenicity which include binding to the estrogen receptors, thereby inducing specific estrogen-responsive gene products leading to stimulation of estrogen receptor(s)-positive breast cancer cell growth [2]. They cause both estrogenic and antiestrogenic effects [3] by binding to estrogen receptors alpha and beta with a preference estrogen receptor [1, 4]. Since estrogen receptors are present in different tissues all over the mammalian body, central nervous system (CNS), reproductive system, mammary gland, bones, lungs, and ovary, phytoestrogens have specific hormonal effects on different tissues [5]. Phytoestrogens act as natural defense against herbivorous animals by controlling their female fertility, attributed to their mimicry and action as antagonists of estrogen [3]. Phytoestrogens exert their effects mediated *via* multiple modes of actions including both genomic and non-genomic mechanisms [6]. Estrogenic and antiestrogenic effects on estrogen receptors (ERs) are some genomic mechanisms of action, while other effects' direct interaction with ERs may not be involved [7]. Phytoestrogens have stable structure and low molecular weight which gains them access through cell membranes [8]. They interact with enzymes and receptors, bind to ERs and provoke specific estrogen-responsive gene products, and stimulate ER-mediated effects in the body. Non-genomic effects exerted by phytoestrogens do not involve direct interaction with ERs. Some of the non-genomic actions include inhibition of tyrosine kinase, antioxidant effects and suppression of angiogenesis, induction of cancer cell differentiation, and DNA topoisomerase activities [9]. Among these, the most well-characterized mode of phytoestrogen action appears to be through estrogen receptor (ER) binding.

A plethora of information related to health benefits of phytoestrogens is available in the literature. The first major attributed health benefits of phytoestrogen are an alternative to hormone replacement therapy (HRT) to alleviate premenopausal symptoms and osteoporosis. Phytoestrogens have also been suggested to have health benefits in cardiovascular diseases, metabolic disorders [10], and breast cancer, but their beneficial role in breast cancer is contradicted by some adverse effects. They have been reported to possess antiviral [11] and antimicrobial properties [12]. In this chapter, we have presented the classification and a detailed description of the role of phytoestrogens in human health along with their positive and negative effects.

15.2 Clinical Importance of Different Classes of Phytoestrogens

Phytoestrogens classification is based on their chemical structures, which resemble estradiol (E2). There are several classes of phytoestrogens found in plants. Broadly, phytoestrogens are classified into two major classes such as flavonoids and non-flavonoids. Flavonoids are further divided into isoflavones, flavones, isoflavans, flavonols, and coumestans. Non-flavonoid consists mainly of lignans, stilbenes, and chalcones. The detailed classification of phytoestrogens along with their clinical significance is presented below.

15.2.1 Flavonoids

15.2.1.1 Isoflavones

The isoflavonoids are 3-phenyl chromones having C6-C3-C6 skeleton in which each C6 moiety is a benzene ring linked *via* a heterocyclic pyrone ring (Fig. 15.1). They can exist as glucosides or as aglycones. The glucosides are further hydrolyzed in the gut to their respective aglycones.

Isoflavones are substituted derivatives of isoflavonoids where two or three hydrogen atoms are replaced by hydroxyl groups. Their structures encompass a planar ring system with a substituted *p*-hydroxy aromatic ring and second in-plane hydroxyl group which are approximately 12Å away from each other. This distance is almost identical to that between the C-3 and C-17 hydroxyls of estradiol [13] (Fig. 15.2). This is why they can bind to receptors. Two ring structures separated with two carbon atoms as well as spacing between two hydroxyls play an important role in binding affinity of isoflavones to ERs [14].

Hydrogen bond interactions with amino acids of the ligand-binding site of the ER are determined by these distances [15]. Isoflavones have gained considerable importance in the areas of clinical nutrition and disease prevention during last few decades. They show beneficial effects on cancer, lipid metabolism, cardiovascular diseases, osteoporosis, and blood coagulation and act as antioxidants [16]. They proved their efficacy as insecticidal, piscicidal, antifungal, and antimicrobial agents. Two of the most extensively studied isoflavones are genistein and daidzein.

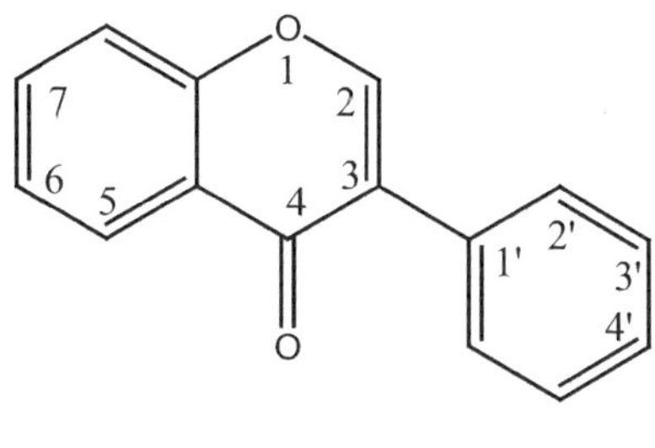

Fig. 15.1 Basic isoflavone skeleton

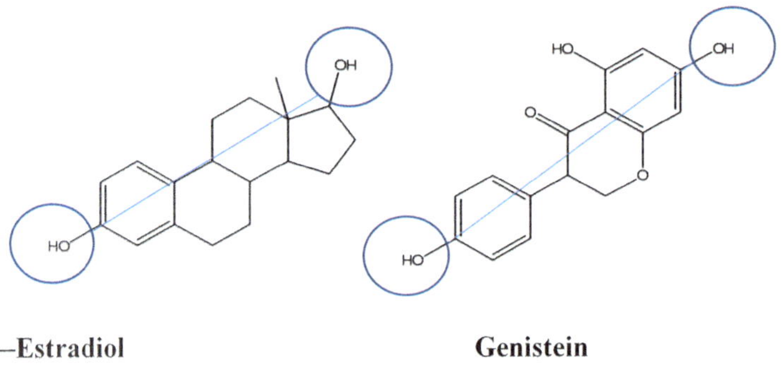

Fig. 15.2 Relative comparison of distances of two hydroxyl groups in estradiol and isoflavones

Fig. 15.3 Genistein

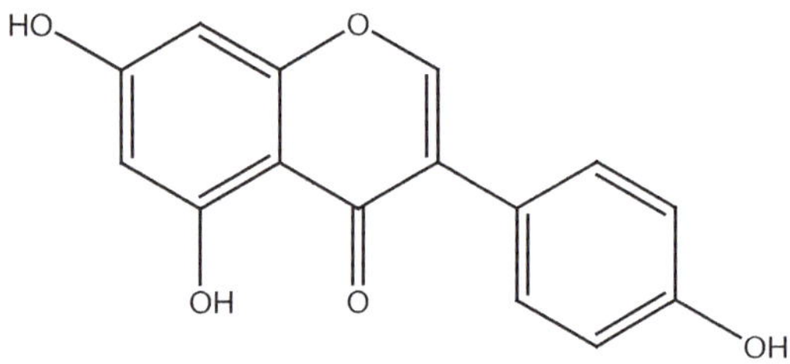

Genistein

Genistein (5,7,4′-trihydroxyisoflavone) (Fig. 15.3) is an isoflavone first isolated from the dyer's broom, *Genista tinctoria*, and is naturally found in a number of plants including soybeans, lupin, fava beans, kudzu, and other legumes and commonly exists as inactive glucoside genistein [17]. It is derived from a precursor molecule, biochanin A, that is converted to genistein after breakdown by intestinal glucosidases.

Genistein is one of the phytoestrogens to show potent receptor binding affinity with the estrogen receptors. It can prop up tissue specific as agonist or antagonist depending on the estrogen receptor and concentration of circulating endogenous estrogens [18]. Despite the similarity, genistein can bind to ERα with relative binding affinity of only 0.05–1% to the binding affinity of 17β-estradiol, while it's greater in case of ERβ [19]. This differential binding affinity of genistein (estrogen receptor-β seven times more than that of estrogen receptor-α) makes it a selective estrogen receptor modulator [20] and thus promotes multiple effects on the target tissue.

Synthetic derivatives have been prepared to test the potency of genistein. Among numerous biological effects displayed by genistein, the most pronounced one is anticancer property. It has been reported to exert antiproliferative properties on different types of cancer. Several mechanisms for the anticancer effects have been proposed including inhibition of angiogenesis, affecting cellular transduction signaling pathways inhibiting carcinogenesis, inactivation of nuclear factor kappa B

(NF-κB), inhibition of metastasis of cancer cells by controlling proteolytic enzymes matrix metalloproteinases (MMPs), topoisomerase, tyrosine kinase activity, and antioxidant properties. Genistein has also been known to cause cell cycle arrest in various human cancer cell lines as described below.

Cancer

The proliferative (estrogenic) and antiproliferative (antiestrogenic) effects in human breast cancer cell lines due to genistein made it the phytoestrogen of greatest interest. It imparts biphasic effect in estrogen receptor (ER)-positive MCF-7 breast cancer cell line which is concentration-dependent, with stimulation of cell growth at lower concentrations between 10 and 1.0 nM but at higher doses between 1 and 10 μM, and above inhibition is observed in both ER-positive and ER-negative cells. At lower concentrations, genistein competes with estradiol for binding to the ER and thus mimic as an estrogen agonist [21]. The antiproliferative effects are clearly non-ER mediated. It induces apoptosis in both ER-positive MCF7 cell lines and ER-negative MDA-MB-468 cancer cell lines. It inhibits the growth of cancer cells by intervening in cell cycle progression (affecting CDKs), which ultimately results in cessation of cell proliferation. PTKs (protein tyrosine kinases) that are involved in tumorigenesis are inhibited by genistein action, and this efficacy has led many laboratories to investigate its therapeutic potential against breast and prostate cancer [22]. In addition to PTKs, other mechanisms by which genistein exhibits anticarcinogenic activity are inhibition of angiogenesis (inhibits TGF-β signaling and the suppression of growth factor receptor VEGF-mediated signaling pathway), BRCA1 modulation, induction of apoptosis through inhibitory effect on NF-κB signaling and Akt pathway, inhibition of aromatase enzyme, and inhibiting DNA replication enzymes DNA topoisomerases I and II.

Androgens are crucial in prostate cancer genesis and act *via* activation of androgen receptor (AR). AR-responsive gene, prostate-specific antigen (PSA) acts as major key factor in prognosis and progression in prostate cancer patients. Genistein, at high concentrations, inhibits the growth of androgen-dependent and independent cells in human prostate cancer cell lines by inhibiting PSA synthesis in prostate cancer cells [23]. It displays anti-metastatic efficacy in prostrate tissue in phase I and phase II clinical trial patients [24]. Genistein inhibit cancer progression in colon cells by inducing apoptosis or inhibiting proliferation, upregulating LDL receptors [25], and affecting epidermal growth factors [26].

Genistein in combination with other chemotherapeutic agents such as centchroman, *cis*platin, gemcitabine, docetaxel, and 5-fluorouracil potentiates their antitumor activity [27–30]. Several other combination therapies such as curcumin-genistein [31], terazosin-genistein [32], and genistein-cetuximab [33] have proved to be effective in various cancer cells. A derivative of genistein, phenoxodiol ((3-(4-hydroxyphenyl)-2H-chromen-7-ol, Fig. 15.4), belonging to the family of drugs called signal transduction inhibitors can arrest several cancer cell lines in the G1 stage of the cell cycle [34]. It stimulates apoptosis *via* multiple pathways. It is a molecule with broad antitumor activity and high specificity for tumor cells. Its biochemical tendencies are predominantly well-matched to reversal of

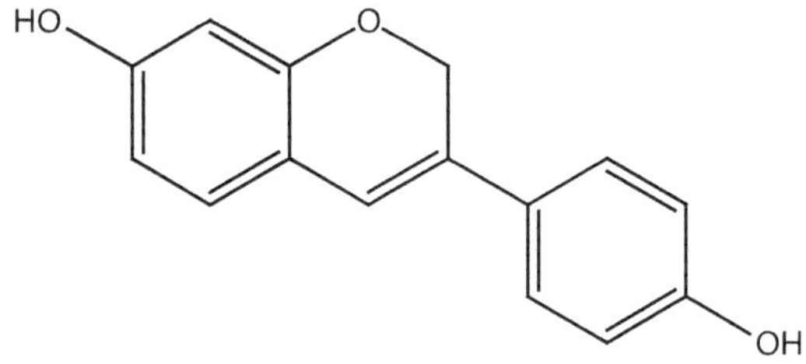

Fig. 15.4 Phenoxodiol [(3-(4-hydroxyphenyl)-2H-chromen-7-ol]

chemoresistance. It is being developed as a chemo-sensitizer of standard chemotherapeutics in solid cancers [35].

Osteoporosis

Osteoporosis is a bone disease causing brittleness in bone due to a reduction in bone tissue. Osteoclasts responsible for bone destruction are dependent on tyrosine kinase enzyme. Estrogen deficiency causes an abundance of the osteoclast. Genistein is a well-known tyrosine kinase inhibitor and can be used as the antiosteoporotic agent. Genistein along with other phytoestrogens play a favorable effect in maintaining a balance between adipogenesis and osteogenesis. It inhibits the expansion of osteoclasts while stimulates the growth of osteoblasts [36, 37]. Antigenotoxic effects of genistein in postmenopausal women leading to positive effects on bone suggest the fact that genistein can be used for the prevention of bone loss in postmenopausal women without getting significant adverse effects [38]. However, later studies suggested that treatment with genistein to ovariectomized rats resulted in improved biomechanical results and enhancement of morphologic parameters but risk involved with the uterus and mammary glands was unclear [39].

Antidiabetic Effects

Diabetes is a metabolic disorder either caused by defective insulin secretion or insulin action or sometimes both play a role. Genistein preserves pancreatic β-cells, ameliorates glucose and lipid metabolisms, and elevates insulin levels [40]. It decreases insulin-induced lipid synthesis from glucose in isolated adipocytes and also inhibits insulin-stimulated glucose oxidation [41]. The antidiabetic potential is mediated by counteracting reactive oxygen species and chelating metals [42]. Since oxidative stress inflammation also plays a role in promoting metabolic disorders, genistein in the dose of 1 mg/kg/day in 30% dimethyl sulfoxide for 45 days ameliorates the inflammatory state to endorse its antidiabetic efficacy [43].

Neuroprotective Effects

Neurodegenerative disorders mainly comprise of Alzheimer's disease (AD) and Parkinson's disease (PD) where oxidative stress-induced neuron cellular apoptosis is believed to be involved in the development of both these diseases. Parkinson's disease (PD) is characterized by the progressive loss of dopaminergic neurons. PD primary pathological events leading to dopaminergic neurons degeneration is supposed to occur *via* oxidative stress. Increased lipid peroxidation and DNA damage are also the result of overproduction of reactive oxygen species (ROS). So, genistein being a potent antioxidant attenuates the neuronal damage and loss through counteracting oxidative stress. Genistein has the potential to suppress PTK expression, which is interpreted to be neuroprotective [44]. It has neuroprotective effects on dopaminergic neurons in the nigrostriatal system, and this effect may be accredited to enhancing Bcl-2 gene expression [45]. Activation of the IGF-I receptor signaling pathway is another neuroprotective effects associated with genistein [46].

Alzheimer's disease (AD) is manifested by different cellular and molecular events and primarily involves deposition of the β-amyloid peptide into senile plaques which initiate oxidative stress and accelerates the degeneration of neurons [47].

Genistein has antioxidant and neuroprotective effects on Alzheimer's disease (AD). It mediates protective effect against Aβ-induced toxicity *via* estrogen receptors. This action stimulates MAP kinases, activates the NF-κB signaling pathway, and thereby induces overexpression of antioxidant manganese superoxide dismutase (MnSOD) [48]. Acute genistein treatment has been suggested to be useful in improving memory deficits in ovariectomized rats [49]. Genistein (10mg/kg) improved the memory of Aβ-injected rats due to the estrogen-like activity instead of antioxidant effect [50]. It has been found that genistein increased PPARg (peroxisome proliferator-activated receptor gamma) levels, which in turn will make overexpressing apolipoprotein E (apoE) thus mediates the degradation of amyloid β (Aβ). Therefore, a clinical trial is undergoing, which studies the role of genistein on treatment of AD *via* activation of PPARg levels [51]. Some of the genistein-polyamine conjugates possessed momentous cholinesterases (ChEs) inhibitory activity with an IC50 value of 2.75 μmol/L which was better than that of rivastigmine (5.60 μmol/L) [52].

Daidzein

Daidzein (7,4'-dihydroxyisoflavone) (Fig. 15.5) is commonly found in *Pueraria mirifica* and Kudzu or *Pueraria lobata*. It commonly exists as inactive glucoside daidzein. It is also obtained from precursor formononetin, which on breakdown by intestinal glucosidases is converted to daidzein. Daidzein is further partially metabolized to equol and *O*-desmethylangiolensin (*O*-DMA).

Daidzein is a comprehensively studied phytoestrogen with respect to its skeletal effects. It can promote various therapeutic effects such as prevention of cancer, cardiovascular diseases, etc. Daidzein inhibits the growth of human prostate, colon,

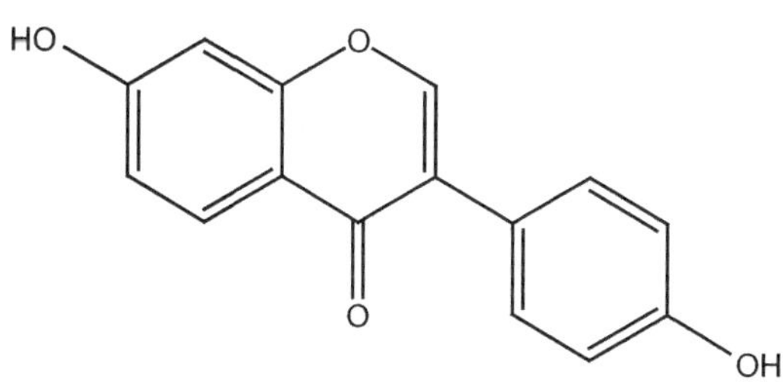

Fig. 15.5 Daidzein (7-hydroxy-3-(4-hydroxyphenyl)-4H-chromen-4-one)

breast, and ovarian carcinomas [53, 54]. It causes cell cycle arrest at the G1 and G2/M phases in human breast cancer cells [40], induces apoptosis [55], and affects angiogenesis/metastasis [56]. Carboxymethyl derivatives of daidzein act as a carrier for daunomycin (anticancer drug used for ovarian cancer) enhancing anticancer potential in human model of ovarian cancer cell lines [57].

Antiosteoporotic functions of daidzein include stimulation of osteoblast and inhibition of osteoclast through ERs and stimulation of protein synthesis in osteoblasts [58] that are further enhanced in synthetic analogs [59].

Ipriflavone, a synthetic isoflavone (prenyl isoflavone derivative) derived from daidzein, has been developed as an oral treatment for acute leukemias [60]. It showed best results for prevention of osteoporosis suggesting that it is a useful and safe alternative to ERT in treating osteoporosis in postmenopausal women [61].

Daidzein displays good antioxidant activity by upregulating the antioxidant enzyme catalase and is thereby linked to chemopreventive potential in many human diseases, including cancer and heart disease [62]. Daidzein along with genistein is among one of the most potent modulators of immune functions [63], but daidzein is also involved in immunosuppressive effects [64]. Daidzein along with genistein can protect against oxidative stress-induced cell apoptosis and cell proliferation inhibition in the diabetic endothelial cell model. This combination therapy may thereby prove to be beneficial for diabetes and chronic lung diseases [65]. Moreover, daidzein anti-oxidative potential imparts neuroprotective effects against Alzheimer's disease [66]. Other beneficial effects of daidzein include decreased LDL oxidation and lipoprotein disorders [67], anti-obesity, and metabolic disorders [68].

Glycitein

Glycitein is commonly found in *Glycine max* (L.) and *Pueraria montana* and accounts for 5–10% of the total isoflavones in soy food products. Glycitein has a structural similarity to those of genistein and daidzein. It has weak estrogenic activity, comparable to that of the other soy isoflavones but much lower than that of diethylstilbestrol (DES) and 17β-estradiol [69]. Glycitein prominently has therapeutic potential for prevention of neurodegenerative disorders [70]. It helps in lowering cholesterol and LDL contents [71] and has preventive measures in cardiovascular diseases [72].

Formononetin

Formononetin is an isoflavonoid found abundantly in *Astragalus mongholicus* Bunge and *Trifolium pratense* L. (red clover). The extracts of these herbs have been used clinically to treat different diseases including cardiovascular diseases since long time. In addition, it has been reported to exert anti-oxidative and estrogenic effects [73] along with hypolipidemic properties. Formononetin has also been found to be beneficial in cardiovascular disorders as it shows vasorelaxation effects [74].

15.2.1.2 Flavones

Flavonoidal chemical entity comprises of 2-phenyl-benzo[α]pyrane or flavane nucleus, which consists of two benzene rings linked through a heterocyclic pyran ring. Flavonoids exert a wide spectrum of biological activities in a multitude of disease states such as cancer, cardiovascular disease, and neurodegenerative disorders. There are reports about their beneficial effects as anti-allergic [75], anti-inflammatory [76], antiviral [77], antioxidant [78], and antitumor activities [79]. Apart from this, they are known to possess hypercholesterolemic, anti-hepatotoxic, and antifertility activities [80] including anti-inflammatory potential [81, 82]. Some of them are also thought to act as natural fungicides and UV protectants. Apigenin and naringenin are some of the extensively studied flavonoids.

Apigenin

Apigenin (Fig. 15.6) is commonly found in *Medicago sativa*, *Allium sativum*, and *Acinos suaveolens*. Apigenin is a weak mutagen showing both anticarcinogenic activity [83] and antigenotoxic effect [84].

Apigenin acts as antiproliferative by inducing cell cycle arrest and apoptosis [85]. It enhances the efficacy of standard chemotherapeutic drug gemcitabine by increasing its antitumor effects [86]. It also possesses antioxidant potential [87] and neuroprotective effects [88]. It enhances the effect of chemotherapeutic drug *cis*platin in prostate cancer cell lines [89]. 8-Prenylapigenin displays anti-inflammatory and vascular protective properties [90].

8-Prenylnaringenin

8-Prenylnaringenin (8-PN) contained in the female flower of hops (*Humulus lupulus* L.). It is one of the strongest plant-derived estrogen receptor ligands and regarded as one of the most potent phytoestrogens [91, 92]. It strongly binds to both estrogen receptor (ER) α and β with a relative binding affinity of about 0.01 (estradiol = 1), ten times higher than that of genistein [93]. 8-PN anticancer effects include induction of apoptosis in hormone-dependent breast cancer cell lines in a dose-dependent manner [94] and intervening in cell signaling pathways [95]. Later studies suggested proliferative effects of 8-PN [96]. Platelet aggregation is involved in a number of vascular diseases such as atherosclerosis, myocardial infarction, coronary artery disease, and thrombosis. 8-PN anti-aggregatory and anti-adhesive potency on human platelets leads to prevention of these diseases [97].

Kaempferol

Kaempferol is one of the most commonly found dietary flavonoids. It is mainly isolated from grapefruit, tea, and broccoli. It exerts its biological effect not only through estrogen receptors but also through some estrogen-related receptors [98]. Its preventive role in postmenopausal conditions, osteoporosis, and cardiovascular diseases has been established [99]. Anticancer effects are observed in colorectal, lung, and prostate cancers. It can induce apoptosis through inhibiting DNA synthesis, inhibiting kinase activities, inducing nuclear DNA degradation, cell cycle arrest, and telomerase inhibition [100–103]. Recent reports suggest its antiviral potential against influenza virus-induced severe lung damage [104].

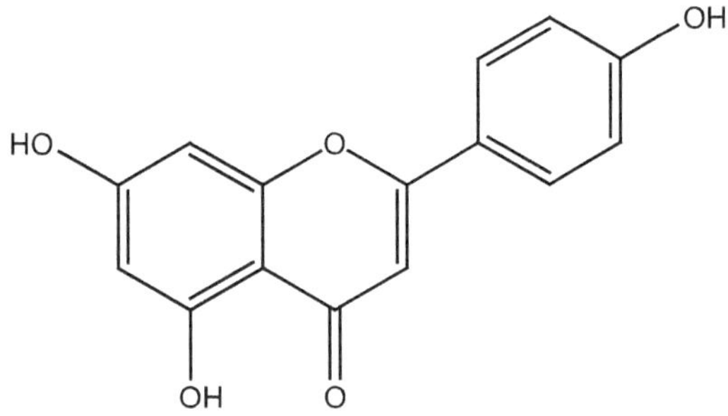

Fig. 15.6 Apigenin (5,7-dihydroxy-2-(4-hydroxyphenyl)-4H-chromen-4-one)

Icariin

Icariin is the main active flavonoid glucoside isolated from *Epimedium pubescens*. It is used as antirheumatics (anti-inflammation) and in tonics (for health promotion) [105]. *Epimedium* flavonoids extract containing prominently icariin have a therapeutic effect on osteoporosis [106, 107], and it also enhances bone healing and thus reduces osteoporosis [108]. Recent studies explored its cardioprotective potential [109], antiangiogenesis against liver fibrosis [110], and neuroprotective efficacies [111].

Silymarin

Silymarin is a purified extract from milk thistle (*Silybum marianum*) comprising of a mixture of four isomeric flavonolignans: silibinin (its main, active component), isosilibinin, silydianin, and silychristin [112]. It has been used for centuries as a natural remedy for liver diseases and claimed for clinical applications in the treatment of viral hepatitis [113]. Additionally, silymarin has also been identified for its anti-inflammatory and cytoprotective effects [114], exhibiting antiosteoporotic [115], immunomodulatory [116], and antitumor activities [117] and antiviral potencies [118]. Silymarin is also effective against neurodegeneration and aging-related disorders [119] and fungal infections [120].

Baicalein

Baicalein is a flavonoid extracted from *Scutellaria baicalensis*. It has been reported to have multiple pharmacological activities including anti-inflammatory, anti-allergic, antiviral, and anti-oxidative effects [121–123]. Antitumor effects of baicalein have been widely explored on various human cancer cell lines [124–127]. Silymarin and baicalein in combination have synergistic anticancer effect on human hepatoma HepG2 cells [128]. It also has beneficial effects in treating memory loss and may have potential as therapeutic lead in combating Parkinson's disease [129]. It is also effective in the prevention of noise-induced hearing loss [130].

15.2.1.3 Isoflavan

The isoflavans are a subclass of flavonoids, composed of C6-C3-C6 skeleton with the central heterocyclic pyran ring. Most common example of this class is equol. Other examples include isoflavans glabridin and hispaglabridins A and B.

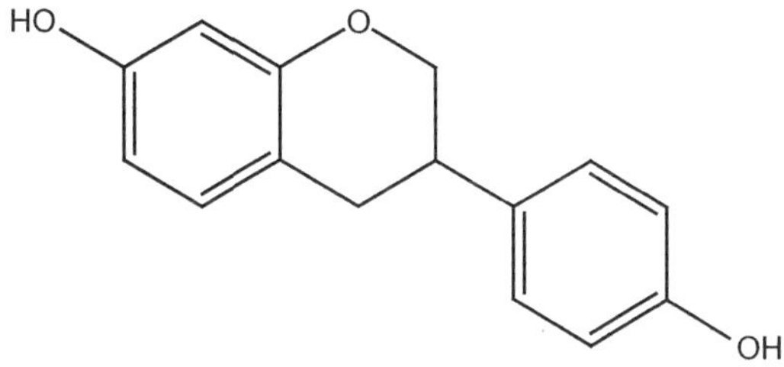

Fig. 15.7 Equol (3,4-dihydro-3-(4-hydroxyphenyl)-2H-chromen-7-ol)

Equol

Equol (Fig. 15.7) is a bioactive metabolite of daidzein that is produced by intestinal bacteria and has estrogenic activity exceeding that of daidzein [131].

It is considered to have therapeutic potential similar to that of the isoflavones genistein and daidzein. Equol being a chiral molecule is profoundly found in nature as enantiomer S-equol. This stereoisomer displays a 13-fold higher relative binding affinity (RBA) for ERβ than ERα which is greater than its parent compound daidzein [132]. Contrary, the R-enantiomer has a stronger RBA for ERα. Equol is known to possess the best antioxidant activity among all isoflavones. It also has anticancerous effects and induces apoptosis in human breast cancer cells [133]. Equol possesses substantial vasodilator [134] and protects cells from H_2O_2-induced cell death to prevent vascular atherosclerosis [135]. To date, few clinical trials have considered equol as a molecule having potential to enhance bone density among postmenopausal women [136, 137]. Studies revealed the fact that ingestion of equol and resveratrol as dietary supplements may improve postmenopausal health problems [138].

15.2.1.4 Flavonol

Any class of flavonoid compounds that are hydroxy derivatives of flavone can be classified as flavonol. One of the most common examples of flavonol is quercetin.

Quercetin (Fig. 15.8) is found in the peel of apples and red onions. Naturally it occurs in form of two glycosides, quercetrin when linked to rhamnose or rutin when linked to rutinose.

Quercetin is presumed to have beneficial health effects such as prevention of cancer and cardiovascular disease [139, 140] and is also a strong antioxidant [141]. Its therapeutic efficacy has been thoroughly investigated for its abilities to express antiproliferative effects by in both prostate cancer and human endometrial adenocarcinoma cell lines [142]. It induces apoptosis in colorectal cancers [143] and in ER-α negative cancer cell lines [144]. Various quercetin derivatives are reported to have DNA topoisomerase activities [145]. Quercetin-amino acid conjugates as safe modulators for multidrug resistance (MDR) [146]. It has protective anti-inflammatory effect against rheumatoid arthritis [147].

15.2.1.5 Coumestans

Coumestans (Fig. 15.9) are plant phenols having estrogenic activity most pronounced among all the phytoestrogens.

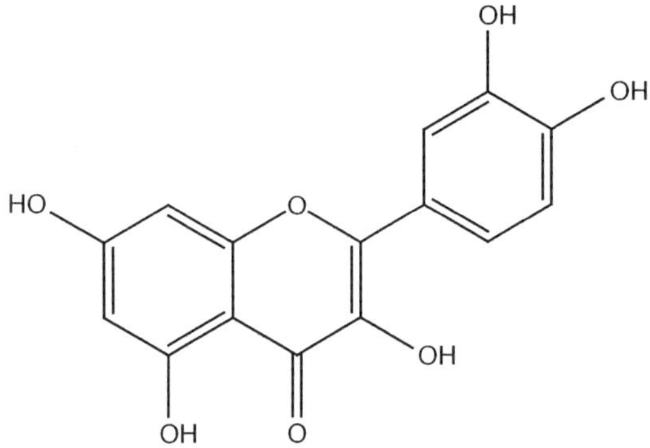

Fig. 15.8 Quercetin (3,5,7-trihydroxy-2-(3,4-dihydroxyphenyl)-4H-chromen-4-one)

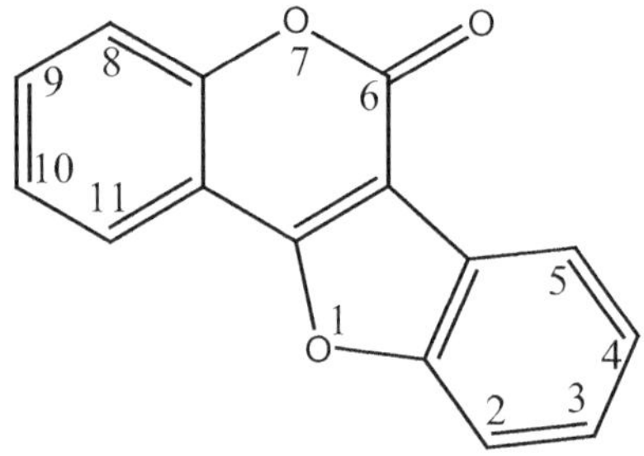

Fig. 15.9 Basic coumestan structure

15.2.1.6 Coumestrol

Coumestrol (Fig. 15.10) is isolated from ladino clover (*Trifolium repens* L.), strawberry clover (*Trifolium fragiferum*), and alfalfa or lucerne (*Medicago sativa* L.). This has a similar structural scaffold to a steroid and thus interacts with estrogen receptors (ERs).

It is reported to have higher binding affinities to ERs than the other phytoestrogen compounds apart from genistein [148], showing similar biphasic effects as genistein on cell growth [149]. Being a phytoestrogen, it displays a wide scale of biological effects on the breast, uterus, and bone. It imparts cytotoxic effects on breast cancer [150] and antitumor effects on prostate cancer [151]. It is a unique molecule as it inhibits resorption of the bone and, at the same time, it stimulates bone mineralization [152]. Since it also has less estrogenic activity than 17β-estradiol, it may prove to be a more potent drug against bone diseases including osteoporosis. Considering the fact, coumestrol and its analogs, coumestrol diacetate and coumestrol dimethyl ether have been evaluated for these effects in vivo with favorable [153]. Psoralidin, a prenylated coumestan and coumestrol, exerts bone-protective effects via osteoblast proliferation and differentiation, thus enhancing bone formation [154]. Coumestrol also displays anticarcinogenic activity [155] and neurobehavioral actions [156] and has neuroprotective effects [157].

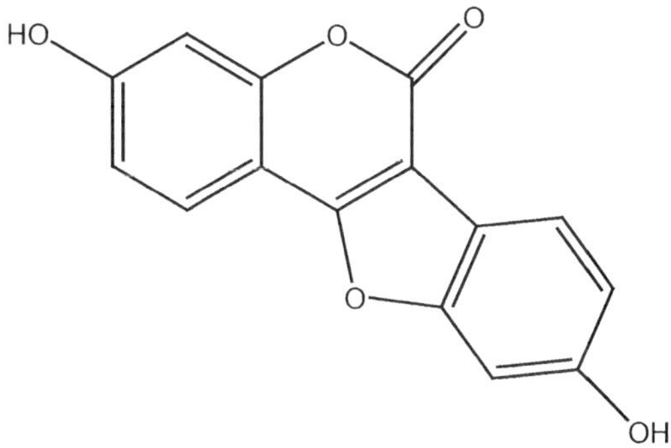

Fig. 15.10 Coumestrol (3,9-dihydroxy-6H-benzofuro[3,2-c] chromen-6-one)

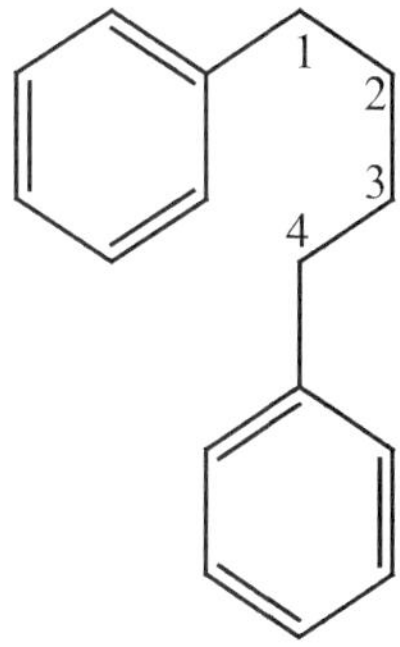

Fig. 15.11 Basic lignan structure

15.2.2 Non-flavonoids

15.2.2.1 Lignans

Lignans (Fig. 15.11) are defined as dimeric phenylpropanoid (C6-C3) compounds containing a dibenzylbutane skeleton and are widely distributed in nature as minor constituents of some plant species.

Lignans occur in high concentration in flaxseed and in lesser concentration in whole grain cereals, vegetables, and fruits. Lignans as they occur in plants are not active estrogens. Such activity is achieved only after the gut flora metabolism of these precursors to the so-called mammalian lignans. They have low molecular weight and with phenolic groups at the *meta* position of the aromatic rings.

As with the isoflavone phytoestrogens, lignans have health-promoting effects particularly in the areas of hormone-dependent cancers and cardiovascular disease [158, 159] and possess estrogenic as well as antiestrogenic activity [160]. Lignans intervene in estrogen metabolism by affecting sex hormone binding globulin (SHBG) [161]. Secoisolariciresinol and matairesinol are the most well-known phytoestrogenic lignans. They are not estrogenic by themselves but on conversion to enterodiol and enterolactone acquire estrogenic action. Enterodiol is further metabolized to enterolactone.

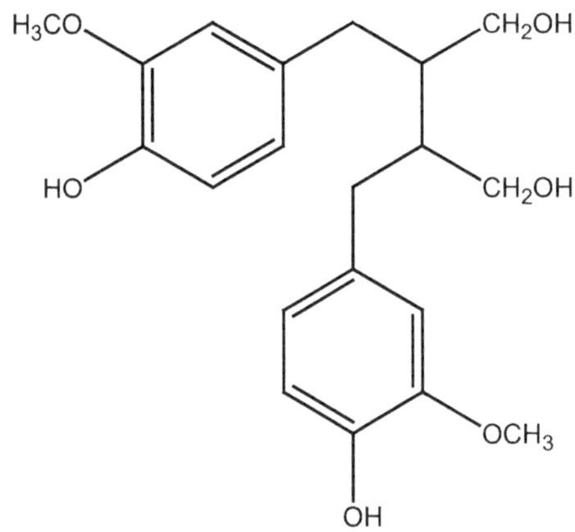

Fig. 15.12 Secoisolariciresinol (4-(4-(4-hydroxy-3-methoxyphenyl)-2,3-bis(hydroxymethyl) butyl)-2-methoxyphenol)

Secoisolariciresinol (Fig. 15.12) is isolated from *Taxus baccata* and *Urtica dioica*. It is effective in lowering serum cholesterol and in preventing the development of diabetes mellitus and has also proved to be good antioxidant along with enterodiol and enterolactone [162]. Enterolactone has a strong protective effect on the breast cancer [163]. It inhibits estradiol proliferative effect on MCF-7 breast cancer cells in culture [164]. Secoisolariciresinol diglucoside (SDG) increases bone mass in offspring in rats when taken during lactation [165]. SDG exerts antidepressant-like effect in mice [166] and is antidiabetic [167].

15.2.2.2 Stilbenes

Stilbenes (Fig. 15.13) are 1,2-diarylethenes. Stilbenoids, hydroxylated derivative of stilbenes, are of pharmaceutical interest. They have been explored for their chemopreventive effects, lipid-lowering and vascular activities, and antidiabetic potential. *Trans*-resveratrol is one the most pronounced and major active compounds of all the monomeric stilbenes.

Resveratrol (Fig. 15.14) is a secondary plant metabolite and exists as *cis* and *trans* isomers. Anticancer properties exhibited by this molecule have drawn considerable attention, since it has a remarkable inhibitory potential in various stages of tumor development.

It may block tumor development at various stages by targeting kinases [168], steroid hormone receptors [169], reactive oxygen species [170], ribonucleotide reductase [171], and DNA polymerases [172]. Resveratrol being structurally similar to the synthetic estrogen diethylstilbestrol (DES: 4,4'-dihydroxy-trans-a,b-diethylstilbene) behaves as a strong estrogen receptor agonist. Unlike other phytoestrogens, it can bind with equal affinity to both the estrogen receptors ERα and ERβ. Among both the enantiomeric forms, the *trans* isomer exerts higher activity in ER-positive cancer cell lines. At lower concentrations (10–25 nM), it displays proliferative effects on MCF-7 cell lines, whereas at concentrations of 0.1 and 1 μM, it

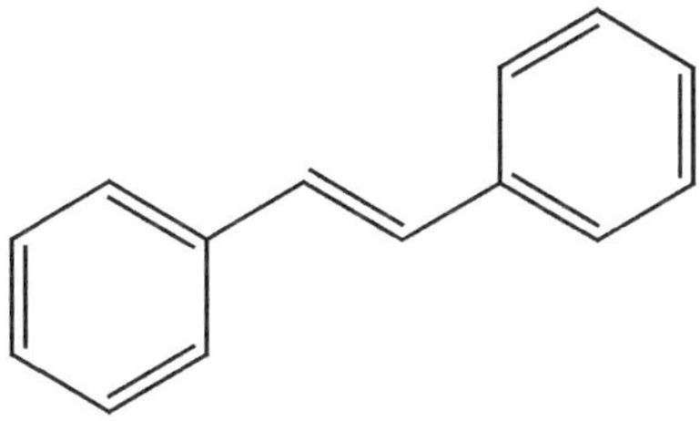

Fig. 15.13 Basic stilbene structure

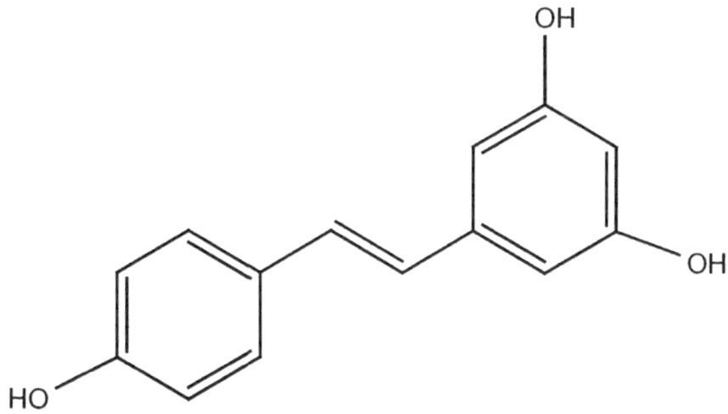

Fig. 15.14 Resveratrol (5-(4-hydroxystyryl) benzene-1,3-diol)

had no effect. At concentrations of 10 μM, resveratrol competes with the endogenous estrogen 17β estradiol and inhibits its binding to ER. This phenomenon can be explained by the fact that in the presence of estrogen resveratrol shows agonist as well as antagonistic effect in ER-positive breast cancer cells, while lack of estradiol leads to induction of ER-dependent transcriptional events by this molecule [173]. Resveratrol anticancer effects include cell cycle arrest (S/G2 phase transition of the cell cycle) and induction of differentiation and apoptosis in several of human cancer cell lines, such as breast, prostate, leukemia, colon, and esophageal cells [174]. Resveratrol in combination therapy with standard chemotherapeutic drugs increases the efficacy of the therapy and concurrently reduces the undesired side effects. Roscovitine (ROSC), a selective cyclin-dependent kinase (CDK) inhibitor in MCF-7 breast cancer cells in combination with resveratrol, enhanced the ROSC-mediated inhibition of cell proliferation and cell cycle arrest [175]. Resveratrol-based aspirin prodrugs exert improved anticancer properties [176]. Apart from combination therapy, resveratrol analogs with improved bioavailability displayed positive results [177]. It has been reviewed extensively as MDR reversion molecule particularly against breast cancer [178]. Considering the above facts, resveratrol might be the most promising candidate for HRT and chemoprevention of breast cancer.

Other activities of resveratrol include antioxidant and neuroprotective properties showing positive effects on Alzheimer's disease [179] and Parkinson's disease [180]. It is in clinical trials phase II to evaluate its neuroprotective potential in patients with AD [181]. Resveratrol also protects the cardiovascular system by a large number of mechanisms [182], possesses anti-inflammatory activity [183] and antimicrobial effects [184], and demonstrates suppression of oxidative DNA damage [185]. Resveratrol also affects in modulation of lipoprotein metabolism by

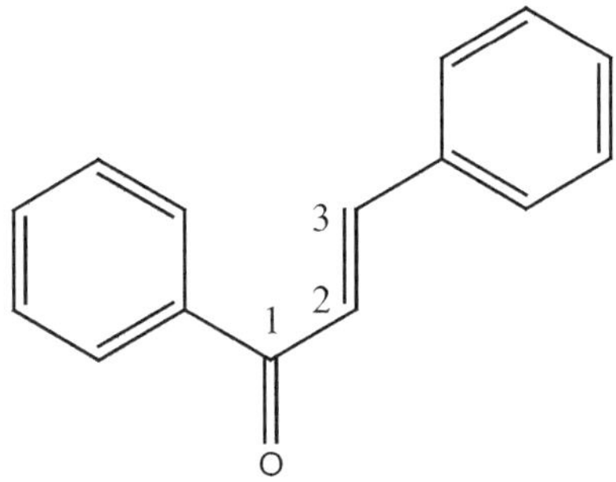

Fig. 15.15 Basic chalcone skeleton

affecting lipogenesis and lipolysis in adipocytes of normal rats [186] and could aid in the treatment of chronic intestinal inflammation [187]. The pure compound is now available in tablets and is recommended as a dietary supplement.

15.2.2.3 Chalcones

Basic chalcone skeleton (Fig. 15.15) comprises of an α, β double bond. Presence of a hydroxyl group at position 2′ and/or 3′ imparts ER-ligand activities, and thus chalcones possess anticarcinogenic potency and antiproliferative effects for different cancer cells [188]. Some chalcone derivatives show biphasic effect in breast cancer cells in a dose-dependent manner. They show proliferative effects at low concentrations but exhibit an ER-independent inhibitory activity at levels ≥10 μM [189]. Two of the commonly found plant chalcones are isoliquiritigenin and licochalcone A.

Isoliquiritigenin

Isoliquiritigenin (Fig. 15.16), a chalcone primarily reported from licorice, has been identified in terms of its anti-oxidative effect [190], anti-inflammatory effect, antiplatelet aggregation [191] and most of all estrogenic properties. Isoliquiritigenin can behave as antitumor agent in several human carcinoma cell lines and induce apoptosis. HL-60 human promyelocytic cell line [192], colon cancer cells [193], human prostate cancer [194], and human hepatoma cells [195] have been found to be inhibited by the action of isoliquiritigenin. Biphasic effects of isoliquiritigenin in MCF-7 breast cancer cells are well-documented. Antitumor mechanism of isoliquiritigenin involves induction of apoptosis [196], inhibition of cell signaling pathways, and inhibition of angiogenesis [197]. Some studies evaluated its antiangiogenetic effects and considered it to be pronounced enough in leukemia cells to be considered as a drug [198]. Not only in cancer, isoliquiritigenin also showed neuroprotection [199], anti-inflammatory potential [200], and bone-protective efficacy [201].

Licochalcone A

Licochalcone A is also a molecule of therapeutic interest and possesses various biological activities including antibacterial [202], anti-parasitic [203], estrogenic,

Fig. 15.16 Isoliquiritigenin ((E)-1-(2,4-dihydroxyphenyl)-3-(4-hydroxyphenyl) prop-2-en-1-one)

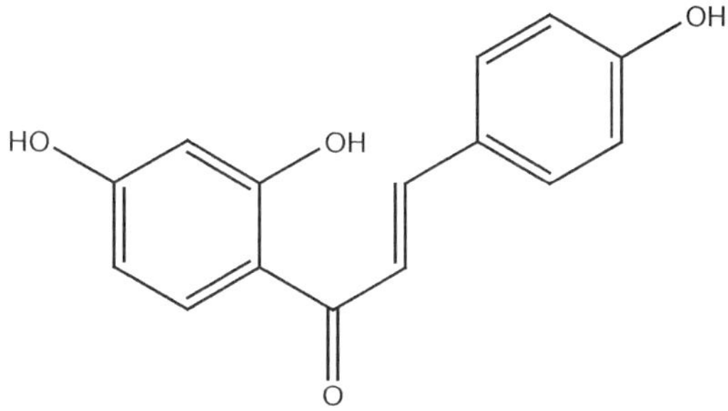

and antitumor effects [204]. Its antiproliferative potency has been found to be very pronounced [205]. Like other phytoestrogens, licochalcone induces apoptosis in MCF-7 breast cancer [206] and leukemia cells [207]. It is also effective in prostate cancer [208] and gastric cancer as it can induce apoptosis by causing cell cycle arrest at G2/M transition [209]. Furthermore, licochalcone A is also effective as an anti-obesity molecule [210] and has bone healing effects [211].

15.3 Phytoestrogens as Therapeutic Agents

The therapeutic potential of phytoestrogens cannot be clearly evaluated as the dose of phytoestrogens administered, their dietary composition, and duration of use come into effect and thus makes them difficult to compare. Data regarding considerable health benefits to cardiovascular, breast, bone, and menopausal symptoms are not strong enough to conclude. Thus the degree to which phytoestrogen consumption confers health benefits excluding side effects still remains unresolved.

15.3.1 Menopausal Symptoms and Prevention of Osteoporosis

Phytoestrogen consumption came into effect as a natural alternative to hormone replacement therapy. They were found to relieve not only from vasomotor perimenopausal symptoms including hot flushes but also played a crucial role in prevention of osteoporosis. Most of the phytoestrogens from all the category including coumestrol, resveratrol, genistein, daidzein, and others are reported to have bone-protective effect in vivo. Although their therapeutic potential seemed dependent on the dose, duration, and route of administration and on the animal model employed, their osteoprotective potential is dependent on their effects on the uterus and mammary glands and the risks involved in using these molecules as osteoprotective agents [39]. Although mixed results have been obtained, still dietary supplements containing numerous phytoestrogens continue to be popular as a natural alternative to hormone replacement.

15.3.2 Cardiovascular Health and Prevention of Heart Disease

Postmenopausal health problems also accompany heart diseases as there is decrease in estradiol levels during menopause [212]. In the current scenario of rise in cardiovascular diseases, soy consumption has been correlated with reduced risk of cardiovascular disorders. Isoflavonoids in particular play a role by increasing plasma level of the good cholesterol (HDL) and deteriorate bad cholesterol (LDL) [213]. Owing to this, phytoestrogens received the great attention and simultaneously stimulated consumption and adoption of soy foods in Western countries. However, there are escalating evidences to suspect the impact of soy consumption on LDL levels, and other cardiovascular diseases might be spurious. Regardless, people at risk for heart disease still consider soy intake as part of their daily meals.

15.3.3 Breast Cancer: Pro or Con?

Role of phytoestrogens in the area of breast cancer as perpetrator or protector has proven to be one of the trickiest questions to be tackled. Although extensive studies regarding this factor have been done, still results have been frustratingly incongruous. The fact that phytoestrogens bind to ERs with relatively high affinity may lead to the idea that high phytoestrogen intake could be the factor behind the increased risk of carcinogenesis. In such a case, breast cancer survivors are put at risk for reoccurrence. For instance, the phytoestrogen genistein enhances multidrug resistance in breast cancer cell [214]. Still, the relatively opposite results with traditionally low cancer rates in Asia stand as evidence of phytoestrogen benefit though this difference could be accounted on the difference in traditional diet of Asian and Western women [215]. On account of these reports, extensive studies have been done taking epidemiological approach to address these concerns, but no concluding results could be obtained due to numerous factors involved. It has been a difficult task to get a clear consensus as to whether or not phytoestrogens are beneficial or harmful particularly in the area of breast cancer. But since the mounting evidences of quantifiable benefits on bone and postmenopausal health, women without grave threat factors for breast cancer or a family history of the same are likely to integrate soy into their diet without momentous concern.

15.3.4 Other Therapeutic Applications

The therapeutic effects of PEs on neurodegenerative disorders have been clearly stated and have extensive effect on CNS. Although recent studies have also suggested negative outcome in case of cognitive functions [216]. Phytoestrogens have benefecial effects in hepatic alterations and have antidiabetic potential as seen with many of the molecules. They display antiaging effect on the skin via estrogen receptors [217]. Their antiviral properties toward a range of exogenous viruses have been clearly established [11]. Phytoestrogens have noteworthy antioxidant potential and are also good anti-inflammatory agents. They display a plethora of therapeutic properties such as antibacterial, antifungal, antidepressant, and anti-obesity to name a few.

15.3.5 Cons: The Endocrine Disrupting Properties of Phytoestrogens

Although phytoestrogens have been established as source of mediating a possible protective role in the areas of certain diseases such as cardiovascular diseases and some hormone-dependent cancers, conversely they also exert a detrimental effect on normal cells. Mutagenic effects, genotoxicity, and cancer cell-promoting properties are some of the perpetrator effect held by them. Isoflavones fit into the category of being an endocrine disruptor, which can be defined as compounds having ability to alter the function(s)/structure of the endocrine system and causing adverse effects. There have been reports implicating that phytoestrogens consumption during critical timings of development may hamper with the organizational role of estrogen and simultaneously causing myriad of adverse health outcome. This has been related to occurrence of malformations in the reproductive system, cancers, and reduced fertility. This also leads to early puberty and disrupted brain organization [218]. Paradoxically, phytoestrogens being effective antioxidants may also cause oxidative DNA damage and may lead to immunosuppressive effects. Phytoestrogens can influence human endogenous retrovirus expression (HERV) causing injurious effects of HERVs on cancer causing genes.

15.4 Conclusions

Phytoestrogens are intriguing, as although their beneficial effects have dominated the research areas, the potentially adverse effects of these compounds have been overlooked. Clearly some of the phytoestrogens have paradoxical effects, and many key factors could be attributed to it. To begin with the ligand-binding affinity of phytoestrogens, both ER could play a role. Also since they act within ER-mediated and ER-independent pathways, this may lead to additive, synergistic, or possible antagonistic interactions. In in vitro and in vivo, studies deliver varying outcome due to contribution of various factors such as involvement of multiple pathways, dosage, and bioavailability. Phytoestrogenic actions are dependent on concentration, varying with the target tissue and upon number and type of ER behaving as selective estrogen receptor modulators (SERM) which makes them difficult to predict the effect. Other key factors are also involved in the discrepancy obtained in the results. Ethnicity of the patient and bioavailability and genetic polymorphism of the PE metabolizing enzyme vary considerably with ethnicity and region of the population involved. Phytoestrogenic content in the supplements and extracts varies reasonably, and the difference in bioavailability of the compound must also be taken into account [219]. The presence or absence of endogenous estrogens also influences the effect of phytoestrogens [220]. All these factors can be attributed for the clinical efficacy of phytoestrogens which is also the necessity of the present scenario in the field of chemotherapy, but the undesired effects cannot be neglected unless critically observed and evaluated. As with many other compounds found in nature, there are many pros and cons associated with phytoestrogens, but overall, phytoestrogen supplements have a safe side-effect profile. To conclude, phytoestrogens are worthy of further investigations to achieve maximum beneficial output with a corresponding minimization of their undesired effects.

Acknowledgments AA is supported by a research fellowship from the Japan Society for the Promotion of Science.

References

1. Rietjens IM, Sotoca AM, Vervoort J, Louisse J (2013) Mechanisms underlying the dualistic mode of action of major soy isoflavones in relation to cell proliferation and cancer risks. Mol Nutr Food Res 57:100–113
2. Knight DC, Eden JA (1996) A review of the clinical effects of phytoestrogens. Obstet Gynecol 87:897–904
3. Yildiz F (2005) Phytoestrogens in Functional Foods. Taylor & Francis Ltd, Boca Raton, pp 3–5:210–211
4. Paterni I., Granchi C., Katzenellenbogen J.A, Minutolo F. (2014) Estrogen receptors alpha (ERα) and beta (ERβ): subtype-selective ligands and clinical potential. Steroids 90 13–29 S0039-128X(14: 00151–00152
5. Böttner M, Thelen P, Jarry H (2013) Estrogen receptor beta: Tissue distribution and the still largely enigmatic physiological function. J Steroid Biochem Mol Biol 139:245–251 pii: S0960-0760(13)00052-6
6. Anderson JJB, Anthony M, Messina M, Garner SC (1999) Effects of phyto-oestrogens on tissues. Nutr Res Rev 12:75–116
7. Messina MJ, Loprinzi CL (2001) Soy for breast cancer survivors: a critical review of the literature. J Nutr 131:3095S–3108S
8. Adlercreutz H (1998b) Evolution, nutrition, intestinal microflora, and prevention of cancer: a hypothesis. Proc Soc Exp Biol Med 217:241–246
9. Kurzer MS, Xu X (1997) Dietary phytoestrogens. Annu Rev Nutr 17:353–381
10. Carlson S, Peng N, Prasain JK, Wyss JM (2008) Effects of botanical dietary supplements on cardiovascular, cognitive, and metabolic function in males and females. Gen Med 5:S76–S90
11. Martin JHJ, Crotty S, Warren P, Nelson PN (2007) Does an apple a day keep the doctor away because a phytoestrogen a day keeps the virus at bay? A review of antiviral properties of phytoestrogen. Phytochemistry 68:266–274
12. Januario AH, Lourenco MV, Domezio LA, Pietro R, Castilho MS, Tomazela DM, da Silva M, Vieira PC, Fernandes JB, Franca SD (2005) Isolation and structure determination of bioactive isoflavones from callus culture of Dipteryx odorata. Chem Pharm Bull 53:740
13. Hu J-Y, Aizawa T (2003) Quantitative structure–activity relationships for estrogen receptor binding affinity of phenolic chemicals. Water Res 37:1213–1222
14. Brzozowski AM, Pike ACW, Dauter Z et al (1997) Molecular basis of agonism and antagonism in the oestrogen receptor. Nature 389:753–758
15. Vaya J, Tamir S (2004) The relation between the chemical structure of flavonoids and their estrogen-like activities. Curr Med Chem 11:1333–1343
16. Cassidy A, Bingham S, Setchell K (1995) Biological effects of isoflavones in young women: importance of the chemical composition of soyabean products. Br J Nutr 74:587–601

17. Fatima A, Singh R (2016) The chemistry and pharmacology of genistein. Nat Prod J 6:3–12
18. Makiewicz L, Garey J, Adlercreutz H, Gurpide E (1993) In vitro bioassays of non-steroidal phytoestrogens. J Steroid Biochem Mol Biol 45:399–405
19. Shutt DA, Cox RI (1972) Steroid and phyto-estrogen binding to sheep Uterine receptors in vitro. J Endocrinol 52:291–310
20. Kuiper GGJM, Lemmen JG, Carlsson B et al (1998) Interaction of estrogenic chemicals and phytoestrogens with estrogen. Endocrinology 139:4252–4263
21. Wang TT, Sathymoorthy N, Phang JM (1996) Molecular effects genistein on estrogen receptor mediated pathways. Carcinogenesis 17:271–275
22. Agarwal R (2000) Cell signaling and regulators of cell cycle as molecular targets for prostate cancer prevention by dietary agents. Biochem Pharmacol 60:1051–1059
23. Lakshman M, Li X, Vijayalakshmi A, Joshua C, Takimoto Chris H, Irene H, Pelling Jill C, Bergan Raymond C (2008) Dietary Genistein inhibits metastasis of human prostate cancer in mice. Cancer Res 68:20–24
24. Pavese JM, Krishn SN, Bergan RC (2014) Genistein inhibits human prostate cancer cell detachment, invasion, and metastasis. Am J Clin Nutr 100:431S–436S
25. Caruso MG, Notarnicola M, Cavallini A, Guerra V, Misciagna G, Di Leo A (1993) Demonstration of low density lipoprotein receptor in human colonic carcinoma and in surrounding mucosa by immunoenzymatic assay. Ital J Gastroenterol 25:361
26. Qi W, Weber CR, Wasland K, Savkovic SD (2011) Genistein inhibits proliferation of colon cancer cells by attenuating a negative effect of epidermal growth factor on tumor suppressor FOXO3 activity. BMC Cancer 11:219
27. Kaushik S, Shyam H, Sharma R, Balapure AK (2016) Genistein synergizes centchroman action in human breast cancer cells. Indian J Pharm 48:637–642
28. Sarkar FH, Li Y (2006) Using chemopreventive agents to enhance the efficacy of cancer therapy. Cancer Res 66:3347–3350
29. Mohammad RM, Banerjee S, Li Y, Aboukameel A, Kucuk O, Sarkar FH (2006) Cisplatin-induced antitumor activity is potentiated by the soy isoflavone genistein in BxPC-3 pancreatic tumor xenografts. Cancer 106:1260–1268
30. Hwang JT, Park IJ, Shin JI, Lee YK, Lee SK, Baik HW, Ha J, Park OJ (2005) Genistein, EGCG, and capsaicin inhibit adipocyte differentiation process via activating AMP-activated protein kinase. Biochem Biophys Res Commun 338:694–699
31. Aditya NP, Shim M, Yang H, Lee YJ, Ko S (2014) Antiangiogenic effect of combined treatment with curcumin and genistein on human prostate cancer cell line. J Funct Foods 8:204–213
32. Chang KL, Cheng HL, Huang LW, Hsieh BS, Hu YC, Chi TT, Shyu HW, Su SJ (2009) Combined effects of terazosin and genistein on a metastatic, hormone-independent human prostate cancer cell line. Cancer Lett 276:14–20
33. Park HJ, Della-Fera MA, Hausman DB, Rayalam S, Ambatia S, Baile CA (2009) Genistein inhibits differentiation of primary human adipocytes. J Nutr Biochem 20:140–148
34. Aguero MF, Fachinetti MM, sheleg Z, Senderowicz AM (2005) phenoxodiol: a novel isoflavone, induces G1 arrest by specific loss in cyclin dependent kinase 2 activity by p- 53 independent induction of p21WAF1/CIP1. Cancer Res 65:3364–3373
35. Choueiri TK, Mekhail T, Hutson TE, Ganapathi R, Kelly GE, Bukowsky RM (2006) Phase I trial of phenoxodiol delivered by continuous intravenous infusion in patients with solid cancer. Ann Oncol 17:860–865
36. Rickard DJ, Monroe DG, Ruesink TJ, Khosla S, Riggs BL, Spelsberg TC (2003) Phytoestrogen genistein acts as an estrogen agonist on human osteoblastic cells through estrogen receptors alpha and beta. J Cell Biochem 89:633–646
37. Yamagishi T, Otsuka E, Hagiwara H (2001) Reciprocal control of expression of mRNAs for osteoclast differentiation factor and OPG in osteogenic stromal cells by genistein: evidence for the involvement of topoisomerase II in osteoclas- togenesis. Endocrinology 142:3632–3637

38. Atteritano M, Pernice F, Mazzaferro S, Mantuano S, Frisina A, D'Anna R, Cannata ML, Bitto A, Squadrito F, Frisina N, Buemi M (2008) Effects of phytoestrogen genistein on cytogenetic biomarkers in postmenopausal women: 1 year randomized, placebo-controlled study. Eur J Pharmacol 589:22–26
39. Sehmisch S, Uffenorde J, Maehlmeyer S, Tezval M, Jarry H, Stuermer KM, Stuermer EK (2010) Evaluation of bone quality and quantity in osteoporotic mice–The effects of genistein and equol. Phytomedicine 17:424–430
40. Choi EJ, Kim G (2008) Daidzein causes cell cycle arrest at the G1 and G2/M phases in human breast cancer MCF-7 and MDA-MB-453 cells. Phytomedicine 15:683–690
41. Bhathena SJ, Velasquez MT (2002) Beneficial role of dietary phytoestrogens in obesity and diabetes. Am J Clin Nutr 76:1191–1201
42. Hamideh S, Marcus BM, Antonio A, Marcelo HL (2011) The antioxidant effect of genistein on the in vitro metal -mediated formation of free radicals. Clin Biochem 44:S226–S226
43. Palanisamy N, Kannappan S, Anuradha CV (2011) Genistein modulates NF-kB-associated renal inflammation, fibrosis and podocyte abnormalities in fructose-fed rats. Eur J Pharmacol 667:355–364
44. Gelinas S, Martinoli MG (2002) Neuroprotective effect of estradiol and phytoestrogens on MPP+-induced cytotoxicity in neuronal PC12 cells. J Neurosci Res 70:90–96
45. Liu LX, Chen W, Xie J, Wong M (2008) Neuroprotective effects of genistein on dopaminergic neurons in the mice model of Parkinson's disease. Neurosci Res 60:156–161
46. Gao Q, Xie J, Wong M, Chen W (2012) IGF-I receptor signaling pathway is involved in the neuroprotective effect of genistein in the neuroblastoma SK-N-SH cells. Eur J Pharmacol 677:39–46
47. Jakob-Roetne R, Jacobsen H (2009) Alzheimer's disease: from pathology to therapeutic approaches. Angew Chem Int Ed Eng 48:3030–3059
48. Bagheri M, Roghanid M, Joghataeib M, Mohsenia S (2012) Genistein inhibits aggregation of exogenous amyloid-beta1-40 and alleviates astrogliosis is in the hippocampus of rats. Brain Res 1429:145–154
49. Alonso A, Gonzalez-Pardo H, Garrido P et al (2010) Acute effects of 17 beta-estradiol and genistein on insulin sensitivity and spatial memory in aged ovariectomized female rats. Age (Dordr) 32(4):421–434
50. Bagheri M, Joghataei M-T, Mohsen S, Roghani M (2011) Genistein ameliorates learning and memory deficits in amyloid β(1-40) rat model of Alzheimer's disease. Neurobiol Learn Mem 95:270–276
51. ClinicalTrials.gov identifier: NCT019825781 Genistein as a Possible Treatment for Alzheimer's Disease. (GENISTEÍNA_2)l juy 2015).
52. Zhang X, Wang J, Hong C, Luon W, Wang C (2015) Design, synthesis and evaluation of genistein-polyamine conjugates as multi-functional anti-Alzheimer agents. Acta Pharm Sin B 5:67–73
53. Sathyamoorthy N, Wang TT (1997) Differential effects of dietary phyto-oestrogens daidzein and equol on human breast cancer MCF-7 cells. Eur J Cancer 33:2384–2389
54. Guo JM, Kang GZ, Xiao BX, Liu DH, Zhang S (2004) Effect of daidzein on cell growth, cell cycle, and telomerase activity of human cervical cancer in vitro. Int J Gynecol Cancer 14:882–888
55. Lee JS, Son KH, Sung MK, Kim YK, Yu R, Kim JS (2003) Anticarcinogenic properties of a daidzein-rich fraction isolated from soybean. J Med Food 6:175–181
56. Rabiau N, Kossaï M, Braud M, Chalabi N, Satih S, Bignon YJ, Bernard-Gallon DJ (2010) Genistein and daidzein act on a panel of genes implicated in cell cycle and angiogenesis by polymerase chain reaction arrays in human prostate cancer cell lines. Cancer Epidemiol 34:200–206
57. Somjen D, Katzburg S, Nevo N, Gayera B, Hodged RP, Reneveyd MD, Kalchenko V, Meshorere A, Sternc N, Kohena F (2008) A daidzein–daunomycin conjugate improves the therapeutic response in an animal model of ovarian carcinoma. J Steroid Biochem Mol Biol 110:144–149

58. Yamaguchi M, Sugimoto E (2000) Stimulatory effect of genistein and daidzein on protein synthesis in osteoblastic MC3T3-E1 cells: activation of aminoacyl-tRNA synthetase. Mol Cell Biochem 214:97
59. Yadav DK, Gautam AK, Kureel J, Srivastava K, Sahai M, Singh D, Chattopadhyay N, Maurya R (2011) Synthetic analogs of daidzein, having more potent osteoblast stimulating effect. Bioorg Med Chem Lett 21:677–681
60. Pagano L, Teofili L, Mele L, Piantelli M, Ranelletti FO et al (1999) Oral ipriflavone (7-isopropoxy-isoflavone) treatment for elderly patients with resistant acute leukemias. Ann Oncol 10:124–125
61. Petilli M, Fiorolli G, Benvenuti S, Frediani U, Gorif BMLIBI, estrogen r (1995) Calcif Tissue Int 56:160–165
62. Kampkötter A, Chovolou A, Kulawik A, Röhrdanz E, Weber N, Proksch P, Wätjen W (2008) Isoflavone daidzein possesses no antioxidant activities in cell-free assays but induces the antioxidant enzyme catalase. Nutr Res 28:620–628
63. Gredel S, Grad C, Rechkemmer G, Watzl B (2008) Phytoestrogens and phytoestrogen metabolites differentially modulate immune parameters in human leukocytes. Food Chem Toxicol 46:3691–3696
64. Yum MK, Jung MY, Cho D, Kim TS (2011) Suppression of dendritic cells' maturation and functions by daidzein, a phytoestrogen. Toxicol Appl Pharmacol 257:174–181
65. Xu SZ, Zhong W, Ghavideldarestani M, Saurabh S, Lindow SW, Atkin SL (2009) Multiple mechanisms of soy isoflavones against oxidative stress-induced endothelium injury. Free Radic Biol Med 47:167–175
66. Hsieh H, Wua W, Hu M (2009) Soy isoflavones attenuate oxidative stress and improve parameters related to aging and Alzheimer's disease in C57BL/6J mice treated with D-galactose. Food Chem Toxicol 47:625–632
67. Borradaile NM, Dedreu LE, Wilcox LJ (2002) Soya phytoestrogens, genistein and daidzein, decrease apolipoprotein B secretion from HepG2 cells through multiple mechanisms. Biochem J 366:531–539
68. He Y, Niu W, Xia C, Cao B (2016) Daidzein reduces the proliferation and adiposeness of 3T3-L1 preadipocytes via regulating adipogenic gene expression. J Funct Foods 22:446–453
69. Song TT, Hendrich S, Murphy PA (1999) Estrogenic activity of glycitein, a soy isoflavone. J Agric Food Chem 47:1607–1610
70. Gutierrez-Zepeda A, Santell R, Wu Z, Brown M, Wu Y, Khan I, Link CD, Zhao B, Luo Y (2005) Soy isoflavone glycitein protects against beta amyloid-induced toxicity and oxidative stress in transgenic *Caenorhabditis elegans. BMC Neurosci* 6:54
71. Chen ZY, Ma KY, Liang Y, Peng C, Zuo Y (2011) Role and classification of cholesterol-lowing functional foods. J Funct Foods 3:61–69
72. Lin CC, Chen TS, Lin YM, Yeh YL, Li YH, Kuo WW, Tsai FJ, Tsai CH, Yen SK, Huang CY (2013) The p38 and NFjB signaling protein activation involved in glycitein protective effects on isoproterenol-treated H9c2 cardiomyoblast cells. J Funct Foods 5:460–465
73. Mu H, Bai YH, Wang ST, Zhu ZM, Zhang YW (2008) Research on antioxidant effects and estrogenic effect of formononetin from Trifolium pratense (red clover). Phytomedicine 16:314–319
74. Wu J, Li Q, Wu M, Guo D, Chen H, Chen S, Seto S, Alice LSA, Poon CCW, Leung GPH, Lee SMY, Kwan Y, Chan S (2010) Formononetin, an isoflavone, relaxes rat isolated aorta through endothelium-dependent and endothelium-independent pathways. J Nutr Biochem 21:613–620
75. Kawai M, Hirano T, Higa S, Arimitsu J, Maruta M, Kuwahara Y, Ohkawara T, Hagihara K, Yamadori T, Shima Y, Ogata A, Kawase I, Tanaka T (2007) Flavonoids and related compounds as anti-allergic substances. Allergol Int 56:113–123
76. Theoharides TC, Alexandrakis M, Kempuraj D, Lytinas M (2001) Anti-inflammatory actions of flavonoids and structural requirements for new design. Int J Immunopathol Pharmacol 14:119–127

77. Arao T, Udayama M, Kinjo J, Nohara T, Funakoshi T, Kojima S (1997) Preventive effects of saponins from Puerariae radix (the root of Pueraria lobata Ohwi) on in vitro immunological injury of rat primary hepatocyte cultures. Biol Pharm Bull 20:988–991
78. Wolfe K, Wu X, Liu RH (2003) Antioxidant activity of apple peels. J Agric Food Chem 51:609–614
79. Joshi KS, Rathos MJ, Joshi RD, Sivakumar M, Mascarenhas M, Kamble S, Lal B, Sharma S (2007) In vitro antitumor properties of a novel cyclin-dependent kinase inhibitor. P276-00. Mol Cancer Ther 6:918–925
80. Formica JV, Regelson W (1995) Review of the biology of quercetin and related bioflavonoids. Food Chem Toxicol 33:1061–1080
81. Middleton E Jr, Kandaswami C, Theoharides TC (2000) The effects of plant flavonoids on mammalian cells: implications for inflammation. Heart disease and cancer. Pharmacol Rev 52:673–751
82. Wang L, Tu YC, Lian TW, Hung JT, Yen JH, Wu MJ (2006) Distinctive antioxidant and anti-inflammatory effects of flavonols. J Agric Food Chem 54:9798–9804
83. Khan TH, Jahangir T, Prasad L et al (2006) Inhibitory effect of apigenin on benzo(a)pyrene mediated genotoxicity in Swiss albino mice. J Pharm Pharmacol 58:1650–1655
84. Siddique YH, Afzal M (2009) Antigenotoxic effect of apigenin against mitomycin C induced genotoxic damage in mice bone marrow cells. Food Chem Toxicol 47:536–539
85. King JC, Lu Q, Li G, Moro A, Takahashi H, Chen M, Go VLW, Reber HA, Eibl G, Hines OJ (2012) Evidence for activation of mutated p53 by apigenin in human pancreatic cancer. Biochim Biophys Acta 1823:593–604
86. Strouch MJ, Milam BM, Melstrom LG, McGill JJ, Salabat MR, Ujiki MB, Ding XZ, Bentrem DJ (2009) The flavonoid apigenin potentiates the growth inhibitory effects of gemcitabine and abrogates gemcitabine resistance in human pancreatic cancer cells. Pancreas 38:409–415
87. Han JY, Ahn SY, Kim CS, Yoo SK, Kim SK et al (2012) Protection of apigenin against kainate-induced excitotoxicity by anti-oxidative effects. Biol Pharm Bull 35:1440–1446
88. Jagan K, Priya CS, Kalpana K, Vidhya R, Anuradha CV (2017) Apigenin attenuates hippocampal oxidative events, inflammation and pathological alterations in rats fed high fat, fructose diet. Biomed Pharmacother 89:323–331
89. Erdogan S, Turkekul K, Serttas R, Erdogan Z (2017) The natural flavonoid apigenin sensitizes human CD44+ prostate cancer stem cells to cisplatin therapy. Biomed Pharmacother 88:210–217
90. Paoletti T, Fallarini S, Gugliesi F, Minassi A, Appendino G, Lombardi G (2009) Anti-inflammatory and vascular protective properties of 8-prenylapigenin. Eur J Pharmacol 620:120–130
91. Milligan SR, Kalita JC, Heyerick A, Rong H, Cooman L, Keukeleire D (1999) Identification of a potent phytoestrogen in hops (Humulus lupulus L.) and beer. J Clin Endocrinol Metab 83:2249–2252
92. Milligan SR, Kalita JC, Pocock V, Van De Kauter V, Stevens JF, Deinzer ML, Rong H, De Keukeleire D (2000) The endocrine activities of 8-prenylnaringenin and related hop (Humulus lupulus L.) flavonoids. J Clin Endocrinol Metab 85:4912–4915
93. Matsumura A, Ghosh A, Pope GS, Darbre PD (2005) Comparative study of oestrogenic properties of eight phytoestrogens in MCF7 human breast cancer cells. J Steroid Biochem Mol Biol 94:431–443
94. Brunelli E, Minassi A, Appendino G, Moro L (2007) 8-Prenylnaringenin, inhibits estrogen receptor-a mediated cell growth and induces apoptosis in MCF-7 breast cancer cells. J Steroid Biochem Mol Biol 107:140–148
95. Brunelli E, Pinton G, Chianale F, Graziani A, Appendino G, Moro L (2009) 8-Prenylnaringenin inhibits epidermal growth factor-induced MCF-7 breast cancer cell proliferation by targeting phosphatidylinositol-3-OH kinase activity. J Steroid Biochem Mol Biol 113:163–170
96. Helle J, Kräker K, Bader MI, Keiler AM, Zierau O, Vollmer G, Welsh JE, Kretzschmar G (2014) Assessment of the proliferative capacity of the flavanones 8-prenylnaringenin,

6-(1.1-dimethylallyl)naringenin and naringenin in MCF-7 cells and the rat mammary gland. Mol Cell Endocrinol 392:125–135
97. Vito CD, Bertoni A, Nalin M, Sampietro S, Zanfa M, Sinigaglia F (2012) The phytoestrogen 8-prenylnaringenin inhibits agonist-dependent activation of human platelets. Biochimica et Biophysica Acta (BBA) 1820:1724–1733
98. Wang J, Fang F, Huang Z, Wang Y, Wong C (2009) Kaempferol is an estrogen-related receptor a and c inverse agonist. FEBS Lett 583:643–647
99. Strauss L, Santti R, Saarinen N, Streng T, Joshi S, Mäkelä S (1998) Dietary phytoestrogens and their role in hormonally dependent disease. Toxicol Lett 102–103:349–354
100. Sahu SC, Gray GC (1994) Kaempferol-induced nuclear DNA damage and lipid peroxidation. Cancer Lett 85:159–164
101. Zhang Y, Chen AY, Li M, Chen C, Yao Q (2008) Ginkgo biloba extract kaempferol inhibits cell proliferation and induces apoptosis in pancreatic cancer cells. J Surg Res 148:17–23
102. Kim S-H, Choi K-C (2013) Anti-cancer effect and underlying mechanism(s) of kaempferol, a phytoestrogen, on the regulation of apoptosis in diverse cancer cell models. Toxicol Res 29:229–234
103. Kashafi E, Moradzadeh M, Mohamadkhani A, Erfaniand S (2017) Kaempferol increases apoptosis in human cervical cancer HeLa cells via PI3K/AKT and telomerase pathways. Biomed Pharmacother 89:573–577
104. Zhang R, Ai X, Duan Y, Xue M, He W, Wang C, Xu T, Xu M, Liu B, Li C, Wang Z, Zhang R, Wang G, Tian S, Liu H (2017) Kaempferol ameliorates H9N2 swine influenza virus-induced acute lung injury by inactivation of TLR4/MyD88-mediated NF-kB and MAPK signaling pathways. Biomed Pharmacother 89:660–672
105. Wu J, Du J, Xu C, Le J, Liu B, Xu Y et al (2010) In vivo and in vitro anti-inflammatory effects of a novel derivative of icariin. Immunopharmacol Immunotoxicol 33:49–54
106. Meng FH, Li YB, Xiong ZL, Jiang ZM, Li FM (2005) Osteoblastic proliferative activity of Epimedium brevicornum Maxim. Phytomedicine 12:189–193
107. Sakai S, Tamura M, Mishima H, Kojima H, Uemura T (2008) Bone regeneration induced by adenoviral vectors carrying til-1/Cbfa1 genes implanted with biodegradable porous materials in animal models of osteonecrosis of the femoral head. J Tissue Eng Regen Med 2:164–167
108. Hsieh T, Sheu S, Sun J, Chen M, Liu M (2010) Icariin isolated from Epimedium pubescens regulates osteoblasts anabolism through BMP-2, SMAD4, and Cbfa1 expression. Phytomedicine 17:414–423
109. Qian ZQ, Wang YW, Li YL, Li YQ, Zhu L, Yanga DL (2017) Icariin prevents hypertension-induced cardiomyocyte apoptosis through the mitochondrial apoptotic pathway. Biomed Pharmacother 88:823–831
110. Algandaby MM, Breikaa RM, Eid BG, Neamatallah TA, Abdel-Naim AB, Ashour OM (2017) Icariin protects against Thioacetamide-induced liver fibrosis in rats: implication of anti-angiogenic and anti-autophagic properties. Pharmacol Rep. https://doi.org/10.1016/j.pharep.2017.02.016
111. Xiong D, Deng Y, Huang B, Yin C, Liu B, Shi J et al (2016) Icariin attenuates cerebral ischemia-reperfusion injury through inhibition of inflammatory response mediated by NF-κB, PPARα and PPARγ in rats. Int Immunopharmacol 30:157–162
112. Wu JW, Lin LC, Tsai TH (2009) Drug–drug interactions of silymarin on the perspective of pharmacokinetics. J Ethnopharmacol 121:185–193
113. Wellington K, Jarvis B (2001) Silymarin: a review of its clinical properties in the management of hepatic disorders. BioDrugs 15:465–489
114. Sonnenbichler J, Zetl I, Cody V, Middleton E, Karbone JB, Berck A (eds) (1988) Specific binding of a flavonolignane derivative to an estradiol receptor. Prog Clin Biol Res 280:369–374
115. El-Shitany NA, Hegazy S, El-desoky K (2010) Evidences for antiosteoporotic and selective estrogen receptor modulator activity of silymarin compared with ethinylestradiol in ovariectomized rats. Phytomedicine 17:116–125

116. Gharagozloo M, Velardi E, Bruscoli E, Agostini M, Di Sante M, Donato V, Amirghofran Z, Riccard C (2010) Silymarin suppress CD4+ T cell activation and proliferation: effects on NF-κB activity and IL-2 production. Pharmacol Res 61:405–409
117. Wu T, Liu W, Guo W, Zhu X (2016) Silymarin suppressed lung cancer growth in mice via inhibiting myeloid-derived suppressor cells. Biomed Pharmacother 81:460–467
118. Song JH, Choi HJ (2011) Silymarin efficacy against influenza A virus replication. Phytomedicine 18:832–835
119. Srivastava S, Sammi SR, Laxman TS, Pant A, Nagar A, Trivedi S, Bhatta RS, Tandon S, Pandey R (2017) Silymarin promotes longevity and alleviates Parkinson's associated pathologies in Caenorhabditis elegans. J Funct Foods 31:32–43
120. Yun DG, Lee DG (2017) Silymarin exerts antifungal effects via membrane-targeted mode of action by increasing permeability and inducing oxidative stress. Biochim Biophys Acta 1859:467–474
121. Kimata M, Inagaki N, Nagai H (2000) Effects of luteolin and other flavonoids on IgE-mediated allergic reactions. Planta Med 66:25–29
122. Hong T, Jin GB, Cho S, Cyong JC (2002) Evaluation of the anti-inflammatory effect of baicalein on dextran sulfate sodium-induced colitis in mice. Planta Med 68:268–271
123. Piao YS, Du YC, Oshima H, Jin JC, Nomura M, Yoshimoto T, Oshima M (2008) Platelet-type 12-lipoxygenase accelerates tumor promotion of mouse epidermal cells through enhancement of cloning efficiency. Carcinogenesis 29:440–447
124. Kuntz S, Wenzel U, Daniel H (1999) Comparative analysis of the effects of flavonoids on proliferation, cytotoxicity, and apoptosis in human colon cancer cell lines. Eur J Nutr 38:133–142
125. Tong WG, Ding XZ, Adrian TE (2002a) The mechanisms of lipoxygenase inhibitor-induced apoptosis in human breast cancer cells. Biochem Biophys Res Commun 296:942–948
126. Tong WG, Ding XZ, Witt RC, Adrian TE (2002b) Lipoxygenase inhibitors attenuate growth of human pancreatic cancer xenografts and induce apoptosis through the mitochondrial pathway. Mol Cancer Ther 1:929–935
127. Chung H, Choi HS, Seo EK, Kang DK, Oh ES (2015) Baicalin and baicalein inhibit transforming growth factor-b1-mediated epithelial-mesenchymal transition in human breast epithelial cells. Biochem Biophys Res Commun 458:707–713
128. Chen C, Huang T, Wong C, Hong C, Tsai Y, Liang C, Lu F, Chang W (2009) Synergistic anti-cancer effect of baicalein and silymarin on human hepatoma HepG2 Cells. Food Chem Toxicol 47:638–644
129. Zhu M, Rajamani S, Kaylor J, Han S, Zhou F, Fink AL (2004) The flavonoid Baicalein inhibits fibrillation of Synuclein and disaggregates existing fibrils. J Biol Chem 279:26846–26857
130. Kang TH, Hong BN, Park C, Kim SY, Park R (2010) Effect of baicalein from Scutellaria baicalensis on prevention of noise-induced hearing loss. Neurosci Lett 469:298–302
131. Setchell KD, Brown NM, Lydeking-Olsen E (2002) The clinical importance of the metabolite equol—a clue to the effectiveness of soy and its isoflavones. J Nutr 132:3577
132. Muthyala RS, Ju YH, Sheng S, Williams LD, Doerge DR, Katzenellenbogen BS, Helferich WG, Katzenellenbogen JA (2004) Equol, a natural estrogenic metabolite from soy isoflavones: convenient preparation and resolution of R- and S-equols and their differing binding and biological activity through estrogen receptors alpha and beta. Bioorg. Med Chem 12:1559–1567
133. Choi EJ, Ahn WS, Bae SM (2009) Equol induces apoptosis through cytochrome c-mediated caspases cascade in human breast cancer MDA-MB-453 cells. Chem Biol Interact 177:7–11
134. Jackman A, Woodman OL, Chrissobolis S, Sobey CG (2007) Vasorelaxant and antioxidant activity of the isoflavone metabolite equol in carotid and cerebral arteries. Brain Res 1141:99–107
135. Chung J, Kim SY, Jo HH, Hwang SJ, Chae B, Kwon DJ, Lew YO, Lim Y, Kim JH, Kim EJ, Kim J, Kim M (2008) Antioxidant effects of equol on bovine aortic endothelial cells. Biochem Biophys Res Commun 375:420–424

136. Setchell KD, Lydeking-Olsen E (2003) Dietary phytoestrogens and their effect on bone: evidence from in vitro and in vivo, human observational, and dietary intervention studies. Am J Clin Nutr 78:593S–609S
137. Lampe JW (2009) Is equol the key to the efficacy of soy foods? Am J Clin Nutr 89:1664S–1667S
138. Davinelli S, Scapagnini G, Marzatico F, Nobile V, Ferrara N, Corbi G (2017) Influence of equol and resveratrol supplementation on health-related quality of life in menopausal women: a randomized, placebo-controlled study. Maturitas 96:77–83
139. Knekt P, Jarvinen R, Seppanen R, Hellovaara M, Teppo L, Pukkala E, Aroma A (1997) Dietary flavonoids and the risk of lung cancer and other malignant neoplasms. Am J Epidemiol 146:223–230
140. Geleijnse JM, Launer LJ, Van der Kuip DA, Hofman A, Witteman JC (2002) Inverse association of tea and flavonoid intakes with incident myocardial infarction: the Rotterdam Study. Am J Clin Nutr 75:880–886
141. Rice-Evans CA, Miller NJ, Paganga G (1996) Structure-antioxidant activity relationships of flavonoids and phenolic acids. Free Radic Biol Med 20:933–956
142. Agullo G, Gamet-Payrastre L, Fernandez Y, Anciaux N, Demigne C, Remesy C (1996) Comparative effects of flavonoids on the growth, viability and metabolism of a colonic adenocarcinoma cell line (HT29 cells). Cancer Lett 105:61–70
143. Russo M, Palumbo R, Tedesco I, Mazzarella G, Russo P, Iacomino G, Russo GL (1999) Quercetin and anti-CD95(Fas/ Apo1) enhance apoptosis in HPB-ALL cell line. FEBS Lett 462:322–328
144. Sotocaa AM, Ratman D, Saag PV, Ström A, Gustafsson JA, Vervoort J, Rietjens IMCM, Murk AJ (2008) Phytoestrogen-mediated inhibition of proliferation of the human T47D breast cancer cells depends on the ER_/ER_ ratio. J Steroid Biochem Mol Biol 112:171–178
145. Bernad FX, Sable S, Cameron B (1997) glycosylated flavones as selective inhibitors of topoisomerases IV. Antimicrob Agents Chemother 41:992–998
146. Kim MK, Choo H, Chong Y (2014) Water-soluble and cleavable Quercetin–amino acid conjugates as safe modulators for P-glycoprotein-based multidrug resistance. J Med Chem 57:7216–7233
147. Haleagrahara N, Miranda-Hernandez S, Alim MA, Hayesa L, Bird G, Ketheesan N (2017) Therapeutic effect of quercetin in collagen-induced arthritis. Biomed Pharmacother 90:38–46
148. Whitten PL, Naftolin F (1998) Reproductive actions of phytoestrogens. Bailliere Clin Endocrinol Metab 12:667–690
149. Saarinen NM, Penttinen PE, Smeds AI, Hurmerinta TT, Makela SI (2005) Structural determinants of plant lignans for growth of mammary tumors and hormonal responses in vivo. J Steroid Biochem Mol Biol 93:209–219
150. Zafar A, Singh S, Naseem I (2016) Cytotoxic activity of soy phytoestrogen coumestrol against human breast cancer MCF-7 cells: insights into the molecular mechanism. Food Chem Toxicol. https://doi.org/10.1016/j.fct.2016.11.034
151. Lim W, Jeong M, Bazer FW, Song G (2017) Coumestrol inhibits proliferation and migration of prostate cancer cells by regulating AKT, ERK1/2, and JNK MAPK cell signaling cascades. J Cell Physiol 232(4):862–871
152. Tsutsumi N (1995) Effect of coumestrol on bone metabolism in organ culture. Biol Pharm Bull 18:1012–1015
153. Wang H, Li H, Moore LB, Johnson MDL, Maglich JM, Goodwin B, Ittoop ORR, Wisely B, Creech K, Parks DJ, Collins JL, Willson TM, Kalpana GV, Venkatesh M, Xie W, Cho SY, Roboz J, Redinbo M, Moore JT, Mani S (2008) The phytoestrogen coumestrol is a naturally occurring antagonist of the human Pregnane X receptor. Mol Endocrinol 22:838–857
154. Zhai Y, Li Y, Wang Y, Cui J, Feng K, Kong X, Chen L (2017) Psoralidin, a prenylated coumestan, as a novel anti-osteoporosis candidate to enhance bone formation of osteoblasts and decrease bone resorption of osteoclasts. Eur J Pharmacol Apr 15(801):62–71

155. Hong Y, Cho M, Yuan Y, Chen S (2008) Molecular basis for the interaction of four different classes of substrates and inhibitors with human aromatase. Biochem Pharmacol 75:1161–1169
156. Whitten PL, Patisaul HB, Younga LJ (2002) Neurobehavioral actions of coumestrol and related isoflavonoids in rodents. Neurotoxicol Teratol 24:47–54
157. Canal Castro C, Pagnussat AS, Orlandi L, Worm P, Moura N, Etgen AM, Alexandre Netto C (2012) Coumestrol has neuroprotective effects before and after global cerebral ischemia in female rats. Brain Res 1474:82–90
158. Dai Q, Franke AA, Jin F, Shu XO, Hebert JR et al (2002) Urinary excretion of phytoestrogens and risk of breast cancer among Chinese women in Shanghai. Cancer Epidemiol Biomark Prev 11:815–821
159. De Kleijn MJJ, Van der Schouw YT, Wilson PWF, Grobbee DE, Jacques PF (2002) Dietary intake of phytoestrogens is associated with a favorable metabolic cardiovascular risk profile in postmenopausal U.S. women: the Framingham study. J Nutr 132:276–282
160. Aehle E, Müller U, Eklund PC, Willför SM, Sippl W, Dräger B (2011) Lignans as food constituents with estrogen and antiestrogen activity. Phytochemistry 72:2396–2405
161. Adlercreutz H, Höckerstedt K, Bannwart C, Bloigu S, Hämäläinen E, Fotsis T, Ollus A (1987) Effect of dietary components, including lignans and phytoestrogens, on enterohepatic circulation and liver metabolism of estrogens and on sex hormone binding globulin (SHBG). Journal of Steroid Biochemistry 27:1135–1144
162. Prasad K (2000) Antioxidant activity of secoisolariciresinol diglucoside- derived metabolites secoisolariciresinol. Enterodiol, and enterolactone. Int J Angiol 9:220–225
163. Boccardo F, Lunardi G, Guglielmini P, Parodi M, Murialdo R, Schettini G, Rubagotti A (2004) Serum enterolactone levels and the risk of breast cancer in women with palpable cysts. Eur J Cancer 40:84–89
164. Mousavi Y, Adlercreutz H (1992) Enterolactone and estradiol inhibit each other's proliferative effect on MCF-7 breast cancer cells in culture. J Steroid Biochem Mol Biol 41:615–619
165. Figueiredo MS, Maia LA, Guarda DS, Lisboa PC, Moura EG (2017) Flaxseed secoisolariciresinol diglucoside (SDG) during lactation improves bone metabolism in offspring at adulthood. J Funct Foods 29:161–171
166. Ma X, Wang R, Zhao X, Zhang C, Sun J, Li J, Zhang L, Shao T, Ruan L, Chen L, Xu Y, Pan J (2013) Antidepressant-like effect of flaxseed secoisolariciresinol diglycoside in ovariectomized mice subjected to unpredictable chronic stress. Metab Brain Dis 28:77–84
167. Hu P, Mei QY, Ma L, Cui WG, Zhou WH, Zhou DS, Zhao Q, Xu DY, Zhao X, Lu Q, Hu ZY (2015) Secoisolariciresinol diglycoside, a flaxseed lignan, exerts analgesic effects in a mouse model of type 1 diabetes: engagement of antioxidant mechanism. Eur J Pharmacol 767(15):183–192
168. Stewart JR, Christman KL, O'Brian CA (2000) Effects of resveratrol on the autophosphorylation of phorbol ester-responsive protein kinases: inhibition of protein kinase D but not protein kinase C isozyme autophosphorylation. Biochem Pharmacol 60:1355–1359
169. Bhat KPL, Lantvit D, Christov K, Mehta RC, Pezzuto JM (2001) Estrogenic and antiestrogenic properties of resveratrol in mammary tumour models. Cancer Res 61:7456–7463
170. Stojanovic S, Sprinz H, Brede O (2001) Efficiency and mechanism of the antioxidant action of trans-resveratrol and its analogues in the radical liposome oxidation. Arch Biochem Biophys 391:79–89
171. Fontecave M, Lepoivre M, Elleingand E, Gerez C, Guittet O (1998) Resveratrol: a remarkable inhibitor of ribonucleotide reductase. FEBS Lett 421:277–279
172. Locatelli GA, Savio M, Forti L, Shevelev I, Ramadan K, Stivala LA, Vannini V, Hubscher U, Spadari S, Maga G (2005) Inhibition of mammalian DNA polymerases by resveratrol: mechanism and structural determinants. Biochem J 389:259–268
173. Bhat KPL, Lantvit D, Christov K, Mehta RC, Pezzuto JM (2001) Estrogenic and antiestrogenic properties of resveratrol in mammary tumour models. Cancer Res 61:7456–7463

174. Sakamoto T, Horiguchi H, Oguma E, Kayama F (2010) Effects of diverse dietary phytoestrogens on cell growth, cell cycle and apoptosis in estrogen-receptor-positive breast cancer cells. J Nutr Biochem 21(9):856–864
175. Wesierska-Gadek J, Kramer MP, Maurer M (2008) Resveratrol modulates roscovitine-mediated cell cycle arrest of human MCF-7 breast cancer cells. Food Chem Toxicol 46:1327–1333
176. Zhu Y, Fu J, Shurlknight KL, Soroka DN, Hu Y, Chen X, Sang S (2015) Novel resveratrol-based aspirin prodrugs: synthesis, metabolism, and anticancer activity. J Med Chem 58(16):6494–6506
177. Du C, Dong MH, Ren YJ, Jin L, Xu C (2016) Design, synthesis and antibreast cancer MCF-7 cells biological evaluation of heterocyclic analogs of resveratrol. J Asian Nat Prod Res 3:1–13
178. Alamolhodaei NS, Tsatsakis AM, Ramezani M, Hayes AW, Karimi G (2017) Resveratrol as MDR reversion molecule in breast cancer: an overview. Food Chem Toxicol 103:223–232
179. Anekonda TS, Reddy PH (2006) Neuronal protection by sirtuins in Alzheimer's disease. J Neurochem 96:305–313
180. Karlsson J, Emgard M, Brundin P, Burkitt MJ (2000) trans-resveratrol protects embryonic mesencephalic cells from tert-butyl hydroperoxide: electron paramagnetic resonance spin trapping evidence for a radical scavenging mechanism. J Neurochem 75:141–150
181. ClinicalTrials.gov Identifier: NCT01504854 Phase II Study to Evaluate the Impact on Biomarkers of Resveratrol Treatment in Patients With Mild to Moderate Alzheimer's Disease April 2016
182. Wallerath T, Deckert G, Ternes T, Anderson H, Li H et al (2002) Resveratrol, a polyphenolic phytoalexin present in red wine, enhances expression and activity of endothelial nitric oxide synthase. Circulation 106:1652–1658
183. Das S, Das DK (2007) Anti-inflammatory responses of resveratrol. Inflamm Allergy Drug Targets 6:168–173
184. Stagos D, Portesis N, Spanou C, Mossialos D, Aligiannis N, Chaita E, Panagoulis C, Reri E, Skaltsounis L, Tsatsakis AM (2012b) Correlation of total polyphenolic content with antioxidant and antibacterial activity of 24 extracts from greek domestic Lamiaceae species. Food Chem Toxicol 50:4115–4124
185. Mizutani K, Keda K, Kawai Y, Yamori Y (2001) Protective effect of resveratrol on oxidative damage in male and female stroke-prone spontaneously hypertensive rats. Clin Exp Pharmacol Physiol 28:55–59
186. Szkudelska K, Nogowski L, Szkudelski T (2009) Resveratrol, a naturally occurring diphenolic compound, affects lipogenesis, lipolysis and the antilipolytic action of insulin in isolated rat adipocytes. J Steroid Biochem Mol Biol 113:17–24
187. Sanchez-Fidalgo S, Cardeno A, Villegas I, Talero E, Alarcon de la Lastra C (2010) Dietary supplementation of resveratrol attenuates chronic colonic inflammation in mice. Eur J Pharmacol 633:78–84
188. De Vincenzo R, Scambia G, Panici PB, Ranelletti FO, Bonanno G, Ercoli A, Delle MF, Ferrari F, Piantelli M, Mancuso S (1995) Effect of synthetic and naturally occurring chalcones on ovarian cancer cell growth: structure–activity relationships. Anticancer Drug Des 10:481–490
189. Tamir S, Eizemberg M, Somjen D, Izrael S, Vaya J (2001) Estrogen-like activity of glabrene and other constituents isolated from licorice root. J Steroid Biochem Mol Biol 78:291–298
190. Haraguchi H, Ishikawa H, Mizutani K, Tamura Y, Kinoshita T (1998) Antioxidative and superoxide scavenging activities of retrochalcones in Glycyrrhiza inflata. Bioorg Med Chem 6:339–347
191. Tawata M, Aida K, Noguchi T, Ozaki Y, Kume S, Sasaki H, Chin M, Onaya T (1992) Antiplatelet action of isoliquiritigenin, an aldose reductase inhibitor in licorice. Eur J Pharmacol 212:87–92

192. Chowdhury SA, Kishino K, Satoh R, Hashimoto K, Kikuchi H, Nishikawa H, Shirataki Y, Sakagami H (2005) Tumor-specificity and apoptosis-inducing activity of stilbenes and flavonoids. Anticancer Res 25:2055–2063
193. Takahashi T, Takasuka N, Iigo M, Baba M, Nishino H, Tsuda H, Okuyama T (2004) Isoliquiritigenin, a flavonoid from licorice, reduces prostaglandin E2 and nitric oxide, causes apoptosis, and suppresses aberrant crypt foci development. Cancer Sci 95:448–453
194. Jung JI, Chung E, Seon MR, Shin HK, Kim EJ, Lim SS, Chung WY, Park KK, Park JH (2006) Isoliquiritigenin (ISL) inhibits ErbB3 signaling in prostate cancer cells. Biofactors 28:159–168
195. Hsu YL, Kuo PL, Lin LT, Lin CC (2005) Isoliquiritigenin inhibits cell proliferation and induces apoptosis in human hepatoma cells. Planta Med 71:130–134
196. Maggiolini M, Statti G, Vivacqua A, Gabriele S, Rago V, Loizzo M, Menichini F, Amdoc S (2002) Estrogenic and antiproliferative activities of isoliquiritigenin in MCF7 breast cancer cells. J Steroid Biochem Mol Biol 82:315–322
197. Kanga S, Choi J, Choi Y, Bae J, Li J, Kim DS, Kim J, Shin S, Lee Y, Kwun I, Kang Y (2010) Licorice isoliquiritigenin dampens angiogenic activity via inhibition of MAPK-responsive signaling pathways leading to induction of matrix metalloproteinases. J Nutr Biochem 21:55–65
198. Li D, Wang Z, Chen H, Wang J, Zheng Q, Shang J, Li J (2009) Isoliquiritigenin induces monocytic differentiation of HL-60 cells. Free Radic Biol Med 46:731–736
199. Zhan C, Yang J (2006) Protective effects of isoliquiritigenin in transient middle cerebral artery occlusion-induced focal cerebral ischemia in rats. Pharmacol Res 53:303–309
200. Chi JH, Seo GS, Cheon JH, Lee SH (2017) Isoliquiritigenin inhibits TNF-α-induced release of high-mobility group box 1 through activation of HDAC in human intestinal epithelial HT-29 cells. Eur J Pharmacol 796:101–109
201. Liu S, Zhu L, Zhang J, Yu J, Cheng X, Peng B (2016) Anti-osteoclastogenic activity of isoliquiritigenin via inhibition of NF-κB-dependent autophagic pathway. Biochem Pharmacol 106:82–93
202. Tsukiyama R, Katsura H, Tokuriki N, Kobayashi M (2002) Antibacterial activity of licochalcone A against spore-forming bacteria. Antimicrob Agents Chemother 46:1226–1230
203. Chen M, Christensen SB, Blom J, Lemmich E, Nadelmann L, Fich K et al (1993) Licochalcone A, a novel antiparasitic agent with potent activity against human pathogenic protozoan species of Leishmania. Antimicrob Agents Chemother 37:2550–2556
204. Iwata S, Nishino T, Nagata N, Satomi Y, Nishino H, Shibata S (1995) Antitumorigenic activities of chalcones I. Inhibitory effects of chalcone derivatives on 32P-Incorporation into phospholipids of Hela cells promoted by 12-O- tetradecanoyl-phorbol 13-acetate (TPA). Biol Pharm Bull 18:1710–1713
205. Chen X, Liu Z, Meng R, Shi C, Guo N (2017) Antioxidative and anticancer properties of Licochalcone A from licorice. J Ethnopharmacol 198:331–337
206. Rafi MM, Rosen RT, Vassil A, Ho CT, Zhang H, Ghai G, Lambert G, Di Paola RS (2000) Modulation of bcl-2 and cytotoxicity by licochalcone-A, a novel estrogenic flavonoid. Anticancer Res 20:2653–2658
207. Rafi M, Vastano BC, Zhu N, Ho CT, Ghai G, Rosen RT, Gallo MA, DiPaola RS (2002) Novel polyphenol molecule isolated from licorice root (Glycyrrhiza glabra) induces apoptosis, G2/M cell cycle arrest, and Bcl-2 phosphorylation in tumor cell lines. J Agric Food Chem 50:677–684
208. Fu Y, Hsieha T, Guo J, Kunicki J, Lee MYWT, Darzynkiewicz Z, Wu JM (2004) Licochalcone-A, a novel flavonoid isolated from licorice root (Glycyrrhiza glabra), causes G2 and late-G1 arrests in androgen-independent PC-3 prostate cancer cells. Biochem Biophys Res Commun 322:263–270
209. Xiao X, Hao M, Yang X, Ba Q, Li M, Ni S, Wang L, Du X (2011) Licochalcone A inhibits growth of gastric cancer cells by arresting cell cycle progression and inducing apoptosis. Cancer Lett 302:69–75

210. Won S, Kim S, Kim Y, Lee P, Ryu J, Kim J, Rhee H (2007) Licochalcone A: a lipase inhibitor from the roots of Glycyrrhiza uralensis. Food Res Int 40:1046–1050
211. Shang F, Ming L, Zhou Z, Yu Y, Sun J, Ding Y, Jin Y (2014) The effect of licochalcone A on cell-aggregates ECM secretion and osteogenic differentiation during bone formation in metaphyseal defects in ovariectomized rats. Biomaterials 35:2789–2797
212. Oliveira PJ, Carvalho RA, Portincasa P, Bonfrate L, Sardao VA Fatty acid oxidation and cardiovascular risk during menopause: a mitochondrial connection? J Lipids 2012, 2012:365798
213. Jenkins DJ, Mirrahimi A, Srichaikul K et al (2010) Soy protein reduces serum cholesterol by both intrinsic and food displacement mechanisms. J Nutr 140(12):2302S–2311S
214. Rigallia JP, Tocchettib GN, Aranab MR, Villanueva SSM, Catania VA, Theile D, Ruiz ML, Weiss J (2016) The phytoestrogen genistein enhances multidrug resistance in breast cancer cell lines by translational regulation of ABC transporters. Cancer Lett 376-1:165–172
215. Henderson BE, Bernstein L (1991) The international variation in breast cancer rates: an epidemiological assessment. Breast Cancer Res Treat 18(1): S11–S117
216. Soni M, White LR, Kridawati A, Bandelow S, Hogervorst E (2016) Phytoestrogen consumption and risk for cognitive decline and dementia: with consideration of thyroid status and other possible mediators. J Steroid Biochem Mol Biol 160:67–77
217. Gopaul R, Knaggs HE, Lephart ED (2012) Biochemical investigation and gene analysis of equol: a plant and soy-derived isoflavonoid with antiaging and antioxidant properties with potential human skin applications. Biofactors 38:44–52
218. Crain DA, Janssen SJ, Edwards TM, Heindel J, Ho SM, Hunt P, Iguchi T, Juul A, McLachlan JA, Schwartz J, Skakkebaek N, Soto AM, Swan S, Walker C, Woodruff TK, Woodruff TJ, Giudice LC, Guillette LJ Jr (2008) Female reproductive disorders: the roles of endocrine-disrupting compounds and developmental timing. Fertil Steril 90:911–940
219. This P, Cremoux PD, Leclercq G, Jacquot Y (2011) A critical view of the effects of phytoestrogens on hot flashes and breast cancer risk. Maturitas 70:222–226
220. Glazier MG, Bowman MA (2001) A review of the evidence for the use of phytoestrogens as a replacement for traditional estrogen replacement therapy. Arch Intern Med 161:1161–1172

Curcumin in Cancer Prevention

16

Akash Sabarwal, Kunal Kumar, Ritis Shyanti, and Rana P. Singh

Abstract

Cancer has become one of the leading causes of death worldwide. Presently, available chemotherapeutic agents have several limitations including severe side effects. Curcumin (diferuloylmethane) is a polyphenol derived from the plant *Curcuma longa*. Curcumin has been used extensively as spice in many Asian countries and in Ayurvedic medicines. It is nontoxic and has shown to possess various medicinal properties including antioxidant, anti-inflammatory, and antibacterial. Recent investigations have shown that curcumin exerts anticancer properties in various cancer cell models and targets variety of biological pathways involved in cell cycle regulation, apoptosis, mutagenesis, angiogenesis, and metastasis. NF-κB, p53, Nrf2, NFAT, MMPs, STATs, and uPA are important molecular targets of curcumin in multiple cancer models. Enzymes involved in redox balance inside the cells including superoxide dismutases, catalase, and glutathione peroxidase are modulated by curcumin. However, bioavailability, water insolubility, short life span, and rapid systemic clearance of curcumin have posed limitations in developing curcumin as an effective chemotherapeutic agent. To address these challenges, curcumin has been used in combinations with many other chemotherapeutic drugs which have shown encouraging results. This chapter deals with the current information available for the cancer chemopreventive activities of curcumin in various in vitro and in vivo cancer models including epidemiological studies and human trials. Also, molecular pathways involved in

A. Sabarwal · K. Kumar · R. Shyanti
School of Life Sciences, Central University of Gujarat, Gandhinagar, Gujarat, India

R. P. Singh (✉)
Cancer Biology Laboratory, School of Life Sciences, Jawaharlal Nehru University, New Delhi, India
e-mail: rana_singh@mail.jnu.ac.in

V. Rani, U. C. S. Yadav (eds.), *Functional Food and Human Health*,
https://doi.org/10.1007/978-981-13-1123-9_16

the manifestation of biological activities of curcumin against various processes of cancer development have been discussed.

Keywords:
Curcumin · Apoptosis · ROS · Cell cycle · Angiogenesis · Metastasis · Radio-sensitizer · Radio-protectant

16.1 Introduction

Curcumin is a yellow-colored compound derived from *Curcuma longa* commonly known as turmeric. Turmeric has been used for many years in several herbal remedies. The compound is one of the most studied component of turmeric with more than 9000 citations in the literature [1]. About two centuries back, curcumin was isolated from the ground rhizome of *Curcuma longa* L. (family Zingiberaceae) by Vogel and Pelletier [2]. Structurally this polyphenolic compound is diferuloylmethane[(1E,6E)-1,7-bis(4-hydroxy-3-methoxyphenyl)-1,6-heptadiene-3,5-dione] and exists in two tautomeric forms (Fig. 16.1).

Curcumin is a multi-targeting compound and has been reported to possess a wide range of preventive or putative therapeutic properties of antioxidant, anti-inflammatory, anticancer, antiarthritic, antiaging, antidiabetic, antimicrobial, antiviral, anti-atherosclerotic, antidepressant, wound healing, and memory-enhancing activities against variety of diseases [3–5]. These nutraceutical properties of curcumin are possibly attributed to methoxy, hydroxyl, α, β-unsaturated carbonyl moiety or diketone groups [6].

Among various pharmacological activities, curcumin had proved its credential as a vital source of chemopreventive agent against varieties of cancer. Curcumin has been studied in multiple human carcinomas including melanoma, head and neck, pancreatic, prostate, breast, colon, and ovarian cancers [7–15]. Many in vitro studies have been conducted on several cancer cell lines showing strong efficacy of curcumin against several human cancers. Curcumin served as a strong inhibitor of cell proliferation in Burkitt's lymphoma cell lines expressing Bax protein [16]. In colon cancer, curcumin enhanced the sensitivity of overexpressed glucose-regulated

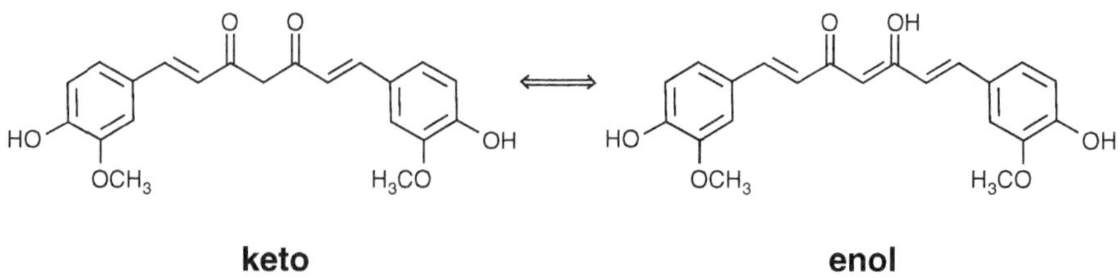

Fig. 16.1 Keto and enol forms of diferuloylmethane [(1E,6E)-1,7-bis(4-hydroxy-3-methoxyphenyl)-1,6-heptadiene-3,5-dione]

protein 78 (the basic regulator of endoplasmic reticulum) in HT-29 cells lines and induced apoptosis in control cells compared to GRP 78 kD DLD-1 cells during cell cycle examination and TUNEL assay [17].

Curcumin prevented progression of colon cancer through inhibiting cell proliferation and cell survival, improving cellular uptake, activating the cascade of caspases, exhibiting controlled release at physiological pH, and promoting intrinsic apoptotic signaling [18]. Another study evaluated the anticancer effect of curcumin in cervical cancer where curcumin controlled the function and expression of P-glycoprotein (Pgp) in multidrug-resistant (MDR) human cervical carcinoma (KB-V1) cells [19]. It also enhanced the cisplatin-induced apoptosis by sensitizing the SiHa cells which are cisplatin-resistant. Thus, curcumin has the potential to reverse the MDR in cervical cancer cells. Goel et al. [3] suggested that 0.5–12 g orally administrated curcumin in a daily dose for 3 months leads to histologic improvement in uterine and cervical intraepithelial carcinoma's precancerous lesions. This improvement in lesions were observed in one out of four patients that have uterine and cervical carcinomas [20]. In other reports, curcuminoids were found to potentiate cytotoxic function of natural killer cells and induce apoptosis [21, 22]. In addition to curcumin, difluorinated curcumin (CDF), an analog of curcumin found in turmeric, has been reported to decrease cell survival, invasive cell migration, formation of pancreatospheres, clonogenicity, and cancer stem cell function in human pancreatic cancer cells.

Bioavailability and tissue distribution of curcumin and CDF can be increased with cyclodextrin [23]. Effect of curcumin on cancer cells was linked with the activation of caspase 3 enzyme in the apoptotic cascade which induces apoptosis in prostate cancer [12]. According to epidemiological studies, consumption of curcumin rich food products greatly reduces the risk of developing cancer of different origins. In one report of breast cancer, effect of curcumin was studied for apoptosis on two MCF-7/TH and MCF-10A cell lines. MCF-7/TH cells displayed a remarkable apoptosis level when treated with certain curcumin concentrations, but the same concentrations of drug showed insignificant level of apoptosis on mammary epithelial cells [24]. Curcumin depicts downregulation of mutant p53 mRNAs, PCNA, and Ki67 as its chemopreventive action in breast cancer cells [25].

In another study the effects of curcumin were investigated in both in vivo and in vitro tests against matrix metalloproteinases, i.e., MMP2 and MMP9, which revealed the range of molecular mechanisms involving the inhibition of NF-κB and nitric oxide production, effects on transcription factors, and intervention in signaling pathways exhibited by curcumin's action to induce apoptosis [26]. Both in vitro and in vivo studies have proven curcumin as an apoptosis inducer against the growth of head and neck cancer through suppression of IKK-mediated NF-κB activation and through NF-κB-regulated gene expressions [8].

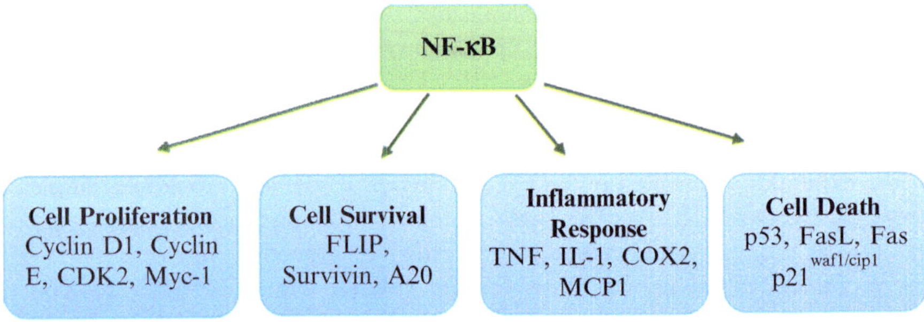

Fig. 16.2 NF-κB contributes to the induction of many genes which are involved in cancer growth and development. These genes regulate many important processes in tumorigenesis including cell proliferation, cell survival, inflammatory responses, and cell death

16.2 Curcumin Chemoprevention by Inhibition of NF-κB

The ability to sense and adjust to external environment is crucial for the survival of all living beings. Regulation through inducible gene expression makes the organisms to acclimatize to the fluctuating outside environment. Different transcription factors play various roles in accomplishing these essential effects. NF-κB was discovered for its role in immunity, which was later found to be involved in various mechanisms that impact cell survival, differentiation, and proliferation. Thus, dysregulation of NF-κB can cause several consequences and diseases including autoimmune diseases, neurodegenerative diseases, cardiovascular disease, diabetes, and cancer [27]. Even though NF-κB target genes that have been extensively studied for their role in inflammation and immune responses, these transcription factors are also involved in the regulation of genes responsible for cell proliferation, survival, migration, and apoptosis (Fig. 16.2). In numerous cancer types, these are found to be constitutively expressed making them important targets for anticancer therapy [28].

16.2.1 Regulation of NF-κB

Nuclear factor κB is a family of protein dimers which bind to the common DNA motif known as κB site. In resting stage or in the absence of stimuli, NF-κB is located inside the cytoplasm associated with IκB proteins. Once the stimulus is triggered, IκB proteins undergo phosphorylation and ubiquitination leading to their proteasomal degradation, thus leaving NF-κB free to translocate into the nucleus resulting in transcription of target genes. IKK complex which comprises of NF-κB dimers and IκB is the crucial part of NF-κB signaling pathway (Fig. 16.3).

Mammalian NF-κB family has five members RelA/p65, RelB, c-Rel, p50 (NF-κB1), and p52 (NF-κB2) [29]. NF-κB binds to the target gene in the form of either homodimers or heterodimers and may cause both positive and negative effects on

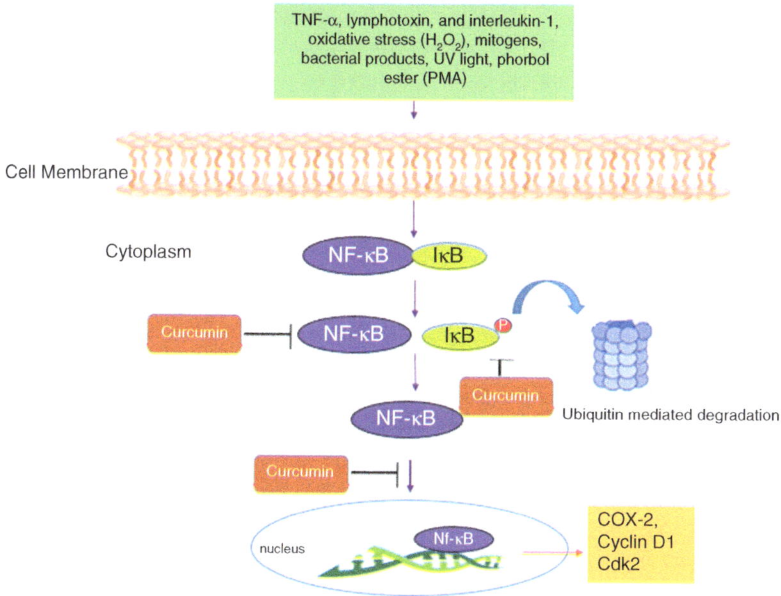

Fig. 16.3 Overview of inhibition of NF-κB pathway mediated through curcumin treatment. Numerous ligands can activate NF-κB by inhibiting its interaction with IκB and increasing its translocation inside the nucleus thus promoting transcription of many genes including cancer promoting factors and cancer. Curcumin blocks IκB phosphorylation and its ubiquitin-mediated proteasomal degradation

the level of transcription. Crystal structures of NF-κB dimers have shown a common RHD domain (Rel Homology Domain) which is present in all the NF-κB family members. This RHD domain promotes the association of NF to the κB sites and also assists in the subunit dimerization process. Only three of the members (RelA/p65, RelB, c-Rel) possess C-terminal transactivation domain (TADs) that have the ability to initiate the process of transcription. The other two proteins, viz., p50 and p52, do not have TADs; however, they still can initiate transcription through heterodimerization with TADs containing proteins or with non-Rel proteins that have transactivation efficiency. The other key regulatory proteins of NF-κB pathway are IκB proteins, which are marked by multiple ankyrin repeat domains. IκB family consists of IκB proteins IκBa, IκBb, IκBe, IκBz, Bcl-3 (B-cell lymphoma 3), and IκBns and the precursor proteins p100 (NF-κB2) and p105 (NF-κB1). The most important role of IκB proteins is to sequester the NF complexes inside the cytoplasm, thus preventing their translocation into the nucleus and binding to κB sites [30].

Phytochemicals have shown effective anticancer potential against different cancer types. Many of them are also found to inhibit NF-κB pathway, thus limiting the transcription of genes involved in cancer progression. Melanoma is one of the deadly cancers which have shown to constitutively express NF-κB, likely playing a central role in disease progression. In melanoma, IκB is also found to be activated through phosphorylation resulting in its degradation, causing increased localization of NF-κB into the nucleus [31]. Curcumin has shown to possess promising inhibitory activity against NF-κB. Curcumin (10–50 μmol/l) inhibited the transactivation of NF-κB in 4046T melanoma cells compared to normal melanocytes. Curcumin has also decreased the phospho-IκBα in these tumor cells, which correlated with the decrease in NF-κB activity. Additionally decrease in the expression of two target genes of NF-κB, i.e., COX-2 and cyclin D1, was also observed. Curcumin caused apoptosis selectively in 4046T cancer cells via cleavage of caspase 3 [32]. In case of colon cancer HCT 116 cells, curcumin triggered apoptosis in time- and dose-dependent fashion and this was linked to NF-κB inhibition. HCT 116 cells were transiently transfected with 3EnhConALuc reporter vector, this vector carried the luciferase reporter gene, and treatment with curcumin caused significant decrease in NF-κB-dependent luciferase activity.

16.2.2 MAPK-NF-κB

Activation of JNK, p38, and ERK through phosphorylation has been investigated in the presence and absence of curcumin, and it was found that curcumin has caused sustained phosphorylation and activation of both JNK and p38 proteins. JNK activation and inhibition of NF-κB promote apoptosis in HCT 116 colorectal cancer cells. However, curcumin did not affect the state of ERK, which is constitutively activated in these cells [33]. One of the most predominant forms found in many cells is the active p65/p50 NF-κB heterodimer [34]. p65/RelA subunit of NF-κB can have effect on cell survival due to activation of NF-κB attributed to upregulation of anti-apoptotic genes including X-linked inhibitor of apoptosis (XIAP), Bcl-xL and A20 and inhibition of sustained activation of c-Jun N-terminal kinase (JNK). However, forced overexpression of p65/Rel A did not affect curcumin-induced JNK activation through phosphorylation but has caused increased apoptosis and cleavage of PARP. This additive effect of p65 upon curcumin-mediated apoptosis was blocked by IκB super-repressor (Delta NIκB) transfection. Additionally, curcumin treatment caused downregulation of anti-apoptotic genes targeted through NF-κB in mock-transfected and p65-overexpressed cells, although expression level was found to be higher in p65-overexpressed cells. Conclusively, the study showed that curcumin-induced apoptosis and JNK activation do not require p65 inhibition but p65 can potentiate the apoptotic effect of curcumin in HCT 116 cells [35]. Many stimuli act as an activator of NF-κB pathway which include inflammatory cytokines (e.g., tumor necrosis factor (TNF-α), lymphotoxin, and interleukin-1), oxidative stress (H_2O_2), mitogens, bacterial products, UV light, and phorbol ester (PMA) [36, 37].

In an in vitro study, TNF-α was employed to activate NF-κB pathway in human myeloid ML-1a cells, and curcumin was employed to block the activation. Blockade of NF-κB activation could be beneficial in many complications including cancer. The results of this study indicated that curcumin (2–60 μM) has significantly blocked the activation of NF-κB. H_2O_2 and PMA are potent activators of NF-κB; thus, curcumin was employed at 50 μM concentration which interestingly completely blocked

NF-κB activation. Curcumin also inhibited TNF-dependent phosphorylation, and proteolysis of IκBα protein, thus, inhibited the translocation of NF-κB inside the nucleus [38]. It has been previously shown that phosphorylation of inducible IκBα is not enough for its dissociation from NF-κB. However, the degradation of IκBα is a necessary event in the activation of NF-κB, and protease inhibitors employed in the study prevented its activation. Protease inhibitors (tosyl phenylalanyl chloromethyl ketone (TPCK), tosyl-L-lysyl-chloromethane hydrochloride (TLCK), N-benzoyl-L-tyrosine ethyl ester (BTEE), or 3,4-dichloroisocoumarin (DCIC)) employed in the study also inhibited inducible phosphorylation of IκBα, indicating a more complex role of proteases in the activation of NF-κB [39]. It was also shown that agents like TPCK can cause alteration in the sulfhydryl group preventing activation of NF-κB; however, this inhibition was prevented in the presence of reducing agents like DTT. DTT can also reverse the inhibitory effect of phenylarsine oxide (a potent protein tyrosine phosphatase inhibitor) on NF-κB activation [39].

To investigate, if presence of reducing agents can reverse the effect of curcumin on the activation of NF-κB by TNF, human myeloid ML-1a cells were exposed to curcumin alone or in combination with DTT. Interestingly, DTT did not reverse the inhibition caused by curcumin, however, significantly reversed the inhibition caused by phenylarsine oxide, indicating that curcumin acts differently than protein tyrosine phosphatase inhibitor in these cells. All compounds (DTT, DMP, phenylarsine oxide) used in this study caused an increase in reactive oxygen intermediates; thus, it is possible that the effect of curcumin is through scavenging these reactive intermediates. Furthermore, inhibitors of respiratory chain complex have also shown to block TNF-mediated NF-κB activation, further suggesting the role of reactive oxygen intermediates [38].

16.2.3 Cancer Stem Cells and NF-κB

Cancer stem cells (CSCs) have important implications in many different cancers including hepatocellular carcinomas (HCC). NF-κB is one of the key pathways which is generally found to be constitutively activated in CSCs of HCC. The characteristic features of HCC are high recurrence rate and multidrug resistance which are found to be associated with CSCs. Thus, drugs which can target CSCs may play pivotal role in managing HCC. The effect of curcumin has been investigated on different HCC-sensitive and HCC-resistant cells [40]. Furthermore, it was also reported that NF-κB is central to inflammation-fibrosis-cancer axis in hepatocellular carcinogenesis [41]. It was found that the differential response to curcumin was directly related to NF-κB inhibition and its downstream signaling including STAT3, cyclin D1, and JNK. In sensitive cells, NF-κB inhibition caused significant decrease in potency of sphere formation and tumorigenicity. In case of resistant cells, curcumin treatment along with histone deacetylase inhibitor, trichostatin A (TSA), made cells susceptible to curcumin treatment [42]. Several studies have shown that HDAC and NF-κB signaling are interrelated [43, 44]. Conclusively, the disruption of NF-κB signaling by curcumin can play an important role in managing prognostically adverse HCCs.

16.3 Chemoprevention by Regulation of p53

Despite numerous genes involved in tumorigenesis, p53 protein (encoded by the human gene TP53) is a master regulator and an established tumor suppressor protein. p53 is a transcription factor that has been reported to play crucial role in cell cycle regulation and apoptosis; thus, dysregulation of p53 can lead to cancer [45, 46]. Additionally, mutations in p53 not only render its tumor-suppressive ability but can also evoke oncogenic functions including invasion, migration, metastasis, cell proliferation, and cell survival [47]. Mdm2 is a p53-specific E3 ubiquitin ligase and cellular antagonist of p53 which does not allow p53 to perform its growth-suppressive function in normal conditions. Under normal conditions, Mdm2 causes ubiquitination and directs proteasomal degradation of p53, thus maintaining low level of p53 protein inside the cells. However, in event of DNA damage, hypoxia, telomerase shortening, and other cellular stresses, p53 is phosphorylated by different proteins including ATM. Phosphorylation of p53 prevents Mdm2 and p53 association, thus, preventing ubiquitination of p53 that results in increased levels of p53 inside the cell. This increased level of p53 causes increased transcription of many proteins that play role in cell cycle arrest, one important target of p53 is p21, a cyclin-dependent kinase inhibitor, which may cause cell cycle arrest [48] (Fig. 16.4).

In case, where DNA damage is so severe and repair machinery is unable to repair the damage, p53 can activate apoptosis pathways, which mainly comprises of its ability to regulate transcription of pro-apoptotic members of the Bcl-2 family. Bcl-2 family members harbor consensus p53 response elements in promoters of their genes [49]. p53 is an important target of curcumin, and various studies have shown that curcumin regulates the expression of p53 in vivo and in vitro models. Curcumin caused apoptosis in human breast cancer cells in a p53-dependent fashion. In MCF-7 cells, in which expression of wild-type p53 can be induced, curcumin (10 μM) treatment caused increase in p53 protein level as early as 4 h of treatment. This increased p53 level started causing increase in the level of Bax protein after 8 h, and the highest level was found after 24 h. To establish the role of p53 in inducing apoptosis, p53-null (MDAH041) and tetracycline-regulated wild-type p53-expressing cell lines (TR9-7) were employed and treated with curcumin. In p53-null (MDAH041) cells, curcumin failed to induce apoptosis. In TR9-7 cells, curcumin moderately induced apoptosis in low p53-expressing condition, i.e., in the presence of tetracycline, which downregulates wild-type p53 expression. However, when TR9-7 cells were released from tetracycline-inhibited p53 expression, curcumin induced a high level of apoptosis. This study indicates clearly that curcumin-induced apoptosis was p53-dependent [50]. Similarly, curcumin induced apoptosis at G2/M phase with deregulated cyclin D1 in mammary epithelial carcinoma cells, but not in their normal counterparts. Curcumin selectively caused increase in p53 levels at G2 phase of

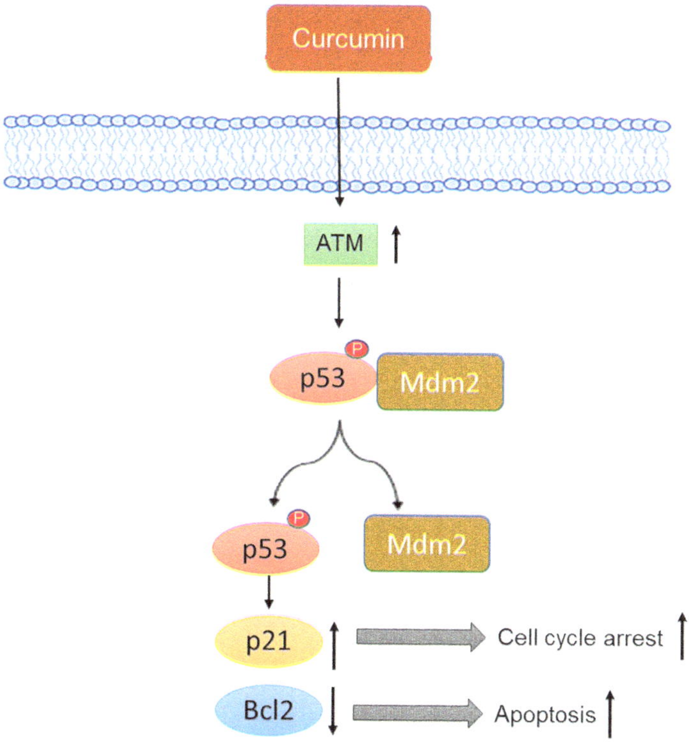

Fig. 16.4 Curcumin induced cell cycle arrest and apoptosis through activation and stabilization of p53. Mdm2 causes ubiquitination and directs proteasomal degradation of p53. Curcumin phosphorylates p53 and inhibits its degradation; increased levels of p53 cause cell cycle arrest and apoptosis

the cell cycle and release of cytochrome c from mitochondria into the cytosol in cancer cells.

Furthermore, experiments conducted with p53-null and dominant-negative and wild-type p53-transfected cells have established that curcumin-induced apoptosis was through p53-dependent pathway. However, curcumin also caused reversible arrest in normal mammary epithelial cells by downregulating cyclin D1 expression and disruption of their association with CDK4/6; curcumin also inhibited phosphorylation and, thus, caused inactivation of retinoblastoma protein. In normal cells, curcumin caused upregulation of p21/Waf1, which lead to arrest of cells in G0 phase of cell cycle and thus escaped from apoptosis in G2 phase. Conversely, in cancer cells cyclin D1 expression remained high, and thus they reached to G2 phase and had undergone apoptosis [51]. In patients with colorectal cancer (CRC), curcumin administration has shown significant effect on weight loss, decreased level of tumor necrosis factor (TNFα) in serum, and induction of apoptosis. Additionally, curcumin administration also increased p53 and Bax levels and decreased Bcl-2 significantly, explaining that p53 is an important target of curcumin in causing apoptosis in CRC [52]. Non-small cell lung cancer (NSCLC) is the leading cause of cancer-related death worldwide. miRNAs play vital roles in numerous steps of carcinogenesis and serve as feasible therapeutic cancer targets. In H460 and A427 cells, curcumin stimulated global miRNA levels. Importantly, miR-192-5p and miR-215 (miR-192-5p/215) are putative anti-oncogenic miRNAs in NSCLC, and curcumin caused significant change in the levels of these two miRNAs. Interestingly, curcumin-induced upregulation of these miRNAs was p53-dependent. Conditional knockdown of wild-type p53 disrupts the miR-192-5p/215 response in H460, A427, and A549 cells. Conversely, conditional expression of wild-type but not R273H mutant p53 caused upregulation of miR-192-5p/215 when treated with curcumin in p53-null H1299 cells. Additionally, miR-192-5p/215 targets XIAP for induction of apoptosis in these cells [53].

So far, we have discussed the studies which reported that curcumin-induced anti-cancer effects in various cancer models were dependent on p53. However, some groups have reported p53-independent pathways in curcumin-induced cell cycle arrest and apoptosis. Curcumin-induced cytotoxicity was investigated in $p53^{+/+}$ and $p53^{-/-}$ HCT116 colon cancer cells, as well as mutant p53 HT-29 colon cancer cells. Curcumin caused significant decrease in cell proliferation and increase in number of dead cells in both the cell lines in a dose (2.5–160 μM)- and time (24–72 h)-dependent fashion. Curcumin also caused chromatin condensation in both $p53^{+/+}$ and $p53^{-/-}$ HCT116 colon cancer cells. Curcumin treatment caused phosphorylation and induction of p53 in wild-type HCT116 cells; however, apoptosis was also observed in $p53^{-/-}$ HCT116 colon cancer cells indicating that the effect was independent of p53 status. A dose-dependent increase in the generation of superoxide anion in both the cell lines was reported, indicating the involvement of oxidative stress for eliciting apoptosis [54]. Similarly, curcumin caused dose-dependent apoptosis in eight melanoma cell lines, four with wild-type p53, while other four with mutant p53. Curcumin did not induce p53, indicating that curcumin-induced apoptosis was independent of p53 and was mediated through Fas receptor/caspase 8 pathway [55]. Curcumin selectively inhibited the proliferation of BKS-2, an immature B-cell lymphoma, and also induced apoptosis in dose- and time-dependent fashion. Curcumin decreased the level of p53, along with egr-1, c-myc, and Bcl-xL, indicating these effects were not dependent on p53 [56, 57].

16.4 Nrf2 and Curcumin

A major defense mechanism against oxidative stress is the activation of Nrf2 antioxidant response element signaling pathway, which regulates the expression of genes involved in removal of reactive oxygen species. Though, the regulatory mechanisms involved are not fully understood, but cytosolic proteins like Keap1 are some of the key players involved in the regulation of Nrf2. In normal conditions, Keap1 binds to Nrf2 which causes ubiquitin-mediated degradation of Nrf2. However, in case of oxidative stress, Nrf2 is phosphorylated that causes dissociation from Keap1 and translocation of Nrf2 into the nucleus. After entering the nucleus,

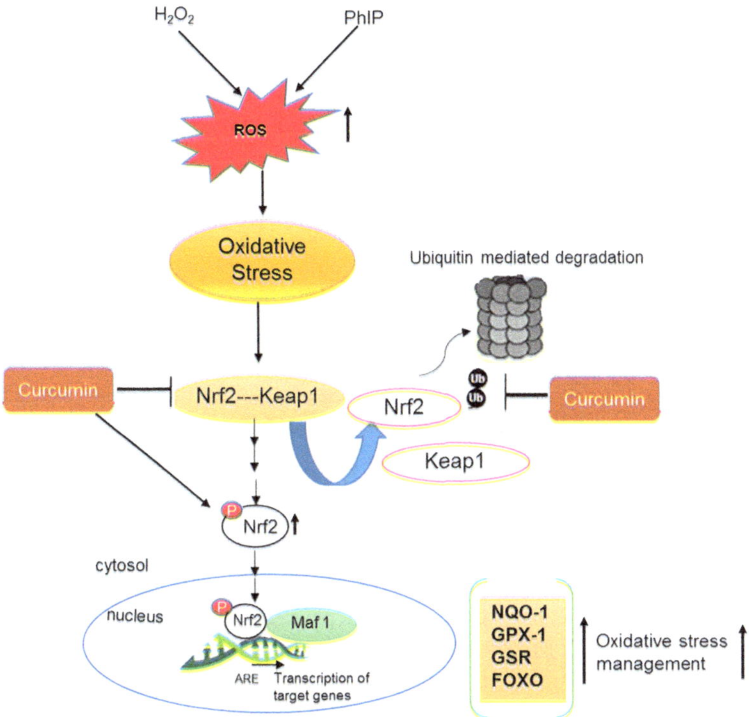

Fig. 16.5 Curcumin activates Nrf2 and causes transcription of genes involved in the management of oxidative stress. PhIP and H_2O_2 treatment causes significant increase in ROS generation; however, curcumin co-treatment blocked the ROS generation by phosphorylating and translocating Nrf2 into the nucleus

Nrf2 may bind to small adaptor proteins such as musculoaponeurotic fibrosarcoma (Maf) and finally targets antioxidant response element (ARE). ARE is a cis-acting response element found in the promoter region of the genes involved in the detoxification of reactive oxygen species and maintaining homeostasis (Fig. 16.5).

Targets of Nrf2 involve GSTA2 (glutathione S-transferase A2), NQO1 (NADPH:quinine oxidoreductase 1), glutathione S-transferase Ya subunit, heme oxygenase1 (HO-1), and ϒ-glutamyl cysteine synthase [58]. Fen1 endonuclease, a DNA repair enzyme, is also closely associated with cell proliferation and is generally found in the development of breast cancer and is found upregulated in many cancers including breast cancer. The link between Fen1 overexpression and Nrf2 is not well established. MCF-7 breast cancer cells were treated with curcumin and a decrease in cell proliferation was observed. Additionally, curcumin treatment caused upregulation of Nrf2 with concomitant downregulation of Fen1 expression. Curcumin treatment also caused nuclear localization of Nrf2 that decreased promoter activity of Fen1 [59].

In case of renal epithelial NRK-52E cells, Nrf2 was upregulated when cells were exposed to low concentrations of curcumin. Curcumin also stimulated Nrf2 activity, and this effect was completely inhibited by the natural repressor Keap1. Curcumin caused translocation of Nrf2 from cytoplasm to nucleus and stimulation of Nrf2 association with ARE sequence and a significant increase in HO-1 expression and heme oxygenase activity levels [60]. PhIP (Amino-1-methyl-6-phenylimidazo [4,5-b]pyridine) is a heterocyclic amine (HCA) and has shown genotoxic and carcinogenic effects. PhIP is formed when meat products are cooked at high temperatures. PhIP generates reactive oxygen species, which in turn may cause DNA damage through formation of DNA adducts and DNA strand breaks. Curcumin blocks the PhIP-induced DNA damage by inhibiting ROS generation. MCF-10A breast epithelial cells were used to screen different dietary phytochemicals that may have potential to inhibit cytotoxicity caused by PhIP. PhIP increased ROS generation in a dose-dependent fashion; however, curcumin co-treatment blocked the ROS generation caused by PhIP. Curcumin also decreased the formation of DNA adducts and inhibited DNA strand breaks caused by PhIP exposure. Oxidative stress marker Nrf2 and its target genes such as NQO-1, GPX-1 (glutathione peroxidase), and GSR (glutathione reductase) and FOXO (forkhead box protein) and their targets such as catalase, GADD-45 (growth arrest and DNA damage-inducible 45), and PRDX-3 (thioredoxin-dependent peroxide reductase) have been studied. DNA damage markers γ-H2A.X and BRCA-1 were induced by PhIP, and curcumin treatment blocked the expression of these genes. p16 (cyclin-dependent kinase inhibitor 2A) expression was inhibited by PhIP treatment; however, curcumin treatment restored its expression. Curcumin targets genes that are involved in the management of oxidative stress. Nrf2 was also found to be an important target of curcumin in MCF-10A breast epithelial cells [60, 61].

16.5 NFAT as a Target of Curcumin

There are many regulatory mechanisms in cells, among them transcription factors are responsible to control the mRNA levels during the transcription process. They have DNA binding sites where they bind at a particular sequence of the genes to control their expression pattern. This regulation could be either ways by positively through the help of activators or negatively through repressors. NFAT (nuclear factor of activated T cells) a transcription factor initially reported in T lymphocytes with its five members (NFAT1–NFAT5); it acts as the transcription activator of IL-2 promoter and regulator of T cells [62, 63]. Recently, it was found that this protein was expressed in both immune and non-immune cells such as adipocytes, cartilage, pancreatic, cardiac and breast cells other than T lymphocytes [64–70]. From the discovery of this molecule, its role in multiple biological processes such as cell cycle, apoptosis, metastasis and angiogenesis was observed in invertebrates and vertebrates [69, 71–73]. NFAT is well known for controlling gene expression at the time of cell activation and differentiation in T lymphocytes [71, 74]. Calcineurin signaling is responsible for the regulation of NFAT in T cells [75].

16.5.1 Structure and Function of NFAT in Normal Physiological Condition

NFAT is composed of two conserved regions, i.e., NFAT homology region (NHR) which shares 22–36% and DNA binding domain (DBD) up to 64–72%, identical sequences among all the five families. NFAT in its inactive form in resting cells is localized in cytosol, mediated by highly NHR phosphorylation. This is activated by dephosphorylation of NHR region by calcium-/calmodulin-dependent serine/threonine phosphatase enzyme calcineurin, whenever there is an elevation in the level of intracellular calcium. Subsequently, the nuclear localization sequence (NLS) gets free after dephosphorylation, NFAT1-4 translocate into the nucleus, binds to the promoter site, and regulates gene expression. NFAT-mediated transcriptional activity is regulated by the reversible activation/inactivation shuttles operated between nucleus and cytosol by the dephosphorylation and phosphorylation enzymatic system [76]. All the NFAT are regulated by calcium-/calcineurin-dependent pathway except NFAT-5, which is regulated during osmotic stress conditions. NFAT-3 is the exception which is not expressed in the immune system, whereas others do [74, 77].

16.5.2 Curcumin Regulates Transcription Factor NFAT

T cells have a major role in organ transplant biology. The immune-suppressive action has immense importance during organ transplant where immune rejection or acceptance plays a convincing role for the successful transplant. From a longer duration, cyclosporin A (CsA) and FK-506 are widely used as key immune suppressor during transplantation, but the rising of CsA-resistant T cells along with toxic effects decreased their efficacy and the implanted organs become more prone to rejection. There are reports of immune-suppressive function of curcumin where its pretreatment enhanced survival of skin-grafted animal and also inhibited CsA-resistant T cells [78, 79]. The mechanism behind the suppressive action of curcumin on T cells is via NFAT where calcium (Ca) mobilization leads to NFAT inactivation and blocked NFAT-dependent cytokine expression [80].

In another report, curcumin's cardioprotective activity was observed to be NFAT-dependent. *Trypanosoma cruzi* infection causes Chagas cardiomyopathy where cascade of endothelin-1 (ET-1) and Ca^{2+}/NFAT activation occurs in cardiomyocytes and induces cyclooxygenase type 2 (COX-2) with release of prostanoids and prohypertrophic peptides. Curcumin hampered Chagas pathogenesis through interfering transcriptional activity of Ca^{2+}-dependent NFAT2/NFATc1 and COX-2 and thereby declining prostaglandin levels in cardiomyocytes. BNP (brain natriuretic peptide) overexpression is a strong indicator of Chagas-mediated cardiac pathogenesis and is reduced by curcumin in both infected and uninfected cells [81]. In human colonic myofibroblasts, canonical transient receptor potential (TRPC) proteins responsible for sensation are inhibited by curcumin *via* TNF-alpha attenuation in NFAT-dependent manner [82]. Overall, the anti-inflammatory properties of curcumin in the form of TNF-alpha, COX-2, and calcium mobilization regulate NFAT-mediated response in cells. Thus, curcumin is an essential polyphenol phytochemical that regulates NFAT transcription factor during normal as well as pathological conditions.

16.6 Regulation of Oxidant and Antioxidant Enzymes

Oxidative stress which results in an imbalance between oxidants and antioxidants is a major player in the regulation of pathogenesis of various diseases such as neuronal cell injury, hemorrhage and shock, myocardial ischemia, cerebral brain ischemia-reperfusion injury, hypoxia, and cancer. Several natural products have been used for the management of oxidative stress and related disease. Curcumin has caught scientific attention as a potential therapeutic agent due to its bifunctional antioxidant properties. It has been reported that curcumin directly reacts with reactive species and induces an upregulation of various cytoprotective and antioxidant proteins. It is able to scavenge hydroxyl radicals (·OH), superoxide anions (O_2·), hydrogen peroxide (H_2O_2), nitric oxide, singlet oxygen, peroxynitrite, and peroxyl radicals (ROO·) [83]. The presence of phenolic groups in the structure of curcumin might explain its ability to react with reactive oxygen species (ROS) and reactive nitrogen species (RNS).

16.6.1 Curcumin Regulates Oxidant Enzymes

The imbalance of oxidation and reduction in tissues during different steps of chronic inflammation contributes to a variety of human cancers [84]. Oxidants may lead to tumor initiation or progression due to overproduction of ROS by the member of NADPH oxidase (NOX) family of protein [85]. However, other potential sources of ROS such as the mitochondrial electron transport chain, the cytochrome P450 system, xanthine oxidase, myeloperoxidase, and uncoupled nitric oxidase synthase can also make substantial contributions to the oxidative atmosphere of the tumor microenvironment [86, 87]. Although curcumin is a powerful antioxidant, it has been shown that it can produce prooxidant effect at higher doses. At higher doses the phenolic hydroxyl group of curcumin and transition metal such as Cu (II) form a complex through chelation. After intermolecular electron transfer reaction, the complex leads to the formation of semiquinone, superoxide and hydroxyl radicals, respectively [88]. Moreover, the activity of thioredoxin reductase can be modified dose dependently by curcumin, leading to oxidation of NADPH by oxygen and subsequently formation of ROS [89].

16.6.2 Curcumin Regulates Antioxidant Enzymes

There are several enzymes expressed in normal and cancer cells that contribute to the production and degradation of ROS. The interaction and regulation of some of these enzymes by curcumin are as follows.

16.6.3 Superoxide Dismutase

The superoxide dismutases (SOD) are the major enzymes that catalyze the dismutation of the superoxide radicals into molecular oxygen or hydrogen peroxide. There are three different isoforms of SOD that have been identified, namely, cytosolic copper-/zinc-containing SOD (SOD-1 or CuZnSOD), mitochondrial manganese-containing SOD (SOD-2 or MnSOD), and the extracellular SOD (SOD-3 or ecSOD). SOD-3 is also a copper-/zinc-containing enzyme, mainly produced and secreted by vascular smooth muscle cells and also reported to bind with glycosaminoglycans present in vascular extracellular matrix on the cell surface of endothelial cells [90]. Single-charged anions such as fluoride and azide are fixed by all types of SOD, but they are distinct in their susceptibilities of Fe (I), Mn (I), or CuZnSODs [91, 92].

The decreased activity of antioxidant enzymes during oxidative stress can result in human cancers. The first line of defense against superoxide radicals is provided by SODs which have been studied in human cancers. The activity of CuZnSOD is found significantly lower in gastric adenocarcinoma samples than normal mucosa [93]. Similarly, the activity of MnSOD is often dysregulated during cancer development. In human esophageal cancers, the decreased MnSOD is associated with increased incidence of esophageal adenocarcinoma [94]. In human oral cancers, a high-level expression of MnSOD was associated with better disease-specific survival, especially in early-stage buccal mucosal squamous cell carcinoma and for patients with moderate or poor cell differentiation [95]. Current studies suggest that ecSOD may play important role in human lung carcinoma. In lung carcinomas the expression level of ecSOD was found downregulated which is the result of ecSOD promoter methylation and loss of heterozygosity [96]. In hepatocellular carcinoma, a virus-mediated tumor suppressor MnSOD gene was delivered which resulted in overexpression of MnSOD and induced cytotoxicity in hepatocellular carcinoma cells [97]. The overexpression of MnSOD has shown radioprotection as well as chemoprotection against antioxidant-based therapeutics [98], while in colorectal cancer, the increased mitochondrial DNA instability and elevated MnSOD expression were also observed [99]. Indeed, the overexpression of MnSOD in human cancer cell lines can cause increased H_2O_2 production and reduce tumor cell proliferation in the absence of anticancer agent [100].

16.6.4 Effect of Curcumin on SOD

A recent study has shown that curcumin oil preserves the expression of MnSOD in esophageal cancer cell and delayed the bile-induced esophageal injury and potential cancer progression in vitro [101]. A herbal medicine *Curcuma aromatica* has shown anticarcinogenic properties in a variety of cancer cell lines and animal models. In further studies, the esophagoduodenal anastomosis (EDA) rats were treated with *Curcuma aromatica* oil and the enzymatic activity and protein level of MnSOD found similar to the control. Hence, curcuma aromatica oil induces MnSOD expression and activity, which is associated with potential protection from incidence of esophageal adenocarcinoma [102]. In human breast epithelial cells which is transformed by the effect of radiation in the presence of estrogen, curcumin enhances the MnSOD protein and catalase expression that protect the cells from oxidative stress damages [103]. Das et al. reported that the expression of SOD in liver of Dalton's lymphoma (DL) and DL+DMSO mice was significantly downregulated compared to normal. In DL and DL+DMSO mice, the expression and activity of both MnSOD and CuZn-SOD were found significantly downregulated. The activity of MnSOD in DL and DL+DMSO mice was 65% and 63% than normal mice, respectively. After treatment of curcumin with different doses such as 50–150 mg/kg body weight, MnSOD activity increased in a dose-dependent manner as up to 1.3–1.6-fold of DL+DMSO mice, respectively. Similarly, the activity of CuZn-SOD was found 61% decreased of normal mice. After treatment with curcumin, the activity of CuZn-SOD increases in a similar dose-dependent manner, up to 1.2–1.5-fold of DL+DMSO mice with 50–150 mg/kg body weight, respectively [104]. Hence, curcumin has potential to elevate the expression and activity of SOD in DL mice.

16.6.5 Glutathione Peroxidase and Catalase

Glutathione peroxidase (GPx) is a family of antioxidant enzymes that catalyze the reduction of H_2O_2 or organic peroxides such as lipid peroxide to water or corresponding alcohols typically using glutathione as a reductant [105]. Glutathione peroxidase has eight variants, out of which four GPx1-4 that have selenoproteins in their catalytic center are found in all mammalian cells. GPx6 is a selenoprotein found in human but not in other mammals [106, 107]. The presence of selenocysteine residue in each of the four identical subunits was suggested to ensure a quick reaction with the hydroperoxide and fast glutathione reduction [108]. GPx have an antioxidant function at different cellular compartments; GPx1 is most abundant in cytosol and mitochondria, GPx2 in gastrointestinal epithelium, and GPx3 abundant in plasma. GPx4 has several isoforms in vertebrates and have varied location and specificity. GPx5 containing cysteine, instead of selenocysteine in its active center, is a protein found in the epididymis [109]. GPx6 is expressed in olfactory epithelium [109]. GPx7 and GPx8 are cysteine GPx with low GPx activity.

Catalase is an antioxidant enzyme which is mainly located in the peroxisome of mammalian cells and, to some extent, in the cytosol. It has two enzymatic activities that depend on the concentration of H_2O_2. It removes H_2O_2 by forming water and molecular oxygen when concentration of H_2O_2 is high; however, at lower concentration of H_2O_2 and in the presence of suitable hydrogen donor (such as phenol, methanol, ethanol, and others), it acts paradoxically by removing H_2O_2 but oxidizing its substrate [110]. By removing H_2O_2, it indirectly detoxifies superoxide radicals which were turned into H_2O_2 by SOD.

It is reported that H_2O_2 induces apoptosis in vitro via activation of caspase cascade in a wide range of cancer cells [111]. Other reports have described a lower level of catalase expression in a large variety of tumor and cancer cell lines compared to normal cells [112, 113]. In healthy cells, i.e., transformed with T-antigen of SV-40 or Ras, catalase was found downregulated, but the underlying mechanism of this downregulation is still unknown [114, 115]. Cancer cell lines resistant to chemotherapeutic agents or hydrogen peroxide, the expression level of catalase was found to be altered [116, 117].

16.6.6 Effect of Curcumin on Glutathione Peroxidase and Catalase

During initial phase of metastasis, expression of various molecules is activated by NF-κB in both liver cells and tumor cells through H_2O_2-mediated pathways. The removal of H_2O_2 is considered to be an effective approach to inhibit metastasis [118]. The antioxidant enzyme GPx and catalase play important role in detoxification of H_2O_2. GPx also detoxify the lipid hydroperoxides using glutathione [119]. Catalase expression is generally suppressed in elevated ROS condition mainly by H_2O_2 at the site of cancer microenvironment and oxygen free radicals. Das et al. reported the low activity of catalase and GPx (GPx1, GPx2, GPx4, and Gpx5) in the liver of cancerous mouse compared to normal mouse, which confirms the depleted expression of antioxidant defense system. They also reported that GPx3 was detected as a faint and thin activity band only in DL- and DMSO-treated DL mice liver. However, in curcumin-treated DL mice and normal mice, these isoforms could not be detected. They also reported that curcumin shifted the activities of all isoforms of GPx and catalase toward normal. Curcumin upregulates the expression of Gpx1 in the liver of lymphoma-bearing mice. Further, the expression of catalase was also found to be upregulated in curcumin-treated DL mice. Thus, curcumin recovers the defected antioxidant defense system by upregulating the activity and expression of antioxidant enzymes. Moreover, curcumin treatment downregulates the p65/p50 and p50/p50 complexes in lymphoma-bearing mice. Also, curcumin treatment decreased the binding of both subunits to DNA in DL mice indicating that curcumin downregulates the NF-κB activation. The overexpression of GPx1 in curcumin-treated DL cancerous mouse was associated with inhibition of NF-κB activation *in vivo*. Hence, curcumin inhibits H_2O_2 signaling-mediated NF-κB activation in the liver of DL-bearing mice by upregulating the expression and activity of endogenous antioxidant enzymes GPx and catalase as well as by decreasing the oxidative status [120].

In a similar study, it was shown that curcumin reduced the DNA damage which was induced by inorganic arsenic (As III) in Swiss albino mice liver cells. Moreover, the As III exposure caused depletion of antioxidant enzyme such as catalase, SOD, GPx, GR and GST, and antioxidants like GSH were significantly increased by the treatment of curcumin in Swiss albino mice liver cells. In conclusion, this study shows the efficacy of curcumin in preventing DNA damage and quenching of ROS induced by As III in Swiss albino mice [121].

16.7 Anti-angiogenic Properties of Curcumin

Angiogenesis is a process of new blood capillary formation from the preexisting blood vessels. Angiogenesis is crucial for the tumor growth and expansion. Tumor cells import nutrients and oxygen for their uncontrolled growth through the tumor angiogenesis process. Tumor cells have capability to form new blood vessel through constitutive expression of proangiogenic factors such as the vascular endothelial growth factor (VEGF) and basic fibroblast growth factor (bFGF). Angiogenesis is strictly controlled by anti-angiogenic molecules in mammals. There are several reports suggesting curcumin has the ability to modulate the tumor angiogenesis process by inducing anti-angiogenic stimulators [122].

Curcumin downregulates the expression of VEGF isomers such as VEGF-A, VEGF-C, and VEGFR2 and decreases microvessel density in breast cancer cell-implanted mice, which leads to inhibition of breast tumor growth. Rodrigues et al. reported that overexpression of osteopontin (OPN), a chemokine-like ECM-associated small integrin-binding ligand N-linked glycoprotein, is one of the major markers of cancer progression in several cancers including breast cancer [123, 124]. Moreover, the targeting of OPN and its downstream signaling pathways has shown promising therapeutic efficacy in various cancers including breast cancer [125]. It is also demonstrated that OPN enhances VEGF expression and hence promotes VEGF-dependent angiogenesis in breast cancer [126]. Philip et al. reported that curcumin suppressed OPN-induced tumor growth in nude mice, by suppressing the pro-MMP2 expression. They demonstrated that curcumin potentially suppressed the OPN-induced cell migration, tumor growth, and NF-κB-mediated pro-MMP2 activation by blocking the IKK/IκBα signaling pathways [127] Chakraborty et al. reported that curcumin inhibits the VEGF expression by abrogating the activity of NF-κB/ATF4 in OPN-induced MDA-MB231 implanted in nude mice. This result clearly suggests that curcumin acts as a potent anti-angiogenic agent in OPN-induced tumor angiogenesis in human breast cancer [128].

Another potent angiogenic molecule is bFGF which plays role in both new blood vessel formation and migration of cancer cells. It is reported that curcumin inhibits the neovascularization even in induced bFGF condition of mouse cornea. Moreover, in TPA-induced transformed human keratinocyte HaCaT cells, curcumin showed no effect on VEGF expression. These results suggest that curcumin acts as an angiogenesis inhibitor [129]. In contrary, Gururaj et al. reported that curcumin downregulates the VEGF mRNA and protein expression level in EAT cells [130]. They also reported the downregulation of Ang1 and Ang2 (newly identified angiogenic ligand) by curcumin in EAT cells.

Perry et al. reported curcumin can cross blood-brain barrier and inhibit angiogenesis in in vivo study. Thus curcumin decreased the gelatinolytic activities of MMP9. Furthermore, curcumin inhibited glioma-induced angiogenesis by decreasing endothelial cell marker such as CD31 and CD105 from newly formed vessels and also by decreasing hemoglobin content in glioma tumor [131]. These results indicate that curcumin may act as an inhibitor of angiogenesis in in vivo.

As reported earlier, the HGF/c-Met pathway promotes tumor angiogenesis and hence cancer progression. When HGF binds to its receptor, c-Met (expressed in endothelial cells) stimulates endothelial cell growth, invasion, and motility [132]. Inhibition of c-Met activity can impair the survival, invasion, and tubulogenesis of HUVECs in vitro and reduces neovascularization and formation of microvessel in tumor model [133, 134]. Jiao et al. reported curcumin suppressed HGF-induced phosphorylation of c-Met/AKT/mTOR in human lung cancer cells. Moreover, curcumin suppressed the EMT in c-Met-overexpressed A549 and PC-9 cells and also reduced the expression of vimentin and increased the expression of E-cadherin [135]. It is reported that HGF promotes angiogenesis by upregulating VEGF expression [136]. Their results show, curcumin significantly decreased the expression of VEGF and microvessel density.

In pancreatic cancer, curcumin has shown to downregulate the NF-κB-regulated gene products such as VEGF, MMP9, cyclin D, and IKKα/β, hence induce apoptosis, and reduce angiogenesis in in vitro and in vivo models [137].

Chen et al. reported that curcumin suppressed the secretion of VEGF in U937 cells and Raji cells that was upregulated by TNFα treatment. The mRNA expression of VEGF165 and VEGF121 were increased by TNFα and reduced by curcumin. They also reported that VEGF and TNFα can induce angiogenesis and curcumin can inhibit angiogenesis in ECV304 cells [138]. Overexpression of COX-2 and VEGF are reported in HepG2 cell line [139]. COX-2 can induce angiogenic factor and can directly take part in angiogenesis. Curcumin decreased the tumor-induced overexpression of VEGF and COX-2 via inhibiting NF-κB signaling pathway [140, 141]. In an in vivo study, Sabrina et al. reported that curcumin inhibited the angiogenesis and tumor growth in a heterotrophic mouse model of breast cancer by downregulating NF-κB-regulated genes such as cyclin D1, p65, and PECAM-1 [142].

Further, hypoxia which is normally detected in the central part of solid tumor can be an important cause of angiogenesis via activating the expression of angiogenic factors [143, 144]. Hypoxia-inducible factor (HIF) is a main factor of this process. HIF consisting HIF-1α and HIF-1β homodimers can regulate the transcription of hypoxia-activated genes. In normal oxygenic condition, α subunit of HIF-1 is rapidly degraded but stable under hypoxia while HIF-1β constitutively expressed [145]. It is reported that curcumin strongly reduces the expression and activity of HIF-1α but did not affect the mRNA of HIF-1α. This suggests that the curcumin reduced the stability of HIF-1α at protein level. To confirming this notion, proteasome inhibitor LLnL was added to the HepG2 cells and checked the protein level of HIF-1α. The treatment of LLnL restored the HIF-1α protein level which was decreased by curcumin. This result indicates curcumin-mediated degradation of HIF-1α is probably proteasome-dependent. Moreover, curcumin reduced the hypoxia-induced mRNA and protein expression of VEGF. These results suggest that curcumin reduces the transactivation function of HIF-1α. These results suggest that hypoxia-induced angiogenesis is inhibited by curcumin via downregulating the activity of HIF-1α [146].

It is reported that COX-2 and COX-derived prostaglandins, induced by VEGF, play important role in HIMEC angiogenesis [147]. Binion et al. reported that the pretreatment of curcumin inhibits the angiogenic effects of VEGF-induced COX-2 mRNA and protein expression as well as PGE2 production in HIMECs. Pretreatment of HIMECs with NS398, a COX-2 inhibitor, reduced the angiogenesis in vitro. These results support the idea that COX-2 and COX-derived prostaglandins play essential role in angiogenesis. They concluded that the inhibitory effect of curcumin could be due to the suppression of PGI2 through the inhibition of COX-2 expression [148]. Curcumin has also shown to inhibit the endothelial nitric oxide synthase (eNOS) expression in endothelial cells which contribute to impaired endothelial tube formation [149]. Taken together, these results suggest that curcumin inhibits angiogenesis by modulating multiple pathways in endothelial cells.

Similarly, Yoysungnoen et al. reported the tumor neocapillary density (NCD) which was significantly increased in HepG2 group of BALB/c nude mice and this increment of neocapillary density was characterized as a time-dependent manner. Moreover, the tissue perfusion on day basis also supported the increased NCD value. The increase NCD and tissue perfusion of HepG2-group was remarkably reduced by curcumin treatment. However, the upregulated COX-2 expression in HepG2 group was suppressed by curcumin treatment in a dose-dependent manner. Interestingly, the increased serum VEGF in HepG2 group was downregulated by the treatment of curcumin in a dose-dependent manner as well [140]. These results also suggest that curcumin could inhibit tumor angiogenesis through reduction of angiogenic biomarkers such as COX-2 and VEGF.

The effect of curcumin can influence the expression profile of micro-RNAs (miRs) in angiogenesis. Recently, miRs came in focus as a powerful mechanism in regulation of gene expression after transcription. In a study, the expression profile of miRs in HUVECs was extensively investigated, and most significantly dysexpressed miRs were selected for their supposed role in the anti-angiogenic effect of curcumin on HUVECs. miR-1246 has been reported to promote angiogenesis by activating Smad signaling in endothelial cells [150], and miR-1275 has been reported to contribute in tissue heterogeneity establishment via epigenetic mechanisms [151]. By bioinformatics means the targets of miR-1275 and miR-1246 were detected VEGF and NKAP, respectively. The NF-κB p65 activation was detected by inhibiting the miR-1246 inhibitor. It also upregulated NF-κB p65 level leading to increased expression of cyclin D1, MMP9, and VEGFB, which finally induced angiogenesis [152]. The anti-angiogenic effect of curcumin with regard to the function of miR-1246 was contrary to the previous report by Yamada et al. [150], in which miR-1246 promotes angiogenesis via activation of Smad pathways.

16.8 Anti-metastatic Properties of Curcumin

16.8.1 Anti-metastatic Effects of Curcumin on Activation of Nuclear Factor-Kappa B (NF-κB)

NF-κB is a transcription factor which regulates the group of genes that control cell proliferation and survival. Tumor invasion-related proteins such as MMPs, urokinase-type plasminogen activator (uPA), interleukin-8 (IL-8), and many other chemokines are regulated by NF-κB [153, 154]. NF-κB activation can be induced by many cancer promoter molecules such as tumor necrosis factor and hydrogen peroxide, and the most common is phorbol ester [155, 156]. Curcumin has the potential to inhibit this activation proved by several works. It is reported that curcumin inhibits the activation of NF-κB through phosphorylation and degradation of IκBα [157, 158]. In estrogen receptor (ER)-negative breast cancer cells and prostate cancer cells, curcumin has shown to inhibit the activation of p65, a member of NF-κB transcription factor family [159, 160]. In breast cancer cells, curcumin downregulates the NF-κB. This downregulation of NF-κB leads to downregulation of many cancer proliferative genes and induces apoptosis, hence inhibiting cancer cell invasion and metastasis [161].

16.8.2 Effect of Curcumin on Signal Transducer and Activator of Transcription 3 (STAT3)

STAT3 is a transcription factor that is constitutively expressed/activated in cancer cells and regulates the cell survival, cell cycle, angiogenesis, and metastasis [162]. In particular, the relation of activated STAT3 with metastasis has been reported in many cancer tumors. Curcumin inhibited the SCLC cell invasion through downregulation of the expression of STAT3-regulated gene products involved in the invasion (e.g., MMP2 and MMP7). Curcumin also downregulates VEGF which explains its anti-metastatic activities. The role of JAK has been concerned in activation of STAT3; curcumin suppressed the phosphorylation of JAK-1, JAK-2, and JAK-3 (upstream proteins of STAT3) in SCLC cells [163]. Curcumin blocked the activation loop of JAK and suppresses the interleukin-6 (IL-6) cytokine signaling pathway and, hence, blocks the signaling required for phosphorylation and activation of STAT3 [163]. As curcumin significantly suppressed the STAT3- and STAT-regulated genes, hence, curcumin treatment reduces the rate of metastasis in SCLC cells. Curcumin also inhibits IL-6-induced phosphorylation of STAT3 in myeloma cells [156, 164] as shown in Fig. 16.6.

16.8.3 Effect of Curcumin on Inflammatory Cytokines

In many cases, cancer development and progression are influenced by chronic inflammation. Inflammatory cytokines have the ability to enhance cancer progression and development by providing growth factors and attracting the mononuclear cells to the cancer site. Angiogenesis is an important process in the cancer progression; the inflammatory cytokines such as CXCL1 and CXCL2 were found to be involved in this process [165]. CXCL1 has been shown to promote migration of breast cancer cells in vitro [166]. It has been identified that CXCL1 and CXCL2 are the major targets of curcumin as anticancer agents in breast cancer cells. In this study, curcumin has shown to downregulate the mRNA and protein expression of

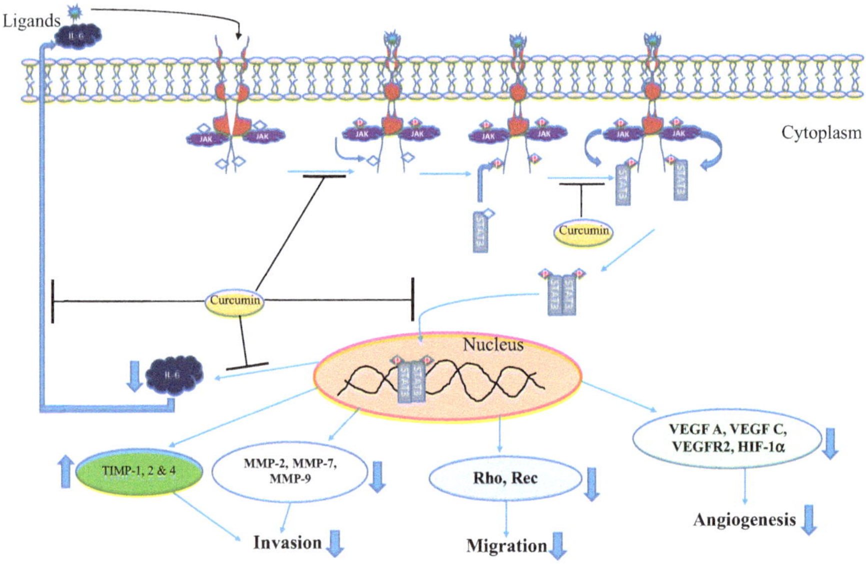

Fig. 16.6 Curcumin inhibits invasion, migration, and angiogenesis. Curcumin blocks the activation loop of JAK and suppresses the interleukin-6 (IL-6)-mediated cytokine signaling pathway and blocks phosphorylation and activation of STAT3. STAT3 modulates transcription of many genes involved in various processes leading to invasion, migration, and angiogenesis

CXCL1 and CXCL2 in a metastatic breast cancer cell line MDA-MB-231. In another study, Helbig et al. has shown that the expression of chemokine receptor CXCR4 is directly induced by NF-κB to promote migration and metastasis in breast cancer cells [167]. Hence, curcumin reduces breast cancer metastasis through decreasing NF-κB-mediated expression of two inflammatory cytokines, namely, CXCL1 and CXCL2, and also reducing the chemotactic receptor CXCR4. It has been reported that the promoter regions of both CXCL1 and CXCL2 genes are identical and in proximity and contain a perfect binding site for NF-κB; hence, curcumin downregulates these inflammatory cytokines by targeting NF-κB in metastatic prostate cancer cells [158].

16.8.4 Effect of Curcumin on Urokinase Plasminogen Activator (uPA) Expression

Cancer cells release a serine-specific protease uPA, which after binding to its receptor uPAR, activates plasmin; this plasmin then mediates the degradation of extracellular matrix (ECM) [168]. The excessive degradation of ECM from primary tumor is a crucial step for cancer invasion and metastasis. uPA secretion is regulated by NF-κB. In breast cancer cells, curcumin significantly reduced the uPA expression and NF-κB DNA binding activity. In colon cancer cell line, curcumin decreased the expression of uPA and MMP9 through NF-κB inhibition. And also the NF-κB inhibition is mediated by activation of AMPK (AMP-activated protein kinase). Moreover, authors have concluded that curcumin suppressed the binding of p65 NF-κB to the uPA and MMP9 promoter region [169]. Based on above results, it can be concluded that curcumin inhibits the invasion and adhesion of breast and colon cancer cells through downregulating the expression of uPA and MMP9 via inhibiting NF-κB activation.

16.8.5 Effect of Curcumin on the Expression of Matrix Metalloproteinase (MMPs)

MMPs are calcium-dependent zinc-containing endopeptidases, capable of degrading all kind of extracellular matrix (ECM) proteins [168]. They play a major role in cell proliferation, migration, angiogenesis, and apoptosis. The expression of MMPs is finely controlled and regulated by tissue inhibitors of metalloproteinase (TIMPs) during normal physiological conditions. However, the overexpression of MMPs is the result of disturbance in balance between MMPs and TIMPs, which further results in various pathogenic processes including tumor invasion and metastasis [170, 171]. Most of the human cancer cells overexpress MMPs, which is related to metastasis. The two main MMPs which degrade the major components of basement membrane are MMP2 and 9 [172]. Hence, these two MMPs are the main therapeutic target for anti-metastatic drugs.

In human lung cancer cells (A549), curcumin inhibited the MMP2 and 9 and VEGF in a dose-dependent manner and hence inhibited the invasion and migration [173]. The anti-invasive properties of curcumin are reported in human lung cancer 801D cells in vitro and in vivo in a mouse model. In this study, curcumin inhibited the Rac-dependent signaling pathways and downregulated the expression of MMP2 and 9, hence inhibiting tumor invasion and migration in both in vivo and in vitro [174]. Overexpression of Ras homolog gene family member A (RhoA) induces the expression of MMPs, which finally promotes the invasion of cancer cells [175]. Breast cancer cells have poor prognosis due to its invasion and migration. Lysophosphatidic acid (LPA) is an inducer of cancer cell invasion and metastasis by activation of Rho/ROCK/MMP signaling pathway in breast cancer. Curcumin inhibits LPA-induced invasion by attenuating Rho/ROCK/MMPs pathway in breast cancer cells [175].

12-O-Tetradecanoylphorbol-13 acetate (TPA) is a potent tumor promoter. TPA promotes tumor invasion of breast cancer cells by inducing MMP9 via protein kinase C pathway [176, 177]. Curcumin has shown inhibition of TPA-induced MMP9 expression and, hence, cell invasion through downregulation of NF-κB and AP1 activation. TPA-induced phosphorylation of p38 and JNK and translocation of PKCα from cytosol to membrane are also strongly inhibited by curcumin [176].

The anti-invasive property of curcumin is also demonstrated in ER-negative breast cancer cell line (MDA-MB231) in a dose-dependent manner. Curcumin has shown to downregulate MMP2 and upregulate TIMP1 and 2 [178]. Additionally, curcumin was shown to inhibit the expression of VEGF and basic fibroblast growth factor in ER-negative breast cancer MDA-MB231 cells [178]. In another study, curcumin was shown to downregulate the expression of MMP2 and 9 and upregulate the expression of TIMP1 and 4 in ER-negative cell lines in a dose- and time-dependent manner. Based on the above findings, it is concluded that curcumin significantly inhibits MMP9 activity by downregulating its expression and upregulating TIMP1 and 4 gene expression in breast cancer cells.

16.8.6 Effect of Curcumin on Epithelial to Mesenchymal Transition (EMT)

Epithelial to mesenchymal transition (EMT) is an important event in the metastatic cascade where the cells acquire migratory and invasive capabilities [179]. The effect of curcumin on EMT and related gene in breast cancer cells has been evaluated. For this study Marcela Gallardo and Gloriam M calaf [180] used triple-positive and triple-negative breast cancer cell lines for ER, PgR, and Erb-3 in relation to EMT process. Curcumin decreased the expression of EMT markers, namely, E-cadherin, Slug, AXL, and Twist1, in MCF-10F (triple-negative) while N-cadherin, β-catenin, Slug, AXL, Twist1, vimentin, fibronectin, ZEB2, EZH2, and STAT3 in tumor 2 (triple-positive) and E-cadherin, N-cadherin, Twist1, AXL, and fibronectin in MDA-MB231 (triple-negative) cells compared to its counterpart. Since MDA-MB231 cells after treatment with curcumin decreased the expression of EMT gene and induced morphological changes, hence, curcumin triggered EMT in MDA-MB231 cells. In this study curcumin inhibited Slug expression which affects E-cadherin and vimentin resulting in decreased cancer cell invasion and metastasis [180].

It is reported earlier that LPS (lipopolysaccharide) mediated the EMT in breast cancer cells by modifying NF-κB signaling [181]. Hunag et al. reported that curcumin inhibits the LPS-induced EMT in breast cancer cells. They have also suggested the ability of curcumin to inhibit tumor invasion is associated with inactivation of NF-κB-Snail signaling and regulate the expression of the important downstream EMT marker, E-cadherin (upregulation) and vimentin (downregulation) [182]. Similarly Chen et al. reported that curcumin suppresses doxorubicin-induced EMT via inhibiting the TGF-β and PI3K/AKT signaling pathways in a triple-negative (ER, PR, and HER2-negative) breast cancer cells [183].

16.9 Synergistic Effect of Curcumin with Other Chemopreventive Agents, Chemotherapeutic Agents, and Radiation Therapy

The rise of cancer burden is huge worldwide, where 14.1 million new cases with 8.2 million deaths occurred in the year 2012 [184]. Despite depending on available cancer treatments including radiotherapy, surgery, cytotoxic therapy, and hormonal and targeted immunotherapies, still there is need of effective treatment modalities to achieve a better cure [185]. Cancer is a complex process operated via variable multiple signaling networks in the cells, which is very difficult to target by any single drug used presently. Thus the success rates of these therapies are not so encouraging and bring secondary complications in patients during and after treatments [186]. Cancer cells developed various escaping methods such as increasing drug efflux pumps, quicker drug metabolism, increasing self-repair pathways, and expressing newer or modified drug targets against the available conventional treatments [187]. Therefore, the use of multidrug phenomena in different combinations could be able to suppress most of the survival mechanisms raised by cancers and may enhance the efficacy. In clinics, combinatorial treatment approach has been adopted to maximize the overall effect on cancer therapeutics [187, 188]. Employing combinatorial therapy will target multiple pathways which will make difficult to survival of cancer cells. Thus the combinatorial therapy may solve the shortcomings of cancer therapeutics. Still this approach is not very easy to apply because of maintenance of proper doses and their unknown interaction and solubility. However, overall, the combinatorial treatment strategy will be a better application than the other conventional treatments if proper combination and safety measures will be taken during the application [189].

Despite curcumin's superior antitumor properties, still its use is restricted because of poor bioavailability, water insolubility, short-life span, rapid systemic clearance, and quick metabolism in the gastrointestinal tract [190]. The reason behind the insolubility in water is due to its molecular structure [(1E,6E)-(1,7-bis(4-hydroxy-3-methoxyphenyl)-1,6-heptadiene-3,5-dione)] which is a keto form in acidic, neutral, and solid conditions. The pKa value which lies at 8.54 makes it difficult to dissolve in aqueous solution and therefore reduces bioavailability [191]. Therefore, to overcome these problems, curcumin was used in combination with chemopreventive agents and radiotherapy. In this section, we have summarized the use and effect of various combinatorial agents along with curcumin for cancer treatments.

16.9.1 Curcumin Used in Combination with Other Chemopreventive Agents

Chemoprevention concept in recent years is found to occupy one of the major places in cancer treatment due to its potential anticancer activity with lesser side effects [192]. Many of them (COX-2 inhibitors, green and black tea derivatives, β-carotene, vitamins, isothiocyanates, and farnesyl transferase inhibitors) have been used as common food stuff or in ancient medicine among different populations worldwide [193].

Presently, we are aware about the chemopreventive potential of curcumin and its limitations against various cancers through several mechanisms. So, in order to improve its anticancer efficacy to a much significant level, combination of other known chemopreventive agents with curcumin is required. Here, we are going to address such combinations used against cancers with their effective anticancer effects.

16.9.1.1 Curcumin-Phospho-sulindac

There were various reports showing the cancer prevention and metastasis by the use of NSAIDs (Nonsteroidal anti-inflammatory drugs), but it was associated with renal and gastrointestinal toxicities. Considering this problem, one group of scientists developed a novel phospho form of NSAIDs, for example, phospho-sulindac (PS, OXT-328), with inhibitory action against breast and colon cancer. PS alone was unable to prevent human lung cancer xenografts but strongly inhibited when curcumin in micellar nanoparticulate formulation was orally supplied to mice prior to PS. This combination showed a promising drug combination which synergistically suppressed tumor growth and improved the pharmacokinetics of PS. Thus, it was observed that curcumin may play a role in enhancing cellular uptake and bioavailability of PS in lung cancer [194].

16.9.1.2 Curcumin-Aspirin-Sulforaphane

In this study, the combination of curcumin, aspirin, and sulforaphane was encapsulated with the help of solid lipid nanoparticle, and the effect was observed on pancreatic cancer cells Panc-1s and MIA PaCa-2. This encapsulation was used to improve the sustained release of drug for longer duration. Synergistic effect was observed both in MTS cell viability and apoptosis studies where the significance rate was comparatively higher than the treatments alone. Thus, curcumin successfully played a major role in reducing pancreatic cancer progression synergistically in the presence of aspirin and sulforaphane [195].

16.9.1.3 Curcumin-Leflunomide-Perindopril

Here, curcumin was combined with two anti-angiogenic drugs leflunomide and perindopril where the combinations were used to target angiogenic pathways in mice, developed with diethylnitrosamine (DEN)-induced hepatocellular carcinoma (HCC). The effect was observed on the expression of CD31-positive endothelial cells and hepatic microvessel density in mice with HCC as compared to the normal liver. The elevated level of hepatic VEGF and HIF-1α in the DEN-treated mice was synergistically abrogated in the presence of the combinations, as compared to the singular treatments of leflunomide and perindopril. Curcumin potentiated the anti-angiogenic activity of leflunomide and perindopril in HCC mice model and improved chemoprevention [196].

16.9.1.4 Curcumin-Catechins

In this study, curcumin and green tea catechin that have anticancer activities against wide varieties of cancers were used together as combinatorial treatment approach against colon cancer. This combination significantly reduced the number of aberrant crypt foci per rat; however, there was no change in foci numbers with curcumin and catechins when treated separately. The colon cancer proliferation index was synergistically inhibited to a significantly higher level in combinatorial treatment as compared to other treatment groups. This showed the improvised efficiency in the chemopreventive value of catechins in the presence of curcumin in colon carcinogenesis [197].

16.9.1.5 Curcumin-(−)-Epigallocatechin-3-gallate (EGCG)

In another combinatorial treatment, EGCG- combined with curcumin wastested in human oral epithelial cells (normal, premalignant, and malignant). Here, the benefits obtained after combining these two agents were the reduction of 4.4–8.5-fold dose of EGCG and 2.2–2.8-fold in case of curcumin. Apart from this, they showed synergistic inhibitory effect on cell proliferation and cell cycle arrest and proved curcumin as a significant partner as a chemopreventive agent [198].

16.9.1.6 Curcumin-Silymarin

Two popular anticancer phytochemicals, silymarin and curcumin, when used in combination to treat colon cancer, synergistic effects were found on cell proliferation inhibition and increased apoptosis with membrane blebbing. The caspase 3/7 activity was almost doubled from threefold to fivefold when combinations were employed compared to curcumin or silymarin treatment alone in colon cancer. Thus, curcumin was found to sensitize colon cancer cells to silymarin treatment and to improve chemopreventive status in colon cancer [199].

16.9.1.7 Curcumin Used in Combination with Other Chemotherapeutic Agents

Chemotherapy used presently has mild to severe toxicity; however, these agents are still in use as major treatments for cancer. To improve the quality of cancer treatment with lesser side effects, it is now used in combination with other treatments such as surgery, radiotherapy, etc. In combination, chemopreventive agents are emerging candidates in cancer therapy that are well suited to support chemotherapeutic agents in an efficient manner at present [200]. Recent studies such as DEN-induced HCC inhibition along with various reports with dose- and time-dependent inhibition of stomach, duodenal, esophageal, colon, and oral cancer support curcumin to be a very good chemopreventive and chemotherapeutic agent against cancers [201]. To further improve the efficacy and to reduce the side effects of chemotherapeutic agents used during cancer treatment, curcumin was used in combinations with doxorubicin, fenretinide, temozolomide, carboplatin, paclitaxel, resveratrol, etc. We have discussed some of the combinatorial studies which showed to be promising aspect in future anticancer treatment regimens.

16.9.1.8 Curcumin-Doxorubicin

Chemosensitizers are small molecules having significant role in chemotherapy where MDR developed by cancer cells gets reversed and enhances the chemotherapeutic potentials of drugs in vitro. But at the same time, chemosensitizers are not much successful candidates with in vivo cancer studies because MDR response was not always effectively reduced. To minimize the limitations and improve the chemosensitizer response of two chemotherapeutic agents, curcumin and doxorubicin were used together with the help of nanotechnology. This strategy was found to be effectively causing sensitization of tumors with proper loading, co-delivery, and co-release of chemotherapeutic agents which solve the problems of conventional chemosensitizers. Subsequently, curcumin augmented the cytotoxicity of doxorubicin against MCF-7/ADR cells. Thus, the combinatorial approach enhanced the antitumor efficacy of these chemotherapeutic agents against MDR cancers [202]. In another study, these drug combinations successfully achieved the synergistic anticancer activity through chemosensitization of MDR tumors and DEN-induced HCC tumor model [203].

16.9.1.9 Curcumin-Fenretinide

Curcumin and fenretinide in combinatorial are extensively used as chemotherapeutic agents against non-small cell lung cancer and have showed significant elevated cytotoxicity in vitro as well as in vivo. This is associated with synergistic overexpression on pro-apoptotic molecule cleaved PARP and also regulates glucose-regulated protein 78 (GRP78), chaperon protein induced by curcumin of endoplasmic reticulum. Therefore, curcumin and fenretinide combination was found to be kpromising in the treatment of lung cancer [204].

16.9.1.10 Curcumin-Temozolomide

Temozolomide is one of the most important DNA alkylating agents that has been used as a chemotherapeutic drug in glioblastoma treatment. However, its use in this cancer has been limited due to the development of resistance. The role of curcumin as a chemosensitizer in various cancers encouraged the combination of these drugs against glioblastomas. It was found that curcumin potentiates the apoptosis effect of temozolomide to a synergistically higher level than the apoptosis with any particular treatments alone. This is based on the pronounced combinatorial effect by ROS generation and p-Akt/mTOR pathway inhibition which leads to an optimized therapy in glioblastoma treatment [205].

16.9.1.11 Curcumin-Carboplatin

Carboplatin is a commonly used chemotherapeutic agent against non-small cell lung cancer (NSCLC), but its efficacy is retarded by the survival signaling pathways along with drug resistance cell development in tumors. Here, the use of curcumin acts as a chemosensitizer when used with carboplatin and produces improved efficacy against lung cancer cells. The synergism of these compounds was found to be efficient against various events of cancer such as on cell proliferation, apoptosis, migration, and invasion in NSCLC cell line, A549. This study again provides support that curcumin can be a strong candidate in the field of cancer treatment with carboplatin [206].

16.9.2 Curcumin Used in Combination with Radiotherapy

Ionizing radiation either from external or internal sources is one of the three major pillars of cancer therapy including surgery and chemotherapy [207]. The cytotoxicity obtained through radiotherapy is an effective approach to treat cancers but the limitations starts with severe side effects. This is very challenging to deliver a maximum dose to the malignant cells without affecting the normal healthy cells surrounding them, which consequently reduces the amount of radiation. Another major limitation with radiotherapy is the development of resistance, i.e., radioresistance, similarly like chemoresistance during chemotherapy [208]. The above major limitations of radiotherapy can be broadly avoidable by considering two objectives such as radiosensitizers and radioprotectors [209]. Curcumin has both these properties established with its significant contributions when combined with radiation therapy against various tumor cells [210]. Here, we have discussed some of the studies where radiotherapy is used in combination with curcumin with maximizing tumor killing efficiency along with minimizing side effects.

16.9.2.1 Curcumin as a Radio-sensitizer

The role of radio-sensitizers is to reduce the irradiation doses which will lower the chance of affecting nearby healthy cells and at the same time will enhance cytotoxicity specifically on the tumors [211]. Curcumin in combination with radiation was found to target receptor of growth factor signaling, cell cycle, and angiogenic and epigenetic modulation pathways in glioblastoma cells, thereby synergistically inducing cell death through apoptosis as well as autophagy. This treatment also hinders the chemo- and radio-resistance phenomena adapted by glioblastoma cells and makes them sensitize toward death [209]. Radio-sensitization of curcumin was also found in nasopharyngeal carcinoma where it was reducing the expression of multidrug resistance gene 1 (MDR1) by increasing miR 593. This radio-sensitization was obtained both in vitro and in vivo studies where 100 mg/kg of curcumin was combined with 4 Gy irradiation [212]. In in vitro study on hepatocellular carcinoma; when treated with curcumin and radiation, cell proliferation, NF-B activity, and metastasis-related pathway were inhibited significantly [213]. In another study, the dual nature of curcumin and trans-resveratrol was found in the form of radioprotector on non-cycling lymphocytes, whereas radio-sensitizer on cycling lymphocytes [214]. These are some of the evidences which support the role of curcumin as a radio-sensitizer in the form of an adjunct small molecule to improve the main stream radiation therapy in cancer treatment.

16.9.2.2 Curcumin as a Radio-protector

Cancer radiation therapy will boost up not only by radiosensitizers but also needs support from radioprotectors. As we have seen in the previous section, curcumin was showing its radio-sensitizing effects on various cancers; it also possess radio-protecting action on non-tumorigenic organs [208]. This section will comprise some studies showing the radio-protective role of curcumin. An initial in vivo study has been performed on mice with head and neck cancer where the mice were categorized into four groups on the basis of treatments before irradiation of 20 Gy. The groups containing curcumin- and lycopene-treated mice had reduced damage significantly on parotid glands as compared to the DMSO or control mice with irradiation [215]. Curcumin analog diarylpentanoid curcumin was found to be a very good radioprotective agent from gamma radiation to the intestinal mucosa of mice [216]. Curcumin and trans-resveratrol are observed to be radio-protective antioxidant polyphenols with protection for non-cycling lymphocytes [212]. There are very few reports available on radio-protective role of curcumin. For the beginning, the above mentioned evidences are sufficient to proceed further for the evaluation of radio-protective role of curcumin and associated mechanisms in various cancer models.

16.10 Curcumin in Clinical Trials

Turmeric which is a readily available source of vital curcumin is a popular spice, food coloring, and flavoring agent in several Asian cuisines. Dose of curcumin estimated in people consuming high levels of turmeric in their diet is nearly about 150 mg/day [217, 218] which is approximately 20 times lower than the recommended. In a short-term human trial of 3–4 months, suggested concentrations of 3600 mg/day induced low level of toxicity and is safe for future cancer chemoprevention clinical trials [219–221]. Inhibitory effect of curcumin on carcinogenesis has been evident in hundreds of scientific research on several animal models of various tumor types of different origin including mammary carcinoma, oral cancer, intestinal tumors, leukemia, prostate, breast, pancreatic, colorectal, cervical and colon cancers. Here in this section, we are going to discuss some information on these clinical trial studies. Numerous preclinical trials, in vitro studies and animal model testing have verified curcumin as a chemopreventive agent which intervene the array of cell signal transduction pathways and molecular mechanisms involved in cancer progression. Safety, tolerability, bioavailability, and mechanism of action of curcumin against multiple human chronic diseases have been claimed through comprehensive clinical trials over the past quarter century. These pre-clinical and clinical trials suggest low bioavailability of curcumin in patients following oral application and readily elimination from body. The key factors which are primarily considered to design cancer chemoprevention clinical trials are (i) the doses at which curcumin should be supplemented, (ii) the selection of the participants for the trials, and (iii) the duration of the trials and respective follow-ups.

16.10.1 Prostate Cancer

A pilot-scaled phase II was designed after successfully completing phase I to assess the efficacy of docetaxel/curcumin in patients with chemotherapy-naive metastatic castration-resistant prostate cancer (CRPC). The study was conducted within a group of 30 patients with progressing CRPC and a rising prostate-specific antigen (PSA) who received docetaxel/prednisone in standard conditions for 6 cycles in combination with orally supplied curcumin, 6000 mg/day (day –4 to day +2 of docetaxel). The result revealed that 26 patients received the scheduled treatment, 2 progressed and 2 died before the end of treatment. Fifty-nine percent of patients responded toward PSA within first three cycles (88% of responders), 40% patient partially responded, and 20 patients were 100% curcumin compliant. The study produced a high response rate, good tolerability, and patient acceptability for randomized trial [222].

16.10.2 Cervical Cancer

A phase II randomized four intervention arm controlled study was performed to evaluate the efficacy of Basant polyherbal vaginal cream that contained extracts of curcumin, reetha, amla, and aloe vera and of curcumin vaginal capsules to eliminate HPV infection from the cervix. Screening of women was done by Pap smear and HPV DNA test. HPV-positive women without high-grade cervical neoplasias (N=287) were treated with vaginal Basant cream, vaginal placebo cream, curcumin vaginal capsules, and placebo vaginal capsules, respectively. The assigned formulation was applied once per day for 1 month except during menstruation and recalled within a week of the last application for repeat HPV test, cytology and colposcopy. HPV clearance rate which was 87.7% in Basant arm was significantly higher than 73.3% of combined placebo arms. Rate of clearance in curcumin application was 81.3% than placebo. Mild to moderate vaginal itching and irritation was slightly higher after Basant application [223].

16.10.3 Breast Cancer

Research on open-label phase I dose trial in 14 patients with advanced and metastatic breast cancer was conducted in order to find maximal tolerated dose of the combination of dose-escalating curcumin and docetaxel. Six cycles of 100 mg/m^2 docetaxel were administered, and curcumin was orally given from 500 mg/d for seven consecutive days by cycle. Three dose-limiting toxicities were observed at the last dose level of curcumin. Two out of three patients at this dose level refused to continue treatment, leading maximal tolerated dose of curcumin at 8000 mg/d.

Eight patients out of 14 had measurable lesions according to RECIST criteria, with 5 PR and 3 SD. The recommended dose of curcumin was 6000 mg/d for seven consecutive days every 3 weeks in combination [224].

16.10.4 Colon Cancer

A pilot-scale study was performed to quantify levels of curcuminoids in colorectal mucosa on 26 patients undergoing colorectal endoscopy or surgical resection and to obtain information on the acceptability and compliance. Patients were treated with curcumin C3 complex (2.35 g) once daily for 14 days prior to endoscopic biopsy or colonic resection. Curcuminoid levels were analyzed in plasma, urine, and colonic mucosa by UPLC-UV (ultra high performance liquid chromatography with UV detection) with characterization by tandem LC-MS/MS. Twenty-four patients commencing curcumin completed the course, out of which six patients reported mild gastrointestinal adverse events. Curcuminoids were detectable in 9/24 plasma samples, in 24/24 urine samples, and in the colonic mucosa of all 23 biopsied participants. Mean tissue levels were found to be 48.4 μg/g (127.8 nmol/g) of parent curcuminoids. Curcumin glucuronide a major conjugate was detectable in 29/35 biopsies. High levels of topical curcumin persisted in the mucosa for up to 40 h postadministration. Sixteen participants agreed to take curcumin in long-term trials should it be of proven benefit. Pharmacologically active levels of curcumin were recovered from colonic mucosa [225].

16.10.5 Colorectal Cancer

In a study of herbal preparation, MB-6 that is composed of curcumin, fermented soybean extract, green tea extract, grape seed extract, *Antrodia camphorata* mycelia and *Spirulina* was evaluated for its effectiveness of leucovorin/5-fluorouracil chemotherapy in patients with colorectal cancer. MB-6, in combination with leucovorin/5-fluorouracil chemotherapy supplied to colon cancer tumor-bearing BALB/c mice, showed increased survival rate and life span compared to chemotherapy alone. In a proof of concept, 16-week randomized clinical study was executed on 72 patients with metastatic colorectal cancer that were treated with leucovorin, 5-fluorouracil and oxaliplatin in combination with either MB-6 or placebo. The follow-up taken after 77 weeks of treatment showed no significant difference in the best overall response rate (primary outcome) and overall survival (secondary outcome) was observed between the two groups. Patients in the MB-6 group had a significantly lower disease progression rate (secondary outcome) than patients in the placebo group. The placebo group had a significantly higher incidence of adverse events (secondary outcome) at least grade 4 compared with the MB-6 group and a significantly higher occurrence of increased serum creatinine compared with the MB-6 group. The result revealed herbal formulation MB-6 as a potential adjunct to chemotherapy in patients with metastatic colorectal cancer [226].

In another study of dose-escalation pilot clinical trial of advanced colorectal human cancer, a proprietary capsule of curcumin with dosage between 440 and 2200 mg/day, containing 36–180 mg of curcumin, was tested on 15 patients for 4 months. Glutathione S-transferase activity showed 59% decrease in blood samples after 440 mg curcuma extract ingestion for 29 days. The effect was not observed for higher-dose concentration. Leukocytic DNA adduct malondialdehyde levels were constant within each patient and unaffected by treatment. For 2–4 months of treatment, disease was found radiologically stable in five patients. The result suggested safe administration of *Curcuma* extract up to 2.2 g daily equivalent to 180 mg of curcumin. Investigation on intestinal metabolism revealed low oral bioavailability of curcumin [227]. Another similar study on the above designed trial 0.45 and 3.6 g of curcumin was orally administered by patients. Three biomarkers of curcumin activity suggested in preclinical trials that are glutathione S-transferase activity, levels of DNA adduct malondialdehyde and inhibition of prostaglandin E2 (PGE2) production induced ex vivo. Chromatography analysis detected curcumin and its glucuronide and sulfate metabolites in plasma and urine. A dose of 3.6 g/day curcumin caused 62% and 57% decreases in inducible PGE2 production in blood samples taken 1 h after dose on days 1 and 29, respectively. A daily oral dose of 3.6 g of curcumin is recommended for phase II trials in the prevention or treatment of cancers ex vivo [227].

16.10.6 Pancreatic Cancer

A clinical trial was evaluated for biological effects of curcumin in advanced pancreatic cancer. In a study of 21 patients, 8 g curcumin was administered orally until disease progression with restaging every 2 months. Serum cytokine levels for interleukin (IL)-6, IL-8, IL-10 and IL-1 receptor antagonists and peripheral blood mononuclear cell expression of curcumin nuclear factor NF-κB and cyclooxygenase-2 were monitored. Curcumin was detectable in glucuronide and sulfate forms at low steady-state levels that suggest poor oral bioavailability. Two patients showed clinical biological activity. One patient had ongoing stable disease for more than 18 months and one patient depicted 73% tumor regression accompanied by serum cytokine levels. Curcumin downregulated expression of NF-κB, COX-2 and phosphorylated STAT3 in blood from patients. There were interpatient variation in plasma curcumin levels and drug levels (22–41 ng/mL) remained constant over the first 4 weeks. Oral curcumin was well tolerated and had limited absorption with biological activity in some patients with pancreatic cancer [228].

In a clinical trial on gemcitabine resistant pancreatic cancer patients, curcumin was studied in combination with gemcitabine-based chemotherapy. No dose-limiting toxicities were observed in the phase I study and oral curcumin 8 g per day was selected as the recommended dose for the phase II study. No dose-limiting toxicities, no intolerability, and 100% compliance rate were observed. Patient's survival time after initiation of curcumin was 161 days and annual survival rate was 19%. Evaluation of phase II resulted in safe and feasible uptake of suggested curcumin dose [229].

In a similar kind of another clinical trial on advanced pancreatic cancer, 17 patients received 8 g curcumin per day along with gemcitabine 1000 mg/m(2) IV weekly × 3 of 4 weeks. Five patients discontinued the trial because of intractable abdominal fullness or pain and in two patients dose was reduced to 4 g/day due to abdominal complaints. Nine percent of patient showed partial response, 36% of patient had stable disease condition, and 55% of patient had tumor progression. The study suggested to prevent the use of high doses of oral curcumin needed to achieve systemic effect [230].

16.10.7 Leukemia

A study was designed to evaluate nitric oxide clearance activity of imatinib along with curcumin in 50 patients of chronic myeloid leukemia. The first group received imatinib alone and the second group received turmeric powder with imatinib for 6 weeks. Significant decrease in nitric oxide level was noted in curcumin-aided imatinib group as compared to imatinib alone and thus demonstrated curcumin as an adjuvant to imatinib which could help in the treatment of chronic myeloid leukemia patients [231] (Table 16.1).

16.11 Conclusions

Currently, the alternative and less toxic chemotherapeutic and chemopreventive agents are essentially desired due to lack of effective cancer therapy. Multiple molecular pathways such as NF-κB activation, Nrf2, NFAT, p53, MMP-mediated metastasis and VEGF-mediated angiogenesis are deregulated in various cancer types and represent potential therapeutic targets. Curcumin has shown to target many mechanisms deregulated in various malignancies. Many encouraging results have been obtained employing curcumin against various cancer models in vitro and in vivo. However, certain limiting biophysical properties such as bioavailability, water insolubility, short life span, rapid systemic clearance and quick metabolism in the gastrointestinal tract have limited the use of curcumin as an effective chemotherapeutics. In order to enhance its anticancer efficacy to a significant level, combination of other known chemotherapeutics and chemopreventive agents with curcumin have shown positive results. Thus, curcumin has cancer-preventive

Table 16.1 List of different cancer clinical trials involving curcumin combination with chemotherapeutic drugs

Type of cancer	Clinical trial	Drug combination(s)	Number of cases	References
Castration-resistant prostate cancer	Phase II	Docetaxel, prednisone, and curcumin	30	[222]
Colorectal cancer	Phase I and phase IIa	Curcumin (C3 complex, Sabinsa) including FOLFOX-based (5-fluorouracil, folinic acid, and oxaliplatin)	33	[232]
Colorectal liver metastases (CRLM)	Phase I	Curcumin, oxaliplatin/5-fluorouracil	12	[233]
Colon cancer	Phase II	MB-6 composition containing curcumin extract with oxaliplatin/5-fluorouracil	72	[226]
Cervical neoplasia	Phase II	Curcumin and curcumin-containing polyherbal cream	287	[223]
Colorectal, gastric, and breast cancer	Phase II	Curcuminoids	80	[234]
Breast cancer	Randomized, double-blind, placebo-controlled clinical trial	Curcumin	30	[235]
Colon cancer	Phase I and II	Curcumin	26	[225]
Pancreatic cancer	Phase I	Highly bioavailable curcumin (Theracurmin)	16	[236]
Monoclonal gammopathy of undetermined significance (MGUS) and smoldering multiple myeloma (SMM)	Randomized, double-blind placebo-controlled crossover study	Curcumin	36	[237]
Chronic myeloid leukemia (CML)	Phase I and phase II	Curcumin as an adjuvant to imatinib	50	[231]
Colorectal neoplasia	Phase IIA	Curcumin	41	[238]
Advanced pancreatic cancer	Phase II	Gemcitabine concurrent with curcumin	17	[230]
Gemcitabine-resistant pancreatic cancer	Phase I/II	Gemcitabine-based chemotherapy plus curcumin	21	[229]
p53-upregulated colorectal cancer	Randomized clinical trials	Curcumin	126	[52]

(continued)

Table 16.1 (continued)

Type of cancer	Clinical trial	Drug combination(s)	Number of cases	References
Advanced or metastatic breast cancer	Phase I	Combination of docetaxel/curcumin	14	[224]
Prostate cancer	Randomized placebo-controlled double-blind study	Isoflavones and curcumin	85	[239]
Advanced pancreatic cancer	Phase II	Curcumin	25	[228]
Monoclonal gammopathy of undetermined significance (MGUS)	Randomized placebo-controlled double-blind study	Curcumin	26	[240]
Colorectal carcinoma	Phase I	Oral curcumin	15	[221]

potential and can be developed as an effective chemotherapeutic agent in coming years which would provide less toxic and cheap medical alternatives for the management of cancer.

References

1. Kunnumakkara AB et al (2016) Curcumin, the golden nutraceutical: multitargeting for multiple chronic diseases. Br J Pharmacol 174:1325–1348
2. Prasad S, Tyagi AK, Aggarwal BB (2014) Recent developments in delivery, bioavailability, absorption and metabolism of curcumin: the golden pigment from golden spice. Cancer Res Treat 46(1):2–18
3. Goel A, Kunnumakkara AB, Aggarwal BB (2008) Curcumin as "Curecumin": from kitchen to clinic. Biochem Pharmacol 75(4):787–809
4. Gupta SC et al (2012) Discovery of curcumin, a component of golden spice, and its miraculous biological activities. Clin Exp Pharmacol Physiol 39(3):283–299
5. Rainey N et al (2015) Curcumin hormesis mediates a cross-talk between autophagy and cell death. Cell Death Dis 6:e2003
6. Aggarwal BB, Gupta SC, Sung B (2013) Curcumin: an orally bioavailable blocker of TNF and other pro-inflammatory biomarkers. Br J Pharmacol 169(8):1672–1692
7. Siwak DR et al (2005) Curcumin-induced antiproliferative and proapoptotic effects in melanoma cells are associated with suppression of IkappaB kinase and nuclear factor kappaB activity and are independent of the B-Raf/mitogen-activated/extracellular signal-regulated protein kinase pathway and the Akt pathway. Cancer 104(4):879–890
8. Aggarwal S et al (2004) Inhibition of growth and survival of human head and neck squamous cell carcinoma cells by curcumin via modulation of nuclear factor-kappaB signaling. Int J Cancer 111(5):679–692
9. LoTempio MM et al (2005) Curcumin suppresses growth of head and neck squamous cell carcinoma. Clin Cancer Res 11(19 Pt 1):6994–7002
10. Wang Z et al (2008) Synergistic effects of multiple natural products in pancreatic cancer cells. Life Sci 83(7-8):293–300

11. Elattar TM, Virji AS (2000) The inhibitory effect of curcumin, genistein, quercetin and cisplatin on the growth of oral cancer cells in vitro. Anticancer Res 20(3a):1733–1738
12. Mukhopadhyay A et al (2001) Curcumin downregulates cell survival mechanisms in human prostate cancer cell lines. Oncogene 20(52):7597–7609
13. Mehta K et al (1997) Antiproliferative effect of curcumin (diferuloylmethane) against human breast tumor cell lines. Anti-Cancer Drugs 8(5):470–481
14. Hanif R et al (1997) Curcumin, a natural plant phenolic food additive, inhibits cell proliferation and induces cell cycle changes in colon adenocarcinoma cell lines by a prostaglandin-independent pathway. J Lab Clin Med 130(6):576–584
15. Lin YG et al (2007) Curcumin inhibits tumor growth and angiogenesis in ovarian carcinoma by targeting the nuclear factor-kappaB pathway. Clin Cancer Res 13(11):3423–3430
16. Hussain AR et al (2008) Curcumin suppresses constitutive activation of nuclear factor-kappa B and requires functional Bax to induce apoptosis in Burkitt's lymphoma cell lines. Mol Cancer Ther 7(10):3318–3329
17. Shakibaei M et al (2014) Curcumin chemosensitizes 5-fluorouracil resistant MMR-deficient human colon cancer cells in high density cultures. PLoS One 9(1):e85397
18. Waghela BN et al (2015) Curcumin conjugated with PLGA potentiates sustainability, anti-proliferative activity and apoptosis in human colon carcinoma cells. PLoS One 10(2):e0117526
19. Chearwae W et al (2004) Biochemical mechanism of modulation of human P-glycoprotein (ABCB1) by curcumin I, II, and III purified from Turmeric powder. Biochem Pharmacol 68(10):2043–2052
20. Singh M, Singh N (2009) Molecular mechanism of curcumin induced cytotoxicity in human cervical carcinoma cells. Mol Cell Biochem 325(1-2):107–119
21. Halder RC et al (2015) Curcuminoids and omega-3 fatty acids with anti-oxidants potentiate cytotoxicity of natural killer cells against pancreatic ductal adenocarcinoma cells and inhibit interferon gamma production. Front Physiol 6:129
22. Fiala M (2015) Curcumin and omega-3 fatty acids enhance NK cell-induced apoptosis of pancreatic cancer cells but curcumin inhibits interferon-gamma production: benefits of omega-3 with curcumin against cancer. Molecules 20(2):3020–3026
23. Imran M et al (2016) Cucurmin; anticancer and antitumor perspectives – a comprehensive review. Crit Rev Food Sci Nutr 58:1271–1293
24. Labbozzetta M et al (2009) Curcumin as a possible lead compound against hormone-independent, multidrug-resistant breast cancer. Ann N Y Acad Sci 1155:278–283
25. Ramachandran C, You W (1999) Differential sensitivity of human mammary epithelial and breast carcinoma cell lines to curcumin. Breast Cancer Res Treat 54(3):269–278
26. Onoda M, Inano H (2000) Effect of curcumin on the production of nitric oxide by cultured rat mammary gland. Nitric Oxide 4(5):505–515
27. Hayden MS, Ghosh S (2012) NF-κB, the first quarter-century: remarkable progress and outstanding questions. Genes Dev 26(3):203–234
28. Karin M (2009) NF-κB as a critical link between inflammation and cancer. Cold Spring Harb Perspect Biol 1(5):a000141
29. Karin M et al (2002) NF-kappaB in cancer: from innocent bystander to major culprit. Nat Rev Cancer 2(4):301–310
30. Karin M (1999) How NF-kappaB is activated: the role of the IkappaB kinase (IKK) complex. Oncogene 18(49):6867–6874
31. Dhawan P, Richmond A (2002) A novel NF-kappa B-inducing kinase-MAPK signaling pathway up-regulates NF-kappa B activity in melanoma cells. J Biol Chem 277(10):7920–7928
32. Marin YE et al (2007) Curcumin downregulates the constitutive activity of NF-kappaB and induces apoptosis in novel mouse melanoma cells. Melanoma Res 17(5):274–283
33. Collett GP, Campbell FC (2004) Curcumin induces c-jun N-terminal kinase-dependent apoptosis in HCT116 human colon cancer cells. Carcinogenesis 25(11):2183–2189
34. Sun Z, Andersson R (2002) NF-kappaB activation and inhibition: a review. Shock 18(2):99–106

35. Collett GP, Campbell FC (2006) Overexpression of p65/RelA potentiates curcumin-induced apoptosis in HCT116 human colon cancer cells. Carcinogenesis 27(6):1285–1291
36. Baeuerle PA (1991) The inducible transcription activator NF-kappa B: regulation by distinct protein subunits. Biochim Biophys Acta 1072(1):63–80
37. Naumann M et al (1997) Neisseria gonorrhoeae epithelial cell interaction leads to the activation of the transcription factors nuclear factor κB and activator protein 1 and the induction of inflammatory cytokines. J Exp Med 186(2):247–258
38. Singh S, Aggarwal BB (1995) Activation of transcription factor NF-kappa B is suppressed by curcumin (diferuloylmethane) [corrected]. J Biol Chem 270(42):24995–25000
39. Finco TS, Beg AA, Baldwin AS (1994) Inducible phosphorylation of I kappa B alpha is not sufficient for its dissociation from NF-kappa B and is inhibited by protease inhibitors. Proc Natl Acad Sci U S A 91(25):11884–11888
40. Shostak K, Chariot A (2011) NF-κB, stem cells and breast cancer: the links get stronger. Breast Cancer Res: BCR 13(4):214–214
41. Luedde T, Schwabe RF (2011) NF-kappaB in the liver–linking injury, fibrosis and hepatocellular carcinoma. Nat Rev Gastroenterol Hepatol 8(2):108–118
42. Marquardt JU et al (2015) Curcumin effectively inhibits oncogenic NF-kappaB signaling and restrains stemness features in liver cancer. J Hepatol 63(3):661–669
43. Anest V et al (2003) A nucleosomal function for IkappaB kinase-alpha in NF-kappaB-dependent gene expression. Nature 423(6940):659–663
44. Kuo HP et al (2013) Epigenetic roles of MLL oncoproteins are dependent on NF-kappaB. Cancer Cell 24(4):423–437
45. Shaw PH (1996) The role of p53 in cell cycle regulation. Pathol Res Pract 192(7):669–675
46. Amaral JD et al (2010) The role of p53 in apoptosis. Discov Med 9(45):145–152
47. Rivlin N et al (2011) Mutations in the p53 tumor suppressor gene: important milestones at the various steps of tumorigenesis. Genes Cancer 2(4):466–474
48. Meek DW (1999) Mechanisms of switching on p53: a role for covalent modification? Oncogene 18(53):7666–7675
49. Fridman JS, Lowe SW (2003) Control of apoptosis by p53. Oncogene 22(56):9030–9040
50. Choudhuri T et al (2002) Curcumin induces apoptosis in human breast cancer cells through p53-dependent Bax induction. FEBS Lett 512(1-3):334–340
51. Choudhuri T et al (2005) Curcumin selectively induces apoptosis in deregulated cyclin D1-expressed cells at G2 phase of cell cycle in a p53-dependent manner. J Biol Chem 280(20):20059–20068
52. He ZY et al (2011) Upregulation of p53 expression in patients with colorectal cancer by administration of curcumin. Cancer Investig 29(3):208–213
53. Ye M et al (2015) Curcumin promotes apoptosis by activating the p53-miR-192-5p/215-XIAP pathway in non-small cell lung cancer. Cancer Lett 357(1):196–205
54. Watson JL et al (2010) Curcumin causes superoxide anion production and p53-independent apoptosis in human colon cancer cells. Cancer Lett 297(1):1–8
55. Bush JA, Cheung KJ Jr, Li G (2001) Curcumin induces apoptosis in human melanoma cells through a Fas receptor/caspase-8 pathway independent of p53. Exp Cell Res 271(2):305–314
56. Han SS et al (1999) Curcumin causes the growth arrest and apoptosis of B cell lymphoma by downregulation of egr-1, c-myc, bcl-XL, NF-kappa B, and p53. Clin Immunol 93(2):152–161
57. Ramakrishnan SK et al (2014) Expression of targeted ribozyme against telomerase RNA causes altered expression of several other genes in tumor cells. Tumour Biol 35(6):5539–5550
58. Kensler TW, Wakabayashi N, Biswal S (2007) Cell survival responses to environmental stresses via the Keap1-Nrf2-ARE pathway. Annu Rev Pharmacol Toxicol 47:89–116
59. Chen B et al (2014) Curcumin inhibits proliferation of breast cancer cells through Nrf2-mediated down-regulation of Fen1 expression. J Steroid Biochem Mol Biol 143:11–18
60. Balogun E et al (2003) Curcumin activates the haem oxygenase-1 gene via regulation of Nrf2 and the antioxidant-responsive element. Biochem J 371(Pt 3):887–895
61. Jain A et al (2015) Curcumin inhibits PhIP induced cytotoxicity in breast epithelial cells through multiple molecular targets. Cancer Lett 365(1):122–131

62. Shaw JP et al (1988) Identification of a putative regulator of early T cell activation genes. Science 241:202–205. J Immunol, 2010. 185(9): p. 4972–4975
63. Durand DB et al (1988) Characterization of antigen receptor response elements within the interleukin-2 enhancer. Mol Cell Biol 8(4):1715–1724
64. Shukla U et al (2009) Tyrosine phosphorylation of 3BP2 regulates B cell receptor-mediated activation of NFAT. J Biol Chem 284(49):33719–33728
65. Zanoni I et al (2009) CD14 regulates the dendritic cell life cycle after LPS exposure through NFAT activation. Nature 460(7252):264–268
66. Ranger AM et al (2000) The nuclear factor of activated T cells (NFAT) transcription factor NFATp (NFATc2) is a repressor of chondrogenesis. J Exp Med 191(1):9–22
67. Ho IC et al (1998) A potential role for the nuclear factor of activated T cells family of transcriptional regulatory proteins in adipogenesis. Proc Natl Acad Sci U S A 95(26):15537–15541
68. Buchholz M et al (2006) Overexpression of c-myc in pancreatic cancer caused by ectopic activation of NFATc1 and the Ca2+/calcineurin signaling pathway. EMBO J 25(15):3714–3724
69. Jauliac S et al (2002) The role of NFAT transcription factors in integrin-mediated carcinoma invasion. Nat Cell Biol 4(7):540–544
70. de la Pompa JL et al (1998) Role of the NF-ATc transcription factor in morphogenesis of cardiac valves and septum. Nature 392(6672):182–186
71. Viola JP et al (2005) NFAT transcription factors: from cell cycle to tumor development. Braz J Med Biol Res 38(3):335–344
72. Baksh S, DeCaprio JA, Burakoff SJ (2000) Calcineurin regulation of the mammalian G0/G1 checkpoint element, cyclin dependent kinase 4. Oncogene 19(24):2820–2827
73. Hernandez GL et al (2001) Selective inhibition of vascular endothelial growth factor-mediated angiogenesis by cyclosporin A: roles of the nuclear factor of activated T cells and cyclooxygenase 2. J Exp Med 193(5):607–620
74. Macian F (2005) NFAT proteins: key regulators of T-cell development and function. Nat Rev Immunol 5(6):472–484
75. Schulz RA, Yutzey KE (2004) Calcineurin signaling and NFAT activation in cardiovascular and skeletal muscle development. Dev Biol 266(1):1–16
76. Mognol GP et al (2016) Cell cycle and apoptosis regulation by NFAT transcription factors: new roles for an old player. Cell Death Dis 7:e2199
77. Neuhofer W (2010) Role of NFAT5 in inflammatory disorders associated with osmotic stress. Curr Genomics 11(8):584–590
78. Chueh SC et al (2003) Curcumin enhances the immunosuppressive activity of cyclosporine in rat cardiac allografts and in mixed lymphocyte reactions. Transplant Proc 35(4):1603–1605
79. Ranjan D et al (2004) Curcumin inhibits mitogen stimulated lymphocyte proliferation, NFkappaB activation, and IL-2 signaling. J Surg Res 121(2):171–177
80. Kliem C et al (2012) Curcumin suppresses T cell activation by blocking Ca2+ mobilization and nuclear factor of activated T cells (NFAT) activation. J Biol Chem 287(13):10200–10209
81. Hernandez M, Wicz S, Corral RS (2016) Cardioprotective actions of curcumin on the pathogenic NFAT/COX-2/prostaglandin E2 pathway induced during Trypanosoma cruzi infection. Phytomedicine 23(12):1392–1400
82. Hai L et al (2011) Counteracting effect of TRPC1-associated Ca2+ influx on TNF-alpha-induced COX-2-dependent prostaglandin E2 production in human colonic myofibroblasts. Am J Physiol Gastrointest Liver Physiol 301(2):G356–G367
83. Sai Krishna Borra PG, Mahendra J, Jayamathi KM, Cherian C, chand NaR (2013) Antioxidant and free radical scavenging activity of curcumin determined by using different in vitro and ex vivo models. J Med Plants Res 7(36):2680–2690
84. Wu Y et al (2014) Molecular mechanisms underlying chronic inflammation-associated cancers. Cancer Lett 345(2):164–173
85. Hussain SP, Hofseth LJ, Harris CC (2003) Radical causes of cancer. Nat Rev Cancer 3(4):276–285
86. Hamanaka RB, Chandel NS (2011) Cell biology. Warburg effect and redox balance. Science 334(6060):1219–1220

87. Giannoni E, Parri M, Chiarugi P (2012) EMT and oxidative stress: a bidirectional interplay affecting tumor malignancy. Antioxid Redox Signal 16(11):1248–1263
88. Chattopadhyay D et al (2014) Dichotomous effect of caffeine, curcumin, and naringenin on genomic DNA of normal and diabetic subjects. Scientifica (Cairo) 2014:649261
89. Cai W et al (2012) Curcumin targeting the thioredoxin system elevates oxidative stress in HeLa cells. Toxicol Appl Pharmacol 262(3):341–348
90. Wassmann S, Wassmann K, Nickenig G (2004) Modulation of oxidant and antioxidant enzyme expression and function in vascular cells. Hypertension 44(4):381–386
91. Leone M et al (1998) Fourier transform infrared analysis of the interaction of azide with the active site of oxidized and reduced bovine Cu, Zn superoxide dismutase. Biochemistry 37(13):4459–4464
92. Vance CK, Miller AF (1998) Spectroscopic comparisons of the pH dependencies of Fe-substituted (Mn)superoxide dismutase and Fe-superoxide dismutase. Biochemistry 37(16):5518–5527
93. Monari M et al (2006) Superoxide dismutase in gastric adenocarcinoma: is it a clinical biomarker in the development of cancer? Biomarkers 11(6):574–584
94. Martin RC et al (2007) Chemoprevention of carcinogenic progression to esophageal adenocarcinoma by the manganese superoxide dismutase supplementation. Clin Cancer Res 13(17):5176–5182
95. Fu TY et al (2011) Manganese superoxide dismutase and glutathione peroxidase as prognostic markers in patients with buccal mucosal squamous cell carcinomas. Head Neck 33(11):1606–1615
96. Teoh-Fitzgerald ML et al (2012) Genetic and epigenetic inactivation of extracellular superoxide dismutase promotes an invasive phenotype in human lung cancer by disrupting ECM homeostasis. Mol Cancer Res 10(1):40–51
97. Huang F et al (2014) Targeting gene-virus-mediated manganese superoxide dismutase effectively suppresses tumor growth in hepatocellular carcinoma in vitro and in vivo. Cancer Biother Radiopharm 29(10):403–411
98. Tarhini AA et al (2011) A phase I study of concurrent chemotherapy (paclitaxel and carboplatin) and thoracic radiotherapy with swallowed manganese superoxide dismutase plasmid liposome protection in patients with locally advanced stage III non-small-cell lung cancer. Hum Gene Ther 22(3):336–342
99. Govatati S et al (2016) Manganese-superoxide dismutase (Mn-SOD) overexpression is a common event in colorectal cancers with mitochondrial microsatellite instability. Tumour Biol 37(8):10357–10364
100. Zhong W et al (1996) Inhibition of cell growth and sensitization to oxidative damage by overexpression of manganese superoxide dismutase in rat glioma cells. Cell Growth Differ 7(9):1175–1186
101. Schiffman SC, Li Y, Martin RC (2012) The association of manganese superoxide dismutase expression in Barrett's esophageal progression with MnTBAP and curcumin oil therapy. J Surg Res 176(2):535–541
102. Li Y et al (2009) Chemoprotective effects of Curcuma aromatica on esophageal carcinogenesis. Ann Surg Oncol 16(2):515–523
103. Calaf GM et al (2011) Protective role of curcumin in oxidative stress of breast cells. Oncol Rep 26(4):1029–1035
104. Das L, Vinayak M (2014) Long term effect of curcumin in regulation of glycolytic pathway and angiogenesis via modulation of stress activated genes in prevention of cancer. PLoS One 9(6):e99583
105. Margis R et al (2008) Glutathione peroxidase family – an evolutionary overview. FEBS J 275(15):3959–3970
106. Kryukov GV et al (2003) Characterization of mammalian selenoproteomes. Science 300(5624):1439–1443
107. Brigelius-Flohe R, Maiorino M (2013) Glutathione peroxidases. Biochim Biophys Acta 1830(5):3289–3303

108. Tappel AL (1978) Glutathione peroxidase and hydroperoxides. Methods Enzymol 52:506–513
109. Ghyselinck NB, Dufaure JP (1990) A mouse cDNA sequence for epididymal androgen-regulated proteins related to glutathione peroxidase. Nucleic Acids Res 18(23):7144
110. Scibior D, Czeczot H (2006) Catalase: structure, properties, functions. Postepy Hig Med Dosw (Online) 60:170–180
111. Fang J, Nakamura H, Iyer AK (2007) Tumor-targeted induction of oxystress for cancer therapy. J Drug Target 15(7-8):475–486
112. Sato K et al (1992) Negative regulation of catalase gene expression in hepatoma cells. Mol Cell Biol 12(6):2525–2533
113. Kwei KA et al (2004) Transcriptional repression of catalase in mouse skin tumor progression. Neoplasia 6(5):440–448
114. Sun Y, Colburn NH, Oberley LW (1993) Depression of catalase gene expression after immortalization and transformation of mouse liver cells. Carcinogenesis 14(8):1505–1510
115. Hoffschir F et al (1993) Decrease in catalase activity and loss of the 11p chromosome arm in the course of SV40 transformation of human fibroblasts. Carcinogenesis 14(8):1569–1572
116. Xu H et al (2005) Concentration-dependent collateral sensitivity of cisplatin-resistant gastric cancer cell sublines. Biochem Biophys Res Commun 328(2):618–622
117. Lee HC et al (2004) Increased expression of antioxidant enzymes in radioresistant variant from U251 human glioblastoma cell line. Int J Mol Med 13(6):883–887
118. Kobayashi Y et al (2008) Hydrogen peroxide-mediated nuclear factor kappaB activation in both liver and tumor cells during initial stages of hepatic metastasis. Cancer Sci 99(8):1546–1552
119. Vina J et al (1996) Exercise causes blood glutathione oxidation in chronic obstructive pulmonary disease: prevention by O_2 therapy. J Appl Physiol (1985) 81(5):2198–2202
120. Das L, Vinayak M (2012) Anti-carcinogenic action of curcumin by activation of antioxidant defence system and inhibition of NF-kappaB signalling in lymphoma-bearing mice. Biosci Rep 32(2):161–170
121. Biswas J et al (2010) Indian spice curcumin may be an effective strategy to combat the genotoxicity of arsenic in Swiss albino mice. Asian Pac J Cancer Prev 11(1):239–247
122. Safe S, Kasiappan R (2016) Natural products as mechanism-based anticancer agents: Sp transcription factors as targets. Phytother Res 30(11):1723–1732
123. Rodrigues LR et al (2007) The role of osteopontin in tumor progression and metastasis in breast cancer. Cancer Epidemiol Biomark Prev 16(6):1087–1097
124. Kasi PD et al (2016) Molecular targets of curcumin for cancer therapy: an updated review. Tumour Biol 37(10):13017–13028
125. Jain S et al (2007) Osteopontin: an emerging therapeutic target for anticancer therapy. Expert Opin Ther Targets 11(1):81–90
126. Chakraborty G, Jain S, Kundu GC (2008) Osteopontin promotes vascular endothelial growth factor-dependent breast tumor growth and angiogenesis via autocrine and paracrine mechanisms. Cancer Res 68(1):152–161
127. Philip S, Kundu GC (2003) Osteopontin induces nuclear factor kappa B-mediated promatrix metalloproteinase-2 activation through I kappa B alpha/IKK signaling pathways, and curcumin (diferulolylmethane) down-regulates these pathways. J Biol Chem 278(16):14487–14497
128. Chakraborty G et al (2008) Curcumin suppresses breast tumor angiogenesis by abrogating osteopontin-induced VEGF expression. Mol Med Rep 1(5):641–646
129. Arbiser JL et al (1998) Curcumin is an in vivo inhibitor of angiogenesis. Mol Med 4(6):376–383
130. Gururaj AE et al (2002) Molecular mechanisms of anti-angiogenic effect of curcumin. Biochem Biophys Res Commun 297(4):934–942
131. Perry MC et al (2010) Curcumin inhibits tumor growth and angiogenesis in glioblastoma xenografts. Mol Nutr Food Res 54(8):1192–1201
132. Ponzo MG et al (2009) Met induces mammary tumors with diverse histologies and is associated with poor outcome and human basal breast cancer. Proc Natl Acad Sci U S A 106(31):12903–12908

133. Heideman DA et al (2004) Suppression of tumor growth, invasion and angiogenesis of human gastric cancer by adenovirus-mediated expression of NK4. J Gene Med 6(3):317–327
134. Zou HY et al (2007) An orally available small-molecule inhibitor of c-Met, PF-2341066, exhibits cytoreductive antitumor efficacy through antiproliferative and antiangiogenic mechanisms. Cancer Res 67(9):4408–4417
135. Jiao D et al (2016) Curcumin inhibited HGF-induced EMT and angiogenesis through regulating c-Met dependent PI3K/Akt/mTOR signaling pathways in lung cancer. Mol Ther Oncolytics 3:16018
136. Zhang YW et al (2003) Hepatocyte growth factor/scatter factor mediates angiogenesis through positive VEGF and negative thrombospondin 1 regulation. Proc Natl Acad Sci U S A 100(22):12718–12723
137. Bimonte S et al (2013) Curcumin inhibits tumor growth and angiogenesis in an orthotopic mouse model of human pancreatic cancer. Biomed Res Int 2013:810423
138. Chen WH, Chen Y, Cui GH (2005) Effects of TNF-alpha and curcumin on the expression of VEGF in Raji and U937 cells and on angiogenesis in ECV304 cells. Chin Med J 118(24):2052–2057
139. Millanta F et al (2006) COX-2 expression in canine and feline invasive mammary carcinomas: correlation with clinicopathological features and prognostic molecular markers. Breast Cancer Res Treat 98(1):115–120
140. Yoysungnoen P et al (2006) Effects of curcumin on tumor angiogenesis and biomarkers, COX-2 and VEGF, in hepatocellular carcinoma cell-implanted nude mice. Clin Hemorheol Microcirc 34(1-2):109–115
141. Li L, Braiteh FS, Kurzrock R (2005) Liposome-encapsulated curcumin: in vitro and in vivo effects on proliferation, apoptosis, signaling, and angiogenesis. Cancer 104(6):1322–1331
142. Bimonte S et al (2015) Dissecting the role of curcumin in tumour growth and angiogenesis in mouse model of human breast cancer. Biomed Res Int 2015:878134
143. Bergers G, Benjamin LE (2003) Tumorigenesis and the angiogenic switch. Nat Rev Cancer 3(6):401–410
144. Vaupel P (2004) The role of hypoxia-induced factors in tumor progression. Oncologist 9(Suppl 5):10–17
145. Jeong JW et al (2002) Regulation and destabilization of HIF-1alpha by ARD1-mediated acetylation. Cell 111(5):709–720
146. Bae MK et al (2006) Curcumin inhibits hypoxia-induced angiogenesis via down-regulation of HIF-1. Oncol Rep 15(6):1557–1562
147. Ogawa H et al (2003) Sodium butyrate inhibits angiogenesis of human intestinal microvascular endothelial cells through COX-2 inhibition. FEBS Lett 554(1-2):88–94
148. Binion DG, Otterson MF, Rafiee P (2008) Curcumin inhibits VEGF-mediated angiogenesis in human intestinal microvascular endothelial cells through COX-2 and MAPK inhibition. Gut 57(11):1509–1517
149. Rossig L et al (2002) Inhibitors of histone deacetylation downregulate the expression of endothelial nitric oxide synthase and compromise endothelial cell function in vasorelaxation and angiogenesis. Circ Res 91(9):837–844
150. Yamada N et al (2014) Colorectal cancer cell-derived microvesicles containing microRNA-1246 promote angiogenesis by activating Smad 1/5/8 signaling elicited by PML down-regulation in endothelial cells. Biochim Biophys Acta 1839(11):1256–1272
151. Katsushima K et al (2012) Contribution of microRNA-1275 to Claudin11 protein suppression via a polycomb-mediated silencing mechanism in human glioma stem-like cells. J Biol Chem 287(33):27396–27406
152. Bai Y et al (2016) Curcumin inhibits angiogenesis by up-regulation of microRNA-1275 and microRNA-1246: a promising therapy for treatment of corneal neovascularization. Cell Prolif 49(6):751–762
153. Fan Y, Mao R, Yang J (2013) NF-kappaB and STAT3 signaling pathways collaboratively link inflammation to cancer. Protein Cell 4(3):176–185
154. Aggarwal BB (2004) Nuclear factor-kappaB: the enemy within. Cancer Cell 6(3):203–208

155. Shishodia S (2013) Molecular mechanisms of curcumin action: gene expression. Biofactors 39(1):37–55
156. Kunnumakkara AB, Anand P, Aggarwal BB (2008) Curcumin inhibits proliferation, invasion, angiogenesis and metastasis of different cancers through interaction with multiple cell signaling proteins. Cancer Lett 269(2):199–225
157. Bachmeier BE et al (2008) Curcumin downregulates the inflammatory cytokines CXCL1 and -2 in breast cancer cells via NFkappaB. Carcinogenesis 29(4):779–789
158. Killian PH et al (2012) Curcumin inhibits prostate cancer metastasis in vivo by targeting the inflammatory cytokines CXCL1 and -2. Carcinogenesis 33(12):2507–2519
159. Jacobs MD, Harrison SC (1998) Structure of an IkappaBalpha/NF-kappaB complex. Cell 95(6):749–758
160. Verma IM et al (1995) Rel/NF-kappa B/I kappa B family: intimate tales of association and dissociation. Genes Dev 9(22):2723–2735
161. Zong H et al (2012) Curcumin inhibits metastatic progression of breast cancer cell through suppression of urokinase-type plasminogen activator by NF-kappa B signaling pathways. Mol Biol Rep 39(4):4803–4808
162. Huang S (2007) Regulation of metastases by signal transducer and activator of transcription 3 signaling pathway: clinical implications. Clin Cancer Res 13(5):1362–1366
163. Yang CL et al (2012) Curcumin blocks small cell lung cancer cells migration, invasion, angiogenesis, cell cycle and neoplasia through Janus kinase-STAT3 signalling pathway. PLoS One 7(5):e37960
164. Bharti AC, Donato N, Aggarwal BB (2003) Curcumin (diferuloylmethane) inhibits constitutive and IL-6-inducible STAT3 phosphorylation in human multiple myeloma cells. J Immunol 171(7):3863–3871
165. Strieter RM et al (1995) The functional role of the ELR motif in CXC chemokine-mediated angiogenesis. J Biol Chem 270(45):27348–27357
166. Youngs SJ et al (1997) Chemokines induce migrational responses in human breast carcinoma cell lines. Int J Cancer 71(2):257–266
167. Helbig G et al (2003) NF-kappaB promotes breast cancer cell migration and metastasis by inducing the expression of the chemokine receptor CXCR4. J Biol Chem 278(24):21631–21638
168. Brooks SA et al (2010) Molecular interactions in cancer cell metastasis. Acta Histochem 112(1):3–25
169. Tong W et al (2016) Curcumin suppresses colon cancer cell invasion via AMPK-induced inhibition of NF–κB, uPA activator and MMP9. Oncol Lett 12(5):4139–4146
170. Chirco R et al (2006) Novel functions of TIMPs in cell signaling. Cancer Metastasis Rev 25(1):99–113
171. Mylona E et al (2006) Expression of tissue inhibitor of matrix metalloproteinases (TIMP)-3 protein in invasive breast carcinoma: relation to tumor phenotype and clinical outcome. Breast Cancer Res 8(5):R57
172. Baldwin AS (2001) Control of oncogenesis and cancer therapy resistance by the transcription factor NF-kappaB. J Clin Invest 107(3):241–246
173. Lin SS et al (2009) Curcumin inhibits the migration and invasion of human A549 lung cancer cells through the inhibition of matrix metalloproteinase-2 and -9 and Vascular Endothelial Growth Factor (VEGF). Cancer Lett 285(2):127–133
174. Chen QY et al (2014) Curcumin inhibits lung cancer cell migration and invasion through Rac1-dependent signaling pathway. J Nutr Biochem 25(2):177–185
175. Sun K et al (2016) Curcumin inhibits LPA-induced invasion by attenuating RhoA/ROCK/MMPs pathway in MCF7 breast cancer cells. Clin Exp Med 16(1):37–47
176. Kim JM et al (2012) Curcumin suppresses the TPA-induced invasion through inhibition of PKCalpha-dependent MMP-expression in MCF-7 human breast cancer cells. Phytomedicine 19(12):1085–1092
177. Lee SO et al (2008) Suppression of PMA-induced tumor cell invasion by capillarisin via the inhibition of NF-kappaB-dependent MMP-9 expression. Biochem Biophys Res Commun 366(4):1019–1024

178. Shao ZM et al (2002) Curcumin exerts multiple suppressive effects on human breast carcinoma cells. Int J Cancer 98(2):234–240
179. Cowin P, Welch DR (2007) Breast cancer progression: controversies and consensus in the molecular mechanisms of metastasis and EMT. J Mammary Gland Biol Neoplasia 12(2-3):99–102
180. Gallardo M, Calaf GM (2016) Curcumin inhibits invasive capabilities through epithelial mesenchymal transition in breast cancer cell lines. Int J Oncol 49(3):1019–1027
181. Chen MC et al (2012) Resveratrol inhibits LPS-induced epithelial-mesenchymal transition in mouse melanoma model. Innate Immun 18(5):685–693
182. Huang T, Chen Z, Fang L (2013) Curcumin inhibits LPS-induced EMT through downregulation of NF-kappaB-Snail signaling in breast cancer cells. Oncol Rep 29(1):117–124
183. Chen WC et al (2013) Curcumin suppresses doxorubicin-induced epithelial-mesenchymal transition via the inhibition of TGF-beta and PI3K/AKT signaling pathways in triple-negative breast cancer cells. J Agric Food Chem 61(48):11817–11824
184. Ferlay J et al (2015) Cancer incidence and mortality worldwide: sources, methods and major patterns in GLOBOCAN 2012. Int J Cancer 136(5):E359–E386
185. Palumbo MO et al (2013) Systemic cancer therapy: achievements and challenges that lie ahead. Front Pharmacol 4:57
186. Mishra J et al (2013) Prospective of colon cancer treatments and scope for combinatorial approach to enhanced cancer cell apoptosis. Crit Rev Oncol Hematol 86(3):232–250
187. Zhang L et al (2008) Nanoparticles in medicine: therapeutic applications and developments. Clin Pharmacol Ther 83(5):761–769
188. Peer D et al (2007) Nanocarriers as an emerging platform for cancer therapy. Nat Nanotechnol 2(12):751–760
189. Matsumura Y, Maeda H (1986) A new concept for macromolecular therapeutics in cancer chemotherapy: mechanism of tumoritropic accumulation of proteins and the antitumor agent smancs. Cancer Res 46(12 Pt 1):6387–6392
190. Lee WH et al (2014) Recent advances in curcumin nanoformulation for cancer therapy. Expert Opin Drug Deliv 11(8):1183–1201
191. Ghosh S, Banerjee S, Sil PC (2015) The beneficial role of curcumin on inflammation, diabetes and neurodegenerative disease: a recent update. Food Chem Toxicol 83:111–124
192. Logsdon CD, Abbruzzese JL (2010) Chemoprevention of pancreatic cancer: ready for the clinic? Cancer Prev Res (Phila) 3(11):1375–1378
193. Cirmi S et al (2016) Chemopreventive agents and inhibitors of cancer hallmarks: may citrus offer new perspectives? Nutrients 8(11):698
194. Cheng KW et al (2013) Curcumin enhances the lung cancer chemopreventive efficacy of phospho-sulindac by improving its pharmacokinetics. Int J Oncol 43(3):895–902
195. Sutaria D et al (2012) Chemoprevention of pancreatic cancer using solid-lipid nanoparticulate delivery of a novel aspirin, curcumin and sulforaphane drug combination regimen. Int J Oncol 41(6):2260–2268
196. Nasr M et al (2014) Targeting different angiogenic pathways with combination of curcumin, leflunomide and perindopril inhibits diethylnitrosamine-induced hepatocellular carcinoma in mice. Eur J Pharmacol 723:267–275
197. Xu G et al (2010) Combination of curcumin and green tea catechins prevents dimethylhydrazine-induced colon carcinogenesis. Food Chem Toxicol 48(1):390–395
198. Khafif A et al (1998) Quantitation of chemopreventive synergism between (-)-epigallocatechin-3-gallate and curcumin in normal, premalignant and malignant human oral epithelial cells. Carcinogenesis 19(3):419–424
199. Montgomery A et al (2016) Curcumin sensitizes Silymarin to exert synergistic anticancer activity in colon cancer cells. J Cancer 7(10):1250–1257
200. Feng S-S, Chien S (2003) Chemotherapeutic engineering: application and further development of chemical engineering principles for chemotherapy of cancer and other diseases. Chem Eng Sci 58(18):4087–4114

201. Chuang SE et al (2000) Curcumin-containing diet inhibits diethylnitrosamine-induced murine hepatocarcinogenesis. Carcinogenesis 21(2):331–335
202. Guo S et al (2016) A nanoparticulate pre-chemosensitizer for efficacious chemotherapy of multidrug resistant breast cancer. Sci Rep 6:21459
203. Zhao X et al (2015) Doxorubicin and curcumin co-delivery by lipid nanoparticles for enhanced treatment of diethylnitrosamine-induced hepatocellular carcinoma in mice. Eur J Pharm Biopharm 93:27–36
204. Chen H et al (2016) Synergistic effect of fenretinide and curcumin for treatment of non-small cell lung cancer. Cancer Biol Ther 17:1–8
205. Yin H et al (2014) Curcumin sensitizes glioblastoma to temozolomide by simultaneously generating ROS and disrupting AKT/mTOR signaling. Oncol Rep 32(4):1610–1616
206. Kang JH et al (2015) Curcumin sensitizes human lung cancer cells to apoptosis and metastasis synergistically combined with carboplatin. Exp Biol Med (Maywood) 240(11):1416–1425
207. Zoller F et al (2009) Endoradiotherapy in cancer treatment—basic concepts and future trends. Eur J Pharmacol 625(1–3):55–62
208. DeNardo SJ, Denardo GL (2006) Targeted radionuclide therapy for solid tumors: an overview. Int J Radiat Oncol Biol Phys 66(2 Suppl):S89–S95
209. Zhao J, Zhou M, Li C (2016) Synthetic nanoparticles for delivery of radioisotopes and radiosensitizers in cancer therapy. Cancer Nanotechnol 7(1):9
210. Verma V (2016) Relationship and interactions of curcumin with radiation therapy. World J Clin Oncol 7(3):275–283
211. Luthra PM, Lal N (2016) Prospective of curcumin, a pleiotropic signalling molecule from Curcuma longa in the treatment of Glioblastoma. Eur J Med Chem 109:23–35
212. Fan H et al (2016) MiR-593 mediates curcumin-induced radiosensitization of nasopharyngeal carcinoma cells via MDR1. Oncol Lett 11(6):3729–3734
213. Hsu F-T et al (2015) Curcumin sensitizes hepatocellular carcinoma cells to radiation via suppression of radiation-induced NF-κB activity. Biomed Res Int 2015:7
214. Sebastia N et al (2014) Curcumin and trans-resveratrol exert cell cycle-dependent radioprotective or radiosensitizing effects as elucidated by the PCC and G2-assay. Mutat Res 766-767:49–55
215. Lopez-Jornet P et al (2016) Radioprotective effects of lycopene and curcumin during local irradiation of parotid glands in Sprague Dawley rats. Br J Oral Maxillofac Surg 54(3):275–279
216. Fukuda K et al (2016) A diarylpentanoid curcumin analog exhibits improved radioprotective potential in the intestinal mucosa. Int J Radiat Biol 92(7):388–394
217. Sharma RA, Gescher AJ, Steward WP (2005) Curcumin: the story so far. Eur J Cancer 41(13):1955–1968
218. Eigner D, Scholz D (1999) Ferula asa-foetida and Curcuma longa in traditional medical treatment and diet in Nepal. J Ethnopharmacol 67(1):1–6
219. Johnson JJ, Mukhtar H (2007) Curcumin for chemoprevention of colon cancer. Cancer Lett 255(2):170–181
220. Hsu CH, Cheng AL (2007) Clinical studies with curcumin. Adv Exp Med Biol 595:471–480
221. Sharma RA et al (2004) Phase I clinical trial of oral curcumin: biomarkers of systemic activity and compliance. Clin Cancer Res 10(20):6847–6854
222. Mahammedi H et al (2016) The new combination Docetaxel, Prednisone and Curcumin in patients with castration-resistant prostate cancer: a pilot phase II study. Oncology 90(2):69–78
223. Basu P et al (2013) Clearance of cervical human papillomavirus infection by topical application of curcumin and curcumin containing polyherbal cream: a phase II randomized controlled study. Asian Pac J Cancer Prev 14(10):5753–5759
224. Bayet-Robert M et al (2010) Phase I dose escalation trial of docetaxel plus curcumin in patients with advanced and metastatic breast cancer. Cancer Biol Ther 9(1):8–14
225. Irving GR et al (2013) Prolonged biologically active colonic tissue levels of curcumin achieved after oral administration–a clinical pilot study including assessment of patient acceptability. Cancer Prev Res (Phila) 6(2):119–128

226. Chen WT et al (2014) Effectiveness of a novel herbal agent MB-6 as a potential adjunct to 5-fluoracil-based chemotherapy in colorectal cancer. Nutr Res 34(7):585–594
227. Sharma RA et al (2001) Pharmacodynamic and pharmacokinetic study of oral Curcuma extract in patients with colorectal cancer. Clin Cancer Res 7(7):1894–1900
228. Dhillon N et al (2008) Phase II trial of curcumin in patients with advanced pancreatic cancer. Clin Cancer Res 14(14):4491–4499
229. Kanai M et al (2011) A phase I/II study of gemcitabine-based chemotherapy plus curcumin for patients with gemcitabine-resistant pancreatic cancer. Cancer Chemother Pharmacol 68(1):157–164
230. Epelbaum R et al (2010) Curcumin and gemcitabine in patients with advanced pancreatic cancer. Nutr Cancer 62(8):1137–1141
231. Ghalaut VS et al (2012) Effect of imatinib therapy with and without turmeric powder on nitric oxide levels in chronic myeloid leukemia. J Oncol Pharm Pract 18(2):186–190
232. Irving GR et al (2015) Combining curcumin (C3-complex, Sabinsa) with standard care FOLFOX chemotherapy in patients with inoperable colorectal cancer (CUFOX): study protocol for a randomised control trial. Trials 16:110
233. James MI et al (2015) Curcumin inhibits cancer stem cell phenotypes in ex vivo models of colorectal liver metastases, and is clinically safe and tolerable in combination with FOLFOX chemotherapy. Cancer Lett 364(2):135–141
234. Panahi Y et al (2014) Adjuvant therapy with bioavailability-boosted curcuminoids suppresses systemic inflammation and improves quality of life in patients with solid tumors: a randomized double-blind placebo-controlled trial. Phytother Res 28(10):1461–1467
235. Ryan JL et al (2013) Curcumin for radiation dermatitis: a randomized, double-blind, placebo-controlled clinical trial of thirty breast cancer patients. Radiat Res 180(1):34–43
236. Kanai M (2014) Therapeutic applications of curcumin for patients with pancreatic cancer. World J Gastroenterol 20(28):9384–9391
237. Golombick T et al (2012) Monoclonal gammopathy of undetermined significance, smoldering multiple myeloma, and curcumin: a randomized, double-blind placebo-controlled cross-over 4g study and an open-label 8g extension study. Am J Hematol 87(5):455–460
238. Carroll RE et al (2011) Phase IIa clinical trial of curcumin for the prevention of colorectal neoplasia. Cancer Prev Res (Phila) 4(3):354–364
239. Ide H et al (2010) Combined inhibitory effects of soy isoflavones and curcumin on the production of prostate-specific antigen. Prostate 70(10):1127–1133
240. Golombick T et al (2009) The potential role of curcumin in patients with monoclonal gammopathy of undefined significance–its effect on paraproteinemia and the urinary N-telopeptide of type I collagen bone turnover marker. Clin Cancer Res 15(18):5917–5922

Nutraceuticals and Their Role in Human Health and Disease

17

Arpita Devi, S. Chennakesavulu, Chava Suresh, and Aramati B. M. Reddy

Abstract

Nutraceuticals, a growing market since the 1990s, is being used for prevention and treatment of various diseases. Nutraceuticals are used as drugs all over the world. While on one hand, researchers are working to discover disease-preventing potential of nutrients found in microorganisms and plants, on the other hand, pharmaceutical industry is trying to enrich food with those compounds.

Nutraceuticals have been found to have anti-inflammat ory, anti-oxidative, antitumor, antidiabetic, and anti-obesity properties. While dietary fibers, probiotics, and prebiotics play a role in digestive system, antioxidants and polyphenols help in relief from ROS and cellular stress. Spices and PUFAs have a number of roles to play in disease prevention.

Keywords

Dietary fibers · Probiotics · Prebiotics · PUFAs · Antioxidant vitamins · Polyphenols

17.1 Introduction

The word nutraceutical is a combination of two words – Latin word "nutrire" which means nourish and English word pharmaceutical [1]. Stephen L. DeFelice coined the term nutraceutical in the year 1990 [1]. Nutraceuticals are often thought to be the same thing as functional foods. FDA has defined nutraceuticals as any specific food used for the prevention or treatment of disease as drugs.

A. Devi · S. Chennakesavulu · C. Suresh · A. B. M. Reddy (✉)
Department of Animal Biology, School of Life Sciences, University of Hyderabad, Hyderabad, India
e-mail: abmsl@uohyd.ernet.in; binduaramati@gmail.com

V. Rani, U. C. S. Yadav (eds.), *Functional Food and Human Health*,
https://doi.org/10.1007/978-981-13-1123-9_17

17.1.1 Difference Between Ancient Indian Ayurveda and Present-Day Nutraceuticals

Ancient Indian Ayurveda believed that a disease can be treated either by nourishment (Brimhana) or by depleting (Langhana) based on the type of disease. But nutraceuticals means treating a disease by total nourishment. Also, Ayurveda uses whole plant or organism or specific parts for treatment, but nutraceuticals also include purified nutrients and nutrient supplement along with whole food.

17.1.2 Difference Between Phytochemicals and Nutraceuticals

Phytochemicals are chemical compounds in plants. These are synthesized by the plants for regulation of their metabolic processes and protection from stress and predators [2]. But nutraceuticals include products obtained from any natural source which also include live microorganisms that have the ability to act as drug. Phytochemicals may not necessarily be beneficial to human health [3]. For example, a compound obtained from *Momordica* has been found to be abortive in all stages of pregnancy [3, 4]. But nutraceuticals have been designed for the improvement of human health. Thus, there are least chances of harm.

17.2 Classification of Nutraceuticals

17.2.1 Nutraceuticals Can Be Divided into Two Broad Categories

(a) *Dietary supplements*: In the year 1994, the Dietary Supplement Health and Education Act (DSHEA), USA, defined the term dietary supplement as: "A product which is consumed orally that contains a 'dietary ingredient' introduced intentionally to supplement the diet". Thus, dietary supplements are food nutrients that can be in the form of extracts, concentrated liquids, capsules, gel caps, soft gel, or powder.

(b) *Functional foods*: Functional foods are whole foods which are enriched with nutrients. In Japan, it is said that all functional foods must fulfill three conditions: foods should be (1) in natural form, (2) can be consumed as often as daily, and (3) should be involved in a biological process so as to prevent or control disease [3].

17.2.2 Based on Food Source, Nutraceuticals Can Be Classified into

(a) *Dietary fibers*: These are naturally occurring substances in food which are not digested by the digestive juices in gastrointestinal tract and are excreted by the digestive system as such. Dietary fiber consists of arabinoxylans, inulin, lignin,

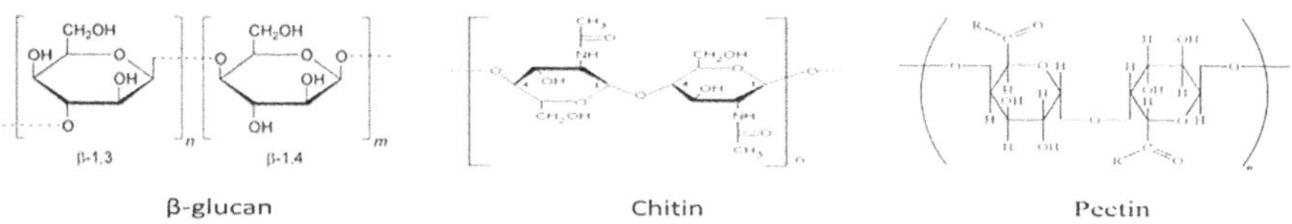

Fig. 17.1 Examples of dietary fibers

Fig. 17.2 Structures of omega-6 fatty acids, viz., α-linoleic acid and arachidonic acid

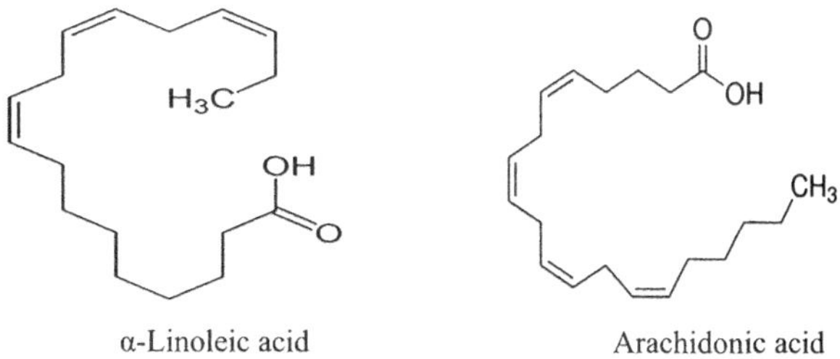

dextrins, chitins, pectins, beta-glucans, cellulose, and oligosaccharides (Fig. 17.1).

(b) *Probiotics*: These are derivatives of live microbial cultures which will be formulated in appropriate doses in order to improve the host's intestinal microbial balance, e.g., *lactobacilli*, *bifidobacteria*, etc. [5].

(c) *Prebiotics*: Roberfroid defines prebiotics as "a selectively fermented ingredient that allows specific changes, both in the composition and/or activity in the gastrointestinal microflora that confers benefits upon host well-being and health" [6]. These are, generally, fructose-based oligosaccharides which are not digested in human digestive tract. Examples of prebiotics are trans-galactooligosaccharide and inulin, larch arabinogalactan (LAG), pectin, resistant starch, beta-glucans, and xylooligosaccharides (XOS) [7–10].

(d) *Polyunsaturated fatty acids (PUFAs)*: The fatty acids that contain more than one double bond in their carbon backbone are called as PUFAs. Though there are many types of PUFAs, but the most common are omega-3 fatty acids, omega-6 fatty acids (Fig. 17.2), eicosapentaenoic acid, and docosahexaenoic acid.

(e) *Antioxidant vitamins*: The three major antioxidant vitamins are beta-carotene, vitamin C (L-ascorbic acid) [11, 12], and vitamin E that include both tocopherols and tocotrienols [13, 14] (Fig. 17.3).

(f) *Polyphenols*: These are phytochemicals produced by plant for protection against reactive oxygen species and photosynthetic stress. Structurally, these are large molecules of multiple phenol structural units (Fig. 17.4).

(g) *Spices*: These are food adjuncts that enhance sensory quality of foods. Minute quantities of dietary spices have antioxidant, chemopreventive, antimutagenic, anti-inflammatory, and immune modulatory effects. Fresh ginger and cumin have highest amount of antioxidant [15].

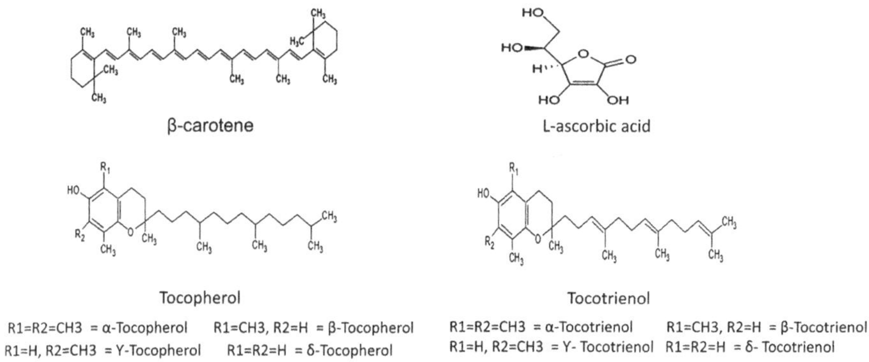

Fig. 17.3 Examples of antioxidant vitamins (β-carotene, vitamin C, and vitamin E)

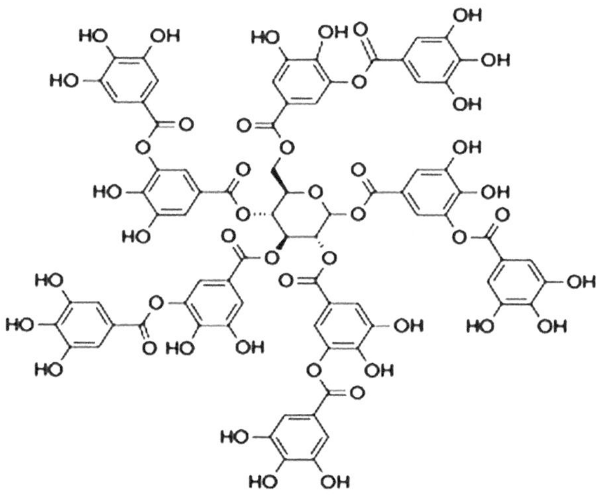

Fig. 17.4 Tannic acid, a polyphenol formed by esterification of ten equivalents of the phenylpropanoid-derived gallic acid

17.3 Potential Health Benefits of Nutraceuticals

As different diseases are spreading vigorously around the world and synthetic drugs have been failing in clinical trials, nutraceuticals have attracted the attention of many scientists due to their therapeutic effects along with potential nutritional property and safety. Over the years, researchers have shown their direct participation in implicating several biological processes, acting as antioxidants, antimutagens, etc. Use of nutraceuticals has shown therapeutic effects in diseases like diabetes, heart disease, arthritis, cancer, hypertension, common cold, dyslipidemia, inflammatory bowel disease, depression, and others.

For ages, vitamins have been known to be a boon for life. These vitamins are chemical compounds which are obtained mostly from plants and microorganisms and have a wide range of health benefits. They can be directly involved in biological processes; for example, vitamin A and niacin are involved in the tryptophan synthesis pathway or can act as coenzymes in different pathways [Table 17.1].

Table 17.1 Mechanism of action of vitamins and deficiency diseases related to them

Vitamins	Disease condition	Mechanism of action
Vitamin A (retinol, retinal, retinoic acid, and beta-carotene)	Night blindness, hyperkeratosis, and keratomalacia	11-*cis*-retinal binds protein "opsin" to form rhodopsin in rods and iodopsinin cones at conserved lysine residue. All-trans-retinoic acid binds to retinoic acid receptors (RARs) and regulates gene transcription [106]
Vitamin B1 (thiamine)	Beriberi, Wernicke-Korsakoff syndrome	Acts as coenzyme in sugars and amino acid catabolism. Its active form, thiamine pyrophosphate (TPP), acts as coenzyme in the conversion of pyruvate to acetyl coenzyme A (CoA). It participates in RNA and DNA synthesis, as well as nerve function [107]
Vitamin B2 (riboflavin)	Ariboflavinosis, glossitis, angular stomatitis	For formation of FAD and FMN, coenzymes for flavoprotein enzymes. Electron transport chain, citric acid cycle, and beta oxidation [108, 109]
Vitamin B3 (niacin)	Pellagra	Needed for the formation of NAD and NADP, coenzymes in many metabolic processes. NAD and NADP act as electron acceptor and donors in many oxidation reduction reactions [108]
Vitamin B5 (pantothenic acid)	Paresthesia	A precursor for the synthesis of coenzyme A. In the synthesis of amino acids, fatty acids, ketones, cholesterol, phospholipids, steroid hormones, neurotransmitters (such as acetylcholine), and antibodies
Vitamin B6 (pyridoxamine)	Anemia, peripheral neuropathy	A coenzyme in the form of pyridoxal 5'-phosphate (PLP) in many enzymatic reactions especially in amino acid metabolism and biosynthesis of neurotransmitters [108]
Vitamin B7 (biotin)	Dermatitis, enteritis	A coenzyme for carboxylase enzymes, mainly acetyl CoA carboxylase [110]
Vitamin B9 (folic acid)	Megaloblastic anemia, birth defects, such as neural tube defects	Involved in synthesis, repair, and methylation of DNA; a cofactor in the form of tetrahydrofolate (THF) [111]. It is often prescribed to pregnant and lactating women because of its role in rapid cell division and growth; aids in erythropoiesis [111]
Vitamin B12 (cyanocobalamine)	Pernicious anemia	A coenzyme involved DNA synthesis and regulation [112], fatty acid and amino acid metabolism, hematopoiesis (bone marrow), and nerve sheath [113]
Vitamin C (ascorbic acid)	Scurvy	Vitamin C plays a major role in collagen synthesis reaction; deficiency of which causes scurvy. In animals, dysregulation of these reactions will delay wound healing and cause bleeding from capillaries. It is an antioxidant vitamin and acts as a reducing agent in several enzymatic and nonenzymatic reactions [114]

(continued)

Table 17.1 (continued)

Vitamins	Disease condition	Mechanism of action
Vitamin D (cholecalciferol, ergocalciferol)	Rickets and osteomalacia	Calcitriol, the active vitamin D metabolite, binds to the vitamin D receptor (VDR) and modulates the gene expression of transport proteins so it can help in calcium absorption in the intestine [115]
Vitamin E (tocopherol, tocotrienol)	Sterility in males and abortions in females	Vitamin E acts as a peroxyl radical scavenger [13]. α-tocopherol brings down the smooth muscle growth by stimulating dephosphorylating enzyme, protein phosphatase 2A, which in turn deactivates protein kinase c by dephosphorylating it. It also plays a role in the functioning of the eye and neurons and inhibition of platelet coagulation and protects lipids and polyunsaturated fatty acids from peroxidation by free radicals [13, 14]
Vitamin K (phylloquinone)	Bleeding diathesis	Blood coagulation, vitamin K carboxylates certain glutamate residues in proteins to facilitate calcium binding especially in Gla domains [115, 116]

High dietary intake of fibers through fruits, vegetables, and whole grains reduces risk of chronic diseases like cancer and cardiovascular diseases [16]. Cancer progression often leads to metastasis, if not treated at early stage. Epidemiological studies have proved that nutrients from natural sources may modify carcinogenesis. A number of phytochemicals possess anticarcinogenic and antimutagenic properties. Phytoestrogens might also be beneficial in prevention of cardiovascular diseases (CVDs) [17]. Phytochemicals can counteract risk factors of CVDs like obesity, hyperlipidemia, hypertension, and diabetes [17, 18]. Phytochemicals help in reducing oxidative stress which is nowadays related to many pathological conditions and one of the main causes of DNA damage and mutation.

Not always chemical compounds act as nutraceuticals. Mineral ions and some plant proteins also have effect on human health. Mineral ions like sodium, calcium, sulfur, zinc, potassium, phosphorous, magnesium, iron, iodine, fluorine, selenium, copper, manganese, etc. are required for many biological processes [Table 17.2]. Some proteins like momordin and polypeptide-p isolated from *Momordica* have been shown to be a protein synthesis inhibitor [19] and have blood glucose-lowering property [18], respectively. Momordins inhibit protein synthesis in eukaryotes by inactivating the ribosomes. Polypeptide-p showed hypoglycemic effect in in vivo experiments on gerbils, langurs, and humans [18, 19].

Therefore, nutraceuticals play a critical role in health improvement, prevent chronic diseases, increase life span by postponing aging, and support smooth functioning and integrity of the body.

Table 17.2 Involvement of minerals in regulation of various cellular processes

Mineral	Involvement in biological processes
Sodium	Involved in functioning of Na + -K+ ATPase enzyme. Also as voltage-gated channel for nerve impulse conduction [117] and helps in maintaining osmotic balance and blood pressure [118, 119]
Potassium	Involved in functioning of Na + -K+ ATPase enzyme [117]. Also as voltage-gated channel for nerve impulse conduction [119]
Calcium	In teeth and bone development. Also plays a role in nerve conductance and blood clotting. It has a major role in muscle movement [120]
Sulfur	Helps in synthesis of methionine and cysteine, thus plays an important role in synthesis of functional protein [121]
Zinc	Helps in functioning of Zn2 + –dependent enzymes like carbonic anhydrase [122]
Phosphorous	Helps in muscle contraction and nerve signaling. It is the major component of nucleic acids [123]
Iron	Important component of hemoglobin [124]
Iodine	Involved in the synthesis of thyroid hormone [125]
Magnesium	Helps in muscle contraction and nerve impulse, stabilizes ATP, activates hexokinase, and thus helps in energy metabolism and many other metalloenzymes [126]
Fluorine	Mineralization of bone and enamel of teeth [127]
Manganese	Helps in the formation of connective tissue, bones, blood clotting factors, and sex hormones, fat and carbohydrate metabolism, calcium absorption. Acts as cofactor for superoxide dismutase, an enzyme involved in removal of free radicals [128]

17.3.1 Nutraceuticals As Anti-inflammatory Agent

Cells are continuously attacked by different stimuli like pathogens, damaged cells, irritants, etc. In response cells undergo inflammation to solve the initial reason of cell injury, to clear out dead cells, and to initiate tissue repair. Chronic inflammation leads to a number of diseases, such as atherosclerosis, rheumatoid arthritis, hay fever, periodontitis, and also cancer. Nutraceuticals with anti-inflammatory property allows inflammation that is required and stops continuous inflammation.

Nutraceuticals act as anti-inflammatory agent by numerous mechanisms like inhibiting the activation of NF-kB, scavenging free radicals, and downregulation or overexpression of cell adhesion molecules and enzymes (cyclooxygenase-2, phospholipase A2, etc.) and pro-inflammatory cytokines (IL-1, IL-6, etc.).

Some of the well-known nutraceuticals with anti-inflammatory properties are phenolic compounds, polyunsaturated fatty acids, phytosterols, tocopherols, and carotenoid molecules present in food sources such as fenugreek, coriander, tomato, carrot, green tea, walnut, fish oil, etc.

17.3.2 Nutraceuticals As Antioxidants

In cells, there is a continuous production of reactive free radicals. These free radicals should be neutralized to protect the cell from various cell damages. Antioxidants are the substances that neutralize the effects of free radicals. Beneficial effects of antioxidants have shown in preventing neurodegenerative diseases like Parkinson's disease, Alzheimer's disease, etc. The chemical constituents that possess antioxidant properties are flavonoid polyphenols, non-flavonoid polyphenols, phenolic acids or phenolic diterpenes, organic sulfur compounds, etc. They exhibit the antioxidant properties either by scavenging free radicals or increasing endogenous cellular redox potential. Some of the regular dietary foods that possess antioxidant properties are citrus fruits, onion, green tea, apple, garlic, turmeric, grapes, rosemary, broccoli, basil, etc.

17.3.3 Nutraceuticals As Anti-obesity Agent

Obesity is a common chronic disease and metabolic syndrome which is due to imbalance between energy intake and energy expenditure. Obesity is associated with various diseases, particularly cardiovascular diseases, type 2 diabetes, obstructive sleep apnea, some cancers, osteoarthritis, and asthma. Scientists have explored the use of nutraceuticals in preventing some of the diseases associated with obesity. There are some nutraceuticals that help to overcome obesity by increasing thyroxine release, metabolism, and lipolysis and inducing secretion of leptin which controls hunger. Some of the widely used nutraceuticals in preventing obesity are curcumin, bottle gourd, black gram, green tea, etc.

17.3.4 Nutraceuticals As Antidiabetic Agent

Diabetes is a metabolic syndrome where a person suffers from long-term persistence of high blood glucose. It is due to low or no insulin production or loss of function of insulin receptors on cell surface or both. There are some nutraceuticals that help in lowering the blood glucose level by increasing insulin production, increasing sensitivity of cells toward insulin by increasing the expression of IRS, GLUT, etc., decreasing the absorption of glucose in blood, increasing hepatic glycogen concentration, and decreasing the concentration of glycogen phosphorylase and gluconeogenic enzymes. Some nutraceuticals that are in use since long time to treat and prevent diabetes are *Emblica officinalis*, citrus fruits, fenugreek, green tea, *Momordica*, etc. [19, 20].

17.4 Mechanism of Action of Different Classes of Nutraceuticals

17.4.1 Role of Dietary Fibers in Disease Prevention

Dietary fibers which are abundant in fruits and vegetables have many health benefits. In an 8-year study demonstrated by Walter C. Willett and his group in 1992, protective effect of fiber consumption on breast cancer in middle-aged women was shown. In another statistical study done by Alicja Wolk et al., in 1999, it was shown that high-fiber intake particularly from cereal may be protective against coronary heart diseases in women [20]. It is known that dietary fiber consumption prevents obesity which inversely correlated with body weight and body fat. Studies have shown that intake of dietary fiber aids in health management through promoting satiation, altering absorption of macronutrients, and secretion of gut hormones [21–23]. But there is no definite evidence as to how dietary fiber aids in weight management. In another study by Chandalia et al., it was shown that increase in the intake of soluble dietary fiber improved glycemic control in patients with type 2 diabetes mellitus and decreased hyperinsulinemia and also lowered plasma lipid concentrations [20]. A significant correlation was found between dietary fiber and C-reactive protein (CRP), a marker of inflammation and predictive marker for cardiovascular events. High-fiber diet decreased the concentration of CRP in blood. In another human study, fiber intake was inversely associated with the risk of gastroesophageal reflux disease symptoms in human subjects [24]. A high-fiber diet, especially from fruit and vegetables, could reduce risk of duodenal ulcers [23].

17.4.2 Role of Probiotics in Disease Prevention

Bifidobacterium produces glutamine from NH4+ and glutamic acid, which maintains mucosal integrity of the intestine. Metabolites of *S. faecalis*, *C. butyricum*, and *B. mesentericus* – butyrate and acetate – create an unfavorable environment for pathogenic *E. coli*, *Salmonella*, and methicillin-resistant *S. aureus* by lowering the intestinal pH [25]. Butyrate also maintains mucosal barrier defenses [25]. Milk fermented by probiotics significantly lowered the level of hydrogen in the breath of lactose-intolerant patients indicating that lactose has been pre-metabolized before entering the large intestine [26]. Reddy et al. in the year 1993 showed that dietary *B. longum* could suppress the expression of total and neonate RAS P-21 in the colon mucosa and tumors [25, 27]. Probiotics can also reduce food allergy with decrease in concentration of serum IgE [28]. Probiotics might inhibit the growth of pathogenic bacteria by secreting bacteriocins such as nisin [29, 30]. In infants, probiotics reduce viral diarrhea by increasing secretory IgA and decreasing viral shedding, indicating an immunological mechanism [31]. Recently, it has also been found that daily intake of heat-killed *Lactobacillus*-137 can decrease the depth of periodontal pockets in patients undergoing supportive periodontal therapy [29].

17.4.3 Role of Prebiotics in Disease Prevention

Prebiotics are the substances which benefits gut microflora and as a result the host is benefitted. Most of the research has been done with fructooligosaccharides – inulin and oligofructose. With the use of the above prebiotics, significant reduction in serum triglycerides has been observed [28, 30]. In the year 2002, Moro et al. fed infants with fructooligosaccharides and galactooligosaccharides and after 28 days observed an increase in the concentration of gut microorganisms – bifidobacteria and lactobacteria [31]. The health benefits of these two bacteria were discussed in the probiotics section. Prebiotics can reduce the activity of microbial enzymes involved in the production of carcinogens, thus reducing the chances of colon cancer [32]. Inulin and fructooligosaccharides have been found to reduce the pre-carcinogen lesions in carcinogen-treated mice [33].

Prebiotics, mostly inulin and fructooligosaccharides, improves the absorption of minerals in the body. In the year 2000, Roberfroid showed increased absorption of calcium and magnesium in ovariectomized rat fed with fructooligosaccharides [28]. Again in 2003, Tahiri, Tressol et al. showed that fructooligosaccharide improved calcium absorption in women who reached menopause. The same effect of fructooligosaccharide in adolescent girls has been shown by Griffin et al. in 2002 [33, 34].

17.4.4 Role of Polyunsaturated Fatty Acids (PUFAs) in Disease Prevention

Fatty acids are the main constituents of cell membrane. Integrity of the cell membrane has to be maintained for survival of the cell. Some fatty acids are synthesized in human body. But some fatty acids mainly polyunsaturated fatty acids cannot be synthesized by animal cell. So, these PUFAs have to be obtained from diet. n-3 PUFAs have an effect on the fluidity of cell membrane. It has been shown by Ma et al., in the year 2004, that n-3 PUFAs can alter lipid raft and caveolae that regulate cholesterol transport, signal transduction, and endocytosis [35, 36]. n-3 PUFAs also decreased expression of adhesion molecules on monocytes to the endothelium, thereby preventing atherosclerosis development [35]. Also, much research is going on to understand the mechanism by which PUFAs can prevent cancer.

17.4.5 Role of Antioxidant Vitamins in Disease Prevention

Beta-carotene, vitamin C, and vitamin E are the three antioxidant vitamins, which help the cells from oxidative stress. Reactive free radicals are always formed inside the cell, and these free radicals always attack low-density lipoproteins and cause their peroxidation. Vitamins C (Fig. 17.5) and E prevent peroxidation by scavenging these free radicals. In a study, it was found that exposure of α-tocopherol inhibits oxidized LDL-induced expression of ICAM-1, VCAM -1, or E-selectin and decreases adhesion of monocytes to human umbilical endothelial cells [37].

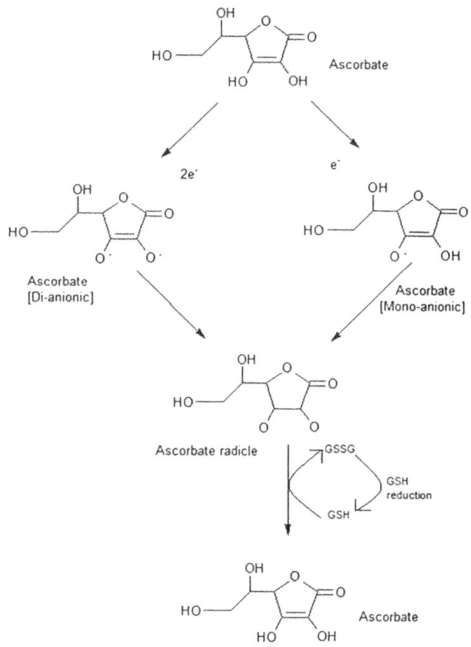

Fig. 17.5 Mechanism of antioxidant activity by vitamin C. (Modified from the original image by Matthias Rath et al., Cellular medicine and natural health, 2015)

Vitamin C also protects the cells from pro-oxidative activity of α-tocopherol (vitamin E) by reducing α-tocopheroxyl radical to α-tocopherol [13, 14]. Antioxidant vitamins have also been found to play a role in cancer prevention, but the exact mechanism by which it is done is not known.

17.4.6 Role of Polyphenols in Disease Prevention

Polyphenols are believed to be anticancer molecules. Among the polyphenols, mechanism of action of epigallocatechin gallate (ECGC) is well elucidated (Fig. 17.6) [38]. Auto-oxidation of EGCG leads to EGFR activation in human esophageal cells which may lead to esophageal cancer. ECGC increases the level of cytoskeletal and structural proteins which may contribute in metastasis. ECGC increases the level of the binding protein 14-3-3 gamma, which is involved in signal transduction and cell survival [39].

Polyphenols stimulate AMPKα1 and –α2 activity and acetyl CoA carboxylase phosphorylation [40]. Phenolic acids exert a direct antiproliferative action at low concentration in T47D human breast cancer cells [41]. ECGC has neuroprotective role via induction of bad degradation and inhibition of caspase-3 and its substrate poly (ADP-ribose) polymerase cleavage [42].

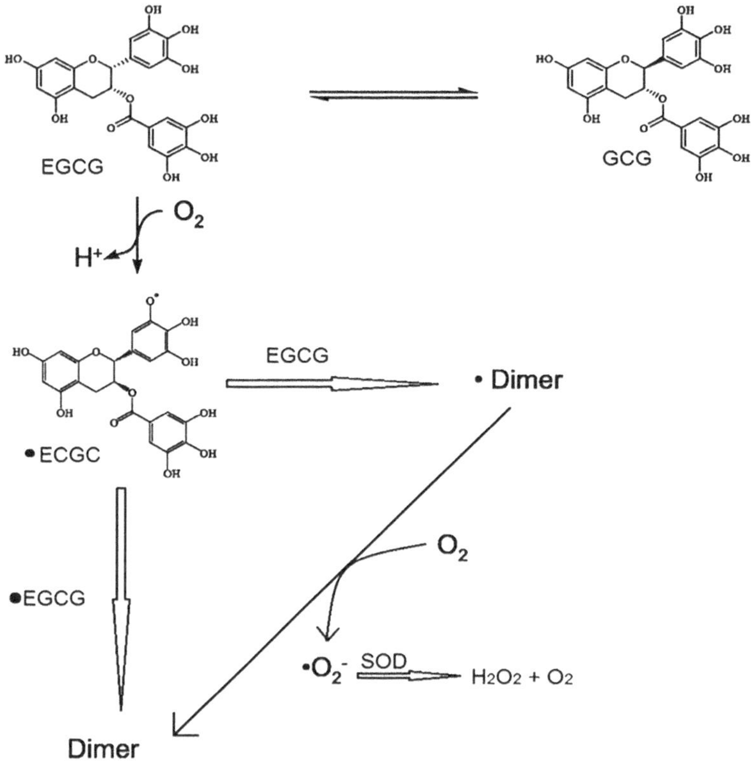

Fig. 17.6 Mechanism of auto-oxidation of EGCG. (Image modified from the original image by Zhe Hou et al., Cancer Research, January 2005)

17.4.7 Role of Spices in Disease Prevention

Spices add taste to our food. But other than taste spices are also highly nutritious. A pinch of spice in our daily meal is sufficient for a healthy living. Chithra and Leelamma in the year 1997 showed that coriander seeds have hypolipidemic properties. The increased activity of plasma lecithin cholesterol acyltransferase, enhanced synthesis of hepatic bile acid and the increased degradation of cholesterol to fecal bile acids and neutral sterols seemed to be responsible for its hypolipidemic effect. Hypoglycemic action of *Murraya koenigii* (curry leaf) and *Brassica juncea* (mustard) was showed by increased concentration of hepatic glycogen by increased activity of glycogen synthetase, decrease in glycogenolysis and gluconeogenesis by decreased activity of glycogen phosphorylase and gluconeogenic enzymes[43]. Essential oil, rich in trans-anethole, from fennel seeds, has antibacterial activity against *Staphylococcus aureus*, *Staphylococcus albus*, *Bacillus subtilis*, *Salmonella typhimurium*, *Pseudomonas aeruginosa*, *Shigella dysenteriae*, and *Escherichia coli* [44]. Oregano seed oil, thymol and carvacrol have antibacterial activity against *Pseudomonas aeruginosa* and *Staphylococcus aureus* [45].

17.5 Some Commonly Used Nutraceuticals and Their Health Benefits

17.5.1 Cinnamon

Cinnamon is a spice in global cuisine. It not only gives taste to the food but also provides us with many health benefits. Cinnamon is full of protective antioxidant like polyphenols, phenolic acid, and flavonoids which protect from the free radical damage [46] and the harm caused by it like DNA damage, mutations, and also aging process. Since free radicals are involved in the formation of advanced glycation products, the compounds present in cinnamon also inhibit the formation of advanced glycation products. Experimentally, it has been shown to decrease the concentration of absorbed glucose in the bloodstream after a high-carbohydrate meal, which is especially important for diabetic patients. Recent studies show that cinnamon increases insulin sensitivity and glucose transport in skeletal muscle cells and adipocytes by enhancing the insulin-stimulated insulin receptor (IR) and IR substrate-1 (IRS-1) tyrosine phosphorylation levels and phosphoinositide 3-kinase (PI3K) and GLUT4 expression [47].

Cinnamon has also been found effective in Alzheimer's disease as it could inhibit Tau aggregation. An A-linked proanthocyanidin trimer molecule was found to be causing this inhibition. Studies with cinnamon extract showed complete disassembly of recombinant Tau filaments and significant change in the morphology of paired helical filaments isolated from the brain of an AD victim [48].

Cinnamon contains a compound called cinnamaldehyde which has the ability to inhibit tumor growth by protecting DNA from damage and inducing apoptosis [94]. Cinnamon regulates vascular endothelial growth factor (VEGF) production and reduces proliferation of cancer cells [49].

Cinnamon also has antimicrobial properties. Ethanol and methylene chloride extracts of cinnamon were studied for their effect on *Helicobacter pylori* growth and urease activity. It was found that methylene chloride extract could inhibit growth of *H. pylori*, and ethanol extract counteracted its urease activity [50](Fig. 17.7).

17.5.2 Apple Cider Vinegar

Apple cider vinegar is the most used of all vinegars because of its benefits to human health. The high concentration of acetic acid helps the body to absorb important minerals from foods. Vinegar also has anti-glycemic property and has a beneficial effect on blood sugar levels [51]. The acetic acid in vinegar helps in lowering the blood sugar by preventing the complete digestion of complex carbohydrates, either by accelerating gastric emptying, by increasing the uptake of glucose by bodily tissues, or by decreasing the gastric pH [52]. Polyphenols such as chlorogenic acid, 4-p-coumarylquinic acid, and caffeic acid are the three major phenols which are present in high levels in apple cider vinegar, could inhibit oxidation of low-density lipoproteins, and improve health by preventing cardiovascular diseases [53]. Recent

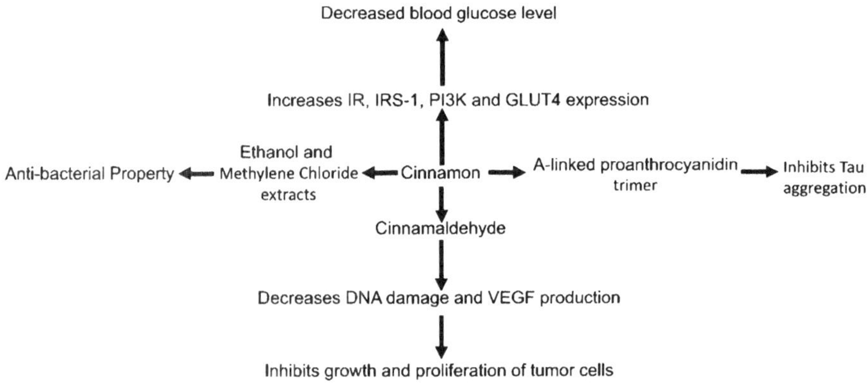

Fig. 17.7 Role of cinnamon in disease prevention

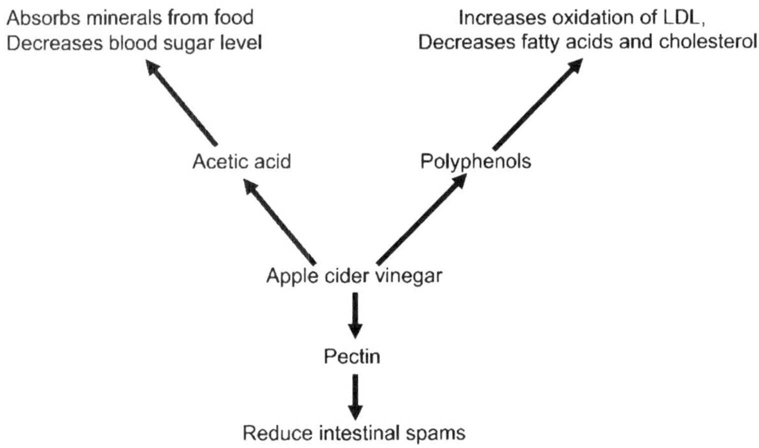

Fig. 17.8 Role of apple cider vinegar in disease prevention

studies showed that consumption of apple cider vinegar can reduce the LDL, triglyceride, and cholesterol levels in patients with hyperlipidemia. Daily consumption of apple cider vinegar improves the acid content of the stomach. The pectin present in apple cider vinegar also helps to relief intestinal spasms [54] (Fig. 17.8).

Chlorogenic acid (CGA) is abundant ester of caffeic acid and (−) quinic acid present in apple cider vinegar. Ohnishi et al. showed that CGA inhibited the formation of conjugated diene from linoleic acid. CGA also has strong inhibitory effect on the hemolysis and peroxidation of H_2O_2-induced mouse erythrocytes [55]. Feng et al. showed the carcinogenesis-protective property of CGA by inhibiting the proliferation of A549 human cancer cells in vitro. CGA blocks UVB or TPA-induced transactivation of AP-1 and NF-κB. CGA also increases the enzymatic activities of glutathione S-transferases (GST) and NAD(P)H:quinone oxidoreductase and also stimulates the nuclear translocation of Nrf2 and induction of GSTA1 antioxidant response element (ARE)-mediated GST activity [56]. Lee and Zhu showed that CGA inhibited *RAR*β gene DNA methylation predominantly through a noncompetitive mechanism in cultured MCF-7 and MAD-MB-231 human breast cancer cells, by demonstrating the increased formation of *S*-adenosyl-l-homocysteine (SAH, a potent inhibitor of DNA methylation) [57]. Ae-Sim Cho et al. showed that CGA inhibits fatty acid and cholesterol biosynthesis by increasing the expression of PPAR-α with a simultaneous increase in β-oxidation of fatty acid in the liver and by efficiently suppressing the activities of fatty acid synthase, HMG-CoA reductase, and ACAT in the liver [58].

17.5.3 Psyllium Husk

Psyllium husk is the seed of the plant called *Plantago ovata*. Psyllium husk has a prominent place in Ayurveda. It has gentle laxative, cooling, as well as diuretic properties [59]. The hygroscopic nature of the soluble fiber in psyllium husk helps in absorbing moisture and swelling up to form soft stools [60]. The insoluble fiber in psyllium husk helps by providing enough roughage to the wastes. Together they make the elimination of the waste easy and painless. Recent study demonstrates that psyllium husk has a gut-stimulatory effect [59] and has natural gelatin substance which decreases the rate of digestion and absorption of glucose in the blood. So it helps to decrease the intake of glucose and control diabetes [61]. The husk of psyllium husk provides a coating layer to the stomach lining cells which helps in protecting the delicate tissue from the direct hazardous effects of stomach acid. The petroleum fraction of the plant caused varying degrees of inhibition in three different species of Entamoeba, i.e., *E. histolytica*, *E. invadens*, and *E. dispar*. Thus, psyllium husk is used in amoebic dysentery [59].

17.5.4 Green Tea

Green tea is one of the most commonly used teas all over the world. It is full of antioxidants and nutrients that have powerful effects on the body. Green tea contains antioxidants called catechins which hinder the growth of many bacterial species. An extract of Japanese green tea was an inhibitor of the growth of dental caries-*causing bacteria*, *Streptococcus mutans*. The analysis of the extract revealed that the presence of polyphenolic compounds such as gallocatechin (GC), epigallocatechin (EGC), and epigallocatechin gallate (EGCG) exhibits antibacterial activity [62]. Chinese green tea was shown to have antioxidant activity by quenching oxygen free radical and H_2O_2-induced cytotoxicity in cultured mouse hepatocytes and human keratinocytes [63]. The polyphenols present in green tea significantly reduce the level of LDL cholesterol in blood and raise the level of HDL cholesterol [64]. The fluoride present in green tea helps in making the bone strong which in turn reduces the risk of osteoporosis in later life.

Gallic acid is also one of the polyphenols found in green tea. In a study conducted by Kim et al., in 2005, gallic acid attenuated immunoglobulin E (IgE)-induced histamine release and PMA induced pro-inflammatory markers from mast cells by the modulation of cAMP and intracellular calcium apart from inhibition of systemic allergic reaction and IgE-mediated local allergic reaction [65]. In 1992, Kroes et al. showed that gallic acid interferes with the functioning of polymorphonuclear leukocytes (PMNs). Inhibition of myeloperoxidase release and activity, scavenging of superoxide ions, as well as a possible interference with the assembly of active NADPH oxidase may result in the inhibition of inflammatory process by gallic acid [66]. In 2015, a unique finding by Hsieh et al. showed teratogenic property of gallic acid by exhibiting characteristic cerebral hemolysis and lipolysis in the musculus longissimus cervicis in mice at doses $\geq$1.02 mg/kg of body weight. Thus the mechanism of action of GA is multidisciplinary. The lipolysis and erythrocyte hemolysis can be related to the downregulation of PPAR-α by GA.; Due to its unique pro-oxidative and autoxidative behavior and the inhibition of carbonic anhydrase properties of GA, upregulates Ras/Raf/JAK/STAT pathway thus leads to inhibition of cell proliferation and differentiation [67]. The similar observation hold good from the studies showing high amount of gallic acid consumed by pregnant women can cause a risk to their fetuses. Gallic acid, at high doses, can cause a reduction in mitochondrial antioxidant activity by increasing miR-17 level, slowing down cell proliferation by increasing miR-21 level. Gallic acid also decreases the ability to repair mitochondrial damage by increasing miR-421 levels [68].

But extensive research on green tea is being carried out in the field of cancer. Data from various laboratories suggests that green tea, particularly its major polyphenolic constituent (−) – epigallocatechin-3-gallate (EGCG) – has tremendous anticancer properties. EGCG can modulate multiple signal transduction pathways that control the proliferation of cells [69]. Epidemiological study in Japan showed that green tea could potentially prevent cancer and cardiovascular diseases [70]. Treatment with epigallocatechin-3-gallate also resulted in apoptosis in human carcinoma keratinocyte, human prostate carcinoma cells, and mouse lymphoma cells, but not in human epidermal keratinocytes. The DNA cell cycle analysis showed that in human epidermoid carcinoma cells, epigallocatechin-3-gallate treatment resulted in arrest in the $G_0 - G_1$ phase of the cell cycle and apoptosis in a dose-dependent manner [71] (Fig. 17.9).

17.5.5 Bitter Melon

Bitter melon belonging to *Momordica* genus has a powerful bitter taste. It is mostly known for its antidiabetic and antitumor effects. Main phytoconstituents present in *Momordica* are gallic acid, gentisic acid, catechin, epicatechin, caffeic acid, chlorogenic acid, polypeptide-p, vicine, charantin, momordicin, momordin, stigmasta-5, 25-diene-3-β-O-glucoside, β-sitosterol-β-D-glucoside, momordicoside (G, F1, F2, I, K, L), etc.

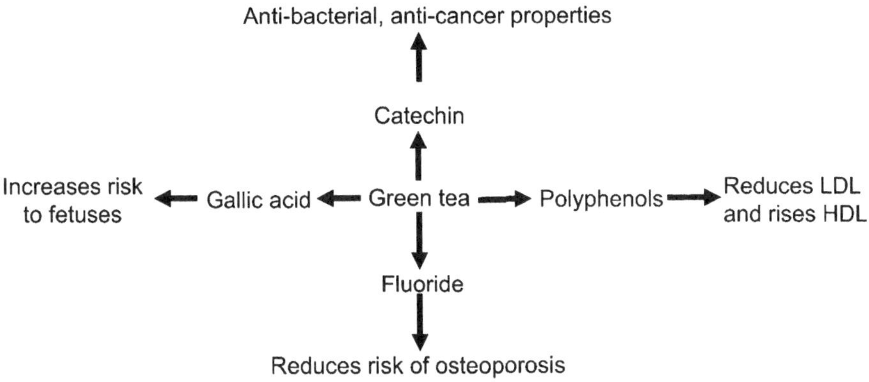

Fig 17.9 Effects of green tea in disease prevention

Momordin is a protein which is found mainly in *Momordica*. It is known as ribosome-binding enzyme and has antitumor properties due to its ability to inhibit protein synthesis by binding to the ribosome. In a study conducted by Lee et al. in 1998, it was found that the anticancer activity of momordin I is by its ability to inhibit protein-DNA interaction. They also found that momordins inhibit AP-1 and its interaction with Jun/Fos. Activated Jun/Fos transmits the tumor-promoting signal from the extracellular environment to nuclear transcription machinery. The final step in this signal transduction is the binding of Jun/Fos to the AP-1 to induce gene transcription. But momordin prevents cell proliferation by preventing this interaction [72]. Again in 2002, Lee et al. found that momordin I is cytotoxic to HL-60 cells with an IC50 of 19.0 μg/ml. These antiproliferative effects of momordin I may be due to induction of apoptosis by modulation of expression of anti-apoptotic protein Bcl-2, pro-apoptotic protein Bax. In addition, momordin I increases the activation of caspase-3 and cleavage of poly (ADP-ribose) polymerase [73]. A recent study by Jing Wang et al. in 2013 showed that momordin Ic induces apoptosis in hepatocellular cells through oxidative stress-regulated mitochondrial dysfunction involving the MAPK and PI3K-mediated iNOS and HO-1 pathways [74].

Charantin is a steroidal glycoside found exclusively in bitter melon and exists as equal mixture of stigmasterol glucoside and β-sitosterol glucoside. It has blood sugar-lowering property similar to insulin. The exact mechanism by which charantin acts to lower blood glucose level is not known, but various reports have shown the hypoglycemic effect of charantin-rich extract *Momordica charantia* (CEMC). In a report submitted in 2014 by Wang et al., it was shown that CEMC improved glucose tolerance by elevating IRS-1 protein levels in mouse liver. Accordingly, CEMC decreased plasma insulin and increased the sensitivity of insulin by increasing the expression of GLUT4 and IRS-1 in the skeletal muscle and liver of KK/HIJ mice. In vitro study showed that CEMC can protect pancreatic β cells from high-glucose-induced cytotoxicity through the expression of IRS-1, GLUT-4, and PPAR-γ in T1D and T2D mice for limited duration. Collectively, the hypoglycemic effects of CEMC

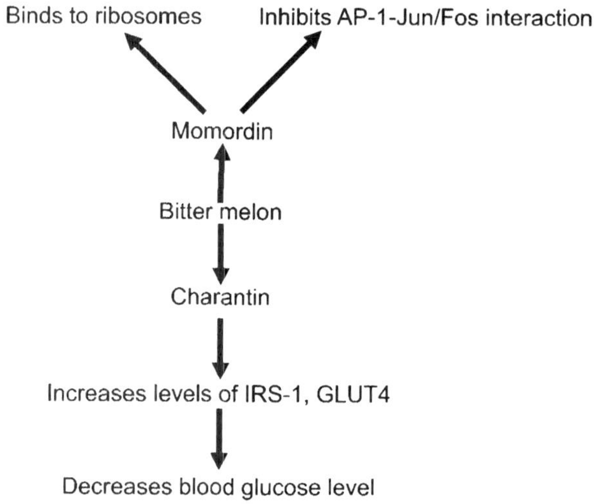

Fig 17.10 Role of bitter melon in disease prevention

suggest that it has potential for increasing insulin sensitivity rather than protection against β-cell dysfunction [75](Fig. 17.10).

17.5.6 Aloe Vera

Aloe vera is a succulent plant species having thick fleshy stems with spiny leaves. The stem is the source of aloe juice which demonstrated to have various benefits from skin to body health. Chemical analysis reveals that *Aloe vera* contains various glucomannans, along with amino acids, vitamins, minerals, and antioxidants. It is rich in vitamins A, B complex, C, and E, folic acid, and niacin. Its juice is also used for skin treatment, minor burns, and sunburns, and so aloe juice is used in a wide range of cosmetics all over the world. *Aloe vera* has inhibitory action on the arachidonic acid pathway via cyclooxygenase [76]. In vitro studies showed that *Aloe vera* increases the collagen content in granulation tissue and the degree of cross-linking of collagen, measured by increased aldehyde content and decreased acid solubility [77]. Lophenol, 24-methyl-lophenol, 24-ethyl-lophenol, cycloartanol, and 24-methylene-cycloartanol are the phytosterols extracted from *Aloe vera* gel, and they have a long-term effect on blood glucose level [77]. *Aloe vera* extracts have been shown to inhibit the growth of microorganisms such as *Streptococcus* and *Shigella* species in vitro [78] (Fig. 17.11).

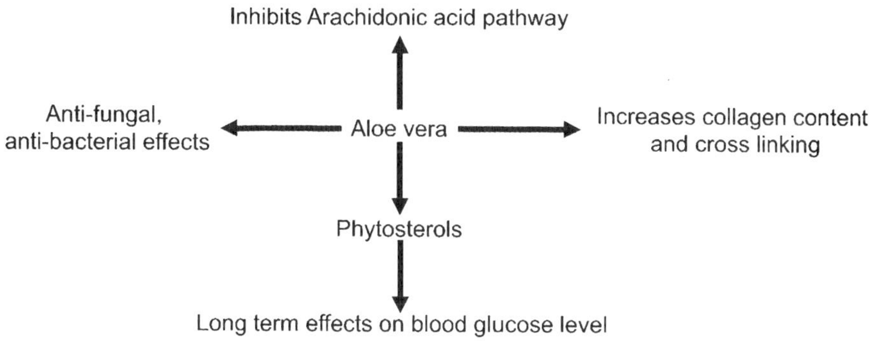

Fig 17.11 Role of *Aloe vera* in disease prevention

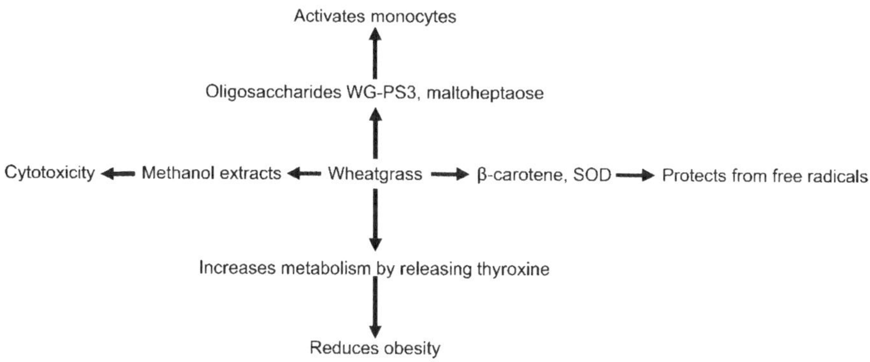

Fig 17.12 Role of wheatgrass in disease prevention

17.5.7 Wheatgrass

Wheatgrass contains 70% chlorophyll and all minerals known to man, and it is rich in vitamins A, B complex, C, E, and K and protein and contains free amino acids. Wheatgrass stimulates metabolism, and its abundance of alkaline minerals helps reduce the acidity in the blood. Wheatgrass stimulates the thyroid glands to produce thyroid hormones which increase the metabolism [79]. This helps to overcome obesity, indigestion, and other problems related to slow metabolism. Wheat grass contains a high amount of beta-carotene and superoxide dismutase, which protects the cell from free radical damage [80]. An oligosaccharide (WG-PS3) found in wheatgrass induces CD69 and Th1 cytokine expression in human PBMCs and CD69, CD80, CD86, IL-12, and TNF-α expression in monocytes [81]. Maltoheptaose, another constituent present in wheatgrass activates monocytes via Toll-like receptor 2 (TLR-2) signaling [132]. Methanol extract of wheatgrass induced the cytotoxic effect by cell cycle arrest at G_1 phase in association with increased p53 and decreased cyclin D1 expression in human laryngeal cells which shows antitumor properties of wheatgrass [82] (Fig. 17.12).

17.5.8 Papaya

Papaya is a magical fruit which can be consumed raw or ripe and has numerous health benefits. It provides a good amount of carbohydrates, potassium, vitamins A, C, and B, and phosphorous as well. Papaya contains an enzyme, papain, which aids in digestion. Because of its laxative property, it is always helpful for people who suffer from gastric problems like diarrhea, gas, constipation, etc. Papaya contains high amount of vitamin C and antioxidants that help in building a stronger immune system. Studies on papaya supplementation improved respiratory burst function and higher abundance of CD68 and CD31 in viable macrophages at the wound site, suggesting effective recruitment of monocytes and an improved proangiogenic response and hepatoprotective roles [83–85].

The seed of papaya has antimicrobial activity against *Trichomonas vaginalis* trophozoites seen in urogenital disorder. The latex of papaya has anti-helminthic activity against *Heligmosomoides polygyrus*, and raw papaya fruit and juice from leaves exhibit antimalarial and anti-dengue activity [83].

17.5.9 Garlic

It is one of the commonly used ingredients in global culinary. It has not only exhibited strong smell and delicious taste but also provided several health benefits. Aged garlic extract (AGE) acts as antioxidant due to rich source of water-soluble and lipid organosulfur compounds and flavonoids, notably allixin and selenium. The AGE quenches ROS, increases glutathione in the cells [86], inhibits platelet aggregation and oxidative modification of LDL, and reduces ischemic/reperfusion damage [87]. AGE also protects against the cardiotoxic effects of cancer chemotherapeutic drugs such as doxorubicin. AGE also protects against liver toxicity by industrial chemicals such as carbon and tetrachloride and an analgesic, acetaminophen [88].

Garlic has an antibiotic property due to high concentration of sulfur, which helps in removing toxins from the digestive system and also helps in building immunity against common colds. Allicin, one of the active principles of garlic, has shown to have selective antimicrobial, antifungal, antiparasitic, and antiviral activities [89, 90].

A sulfur containing amino acid of garlic, S-allyl cysteine sulphoxide (SACS), which is the precursor of allicin and garlic oil, has shown significant antidiabetic effects in alloxan diabetic rats [91](Fig. 17.13).

17.5.10 Indian Gooseberry

Indian gooseberry is a citrus fruit. Its scientific name is *Emblica officinalis*. It is a rich source of vitamin C. In Ayurveda, all parts of the plant are used as medicines. It may be used to promote longevity, alleviate asthma, enhance digestion, treat constipation, stimulate hair growth, reduce fever, purify the blood, reduce cough, strengthen the heart, benefit the eyes, and enhance intellect. Indian gooseberry is very effective in diabetic patients. Ellagic acid, found in Indian gooseberry, acts on

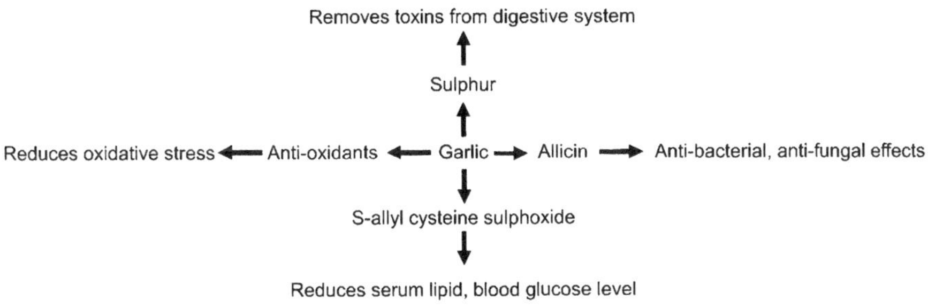

Fig 17.13 Role of garlic in disease prevention

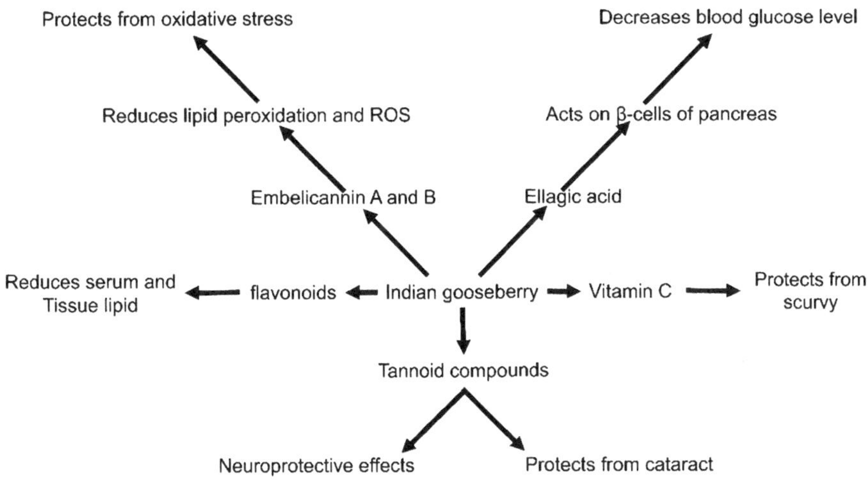

Fig 17.14 Role of Indian gooseberry in disease prevention

β cells of pancreas, stimulates insulin secretion, and thus has antidiabetic property [92]. Recent study shows the neuroprotective effect of tannoid principles of *Indian gooseberry* against memory loss caused by aluminum chloride ($AlCl_3$) intoxication [93]. The tannoid principles were found to play a role in prevention of sugar cataract by inhibiting aldose reductase. Its extracts were found to inhibit the CCl_4-administered hepatotoxicity [94].

Recent studies show that flavonoids from *Emblica officinalis* effectively reduce serum and tissue lipid of rats. Hepatic HMG-CoA reductase activity was significantly inhibited in rats fed with *E. officinalis* flavonoids [95]. It has also been found that ethyl acetate (EtOAc) extract of amla significantly inhibited thiobarbituric acid (TBA)-reactive substance level in the Cu^{2+}-induced LDL oxidation [96]. Aqueous extract of Indian gooseberry was found to be a potent inhibitor of lipid peroxidation and scavenger of hydroxyl and superoxide radicals in vitro. The extract rich in emblicanin A and B was found to increase the cortical and striatal concentrations of antioxidant enzymes superoxide dismutase, catalase, and glutathione peroxidase (Fig. 17.14).

Recent findings show that Indian gooseberry extract modulates tumor cells and its microenvironment in ovarian cancer cells. It decreased cell proliferation in p53-independent manner in human colon cancer stem cells and induced apoptosis [97]. The suppression of cell proliferation was by downregulation of c-Myc and cyclin D1 by targeting Wnt/β-catenin signaling. It also induces mitochondrial-mediated apoptosis as seen by elevated level of Bax/Bcl-2 ratio [97].

17.5.11 Guggul

Guggul is the common name for the flowering mukul myrrh tree (*Commiphora wightii*). It has been used as an Ayurvedic medicine over 2000 years. In Sanskrit the word "guggulu," means "protects from disease." Chemical analysis shows that guggul contains diterpenes, sterols, steroids, esters, and higher alcohols. Its active constituents, the Z- and E-guggulsterones, exhibit their activities by binding to nuclear receptors [98]. Guggulsterones have been found to control transcription factors such as NF-kB [99], STAT, and steroid receptors [100] and thus regulate gene expression. Guggulsterone is a potent antagonist of the farnesoid X receptor or FRX, a nuclear hormone receptor which is activated by bile acids [101]. In mouse macrophage cells, myrrh methanol extract of guggul inhibits the production of NO, PGE2, IL-1β, and TNF-α by downregulating iNOS and COX-2 gene expression. Expression of Cyp3a11 mRNA is also increased in cultured mouse hepatocytes by E-guggulsterone and gugulipid [101].

Guggul contains certain compounds which activate thyroid hormones. Thyroid hormone plays an important role in the regulation of normal metabolism. Administration of guggulsterone increases in iodine uptake by the thyroid gland and enhanced the activities of thyroid peroxidase and protease as well as oxygen consumption [102].

The resin of guggul contains guggulsterones, ketonic steroid compounds, which possess the ability to lower blood levels of LDL cholesterol in the body [103]. Thus there is decrease in plaque buildup on the arterial walls thereby preventing atherosclerosis [104, 105].

The presence of anti-inflammatory compounds such as phenolics, alkaloids, tannins, and steroids in guggul helps to fight infections. The two powerful antioxidants (gugulipid and guggulsterone) present in guggulu help in preventing inflammation by stimulating the immune system [98] (Fig. 17.15).

17.6 Modern Nutraceutical Market

The modern nutraceutical market started in Japan during the 1980s. With the knowledge from Ayurveda and ancient folk medicinal uses of natural herbs and spices, the nutraceutical industry has grown with the expansion and exploration of modern technology. According to Transparency Market Research, the global nutraceuticals market is expanding at a compound annual growth rate of 7.3% from 2015 to 2021.

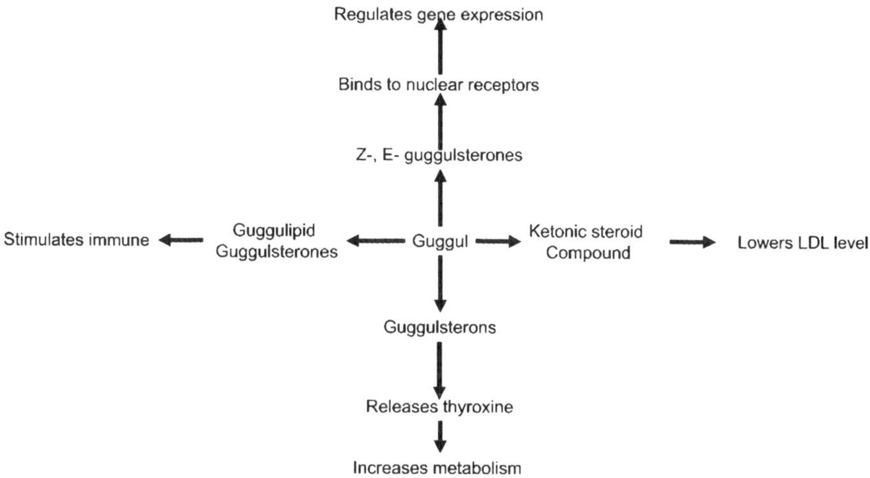

Fig 17.15 Role of guggul in disease prevention

This market has already earned US$182.6 billion in 2015, indicating that the earning will be approximately US$278.95 billion by the end of 2021. These figures indicate that nutraceuticals have potential health benefits that lead to demand among consumers. Now, the global nutraceuticals market is divided into different segments such as functional foods, beverages, dietary supplements, and personal care product. However only functional foods are in demand, whereas other segments are at slower pace in attracting the consumer market.

17.7 Conclusion and Future Prospects

Although nutraceuticals market is flourishing, but there are certain loopholes in the area. In case of research, researchers should study the effect of nutrients on model organisms after studying on cells to come to a definite conclusion. Although nutraceutical market is not regulated by the FDA, the importance of clinical trials cannot be denied. For the pharmaceutical industry, as they are successful in enriching food with beneficial nutrients, they should also try to deplete the levels of compounds which may be harmful. Also, it is very important to understand the effect of storage, temperature, climate, humidity, ripening, cooking, etc. on the nutrients so that the nutrients are consumed in active form.

Acknowledgment Arpita Devi, S. Chennakesavulu, and Chava Suresh acknowledge CSIR for fellowship for doctoral studies and Dr. ABM Reddy acknowledges UoH-UPE-II, DBT-RGYI, DBT-RNAi, and DAE-BRNS for funding laboratory.

References

1. Kalra EK (2003) Nutraceutical--definition and introduction. AAPS PharmSci 5(3):E25
2. Molyneux RJ et al (2007) Phytochemicals: the good, the bad and the ugly? Phytochemistry 68(22–24):2973–2985
3. Hardy G (2000) Nutraceuticals and functional foods: introduction and meaning. Nutrition 16(7–8):688–689
4. Koneri R, Balaraman R, Saraswati C (2006) Antiovulatory and abortifacient potential of the ethanolic extract of roots of *Momordica cymbalaria* Fenzl in rats. Indian J Pharmacol 38(2):111–114
5. Arena MP et al (2014) Barley beta-glucans-containing food enhances probiotic performances of beneficial bacteria. Int J Mol Sci 15(2):3025–3039
6. Roberfroid M (2007) Prebiotics: the concept revisited. J Nutr 137(3 Suppl 2):830S–837S
7. Gomez B et al (2014) Purification, characterization, and prebiotic properties of pectic oligosaccharides from orange peel wastes. J Agric Food Chem 62(40):9769–9782
8. Kelly GS (1999) Larch arabinogalactan: clinical relevance of a novel immune-enhancing polysaccharide. Altern Med Rev 4(2):96–103
9. Zaman SA, Sarbini SR (2016) The potential of resistant starch as a prebiotic. Crit Rev Biotechnol 36(3):578–584
10. Jain I, Kumar V, Satyanarayana T (2015) Xylooligosaccharides: an economical prebiotic from agroresidues and their health benefits. Indian J Exp Biol 53(3):131–142
11. Meister A (1994) Glutathione-ascorbic acid antioxidant system in animals. J Biol Chem 269(13):9397–9400
12. Wells WW et al (1990) Mammalian thioltransferase (glutaredoxin) and protein disulfide isomerase have dehydroascorbate reductase activity. J Biol Chem 265(26):15361–15364
13. Herrera E, Barbas C (2001) Vitamin E: action, metabolism and perspectives. J Physiol Biochem 57(2):43–56
14. Packer L, Weber SU, Rimbach G (2001) Molecular aspects of alpha-tocotrienol antioxidant action and cell signalling. J Nutr 131(2):369S–373S
15. Ninfali P et al (2005) Antioxidant capacity of vegetables, spices and dressings relevant to nutrition. Br J Nutr 93(2):257–266
16. Salas-Salvado J et al (2006) Dietary fibre, nuts and cardiovascular diseases. Br J Nutr 96(Suppl 2):S46–S51
17. Liu RH (2004) Potential synergy of phytochemicals in cancer prevention: mechanism of action. J Nutr 134(12 Suppl):3479S–3485S
18. Khanna P et al (1981) Hypoglycemic activity of polypeptide-p from a plant source. J Nat Prod 44(6):648–655
19. Husain J, Tickle IJ, Wood SP (1994) Crystal structure of momordin, a type I ribosome inactivating protein from the seeds of Momordica charantia. FEBS Lett 342(2):154–158
20. Chandalia M et al (2000) Beneficial effects of high dietary fiber intake in patients with type 2 diabetes mellitus. N Engl J Med 342(19):1392–1398
21. Ajani UA, Ford ES, Mokdad AH (2004) Dietary fiber and C-reactive protein: findings from national health and nutrition examination survey data. J Nutr 134(5):1181–1185
22. El-Serag HB, Satia JA, Rabeneck L (2005) Dietary intake and the risk of gastro-oesophageal reflux disease: a cross sectional study in volunteers. Gut 54(1):11–17
23. Ryan-Harshman M, Aldoori W (2004) How diet and lifestyle affect duodenal ulcers. Review of the evidence. Can Fam Physician 50:727–732
24. Seo G (1989) Inhibition of growth of some enteropathogenic strains in mixed cultures of Streptococcus faecalis and Clostridium butyricum. Microb Lett 40:151–160
25. Reddy BS, Rivenson A (1993) Inhibitory effect of Bifidobacterium longum on colon, mammary, and liver carcinogenesis induced by 2-amino-3-methylimidazo[4,5-f]quinoline, a food mutagen. Cancer Res 53(17):3914–3918

26. Van de Water J, Keen CL, Gershwin ME (1999) The influence of chronic yogurt consumption on immunity. J Nutr 129(7 Suppl):1492S–1495S
27. Alvarez-Sieiro P et al (2016) Bacteriocins of lactic acid bacteria: extending the family. Appl Microbiol Biotechnol 100(7):2939–2951
28. Roberfroid MB (2000) Prebiotics and probiotics: are they functional foods? Am J Clin Nutr 71(6 Suppl):1682S–1687S Discussion 1688S–90S
29. Iwasaki K et al (2016) Daily intake of heat-killed lactobacillus plantarum L-137 decreases the probing depth in patients undergoing supportive periodontal therapy. Oral Health Prev Dent 14(3):207–214
30. Pereira DI, Gibson GR (2002) Effects of consumption of probiotics and prebiotics on serum lipid levels in humans. Crit Rev Biochem Mol Biol 37(4):259–281
31. Moro G et al (2002) Dosage-related bifidogenic effects of galacto- and fructooligosaccharides in formula-fed term infants. J Pediatr Gastroenterol Nutr 34(3):291–295
32. Pool-Zobel B et al (2002) Experimental evidences on the potential of prebiotic fructans to reduce the risk of colon cancer. Br J Nutr 87(Suppl 2):S273–S281
33. Griffin IJ, Davila PM, Abrams SA (2002) Non-digestible oligosaccharides and calcium absorption in girls with adequate calcium intakes. Br J Nutr 87(Suppl 2):S187–S191
34. Tahiri M et al (2003) Effect of short-chain fructooligosaccharides on intestinal calcium absorption and calcium status in postmenopausal women: a stable-isotope study. Am J Clin Nutr 77(2):449–457
35. Ma DW et al (2004) n-3 PUFA alter caveolae lipid composition and resident protein localization in mouse colon. FASEB J 18(9):1040–1042
36. Hughes DA, Southon S, Pinder AC (1996) (n-3) polyunsaturated fatty acids modulate the expression of functionally associated molecules on human monocytes in vitro. J Nutr 126(3):603–610
37. Cominacini L et al (1997) Antioxidants inhibit the expression of intercellular cell adhesion molecule-1 and vascular cell adhesion molecule-1 induced by oxidized LDL on human umbilical vein endothelial cells. Free Radic Biol Med 22(1–2):117–127
38. Kim HS, Quon MJ, Kim JA (2014) New insights into the mechanisms of polyphenols beyond antioxidant properties; lessons from the green tea polyphenol, epigallocatechin 3-gallate. Redox Biol 2:187–195
39. Hou Z et al (2005) Mechanism of action of (−)-epigallocatechin-3-gallate: auto-oxidation-dependent inactivation of epidermal growth factor receptor and direct effects on growth inhibition in human esophageal cancer KYSE 150 cells. Cancer Res 65(17):8049–8056
40. Zang M et al (2006) Polyphenols stimulate AMP-activated protein kinase, lower lipids, and inhibit accelerated atherosclerosis in diabetic LDL receptor-deficient mice. Diabetes 55(8):2180–2191
41. Kampa M et al (2004) Antiproliferative and apoptotic effects of selective phenolic acids on T47D human breast cancer cells: potential mechanisms of action. Breast Cancer Res 6(2):R63–R74
42. Weinreb O et al (2009) Neuroprotective molecular mechanisms of (−)-epigallocatechin-3-gallate: a reflective outcome of its antioxidant, iron chelating and neuritogenic properties. Genes Nutr 4(4):283–296
43. Khan BA, Abraham A, Leelamma S (1995) Hypoglycemic action of Murraya koenigii (curry leaf) and Brassica juncea (mustard): mechanism of action. Indian J Biochem Biophys 32(2):106–108
44. Diao W-R et al (2014) Chemical composition, antibacterial activity and mechanism of action of essential oil from seeds of fennel (Foeniculum vulgare Mill.). Food Control 35(1):109–116
45. Lambert RJ et al (2001) A study of the minimum inhibitory concentration and mode of action of oregano essential oil, thymol and carvacrol. J Appl Microbiol 91(3):453–462
46. Prakash D et al (2007) Total phenol, antioxidant and free radical scavenging activities of some medicinal plants. Int J Food Sci Nutr 58(1):18–28
47. Qin B, Panickar KS, Anderson RA (2010) Cinnamon: potential role in the prevention of insulin resistance, metabolic syndrome, and type 2 diabetes. J Diabetes Sci Technol 4(3):685–693

48. Peterson DW et al (2009) Cinnamon extract inhibits tau aggregation associated with Alzheimer's disease in vitro. J Alzheimers Dis 17(3):585–597
49. Zhang K et al (2017) Cinnamon extract reduces VEGF expression via suppressing HIF-1alpha gene expression and inhibits tumor growth in mice. Mol Carcinog 56(2):436–446
50. Tabak M, Armon R, Neeman I (1999) Cinnamon extracts' inhibitory effect on Helicobacter pylori. J Ethnopharmacol 67(3):269–277
51. Johnston CS et al (2010) Examination of the antiglycemic properties of vinegar in healthy adults. Ann Nutr Metab 56(1):74–79
52. Liatis S et al (2010) Vinegar reduces postprandial hyperglycaemia in patients with type II diabetes when added to a high, but not to a low, glycaemic index meal. Eur J Clin Nutr 64(7):727–732
53. Naziroglu M et al (2014) Apple cider vinegar modulates serum lipid profile, erythrocyte, kidney, and liver membrane oxidative stress in ovariectomized mice fed high cholesterol. J Membr Biol 247(8):667–673
54. Shishehbor F et al (2008) Apple cider vinegar attenuates lipid profile in normal and diabetic rats. Pak J Biol Sci 11(23):2634–2638
55. Ohnishi M et al (1994) Inhibitory effects of chlorogenic acids on linoleic acid peroxidation and haemolysis. Phytochemistry 36(3):579–583
56. Feng R et al (2005) Inhibition of activator protein-1, NF-kappaB, and MAPKs and induction of phase 2 detoxifying enzyme activity by chlorogenic acid. J Biol Chem 280(30):27888–27895
57. Lee WJ, Zhu BT (2006) Inhibition of DNA methylation by caffeic acid and chlorogenic acid, two common catechol-containing coffee polyphenols. Carcinogenesis 27(2):269–277
58. Cho AS et al (2010) Chlorogenic acid exhibits anti-obesity property and improves lipid metabolism in high-fat diet-induced-obese mice. Food Chem Toxicol 48(3):937–943
59. Mehmood MH et al (2011) Pharmacological basis for the medicinal use of psyllium husk (Ispaghula) in constipation and diarrhea. Dig Dis Sci 56(5):1460–1471
60. Marlett JA, Kajs TM, Fischer MH (2000) An unfermented gel component of psyllium seed husk promotes laxation as a lubricant in humans. Am J Clin Nutr 72(3):784–789
61. Zaman V et al (2002) The presence of antiamoebic constituents in psyllium husk. Phytother Res 16(1):78–79
62. Sakanaka S et al (1989) Antibacterial substances in Japanese green tea extract against Streptococcus mutans, a cariogenic bacterium. Agric Biol Chem 53(9):2307–2311
63. Ruch RJ, Cheng SJ, Klaunig JE (1989) Prevention of cytotoxicity and inhibition of intercellular communication by antioxidant catechins isolated from Chinese green tea. Carcinogenesis 10(6):1003–1008
64. Graham HN (1992) Green tea composition, consumption, and polyphenol chemistry. Prev Med 21(3):334–350
65. Kim SH et al (2006) Gallic acid inhibits histamine release and pro-inflammatory cytokine production in mast cells. Toxicol Sci 91(1):123–131
66. Kroes BH et al (1992) Anti-inflammatory activity of gallic acid. Planta Med 58(6):499–504
67. Hsieh CL et al (2015) The teratogenicity and the action mechanism of gallic acid relating with brain and cervical muscles. PLoS One 10(6):e0119516
68. Paolini A et al (2015) Gallic acid exerts a protective or an anti-proliferative effect on glioma T98G cells via dose-dependent epigenetic regulation mediated by miRNAs. Int J Oncol 46(4):1491–1497
69. Khan N et al (2006) Targeting multiple signaling pathways by green tea polyphenol (−)-epigallocatechin-3-gallate. Cancer Res 66(5):2500–2505
70. Kuriyama S et al (2006) Green tea consumption and mortality due to cardiovascular disease, cancer, and all causes in Japan: the Ohsaki study. JAMA 296(10):1255–1265
71. Ahmad N et al (1997) Green tea constituent epigallocatechin-3-gallate and induction of apoptosis and cell cycle arrest in human carcinoma cells. J Natl Cancer Inst 89(24):1881–1886
72. Lee DK et al (1998) Momordins inhibit both AP-1 function and cell proliferation. Anticancer Res 18(1A):119–124

73. Kim JH et al (2002) Induction of apoptosis by momordin I in promyelocytic leukemia (HL-60) cells. Anticancer Res 22(3):1885–1889
74. Wang J et al (2013) Momordin Ic induces HepG2 cell apoptosis through MAPK and PI3K/Akt-mediated mitochondrial pathways. Apoptosis 18(6):751–765
75. Wang HY et al (2014) Differential anti-diabetic effects and mechanism of action of charantin-rich extract of Taiwanese Momordica charantia between type 1 and type 2 diabetic mice. Food Chem Toxicol 69:347–356
76. Chithra P, Sajithlal GB, Chandrakasan G (1998) Influence of Aloe vera on collagen characteristics in healing dermal wounds in rats. Mol Cell Biochem 181(1–2):71–76
77. Tanaka M et al (2006) Identification of five phytosterols from Aloe vera gel as anti-diabetic compounds. Biol Pharm Bull 29(7):1418–1422
78. Arunkumar S, Muthuselvam M (2009) Analysis of phytochemical constituents and antimicrobial activities of Aloe vera L. against clinical pathogens. World J Agric Sci 5(5):572–576
79. Jorige A, Akula A (2015) Neuroprotective role of wheatgrass powder in experimental diabetic neuropathy via modulating oxidative stress markers in rat sciatic nerves. Am J Phytomed Clin Ther 3(7):529–540
80. Kashudhan H, Dixit A, Upadhyay A (2017) Optimization of ingredients for the development of wheatgrass based therapeutical juice using response surface methodology (RSM). J Pharmacogn Phytochem 6(2):338–345
81. Tsai CC et al (2013) The immunologically active oligosaccharides isolated from wheatgrass modulate monocytes via Toll-like receptor-2 signaling. J Biol Chem 288(24):17689–17697
82. Shakya G, Balasubramanian S, Rajagopalan R (2015) Methanol extract of wheatgrass induces G1 cell cycle arrest in a p53-dependent manner and down regulates the expression of cyclin D1 in human laryngeal cancer cells-an in vitro and in silico approach. Pharmacogn Mag 11(Suppl 1):S139
83. Satrija F et al (1995) Anthelmintic activity of papaya latex against patent Heligmosomoides polygyrus infections in mice. J Ethnopharmacol 48(3):161–164
84. Collard E, Roy S (2010) Improved function of diabetic wound-site macrophages and accelerated wound closure in response to oral supplementation of a fermented papaya preparation. Antioxid Redox Signal 13(5):599–606
85. Adeneye A et al (2009) The aqueous seed extract of Carica papaya Linn. Prevents carbon tetrachloride induced hepatotoxicity in rats. Int J Appl Res Nat Prod 2(2):19–32
86. Borek C (2001) Antioxidant health effects of aged garlic extract. J Nutr 131(3):1010S–1015S
87. Rahman K, Lowe GM, Smith S (2016) Aged garlic extract inhibits human platelet aggregation by altering intracellular signaling and platelet shape change. J Nutr 146(2):410S–415S
88. Alkreathy H et al (2010) Aged garlic extract protects against doxorubicin-induced cardiotoxicity in rats. Food Chem Toxicol 48(3):951–956
89. Ankri S, Mirelman D (1999) Antimicrobial properties of allicin from garlic. Microbes Infect 1(2):125–129
90. Ross Z et al (2001) Antimicrobial properties of garlic oil against human enteric bacteria: evaluation of methodologies and comparisons with garlic oil sulfides and garlic powder. Appl Environ Microbiol 67(1):475–480
91. Sheela C, Augusti K (1992) Antidiabetic effects of S-allyl cysteine sulphoxide isolated from garlic Allium sativum Linn. Indian J Exp Biol 30(6):523–526
92. Fatima N et al (2017) Ellagic acid in Emblica officinalis exerts anti-diabetic activity through the action on β-cells of pancreas. Eur J Nutr 56(2):591–601
93. Justin Thenmozhi A et al (2016) Tannoid principles of Emblica officinalis renovate cognitive deficits and attenuate amyloid pathologies against aluminum chloride induced rat model of Alzheimer's disease. Nutr Neurosci 19(6):269–278
94. Suryanarayana P et al (2004) Inhibition of aldose reductase by tannoid principles of Emblica officinalis: implications for the prevention of sugar cataract. Mol Vis 10:148
95. Bhuvaneswari R, Chidambaranathan N, Jegatheesan K (2014) Hepatoprotective effect of embilica officinalis and its silver nanoparticles against CCl4 induced hepatotoxicity in wistar albino rats. Dig J Nanomater Biostruct (DJNB) 9(1):223

96. Kim HJ et al (2005) Influence of amla (Emblica officinalis Gaertn.) on hypercholesterolemia and lipid peroxidation in cholesterol-fed rats. J Nutr Sci Vitaminol 51(6):413–418
97. Vadde R et al (2016) Indian gooseberry (Emblica officinalis Gaertn.) suppresses cell proliferation and induces apoptosis in human colon cancer stem cells independent of p53 status via suppression of c-Myc and cyclin D1. J Funct Foods 25:267–278
98. Shah R, Gulati V, Palombo EA (2012) Pharmacological properties of guggulsterones, the major active components of gum guggul. Phytother Res 26(11):1594–1605
99. Lv N et al (2008) Guggulsterone, a plant sterol, inhibits NF-κB activation and protects pancreatic β cells from cytokine toxicity. Mol Cell Endocrinol 289(1):49–59
100. Macha MA et al (2013) Guggulsterone decreases proliferation and metastatic behavior of pancreatic cancer cells by modulating JAK/STAT and Src/FAK signaling. Cancer Lett 341(2):166–177
101. Brobst DE et al (2004) Guggulsterone activates multiple nuclear receptors and induces CYP3A gene expression through the pregnane X receptor. J Pharmacol Exp Ther 310(2):528–535
102. Tripathi YB, Malhotra O, Tripathi S (1984) Thyroid stimulating action of Z-guggulsterone obtained from Commiphora mukul. Planta Med 50(01):78–80
103. Wu J et al (2002) The hypolipidemic natural product guggulsterone acts as an antagonist of the bile acid receptor. Mol Endocrinol 16(7):1590–1597
104. Chander R, Khanna A, Kapoor N (1996) Lipid lowering activity of guggulsterone from Commiphora mukul in hyperlipaemic rats. Phytother Res 10(6):508–511
105. Baldwa VS et al (1981) Effects of Commiphora Mukul (Guggul) in experimentally induced hyperlipemia and atherosclerosis. J Assoc Physicians India 29(1):13–17
106. Duester G (2008) Retinoic acid synthesis and signaling during early organogenesis. Cell 134(6):921–931
107. Frederich M et al (2009) Thiaminylated adenine nucleotides. Chemical synthesis, structural characterization and natural occurrence. FEBS J 276(12):3256–3268
108. Snell EE (1953) Summary of known metabolic functions of nicotinic acid, riboflavin and vitamin B6. Physiol Rev 33(4):509–524
109. Mandula B, Beutler E (1970) Synthesis of riboflavin nucleotides by mature human erythrocytes. Blood 36(4):491–499
110. Gravel RA, Robinson BH (1985) Biotin-dependent carboxylase deficiencies (propionyl-CoA and pyruvate carboxylases). Ann N Y Acad Sci 447:225–234
111. Goh YI, Koren G (2008) Folic acid in pregnancy and fetal outcomes. J Obstet Gynaecol 28(1):3–13
112. Mierzejewski K, Rozengurt E (1976) Stimulation of DNA synthesis and cell division in a chemically defined medium: effect of epidermal growth factor, insulin and vitamin B12 on resting cultures of 3T6 cells. Biochem Biophys Res Commun 73(2):271–278
113. Zittoun J (1993) Anemias due to disorder of folate, vitamin B12 and transcobalamin metabolism. Rev Prat 43(11):1358–1363
114. Peterkofsky B (1991) Ascorbate requirement for hydroxylation and secretion of procollagen: relationship to inhibition of collagen synthesis in scurvy. Am J Clin Nutr 54(6 Suppl):1135S–1140S
115. Yamada SSMYK (2004) Vitamin D and rickets, vol 6. Karger Publishers, pp 50–68
116. Price PA, Williamson MK (1985) Primary structure of bovine matrix Gla protein, a new vitamin K-dependent bone protein. J Biol Chem 260(28):14971–14975
117. Rose CR (2002) Book review: Na+ signals at central synapses. Neuroscientist 8(6):532–539
118. Titze J et al (2006) Extrarenal Na+ balance, volume, and blood pressure homeostasis in intact and ovariectomized deoxycorticosterone-acetate salt rats. Hypertension 47(6):1101–1107
119. Ransom CB, Ransom BR, Sontheimer H (2000) Activity-dependent extracellular K+ accumulation in rat optic nerve: the role of glial and axonal Na+ pumps. J Physiol 522(Pt 3):427–442
120. Ebashi S, Endo M (1968) Calcium ion and muscle contraction. Prog Biophys Mol Biol 18:123–183
121. Stipanuk MH (2004) Sulfur amino acid metabolism: pathways for production and removal of homocysteine and cysteine. Annu Rev Nutr 24:539–577

122. Thorslund A, Lindskog S (1967) Studies of the esterase activity and the anion inhibition of bovine zinc and cobalt carbonic anhydrases. Eur J Biochem 3(1):117–123
123. Schneider WC (1946) Phosphorus compounds in animal tissues; a comparison of methods for the estimation of nucleic acids. J Biol Chem 164(2):747–751
124. Morgan EH (1961) Plasma-iron and haemoglobin levels in pregnancy. The effect of oral iron. Lancet 1(7167):9–12
125. Sun W (2014) Iodine-131 and thyroid function. Environ Health Perspect 122(2):A40
126. DelaFuente G, Lagunas R, Sols A (1970) Induced fit in yeast hexokinase. Eur J Biochem 16(2):226–233
127. Zipkin I, Mc CF (1952) Deposition of fluorine in the bones and teeth of the growing rat. J Nutr 47(4):611–620
128. Ravindranath SD, Fridovich I (1975) Isolation and characterization of a manganese-containing superoxide dismutase from yeast. J Biol Chem 250(15):6107–6112

Part III

Functional Food and Human Health: Effective Delivery Methods, Product Development and Toxicity

Extraction and Characterization of Phytochemicals

18

Aditi Khare, Gauransh Jain, and Vibha Rani

Abstract

Recent advances in phytochemical uses have shown its activity as protective role against a variety of diseases, which marks its importance in developing functional food and even in nutraceuticals. Extraction from plant materials is a crucial step to the food processing industries. The whole process revolves around eliminating the less concerned part to maximize benefits out of herbs. The main aim of standardizing the extraction procedures is to eliminate unwanted material and to obtain the most enriched bioactive portion of the herbs, with maximal therapeutic potential using different solvent systems. Recent advances have developed a wide range of technologies in extraction of active components and other aromatic essential oils from plants and herbs having a known therapeutic value. Adoption of a certain procedure depends upon the yield and economic feasibility of the process concerned with the plant. This chapter gives a complete insight into the various available methods for phytochemical extraction, their purification, and finally its characterization.

Keywords

Phytochemicals · Functional food · Extraction · HPLC · OPLC · HPTLC · Chromatography · NMR · MALDI-TOF

Authors Aditi Khare and Gauransh Jain have equally contributed to this chapter.

A. Khare · G. Jain · V. Rani (✉)
Department of Biotechnology, Jaypee Institute of Information Technology, Noida, Uttar Pradesh, India
e-mail: Vibha.rani@jiit.ac.in

V. Rani, U. C. S. Yadav (eds.), *Functional Food and Human Health*,
https://doi.org/10.1007/978-981-13-1123-9_18

18.1 Introduction

Phytochemicals are biologically active, plant-based secondary metabolites which aid plants with color and play an important role in exerting defense response against pest and other predators.

Traditional use of plants for medicinal purposes is practiced across the globe since the very first civilization. The awareness of these plants was created by years of struggles to cure the illness for which the man moved toward the barks, fruits, roots, and other parts of plant. Contemporary science acknowledged this plant-based therapeutic action and became a base for modern pharmacotherapy, where a range of drug belongs to the plant origin. Even in modern times, majority of the pharmacotherapy relies on plants and plant-based molecules [1].

According to the World Health Organization (WHO), "medicinal plant" is the one which has more than one of its parts rich in content depicting a therapeutic value [2]. Therapeutic efficiency should also include triggering of a chemo-pharmaceutical-based action in the body. The extract must contain an active constituent having a therapeutic effect. Such chemical components are called phytochemicals or phytoconstituents [3]. Though their action doesn't play a fair role in nutrition, however, their active role in protection against invading pathogens has been reported. This property imparts an antimicrobial action and thus forms a backbone in therapeutic role of the plant.

18.2 Uses of Phytomolecules

There exist different modes of action for different phytochemicals, thus imparting their varied uses. The basis of action of a phytomolecule depends majorly on its antioxidant activity, which helps in protection of our cells against different types of oxidative damage, which also increases the risk of developing tumors [4]. The active constituent mainly contains phytochemicals like allyl sulfides, which are found abundantly in onions and garlic, carotenoids and flavonoids obtained majorly from fruits and vegetables, and polyphenols which are present in tea and grapes [5]. Other phytomolecules like isoflavones, which are found in soy, can imitate the human estrogens and thus prove its action in reducing menopausal symptoms. It also has positive effects on osteoporosis [6].

Indoles, from cabbages, also play an important role in maintaining fertility. They stimulate the enzymes which are responsible for maintaining the estrogen levels and also reduce the risk of development of breast cancer [7]. Similar effects have been shown by different protease inhibitors obtained mainly from soy and terpenes from citrus fruits and cherries. Beans rich in saponins regulate the process of replication of DNA and thus prevent malignancy. Capsaicin generally found in black pepper is also anti-carcinogenic. Allicin obtained from garlic is well known for its antibacterial properties [8]. Some phytochemicals also play an important role in defense mechanism. They can bind to the cell walls and hence compete with the foreign-invading pathogens for adherence. Proanthocyanidin is one such molecule with anti-adhesion properties found in cranberry [9]. Its consumption also reduces the risk of the urinary tract infections and improves the dental health.

18.3 Importance of Extraction Procedures

Extraction is a crucial step in the analysis of plants having medicinal properties. It becomes necessary to separate the desired constituent from plant materials for its further analysis. The term extraction basically involves separation of the main constituent having pharmaceutical response from a medicinal plant or animal tissues with the involvement of a basic solvent while following a specific extraction procedure [10]. The extract obtained directly from plants is relatively impure and needs further refining. These extracted crude products may exist as impure mixed liquids or semisolids or in form of a fine powder [11].

There are different methods involved in extraction process which are primarily aimed in order to separate the pharmaceutically useful substance from the whole part neglecting the inert portion. Next to the extraction process, the available crude extract is subjected to a variety of refining, which is finally followed by its characterization [12]. The finalized product can either be taken wholly as in form of a tincture or fluid extract or can be formulated in a specific dose in form of capsules or oral suspensions. Sometimes certain agents like an excipient molecules are also added in the formulation so as to increase the therapeutic efficiency of drug [13]. Thus standardization forms the basis of drug formulation and enhances the quality of drug molecule.

Prior to extraction exists certain basis of treatment and operations like washing and drying of plant materials and powdering or grinding to obtain a homogenous extract. This is done in order to increase the surface area of particles for interaction with solvent for efficient removal of active constituent [14].

The process of extraction and pretreatment are highly regulated and specific for plants. This ensures that the potentially active constituent molecules are not degraded or distorted due to high temperature [15]. Forthcoming sections focus on the different extraction techniques and methods involved in characterization of these phytomolecules.

18.4 Solvent System for Extraction

The selection of a solvent for dissolving the phytoconstituents primarily depends upon the nature of desired compound. Solvent selection basically depends on the concept of "like dissolves like." There exists a varied range of solvent systems for efficient extraction of the bioactive component [16].

For the extraction of a hydrophilic molecule, solvent solution should comprise of a polar molecule like ethanol, methanol, and ethyl acetate. Similarly for extraction of a lipophilic, i.e., a more hydrophobic compound, a mixture of dichloromethane and methanol can be used [17]. Dichloromethane can also be used singly as a solvent. Nature of the bioactive component describes the solvent system to be used. Target molecule can be heat tolerant to heat labile; this variation decides the solvent and extraction procedures to be used further [18].

18.5 Steps Involved in Phytochemical Extraction

Extraction refers to the sequential steps involved in separation of the pharmaceutically active component from plant/animal portion by the use of a specific solvent. This also includes post-separation refining and characterization of the separated crude component that can be then validated quantitatively [19]. This extract can further be further processed and standardized and availed for commercial benefits. Thus, standardization of extraction procedures contributes significantly to the final quality of the herbal drug.

So the whole process of phytochemical extraction involves three basic steps:

1. Procurement of plants
2. Extraction of bioactive constituents
3. Characterization of derived constituents

18.5.1 Procurement of Plants

Procurement or the collection of plants/herbs for extraction process marks an important step which decides the yield and the quality of extract. Plants obtained for extract preparation must be of same variety and almost same size and should be chosen from a particular region, for enhanced separation and less variability. Steps involved in the process of selecting the plants for extract preparation are given below.

18.5.1.1 Collection of Plants

Plants as a part are collected from forests or herbariums and are screened for any deformities and differences. Generally samples of the same varieties, having similar characteristics, are obtained. Plants grown in forests are generally free from chemical pesticide presence but may be contaminated with pathogens and other toxins; therefore a proper washing and decontamination process of the gathered batch are necessary [20].

18.5.1.2 Cleaning of Collected Plants and Their Parts

Soon after the collection procedure is over, the prime focus shifts toward making the batch free from any pathogen and insects. Also presence of dirt and other toxins (like animal excreta, bird droppings, and other plant excreta) may hinder the characterization techniques and also reduces the yield of extracts [20]. Cleaning process involves a series of sequential steps like washing, cutting, and peeling leaves from stems. Process of cleaning could be by hands or is mechanized involving different mechanical processes and detergents.

18.5.1.3 Drying of Plant Parts

The aim of drying is to reduce the water content from the plants completely for their storage. Plant parts in dried forms extend their shelf life and avoid any fungal and other bacterial growth. Drying of material can also be done either naturally, i.e., under the sun, or with the aid of a hot air oven [21]. The use of an artificial source of drying reduces the time and maintains homogeneity in the batches.

Natural Process of Drying

Sun drying is the most common natural process. After proper washing and cleaning, the plant parts are spread onto huge drying panels and left under the sun for some days, until a dry product is obtained. Sun drying proves to be an inexpensive method and has been widely followed [22]. Still the process is time taking and batch-to-batch variation may arise in the product. Other natural methods of drying include placing plants on drying stands and frames, which are left for air drying under the sheds. Both the processes are highly variable and depend on humidity and temperature [23].

Mechanized Drying (Artificial Drying)

This includes the use of air dryers, which are energy driven. Time of drying reduces significantly making the product homogenously dried. Commonly for medicinal plants, warm air drying is preferred. Plant parts are spread onto plates in air dryer and warm air is blown using a blower [24]. The temperature and time of the whole process can be varied and monitored. Time of preparation drastically reduces but enhances the cost of production. These methods also include a large number of workers for manual loading and unloading of sample. Time of drying also varies depending on the fragility of the product [25]. Time varies from minutes to hours for flowers and leaves, whereas the same is comparatively higher for roots and barks.

18.5.1.4 Powdering

Prior to solvent extraction, the dried products are crushed and grinded forming a homogenous powdered product. This grinding increases the surface area of the particles for efficient extraction and makes the product fine. This process of powdering also can be mechanical or can include use of automated grinders and shredders.

18.5.2 Methods of Extraction

After obtaining plants/herbs, proper methods of extraction should be followed for increasing the separation of active components from the plant. This generally requires different physical as well as chemical approaches. Different techniques for efficient extraction are explained in the upcoming sections.

18.5.2.1 Homogenization

Several researches have discussed homogenization of plant extract in different solvent. The plant parts are grinded in blender and put into a specific volume of solvent so as to create a specific concentration solution. Extract can be both dried and in liquid form. The resulting mixture is shaken vigorously or left for 24 h for efficient molecule movement, until the active constituent evades out of the extract into the solution. This process is followed by filtration of extract. The filtrate obtained is rich in the phytoconstituent, and now this extract can be put into various uses. It can also be dried under elevated pressure, process facilitated by a lyophilizer [24]. This drying increases its shelf life and extracts can be stored for a longer time. The dried extract can also be redissolved in a solvent to determine its concentration. Filtered extract can also be centrifuged so as to obtain a clear solution.

18.5.2.2 Serial Exhaustive Extraction

It is one of the most used methods of extraction which involves successive extraction with some polar solvents. This ensures a wide variety of phytomolecules of different polarities to be extracted. Some studies involve the use of Soxhlet extraction of a dried plant part with a polar solvent [25]. Since this method involves the use of heat, it becomes unsuitable for heat labile compounds.

18.5.2.3 Soxhlet Extraction

Soxhlet apparatus works on the basis of solvent extraction with the aid of polar solvents. It proves to be useful when the desired component has a limited solubility in a solvent and the impurities are relatively insoluble. Thus it gives an efficient separation of components based on temp regulation. If the desired compound has a high solubility in a solvent, then a simple filtration can be used to separate the compound from the insoluble substance. The main advantage of this process relies on the fact that solvent is recycled and can be reused in the next batch [25]. Another advantage of this process is the reduced time of phytoextraction, as compared to the other methods. However this method is not useful for components which are heat sensitive, since prolonged heating may denature the proteins and thus degrade the compounds.

18.5.2.4 Maceration

The process of maceration involves processing of dried and fine-powdered components that were obtained during extraction and are mixed with a solvent. The resultant suspension is transferred to a stoppered container and is agitated until a homogenous solution is obtained. This generally involves soaking and agitating of the plant-based extract with the solvent and then allowing the solvent to drain off, leaving a homogenous solution [26]. Micelles are further removed by centrifugation. This is one of the most followed methods which are best suitable for extraction of thermolabile components. However, this does not completely extract the phyto ingredients from plant material.

18.5.2.5 Decoction

Decoction is one of the main methods for extraction of a water-soluble component. The procedure involves boiling of a crude extract in water for 15 min and then cooling it [27]. This is followed by straining and passing cold water through the drug to increase the volume.

18.5.2.6 Digestion

This is a customized maceration process which involves application of heat during the maceration extraction step. Gentle heat is applied during the process and thereby helps in increasing the solvent efficiency of extract [28]. This process can be proved useful for extraction of components which are heat tolerant and also takes less time than the maceration process.

18.5.2.7 Percolation

Percolation is the most widely used method for extraction of phytoconstituents in the preparation of tinctures and fluid extracts. A device percolator structurally comprises of a narrow, cone-shaped vessel which is open from both the ends. There is a limited exposure of solvent to the solid extract; therefore in a closed container, extract is allowed to stand for about 4 h, and the percolator is closed [28]. This step is called "packing" of percolator. Then repetitive steps of rinsing of plant material with solvent are followed, until the complete removal of phytocompounds. Similarly the solvent is re-pumped and contains active constituent. Solvent is reused until a saturation is observed. The outlet of the instrument is opened for the solvent containing phyto ingredient to drip slowly, and this process is further enhanced by pressing the extract [29]. The solvent obtained is then subjected to different processes like filtration, standing, clearing, and decanting, in order to efficiently separate the desired component. This process excels over maceration process due to better extraction efficiency and less time requirement. Percolation can also be useful for heat labile drugs.

18.5.2.8 Sonication

Sonication involves the use of sound energy in order to agitate molecules. Ultrasound waves of a range of frequencies from 20 kHz to 2000 kHz are used in the procedure so as to increase the reach of solvent molecules in plant product and thereby increase permeation in cell wall. This is an efficient process as it enhances the rate of solvent distribution but, on the other hand, it also increases the cost of production [29]. Another disadvantage of the process is relative deterioration of the plant active molecules due to high-energy ultrasound waves, leading to formation of free radicals and loss of the pharmacological action.

18.5.3 Characterization of Derived Phytoconstituents

Plants have been found to possess wide spectra of phytochemicals such as flavonoids, alkaloids, saponins, tannins, glycosides, terpenoids, phenols, polyphenols,

Table 18.1 Different tests for alkaloids present in phytochemical extract

Reagent Test	Composition	Colour obtained
Meyer's test	Potassiomercuric iodide sol.	Cream ppt.
Wagner's test	12 + KI	Reddish-Brown ppt.
Murexide test	Ammonium vapor	Pink solution
Tannic acid	Tannic acid	Ppt. (no colour change)
Hager's reagent	Picric acid	Yellow ppt.
Dragendoeff's reagent	Potassium bismuth iodide, potassium chlorate	Orange ppt.

phlobatannins, steroids, etc. Therefore there is a need of identifying, isolating, and purifying the active phytoconstituent in the crude extracts.

18.5.3.1 Qualitative Estimation of Metabolites

Several qualitative methods are employed on extracts to evaluate the presence of various majorly expressed phytochemicals (secondary metabolites) in plant extracts. These metabolites are said to have various supportive roles other than normal expression and are the main responsible agents for plant's therapeutic efficiency. Different qualitative tests can be done to indicate the presence of such compounds.

Qualitative Test for Presence of Alkaloids

Mayer's Test

Mayer's reagent typically consists of a solution of HgI_4. Reaction of lone pair of N present in sample with Mayer's reagent yields a white precipitate. This indicates the presence of free N in the sample and thereby determines the concentration qualitatively. However alkaloids lacking a lone pair give negative results, so they should be validated quantitatively.

Addition of two drops of Mayer's reagent to a few ml plant extract in a test tube yields a white precipitate. Appearance of a precipitate is the indication of presence of alkaloids [30]. Various other methods are also reported to analyze the alkaloid presence (Table 18.1).

Qualitative Test for Presence of Amino Acids

Ninhydrin Test

Ninhydrin (2,2-dihydroxyindane-1,3-dione) is used to detect presence of a primary and secondary amine. On reaction with a free amine, the solution turns blue due to formation of Ruhemann's purple. This indicates the presence of amines or lysine residues in proteins [30]. Reaction involving the Ninhydrin test is shown in the following Fig. 18.1.

Benedict's Test

Reducing sugars upon heating in the presence of an alkali reduces the Cu^{2+} ion present in solution into insoluble red-colored Cu_2O adduct. Reducing sugars first gets converted into enediols, which further reduces the free Cu^{2+} ions. The insoluble red product is chemically copper (I) oxide (Cu_2O). Analysis can be done spectrophotometrically

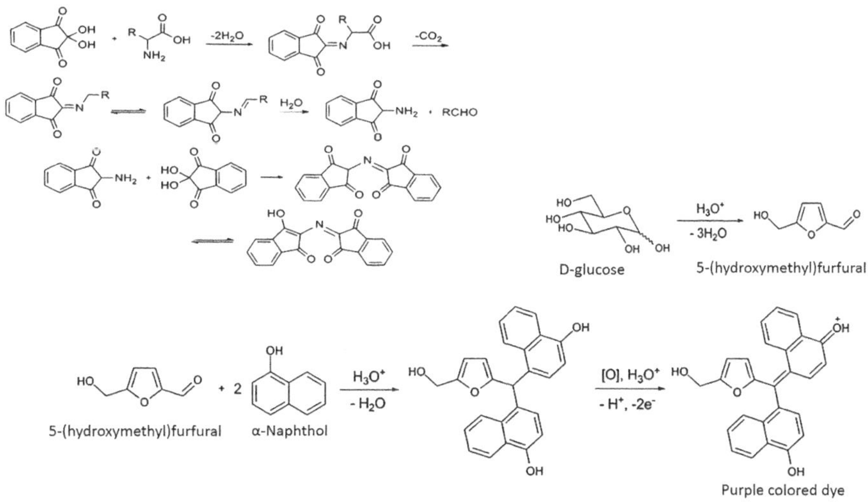

Fig. 18.1 Reactions taking place during Ninhydrin test

to quantitate the concentration of carbohydrates present as the bioactive agents in the extracts 0.5 ml of the crude extract is mixed in equimolar concentration with Benedict's reagent, and heated for 2 min in a boiling water bath. Appearance of a reddish brown color indicates the carbohydrates [31].

Qualitative Test for Presence of Phenolic Compounds and Tannins

Gelatin Hydrolysis Test

Gelatin hydrolysis test evaluates the presence of phenolic compounds in the bioactive constituent. 50 mg dried extract is dissolved in 5 ml of distilled water. To the resulting solution, 10% NaCl and 1% gelatin are added. Gelatin acts as a substrate, and formation of a white insoluble precipitate indicates the presence of phenols. It is a semiquantitative test and can be evaluated spectrophotometrically [32].

Ferric Chloride Test

Ferric chloride test is used to evaluate the presence of phenols and polyphenols in a plant extract sample. However certain enols and oximes also give positive results, which can create a sense of false-positive results. This test only gives an indication of presence of phenols, which can be further evaluated using techniques like spectrophotometer.

50 mg dried extract is dissolved in 5 ml of distilled water, and to this, a few drops of 5% ferric chloride solution were added; soon development of a dark green color indicates the presence of a phenolic compound. This method also relies on spectrophotometric analysis for quantitative analysis [33].

Alkaline Reagent Test

Addition of 10% ammonium hydroxide solution to aqueous extract of plant yields a yellow fluorescence, which clearly mentions the presence of flavonoids in the extract.

Qualitative Test for Presence of Phytosterols in Plant-Based Extract

Liebermann–Burchard's Test

The Liebermann–Burchard or acetic anhydride test is a primary test for the detection of polysterols and cholesterol in the bioactive agent. This colorimetric assay leads to formation of deep green color, which validates the result quantitatively. Conjugation between the –OH group of polysterols and reagents results in color development.

50 mg of crude extract is dissolved in 2 ml acetic anhydride. In the resulting solution, few drops of concentrated sulfuric acid were added along the walls, which results in color development. Intensity of color in an array clearly depicts the concentration of polysterols, which can be further quantified spectrophotometrically [31].

Qualitative Test for Presence of Proteins in Plant-Based Extract

Millon's Test

Millon's reagent is chemically mercurous and mercuric nitrate solutions. On reaction with a hydroxybenzene radical, a red- or pink-colored compound is observed. Product named mercuric fumarate is formed after the reaction giving a pinkish or red color [32].

Biuret Test

Also known as Piotrowski's test, the biuret test detects the presence of peptide linkages in the solution, which indirectly indicates the presence of protein component in the solution. The resultant solution gives a red color intensity, and intensity of a colored product depicts the concentration quantitatively. Absorption at 540 nm quantitates the concentration of proteins.

2 ml of filtrate is treated with 1 drop of 2% copper sulfate solution. To this 1 ml of ethanol (95%) is added, followed by excess of potassium hydroxide pellets. Pink color ethanolic layer indicates the presence of protein [33]. A complete and comprehensive collection of tests along with their tests has been documented in Table 18.2.

18.6 Qualitative and Quantitative Estimation of Polyphenols and Other Functional Groups

Phytochemicals present in the crude plant extract consist of a large number of secondary metabolites found in plants. Phytochemicals extracted from the crude extract are subjected to various qualitative and quantitative analyses for analyzing the levels of bioactive compounds in the extract and working out on the probable composition of the bioactive component.

Table 18.2 Qualitative tests for major secondary metabolites present in phytoextracts

Phyto compounds	Tests
Carbohydrates	Molisch test
Reducing sugars	Fehling's test, Benedict's test
Alkaloids	Hager's test, Wagner test
Phenols	Ferric chloride test, lead acetate test
Phytosteroids	Salkowski's test
Terpenoids	Salkowski's test
Gums/mucilage	Ruthenium red solution

18.6.1 Chromatography Technique

Chromatography involves the separation of components from the extract based on either molecular weight or charge. The separation indicates the presence and composition of various compounds in the extract and helps in its characterization.

18.6.1.1 Gas Chromatography

Gas chromatography can be used for efficient qualitative analysis of compounds and can be applied on all forms of agents – solid, liquid, and gaseous samples. Both soluble and insoluble compounds can be characterized by the use of gas chromatography, which involves two basic steps. The first involves the conversion of compound into gaseous form and secondly analyzing it on the basis of mass to charge ratio. This technique can further be used to detect, separate, and characterize the components and assess it on the basis of chemical groups present in the sample.

GC chromatography can be used for volatile components of an extract, which can be vaporized without decomposition. This method involves converting the specie into an equilibrium phase lying between both gas and liquid. This conversion into gaseous phase enables the specie to flow while keeping the sample molecules immobile as they lie in liquid phase. The extent of specie distribution in gaseous phase decides the exposure and migration and hence increases the accuracy in detection. This extent of dissolution of specie in both phases decides the rate of specie migration. Gas chromatography hence based upon the rate of specie migration detects the sample. Thus it can be adopted as a primary quantitative analysis method for phytocompounds [34]. A simple figure depicting the process involved in gas chromatography for phytochemical analysis is shown in Fig. 18.2.

18.6.1.2 High-Performance Liquid Chromatography (HPLC)

Further advances in the characterization procedure involve the use of high-performance liquid chromatography. HPLC is applicable for the separation, detection, and qualitative and quantitative analysis of phytochemicals. Also known as high-pressure liquid chromatography, it is the advancement in specie detection and evaluating the chemical makeup of phytocompounds. Having the basis of sample-based chromatographic separation, the separation of phytomolecules depends upon

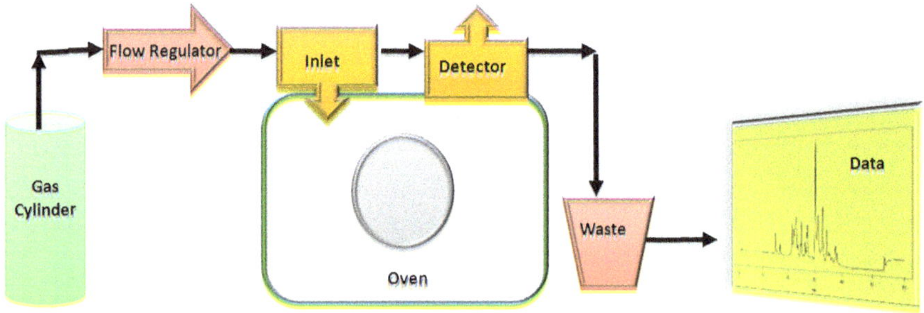

Fig. 18.2 Overview of gas chromatography

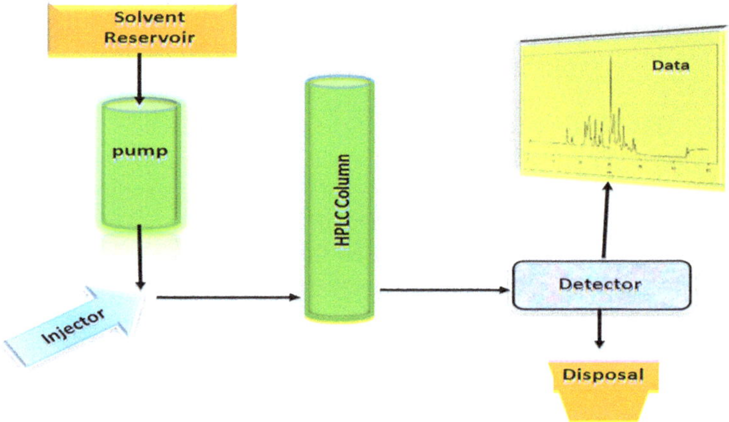

Fig. 18.3 Depicts a simple outlay describing the process of HPLC along with the mechanism of processing and signal generation

its affinity for binding to solid-based tight column. Mechanism involved in the process of HPLC has been shown in the following Fig. 18.3.

Size and thickness of column varies with equipment, and hence it also determines the nature of sample to be detected. The solvent molecules make the mobile phase, and separation is aided with the help of pressure. High pressure of about 400 bars is required to elute out the analyte out of the thin column and hence can be detected for chemical groups. HPLC-based separation also gives an insight into the chemical makeup of the molecules. But only a certain amount of knowledge can be inferred using this technique regarding compounds. Still HPLC can be used in both quantitative and qualitative analyses for our phytocompounds [33, 34].

18.6.1.3 High-Performance Thin Layer Chromatography (HPTLC)

Advancements in the instrument are aimed at increasing the sensitivity of molecular detection, along with reducing time of operation, and at the same time maintaining the accuracy of results. High-performance thin layer chromatography is a modified

Table 18.3 Difference between HPTLC and HPLC

Criteria	HPTLC	HPLC
Sorbent layer	100 μm	250 μm
Efficiency	+++	+
Separation	3–5 cm	10–15 cm
Analysis time	Less	Less
Solid support	C8 and C18	Silica gel
Spotting	Auto	Manual
Scanning	+	–

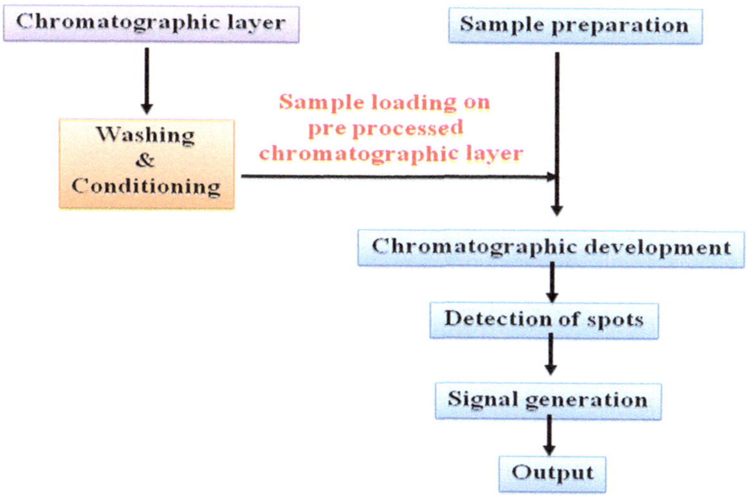

Fig. 18.4 Steps involved in HPTLC process

version of thin layer chromatography. Table 18.3 differentiates between TLC and HPTLC. It can be conceptually described as a planer chromatography involving use of high-end detectors and column separators, also known as layers. These high-performance layers are pre-coated with a sorbent of particle size of 5–7 microns and possess a thickness of 150–200 microns.

The reduction in the thickness of the layer and the particle size result in increasing the plate efficiency along with nature of separation. HPTLC is suitable for qualitative and quantitative estimation for samples with minute quantity of volatile phytocompounds. Figure 18.4 depicts a simple flowchart describing the steps involved in HPTLC for phytochemical analysis.

18.6.1.4 Optimum Performance Laminar Chromatography (OPLC)

OPLC is one of the most used chromatographic techniques for phytomolecule drug distribution till date. This method can be also useful for further analysis and characterization of different chemical groups marking therapeutic effect in the extract.

OPLC is one of the major advancements when it comes to specie detection because it combines the advantages of both HPLC and TLC. All the analyses are carried out directly on crude plant extracts keeping one reference substance. The system is capable of detecting and separating substances of quantity even as low as 10–15 mg and is capable of processing about 4–8 samples at a time. The latter feature depends on the make and differs in different models. In OPLC a pump forces the liquid mobile phase over a stationary phase load comprising of silica and separates the compounds [32]. OPLC is another type of planar chromatography, and columns are generally made up of cylindrical glass or stainless steel. Pressure of over 50 bars is applied over the columns for efficient separation, and the mobile phase is forced to drip at a constant speed through a solvent delivery pump. Sample contained in a 96-well plate. System also consists of a sample holder, an OPLC purification unit, and a four-channel diode array detector to monitor the eluent. The effluents are finally collected in a fraction collector and are thereby separated.

18.6.2 Non-chromatographic Techniques

Advancements in the characterization of phytomolecules involve the use of other high-end non-chromatographic techniques that are aimed to increase the accuracy of detection and reduce the time of detection. Some other techniques, which do not involve the use of chromatography, are being generally employed for characterization purpose. Later part of the chapter deals with certain instrumentation techniques widely used.

18.6.2.1 Immunoassay

Immunoassays involve the use of monoclonal antibodies for separation of low molecular weight compounds based on enzyme-linked receptor binding and its further analysis. These techniques serve as fast and accurate and require less time for detection of compounds. Since it requires receptor binding, therefore it shows high specificity and sensitivity than the conventional chromatographic techniques [31].

18.6.2.2 Fourier-Transform Infrared Spectroscopy (FTIR)

FTIR is one of the most promising tools when it comes to characterization of compounds or functional groups (chemical bonds) present in plant extract. The process involves comparison of the spectra obtained because of different samples with the "fingerprints" of pure compounds taken as reference. A library is constructed of the spectra obtained by FTIR analysis of pure compounds and serves as the reference for other phytomolecules [33]. For most common plant compounds, the spectrum of an unknown compound can be identified by comparison to a library of known compounds. For FTIR analysis, samples are to be prepared following various protocols present in the literature. Liquid extracts can be easily and directly poured between plates. Plates are composed of sodium chloride onto which the liquid forms a layer. Solid samples can either be milled with potassium bromide before using and then can be formed into a pellet or can be dissolved in methylene chloride before placing them onto the sample injector.

18.6.2.3 Nuclear Magnetic Resonance (NMR) Spectroscopy

Most recent developments in the phytochemical extraction deal with improved accuracy in molecular detection and predicting the chemical makeup of molecules in the extract. Nuclear magnetic resonance (NMR) spectroscopy is helpful in predicting the 2D structure of molecules, thereby unfolding the chemical properties of extract. The process of NMR improves detection and is achieved using solid-state NMR. Compounds/groups to be detected share the name with the technique, i.e., C-NMR is used for carbon detection. Similarly, H-NMR is used for hydrogen detection in the given sample.

18.6.2.4 MALDI-TOF

Matrix-assisted laser desorption ionization-time of flight mass spectrometry (MALDI-TOF) technique has made it possible of creating insights into the molecular analysis and makeup of phytocompounds by predicting spatial arrangement of proteins and other phytoconstituents. MALDI is one of the common MSI techniques, which enabled direct measurement of molecular ions via soft imaging technique. MALDI involves the absorption of harsh laser onto a matrix, which is generally made up of small organic compounds. Mechanism of this detection process involves initial desorption of analytes from solid to gas phase and final propagation of ionization within the plume. Depending on the make of the instrument, MALDI instruments can be used to measure a varied range of mass for the biomolecules. MALDI instrument consists of different laser sources with either UV (ultraviolet) or IR (infrared) ranges of varied wavelengths, which promotes desorption. The out throw of electrons forming a plume is generally detected and converted into signal, and results are obtained. Laser desorption ionization (LDI) involves volatilization of molecules initiated by absorption of laser energy (Fig. 18.5) [34, 35].

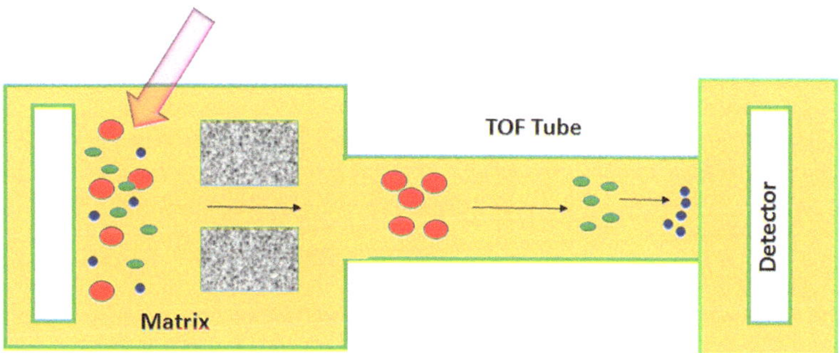

Fig. 18.5 Process involved in MALDI-TOF

18.7 Conclusion

This extensive research over the various phytochemical techniques and their extraction paves the path for its use in functional food. With the emergence of modern era, it marked the beginning of health consciousness in consumers. Functional food and nutraceutical market alone holds a large share in food processing industry. Phytochemical extraction revisiting from starting comes up with involvement of various physical and chemical processes of removing out phytoconstituents safely from the extract along with maintaining the functional viability and structure of groups. Then it is further characterized using different chromatography techniques and even involves use of modern techniques like MALDI and NMR for phytochemical derivation. This advancement further increased accuracy and detection capability with compounds.

References

1. Aladejimokun A, Daramola K, Osabiya O, Arije O (2017) Comparative study of phytochemical constituents and antimicrobial activities of *Acalypha wilkesiana* and *Acalypha godseffiana* extracts. J Adv Microbiol 4(1):1–7
2. Babar M (2013) Antiviral drug therapy- exploiting medicinal plants. J Antivir Antiretrovir 05(02)
3. Colombo M (2012) Herbal preparations and homemade herbal teas for children. Forum Nutr 11(1):19–24
4. Di Fabio G, Romanucci V, Zarrelli M, Giordano M, Zarrelli A (2013) C-4 gem-dimethylated oleanes *of Gymnema sylvestre* and their pharmacological activities. Molecules 18(12):14892–14919
5. Djouonzo P, Ladoh-Yemeda C, Tchinda A, Mbafor J (2016) Antiradical activity, total phenolic and flavonoid content of extracts from the stem bark of *Pterocarpus erinaceus*. Schol Acad J Biosci 4(6)
6. Duodu K (2014) Effects of processing on phenolic phytochemicals in cereals and legumes. Cereal Foods World 59(2):64–70
7. El-Khateeb A, Azzaz N, Mahmoud H (2014) Phytochemical constituents, hypoglycemic and haematological effects of methanolic *Acalypha wilkesiana* leave extract on streptozotocin-induced diabetic rats. Eur J Chem 5(3):430–438
8. Ganjhu R, Mudgal P, Maity H, Dowarha D, Devadiga S, Nag S, Arun kumar G (2015) Herbal plants and plant preparations as remedial approach for viral diseases. Virus Dis 26(4):225–236
9. Geerkens C, Matejka A, Carle R, Schweiggert R (2015) Development and validation of an HPLC method for the determination of alk(en)ylresorcinols using rapid ultrasound-assisted extraction of mango peels and rye grains. Food Chem 169:261–269
10. Giambanelli E, Ferioli F, Koçaoglu B, Jorjadze M, Alexieva I, Darbinyan N, D'Antuono L (2013) A comparative study of bioactive compounds in primitive wheat populations from Italy, Turkey, Georgia, Bulgaria and Armenia. J Sci Food Agric 93(14):3490–3501
11. Hole A, Grimmer S, Jensen M, Sahlstrøm S (2012) Synergistic and suppressive effects of dietary phenolic acids and other phytochemicals from cereal extract on nuclear factor kappa B activity. Food Chem 133(3):969–977
12. Kumar G, Krishna A (2013) Studies on the nutraceuticals composition of wheat derived oils wheat bran oil and wheat germ oil. J Food Sci Technol 52(2):1145–1151
13. Lin L, Chen T, Lin S, Chung C, Lin T, Wang G, Anderson R, Lin C, Richardson C (2013) Broad-spectrum antiviral activity of chebulagic acid and punicalagin against viruses that use glycosaminoglycans for entry. BMC Microbiol 13(1):187

14. Mithat Kara G (2016) Some chemical and physical properties, fatty acid composition and bioactive compounds of wheat germ oils extracted from different wheat cultivars. Tarım Bilim Derg 22(3):433–443
15. Sen S, Chakraborty R (2015) Toward the integration and advancement of herbal medicine: a focus on traditional Indian medicine. Bot: Targets Ther:33
16. Singh I, Rani P, Kumar P (2017) Microwave assisted grafting of gums and extraction of natural materials. Mini-Rev Med Chem 17:1573
17. Tiwari P, Mishra B, Sangwan N (2014) Phytochemical and pharmacological properties of Gymnemasylvestre: an important medicinal plant. Biomed Res Int 2014:1–18
18. Tunchaiyaphum S, Eshtiaghi M, Yoswathana N (2013) Extraction of bioactive compounds from mango peels using green technology. Int J Chem Eng Appl:194–198
19. Vats S, Kamal R (2016) Identification of flavonoids from plant parts and callus culture of *Gymnema sylvestre* R. Br.: an antidiabetic plant. Curr Bioact Compd 12(4):264–268
20. Asma E, Naiima B, Fethia H (2015) Antifungal activity of *Citharexylum quadrangulare Jacq.* Extracts against phytopathogenic fungi. Afr J Microbiol Res 9(29):1764–1769
21. Addie R, Balluff B, Bovée J, Morreau H, McDonnell L (2015) Current state and future challenges of mass spectrometry imaging for clinical research. Anal Chem 87(13):6426–6433
22. Cvetanović A, Švarc-Gajić J, Mašković P, Savić S, Nikolić L (2015) Antioxidant and biological activity of chamomile extracts obtained by different techniques: perspective of using superheated water for isolation of biologically active compounds. Ind Crop Prod 65:582–591
23. Eriksson C, Masaki N, Yao I, Hayasaka T, Setou M (2013) MALDI imaging mass spectrometry—a mini review of methods and recent developments. Mass Spectrom 2:S0022–S0022
24. Jamila N, Khairuddean M, Khan S, Khan N (2014) Complete NMR assignments of bioactive rotameric (3 → 8) biflavonoids from the bark of *Garcinia hombroniana*. Magn Reson Chem 52(7):345–352
25. Maksimovic S, Tadic V, Skala D, Zizovic I (2017) Separation of phytochemicals from *Helichrysum italicum*: an analysis of different isolation techniques and biological activity of prepared extracts. Phytochemistry 138:9–28
26. Neubert P, Walch A (2013) Current frontiers in clinical research application of MALDI imaging mass spectrometry. Expert Rev Proteomics 10(3):259–273
27. Ramak A, Ezatpour B, Talei G (2017) Investigation on chemical composition, antimicrobial, antioxidant, and cytotoxic properties of essential oil from *Dracocephalum kotschyi boiss*. Afr J Tradit Complement Altern Med 14(3):209–217
28. Rochman J, Siswoyo T, Ratnadewi A (2017) Antioxidant activity study and A-glucosidase inhibitor phenolic leaf extract Bungur (*Lagerstroemia speciosa*) from Meru National Park Betiri. J ILMU DASAR 17(1):39
29. Sabine Becker J (2013) Imaging of metals in biological tissue by laser ablation inductively coupled plasma mass spectrometry (LA-ICP-MS): state of the art and future developments. J Mass Spectrom 48(2):255–268
30. SaiSaraswathi V, Saravanan D, Santhakumar K (2017) Isolation of quercetin from the methanolic extract of Lagerstroemia speciosa by HPLC technique, its cytotoxicity against MCF-7 cells and photocatalytic activity. J Photochem Photobiol B Biol 171:20–26
31. Schmitz C, Fritsch L, Fischer R, Schillberg S, Rasche S (2016) Statistical experimental designs for the production of secondary metabolites in plant cell suspension cultures. Biotechnol Lett 38(12):2007–2014
32. Szydłowska-Czerniak A, Tułodziecka A (2014) Optimization of ultrasound-assisted extraction procedure to determine antioxidant capacity of rapeseed cultivars. Food Anal Methods 8(3):778–789
33. Wocheslander S, Groß F, Scholz B, Engel K (2017) Quantitation of acyl chain oxidation products formed upon thermo-oxidation of phytosteryl/-stanylOleates and linoleates. J Agric Food Chem 65(11):2435–2442
34. Zhu H, Lu Z, Cai J, Li J, Gao L (2009) Development of a headspace–SPME–GC/MS method to determine volatile organic compounds released from textiles. Polym Test 28(5):521–527
35. Tetik S (2016) Mass spectrometry technology for protein biomarker discovery. Imaging J Clin Med Sci 2016:14–16

Nanotechnology in the Food Industry: Perspectives and Prospects

19

Himanshu Sukhpal, Stuti Awasthy, and Indira P. Sarethy

Abstract

In the present scenario, it has been estimated that out of the 7 billion people inhabiting the Earth, 0.9 billion face inadequate food supply (food insecurity) and 2 billion are malnourished, out of which over 800 million suffer from severe malnutrition and 36 million more die from lack of food. The projected rise in population curves indicates that by the year 2020, the Earth will be populated by 8 billion people and by 2050, the global population is expected to hit 9.6 billion. Wastage of food in various forms (lack of efficient storage methods, delayed transportation to market, pathogen and pest attacks, natural calamities, poor bioavailability, short shelf life) leads to ineffective utilization of the produced food. While various strategies exist to counter these issues, it is postulated that nanotechnology can also play a significant role in mitigating many of the contributing factors. Food supplements can be incorporated in the form of nanoparticles to enhance nutritional value or to improve the taste, texture and consistency attributes of the food. Nano-textured foodstuffs enable reduction of fat usage thereby contributing to healthier food. Nanotechnology has application in packaging food materials as it can provide resistance to physically stressful conditions, from agents that cause degradation of the packaging material and from pathogens, and can also facilitate in increasing the shelf life of the food items. Silver and gold nanoparticles in biosensor-incorporated packaging material can detect spoilage or contamination of food. In agriculture, nano-particulate fertilizers can be more effective with their increased surface area. Nano-pesticides can be enabled for controlled release under desired conditions. Despite these promising

H. Sukhpal · S. Awasthy · I. P. Sarethy (✉)
Department of Biotechnology, Jaypee Institute of Information Technology, Noida, India
e-mail: indirap.sarethy@jiit.ac.in

V. Rani, U. C. S. Yadav (eds.), *Functional Food and Human Health*,
https://doi.org/10.1007/978-981-13-1123-9_19

aspects, concerns remain about toxicity of nanoparticles inside the body. Nevertheless, rapid advances have resulted in many countries designing specific regulations focusing on nanotechnology aspects.

Keywords
Nanopesticide · Nanofertilizer · Smart packaging · Functional food · Nano-textured food · Nanoencapsulation

Freedom from want of food therefore must mean making available for every citizen in every country sufficient of the right kind of food for health. If we are planning food for the people no lower standard can be accepted. – Sir John Boyd Orr (1943)

19.1 Introduction

To meet the growing food needs of increasing population worldwide, large tracts of land, especially forest land, are continuously being converted into agricultural areas.

'To boost the economy, we are demolishing the ecology'.

Scientists have been making efforts to increase the yield and production on the same amount of land by intensely using fertilizers, adopting genetically improved plants in many countries and using improved technology for cultivation, but which is not enough to meet the world's demand.

19.2 The Growing World and Its Demands

It has been estimated that out of the 7 billion people inhabiting the Earth, 0.9 billion have an inadequate supply of food (food insecure). Amongst these, 2 billion are malnourished [48], over 800 million people suffer from various degrees of malnutrition, and 36 million die from lack of access to food, implying that the available food supply cannot sustain current population increases. As per Food and Agriculture Organization [26] data of the Food Security Information Network, 108 million people face food insecurity.

A UNESCO report predicts that global population is likely to increase by 2–3 billion people over the next 40 years. Along with changing diets, the food demand can increase 70% by 2050 (http://www.unesco.org). A recent report in *Financial Times* [30] estimates that 795 million people globally suffer from chronic undernourishment. Of this 780 million are in developing countries. The sub-Saharan countries show dismal statistics with one in four being undernourished. Further, there are 11 million people undernourished in developed countries [26].

The FAO [25] has proposed four dimensions of food security: (1) food physically available to people, (2) access to food (affordable as well as physical), (3) food utilization and (4) stability of the other dimensions over time. All these dimensions must be satisfied to avoid or overcome food insecurity.

With all these statistics available, many questions need to be addressed to avoid a catastrophe:

- Are we producing enough food to meet the demand of today's world? If yes, where is it all going?
- Do we have the capability to feed that many people?
- Are the current approaches towards increasing the food production environmentally sustainable and safe for us?

Surprisingly, the answer is yes to the first two questions, suggesting that malnutrition and lack of access to food can be addressed with a few changes in food operations – from farm to fork – as well as adoption of new technologies. The answer to the third question is ambivalent since different countries have different standards and accurate estimates are lacking on actual application of technologies.

19.3 The Issue

One of the major arguments today is that are we producing enough food or are we just wasting too much? A report by FAO (http://www.fao.org/food-loss-and-food-waste/en/) [22] suggests that 1.3 billion tons of food is wasted every year implying that one-third of all food produced worldwide (worth around US$1 trillion) is wasted or lost during food production and/or consumption. Expressing this in terms of calories, Lipinski et al. [52] show that one in four calories is never consumed. Climate change patterns, in addition, may contribute to increasing global food prices [60].

The food produced is wasted at many stages (Fig. 19.1). A big share can be wasted during harvesting/in fields due to other reasons, improper conditions during transportation leading to pest and pathogen attack, unequal distribution amongst people and reduced bioavailability in the body (due to various reasons such as degradation of nutrients, lack of solubility of food in blood, unstable nutrients, adverse interactions with other food particles, low retention time in blood, etc.).

19.4 Nanotechnology for the Food Industry

There are many ways in which such food shortage due to various reasons can be addressed and, indeed, is being implemented in varying scales with different degrees of successes. Amongst the many solutions, nanotechnology can have a considerably big role in the coming years. This chapter provides a broad overview of some areas in the food industry where nanotechnology is expected to contribute, to improve products and processes.

An article in www.nutraingredients.com suggests that the global nanofood market is expected to reach €19 billion by 2020 [16], a huge jump from €6.6 billion now. This reflects a quantum increase in potential usage of nanotechnology-based

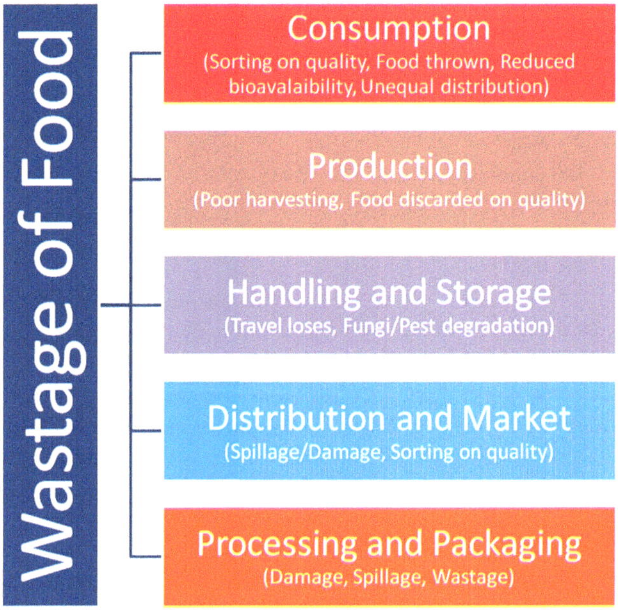

Fig. 19.1 The different reasons resulting in wastage of food

products, but not surprising considering the various fields where the applications exist.

The technology dealing with the manipulation of matter at the nanometre scale (1 nm = 10^{-9} m) is known as nanotechnology. In recent times, nanotechnology has been applied in various sectors such as in electronics, medicine, diagnostics and military [94]. The application of nanotechnology in the food industry, while not being a recently explored concept, has been gaining ground in the last few years. There are many ways in which nanotechnology is being applied, such as to improve food by improving their dissolution, bioavailability and retention time inside the body, enhancing the shelf life of processed foods, supplementing and complementing the nutritional value of food, delivering food particles to their correct sites in the body and facilitating in detecting spoilage, food processing and food packaging.

Briefly, the various sectors where nanotechnology can play a role and a few examples where it can be applied are shown below:

- Agriculture – enhancing the efficiency of fertilizers and pesticides [58]
- Handling and Storage – pest-resistant nanomaterial coatings
- Processing – nanoencapsulation, nanoemulsions, etc. [59]
- Packaging – biodegradable, edible packaging material, and antimicrobial packaging [6, 63, 79]
- Distribution and Marketing – enhanced shelf lives using nanocoatings of oxidative barriers
- Consumption – nutraceuticals, food additives, etc. [1]

- Site-Specific Delivery – specially of functional foods [29]
- Production and Processing – enhancing security using nanosensors that can detect microbial and/or chemical attack [103]
- Tracking Production and Logistics – gadgets to document records of articles and tracking specific shipments [69]

19.5 Applications and Techniques of Nanotechnology in the Food Industry

New breakthroughs and innovations are being announced everyday in the field of nanotechnology, some of them related to the food sector. Figure 19.2 provides an idea about the various sectors in the food industry where nanotechnology can be applied. Starting from the point of production, namely, agriculture, including food and feed sectors, and leading to processing and packaging of food, many examples exist of nanotechnology-based inventions and interventions.

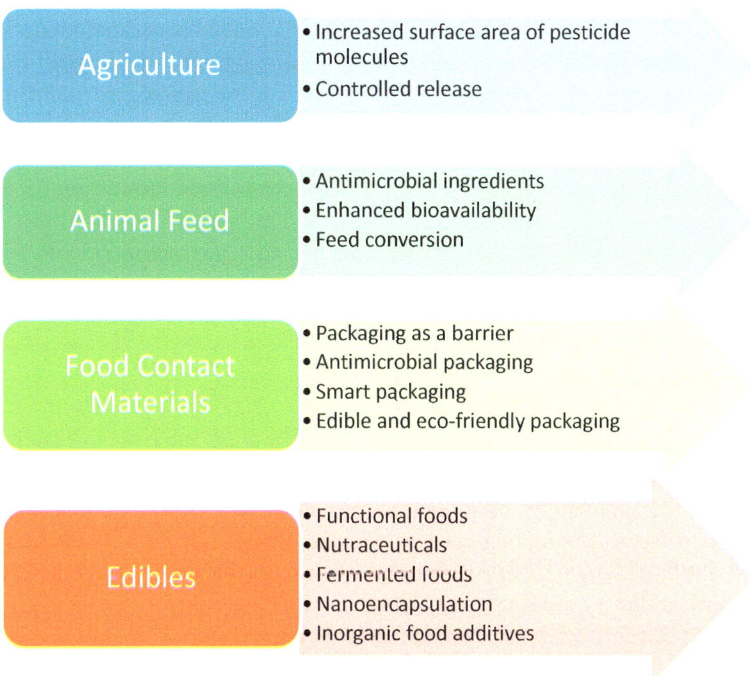

Fig. 19.2 Applications of nanotechnology in food industry

19.5.1 Agricultural Practices

Pesticides and fertilizers are extensively used for protection and better growth of crops. Globally, it has been estimated that pesticide consumption accounts for about 2 million tonnes per year [18], amongst which Europe uses 45% and the USA 25%. The FAO has estimated that the global demand for fertilizer nutrients will grow at 1.6 percent per annum from 2014 to 2019 [27]. Extensive and sometimes indiscriminate use of these essentially synthetic compounds has resulted in soil degradation in many regions. There has been a lack of effective alternatives.

It has been suggested that fertilizers and pesticides can be made more effective by breaking them up into nanoparticles. Thereby, the total amount of these compounds sprayed on agricultural fields could be significantly reduced. This is essential, because the nanoscale size of these particles enables a larger surface area to be available for action. The nano-sized fertilizers can facilitate increased uptake by roots [57]. In the same way, nano-sized pesticides with larger surface areas can provide for better action on pests. There can also be a control on the release of pesticides in desired conditions [15] by facilitating adsorbing of the pesticide molecules onto porous hollow structures, such as the porous hollow silica nanoparticles (PHSNs). Wang et al. [101] showed that abamectin, though a biopesticide, still suffers from many drawbacks such as poor water solubility, dispersity and poor controlled release. Loading abamectin with porous silica nanoparticles (Abam-PSNs) improved these features. A similar study with avermectin showed that its UV-protecting property increased and provided for sustained controlled release with increasing shell thickness of PHSNs [51]. The authors proposed that shell thickness could be the main factor deciding the release of avermectin.

A search for the term 'nanopesticide' in Google Patents showed up 65 results (https://patents.google.com). Around 110 results show up for 'nano fertilizer'. This suggests that such filings are likely to increase, considering the importance of reducing residual effects and persistence of pesticides and fertilizers in soil.

19.5.2 Animal Feed

Usage of nanotechnology in animal feed shows around 3000 results in Google Patents, some pertaining to coating and some with designing and composition of feed. While the applications of animal feed additives in nano-form are very limited and many are in the development phase, many applications focus on testing replacement of antibiotics by nanomaterials which can adsorb/absorb the bacteria/toxins. Some applications try to improve digestibility of the feed.

A global concern amongst farmers is of mycotoxin contamination in animal feed. Nanoclays (such as montmorillonite-based nanoclays) and various nanoparticles (such as polystyrene-based particles) have been shown to selectively take up mycotoxins and prevent them from reaching the animal gut and thereby preclude mycotoxicosis. Yuan et al. [111] studied the efficacy of nanosilicate clay platelets (NSCP) as a feed additive for ameliorating mycotoxicosis caused by fumonisin B1

(FB1) in chicken using proteomic and biochemical analyses. When NSCP was supplemented in diet (40 mg/kg of feed), growth performances in chickens were shown to be improved with respect to FB1 toxicosis. Nanosilver composites (nanocid) were fed to broiler chicken, which were earlier induced with aflatoxicosis, and their blood parameters (mean corpuscular volume, mean corpuscular haemoglobin, mean corpuscular haemoglobin concentration lymphocytes, neutrophils, basophils, monocytes and eosinophils percentage) were analysed [28]. It was seen that haemoglobin percentage and white blood cell count in all the groups fed with 3 ppm aflatoxin decreased, except those fed with nanocid + aflatoxin. Hence nanocid was suggested to be useful for bringing down the adverse effects of aflatoxicosis in chicken.

Xun et al. [108] studied the effect of nano-selenium (NS) and yeast selenium (YS) supplementation in sheep feed and how it affects feed digestibility, rumen fermentation and urinary purine derivatives. They observed that ruminal fermentation and feed conversion efficiency were improved by feeding NS, suggesting that NS could be used as a selenium source in ruminants. Nano-forms of selenium have been proposed as an important formulation, considering its anticarcinogenic and antioxidant, antimicrobial and immunomodulatory properties. It has been suggested that it could improve stress resilience in fish [81] and in livestock [84].

19.5.3 Food Contact Material

Food packaging materials have to be resistant to physical stress, chemical degrading agents and biological pathogens, must provide sufficient protection to the food items in terms of maintaining the shelf life and must not adversely affect the food articles. Continuing research in the packaging industry has led to improved packaging quality and enhanced shelf life of the food packaged and biodegradable packing. The nanotechnology-based interventions used in packaging are briefly described in the following sections.

19.5.3.1 Packaging as a Barrier

Packaging material must preferably not allow any exchange of gases, thus maintaining an inert environment. This can prevent spoilage and/or microbial growth. Nanocomposites, as part of the packaging material, can facilitate to increase the surface area of the polymer matrix, thus resulting in a complicated path precluding the entry of gas molecules. Nanoclays are naturally obtained from volcanic ash. Precipitation as crystals from water results in a phyllosilicate group of minerals called montmorillonite, which provides a barrier to permeation of gases. Clays and silicates have a layered structure, are relatively inexpensive and can be processed easily [7]. Honarvar et al. [33] suggest that nanocomposites in packaging material need to be thoroughly assessed for their risks since their physical and chemical characteristics are different and their migration into food matrices can pose potential risk to consumers. Commercially developed polyamide-based nanoclays (sold under the trade names Durethan, Imperm and Aegis) have shown better durability and protection. Readers are requested to read the review by Farhoodi [23], where the author has comprehensively reviewed the utility, risks and safety evaluations of nanocomposites as food packaging material.

19.5.3.2 Packaging with Antimicrobial Nanoparticles

Silver has been known, since ancient times, for its antimicrobial action and is attractive from the viewpoint as part of packaging material [53]. Silver as nanoparticles, in ion form, inhibits a wide range of biological processes in bacteria [107]. It can inhibit aquatic organisms, if leached. However, Fabrega et al. [20] showed that the effect was not significant. Silver nanoparticles are used in all forms such as biotexiles, electrical appliances, refrigerators, kitchenware, etc. [72].

Titanium dioxide (TiO2) nanopowder is being studied as a filler material due to its inertness, its UV-filtering property and its antibacterial nature [88]. Many studies, such as that of Charpentier et al. [14], have focussed on its antibacterial property. However, the antibacterial nature depends on UV irradiation and thus is a limitation [88].

Chitosan is a partially deacetylated polymer of N-acetyl glucosamine. It is obtained by alkaline deacetylation of chitin. Chitin is present in exoskeleton of arthropods and cell wall of many fungi. Dutta et al. [19] have extensively discussed the potential advantages and many applications of chitosan-based packaging material for food. Its antimicrobial properties are an additional advantage as it can be used as edible film coating, as sprays and as dips.

Azeredo's group [8] studied the effect of cellulose nanofibres on physical properties of chitosan films. They found that addition of cellulose nanofibres to chitosan films improved mechanical characteristics and, thereby, the water vapour barrier properties, comparable to many synthetic polymers, but not enough flexibility, and hence was suitable for limited applications [8]. However, an important and significant characteristic would be that they are biodegradable and hence eco-friendly.

Interestingly, vitamin-based chitosan nanoparticles (CSNPs), comprising of Vitamin C and E, were proposed as novel food packaging material [4]. The vitamins, being antioxidants, could help enhance the shelf life of the packaged food. Such incorporation of vitamins was seen to provide better storage of hydrophilic and lipophilic food material.

19.5.3.3 Smart Packaging Materials

Helmke and Minerick [32] have hypothesized about the controlling factors in cell-biomaterial interactions. They have proposed that traditional approaches characterize behaviour of millions of cells as a single unit, which may not be reflective of capability of detecting trace quantities of toxins or small increases in pathogen numbers. Modern approaches at the nanoscale level can potentially overcome this gap and help to understand events at the unicellular level. Bülbül et al. [13] have elegantly summarized the recent developments in development of inexpensive and portable nanoparticle-based biosensors.

An excellent review by Biji and co-workers [9] summarizes the evolution of packaging material for food. Traditional food packaging material primarily focused on protection of the food from external deteriorants (such as heat, light, moisture, microbial contaminants, gaseous emissions, pressure, etc.). Shapes and sizes of the packing material are evolving to provide greater comfort of use [56, 109]. The current thinking is to design and develop active and intelligent packing materials. The active materials contain additives to prolong shelf life of the food, and intelligent systems monitor the packaged food in real time as to their quality during packing, transportation and storage [9].

In a recent review, Inbaraj and Chen [43] provide examples of how nanomaterials are increasingly being used as sensors and biosensors for detecting microbial pathogens and toxins. Many nanoparticles such as gold, in the form of nanoparticles and nanorods, quantum dots, silver and silica nanoparticles have been detailed in various reports [50, 96]. The optical and electrochemical properties of the nanomaterial play a role in detecting pathogens and toxins [70, 77]. Some examples of nanoparticle-based biosensors for the food industry are briefly described below:

Gold Nanoparticles

These nanoparticles provide for efficient and visible colour change from red to blue based on aggregation-induced interparticle surface plasmon coupling of AuNPs with antibodies or aptamers [2]. Gold nanoparticles can increase the surface area and conductivity in sensors and have been reported to be used for detection of many chemical and biological contaminants ([13] and references therein; [61, 71]).

Wang and Alocilja [100] developed an electrochemical biosensor for detecting *E. coli* O157:H7, using antibody-modified nanoparticles. Usage of polymer-coated magnetic nanoparticles (MNPs) and carbohydrate-capped AuNPs resulted in rapid pathogen detection. Storhoff et al. [93] showed that aggregate size determines the optical properties of DNA-linked nanoparticle assemblies. The length of the oligonucleotide linker does not play a significant role. Hence colorimetric detection using these nanoparticles should focus on the size.

Silver Nanoparticles

Various sensing techniques based on the optical properties of silver nanoparticles have been reported [87]. The switch in the light reflecting properties from dispersed phase silver nanoparticles and aggregated nanoparticles from yellow to brown is directly linked to the change in concentration of a target molecule, such as a pollutant gas released on spoilage [98]. This has resulted in silver nanoparticle-based assays for identification of pesticides, metals and proteins [13, 92, 106]. However, these have certain limitations such as chemical degradation to silver ions on functionalizing and oxidation of surfaces [102].

Cerium Oxide Nanoparticles

Cerium nanoparticles (nanoceria) are optically active and, depending on specific interactions on their surface, can reflect different and multiple colour patterns, thus making them attractive for detecting multiple targets [13]. Portable colorimetric sensors were designed to detect food antioxidants and glucose [67, 80, 86] and mycotoxins [13].

19.5.3.4 Edible and Eco-friendly Packaging Materials

One of the major factors concerning every industry and overall human life is the continuous degradation of the environment, in which plastics used in the food packaging industry play a dominant role. These plastics do not get decomposed when dumped in soil and, rather, alter the nature of soil. When burnt they cause accumulation of toxic gases in the environment. Therefore biodegradable plastics were proposed and came into use to a limited extent, but they lack mechanical strength and more.

To overcome these problems, nanoparticles obtained as proteins, carbohydrates and lipids from animal and plant sources can be used to make packaging material which is biodegradable and renewable, has a high mechanical strength and is even edible. Cruz-Romero et al. [17] showed that low and medium molecular weight chitosan showed the highest antimicrobial activity. In addition, nano-sized solubilizates of benzoic acid and sorbic acid had better antimicrobial activity than the non-nano ones. The authors suggested that such nano-sized solubilisates, which are food compatible too, can have a double use: they could be utilized as smart antimicrobial packaging solution, as also eliciting a similar antimicrobial effect with lesser quantity of the antimicrobial substance.

In a recent review, Ramos et al. [78] have discussed the advances in usage of gelatin-based films as food coatings. To address consumer demands of reducing petroleum-based products for coating food and substitute with edible ones from natural sources [55], many biodegradable polymers have been reported to be effective such as corn zein, soy protein, collagen, gelatin [10], casein [5] and whey protein. Metallic nanofillers, with silver [45] or zinc [85], incorporated in gelatin films, have shown antimicrobial effects as well as improved shelf life. Chitosan nanoparticles have also been added as reinforcement agents in a gelatin matrix, resulting in increased chain mobility and improved thermal stability [35]. Nanoclay as reinforcement for gelatin has also been reported [46].

19.6 Functional Foods

A functional food is a food which not only has nutrients but also several non-nutrients essential for the performance of a variety of functions inside the body, keeping a person healthy and fit. Functional foods might be found naturally like cultivated fruits/veggies, etc. or fishes which have a number of naturally present compounds beneficial apart from nutrition, such as the lycopene present in tomatoes, the omega-3 fatty acids present in cod livers, etc. The quantity of functional compounds in a food product can be increased using specific cultivation/farming practices or breeding methods. Examples are vitamin-enriched broccoli and soybeans or β-carotene-rich rice. Animal-derived food products can have their compositions enhanced through the animal's feed, like improving the levels of conjugated linoleic acids or omega-3 in dairy and poultry products.

The efforts to increase the nutritional value of a food product can be futile if the nutrients are not taken up by the body or not delivered to their target sites. Singh [90] has provided a bird's eye view of various nanotechnology-based delivery solutions such as protein-polysaccharide coacervates, multiple emulsions, liposomes and cochleates. Understanding the component to be introduced as functional ingredient in the food, delivering it to the correct target site and, at the same time, maintaining it in the proper structural form in the food matrix are some challenges associated with many of these nanodelivery systems.

19.6.1 Nutraceuticals

The terms functional foods and nutraceuticals are sometimes used interchangeably. Nutraceutical refers to a component specifically added for purposes of imparting additional nutritional value to a food. Altering the composition, or the nutritional value of a conventional food product by supplementing it with added vitamins/minerals or other essential nutrients, is not a recent concept. Some approaches were more focused on genetic modifications, which faced public opposition and non-acceptance. Recent approaches look towards addition of nutrient nanoparticles as food supplements to enhance nutritional value and improve health and well-being. Nano-sized structures are processed as stable emulsions/mixtures to improve the taste, texture and consistency attributes of the food.

Nano-textured foodstuffs can, for example, cut down the usage of fats, thus making the food healthier. An example of such a product is nano-textured mayonnaise, ice cream or butter, which will be low in fat but creamy as the original food. So they would be considered healthy but retain the taste of the original food product [94]. Micronized cellulose, starch, rice flour, wheat and spices are still in development stages but hold promise [42, 95].

19.6.2 Fermented Products

In wine production, nanomaterials are used for degradation of pollutants, as antimicrobial agents, especially silver nanoparticles, as nanosensors to address quality control issues and as nanobiosensors to detect polyphenols, ethanol, glucose, any mycotoxins and pesticide residues [62].

19.6.3 Nanoencapsulation

Nanoencapsulation refers to techniques that result in enclosing food substances in miniature form. Techniques such as nanocomposite, nanoemulsification and nanoestructuration are used for protection of vitamins, antioxidants, lipids, proteins and other important components. Quintanilla-Carvajal et al. [73] have exhaustively commented on the many such techniques and applications.

An encapsulation over a food material prevents instability due to unfavourable conditions, increases/decreases solubilization properties or eliminates foul odour. It involves enclosing substances in miniaturized capsules allowing for the controlled release of the core active molecule [83]. Nanostructured encapsulation systems such as liposomes, nanoemulsions, microemulsions, solid lipid nanoparticles (SLNs) and polymeric nanoparticles have been studied, and many have been found to improve the dispersibility, stability and bioavailability of bioactive compounds [101].

Micelles of starch have been reported to encapsulate the core active component of turmeric, known as curcumin, which is hydrophobic in nature [111]. The action of antioxidants can be augmented using the lipid-principled nanoencapsulation techniques, thus enhancing their dissolution ratio and bioavailability, reducing instability and avoiding interactions with other food components [94]. Degradation of essential nutrients and components in food, such as antioxidants, amino acids and carbohydrates, can be prevented by the use of octenyl succinic anhydride-ε-polylysine, which functions as surfactants or emulsifiers in the encapsulation of nutraceuticals [110].

Lipid-based nanoencapsulation techniques are specially suited for antioxidants, as their solubility, stability and bioavailability are enhanced. Nanoliposomes, nanocochleates, colloidosomes and archaeosomes are some of the nanoencapsulation systems for lipids. Nanoliposomes and their huge potential for encapsulation of hydrophobic molecules have been reviewed by Mozafari et al. [64]. Nanocochleates, nanocoiled particles enclosing nutrients, have been postulated to be useful for micronutrients [91]. Hydrophobic molecules such as fat blockers, vitamins, etc. can be placed inside colloidosomes and be effective in their action in smaller quantities [105].

Nanoencapsulation of probiotics can be very desirable to prevent their degradation during their transport through the gastrointestinal tract. These protected probiotics can also serve as 'de novo vaccines' [97].

19.6.4 Inorganic Food Additives

Approximately 45,000 patents have been registered as nanofood additives (https://patents.google.com/?q=nano+food+additives). Engineered nanomaterials of silver, iron, calcium, magnesium, selenium and silicates find wide applications as food additives. Nanosilver, apart from its use as an antimicrobial agent, is also finding use as antiodorant and health supplement. Nanosilica has use in food contact surfaces and nano-selenium as additive in green tea products. Nanocalcium is being considered as an ingredient in chewing gums and nanomagnesium and nano-iron as health supplements. Nanosalt is being proposed as an additive, which can help in less intake of salt [83].

19.7 But, What's the Catch?

While nanotechnology offers promise for many aspects in the food industry, correspondingly the risk factors (perceived and real) have also increased. Primarily, concerns around the risk assessment of nanoparticles and nanostructures exist. Many physiochemical properties of nanoparticles such as size distribution, electrostatics, surface area, general morphology and aggregation can bring about different types of interactions between nanomaterials and the body tissue [89]. The potential routes of human exposure to nanoparticles are the oral cavity, lungs, skin and the gastrointestinal tract. Each part of the body from the oral cavity to the colon presents a completely different environment, in terms of tissue type, pH, fluids and microflora present. The development stage – baby, children, adult, geriatric – can also play a determinative part in triggering possible immune responses [99].

19.7.1 The Size

The size of the nanoparticles determines the surface area, and the surface area plays a crucial role in several of the responses of the body, like respiratory disorders, especially those with cardiovascular problems [65]. There are in vitro evidences that the scaling of particles is a crucial aspect in dermal-cell cytotoxicity [3]. Photodynamic therapy efficiency of 5-aminolevolenic acid as photosensitizer increased in the presence of 4 nm gold nanoparticles. Adsorbed nanoparticles of various sizes can elicit different immune responses [47]. In mouse keratinocyes, amorphous silica nanoparticles below 100 nm induced cytotoxicity. However, all sizes of nanoparticles were taken up into the cytoplasm.

Borm and Kreyling [11] have extensively discussed the various modes by which nanoparticles exert their (adverse) effects – on the central nervous system, on translocation from lungs into the bloodstream and on inflammatory responses in the lungs. The evidences point to risks on inhalation of nanoparticles. The authors hypothesize that nanoparticles less than 40 nm in size are similar to large proteins and therefore, pose potential issues of forming a complex with the latter. They also suggest that since increasing research is being carried out on nanoparticles as carriers for drugs, it would be prudent to cover the gap existing on studies between toxicity of nanoparticles and their utility as drug carriers.

Exposure to polystyrene nanoparticles affected iron uptake and transport in an in vitro model of the intestinal epithelium and an in vivo chicken intestinal loop model [54]. Disruption of intestinal membrane by nanoparticles resulted in increased iron transport, while chicken cells presented lower iron absorption.

19.7.2 The Chemical Composition

Many reagents used during the production of nanoparticles can be toxic themselves. Some quantities of such reagents, such as sodium borohydride, hydrazine, cetyltrimethyl ammonium bromide and various polymers, if they remain in the final product, can result in exposure to toxins that are unrelated to the nanomaterials themselves (Lee and references therein [34, 44, 49, 74, 104]). Titanium oxide has different crystalline forms, each with a unique cytotoxic effect [12].

19.7.3 The Surface Properties

There are certain aspects associated with surface structure of the nanoparticles, such as hydrophobicity, charge, texture, etc. that could result in a varied toxicological response in the human body [68]. Coatings on the surfaces of zinc oxide (ZnO) nanoparticles result in lowered capacity of the nanoparticles to generate free radicals and, thereby, reduced toxicity. While ionic zinc did not exert toxicity, uncoated nanoparticles exhibited toxicity, but surface-coated ones did not show cytotoxicity in human hepatic stellate cells (hHSCs). However, more studies are required in this aspect, especially with standardization of the coating material.

19.8 Safety

Falkner and Jaspers [21] emphasize that nanoscience is increasingly being implemented globally and hence international harmonization of regulations is essential. The National Institute of Occupational Safety and Health (NIOSH), a US-based federal agency, provides guidance on the occupational health and safety issues arising out of nanotechnology applications (https://www.nano.gov//node/135). It carries out research on how workers might be affected in industries involved in manufacturing or use of nanomaterials and how the nanomaterials interact with body systems. Its interactive site https://nanohub.org/groups/gng hosts the Good NanoGuide, which provides a collaborative platform for scientists to share ideas on how best to handle nanomaterials. It also comes out with periodically updated publication documents on toxicity assessment of nanoparticles, assessment of exposures of workers, process controls and protective equipment for personnel. Schulte and Salamanca-Buentello [82] have provided a broad framework for implementing safe practices in workplaces dealing with nanomaterials. They suggest unbiased and critical analyses of hazards, implementation of effective controls and emphasis on toxicological research.

When working with nanomaterials, appropriate laboratory safety practices should be followed. A few of the suggestions are that Personnel Protective Equipment (PPE) should be used which includes wearing gloves, lab coats, safety glasses, face shields and closed-toed shoes to avoid skin contact with nanoparticles [75]. Powder exhaust laminar flow hoods and appropriate respiratory protection should be used for safe handling [24]. Proper training should be given to laboratory personnel about the risks associated with nanoparticles and workplace hazards.

As with every new technology, public perception plays a key role in acceptance of a product [31]. Public perception for a product or technology can be largely improved by making risk assessment of worker's exposure to nanomaterials transparent. Appropriate research on toxicity aspects of nanotechnology and conveying it to the public with facts and figures can make nanotechnology-based products more acceptable.

19.9 Regulatory Aspects

Nanotechnology, being a new and emerging field, regulations governing approval after appropriate toxicity studies are not very structured in many countries. Table 19.1 provides an overview of the regulations existing in countries pertaining to nanotechnology-based (food) products.

While some major countries including the USA, Canada, the European Union, Australia, New Zealand and China have not yet formulated any specific legislation on the regulation of nanofood, in general, it is subject to the same public health and food safety regulations that apply to other food types. Findings on toxicity have been inconclusive or not sufficient to warrant exclusive laws and regulations on nanotechnology-based products. Hence, many countries have preferred to take a cautious approach and included aspects of nanotechnology that need regulation in their existing regulations. The same applies to the food sector as well. The Food Standards Australia New Zealand (FSANZ) (Table 19.1) mentions that adequate precautions need to be taken with respect to nanotechnology-based products; the US Food and Drug Administration (USFDA) addresses nanotechnology in food, feed and cosmetics by means of appropriate guidance documents; and the European Union regulations

Table 19.1 Regulatory law aspects related to nanotechnology in New Zealand, Australia, the USA and the European Union

Regulatory body	Aspects of regulation
Food Standards Australia New Zealand (FSANZ) http://www.foodstandards.gov.au/consumer/foodtech/nanotech/Pages/default.aspx	Any new food manufactured using nanotechnology, which may have potential safety concerns, will be required to undergo safety assessment
	Information is mandatory about particle size, size distribution, morphology and size-dependent properties
	Strategies to manage any potential risks associated with nanotechnologies in foods to ensure protection of public health
US Food and Drug Administration (USFDA) https://www.fda.gov/scienceresearch/specialtopics/nanotechnology/default.htm	Regulates many products (foods, cosmetics, drugs, devices, veterinary and tobacco products), which may utilize nanotechnology/contain nanomaterials
	Considering Whether an FDA-Regulated Product Involves the Application of Nanotechnology: Federal Register notice: 79 FR 36534 (June 27, 2014)
	Safety of Nanomaterials in Cosmetic Products: Federal Register notice: 79 FR 36532 (June 27, 2014)
	Use of Nanomaterials in Food for Animals: Federal Register notice: 80 FR 46587 (August 5, 2015)
European Union https://www.efsa.europa.eu/en/topics/topic/nanotechnology	'Integrated, safe and responsible approach' to the development of nanotechnologies, such as reviewing existing laws and modifying them as required, monitoring safety issues, engaging in dialogues with stakeholders and citizens
	Regulatory aspects of nanomaterials [Sec(2008) 2036]

mention about regulating nanotechnology products in the same way as other products and involving stakeholders in the evolving nanotechnology field.

There are separate regulations for novel foods and novel foods ingredients, food additives, food supplements, vitamins, minerals and food contact materials. A regulation on the provision of food information to consumers exists. Depending on the type of nanomaterial or the purpose of its use as food or part of food, it would fall under its respective legislation.

19.10 Conclusion

Nanotechnology has promising applications in the food industry, through systems which possess efficient barrier and antimicrobial properties, as well as in food processing through improvement of sensory characteristics and nanoencapsulation of bioactive compounds. Other prospects include improvements in the nutrition and functional properties of food products. However, the actual application of nanotechnology in food remains a huge challenge. Different nanomaterials have different preparation methodologies depending on their type, and hence they lack scalability which can lead to lower performance efficiency and be less cost effective [76].

Rapid increase in development of high-throughput technologies is expected to lead to a rapid rise in usage of nanomaterials in other industry sectors. One solution that is expected to counter the perceived toxicity and non-acceptance of nanoproducts is the use of phytochemical-treated nanoparticles as substitutes for the existing ones [49]. Transcriptomic studies, on application of nanoparticles, can provide real-time data points on the effect on proteome.

Experimental studies are being conducted on mammalian cell types. Computational models can help in prediction of potential toxicities of new and modified nanomaterials. However, many experimentally conducted toxicity studies use excessive and unrealistically high doses, which cannot be extrapolated to real-life situations. Hence, there may be a requirement for more studies using realistically applied doses. Therefore there is a need to study more realistic lower exposure scenarios, both acute and chronic, to have better relevance to the real-world scenarios [66]. Further studies are necessary to assess the health effects of chronic exposure to nanomaterials, as nanoparticle toxicology is still a very young science. Associated health risks depend on the interaction of nanoparticles with the body.

References

1. Acosta E (2009) Bioavailability of nanoparticles in nutrient and nutraceutical delivery. Curr Opin Colloid Interface Sci 14(1):3–15
2. Ambrosi A, Airo F, Merkoci A (2010) Enhanced gold nanoparticle based ELISA for a breast cancer biomarker. Anal Chem 82(3):1151–1156
3. Amini SM, Kharrazi S, Hadizadeh M, Fateh M, Saber R (2013) Effect of gold nanoparticles on photodynamic efficiency of 5-aminolevolenic acid photosensitiser in epidermal carcinoma cell line: an in vitro study. IET Nanobiotechnol 7(4):151–156

4. Aresta A, Calvano CD, Trapani A, Cellamare S, Zambonin CG, Giglio ED (2013) Development and analytical characterization of vitamin (s)-loaded chitosan nanoparticles for potential food packaging applications. J Nanopart Res 15(4):1592
5. Arrieta MP, Peltzer MA, López J, del Carmen Garrigós M, Valente AJ, Jiménez A (2014) Functional properties of sodium and calcium caseinate antimicrobial active films containing carvacrol. J Food Eng 121:94–101
6. Avella M, De Vlieger JJ, Errico ME, Fischer S, Vacca P, Volpe MG (2005) Biodegradable starch/clay nanocomposite films for food packaging applications. Food Chem 93(3):467–447
7. Azeredo H, Mattoso LHC, Wood D, Williams TG, Avena-Bustillos RJ, McHugh TH (2009) Nanocomposite edible films from mango puree reinforced with cellulose nanofibers. J Food Sci 74(5):N31–N35
8. Azeredo H, Mattoso LHC, Avena-Bustillos RJ, Munford ML, Wood D, McHugh TH (2010) Nanocellulose reinforced chitosan composite films as affected by nanofiller loading and plasticizer content. J Food Sci 75(1):N1–N7
9. Biji KB, Ravishankar CN, Mohan CO, Gopal TS (2015) Smart packaging systems for food applications: a review. J Food Sci Technol 52(10):6125–6135
10. Biscarat J, Charmette C, Sanchez J, Pochat-Bohatier C (2015) Development of a new family of food packaging bioplastics from cross-linked gelatin based films. Can J Chem Eng 93(2):176–182
11. Borm PJ, Kreyling W (2004) Toxicological hazards of inhaled nanoparticles—potential implications for drug delivery. J Nanosci Nanotechnol 4(5):521–531
12. Braydich-Stolle LK, Schaeublin NM, Murdock RC, Jiang J, Biswas P, Schlager JJ, Hussain SM (2009) Crystal structure mediates mode of cell death in TiO2 nanotoxicity. J Nanopart Res 11(6):1361–1374
13. Bülbül G, Hayat A, Andreescu S (2015) Portable nanoparticle-based sensors for food safety assessment. Sensors (Basel) 15(12):30736–30758
14. Charpentier PA, Burgess K, Wang L, Chowdhury RR, Lotus AF, Moula G (2012) Nano-TiO2/polyurethane composites for antibacterial and self-cleaning coatings. Nanotechnology 23(42):425606
15. Chaudhry Q, Castle L (2011) Food applications of nanotechnologies: an overview of opportunities and challenges for developing countries. Trends Food Sci Technol 22(11):595–603
16. Chu W (2017) Special delivery: nano's nutrient gift presents huge potential. Is nanotechnology finally walking the talk when it comes to nutrient enhancement in food and supplements? http://www.nutraingredients.com/Research/Special-delivery-Nano-s-nutrient-gift-presents-huge-potential. Accessed on 25 Aug 2017
17. Cruz-Romero MC, Murphy T, Morris M, Cummins E, Kerry JP (2013) Antimicrobial activity of chitosan, organic acids and nano-sized solubilisates for potential use in smart antimicrobially-active packaging for potential food applications. Food Control 34(2):393–397
18. De A, Bose R, Kumar A, Mozumdar S (2014) Targeted delivery of pesticides using biodegradable polymeric nanoparticles. Springer Briefs in Molecular Science. Springer, New Delhi. https://doi.org/10.1007/978-81-322-1689-6
19. Dutta PK, Tripathi S, Mehrotra GK, Dutta J (2009) Perspectives for chitosan based antimicrobial films in food applications. Food Chem 114(4):1173–1182
20. Fabrega J, Fawcett SR, Renshaw JC, Lead JR (2009) Silver nanoparticle impact on bacterial growth: effect of pH concentration and organic matter. Environ Sci Technol 43(19):7285–7290
21. Falkner R, Jaspers N (2012) Regulating nanotechnologies: risk, uncertainty and the global governance gap. Global Environ Politics 12(1):30–55
22. FAO. http://www.fao.org/food-loss-and-food-waste/en/. Accessed on 25 Aug 2017
23. Farhoodi M (2016) Nanocomposite materials for food packaging applications: characterization and safety evaluation. Food Eng Rev 8(1):35–51
24. Feder BJ (2004) Study raises concerns about carbon particles. New York Times, p 153
25. Food and Agriculture Organization (2012) FAO Statistical Yearbook 2012. http://www.fao.org/docrep/015/i2490e/i2490e03a.pdf

26. Food and Agriculture Organization (2015) International Fund for Agricultural Development, World Food Program. "The State of Food Insecurity in the World 2015. Strengthening the enabling environment for food security and nutrition". FAO, Rome
27. Food and Agriculture Organization (2016) – Rome, 2016, World fertilizer trends and outlook to 2019 summary report
28. Gholami-Ahangaran M, Zia-Jahromi N (2014) Effect of nanosilver on blood parameters in chickens having aflatoxicosis. Toxicol Ind Health 30(2):192–196
29. Gibbs BF, Kermasha S, Alli I, Mulligan CN (1999) Encapsulation in the food industry: a review. Int J Food Sci Nutr 50(3):213–224
30. Greaves S, Faunce L, Cocco F (2017) https://www.ft.com/content/9d2b0b36-f784-11e6-9516-2d969e0d3b65. Accessed on 25 August 2017
31. Gutiérrez FJ, Mussons ML, Gatón P and Rojo R (2012) Nanotechnology and food industry, scientific, health and social aspects of the food industry. Dr. Benjamin Valdez (Ed.). InTech. https://doi.org/10.5772/33458. Available from: https://www.intechopen.com/books/scientific-health-and-social-aspects-of-the-food-industry/nanotechnology-and-food-industry. Accessed on 25 August 2017
32. Helmke BP, Minerick AR (2006) Designing a nano-interface in a microfluidic chip to probe living cells: challenges and perspectives. Proc Natl Acad Sci 103(17):6419–6424
33. Honarvar Z, Hadian Z, Mashayekh M (2016) Nanocomposites in food packaging applications and their risk assessment for health. Electron Physician 8(6):2531–2538
34. Hone DC, Walker PI, Evans-Gowing R, FitzGerald S, Beeby A, Chambrier I, Cook MJ, Russell DA (2002) Generation of cytotoxic singlet oxygen via phthalocyanine-stabilized gold nanoparticles: a potential delivery vehicle for photodynamic therapy. Langmuir 18(8):2985–2987
35. Hosseini SF, Rezaei M, Zandi M, Farahmandghavi F (2015) Fabrication of bio-nanocomposite films based on fish gelatin reinforced with chitosan nanoparticles. Food Hydrocoll 44:172–182
36. http://www.foodstandards.gov.au/consumer/foodtech/nanotech/Pages/default.aspx
37. https://nanohub.org/groups/gng. Accessed on 25 August 2017
38. https://patents.google.com. Accessed on 25 August 2017
39. https://www.efsa.europa.eu/en/topics/topic/nanotechnology. Accessed on 25 August 2017
40. https://www.fda.gov/scienceresearch/specialtopics/nanotechnology/default.htm. Accessed on 25 August 2017
41. https://www.nano.gov//node/135. Accessed on 25 August 2017
42. Hwang LS, Yeh AI (2010) Applying nanotechnology in food in Taiwan. International conference on food applications of nanoscale Science
43. Inbaraj BS, Chen BH (2016) Nanomaterial-based sensors for detection of foodborne bacterial pathogens and toxins as well as pork adulteration in meat products. J Food Drug Anal 24(1):15–28
44. Kajihara Y, Murakami M, Imagawa T, Otsuguro K, Ito S, Ohta T (2010) Histamine potentiates acid-induced responses mediating transient receptor potential V1 in mouse primary sensory neurons. Neuroscience 166(1):292–304
45. Kanmani P, Rhim JW (2014) Physical, mechanical and antimicrobial properties of gelatin based active nanocomposite films containing AgNPs and nanoclay. Food Hydrocoll 35:644–652
46. Kanmani P, Rhim JW (2014) Physicochemical properties of gelatin/silver nanoparticle antimicrobial composite films. Food Chem 148:162–169
47. Kyung OY, Grabinski CM, Schrand AM, Murdock RC, Wang W, Gu B, Schlager JJ, Hussain SM (2009) Toxicity of amorphous silica nanoparticles in mouse keratinocytes. J Nanopart Res 11(1):15–24
48. Lang T, Heasman M (2015) Food wars: the global battle for mouths, minds and markets. Routledge, Abingdon
49. Lee J, Park EY, Lee J (2014) Non-toxic nanoparticles from phytochemicals: preparation and biomedical application. Bioprocess Biosyst Eng 37(6):983–989

50. Leonard P, Hearty S, Brennan J, Dunne L, Quinn J, Chakraborty T, O'Kennedy R (2003) Advances in biosensors for detection of pathogens in food and water. Enzyme Microb Technol 32(1):3–13
51. Li ZZ, Xu SA, Wen LX, Liu F, Liu AQ, Wang Q, Sun HY, Yu W, Chen JF (2006) Controlled release of avermectin from porous hollow silica nanoparticles: influence of shell thickness on loading efficiency, UV-shielding property and release. J Control Release 111(1):81–88
52. Lipinski B, Hanson C, Lomax J, Kitinoja L, Waite R, Searchinger T (2013) Reducing food loss and waste. World Resources Institute Working Paper June
53. Loher S, Schneider OD, Maienfisch T, Bokorny S, Stark WJ (2008) Microorganism-triggered release of silver nanoparticles from biodegradable oxide carriers allows preparation of self-sterilizing polymer surfaces. Small 4(6):824–832
54. Mahler GJ, Esch MB, Tako E, Southard TL, Archer SD, Glahn RP, Shuler ML (2012) Oral exposure to polystyrene nanoparticles affects iron absorption. Nat Nanotechnol 7(4):264–271
55. Malhotra B, Keshwani A, Kharkwal H (2015) Natural polymer based cling films for food packaging. Int J Pharm Pharm Sci 7:10–18
56. Marsh K, Bugusu B (2007) Food packaging—roles materials and Environ issues. J food Sci 72(3):R39–R55
57. Martínez-Fernández D, Barroso D, Komárek M (2016) Root water transport of Helianthus annuus L under iron oxide nanoparticle exposure. Environ Sci Pollut Res 23(2):1732–1741
58. Martirosyan A, Schneider YJ (2014) Engineered nanomaterials in food: implications for food safety and consumer health. Int J Environ Res Public Health 11(6):5720–5750
59. McClements DJ, Decker EA (2000) Lipid oxidation in oil-in-water emulsions: impact of molecular environment on chemical reactions in heterogeneous food systems. J Food Sci 65(8):1270–1282
60. Mills E, Roth RJ Jr, Lecomte E (2006) Availability and affordability of insurance under climate change: a growing challenge for the United States. J Insur Regul 25(2):109
61. Miranda OR, Li X, Garcia-Gonzalez L, Zhu ZJ, Yan B, Bunz UH, Rotello VM (2011) Colorimetric bacteria sensing using a supramolecular enzyme–nanoparticle biosensor. J Am Chem Soc 133(25):9650–9653
62. Monge M, Moreno-Arribas MV (2016) Applications of nanotechnology in wine production and quality and safety control. In: Wine safety, consumer preference, and human health. Springer, Berlin, pp 51–69
63. Morones JR, Elechiguerra JL, Camacho A, Holt K, Kouri JB, Ramírez JT, Yacaman MJ (2005) The bactericidal effect of silver nanoparticles. Nanotechnology 16(10):2346
64. Mozafari RM, Johnson C, Hatziantoniou S, Demetzos C (2008) Nanoliposomes and their applications in food nanotechnology. J Lipos Res 18(4):309–327
65. Oberdörster G (2000) Pulmonary effects of inhaled ultrafine particles. Int Arch Occup Environ Health 74(1):1–8
66. Oberdörster G (2010) Safety assessment for nanotechnology and nanomedicine: concepts of nanotoxicology. J Internal Med 267(1):89–105
67. Ornatska M, Sharpe E, Andreescu D, Andreescu S (2011) Paper bioassay based on ceria nanoparticles as colorimetric probes. Anal Chem 83(11):4273–4280
68. Osmond-McLeod MJ, Oytam Y, Osmond RIW, Sobhanmanesh F, McCall MJ (2014) Surface coatings protect against the in vitro toxicity of zinc oxide nanoparticles in human hepatic stellate cells. J Nanomed Nanotechnol 5:232. https://doi.org/10.4172/2157-7439.1000232
69. Ozimek L, Pospiech E, Narine S (2010) Nanotechnologies in food and meat processing. Acta Sci Pol Technol Alimentaria 9(4):401–412
70. Pérez-López B, Merkoçi A (2011) Nanomaterials based biosensors for food analysis applications. Trends Food Sci Technol 22(11):625–639
71. Phillips RL, Miranda OR, You CC, Rotello VM, Bunz UH (2008) Rapid and efficient identification of bacteria using gold-nanoparticle–poly (para-phenyleneethynylene) constructs. Angew Chem Int Ed 47(14):2590–2594
72. Prabhu S, Poulose EK (2012) Silver nanoparticles: mechanism of antimicrobial action, synthesis, medical applications, and toxicity effects. Int Nano Lett 2(1):32

73. Quintanilla-Carvajal MX, Camacho-Dıaz BH, Meraz-Torres LS, Chanona-Perez JJ, Alamilla-Beltran L, Gutierrez-Lopez GF (2009) Nanoencapsulation: a new trend in food engineering processing. Food Eng Rev 2(1):39–50. https://doi.org/10.1007/s12393-009-9012-6
74. Radi RBJS, Beckman JS, Bush KM, Freeman BA (1991) Peroxynitrite oxidation of sulfhydryls. The cytotoxic potential of superoxide and nitric oxide. J Biol Chem 266(7):4244–4250
75. Raja P, Huynh M and Khabashesku VN (2015) Safe handling and disposal of nanostructured materials. Offshore Technology Conference May
76. Ramachandraiah K, Han SG, Chin KB (2015) Nanotechnology in meat processing and packaging: potential applications—a review. Asian-Australas J Anim Sci 28(2):290
77. Ramon-Marquez T, Medina-Castillo AL, Fernandez-Gutierrez A, Fernandez-Sanchez JF (2016) Novel optical sensing film based on a functional nonwoven nanofibre mat for an easy, fast highly selective sensitive detection of tryptamine in beer. Biosens Bioelectron 79:600–607
78. Ramos M, Valdés A, Beltrán A, Garrigós MC (2016) Gelatin-based films and coatings for food packaging applications. Coatings 6: 41
79. Ren G, Hu D, Cheng EW, Vargas-Reus MA, Reip P, Allaker RP (2009) Characterisation of copper oxide nanoparticles for antimicrobial applications. Int J Antimicrob Agents 33(6):587–590
80. Saha B, Evers TH, Prins MW (2014) How antibody surface coverage on nanoparticles determines the activity and kinetics of antigen capturing for biosensing. Anal Chem 86(16):8158–8166
81. Sarkar B, Bhattacharjee S, Daware A, Tribedi P, Krishnani KK, Minhas PS (2015) Selenium nanoparticles for stress-resilient fish and livestock. Nanoscale Res Lett 10(1):371
82. Schulte PA, Salamanca-Buentello F (2007) Ethical and scientific issues of nanotechnology in the workplace. Ciên Saúde Colet 12(5):1319–1332
83. Sekhon BS (2010) Food nanotechnology–an overview. Nanotechnol Sci Appl 3(1):1–15
84. Selim NA, Radwan NL, Youssef SF, Eldin TAS, Elwafa SA (2015) Effect of inclusion inorganic, organic or nano selenium forms in broiler diets on: 2-Physiological, immunological and toxicity statuses of broiler chicks. Int J Poult Sci 14(3):144
85. Shankar S, Teng X, Li G, Rhim JW (2015) Preparation characterization, and antimicrobial activity of gelatin/ZnO nanocomposite films. Food Hydrocoll 45:264–271
86. Sharpe E, Frasco T, Andreescu D, Andreescu S (2013) Portable ceria nanoparticle-based assay for rapid detection of food antioxidants (NanoCerac). Analyst 138(1):249–262
87. Sherry LJ, Jin R, Mirkin CA, Schatz GC, Van Duyne RP (2006) Localized surface plasmon resonance spectroscopy of single silver triangular nanoprisms. Nano Lett 6(9):2060–2065
88. Shi L, Zhao Y, Zhang X, Su H, Tan T (2008) Antibacterial and anti-mildew behavior of chitosan/nano-TiO2 composite emulsion. Korean J Chem Eng 25(6):1434–1438
89. Shin SW, Song IH, Um SH (2015) Role of physicochemical properties in nanoparticle toxicity. Nanomaterials 5(3):1351–1365
90. Singh H (2016) Nanotechnology applications in functional foods; opportunities and challenges. Prev Nutr Food Sci 21(1):1
91. Sinha VK, Vinay A, Bhinge JR (2008) Nanocochleates: a novel drug delivery technology. Pharmainfo Net 6(5):28
92. Song J, Wu F, Wan Y, Ma L (2015) Colorimetric detection of melamine in pretreated milk using silver nanoparticles functionalized with sulfanilic acid. Food Control 50:356–361
93. Storhoff JJ, Lazarides AA, Mucic RC, Mirkin CA, Letsinger RL, Schatz GC (2000) What controls the optical properties of DNA-linked gold Nanopart assemblies? J Am Chem Soc 122(19):4640–4650
94. Thangavel G, Thiruvengadam S (2014) Nanotechnology in food industry – a review. Int J Chem Tech Res 16(9):4096–4101
95. Tsukamoto K, Wakayama J and Sugiyama S (2010) Nanobiotechnology approach for food and food related fields. Poster presented at the International Conference on Food Applications of Nanoscale Science (ICOFANS), Tokyo, Japan, pp 9–11

96. Valdés MG, González ACV, Calzón JAG, Díaz-García ME (2009) Analytical nanotechnology for food analysis. Microchim Acta 166(1-2):1–19
97. Vidhyalakshmi R, Bhakyaraj R, Subhasree RS (2009) Encapsulation "the future of probiotics"-a review. Adv Biol Res 3(3-4):96–103
98. Vilela D, González MC, Escarpa A (2012) Sensing colorimetric approaches based on gold and silver nanoparticles aggregation: chemical creativity behind the assay A review. Anal Chim Acta 751:24–43
99. Vishwakarma V, Samal SS, Manoharan N (2010) Safety and risk associated with nanoparticles-a review. J Miner Mater Charact Eng 9(05):455
100. Wang Y, Alocilja EC (2015) Gold nanoparticle-labeled biosensor for rapid and sensitive detection of bacterial pathogens. J Biol Eng 9(1):16
101. Wang S, Su R, Nie S, Sun M, Zhang J, Wu D, Moustaid-Moussa N (2014) Application of nanotechnology in improving bioavailability and bioactivity of diet-derived phytochemicals. J Nutr Biochem 25(4):363–376
102. Wei H, Chen C, Han B, Wang E (2008) Enzyme colorimetric assay using unmodified silver nanoparticles. Anal Chem 80(18):7051–7055
103. Weiss J, Takhistov P, McClements DJ (2006) Functional materials in food nanotechnology. J Food Sci 71(9):R107–R116
104. Wickramasinghe SN, Gardner B, Barden G (1987) Circulating cytotoxic protein generated after ethanol consumption: identification and mechanism of reaction with cells. Lancet 330(8551):122–126
105. Xia S, Xu S, Zhang X (2006) Optimization in the preparation of coenzyme Q10 nanoliposomes. J Agric Food Chem 54(17):6358–6366
106. Xiong D, Li H (2008) Colorimetric detection of pesticides based on calixarene modified silver nanoparticles in water. Nanotechnology 19(46):465502
107. Xiu ZM, Zhang QB, Puppala HL, Colvin VL, Alvarez PJ (2012) Negligible particle-specific antibacterial activity of silver nanoparticles. Nano Lett 12(8):4271–4275
108. Xun W, Shi L, Yue W, Zhang C, Ren Y and Liu Q (2012) Effect of high-dose nano-selenium and selenium–yeast on feed digestibility, rumen fermentation, and purine derivatives in sheep. Biol Trace Elem Res 150(1–3): 130–136
109. Yam KL, Takhistov PT, Miltz J (2005) Intelligent packaging: concepts and applications. J Food Sci 70(1):R1–R10
110. Yu H, Huang Y, Huang Q (2009) Synthesis and characterization of novel antimicrobial emulsifiers from ε-polylysine. J Agric food Chem 58(2):1290–1295
111. Yu H and Huang Q (2010) Enhanced in vitro anti-cancer activity of curcumin encapsulated in hydrophobically modified starch. Food Chem, 119(2):669-674
112. Yuan CW, Huang JT, Chen CC, Tang PC, Huang JW, Lin JJ, Huang SY, Chen SE (2017) Evaluation of efficacy and toxicity of exfoliated silicate nanoclays as a feed additive for fumonisin detoxification. J Agric Food Chem 65(31):6564–6571

20 Nano-delivery of Food-Derived Biomolecules: An Overview

Dhwani Jhala, Hilal Rather, and Rajesh Vasita

Abstract

Certain bioactive molecules present in food have therapeutic effects such as anti-oxidant, anti-carcinogenic, anti-mutagenic, anti-ageing and anti-inflammatory activity. Although these molecules can be consumed via food, their effects are often subordinated due to degradation during digestion and lesser bioavailability at target site. Nanocarriers such as nanoparticles, nanoliposomes and nanopoly-mersomes have been developed for the food-derived biomolecules that can overcome traditional drug delivery limitations and provide advantages such as increased solubility and stability, protection against degradation, target specificity, increased bioavailability and regulated release. This chapter discusses different types of nanocarriers for delivery of food-derived biomolecules along with their advantages over conventional drug delivery methods. It also addresses the clearance of these nanocarriers from the body, their toxicity and ethics-related concerns. Overall, it provides an overview of nanocarriers used for delivery of food-derived bioactive molecules.

Keywords

Drug delivery · Nanocarriers · Nanoparticles · Nanoliposomes · Polymersomes · Nanotoxicity · Nutraceuticals

Authors Dhwani Jhala and Hilal Rather have been contributed equally to this chapter.

D. Jhala · H. Rather · R. Vasita (✉)
Biomaterials and Biomimetics Laboratory, School of Life Sciences,
Central University of Gujarat, Gandhinagar, Gujarat, India

V. Rani, U. C. S. Yadav (eds.), *Functional Food and Human Health*,
https://doi.org/10.1007/978-981-13-1123-9_20

20.1 Introduction

The concept of controlled, sustained and/or targeted drug delivery is well established and clinically practised for the last two decades. However, the introduction of nanotechnology-based delivery devices has revolutionized the concept of drug delivery due to their higher loading capacity, target specificity and lesser toxicity. The targeted delivery ensures the adequate bioavailability of drug and reduces the frequent dosage requirement. Furthermore, their smaller size enhances the blood circulation duration, cell internalization and removal from the body. A variety of nano-delivery vehicles such as nanoparticles, nanoliposomes, nanopolymersomes and nanoclays have been developed for delivering bioactive molecules. Today, nano-drug delivery systems for hundreds of drugs have been researched and developed, and more than 20 FDA-approved systems are commercially available in pharmaceutical market; however, it is still in the developing stage for food-derived drugs.

Research on the medical benefits of food-derived compounds has a very long history; for example, turmeric-containing curcumin has been used for health benefits dating back nearly 4000 years. However, delivery of such food-derived drugs using specialized delivery systems was not much focused due to limited understanding on mechanism and natural availability. In recent years, the advantages of nano-drug delivery systems were extended for delivery of food-derived drugs too. The nanocarriers can provide increased bioavailability, target specificity, better distribution and protection to encapsulated bioactive molecules. Some of the key advantages are discussed below.

20.1.1 Solubility

Solubility is a key property for any therapeutic substance as it governs its dissolution and absorption. However, most of the food-derived bioactive molecules are either insoluble or partially soluble in water. Methods and materials used for nanocarrier's fabrication provide an opportunity to encapsulate water-insoluble bioactive molecules, which effectively enables them to overcome their solubility limitations impacting their efficient delivery [1]. Solubility affects the applicability of bioactive molecules in variety of ways as insoluble nutrients are not useful for enhancing food utilization [2]. Poorly soluble compounds have limited bioavailability because they do not remain at the site of action for a longer time which results in lesser absorption by target tissue. Therefore, higher and repeated doses are required to enhance their bioavailability. However, nanocarriers have been used to solubilize a number of bioactive molecules to increase their bioavailability. For example, Ubisol-Aqua™, a commercially available carrier for water-insoluble nutraceuticals, has been used to solubilize vitamin A acetate, vitamin A palmitate, vitamin B pentapalmitate, coenzyme Q10, squalene, α-tocopheryl acetate and β-carotene by reducing their particle size [3, 4].

20.1.2 Stability

Stability of bioactive molecules is another critical factor to be looked into which includes mechanical, thermal and pH stability. Several commercial products use the advanced drug delivery system which improvises the stability of cargo [1]; for example, Aquanova is used as a carrier for water-insoluble compounds. The solubilized complex is stable through a wide range of temperatures (−80 to 120 °C), pH range (2.0–8.0) and phase changes (freezing, melting). Furthermore, Aquanova creates a liquid carrier solution and transports the material in the smallest capsules called product micelles. These micelles are of 30 nm diameter and are mechanically, thermally and pH stable [5, 6]. This is important for making the formulations and ensuring compound stability in the gastrointestinal track. The product micelles bypass the physiological mechanism of intestinal micelle formation which otherwise enhances the absorption of poorly soluble bioactive molecules and consequently, higher bioavailability is achieved. In Bioral® nanocochleate delivery system, lipid layers around the cargo potentially protect it from degradation by acid or digestive enzymes in the stomach [1].

20.1.3 Bioavailability

The nanosized self-assembled structured liquids improve bioavailability of food-derived drugs because of its potential to release the drugs into the membrane between the digestive system and the blood. Curcumin, the most exploited food-derived drug, has shown to modulate several targets that have been linked with cancer and various other chronic diseases. However, poor bioavailability of curcumin limits its applications [7–9]. Therefore, various drug delivery approaches including nanoemulsions, microemulsions, micelles and self-microemulsifying systems have been used to enhance the bioavailability of curcumin [10]. Kakkar et al. reported that lipid-based nanoformulations improve the bioavailability of curcumin [11]. Furthermore, it was demonstrated that the lipid nanoparticles could cross the blood-brain barrier and enhance the bioavailability of curcumin in the brain [12].

As the digested food leaves the stomach and enters the duodenum, it mixes with bile salts released by the gall bladder which helps in the emulsification of hydrophobic compounds like fats. The average size of bile salt aggregates ranges from 4 nm to 60 nm [13]. Thus, these bile salt micelles act as nature's nanoparticle delivery system. Faulks and Southon reported that 70 mg of apolar and 44 mg of polar carotenoids could be absorbed in a single meal without the use of nanocarrier systems [14]. However, authors indicated that bile salts can emulsify carotenoids dissolved in triglycerides, but cannot solubilize pure carotenoids. Therefore, triglyceride-based nanocarrier systems could provide the alternative for nanoemulsifying systems and hence enhance their digestion. After the duodenum, the suspension enters into the small intestine, where epithelial cells are covered with microvilli which are in turn covered by a mucous layer of an anionic glycoprotein which acts as a key factor in the uptake of nanoparticles. Therefore, direct nanocarrier-based delivery of the active ingredients is a viable route to enhance their bioavailability [15–18].

20.2 Types of Delivery Vehicles

As mentioned in introduction, there are various types of drug-delivering nanocarriers; however, only those associated with delivery of food-derived bioactive molecules are discussed here.

20.2.1 Nanoparticles

Different types of nanoscale drug delivery vehicles have been reported, viz. nanocapsules, nanogels, nanospheres and nanocrystals which can be commonly referred as 'nanoparticles' (NPs). Nanoparticles used for drug delivery have been traditionally defined as 'submicron-sized polymeric colloidal particles with a therapeutic agent of interest encapsulated within their polymeric matrix or adsorbed or conjugated onto the surface' [19]. However, nanoparticles from other materials than polymer, such as metal, lipid, protein and polysaccharides, have also been fabricated.

Drugs have been classified in four major classes by Biopharmaceutics Classification System (BCS) based on their aqueous solubility and intestinal permeability (Table 20.1). Nanoparticle-based systems can be useful for delivery of all the four classes of drugs. Nanoparticles provide several advantages over conventional carriers due to their smaller size. They can improve drug solubility and increase permeation rate. They can enhance penetration into blood capillaries and ultimately, increase cellular internalization. Moreover, nanosized particles (100 nm) have shown 15–250-fold higher uptake efficiency compared to micro-sized particles (1 and 10 μm) [15]. They have also been reported to cross blood-brain barrier which is limitation of many traditional drugs and their delivery systems [20]. NPs also facilitate cellular internalization of drugs. They are internalized by early endosomes which are then transferred to sorting endosomes. Then, most of them are transported to lysosomes, whereas some are exported out of the cells. From lysosomes, NPs escape into cytoplasm where they release their drug load [19]. They protect drug

Table 20.1 Benefits of nanoparticles for different classes of drugs

Drug class	Solubility	Permeability	Benefits of nanoparticles
Class I	High	High	Sustained or controlled release
Class II	Low	High	Increase solubility or bypass the need of solubility for absorption
Class III	High	Low	Increase permeation rate
Class IV	Low	Low	Protect from enzymatic degradation, efflux and first-pass effect

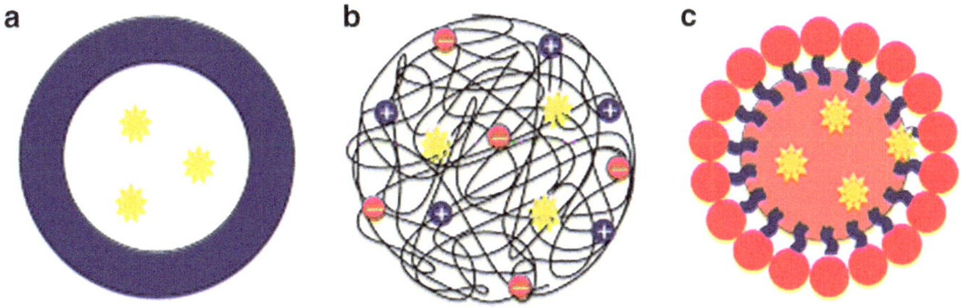

Fig. 20.1 Illustrations of (**a**) metal, (**b**) polymeric and (**c**) solid lipid nanoparticles. Encapsulated drugs are shown as yellow stars inside nanoparticles

from enzymatic degradation, efflux and first-pass effect. NPs can be used for controlled, sustained or targeted drug delivery based upon specific application. NPs also help in reducing immunogenicity of specific peptide- or protein-based biological drugs which can be otherwise quickly metabolized in vivo [21].

Nanoparticles can be synthesized by two approaches: (1) top-down approach where large molecules are broken down into smaller particles via different size reduction methods and (2) bottom-up approach where atoms self-assemble together to make nuclei and then nanosized particles. This is facilitated by chemical or biological reduction methods. Both these methods are currently in use and have been used for developing different drug delivery vehicles.

Here, nanoparticles have been classified as metal nanoparticles, polymeric NPs and solid lipid NPs based on their composition. Schematic representation of all three types of nanoparticles is shown in Fig. 20.1. Apart from this, other materials such as proteins and polysaccharides have also been used for nanoparticle synthesis.

20.2.1.1 Metal Nanoparticles

Metal nanoparticles are suspension of nanosized particles of metal within the size range of 1–100 nm. Metal NPs were first recognized by Faraday in 1857, whereas they were quantitatively explained through 'Mie theory' later on. Nanoparticles of noble metals such as gold, silver and platinum are widely used for electronic, magnetic and information storage applications due to their special intrinsic properties [22–24]. Gold has a long medicinal history for use against cancer and arthritis, making it most commonly used metal for drug delivery applications [25]. Further, other metals such as silver, platinum, titanium, palladium, iron, aluminium and copper are also used for nanoparticle-based drug delivery (Fig. 20.1a).

Metal nanoparticles can be synthesized by three different methods: (1) by using dendrimers, (2) by using microemulsions and (3) by direct reduction of metal salts. In case of food-derived drugs, the third method is the most commonly used since the food molecules act as natural biological reductants, making the nanoparticle fabrication a one-step reaction and 'green synthesis' by avoiding the use of harsh chemicals. Variety of plant extracts and microorganisms have been used in order to synthesize gold nanoparticles. The plant extract works as a reducing agent during nanoparticle synthesis as well as provides phytochemicals with health-benefiting properties. Thus, synthesizing metallic nanoparticles with drugs is a one-step and straightforward process. Crude extract of broccoli has been used to make gold nanoparticles which showed anticancer activity [26]. The extract contained mixture of phytochemicals, viz. glucoraphanin, phenethyl glucosinolates, quercetin, folic acid, vitamin C, allyl isothiocyanates, 2-phenylethyl isothiocyanates and sulforaphane which has been effective against prostate cancer, breast cancer and myeloma. Cumin has also been used to make biocompatible gold NPs (along with gum Arabic used as a stabilizer) with antioxidant properties [27]. Moreover, metallic NPs have advantage of antibacterial properties which is an intrinsic property of metal and owes to large surface area to volume ratio. Basil has been used for synthesis of silver NPs [28]. The intrinsic drug properties of basil and silver both render antibacterial activity to NPs. Overall, various food extracts such as grape, banana, egg, honey, citrus fruits, olive, tea leaves and garlic have been widely used for green synthesis of metallic nanoparticles [29–36], but mainly due to their reducing nature. To the best of our knowledge, metallic nanoparticles have not been reported for delivery of encapsulated food-derived drugs.

20.2.1.2 Polymeric Nanoparticles

Polymeric nanoparticles are solid colloidal particles, often in submicron size range (<1000 nm). Biodegradable synthetic polymers such as polylactic acid (PLA), poly(lactic-co-glycolic acid) and polyethylene glycol (PEG) and natural polymers such as chitosan, alginate and gelatin have been used to synthesize nanoparticles. They are all biocompatible with tunable biodegradable properties. They can be either anionic or cationic depending upon the nature of polymeric chains and can be used to encapsulate drug within their matrix (Fig. 20.1b). The drug loading, release duration and pattern can be precisely controlled with polymeric NPs as compared to non-polymeric NPs. Cytotoxic studies on PLA/PLGA NPs showed that mononuclear phagocytic system in the body rapidly clears these NPs by endocytosis and therefore they do not cause any cytotoxicity [37]. Polymeric NPs can be functionalized with specific antibodies or other biomolecules for targeted drug delivery. Moreover, surface modification of the polymeric NPs such as PEGylation can provide further advantages of increased time in circulatory system by improving their pharmacokinetic and pharmacodynamic properties [38].

Several studies of food-derived drug delivery have been reported using polymeric NPs. Ellagic acid is one such drug which is found in berries and known for its antioxidant, anti-carcinogenic and anti-mutagenic activities [39, 40]. However, it has several limitations such as low solubility, rapid metabolism in GI tract, first-pass effect and binding to cellular DNA and proteins [41]; and therefore, it comes under class IV of the BCS. To overcome the disadvantages of ellagic acid, it has been successfully encapsulated in PLGA-PEG NPs by a novel emulsion-diffusion-evaporation method [41]. These NPs can be further studied for applications related to drug's properties. Cancer is one of the leading causes of death worldwide and therefore prime area for pharmaceutical research. Various studies using

Table 20.2 Examples of curcumin-loaded polymeric nanoparticles

Sr. no.	Polymer	Fabrication method	Size	Application	References
1.	Poly(maleic anhydride-alt-1-octadecene)-polyethylene glycol (C_{18}PMH-PEG)	Reprecipitation and anchored to polymer by ultrasonication	80–90 nm	Colon cancer (CT-26 cells)	[45]
2.	Poly(lactic-co-glycolic acid) (PLGA)	Nano-precipitation	560 nm	Ovarian cancer (A2780CP cells)	[46]
3.	2-Hydroxyethyl methacrylate (PHEMA)	Nano-precipitation	300 nm	Ovarian cancer (SKOV-3 cells)	[47]
4.	Poly(lactic-co-glycolic acid) (PLGA)	Solid/oil/water emulsion solvent evaporation method	45 nm	Prostate cancer (LNCaP, PC3 and DU-145 cells)	[43]
5.	Alginate-chitosan-pluronic	Ionotropic pre-gelation and polycationic cross-linking	100 nm	Cervical cancer (HeLa cells)	[48]
6.	Poly(lactide-co-glycolide) and polyethylene glycol (PEG)	Nano-precipitation	80 nm	Leukaemia (KBM-5 cells)	[49]
7.	Chitosan and silk fibroin	Capillary-microdot technique	<100 nm	Breast cancer (MCF-7 and MDA-MB-453 cells)	[44]

food-derived drugs loaded in polymeric NPs have been done that can enhance properties of drugs for anticancer activity. Ferulic acid, found in apples, artichokes, coffee, peanuts and oranges, has been loaded in PLGA NPs for lung cancer treatment. The NPs were synthesized by double emulsion method and have been effective against NCI-H460 lung cancer cells [42].

Curcumin, a well-known phytochemical found in turmeric, has variety of benefits. It has anti-inflammatory, anti-carcinogenic, antimalarial, anti-HIV and antioxidant activity; it improves brain function, reduces risk of heart diseases and is useful in Alzheimer's disease, arthritis, depression and age-related issues. Despite of all these properties, it suffers from poor aqueous solubility, chemical instability and rapid metabolism, and therefore it requires delivery vehicles for its optimum benefit. Curcumin has been loaded in PLGA nanospheres for use against prostate cancer [43]. Apart from synthetic polymers, natural polymers such as chitosan and silk fibroin NPs have also been synthesized using curcumin against breast cancer [44]. Thus, same drug can be used with different polymers and different fabrication methods to synthesize nanoparticles with ranges of sizes and applications. Table 20.2 shows how curcumin has been loaded in different polymeric NPs for specific properties and applications against different cancer cells.

20.2.1.3 Solid Lipid Nanoparticles

Solid lipid nanoparticles have a solid lipid matrix, stabilized by aqueous surfactants that can be loaded with the drug in core, shell or entire matrix (Fig. 20.1c). The lipids used are triglycerides, fatty acids, waxes, glycerols and hard fats. They are mainly used for BCS class II and IV drugs which suffer from poor solubility. Use of SLNs provides advantages of improved drug stability, biocompatibility and controlled and sustained release compared to traditional drug carrier systems such as emulsions, liposomes and polymeric nanoparticles. However, uncontrolled gelation tendency and particle growth are some of the disadvantages associated with SLNs.

SLNs have been used for few food-derived drug delivery. As discussed in previous section, curcumin suffers from poor water solubility, stability and oral bioavailability, despite having multiple health benefits. SLNs have also been used to overcome these limitations. SLNs made up of Smoothex STD, Rikemal PS-100 and palmitic acid as a lipid phase and Tween 80 as a surfactant were developed for oral delivery of curcumin. These SLNs could enhance feasibility and bioavailability of curcumin. Furthermore, to improve its absorption and oral bioavailability, a mucoadhesive natural polymer, chitosan, was coated on these SLNs [50]. Another drug, known as epigallocatechin gallate (EGCG), which is found in different teas, apple skin, plums, onions and hazelnuts has shown promising effect on melanoma cancer and Alzheimer's disease. However, it has low bioavailability and ineffective delivery to central nervous system. Smith et al. fabricated nanolipidic particles with encapsulated EGCG which showed double oral bioavailability in rat models and increased effectiveness against Alzheimer's disease [51].

20.2.1.4 Other Nanoparticles

Widely used nanoparticles made up of metals, polymers or lipids have been discussed above. However, there are few studies where mixture of these materials or some other classes of materials has been used to synthesize NPs for food-derived drug delivery. Huang et al. developed core-shell protein-polysaccharide NPs for curcumin delivery. They have used zein (primary protein of corn) as a core and mixture of alginate and pectin as shell to develop chemically stable NPs that can incorporate hydrophobic drugs such as curcumin and carry out their delivery [52].

Quercetin, a drug commonly found in many fruits, vegetables, grains and leaves, has been effective against cancer; however, low bioavailability and poor water solubility have confined its application. Lecithin (lipid)-chitosan (natural polymer) NPs have been prepared for topical delivery of quercetin against skin cancer [53]. Recently, targeted delivery of quercetin has been explored for breast cancer using mesoporous silica nanoparticles (MSNs) [49]. MSNs can provide tunable porosity, chemical and mechanical stability, more biocompatibility and less toxicity compared to organic drug carriers [28]. For targeted delivery, the MSNs were coated with folic acid, as cancer cells tend to have more receptors for folic acids. Folate-armed MSNs showed higher bioavailability and anticancer effect in MDA-MB-231 breast cancer cells [49].

Recently, Zhang et al. developed a novel method of drug delivery for treatment of inflammatory bowel disease. They have derived NPs from edible ginger with

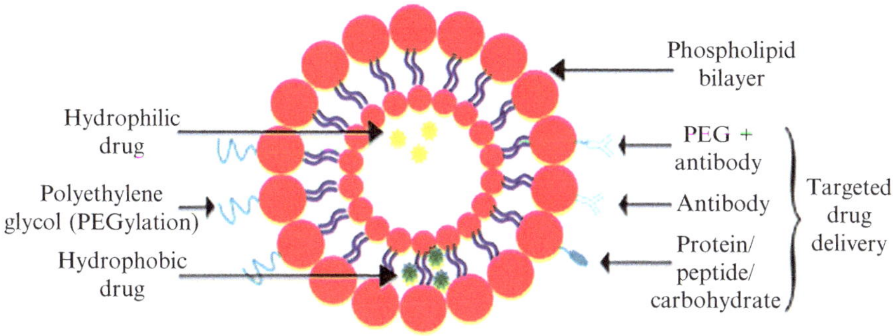

Fig. 20.2 Multifunctional nanoliposomes for drug delivery

~230 nm size and large number of bioactive molecules from ginger such as 6-gingerol and 6-shagol. These NPs do not consist of any supporting matrix and therefore are very natural and unique in their composition. Their oral administration has resulted in increased proliferation of intestinal epithelial cells and increased anti-inflammatory activity in mouse models which confirmed their therapeutic potential [54].

To summarize, nanoparticles are one of the most widely studied classes of drug delivery vehicles due to their variety in composition, size and structure. They also provide range of variations in biodegradability, target specificity and release kinetics and therefore have been successfully used for delivery of food-derived biomolecules for different target-specific and sustained drug delivery applications such as cancer and Alzheimer's disease.

20.2.2 Liposomes

Liposomes can be defined as hollow lipid vesicles composed of bilayered membrane of phospholipids (Fig. 20.2). Unlike SLNs, liposomal core is aqueous and therefore comparatively less stable. Liposomes have been used in variety of fields including drug delivery, diagnosis, food industry and cosmetic products [55–57]. Nanoliposome term has been recently coined specifically for the liposomes in nanometre range [58]. Due to smaller size, nanoliposomes can provide added advantage of larger surface area, thus increased bioavailability of encapsulated drug [58]. They can be synthesized using variety of methods such as extrusion, sonication, microfluidization and freeze-thawing. Nanoliposomal drug delivery can be done by oral, topical, nasal or parenteral route [58–60].

Liposomes are the most commonly studied drug delivery vehicles due to their ability to encapsulate both hydrophilic and hydrophobic drugs, biocompatibility, controllable physicochemical properties, self-assembling ability and target-specific delivery [61]. Hydrophilic drugs can be encapsulated in central aqueous compartment, whereas hydrophobic drugs can be loaded in lipid membrane (Fig. 20.2) [56]. Inside the body, external forces can disintegrate nanoliposomes;

therefore, sometimes, they are additionally stabilized by coating with inert polymers such as polyethylene glycol (PEG). Moreover, for targeted drug delivery, specific antibody, protein, peptide or carbohydrates can be attached to nanoliposome structures (Fig. 20.2).

Curcumin is the most widely used drug for nanoliposome-based delivery. Different methods of liposomal formulations such as vortexing, extrusion and sonication have been used to synthesize liposomes for curcumin drug delivery [55, 62–65]. Thangapazham et al. developed nanoliposomal system for targeted delivery of curcumin against prostate cancer by coating liposomes with prostate-specific antigen (PSA) antibodies, and these nanoliposomes showed 70–80% inhibition of cellular proliferation with ~10-fold lesser curcumin compared to free curcumin [62]. Curcumin has also been used as a potential drug for Alzheimer's disease [66]. Anti-transferrin antibody, which is specific against blood-brain barrier transport mediator protein, was attached to nanoliposomes for targeted delivery of curcumin to the brain [66].

Nanoliposome-based drug delivery system has been also used for β-cryptoxanthin drug known for its anticancer effects [67]. β-Cryptoxanthin is a carotenoid commonly present in pumpkin, orange, corn, peas and sweet potatoes [68]. It has shown anticancer effects in lung, breast, bladder and colon cancer cells [69, 70]; however, its insolubility in water restricts its use in therapeutics. Therefore, nanoliposomes with average diameter of 96.2 nm were developed by extrusion method. They showed strong apoptotic effect in K562 human leukaemia cells [67]. Another drug, resveratrol, found in nuts, berries, cocoa and wine, is known for its anticancer, anti-inflammatory, antidiabetic and cardio- and neuroprotective effects. However, it suffers from drawbacks of low bioavailability and rapid metabolism. Mohan et al. used nanoliposomes to deliver two drugs – resveratrol and 5-fluorouracil – simultaneously for the synergistic effect against head and neck carcinoma [71]. Both these hydrophobic drugs were encapsulated in lipid bilayer, and nanoliposomes were PEGylated for stabilization. The liposomal carrier showed successful delivery of both drugs and combination effect of cytotoxicity. Thus, liposomal nanocarriers have been used for different types of food-derived drugs for specific and targeted delivery.

20.2.3 Polymersomes and Polymeric Micelles

Liposomes are widely used for drug delivery applications, mainly because of its ability to encapsulate hydrophilic as well as hydrophobic drugs. However, lipid bilayers are usually very thin (4–5 nm) and are subject to instability [72]. Therefore, polymersomes have been developed which are liposomes like structure made up of polymers. Polymersomes are artificial vesicles synthesized from amphiphilic block copolymers arranged in a bilayer micelle structure having a hydrophilic core and hydrophobic shell (Fig. 20.3a). Unlike liposomes, polymersome membrane is comparatively thick (5–30 nm), and therefore they have more robustness [72]. Properties and thickness of membrane can be controlled by composition and

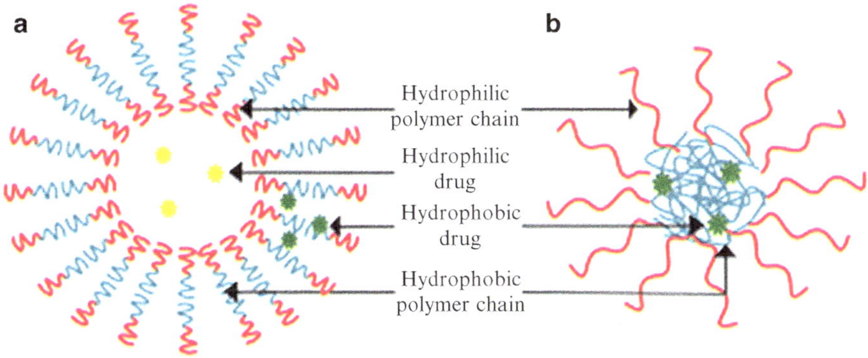

Fig. 20.3 Illustrations of (**a**) polymersome and (**b**) polymeric micelle

molecular weight of polymers [73, 74]. In most cases, PEG is used as a hydrophilic polymer because it increases overall blood circulation time of polymersomes [75, 76]. The relation between membrane thickness (d) and polymer molecular weight (MW) has been found as d α $MW^{0.55}$ [77]. Different methods have been used to form polymersomes such as film hydration, solvent exchange, electroformation and double emulsion strategy.

The hydrophilic part within total mass of copolymers is known as f value which plays important role in their self-assembly. It is found that if f value is >50%, it favours micelle formation, whereas f value between 25 and 40% favours formation of polymersomes. Polymeric micelles are another type of self-assembly structures made by amphiphilic block copolymers (Fig. 20.3b). These structures are generally 5–100 nm in size; however, they are not as stable as polymersomes.

As discussed in liposomal drug delivery, curcumin delivery across blood-brain barrier is difficult and needs a special carrier. Similar to liposomes, brain-targeting polymersomes have been developed, functionalized by transferring and Tet-1 peptide and loaded with curcumin [78]. The study showed successful drug delivery to the brain and showed potential of the device for targeting therapeutics. Curcumin along with angiostatin drug has been also used for anti-angiogenic therapy using PEG-polycaprolactone (PCL) polymersomes [79]. Polymersomes from hydrophilic polymer PEG and hydrophobic polymer PCL were prepared by film hydration method with average size of 153 nm. They showed significant decrease in proliferation, migration and angiogenesis in vitro and in vivo. Polymeric micelles have also been used for curcumin drug delivery for its anticancer activity [80]. Alginate/Pluronic tri-block copolymer micelles (<100 nm diameter) were successfully used to deliver curcumin and tested against L929 fibrosarcoma cells.

Thus, food-derived drug delivery has been studied using different nanocarriers such as nanoparticles, polymersomes, liposomes and polymeric micelles. Apart from this, some other types of carriers such as phytosomes, nanoclays and nanoemulsion have also been researched and may be developed in future. All these nanocarriers used as delivery systems have shown more therapeutic effect against a particular disease compared to the drug alone. However, as these nanocarriers are foreign materials for the body, it is essential to understand their distribution, clearance and toxicity, which have been discussed further.

20.3 Clearance of Nanocarriers

As a standard practice, the US FDA requires that all carrier agents administered for detection and treatment purpose be completely cleared from the human body in a reasonable time period [81]. Thus, while making nanocarriers for nutraceutical and therapeutic delivery, their biodistribution and clearance are primary requirement for their future success in clinical practices. In this section of the chapter, we we have discussed the clearance of nanocarriers from the body via two principle systems renal and hepatobilary.

20.3.1 Renal Clearance

Small molecules are generally cleared by renal system because of their smaller size than the kidney filtration threshold (KFT ~ 5.5 nm) and are rapidly eliminated from the body [82, 83]. Renal clearance is desirable as potential health hazards resulting from accumulation of nanoparticles in the body can be reduced [84]. However, passage of molecules through the kidney is dependent on their size, shape and charge due to the structure of the glomerular capillary wall [85]. To escape from the blood, a nanocarrier has to pass through the 70–90 nm aperture of endothelium, glomerular basement membrane meshwork structure with 2–8 nm pores and the epithelium with 4–11 nm filtration slits between the podocyte extensions (Fig. 20.4b). The factors affecting distribution and clearance of nanocarriers are listed below:

20.3.1.1 Hydrodynamic Diameter

Hydrodynamic diameter (HD) is a critical parameter for nanocarrier design, their distribution and renal clearance. The mammalian vasculature has an average pore size of around 5–6 nm; therefore globular NPs/proteins having hydrodynamic diameter of <6 nm can easily pass through glomerular capillary wall and are rapidly cleared from the body by renal filtration and urinary excretion [81], whereas it is difficult for larger NPs (HD > 8 nm) to cross through glomerular capillary wall (Fig. 20.4a, b) [85]. Although, with increase in HD of the therapeutic agents, their filtration through glomeruli decreases, whereas their blood and whole-body half-life increases.

20.3.1.2 Surface Charge

Surface charge of the nanocarriers plays an important role in the kidney filtration because the glomerular capillary wall is negatively charged. In case of NPs with an HD (6–8 nm) larger than KFT, only positively charged NPs can pass the filtration barrier of the kidney due to the charge interactions [86].

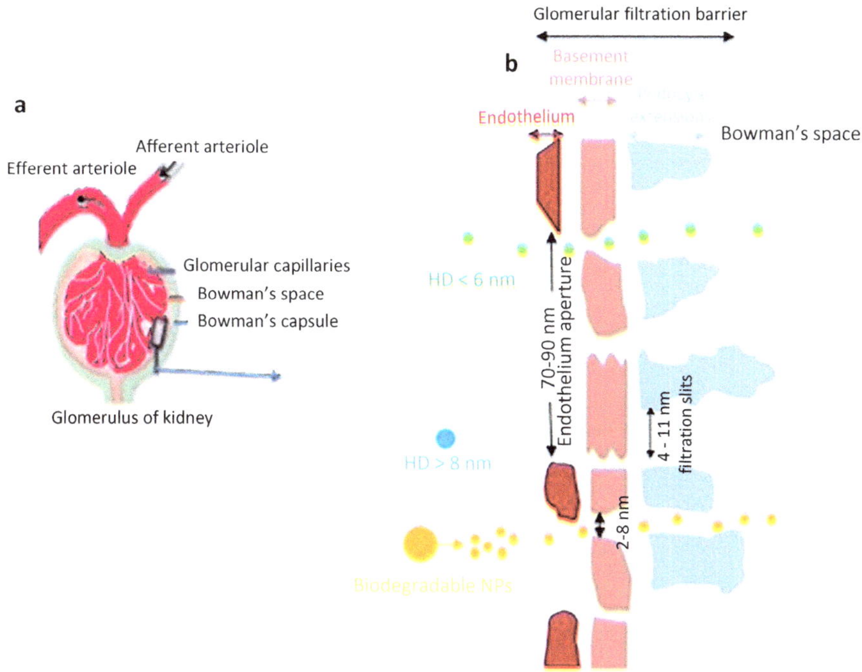

Fig. 20.4 (**a**) Glomerulus and (**b**) glomerular filtration as function of nanoparticle hydrodynamic diameter

20.3.1.3 Shape

Shape of nanocarriers is also an important factor for kidney filtration because of the rectangular shape of the glomerular basement membrane (GBM) slit [87]. Spherical molecules are generally renal clearable. However, molecules with high aspect ratio but smaller width or diameter than the KFT were also found to be cleared by the kidney. For example, large linear-structured carbon nanotubes with a width smaller than KFT can filter through the kidney.

20.3.2 Hepatobilary Clearance

The hepatobiliary system is the primary route of excretion for nanocarriers that do not biodegrade in vivo and do not undergo renal clearance. Excretion by this system is through the liver, into bile and into faeces. The liver accumulates and sequesters 30–99% of administered nanocarriers from the bloodstream and acts as a biological filtration system [88]. Hepatic clearance is more complex than renal clearance due to prolonged retention of nanocarriers within the liver parenchyma which prevents their elimination from the body [89]. Nanocarriers enter the liver via portal triads. To successfully transit through the hepatobiliary system, nanocarriers must evade sequestration in Kupffer cells. Kupffer cells and hepatocytes play

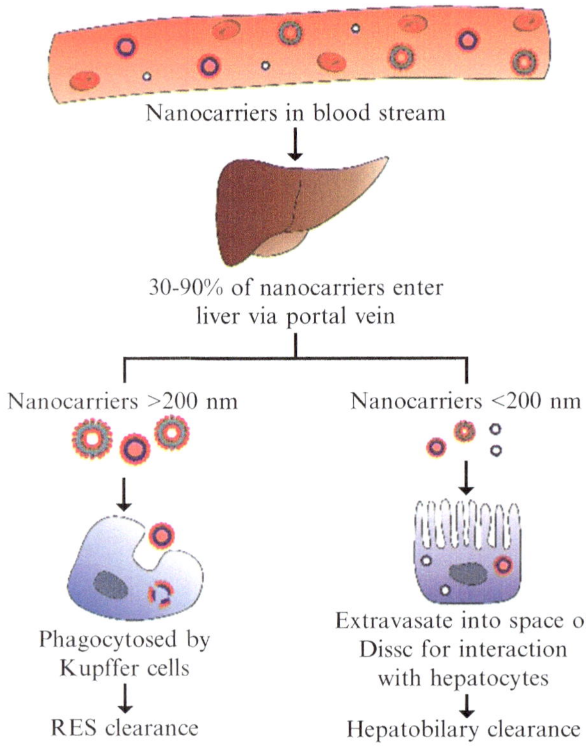

Fig. 20.5 Flow diagram showing the hepatobiliary clearance of nanocarriers

the major role to catabolize and eliminate the nanocarriers from the body [90, 91]. Phagocytic Kupffer cells have ciliated borders and stellate branches that serve as highly adapted mechanical traps for the removal of unwanted substances from the blood. Kupffer cells possess numerous receptors for endocytosis of opsonized nanocarriers (Fig. 20.5) [92].

Hepatocytes follow enzymatic breakdown, endocytosis and excretion of nanocarriers into the bile via biliary system (Fig. 20.5). Nanocarriers with diameter smaller than liver sinusoidal fenestrations (up to 200 nm) can cross to the perisinusoidal space/space of Disse and interact with hepatocytes [93]. Furthermore, interaction of nano devices with liver sinusoidal endothelial cells and transcytosis to the hepatocytes is the alternative pathway for their hepatocyte interaction. Although both hepatocytes and Kupffer cells are equally important for hepatic clearance, however an important difference exists between them. Excretion of nanocarriers processed by the hepatocytes is potentially excreted into the bile, whereas Kupffer cells are part of the RES and rely exclusively on intracellular degradation for nanocarrier's removal. Particles that are not broken down by intracellular processes will remain within the Kupffer cells and will therefore be retained by the body. Kupffer cells express scavenger receptors and hence efficiently take up negatively charged nanocarriers [93–95]. On the other hand, hepatocytes take up positively charged nanocarriers [93, 94].

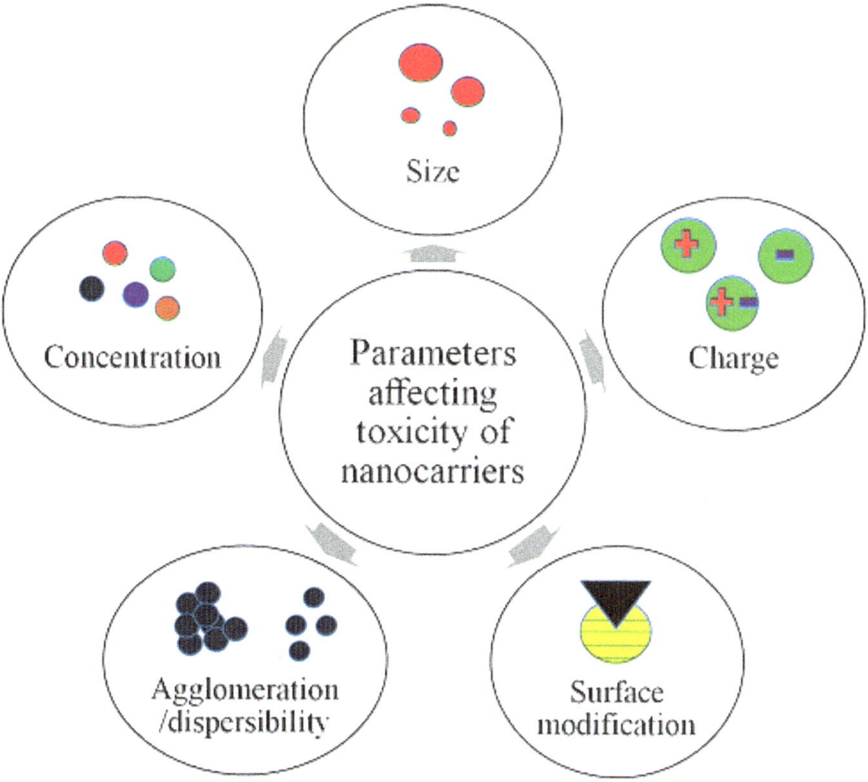

Fig. 20.6 Schematic shows the parameters affecting the toxicity of nanocarriers

20.4 Toxicity of Nanocarriers

The goal of nanocarrier-based delivery is to enhance either delivery or uptake of different agents into target cells or tissues and to reduce side effects associated with particular drugs [96]. However, several findings have reported unexpected toxicities of nanostructured systems, which lead to the origin of nanotoxicology. Nanotoxicology is an emerging branch of nanotechnology to study the interactions of nanostructures with biological systems and to explain the relationship between properties of nanocarriers such as size, shape, surface, composition and aggregation with induction of toxic responses [97]. All these parameters can interfere with cellular membrane stability by promoting an imbalance between intra- and extracellular ions, proteins and other vital molecules required to maintain normal cellular functions and properties, leading to cytotoxicity [98]. Furthermore, the nano-biointerfaces could also be influenced by the composition of intra and extra microenvironment of cell (DNA, proteins, phospholipids, organelles, endocytic vesicles and biological fluids) [99]. Some of the critical parameters are discussed in detail here and shown in Fig. 20.6.

20.4.1 Size

The size of a nanocarrier impacts its potential to cross various biological barriers which may cause undesirable effects such as damaging cell membranes and triggering immune reactions [100]. For example, in vitro cytotoxicity studies using gold NPs at different concentrations showed that 1.4 nm gold NPs inhibited the growth of fibroblasts, epithelial cells, melanoma cells and macrophages; however, NPs of 15 nm size were less toxic at similar concentrations [101]. Furthermore, Szoka et al. have reported liposome-intercalated amphotericin B complex as a fungicide and showed that toxicity of the liposomes increased with size [102].

20.4.2 Charge

Charge of nanocarriers influences their accumulation in nontargeted organs [103]. At cellular level, cationic macromolecules cause cell lysis by interacting with the negatively charged cell membranes [104]. Similarly, cationic nanocarriers can interact with the cell membrane, organelles and proteins and disrupt their integrity and cause cell lysis [105]. For example, cationic dendrimers having surface groups such as amine, guanidine, sulfonate, carboxylate or phosphonate were reported to be more cytotoxic than PEGylated or anionic dendrimers [106]. Dokka et al. reported that cationic liposomes cause reactive oxygen species-mediated pulmonary toxicity in mice [107]. Cationic nanocarriers cause acute inflammatory response by induction of cell necrosis [108]. Furthermore, exposure of cationic micelles and liposomes increased levels of DNA strand breaks in the lung and spleen [109]. When NPs with different surface properties were evaluated, low concentrations of neutral and anionic NPs were found to have no effect on blood-brain barrier integrity, whereas high concentrations of anionic and cationic NPs were toxic. Therefore, surface charges of NPs must be considered for their toxicity [110].

20.4.3 Surface Modification

Nanocarriers possess large surface area due to small particle sizes. Even though increased surface area of nano-delivery system has the advantage of providing opportunities for material delivery, there are inherent disadvantages which make them toxic if not rationally designed. For example, the stability of surface coatings of NPs is a critical factor regulating their toxicity inside a biological environment [100, 111]. While surface modifications allow nanocarriers to be stabilized and avoid agglomeration, they may shield or influence the effects of nanomaterials on biological systems. For example, PEGylation has been shown to stabilize nanocarriers by protecting them from opsonization; however, recent findings demonstrated that confirmations of PEG on NPs are key factor influencing the opsonin functionality; for example, 'brush-like' and 'intermediate' configurations of PEG surface reduced phagocytosis, whereas 'mushroom-like' configuration activated the complement system [112]. Hydrophilicity of the surface molecules also plays a major role in the toxicity of the nanocarriers. Andrea Beyerle et al. have toxicologically evaluated the poly(ethylene imine)-based nanocarriers, with hydrophilic and hydrophobic PEG modifications for pulmonary application in mice. They observed that the hydrophobic modifications caused severe inflammatory response accompanied by an acute antioxidant response, whereas hydrophilic modifications, with high PEG-grafting degrees, induced less pro-inflammatory response without depleting macrophages while showing only a minor oxidative stress response in the lungs [113]. Furthermore, Doxil®, a PEGylated liposome, was found to induce complement activation-related pseudo-allergy [114–116].

20.4.4 Agglomeration and Dispersibility

Agglomeration alters the physicochemical properties of nanocarriers such as size distribution, surface-to-volume ratio, surface activity and concentration, thereby affecting their toxicity due to lymphatic and vascular blockage. Therefore, prevention of agglomeration is a key factor while considering nanocarriers for clinical use [96]. Nanocarriers tend to agglomerate via Van der Waal's forces and Brownian motion due to increased surface area to volume ratio. Agglomeration can occur during synthesis or once the nanocarriers are introduced into a solution where chemical bonding or electrostatic interactions may lead to their aggregation [100]. Several factors such as nanocarrier size, composition, concentration, surface coating, dispersant characteristics and zeta potential can influence agglomeration [100, 117, 118].

20.4.5 Concentration

Concentration of any therapeutic compound is critical factor for its effectiveness. Therefore, it is important to consider the concentration of nanocarriers for their use. To understand the effect of concentration, three different nanocarriers nanospheres, nanocapsules and nanostructured lipid carriers were studied by Mendes et al. [119]. Human lymphocytes and Balb/c 3T3 fibroblasts were incubated with each particle in eight different dilutions varying from 2.1×10^4 to 2.1×10^{11} particles/mL. The results suggested that changes in their population density (number/mL) resulted in cytotoxicity responses. Results also suggested that depending on the kind of NP, the number of particles in the dispersion can negatively influence cell viability in preclinical drug development and delivery studies [119]. Mayhew et al. reported that phosphatidylcholine or dipalmitoylphosphatidyl choline liposomes were non-toxic (ID_{50}) at 3000–4000 μM, whereas phosphatidylglycerol- and phosphatidyl serine-containing liposomes were toxic in the range 130–3000 μM and stearylamine- and cardiolipin-containing liposomes were toxic at 200 μM liposomal lipid concentrations or less [120].

20.5 Ethical Concerns of Nanocarriers

Nanocarriers show tremendous potential for improving medical diagnosis and treatment; however, it also raises a variety of ethical, social and legal concerns because of the lack of common properties other than size and the paucity of data on their physicochemical properties. Nanocarriers are not a unified class of compounds; hence each of its types must be evaluated on its own terms. The variation in size (1–100 nm) and shape can have unpredictable effects. For example, NP that is non-toxic at 50 nm may be toxic at 1 nm or vice versa. Furthermore, NPs have potential to diffuse into surrounding tissues and cause adverse effect on the cells by inducing the oxidative damage [121]. The production method of nanocarriers is also an ethical issue, for example, the production methods of engineered nanoparticles may expose workers by dermal absorption, inhalation and ingestion [122]. People are more concerned about the undesirable effects which could increase by using nanocarriers for food-derived drug delivery. Hence, labelling of nanonutraceutical products is one of the major ethical issues which could increase the value of the products. Nanonutraceutical companies take advantage of improper labelling by only providing the details of desirable effects and not the undesirable effects of the NPs, irrespective of how least the effect is. Therefore, it is a major ethical issue as most of the people are unaware about the undesirable effects of the NPs used in the formulations. Unless there is cooperation between the nanonutraceutical industry and public, use of NPs will always raise the ethical concerns.

20.6 Conclusion and Future Perspective

Bioactive molecules in food often have therapeutic values in addition to their nutritional values. Curcumin, ellagic acid, ferulic acid, epigallocatechin gallate, quercetin and β-cryptoxanthin are some of the examples of such bioactive molecules found in different food items. Although they have been practised as a remedy for diseases since ancient time, they may not work at their optimum capacity due to lesser bioavailability, degradation during digestion, non-specificity and rapid clearance. Recent developments in use of nanotechnology for pharmaceutics have promoted delivery of food-derived biomolecules by developing nanocarriers. A variety of nanocarriers such as nanoparticles, nanoliposomes, polymersomes and polymeric micelles has been fabricated from metals, natural and synthetic polymers, lipids, polysaccharides and proteins. These nanocarriers can encapsulate both hydrophilic and hydrophobic drugs and have tunable properties that can be tailored depending upon target application. Once inside the body, nanocarriers are cleared by two main mechanisms, renal clearance and hepatobilary clearance. Further, their toxicity depends upon several factors such as their size, charge, concentration and surface properties. However, since properties of each class of nanocarriers are drastically different from each other, their drug release pattern and clearance from the body has to be studied individually. Lack of common physicochemical properties and proper guidelines has raised some ethical concerns related to nanocarriers. With nanotechnological advancements in pharmaceutics, more food-derived biomolecules can be delivered via nanocarriers in more regulated manner.

Acknowledgement Authors DJ and HR appreciatively acknowledge the Council of Scientific and Industrial Research and University Grant Commission, India, for fellowship support, respectively, whereas RV would like to acknowledge SERB, Department of Science and Technology, Govt. of India, and Gujarat State Biotechnology Mission, Govt. of Gujarat, India, for their financial support.

References

1. Aklakur M, Asharf Rather M, Kumar N (2016) Nanodelivery: an emerging avenue for nutraceuticals and drug delivery. Crit Rev Food Sci Nutr 56:2352–2361
2. Jani P, Halbert GW, Langridge J et al (1990) Nanoparticle uptake by the rat gastrointestinal mucosa: quantitation and particle size dependency. J Pharm Pharmacol 42:821–826
3. Sikorska M, Lanthier P, Miller H et al (2014) Nanomicellar formulation of coenzyme Q(10) (Ubisol-Q(10)) effectively blocks ongoing neurodegeneration in the mouse 1-methyl-4-phenyl-1,2,3,6-tetrahydropyridine model: potential use as an adjuvant treatment in Parkinson's disease. Neurobiol Aging 35:2329–2346
4. Muthukumaran K, Leahy S, Harrison K et al (2014) Orally delivered water soluble Coenzyme Q10 (Ubisol-Q10) blocks on-going neurodegeneration in rats exposed to paraquat: potential for therapeutic application in Parkinson's disease. BMC Neurosci 15:21–31
5. Silva HD, Cerqueira MÂ, Vicente AA (2012) Nanoemulsions for food applications: development and characterization. Food Bioprocess Technol 5:854–867
6. Ranjan S, Dasgupta N, Chakraborty AR et al (2014) Nanoscience and nanotechnologies in food industries: opportunities and research trends. J Nanopart Res 16:2464–2489
7. Tonnesen HH, Masson M, Loftsson T (2002) Studies of curcumin and curcuminoids. XXVII. Cyclodextrin complexation: solubility, chemical and photochemical stability. Int J Pharm 244:127–135
8. Aggarwal BB, Kumar A, Bharti AC (2003) Anticancer potential of curcumin: preclinical and clinical studies. Anticancer Res 23:363–398
9. Anand P, Thomas SG, Kunnumakkara AB et al (2008) Biological activities of curcumin and its analogues (Congeners) made by man and mother nature. Biochem Pharmacol 76:1590–1611
10. Yallapu MM, Jaggi M, Chauhan SC (2012) Curcumin nanoformulations: a future nanomedicine for cancer. Drug Discov Today 17:71–80
11. Kakkar V, Singh S, Singla D et al (2010) Pharmacokinetic applicability of a validated liquid chromatography tandem mass spectroscopy method for orally administered curcumin loaded solid lipid nanoparticles to rats. J Chromatogr B Anal Technol Biomed Life Sci 878:3427–3431
12. Kundu P, Mohanty C, Sahoo SK (2012) Antiglioma activity of curcumin-loaded lipid nanoparticles and its enhanced bioavailability in brain tissue for effective glioblastoma therapy. Acta Biomater 8:2670–2687
13. Hernell O, Staggers JE, Carey MC (1990) Physical-chemical behavior of dietary and biliary lipids during intestinal digestion and absorption. 2. Phase analysis and aggregation states of luminal lipids during duodenal fat digestion in healthy adult human beings. Biochemistry 29:2041–2056
14. Faulks RM, Southon S (2005) Challenges to understanding and measuring carotenoid bioavailability. Biochim Biophys Acta 1740:95–100
15. Desai MP, Labhasetwar V, Amidon GL et al (1996) Gastrointestinal uptake of biodegradable microparticles: effect of particle size. Pharm Res 13:1838–1845
16. Hussain N, Jaitley V, Florence AT (2001) Recent advances in the understanding of uptake of microparticulates across the gastrointestinal lymphatics. Adv Drug Deliv Rev 50:107–142

17. Delie F (1998) Evaluation of nano- and microparticle uptake by the gastrointestinal tract. Adv Drug Deliv Rev 34:221–233
18. des Rieux A, Fievez V, Garinot M et al (2006) Nanoparticles as potential oral delivery systems of proteins and vaccines: a mechanistic approach. J Control Release 116:1–27
19. Panyam J, Labhasetwar V (2003) Biodegradable nanoparticles for drug and gene delivery to cells and tissue. Adv Drug Deliv Rev 55:329–347
20. Roney C, Kulkarni P, Arora V et al (2005) Targeted nanoparticles for drug delivery through the blood-brain barrier for Alzheimer's disease. J Control Release 108:193–214
21. Duncan R (2006) Polymer conjugates as anticancer nanomedicines. Nat Rev Cancer 6:688–701
22. Gratzel M (2001) Photoelectrochemical cells. Nature 414:338–344
23. Dai J, Bruening ML (2002) Catalytic nanoparticles formed by reduction of metal ions in multilayered polyelectrolyte films. Nano Lett 2:497–501
24. Murray CB, Sun S, Doyle H et al (2001) Monodisperse 3d transition-metal (Co,Ni,Fe) nanoparticles and their assembly into nanoparticle superlattices. MRS Bull 26:985–991
25. Bhattacharya R, Mukherjee P (2008) Biological properties of "naked" metal nanoparticles. Adv Drug Deliv Rev 60:1289–1306
26. Khoobchandani M, Zambre A, Katti K et al (2013) Green nanotechnology from Brassicaceae: development of broccoli phytochemicals–encapsulated gold nanoparticles and their applications in nanomedicine. Int J Green Nanotechnol 1:1–15
27. Katti K, Chanda N, Shukla R et al (2009) Green nanotechnology from cumin phytochemicals: generation of biocompatible gold nanoparticles. Int J Green Nanotechnol Biomed 1:39–52
28. Ahmad N, Sharma S, Alam Md K et al (2010) Rapid synthesis of silver nanoparticles using dried medicinal plant of basil. Colloids Surf B Biointerfaces 81:81–86
29. Krishnaswamy K, Vali H, Orsat V (2014) Value-adding to grape waste: green synthesis of gold nanoparticles. J Food Eng 142:210–220
30. Bankar A, Joshi B, Kumar AR et al (2010) Banana peel extract mediated synthesis of gold nanoparticles. Colloids Surf B, Biointerfaces 80:45–50
31. Li L, Xi L, Changhui F et al (2015) Biosynthesis of fluorescent gold nanoclusters for in vitro and in vivo tumor imaging. Opt Commun 355:567–574
32. Philip D (2009) Honey mediated green synthesis of gold nanoparticles. Spectrochim Acta Mol 73:650–653
33. Sujitha MV, Kannan S (2013) Green synthesis of gold nanoparticles using Citrus fruits (Citrus limon, Citrus reticulata and Citrus sinensis) aqueous extract and its characterization. Spectrochim Acta Mol 102:15–23
34. Khalil MMH, Ismail Eman H, Fatma E-M (2012) Biosynthesis of Au nanoparticles using olive leaf extract: 1st Nano Updates. Arab J Chem 5:431–437
35. Nune SK, Nripen C, Ravi S et al (2009) Green nanotechnology from tea: phytochemicals in tea as building blocks for production of biocompatible gold nanoparticles. J Mater Chem 19:2912–2920
36. Menon D, Basanth A, Retnakumari A et al (2012) Green synthesis of biocompatible gold nanocrystals with tunable surface plasmon resonance using garlic phytochemicals. J Biomed Nanotechnol 8:901–911
37. Nie S, Xing Y, Kim GJ et al (2007) Nanotechnology applications in cancer. Annu Rev Biomed Eng 9:257–288
38. Gref R, Domb A, Quellec P et al (1995) The controlled intravenous delivery of drugs using PEG-coated sterically stabilized nanospheres. Adv Drug Deliv Rev 16:215–233
39. Han DH, Lee MJ, Kim JH (2006) Antioxidant and apoptosis-inducing activities of ellagic acid. Anticancer Res 26:3601–3606
40. Zhang H-M, Lei Z, Hao L et al (2014) Research progress on the anticarcinogenic actions and mechanisms of ellagic acid. Cancer Biol Med 11:92–100
41. Bala I, Bhardwaj V, Hariharan S et al (2005) Design of biodegradable nanoparticles: a novel approach to encapsulating poorly soluble phytochemical ellagic acid. Nanotechnology 16:2819–2822

42. Jose Merlin JP, Rajendra Prasad N, Shibli SMA et al (2012) Ferulic acid loaded Poly-d,l-lactide-co-glycolide nanoparticles: systematic study of particle size, drug encapsulation efficiency and anticancer effect in non-small cell lung carcinoma cell line in vitro. Biomed Prev Nutr 2:69–76
43. Mukerjee A, Vishwanatha JK (2009) Formulation, characterization and evaluation of curcumin-loaded PLGA nanospheres for cancer therapy. Anticancer Res 29:3867–3875
44. Gupta V, Aseh A, Ríos CN et al (2009) Fabrication and characterization of silk fibroin-derived curcumin nanoparticles for cancer therapy. Int J Nanomed 4:115–122
45. Zhang J, Li S, An FF et al (2015) Self-carried curcumin nanoparticles for in vitro and in vivo cancer therapy with real-time monitoring of drug release. Nanoscale 7:13503–13510
46. Yallapu MM, Gupta BK, Jaggi M et al (2010) Fabrication of curcumin encapsulated PLGA nanoparticles for improved therapeutic effects in metastatic cancer cells. J Colloid Interface Sci 351:19–29
47. Kumar SS, Surianarayanan M, Vijayaraghavan R et al (2014) Curcumin loaded poly(2-hydroxyethyl methacrylate) nanoparticles from gelled ionic liquid – in vitro cytotoxicity and anti-cancer activity in SKOV-3 cells. Eur J Pharm Sci 51:34–44
48. Kumar Das R, Kasoju N, Bora U (2010) Encapsulation of curcumin in alginate-chitosan-pluronic composite nanoparticles for delivery to cancer cells. Nanomed Nanotech Biol Med 6:153–160
49. Anand P, Nair HB, Sung B et al (2010) Design of curcumin-loaded PLGA nanoparticles formulation with enhanced cellular uptake, and increased bioactivity in vitro and superior bioavailability in vivo. Biochem Pharmacol 79:330–338
50. Ramalingam P, Yoo SW, Ko YT (2016) Nanodelivery systems based on mucoadhesive polymer coated solid lipid nanoparticles to improve the oral intake of food curcumin. Food Res Int 84:113–119
51. Smith A, Giunta B, Bickford PC et al (2010) Nanolipidic particles improve the bioavailability and alpha-secretase inducing ability of epigallocatechin-3-gallate (EGCG) for the treatment of Alzheimer's disease. Int J Pharm 389:207–212
52. Huang X, Huang X, Gong Y et al (2016) Enhancement of curcumin water dispersibility and antioxidant activity using core–shell protein–polysaccharide nanoparticles. Food Res Int 87:1–9
53. Tan Q, Liu W, Guo C et al (2011) Preparation and evaluation of quercetin-loaded lecithin-chitosan nanoparticles for topical delivery. Int J Nanomed 6:1621–1630
54. Zhang M, Viennois E, Prasad M et al (2016) Edible ginger-derived nanoparticles: a novel therapeutic approach for the prevention and treatment of inflammatory bowel disease and colitis-associated cancer. Biomaterials 101:321–340
55. Anh NT, Quan DT, Duc CTD et al (2016) Micro and nano liposome vesicles containing curcumin for a drug delivery system. Adv Nat Sci, Nanosci Nanotechnol 7:035003–035008
56. Langer R (1990) New methods of drug delivery. Science 249:1527–1533
57. Mozafari MR, Johnson C, Hatziantoniou S et al (2008) Nanoliposomes and their applications in food nanotechnology. J Liposome Res 18:309–327
58. Mozafari MR, Pardakhty A, Azarmi S et al (2009) Role of nanocarrier systems in cancer nanotherapy. J Liposome Res 19:310–321
59. Shoji Y, Nakashima H (2004) Nutraceutics and delivery systems. J Drug Target 12:385–391
60. Li C, Zhang X, Huang X et al (2013) Preparation and characterization of flexible nanoliposomes loaded with daptomycin, a novel antibiotic, for topical skin therapy. Int J Nanomed 8:1285–1292
61. Sercombe L, Veerati T, Moheimani F et al (2015) Advances and challenges of liposome assisted drug delivery. Front Pharmacol 6:286–298
62. Thangapazham RL, Puri A, Tele S et al (2008) Evaluation of a nanotechnology-based carrier for delivery of curcumin in prostate cancer cells. Int J Oncol 32:1119–1123
63. Li L, Braiteh FS, Kurzrock R (2005) Liposome-encapsulated curcumin: in vitro and in vivo effects on proliferation, apoptosis, signaling, and angiogenesis. Cancer 104:1322–1331

64. Chen C, Johnston TD, Jeon H et al (2009) An in vitro study of liposomal curcumin: stability, toxicity and biological activity in human lymphocytes and Epstein-Barr virus-transformed human B-cells. Int J Pharm 366:133–139
65. Wang D, Veena MS, Stevenson K et al (2008) Liposome-encapsulated curcumin suppresses growth of head and neck squamous cell carcinoma in vitro and in xenografts through the inhibition of nuclear factor kappaB by an AKT-independent pathway. Clin Cancer Res 14:6228–6236
66. Mourtas S, Lazar AN, Markoutsa E et al (2014) Multifunctional nanoliposomes with curcumin-lipid derivative and brain targeting functionality with potential applications for Alzheimer disease. Eur J Med Chem 80:175–183
67. Gharib A, Faezizadeh Z, Godarzee M (2015) Preparation and characterization of nanoliposomal beta-cryptoxanthin and its effect on proliferation and apoptosis in human leukemia cell line K562. Trop J Pharm Res 14:187–194
68. Tanaka T, Tanaka T, Tanaka M et al (2012) Cancer chemoprevention by citrus pulp and juices containing high amounts of beta-cryptoxanthin and hesperidin. J Biomed Biotechnol 2012:516981
69. Wu C, Han L, Riaz H et al (2013) The chemopreventive effect of beta-cryptoxanthin from mandarin on human stomach cells (BGC-823). Food Chem 136:1122–1129
70. Shimoda H, Shan SJ, Tanaka J et al (2012) Beta-Cryptoxanthin suppresses UVB-induced melanogenesis in mouse: involvement of the inhibition of prostaglandin E2 and melanocyte-stimulating hormone pathways. J Pharm Pharmacol 64:1165–1176
71. Mohan A, Narayanan S, Sethuraman S et al (2014) Novel resveratrol and 5-fluorouracil coencapsulated in PEGylated nanoliposomes improve chemotherapeutic efficacy of combination against head and neck squamous cell carcinoma. Biomed Res Int 2014:424239–424252
72. Meng F, Zhiyuan Z (2011) Polymersomes spanning from nano- to microscales: advanced vehicles for controlled drug delivery and robust vesicles for virus and cell mimicking. J Phys Chem Lett 2:1533–1539
73. Bermudez H, Brannan AK, Hammer DA et al (2002) Molecular weight dependence of polymersome membrane structure, elasticity, and stability. Macromolecules 35:8203–8208
74. Battaglia G, Ryan AJ, Tomas S (2006) Polymeric vesicle permeability: a facile chemical assay. Langmuir 22:4910–4913
75. Photos PJ, Lucie B, Bohdana D et al (2003) Polymer vesicles in vivo: correlations with PEG molecular weight. J Control Release 90:323–334
76. Awasthi VD, Garcia D, Klipper R et al (2004) Neutral and anionic liposome-encapsulated hemoglobin: effect of postinserted poly(ethylene glycol)-distearoylphosphatidylethanolamine on distribution and circulation kinetics. J Pharmacol Exp Ther 309:241–248
77. Anajafi T, Mallik S (2015) Polymersome-based drug-delivery strategies for cancer therapeutics. Ther Deliv 6:521–534
78. Jia T, Sun Z, Lu Y et al (2016) A dual brain-targeting curcumin-loaded polymersomes ameliorated cognitive dysfunction in intrahippocampal amyloid-beta1-42-injected mice. Int J Nanomed 11:3765–3775
79. Cao Y, Li Y, Wu Y et al (2016) Co-delivery of angiostatin and curcumin by a biodegradable polymersome for antiangiogenic therapy. RSC Adv 6:105442–105448
80. Franklin J, Jinu G (2014) Curcumin encapsulated alginate/pluronic block copolymer micelles as a promising therapeutic agent. UKJPB 2:6–12
81. Choi HS, Liu W, Misra P et al (2007) Renal clearance of quantum dots. Nat Biotechnol 25:1165–1170
82. Chan KW, Wong WT (2007) Small molecular gadolinium(III) complexes as MRI contrast agents for diagnostic imaging. Coord Chem Rev 251:2428–2451
83. Miller JC, Thrall JH (2004) Clinical molecular imaging. J Am Coll Radiol 1:4–23
84. Barstow L, Small RE (1990) Liver function assessment by drug metabolism. Pharmacotherapy 10:280–288
85. Longmire M, Choyke PL, Kobayashi H (2008) Clearance properties of nano-sized particles and molecules as imaging agents: considerations and caveats. Nanomedicine 3:703–717

86. Ohlson M, Sorensson J, Haraldsson B (2001) A gel-membrane model of glomerular charge and size selectivity in series. Am J Physiol Renal Physiol 280:396–405
87. Longmire MR, Ogawa M, Choyke PL et al (2011) Biologically optimized nanosized molecules and particles: more than just size. Bioconjug Chem 22:993–1000
88. Zhang YN, Poon W, Tavares AJ et al (2016) Nanoparticle–liver interactions: cellular uptake and hepatobiliary elimination. J Controll Release 240:332–348
89. Choi HS, Liu W, Misra P et al (2007) Renal clearance of nanoparticles. Nat Biotechnol 25:1165–1170
90. He C, Hu Y, Yin L et al (2010) Effects of particle size and surface charge on cellular uptake and biodistribution of polymeric nanoparticles. Biomaterials 31:3657–3666
91. Ogawara K, Yoshida M, Higaki K et al (1999) Hepatic uptake of polystyrene microspheres in rats: effect of particle size on intrahepatic distribution. J Controll Release 59:15–22
92. Kuntz E, Hans-Dieter K (2006) Hepatology: principles and practice: history, morphology, biochemistry, diagnostics, clinic, therapy, vol XIII, 2nd edn. Springer, Heidelberg, 906 p
93. Wang H, Thorling CA, Liang X et al (2015) Diagnostic imaging and therapeutic application of nanoparticles targeting the liver. J Mater Chem B 3:939–958
94. Poelstra K, Prakash J, Beljaars L (2012) Drug targeting to the diseased liver. J Controll Release 161:188–197
95. Bartsch M, Weeke-Klimp AH, Meijer DK et al (2002) Massive and selective delivery of lipid-coated cationic lipoplexes of oligonucleotides targeted in vivo to hepatic endothelial cells. Pharm Res 19:676–680
96. Oberdorster G (2010) Safety assessment for nanotechnology and nanomedicine: concepts of nanotoxicology. J Intern Med 267:89–105
97. Puri A, Loomis K, Smith B et al (2009) Lipid-based nanoparticles as pharmaceutical drug carriers: from concepts to clinic. Crit Rev Ther Drug Carrier Syst 26:523–580
98. Love SA, Maurer-Jones MA, Thompson JW et al (2012) Assessing nanoparticle toxicity. Annu Rev Anal Chem 5:181–205
99. Verma A, Stellacci F (2010) Effect of surface properties on nanoparticle-cell interactions. Small 6:12–21
100. Dhawan A, Sharma V (2010) Toxicity assessment of nanomaterials: methods and challenges. Anal Bioanal Chem 398:589–605
101. Pan Y, Neuss S, Leifert A et al (2007) Size-dependent cytotoxicity of gold nanoparticles. Small 3:1941–1949
102. Szoka FC, Milholland D, Barza M (1987) Effect of lipid composition and liposome size on toxicity and in vitro fungicidal activity of liposome-intercalated amphotericin B. Antimicrob Agents Chemother 31:421–429
103. De Jong WH, Borm PJ (2008) Drug delivery and nanoparticles: applications and hazards. Int J Nanomed 3:133–149
104. Fischer D, Li Y, Ahlemeyer B et al (2003) In vitro cytotoxicity testing of polycations: influence of polymer structure on cell viability and hemolysis. Biomaterials 24:1121–1131
105. Mecke A, Majoros IJ, Patri AK et al (2005) Lipid bilayer disruption by polycationic polymers: the roles of size and chemical functional group. Langmuir 21:10348–10354
106. Chen HT, Neerman MF, Parrish AR et al (2004) Cytotoxicity, hemolysis, and acute in vivo toxicity of dendrimers based on melamine, candidate vehicles for drug delivery. J Am Chem Soc 126:10044–10048
107. Dokka S, Toledo D, Shi X et al (2000) Oxygen radical-mediated pulmonary toxicity induced by some cationic liposomes. Pharm Res 17:521–525
108. Wei X, Bin S, Zhiyao H et al (2015) Cationic nanocarriers induce cell necrosis through impairment of Na+/K+ -ATPase and cause subsequent inflammatory response. Cell Res 25:237–253
109. Knudsen KB, Northeved H, Kumar PE et al (2015) In vivo toxicity of cationic micelles and liposomes. Nanomedicine 11:467–477
110. Lockman PR, Koziara JM, Mumper RJ et al (2004) Nanoparticle surface charges alter blood-brain barrier integrity and permeability. J Drug Target 12:635–641

111. Powers KW, Brown SC, Krishna VB et al (2006) Research strategies for safety evaluation of nanomaterials. Part VI. Characterization of nanoscale particles for toxicological evaluation. Toxicol Sci 90:296–303
112. Manocha B, Margaritis A (2008) Production and characterization of gamma-polyglutamic acid nanoparticles for controlled anticancer drug release. Crit Rev Biotechnol 28:83–99
113. Beyerle A, Andrea B, Atrayee B et al (2011) Inflammatory responses to pulmonary application of PEI-based siRNA nanocarriers in mice. Biomaterials 32:8694–8701
114. Moghimi SM, Andersen AJ, Hashemi SH et al (2010) Complement activation cascade triggered by PEG-PL engineered nanomedicines and carbon nanotubes: the challenges ahead. J Control Release 146:175–181
115. Szebeni J, Muggia F, Gabizon A et al (2011) Activation of complement by therapeutic liposomes and other lipid excipient-based therapeutic products: prediction and prevention. Adv Drug Deliv Rev 63:1020–1030
116. Szebeni J (2005) Complement activation-related pseudoallergy: a new class of drug-induced acute immune toxicity. Toxicology 216:106–121
117. Borm P, Klaessig FC, Landry TD et al (2006) Research strategies for safety evaluation of nanomaterials. Part V: Role of dissolution in biological fate and effects of nanoscale particles. Toxicol Sci 90:23–32
118. Hall JB, Dobrovolskaia MA, Patri AK et al (2007) Characterization of nanoparticles for therapeutics. Nanomedicine 2:789–803
119. Mendes LP, Delgado JM, Costa AD et al (2015) Biodegradable nanoparticles designed for drug delivery: the number of nanoparticles impacts on cytotoxicity. Toxicol In Vitro 29:1268–1274
120. Mayhew E, Ito M, Lazo R (1987) Toxicity of non-drug-containing liposomes for cultured human cells. Exp Cell Res 171:195–202
121. Zhang Y, Wan-Xi Y (2016) Tight junction between endothelial cells: the interaction between nanoparticles and blood vessels. Beilstein J Nanotechnol 7:675–684
122. Aitken RJ, Creely KS, Tran CL (2004) Nanoparticles: an occupational hygiene review. Health safety executive, research report, vol 274. HSE Books, London

Phytochemicals in Clinical Studies: Current Perspective

21

Shashank Kumar, Deepak Kumar, Audesh Bhat, and Ajay Kumar

Abstract

Natural plants and their derived products are effective against various diseases for their prevention and treatment for several years. Availability of synthetic drugs with faster and selective responses against various diseases decreased the use of herbal products. However, due to adverse side effects of synthetic drugs, in the last few years, the clinical use of plant-derived medicines has been exponentially increased in developing as well as developed countries due to their easy availability, low cost, and least toxicity. Therefore, in recent years, several human trials have been done to analyze the toxicology, pharmacokinetics, and biologically effective dose of phytochemicals against cancer, diabetes, and bacterial and viral diseases for their clinical application. In the present book chapter, we have discussed the clinical studies on phytochemicals against several diseases including diabetes, cancer, and diseases associated with microbial infections including women health problems.

Keywords

Phytochemicals · Drug discovery · Clinical trial · Women health · Cancer · Bacterial diseases · Viral diseases

S. Kumar
Centre of Biochemistry and Microbial Sciences, Central University of Punjab, Bathinda, India

D. Kumar
Department of Botany, Central University of Jammu, Jammu, India

A. Bhat
Centre for Molecular Biology, Central University of Jammu, Jammu, India

A. Kumar (✉)
Department of Zoology, Banaras Hindu University, Varanasi, India
e-mail: ajayzoo@bhu.ac.in

V. Rani, U. C. S. Yadav (eds.), *Functional Food and Human Health*,
https://doi.org/10.1007/978-981-13-1123-9_21

21.1 An Introduction to Drug Discovery and Clinical Trial

Phytotherapy deals with the prevention and treatment of diseases by plant material or natural products [1]. A large number of modern drugs owed their origin from ethnobotanical remedies. Plants synthesize various compounds known as secondary metabolites from few biosynthetic pathways. These secondary metabolites are known as phytochemicals such as phenolics, flavonoids, terpenes, anthraquinone, etc. Secondary metabolites are the products/intermediate products of metabolic pathways which are indirectly involved in the growth and reproduction of plants. It has been reported that primary healthcare of 80–90% of some populations depends entirely on traditional medicines [2, 3]. Drug development is an extremely regulated and complex procedure that comprises academic research institutions, the pharmaceutical industry, and government agencies. Current research dealing with the pathogenesis of various diseases has potential to produce recurrent drug targets resulting in the gathering of several molecules under the varying developmental stage, and thus, there is a need for clinical testing. Mechanism of action of botanicals involved different factors such as phytochemical, biological, and analytical techniques [4]. A reliable source for the procurement of plant material is very crucial for a robust scientific study. Different parameters such as the location of the plant, the temperature of the environment, harvest time, etc. alter the phytochemical profile of the plant [5]. Bioactive phytoconstituents and their mode of action should be identified to obtain efficacious and safe plant-based remedies. Phytochemicals have diverse solubility index in different solvents. The dissolution of plant extracts in different solvents in different clinical studies creates unsatisfactory conclusions [4]. The complex nature of botanical extracts requires the appropriate analysis methods for pharmacological/clinical studies of individual phytoconstituents [6, 7]. Some phytochemicals may give false-positive results because of different interactions with cellular moieties [8, 9]. Botanical extracts mostly show additive, synergistic, and/or antagonistic effects [10, 11]. Synergistic or antagonistic activities of natural products can be estimated by utilizing isobol method or the Chou-Talalay combination index [12]. After the isolation and characterization of phytochemicals, in vivo and clinical studies have to be performed to examine the several parameters such as bioavailability, safety, toxicity, and effectiveness for their clinical use. The focus of the present book chapter is to update the knowledge of scientists, researchers, and doctors who are either working in this area or having a keen interest in this field. The chapter provides the information about the clinical data obtained from several clinical studies using phytochemicals on the diverse group of patients suffering from cancer, diabetes, and microbial infections including women health-related problems. The chapter also deals with the drug discovery of botanical origin.

21.1.1 Innovations in Drug Discovery

The broad spectrum of medical biology defines drug discovery as "a process by which novel medicine are exposed to worldwide to treat diseases." Drug discovery research is not much older than a century. The eighteenth-century discoveries define some of the indispensable foundations of the chemical theory. Avogadro's atomic hypothesis as well as the periodic table of elements was established. A model was developed that established the elements according to their valences and atomic weight. Subsequently, acid and base theory was also established. August Kekule communicated his pioneering benzene theory based on aromatic organic molecules which were the birth time of coal-tar derivatives and dyes [13]. Dye chemistry evolution had a reflective influence on medical biology. Discriminating potential of the dyes for biological sample hypothesized the presence of chemoreceptors. It was the time of chemotherapy. Chemotherapy generally involves drugs in combinatorial form and aims at prolonging life or to reduce symptoms. It was the extraordinary therapeutic triumphs on the life-threatening diseases [13]. In 1815, Friedrich Wilhelm Adam Serturner isolated morphine from the opium extract. In 1848, papaverine was isolated, but its antispasmodic activity was not discovered until 1917 [13]. Analytical chemistry demonstrated the isolation as well as purification of the bioactive ingredients of medicinal plants and also verified its medicinal value in the nineteenth century. As bioactive properties from plants became exposed, most of the pharmaceutical companies addressed the difficulties of impure drugs for providing standardized preparations. Finally, pharmacology demanded its valuable place in medical practices. During 1871–1918, Oswald Schmiedeberg, from University of Strasbourg, laid several of the important foundations of pharmacology [13]. Innovative organizations to support interdisciplinary drug development and research had to be created. In the nineteenth century, one-third deaths in the USA were from three common causes such as pneumonia, tuberculosis, and diarrhea which are very rare today because they are treatable now. In 1940, the death rate from these diseases was 1 in 11. In 2000, the odds were down to 1 in 25. Pneumonia remains in the top ten list which is one of the main death-threatening diseases among the three diseases mentioned above. It is now led by additional complex circumstances such as CVDs (cardiovascular diseases) as well as cancer. Various factors (improved sanitation and vaccination) undoubtedly act as key players in increasing the life expectancy during the twentieth century.

21.1.2 Modern Molecular Approaches for Drug Discovery

Current age is very exciting for medical science because chemists are involved in all biomedical research fields. Chemistry is thus fundamentally well-situated to have a chief impact on drug discovery since other fields cannot produce novel molecules. Classical plant-based drug discovery involved extraction, assay-based functional fractionation, isolation and characterization, and target validation which have been steadily substituted by molecular target-based drug discovery. Computer-based in silico high-throughput screening of huge libraries leads to identification and optimization from hits and has been in the convention for the last two decades [14]. Natural products-based constitute about 60% of the FDA-approved drugs. Combinatorial chemistry in the form of equivalent synthesis or diversity-oriented synthesis (DOS) for the optimization of promising lead compounds has been fruitful in numerous drug discovery and developmental cases. The library approach is particularly beneficial for clearing ADME/T requirements. Structural biology, computational biology, chemical biology, and genetics are the four key pillars of modern drug discovery. With these advanced tools in hand, exploiting combinatorial chemistry for focused libraries and rational drug design becomes possible. Designing hybrid molecules with a dual mode of action is a decent example to deliver a hopeful approach toward modern drug discovery. New antimicrobial drug in contradiction of multidrug-resistant (MDR) strains of bacteria like tuberculosis is evolving by comprehensive usage of modern tools [14]. The terms "biology-oriented synthesis (BIOS)" and "function-oriented synthesis (FOS)" have recently arisen as a reasonable development in chemical genetics for discovering the intrinsic diversity and complication of the structures of natural products. By taking benefit of the specific gene, use of combinatorial biosynthesis offers another attractive approach to modern drug discovery [14]. Chemistry is also playing a significant role in targeted drug delivery. It is accepted that organic, medicinal, materials, as well as nano-chemistry, have now become extremely involved in the drug delivery development systems. Carbon nanotubes and polymers are used as a vehicle for drug delivery, which is a novel practise in drug discovery. Tumor-specific monoclonal antibodies, omega-3 fatty acids, vitamins, and aptamers have been emerging as extremely promising methods to successful chemotherapy with a minimum of adverse side effects in the tumor-targeted drug delivery systems. Aptamers are oligonucleotide/peptide units that bind to a particular target molecule typically created and selected from a big random sequence pool.

21.1.3 In Silico Technology

In silico approaches can help in recognizing drug targets through bioinformatic tools. The approach has potential to predict the target structures for possible binding, generation of candidate molecules, specific active sites detection, drug-likeness prediction, drug-protein docking, etc. All aspects of computational method permeating with supercomputer workstations form the core of structure-based drug design [15]. Extraordinary performance computing data management software and Internet are enabling access of large amount of data and transforming the enormous complex biological data into feasible data for knowledge in current-day drug discovery process. In silico study also provides novel targets and their interaction with the lead compounds with desired properties. Computational techniques provide the benefit of bringing new drug candidates more rapidly at a lesser cost. Foremost roles of computation in drug discovery comprise virtual screening, de novo design, in silico ADME/T prediction, and delivering progressive methods for shaping protein-ligand binding [15]. In the current scenario, additional protein structure targets become accessible via nuclear magnetic resonance, crystallography, and bioinformatic methods, and there is an urgent need of computational tools that can recognize and investigate the active sites and propose the potential drug molecules which can interact with these specific active sites. Time and money necessary for planning a new drug are massive and intolerable. Some data interprets that, it takes about $880 million cost and 14 years of research time to develop a new drug before it is announced in the market. Interference of computers at approximately rational steps is authoritative to lower the cost and time needed in the drug discovery events [15]. Computer simulations are fast and comparatively cheap option for preliminary screening of both active targets and potential drugs. De novo pharmaceuticals in Cambridge, UK, have made a set of software for virtual screening, for docking, as well as for ligand-based design. If the structure of the active target is known, then their site explorer can predict the potential of drug-binding sites and can estimate the interactions between these active sites and the drug. If the structure of the active target is unknown, then its Quasi2 software will create a virtual protein known to be significant in binding [15]. The SkelGen suite of the packages can then use their own data to create a new chemical structure enhanced for interaction with a target's active site. The pharmaceuticals company is cooperating with GeneFormatics of San Diego, California, in an agenda dedicated on inhibitors of the M10 family of matrix metalloproteinases, enzymes which are critical players of inflammation and cancer. GeneFormatics is using proteomics to recognize the target enzymes and illustrate their active sites while de novo is docking/virtual screen model of small molecules against the proteins. Software that can model ligand-receptor communications is obtainable from a number of firms including Accelrys in San Diego, California; Tripos of St Louis, Missouri; and Metaphorics in Mission Viejo, California. A number of software is accessible free to researchers at nonprofit organizations like GOLD made by the Cambridge Crystallographic Data Centre, UK, and AutoDock 3 from the Scripps Research Institute in La Jolla, California. Molsoft in La Jolla, California, designs ICM molecular modelling software and releases an ICM browser for the Apple Macintosh [15]. Another suite for structural homology programs, Accelrys, identifies the potential function, fold family, and 3D structure of target proteins by matching them with sequences and structural homologues of their known function. When the protein's structure is identified, then functional info can be collected using different modules within Accelrys's Insight II program which supports various processes such as NMR studies, X-ray crystallography, and protein modelling. LION bioscience target engine in Heidelberg, Germany, supports target prediction by subscription of the potential to examine the gene sequence and expression data, map potential functional features onto protein structure, find homologous structures, view related gene annotation and protein pathway info, and use text mining to find functional interactions. Proteins themselves established as vigorous drugs in biotherapeutics. One software suite intended to support protein optimization function is protein design automation (PDA) created by Xencor. PDA computationally screens huge numbers of amino acid changes in an acknowledged protein structure. It then initiates functional information from the 3D protein structure and designs novel features into the protein to enhance its function [15].

21.1.4 Clinical Trials

Preclinical improvement begins before clinical trials. The main aims are to govern care and efficiency of the intervention. If preclinical studies display that the treatment is safe and functioning, clinical trials are ongoing. Clinical trial phases agree if an involvement would be valuable or injurious to humans. It comprises different phases named as 0, I, II, III, IV, and V.

21.1.4.1 Phase 0 Clinical Trial

The investigational new drug (IND) (also recognized as phase 0 studies) studies can regulate a certain mode of action that can be detected in humans to offer evidence on pharmacokinetics, select hopeful yields from a group of candidates, and assess bio-distribution. The determination of these studies is to profit in the go versus no-go conclusion making the process of a drug's destiny early in the development process by means of human models rather than trusting on animal data. Limitations of this type of trials comprise lack of therapeutic commitment, motivation of patients to join, may delay or exclude patients from other clinical trials that may have therapeutic commitment, micro-dosing pharmacokinetics, and association with therapeutic dose and obtainability of sensitive analytical methods [16].

21.1.4.2 Phase I Clinical Trial

A phase I clinical trial assesses the finest way to manage a drug and its occurrence, dose (maximum tolerated dose – MTD), and side effects. Pharmacokinetics, tolerability, and pharmacodynamics are assessed. These studies regulate most importantly for the safe treatment. Trials normally comprise 20–100 patients and are checked by the clinical investigator. Doses are amplified if there are no severe side effects, and patients are tested to determine if he or she is answering to the treatment. Dose-toxicity as well as dose-efficacy curves are determined during this phase and involve single arising dose trials (phase IA), multiple arising dose trials (phase IB), and food effect studies. They are easy to implement and do not need exceptional software (Fig. 21.1).

21.1.4.3 Phase II Clinical Trial

Phase I clinical study evaluates MTD, while phase II studies assess the possible effectiveness and symbolize treatment advantage for the disease. These studies are accomplished on larger groups (100–300 subjects) and are intended to evaluate how healthy the drug works and to continue safety valuations. Trials are frequently accompanied in a multi-institution setting. Phase II may be distributed into phase IIA which is preliminary clinical trials to assess effectiveness and care in designated populations with the disease or disorder to be treated diagnosed or banned and phase IIB which are the toughest trials designed to validate effectiveness.

21.1.4.4 Phase III Clinical Trial

Phase III trials include full-scale assessment of treatment and are aimed to compare the effectiveness of the novel treatment with the typical treatment. These are the

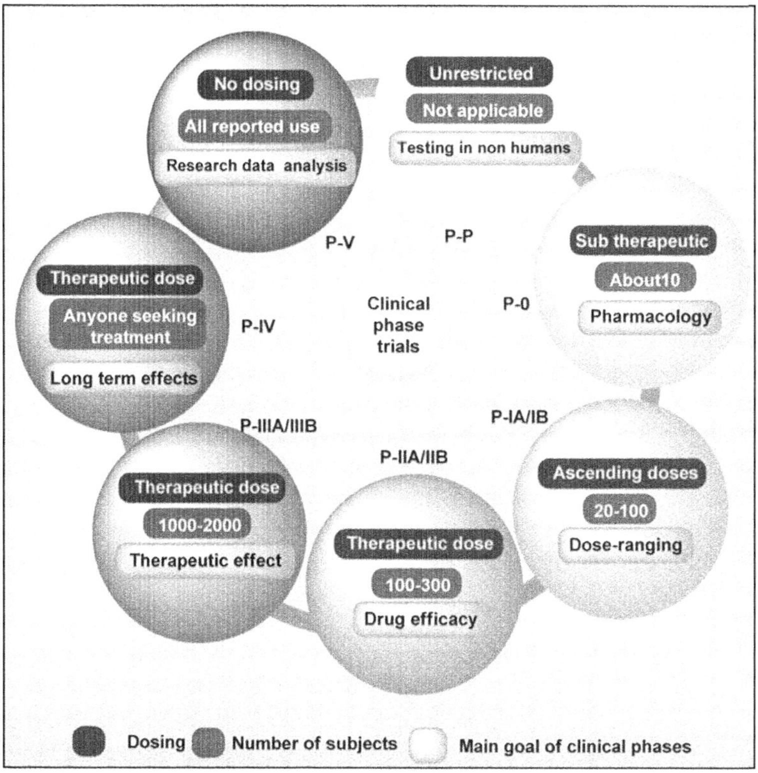

Fig. 21.1 Clinical phase trials: stage of drug development

most laborious and extensive type of technical clinical study of a novel treatment. This is the premarketing phase of clinical trials. These are typically the most luxurious and time-consuming clinical trials. The trials may be problematic to design and run. Large groups (around 100–3000 subjects) are recruited, and trial strategies have comprised randomized skillful trials, uncontrolled trials, historical controls, non-randomized concurrent trials, factorial designs, and group consecutive designs. Patients are checked by the clinical investigator and individual physician.

21.1.4.5 Phase IV Clinical Trial

Phase IV trials include all studies achieved after drug endorsement and linked to the accepted indication [17]. These are post advertising investigation studies. The focus of the trial is on how drugs work in the definite world. Effectiveness and discovery of rare or long-term opposing effects on a greater patient population and lengthier time period are measured, healthcare burdens and results are determined, and pharmacogenetics is studied. New clinical suggestions for a drug may be recognized, and large number of patients and doctors are involved [18].

21.1.4.6 Phase V Clinical Trial

This step of clinical trials denotes relative efficacy and community-based investigation. The investigation is done on the collected data. All stated uses are assessed. Patients are not observed. Its chief attention is to determine incorporation of a new therapy into widespread clinical training.

21.2 Phytochemicals and Women Health

In women, hormone levels play a significant role in health and their well-being, which can differ intensely with age. Mature women are extra vulnerable to cancer, cardiac, and neurological disorders/diseases [19]. These pathological manifestations are stimulated by inflammation as well as oxidative stress [20, 21]. Due to confusion in references and doubt in the medical system, nowadays, women have missing confidence in conventional pharmaceuticals [22]. In this regard, botanicals can act as natural replacements to traditional pharmaceutical treatments used in menopausal (including pre- and post-conditions also) women. The following problems are very sensitive regarding women's health and well-being.

21.2.1 Premenstrual Syndrome

Oral contraceptives, NSAIDs (nonsteroidal anti-inflammatory drugs), SSRIs (selective serotonin reuptake inhibitors), dopamine agonists, and diuretic agents are different pharmacological approaches available to treat premenstrual syndrome (PMS) [23]. Different complications such as nausea, water retention, and weight gain arise as question marks on these pharmaceutical treatments.

21.2.2 Urinary Tract Infections

Women tend to have more frequent urinary tract infections (UTIs). The major problem with excessive antibiotic use, particularly to treat recurring UTIs, is development of drug-resistant bacteria [24].

21.2.3 Pregnancy

Morning sickness is the most shared adverse indication associated with pregnancy. Adverse drug reactions associated with pregnancy medications have been reported. Thus, women often look for natural alternatives to control symptoms.

21.2.4 Lactation

The most common reason for breastfeeding failure is decreased milk production. There are several prescribed drugs that stimulate lactation, but side effects have been stated with the prescription of galactagogues. Due to this, many women prefer natural galactagogues, which avoid the side effects including severe depression, nausea, gastric upset, and diarrhea [25, 26].

21.2.5 Menopausal Symptoms

The most common menopausal sign contributing to the adverse quality of life in women includes hot flashes and night sweats [27]. Women have been disheartened from consuming traditional hormone treatment because of the augmented hazard including breast cancer, CVDs, and stroke [28, 29] (Table 21.1).

21.3 Botanicals in Antidiabetic Clinical Trials

Diabetes mellitus, a chronic metabolic disorder that leads to high level of glucose in the blood, is one of the major health concerns of modern times across the globe. Around 422 million people were affected with diabetes globally in 2014, and the number is expected to increase further due to rapid changes in lifestyle and food habits [56]. Diabetes-associated microvascular (nephropathy, neuropathy, retinopathy) and macrovascular (cerebrovascular accidents, peripheral vascular disease, and cardiovascular disease) complications are the main reasons of morbidity and mortality in patients with diabetes [57]. There are two major types of diabetes mellitus: type I and type II. Patients with type I diabetes depend on the exogenous supply of insulin for controlling their blood glucose levels as the pancreatic β-cells fail to produce any insulin. Type I diabetes is less common (5–10%) and is caused by genetic and environmental factors. On the other hand, type II diabetic patients are not always dependent on exogenous insulin and can control their blood glucose levels through exercise, dietary changes, and medication. The increased blood glucose in type II is primarily because the cells become unresponsive to insulin signaling, and over the time, the production of insulin progressively decreases [58]. Type II diabetes is more common, constituting 90–95% of the patients with diabetes. Being a complex metabolic disorder that involves both genetic and environmental factors, a complete cure for diabetes mellitus has not been developed yet, despite several decades of intensive research. Although several allopathic formulations are currently in use, plant products are becoming more popular the world over to manage type II diabetes due to their fewer side effects and natural origin. Botanical products with antidiabetic properties can improve not only the glucose metabolism but also health of the patients by hypoglycemic effects and by improving capillary function, antioxidant status, and lipid metabolism. Therefore, botanical products are important for both controlling the type II diabetes and preventing it, especially for

Table 21.1 List of botanicals and their manner of action against women problems

Women problem	Botanicals used	Mode of action	References
Premenstrual syndrome	*Vitex agnus-castus*	Phytochemicals act through the dopamine receptors, which decrease thyrotropin-releasing hormone, prolactin levels	[30, 31]
	Angelica sinensis	Inhibits the contraction of the uterus and improves microcirculation	[32]
	Viburnum opulus	Relax uterine	[33–35]
	V. prunifolium	Antispasmodics	
	Zingiber officinale	Antispasmodic and estrogenic activities	[36, 37]
	Valeriana officinalis	Antispasmodic effects and reduction in pain severity	[38, 39]
	Oenothera biennis	Relieves PMS symptoms	[40]
Urinary tract infections	*Vaccinium macrocarpon*	Inhibition of adhesion of P-fimbriated *E. col*	[41–43]
	Arctostaphylos uva-ursi	Astringent effects	[44]
Pregnancy	*Zingiber officinale*	Accelerates gastric emptying (promotility, prokinetic) and stimulates gastric antral contractions	[45, 46]
Lactation	*Trigonella foenum-graecum*	Stimulates sweat production stimulates breastmilk production through this mechanism	[25]
	Silybum marianum	Liver detoxification and galactagogue activity	[47, 48]
Menopausal symptoms	*Actaea/Cimicifuga racemosa*	Modulates serotonergic system similar to the effects of antidepressants	[49–51]
	Valeriana officinalis	Serotonergic activity	[52]
	Glycine max	Relieves menopausal symptoms such as reduced menopausal hot flashes	[53]
	Trifolium pratense	Reduces hot flashes and positive effects on endometrial thickness and breast density	[53]
	Humulus lupulus	Reduces hot flashes	[54]
	Glycyrrhiza species	Selective estrogen receptor β agonist, i.e., estrogenic effects and drop in hot flashes	[55]

people who are at a greater risk of developing diabetes [59, 60]. Reports indicate that a large number of plants have hypoglycemic effects in patients with diabetes. However, only some of them such as cinnamon (*Cinnamomum zeylanicum*, *Cinnamomum cassia*), fenugreek (*Trigonella foenum-graecum*), gymnema (*Gymnema sylvestre*), aloe (*Aloe vera L*), Asian ginseng (*Panax ginseng C. A. Meyer*), bitter melon (*Momordica charantia*), American ginseng (*Panax quinquefolius L*), milk thistle (*Silybum marianum*), salacia (*Salacia oblonga*; *Salacia reticulata*), and nopal (*Opuntia streptacantha*) have been tested clinically for their efficacy and safety [61].

21.3.1 History of Natural Antidiabetic Drugs

The earliest recorded use of medicinal plants for the treatment of human diseases dates back to 2600 B.C. from Mesopotamia. Although plant-based medicine is not presently in extensive use since the advent of pharmaceutical compounds, especially in the Western world, plants or plant extracts have been in extensive use for thousands of years to treat human diseases including diabetes. Botanicals like gymnema, fenugreek, bitter melon, ginseng, and nopal are used by patients with diabetes around the world as they are claimed to decline blood glucose level or decrease complications of diabetes. Similar claims have also been observed for milk thistle, aloe, and bilberry; however, the evidence to support these claims is not substantial. Gymnema (*Gymnema sylvestre*), a member of the milkweed family found in the tropical forests of Africa and India, has been in use to treat diabetes for more than 2000 years. Bilberry, a closely related plant to the American cranberry, huckleberry, and blueberry are used in two forms for medicinal use: the dried fruit and the leaf. The consumption of bilberry dried improves visual acuity and night vision and treats diarrhea. The leaf has several medicinal benefits such as in diabetes, arthritis, circulatory disorders, cataract, and varicose veins. The bilberry leaf is a common constituent of the "antidiabetic" tea, and it is the leaf which is used as a "blood sugar-reducing" drug in the folk medicine.

21.3.2 Phytochemistry of Antidiabetic Drugs

More than three decades of research has revealed that plants are a potential source for the development of new drugs including antidiabetic drugs [62]. Among the various plant products, only some, such as fenugreek, gymnema, aloe, bitter melon, bilberry, ginseng, and milk thistle, have been clinically evaluated. The phytochemistry and mechanism of action of these clinically tested botanicals are given in Table 21.2.

21.3.3 Clinical Investigations of Botanicals on Diabetes

21.3.3.1 Cinnamon

Several studies indicate the hypoglycemic activity of cinnamon in diabetic patients. In a randomized 40-day trial, administration of 1, 3, and 6 gm per day of cinnamon in type II diabetic patients improved blood glucose and lipid profile and reduced incidence of risk factors associated with diabetes [80]. Following the administration of 3 gms of aqueous cinnamon extract per day in patients with type II diabetes who were receiving oral hypoglycemic treatment, a significant reduction of the initial fasting plasma glucose (FPG) was observed [81]. In a randomized trial, a reduction of 0.83% in HbA1c was observed following the administration of 1 gm of cinnamon for 90 days in diabetic patients as compared to 0.37% in diabetic patients who received usual care alone [82]. A double-blind, randomized clinical trial conducted

Table 21.2 Phytochemical profile of some commonly used antidiabetic medicinal plants and their mechanism of action

	Botanical	Phytochemicals	Mechanism of action	References
1	Cinnamon (*Cinnamomum zeylanicum*, *Cinnamomum cassia*)	Polyphenolics	1. Increases glucose uptake by cells through enhanced insulin receptor phosphorylation and the transport of GLUT-4 to the plasma membrane	[63–65]
			2. Increases insulin sensitivity by increasing the expression of PARPα and PARPβ receptors	
			3. Inhibits intestinal glucosidases and pancreatic amylase	
2	Fenugreek (*Trigonella foenum-graecum*)	Saponins, trigonelline alkaloids, gentianine, carpaine compounds, C-glycosides, 4-hydroxyisoleucine, and fenugreekine	1. Delays gastric emptying, slows carbohydrate absorption, and inhibits glucose transport	[66, 67]
			2. Increases erythrocytic insulin receptors and improves peripheral glucose utilization	
3	Gymnema (*Gymnema sylvestre*)	Gymnemic acids (gymnemosides), saponins, stigmasterol, quercitol, the amino acid derivatives choline, betaine, and trimethylamine	1. Increases activity of enzymes responsible for glucose uptake and utilization	[68–70]
			2. May increase β-cell number, stimulate β-cell function, and/or increase insulin release by increasing cell permeability	
4	Bitter melon (*Momordica charantia*)	Glycosides mormordin, charantin, alkaloid mormordicine, polypeptide-P. Seeds contain the β- and α-mormorcharin, abortifacients, and pyrimidine nucleoside vicine	1. The polypeptide-P also known as p-insulin works by mimicking the action of human insulin in the body	[71]
			2. The other potential mechanisms are possibly through stimulation of peripheral and skeletal muscle to utilize glucose, suppression of key gluconeogenic enzymes, enhanced glucose oxidation by enzyme G6PDH pathway, inhibition of intestinal glucose uptake, preservation of islet β cells and their function	

5	Ginseng (*Panax quinquefolius L*, *Panax ginseng* C.A. Meyer, *Eleutherococcus senticosus* Maximum)	Ginsenosides (a family of steroid-like compounds) Ginsenosides are tetracyclic triterpenoid saponin glycosides	Various studies conducted in animal and cells suggest four mechanisms that might account for the blood glucose modulation effect of ginseng:	[72]
			1. Modulation of glucose metabolism	
			2. Modulation of insulin production and secretion	
			3. Modulation of inflammatory pathway	
			4. Modulation of glucose uptake	
6	Nopal (*Opuntia streptacantha Lemaire*)	Soluble fibers (pectin and mucilage), polysaccharides	1. The dietary fiber modifies the absorption of glucose	[73, 74]
			2. Increases insulin sensitivity	
7	Garlic (*Allium sativum*)	Allicin, allyl propyl disulfide, and S-allyl cysteine sulfoxide	Prevents inactivation of insulin by the liver and thus increases insulin levels in the blood	[75, 76]
8	Aloe vera	Juice contains barbaloin, cathartic anthraquinone, and other substances	Aloe gel taken internally may cause a mild reduction in mean glucose levels	
		Gel obtained from the inner portion of the leaves contains glucomannan polysaccharide that is similar to guar gum		
9	Bilberry (*Vaccinium myrtillus*)	Anthocyanosides (bioflavonoids chemicals)	The powerful antioxidant property of anthocyanosides is thought to protect the pancreatic β-cells from the glucose-induced oxidative stress. Flavonoids are also known to prevent diabetic cataracts	[77]
10	Milk thistle (*Silybum marianum*)	Silymarin (composed of three main constituents, silybin A and B, silychristin, and silidianin). Silybin is the most potent biologically active component	1. Protects against toxic free radicals via its antioxidant activity	[78, 79]
			2. silybin decreases DHA gluconeogenesis and glucolysis	
			3. silybin increases β-cell viability by decreasing β-cell cytotoxicity, inhibiting fibrillation, and improving oligomerization	

by Akilen et al. (2010) in diabetic patients (type II) on oral hypoglycemic treatment established a significant reduction in HbA1c levels along with reduced body mass index, FPG, blood pressure, and waist circumference following the administration of 2 gms of cinnamon daily for 12 weeks [83]. In contrast, in an earlier randomized, placebo-controlled trial, conducted by Suppapitiporn et al. (2006), reduction of HbA1c levels in diabetic patients (type II) on oral therapy containing metformin or sulfonylurea when treated with cinnamon *cassia* powder for 12 weeks was nonsignificant [84]. In two recent independent studies, a significant reduction in fasting blood glucose (FBG) levels was reported following the administration of 500 mg/day cinnamon for 2 months, [59] and 6 and 12 weeks [85] when compared with the placebo control group.

21.3.3.2 Fenugreek

Considering its antidiabetic properties as claimed in various cultural practices, several clinical trials were conducted to evaluate the safety and efficacy of fenugreek seeds in diabetic patients. However, the majority of these trials were short term involving fewer patients and did not report study design details adequately. A significant reduction in blood glucose parameters has only been reported in clinical trials that administered fenugreek seed powder (≥5 gm) [86]. Medium to high doses of fenugreek seed powder (5–25 gm) also declined postprandial glucose levels in some studies [86]. In a randomized clinical trial, one group of type I diabetic patients were given 50 gm of defatted fenugreek seed powder in chapati (unleavened bread) twice daily and another group placebo. The mean FBG decreased significantly from the baseline of 272 mg/dl to 196 mg/dl along with a significant decline in total serum triglycerides, cholesterol, and LDL cholesterol without changing HDL cholesterol as compared with the placebo group [87]. In a large 6-month clinical trial involving 60 inadequately controlled type II diabetic patients, administration of 25 gm/day of fenugreek powder with two equal doses at lunch and dinner caused a significant mean decrease of 39 mg/dl FBG from baseline after 24 weeks. The mean HbA1c significantly decreased from 9.6% to 8.4% after 8 weeks. A meta-analysis of ten clinical trials revealed a significant decline in the fasting glucose levels and HbA1c levels in type II diabetic patients following the intake of fenugreek seeds for 6 months [86].

21.3.3.3 Gymnema

Only a few clinical trials have been carried out to examine the antidiabetic properties of gymnema. When type I diabetic patients, who were on insulin, were given 200 mg gymnema capsules after breakfast and supper, the mean FBG of 232 mg/dl at baseline decreased to 177 mg/dl after 6–8 months, 150 mg/dl after 16–18 months, and 152 mg/dl after 20–24 months. The levels of HbA1c also showed a steady decrease with time. After 16–18 months, the mean HbA1c in the 22 patients on gymnema was 9% and after 26–30 months 8.2% in the 6 remaining patients. The mean insulin dose also decreased to 45 units/day after 6–8 months and then to 30 units/day at 26–30 months from 60 units/day at baseline. In another trial, 22 type II diabetic patients on sulfonylureas were given 400 mg/day of gymnema, and 25 patients were given placebo for a period of 18–20 months. The mean HbA1c decreased to 8.48% in the patients who received gymnema capsules from a baseline of 11.9%. In the same treatment group, the mean FBG decreased from 174 to 124 mg/dl after 18–20 months. No significant changes in HbA1c, FBG, or lipids levels were reported in patients who received placebo.

21.3.3.4 Bitter Melon

Although sold under the brand names Karela, Gourdin, and Glucobetic in India, Canada, the USA, the UK, and many Asian countries in the form of capsules or tablets, the human clinical evidence about the effectiveness of bitter melon to treat diabetes is inconclusive. The first documented therapeutic effect of bitter melon was reported in 1956 by Lakholia [88]. More than seven controlled or uncontrolled clinical trials have been conducted so far with most showing a significant decline in blood glucose levels; however, the majority of them lacked proper controls. Following the administration of 4.8 gm of bitter gourd powder for 3 months in an open-label uncontrolled trial, the level of blood glucose decreased significantly [71]. A similar effect was observed in a multicenter, randomized, double-blind, active-control trial following the administration of 500 mg capsule of dried powder for 4 weeks [89].

21.3.3.5 Garlic

Garlic, a common item in many kitchens, is believed to have several medicinal properties including antidiabetic properties. However, the majority of contemporary medical research on garlic has focused on the treatment of cardiovascular-related diseases, and there is less evidence to support its antidiabetic use. Evidence from animal studies suggests that there may be an enhanced insulin secretion or less insulin degradation due to the chemical components of garlic [75, 76]. Although preclinical data support a hypoglycemic property of garlic in diabetic rats [90], few studies done on patients of type II diabetes have not demonstrated any significant changes in blood glucose or insulin levels following oral administration of garlic [91].

21.3.3.6 Aloe vera

Two independent controlled, non-randomized trials have reported decrease in FBG in patients with type II diabetes when given aloe gel juice for 6 weeks; however, the evidence is inconclusive due to insufficient details provided by the authors about these studies [92, 93].

21.3.3.7 Nopal

The evidence collected from various clinical trials evaluating the hypoglycemic effect of nopal suggests that use of broiled or fresh nopal stems significantly reduces FBG in patients with diabetes. In a randomized trial, with type II diabetic patients either administered with diet alone or along with sulfonylurea, in the group of patients who received 500 gms of broiled nopal, the blood glucose decreased from 222 mg/dl fasting to 203 mg/dl after 60 min, 198 mg/dl after 120 min, and 183 mg/

dl after 180 min of taking nopal [94]. On the other hand, in the group which was only administered with water, there was no significant drop in blood glucose, but there was a slight drop in the group which received broiled zucchini (from mean baseline of 246 mg/dl fasting to 226 mg/dl after 60 min, 219 mg/dl after 120 min, and 206 mg/dl after 180 min). In another trial, type II diabetic patients who received 500 gm of broiled nopal showed a significant mean decrease of 21, 28, and 41 mg/dl glucose concentrations at 60, 120, and 180 min, respectively, from baseline [95]. No significant difference was observed in the group administered with water alone. In nondiabetic patients, the glucose levels stayed stable post nopal consumption.

21.3.3.8 Milk Thistle

A daily dose of 600 mg silimarin, a derivative of milk thistle, for 6 months resulted in a significant reduction in FBG and mean daily glucose levels in diabetic patients and alcoholic liver cirrhosis patients from the second month onward of the treatment. A 20% decrease in the insulin requirement suggested an alleviation of insulin resistance after silimarin treatment [96]. In another trial, diabetic patients with the end-stage renal disease were administered with 350 mg intravenous bolus dose of silybin for more than 24 h. The cellular surface thiol status in peripheral blood lymphocytes was restored in these patients along with improved T-cell activation and reduction in TNFα levels and inflammation. In another randomized, controlled trial, half of the patients with type II diabetes who presented at baseline with diabetic macroalbuminuria treated with a dose of 140 mg of silimarin thrice a day reported a 50% reduction in urinary and serum MDA levels, urinary albumin excretion, and urinary TNFα levels [97]. A reduction from 156 ± 46 mg/dl to 133 ± 39 mg/dl blood glucose was seen when patients were administered with 200 mg of silimarin thrice a day before meals for 4 months [98]. A decrease from 7.82% ± 2.01 to 6.78% ± 1.05 HbA1c levels was also reported [98]. When administered as a fixed combination along with an alkaloid derived from *Berberis aristata*, silimarin had a synergistic hypoglycemic effect in patients with type II diabetes [99].

Even though the evidence available from numerous clinical trials is promising and does support a positive impact of various botanicals on the management of glucose levels in patients with diabetes, more clinical trials with large sample size and longer durations are needed to authenticate the use of various botanical products against diabetes, especially type II diabetes. The evidence also suggests that if taken in combination with the allopathic antidiabetic medicine, the botanical products can result in the better management of blood glucose levels in diabetic patients.

21.4 Cancer Preventive and Therapeutic Potential of Phytochemicals

Cancer is a noncommunicable disease, and it is the second leading cause of death in developing as well as developed countries. Several population-based studies suggest that high consumption of vegetables and fruits helps in the reduction of cancer risk [100–102]. Further, phytochemicals have been used for the treatment of cancer for the past several years due to their easy availability, low cost, less toxicity, and few or no side effects. In the USA, approximately 50–60% of cancer patients are either exclusively treated with the plant- or nutrient-derived components or along with conventional therapeutic regimen such as chemotherapy and/or radiotherapy [103]. Therefore, recently, phytochemicals are emerging as promising chemopreventive and therapeutic agents for cancer treatment. There are a number of phytochemicals whose anticancerous and chemopreventive properties have been examined in in vitro and in vivo conditions, but this chapter only discusses about some widely known phytochemicals on which clinical studies have been conducted or still going on such as curcumin, epigallocatechin gallate, sulforaphane, and resveratrol.

21.4.1 Clinical Studies of Curcumin on Cancer Patients

Curcumin (diferuloylmethane) is a polyphenol, derived from the *Curcuma longa* (turmeric). It acts as a pleiotropic molecule due to its anti-inflammatory, hypoglycemic, antioxidant, wound-healing, and antimicrobial properties [104]. Several preclinical studies indicate the therapeutic value of curcumin against various human diseases [105]. Curcumin does not cause any harmful or side effects on animals [106] or humans [107] even at its higher doses. The first evidence of curcumin's anticancerous activities in human was shown by Kuttan et al. (1987) [108]. In Taiwan, Cheng and co-workers performed a phase I trial of curcumin on 24 cancer patients including intestinal metaplasia of the stomach, uterine cervical intraepithelial neoplasm, resected urinary bladder cancer, oral leukoplakia, and arsenic Bowen's disease of the skin. In this investigation, a single dose of curcumin ranging from 500 to 8000 mg/day was administered daily through orally up to 3 months, and interestingly, curcumin did not cause any toxicity at any of the administered doses [109]. Moreover, this study showed histologic improvement of precancerous lesions in some of the patients who underwent this trial [109].

In another independent study, Sharma et al. (2001) conducted a study on advanced colorectal cancer (CRC) patients and revealed that daily consumption of 440–2200 mg single dose of curcuma extract for 4 months does not exert dose-limiting toxicity [110]. Further, a study recommended that 3.6 gm of curcumin can be taken orally daily for the prevention or treatment of malignancies located other than gastrointestinal tract [111]. In addition, an investigation also suggested 3.6 gm daily dose of curcumin as the pharmacologically effective dose for colorectal patients [112]. Further, He et al. (2011) explored that curcumin can also help in the improvement of CRC patient's health as it increases the body weight and number of tumor apoptotic cells, along with a declined serum TNF-α level probably via increasing the expression of p53 in tumor cells [113], but further studies are needed for exploring the exact mechanisms behind the health improvement of CRC patients. Curcumin along with quercetin suppressed colorectal adenomas in patients having FAP (familial adenomatous polyposis, an autosomal-dominant disorder) [114], and thus curcumin can also be used as combinational therapy for the treatment of CRC. Irving et al. (2013) conducted a clinical pilot study to evaluate the acceptable daily dose of curcumin by quantifying the level of curcuminoids in colorectal mucosa of 26 patients who underwent colorectal endoscopy or surgical resection [115]. In this study, 24 of 26 patients completed the course of curcumin (Curcumin C3 complex (2.35 g)) before endoscopic biopsy or colonic resection. The results of this study interestingly showed that curcuminoids were detectable in the plasma of 9 of 24 patients, urine of all participated patients, and in the colon mucosal biopsies of all 23 patients [115]. Thus, authors of this study suggested that biologically active levels of curcumin recovered in colon mucosal biopsies of patients and therefore the regimen used in this study can be utilized for long-term trials. Moreover, recently Irving et al. (2015) designed a phase I randomized control trial to evaluate the acceptable and safe dose of curcumin in colorectal cancer patients receiving standard care FOLFOX-based (5-fluorouracil, folinic acid, and oxaliplatin) chemotherapy [116]. In addition, in vitro and preclinical studies carried out by James et al. (2015) demonstrated that curcumin alone or with FOLFOX chemotherapy inhibited the number of cancer stem cells derived from colorectal cancer patients with liver metastases and curcumin dose up to 2 g/day is safe and well tolerable along with FOLFOX chemotherapy in colorectal liver metastatic (CRLM) cancer patients. Therefore, this study suggested that curcumin can be used as an adjuvant therapeutic drug for the treatment of CRLM cancer patients [117].

In a phase II clinical trial, curcumin did not exert any toxicity and was well-tolerated in advanced pancreatic cancer (APC) patients [118]. In another independent phase II clinical trial, APC patients were ingested with 8 g/day oral dose of curcumin for 4 weeks along with an intravenous dose of gemcitabine (1000 mg/m^2) three times a week. The study suggested that in combination with gemcitabine, 8 g/day oral dose of curcumin was above the maximum tolerated dose [119]. However, a study on 21 patients with gemcitabine-resistant pancreatic cancer suggested that 8 g/day dose of curcumin along with gemcitabine was well-tolerated with no toxicities [120].

A study conducted on 29 multiple myeloma patients for the evaluation of safety, tolerability, and clinical efficacy of curcumin suggested that curcumin might be a potent therapeutic candidate against the multiple myeloma [121]. In addition, the therapeutic value of curcumin was observed in patients with monoclonal gammopathy of undefined significance (MGUS), a model for studying multiple myeloma precursor disease [122]. In this investigation, the impact of curcumin on plasma cells and osteoclasts was evaluated on 26 MGUS patients. This study showed 12%–30% decline paraprotein level in five of ten patients with paraprotein >20 g/L [122]. Moreover, 27% of total patients (7 of 26 patients) showed a reduction in urinary N-telopeptide of type I collagen [122]. Thus, this investigation suggested that curcumin can be used as an early intervention agent against multiple myeloma. Further, for developing curcumin therapy as an early intervention strategy against multiple myeloma, Golombick et al. (2012) evaluated free light chain (FLC) response and bone turnover in curcumin-administered patients with MGUS and smoldering multiple myeloma (SMM) and suggested a chemopreventive role of curcumin in multiple myeloma patients [123].

An open-label phase I trial on advanced and metastatic breast cancer patients indicated 8 g/day dose of curcumin as maximum tolerable dose when administered along with a standard dose of docetaxel, whereas 6 g/day of curcumin was reported as the recommended dose along with the docetaxel [124]. Further, a clinical trial of curcumin was conducted on 30 patients with breast cancer to estimate the radiation dermatitis severity reduction ability of curcumin [125]. In this study, randomized patients were taken 6 g/day of curcumin or placebo orally thrice during the course of radiation therapy. This study suggested that curcumin declines radiation dermatitis severity in breast cancer patients and thus can be used during the radiation therapy for the reduction of radiation dermatitis severity [125].

Curcumin in combination with soy isoflavones synergistically suppressed PSA (prostate-specific antigen) production in a randomized, double-blind controlled study on prostate cancer negative males with high PSA level [126]. Thus, this study indicated that curcumin might be helpful for the prevention of prostate cancer as the risk of prostate cancer is directly related to an increased level of PSA. Further, in a phase II pilot study, Mahammedi et al. (2016) evaluated the efficacy of curcumin against chemotherapy-naive metastatic castration-resistant prostate cancer (CRPC) patients and a rising prostate-specific antigen (PSA) [127]. In this study, 30 patients received docetaxel or prednisone in combination with oral curcumin (6000 mg/day) for 6 cycles. After analyzing the results of this study, they concluded that the regimen used in this study was well tolerable with no severe side effects [127], and thus the regimen can be used for metastatic prostate cancer treatment.

Curcumin has also an antitumor potential against the head and neck squamous cell carcinoma (HNSCC), and its antitumor potential is due to inhibition of IKKβ kinase activity as observed in HNSCC patients [128]. An investigation by Ghalaut et al. (2012) indicated that imatinib in combination with curcumin significantly decreased the level of nitric oxide in chronic myeloid leukemia (CML) patients as compared to patients those were receiving only imatinib [129]. Therefore, curcumin can be used as an adjuvant therapeutic drug for CML patient treatment.

Panahi et al. (2014) conducted a clinical trial of curcuminoid preparations (180 mg/day) on patients with solid tumors who underwent standard chemotherapy regimens for the evaluation of curcuminoids efficacy as adjuvant therapy in cancer patients [130]. After getting the results of this trial, they concluded that curcuminoid preparation significantly improves quality of life and reduces the systemic inflammation in solid tumor patients who underwent the standard chemotherapy regimens [130].

In summary, research and clinical studies discussed above indicate that curcumin could be a better therapeutic option for cancers of different origin, yet larger and well-controlled clinical trials are required which can help in determining the clinical therapeutic value of curcumin against a broader range of cancers.

21.4.2 Cancer Clinical Investigations of Epigallocatechin Gallate

Epigallocatechin gallate (EGCG) is the most abundant catechin compounds in green tea which possesses the most potent antioxidative activity, and recently, it has emerged as an adjunct therapeutic option for cancer treatment. In a recent clinical study, the oral administration of EGCG in combination with concurrent chemotherapy and radiotherapy to 24 patients of unresectable stage III non-small cell lung cancer caused regression of esophagitis to grade 0/1 in 22 out of 24 patients, whereas a persisted grade 2 esophagitis was observed in the rest of two patients at the end of radiotherapy [131]. The results of this study also suggested that the oral administration of EGCG was effective and recommended 440 μmol/L for phase II clinical trial. Recently, in a phase II trial, Zhao et al. (2015) showed the potential role of EGCG in suppressing acute radiation-induced esophagitis (ARIE) in stage III lung cancer patients who underwent radiotherapy [132]. Further, administration of EGCG in combination with radiotherapy to breast cancer patients suggested that administration of EGCG is well-tolerated and may be helpful in treating the radiation dermatitis which develops during or after the radiotherapy [133].

21.4.3 Sulforaphane in Cancer Clinical Studies

Sulforaphane is an organosulfur compound, produced by the myrosinase in GI tract due to the hydrolytic conversion of glucoraphanin. Some of the cruciferous vegetables, namely, broccoli sprouts and cauliflower sprouts, are rich in glucoraphanin. A research group of Alumakal et al. (2015) did the first clinical trial of sulforaphane-rich extracts on the patients with prostate cancer and found less than 50% PSA declines in most of the patients administered with sulforaphane-rich extracts and thus suggested further investigations with higher dose to confirm the role of sulforaphane as a preventive or anticancer agent [134]. A recent report indicates that daily oral administration of sulforaphane to prostate cancer patients with biochemical recurrence (a state of enhanced PSA level in serum of prostate cancer patients after radical prostatectomy) after radical prostatectomy effectively manages the biochemical recurrences [135]. The chemopreventive nature of sulforaphane is also evaluated in women with abnormal mammograms and scheduled for breast biopsy [136]. But still further studies on are needed to analyze the clinical efficacy of sulforaphane.

21.4.4 Cancer Clinical Studies on Genistein

Genistein is a predominant isoflavone isolated from dietary soybean and acts as a tyrosine kinase inhibitor by inhibiting DNA topoisomerase II [137, 138]. The anticancerous activity of genistein has been reported against different types of cancer in several in vitro and in vivo studies, mainly via modulating cell cycle, apoptosis, angiogenesis, and metastasis [139]. Till date, only limited clinical studies have been conducted to examine the clinical efficacy of genistein. A clinical study by deVere White et al. (2004) on prostate cancer patients explored that genistein did not decline the level of PSA ≥50% in all investigated patients except one (total number of patients used in the investigation were 52) and therefore it was concluded that genistein alone may not be a better therapeutic agent for prostate cancer treatment [140]. However, a placebo-controlled, block-randomized double-blind phase II investigation explored that a dose of genistein equivalent to diet rich soy declined the serum level of PSA in the serum of localized prostate cancer patients [141]. Further, genistein was well-tolerated and did not alter the hormonal level. In a phase II randomized, placebo-controlled trial study on urothelial bladder cancer, genistein significantly decreased the bladder cancer tissue p-EGFR staining at its lower dose rather than its higher dose [142]. However, no significant differences had been observed in tumor tissue staining of other biomarkers such as COX-2, Ki-67, activated caspase-3, Akt, p-Akt, mitogen-activated protein kinase (MAPK), or p-MAPK between treatment groups [142]. Therefore, this study suggested that genistein or oral soy products are not an effective preventive or therapeutic agent when given alone. Another phase II placebo-controlled, randomized, double-blind clinical trial investigation on patients with localized prostate cancer indicated that genistein intervention did not affect the expression of proliferation, cell cycle, apoptosis, or neuroendocrine-associated molecules such as p21 Waf1/Cip1, p53, Ki67 nuclear antigen, BCL-2, BAX, neuron-specific enolase, and cytoplasmic chromogranin A, whereas it significantly declined the expression of KLK4 (androgen-related biomarker) in tumor cells at gene level [143]. Further, genistein intervention upregulated the expression of nuclear p27Kip1 (cell cycle-related gene), an inhibitor of the cyclin-dependent kinases [143]. Therefore, genistein could be a chemopreventive agent for prostate cancer as its intervention modulates the expression biomarkers associated with prostate cancer. More detailed study on larger sample size as well as for longer duration is warranted.

21.4.5 Clinical Studies of Resveratrol on Cancer Patients

Resveratrol (3,5,4′-trihydroxy-trans-stilbene) is a plant-derived non-flavonoid polyphenol present in grapes, berries, peanuts, and other plant sources. Most of the reported trials are basically associated with the evaluation of its safety, tolerability, bioavailability, and pharmacokinetics. A dose escalation study by Brown et al. (2010) was done to evaluate the parameters such as safety, pharmacokinetics, and effects on the level of circulating IGF-I and IGF-binding protein-3 (IGFBP 3) after repeated ingestion of resveratrol [144]. In this study, 40 healthy volunteers were administered with different doses (0.5, 1.0, 2.5, or 5.0 g) of resveratrol for 29 days, and the plasma level of IGF-I and IGFBP-3 was measured. Further, this investigation indicated that administration of resveratrol declines the level of circulating IGF-I and IGFBP-3 in all volunteers, but a maximum inhibitory effect was observed at 2.5 g/day dose of resveratrol [144]. Therefore, authors concluded that the level of circulating IGF-I and IGFBP-3 may be emerged as future potential markers for evaluating the chemopreventive activity of resveratrol during its clinical trials on cancer patients.

The efficacy of resveratrol in cancers such as multiple myeloma, breast cancer, follicular lymphoma, neuroendocrine tumor, and colorectal cancer is only reported in few studies [145]. But most of the cancer-related clinical trials have been done on colorectal cancer, and these investigations indicate that resveratrol inhibits the progression of colorectal cancer, possibly due to direct contact and exposure of colonic tissues with resveratrol [145].

In 2009, the first cancer clinical trial of resveratrol was conducted by Nguyen et al. on colon cancer [146]. In this study, they were interested to see the effect of a low dose of resveratrol and resveratrol-containing grape powder (GP) on the level of Wnt target genes in normal and colon cancer tissues obtained from eight colon cancer patients and revealed that resveratrol or grape powder significantly inhibited the expression of Wnt pathway target genes such as cyclin D1, myc, axin II, TCF7, and jun in normal colonic mucosa but not in the cancer tissue [146]. Moreover, the maximum inhibitory effect on the expression of target genes of Wnt signaling pathway was observed in patients administered with 80 g GP per day. Therefore, this study emphasizes the need for further clinical trials to confirm the cancer preventive role of resveratrol. Further, in a study, Patel et al. (2010) detected resveratrol and its metabolites such as resveratrol-3-O-glucuronide, resveratrol-4′-O-sulfate, resveratrol-4′-O-glucuronide, resveratrol-3-O-sulfate, resveratrol disulfate, and resveratrol sulfate glucuronide in the human colorectal tissue [147]. In addition, they also reported that resveratrol reduces the proliferation of tumor cells by 5% and oral administration of 0.5 and 1.0 g doses of resveratrol is sufficient to elicit its chemopreventive activity [147]. In a phase I study, safety, pharmacokinetics, and pharmacodynamics of micronized resveratrol (SRT501) were evaluated in metastatic colorectal cancer patients who underwent hepatectomy following administration of 5 g of SRT501 [148]. This study indicated that resveratrol was well tolerable and elicited pharmacologic effect as a significant increase in cleaved caspase-3 positive cells was observed in SRT501-treated malignant hepatic tissue as compared to tissue of placebo-treated patients [148]. Therefore, authors suggested that micronized resveratrol is well tolerable and has better bioavailability than nonmicronized resveratrol. Zhu et al. (2012) evaluated the dose-related effect (5 or 50 mg trans-resveratrol twice daily for 12 weeks) of trans-resveratrol on DNA methylation and prostaglandin expression in 39 adult women at high risk of breast cancer [149]. Interestingly, they found a declined level of RASSF-1α methylation with increased circulating levels of trans-resveratrol and its metabolite resveratrol-glucuronide in serum. Further, they also reported a decreased RASSF-1α methylation with decreasing expression of PGE_2 in nipple aspirate fluid [149]. The authors of this investigation suggested further validation of this study on a larger sample size.

In another study, Popat et al. (2013) evaluated the safety, pharmacokinetics, and efficacy of SRT501 or its combination with bortezomib in patients with relapsed or refractory multiple myeloma [150]. In this study, they found several severe side effects like nausea, diarrhea, vomiting, fatigue, anemia, infections, and renal failure in patients, and thus the trial was prematurely closed [150]. Therefore, this study indicates the risk of developing resveratrol as a novel therapeutic agent in multiple myeloma patients. It is important to decipher the mechanisms behind the severe effects of resveratrol as it will help in reducing its side effect and enhancement of its therapeutic potential in multiple myeloma patients. The detailed information regarding the clinical trials of phytochemical on cancers of different origin is listed in Table 21.3.

21.5 Phytochemicals in Treatment of Bacterial Diseases

"Diet cures more than doctors!" This age-old proverb illustrates the importance of diet in the health and well-being of humans. Nature is the excellent combinatorial chemist and has the key to all the diseases of humans. Many of the plant species growing all around the world have pharmacological action directly to human beings.

The development of bacterial resistance to the present-day available antibiotics has fascinated the need to search for new antibacterial agents. The Gram-positive bacteria like *Staphylococcus aureus* are responsible mainly for the food poisoning, osteomyelitis, endocarditis, toxic shock syndrome, and postoperative wound infections [151], whereas the Gram-negative bacterium like *Escherichia colitis* present in the human intestine causes infection of the lower urinary tract, coleocystis, or septicemia [152]. It has been observed that due to the indiscriminate use of commercial antibiotic drugs which are used on a routine basis for treatment of infectious diseases, the multiple drug resistance in human pathogenic microorganisms developed. There are several causes of the development of resistance to the antibiotic which includes the specific nature of the relationship of bacteria to antibiotics along with the usage of the antibacterial agent, host characteristics, and the changing environment. To overcome these situations, scientists are in the search of new antimicrobial substances from various natural sources which could also be cost-effective, and considering this into account, the sudden increased demand for natural ingredients might be as components of functional foods, food additives, preventing diseases, nutraceuticals, and other applications. The natural components identified from diverse medicinal plants are the important source of biologically active compounds, of which many of them have been the basis for the development of new lead chemicals for biopharmaceuticals.

21.5.1 Molecular Mechanisms of Antibacterial Nature of Herbal Components

Herbal components have been reported to be important alternatives for multiple drug-resistant strains, and it is scientifically proven and referred that bacteria cannot develop resistance to herbal medicines [153, 154], but there is also no any clear understanding about the survival mechanisms of microorganisms against the naturally occurring antibacterial compounds. The six possible mechanisms of antibacterial resistance/action that have been categorized and reported are as follows: (1) by

Table 21.3 Phytochemicals and their clinical trials on cancer patients

Trial	Type of participants	Aim	References
Clinical trials of curcumin			
Ethanolic extract of turmeric and an ointment of curcumin	Patients with external cancerous lesions	Anticancerous activities of turmeric and curcumin	[108]
Phase I clinical trial of curcumin	Patients with urinary bladder cancer, arsenic Bowen's disease of the skin, uterine cervical intraepithelial neoplasm, oral leucoplakia, and intestinal metaplasia of the stomach	Toxicology, pharmacokinetics, and biologically effective dose of curcumin, phase I trial	[109]
Phase I clinical trial of oral curcumin	Advanced colorectal cancer patients	Pharmacology of curcumin in humans	[111]
Pharmacologically effective dose of curcumin	Colorectal cancer patients	Curcumin levels in the colorectal and their pharmacodynamic consequences	[112]
Combinatorial effect of curcumin and quercetin	Familial adenomatous polyposis patients	Efficacy of combination of curcumin and quercetin on adenomas in patients with familial adenomatous polyposis	[114]
Phase I/II study of curcumin on multiple myeloma	Multiple myeloma patients	Clinical safety and biological effects of curcumin	[121]
Phase II clinical trial of curcumin in advanced pancreatic cancer	Patients with advanced pancreatic cancer	Clinical biological effects of curcumin and tumor inhibitory properties, against advanced pancreatic cancer	[118]
Curcumin as a therapeutic intervention for MGUS patients	Patients with Monoclonal gammopathy of undetermined significance (MGUS): a model for studying multiple myeloma precursor disease	Biological effects of curcumin	[122]
Gemcitabine with Curcumin for pancreatic cancer	Patients with advanced pancreatic cancer	Activity and feasibility of gemcitabine in combination with curcumin in advanced pancreatic cancer patients	[119]

(continued)

Table 21.3 (continued)

Trial	Type of participants	Aim	References
Phase I dose escalation trial of docetaxel and curcumin	Patients with advanced and metastatic breast cancer	Feasibility and tolerability of docetaxel and curcumin combination	[124]
Combinatorial effect of soy isoflavones and curcumin	Prostate cancer negative men with high prostate-specific antigen (PSA) level	Effects of soy isoflavones and curcumin on serum PSA levels	[126]
A pilot study of curcumin on head and neck cancer	Patients with head and neck cancer	Biological effects of curcumin	[128]
Antitumor potential of curcumin	Colorectal cancer patients	Inhibitory mechanism of curcumin on colorectal cancer	[113]
A phase I/II study of curcumin and gemcitabine	Gemcitabine-resistant pancreatic cancer patients	Safety and feasibility of curcumin with gemcitabine combination therapy	[120]
A randomized, double-blind placebo-controlled crossover study on curcumin	Patients with monoclonal gammopathy of undetermined significance (MGUS) and smoldering multiple myeloma (SMM)	Assessment of curcumin effect on free light chain (FLC) response and bone turnover in patients with MGUS and SMM	[123]
Effectiveness of turmeric powder in combination with imatinib	Chronic myeloid leukemia patients	Adjunct therapeutic potential of curcumin in combination with imatinib	[119]
A clinical pilot study on oral curcumin	Colorectal cancer patients	Quantification of curcumin level in colorectal mucosa of patients	[115]
A randomized double-blind placebo-controlled clinical trial of curcumin for radiation dermatitis	Breast cancer patients	Modulatory action of curcumin on radiation dermatitis in patients receiving radiotherapy (RT) for breast cancer	[125]
A randomized double-blind placebo-controlled trial of curcumin for adjunct therapy	Patients with solid tumors	Curcuminoids efficacy in adjuvant therapy of cancer patients	[130]

(continued)

Table 21.3 (continued)

Trial	Type of participants	Aim	References
Phase I study on combinational therapy of curcumin and FOLFOX	Patients with colorectal liver metastasis	Safety and tolerability of curcumin and FOLFOX in combinational therapy	[117]
Phase I/IIa study of oral curcumin with standard care FOLFOX-based (5-fluorouracil, folinic acid, and oxaliplatin) chemotherapy	Colorectal cancer patients with inoperable liver metastases	Safety and tolerability of curcumin in combinational treatment with FOLFOX-based chemotherapy	[116]
Phase II trial of docetaxel, prednisone, and curcumin	Metastatic castration-resistant prostate cancer patients	Efficacy of curcumin in combination with docetaxel or prednisone	[127]
Curcumin with or without bioperine	Patients with multiple myeloma	Pharmacokinetics, pharmacodynamics, tolerance and biological effects of curcumin *vs* curcumin plus bioperine	NCT00113841 (started in 2005 and completed in 2011)
Phase II A trial of curcumin on colon cancer in smokers with aberrant crypt foci (ACF)	Colon cancer patients	Colon cancer inhibitory mechanism of curcumin in smokers with ACF	NCT00365209 (started in 2006 and completed in completed in 2015)
Chemopreventive nature of curcumin in colorectal cancer	Colorectal cancer patients	Antiproliferative action of curcumin in the colonic mucosa of patient with sporadic adenomatous polyps	NCT00118989 (started in 2005 and completed in 2014)
Phase III trial of gemcitabine, curcumin, and celebrex	Advance pancreatic cancer patients	Clinical efficacy of gemcitabine, curcumin, and celebrex in pancreatic cancer	NCT00486460 (started in 2007 but not completed)
Clinical trials of epigallocatechin gallate			
A phase I study of EGCG in combination with standard chemoradiation	Locally advanced stage III non-small cell lung cancer patients	Feasibility, safety, and effectiveness of oral administration of EGCG in combination with standard chemoradiation	[131]
A phase II trial of EGCG	Stage IIIA and IIIB lung cancer patients	Safety and effectiveness of EGCG in treatment of esophagitis induced by the radiation therapy	[132]
Clinical trials of sulforaphane			
A phase II study of sulforaphane-rich broccoli sprout extracts	Patients with recurrent prostate cancer	Biologic effects of sulforaphane-rich extracts	[134]

(continued)

Table 21.3 (continued)

Trial	Type of participants	Aim	References
A double-blinded, randomized, placebo-controlled multicenter trial of sulforaphane	Patients with high level of PSA after radical prostatectomy	Effects of sulforaphane on biochemical recurrences in prostate cancer after radical prostatectomy	[135]
A double-blinded, randomized controlled trial of sulforaphane	Women with abnormal mammograms and scheduled for breast biopsy	Bioavailability and chemopreventive activity of sulforaphane	[136]
Clinical trials of genistein			
An open-label pilot study of genistein	Prostate cancer patients	Biologic effects of genistein-rich extracts	[140]
A randomized, placebo-controlled, double-blind phase II clinical trial of genistein	Patients with localized prostate cancer	Efficacy and safety of genistein	[141]
A phase II randomized, placebo-controlled trial of genistein	Patients urothelial bladder cancer	Biological effects of genistein	[142]
A phase II placebo-controlled, randomized, double-blind clinical trial of genistein	Localized prostate cancer patients	Biological effects of genistein	[143]
Clinical trials of resveratrol			
A phase I pilot clinical trial of resveratrol and grape powder	Patients with colon cancer	Biological effects of resveratrol	[146]
Clinical pharmacology of resveratrol	Patients with colorectal cancer	Anticarcinogenic effect of resveratrol and estimation of resveratrol and its metabolite concentration in the colorectal tissue	[147]
Phase I randomized, double-blind pilot study of micronized resveratrol (SRT501)	Colorectal cancer patients with hepatic metastases	Safety, pharmacokinetics, and pharmacodynamics of micronized resveratrol (SRT501)	[148]
A randomized, double-blind placebo-controlled trial of trans-resveratrol	Women with high breast cancer risk	Biological effects of trans resveratrol	[149]
A phase II study of SRT501 (resveratrol) with bortezomib	Relapsed and or refractory multiple myeloma patients	Efficacy and safety of resveratrol	[150]

interaction with membrane proteins (ATPases and others), (2) by disintegrating the cytoplasmic membrane, (3) by destruction of the proton motive force with leakage of ions, (4) by disturbing the Gram-negative bacteria outer membrane and also release of lipopolysaccharides, (5) by inhibiting the synthesis of enzyme, and (6) by coagulation of the cell components [155].

21.5.2 Clinical Studies on Herbal Extract for Bacterial Infection

To date, a few selected clinical trials with plant antimicrobials have been done. Giron et al. (1988) evaluated *S. nigrescens* against intravaginal suppositories, in *C. albicans* vaginitis confirmed women [156]. From the ancient times, people use pomegranates (*Punica granatum* L.) as an antibacterial agent with extracts obtained from its fruit and bark that are effective against diarrhea and dysentery [157]. Over the years, many scientists have focused on studying the bactericidal activity of pomegranates against various bacterial strains. Fruit methanolic extract components contain higher amount of hydrolysable tannins, gallic acid and ellagic acid, a component of ellagitannins exhibiting the broad range antibacterial activity [158]. However, other compounds also contain biological activity and may contribute synergistically as mixtures to bring about the effects, including flavonols (quercetin and myricetin) and anthocyanins (pelargonidin-3-galactose and cyanidin-3-glucose) [159]. Using these compounds, several clinical trials have done and showed the effectiveness of extracts of pomegranate in reductions of oral plaque [160], gingivitis [161], periodontitis [162], and stomach ulcers caused by devastating pathogen *Helicobacter pylori* [163].

The antibacterial effect of cranberry juice on urinary tract infection was undertaken on elderly women in 1996, and no possible authentic outcome has been reported; however, very less number of bacteria have been survived in the treated women compared to the untreated ones [164]. At least four ayurvedic formulations have been tested against a placebo for their performance against acne vulgaris, out of which one formulation, namely, Sunder Vati displayed a drastic reduction in lesion count compared to others and the placebo without resembling any toxic side effect [165].

The different combinations of garlic (*Allium sativum*) extracts were also tested for the effectiveness of herbal products for the complete destruction of *H. pylori*. The patients infected with *H. pylori* were treated with garlic oil capsules (275 mg, for three times in a day) and compared with the garlic and omeprazole combined treatment (20 mg, twice a day). After treatment for 1 month, *H. pylori* presence in biopsy specimens was confirmed by the urease test and microscopy. From these results, it was concluded that lower level of the allicin, an antibacterial component in the garlic capsules, might be beneficial for the lack of effect [166].

Curcumin is another important highly effective molecule that was clinically proven to display antibacterial activity against a range of human diseases. In pre-clinical studies, the effectiveness of a 7-day non-antibiotic therapy (including lactoferrin, N-acetylcysteine, curcumin, and pantoprazole) for the abolition of *H. pylori* infection and decrease of gastric inflammation was investigated in 25 patients. It was found that these combinations were not effective for the *H. pylori* eradication; however, after 2 months of treatment, significant improvement in dyspeptic symptoms and a reduction in serologic marks of gastric inflammation were observed. This study recommends that curcumin might have a limited anti-bactericidal effect on *H. pylori* [167].

Together, these findings indicate sufficient evidence that plant extracts or purified phytochemicals have the prospect to be developed into agents that might be used as preventative or treatment therapies for bacterial diseases. However, establishing the efficiency of these agents will be important to ensure whether they serve therapeutic benefits, either alone or in combination with conventional therapies. Such efforts can help reduce the overall load of bacterial diseases worldwide. However, the safety and possible side effects of such products must also be considered.

21.6 Role of Phytochemicals in Viral Diseases

Infectious diseases are major threats to public health throughout the world. Several medicinal plants have widely been used against various infectious diseases, and about 25% of current medicines of infectious diseases are isolated from medicinal plants [168]. Further, viral diseases are important causes of human deaths worldwide due to high mortality and morbidity rate. Phytochemicals could be exploited as antiviral agents as viruses are becoming drug resistance to several available synthetic drugs due to their high specificity. In this section, we shall discuss the clinical studies conducted on patients with viral infection.

21.6.1 Clinical Studies on Curcumin in Virally Infected Patients

The antiviral activity of curcumin was examined on 40 AIDS patients in a clinical trial from New England [169]. In this study, viral load was tested along with the baseline testing in the fourth and eighth week following administration of the high and low dose of curcumin. The results of this investigation suggested no evidence of curcumin-associated viral load reduction or increased CD4 counts [169]. However, patients showed their interest in taking the curcumin as they felt better. Therefore, this clinical trial indicated that curcumin does not have any anti-HIV properties.

In a phase II study, Basu et al. (2013) inspected the efficiency of Basant polyherbal vaginal cream (prepared from extracts of curcumin, reetha, amla, and aloe vera) and curcumin vaginal capsules in clearing human papillomavirus (HPV) from the cervix of HPV-positive women [170]. This study interestingly showed that Basant arm significantly cleared HPV with a higher rate (87.7%) than the combined placebo arms (73.3%). Moreover, curcumin also showed a high rate of HPV clearance than placebo but with a statistically insignificant difference [170].

Overall, both these studies indicated that curcumin may have a therapeutic value against patients with the viral infection but not in all cases.

21.6.2 Clinical Studies on Epigallocatechin Gallate in Patients Infected with Viruses

The chemopreventive efficacy of green tea extracts containing polyphenon E and EGCG was evaluated in patients with HPV-infected cervical lesions [171]. In this study, the investigators delivered green tea extracts to the patients either in ointment or capsule form. The study demonstrated that both forms of green tea extracts (GTE) (ointment and capsule) showed effectiveness against cervical lesions treatment [171]. The authors concluded that GTE can be exploited as an effective therapeutic agent for the treatment of patients with cervical lesions. Halegoua-De Marzio et al. (2012) examined the safety, admissibility, and disposition of a single oral dose of epigallocatechin-3-gallate (EGCG) in cirrhotic and noncirrhotic hepatitis C patients [172]. They suggested that 400 mg oral dose of EGCG was well tolerable and safe in case of all studied patients and recommended the investigation related to long-term safety and anticancer efficacy of EGCG in cirrhotic patients [172]. In a phase II clinical trial, Zhao et al. (2015) reported that topical application of lipophilic epigallocatechin-3-gallate (AverTeaX formula) effectively inhibited the infection of herpes simplex labialis, and thus they showed the clinical efficacy of epigallocatechin-3-gallate against herpes simplex labialis infection [173].

21.6.3 A Clinical Study on Antiviral Immune Responses of Sulforaphane

A recent study by Muller at al. (2016) indicated that sulforaphane-containing broccoli sprout homogenate (BSH) ingestion to healthy volunteers inoculated with attenuated influenza virus (LAIV) further augmented their respiratory antiviral immune responses [174]. They observed a further increased production of granzyme B in NK cells isolated from BSH-ingested LAIV-inoculated volunteers. Thus, this study suggests that sulforaphane can be used as an immunostimulating agent against influenza virus infection.

A summary of clinical trials of phytochemicals on virally infected patients is shown in Table 21.4.

Table 21.4 Phytochemicals and their clinical trials in virally infected patients

Trial	Type of participants	Aim	References
Clinical trials of curcumin			
Clinical trial on antiviral effect of curcumin	AIDS patients	Evaluation of viral load and CD4 counts	[169]
A phase II randomized controlled study on curcumin	HPV-positive women	Antiviral efficacy of curcumin	[170]
Clinical trials of epigallocatechin gallate			
Effectiveness of green tea extracts on patients with cervical lesions	Patients with human papillomavirus (HPV) infection and cervical lesions	Efficacy of green tea extracts	[171]
Effect of cirrhosis on EGCG disposition	Cirrhotic and noncirrhotic patients with hepatitis C virus (HCV)	Safety, tolerability, and disposition of a single oral dose of EGCG in cirrhotic patients with hepatitis C virus (HCV)	[172]
A phase II clinical trial of topical application of lipophilic epigallocatechin-3-gallate (AverTeaX)	Patients infected with recurrent herpes labialis (RHL)	Clinical efficacy of topical proprietary formulation containing lipophilic catechins (AverTeaX) on recurrent herpes labialis	[173]
Clinical trials of sulforaphane			
A randomized, double-blind study sulforaphane-containing broccoli sprout homogenates (BSH)	Healthy volunteers	Evaluation of peripheral blood immune cell populations	[174]

21.7 Conclusion

Several in vitro and in vivo reports indicate the medicinal properties of phytochemicals against myriads of diseases. However, till date, only a few clinical studies have been carried out only on a few selected phytochemicals. The clinical investigations help in the determination of bioavailability, toxicity, and therapeutic value of phytochemicals alone or in combinational therapy, which decides their future therapeutic application. These investigations also provide a new platform for conducting the clinical studies on other reported potential phytochemicals. A number of clinical studies discussed in this chapter support the clinical use of phytochemicals against different health-related problems due to their bioavailability and least toxicity. Therefore, it may be safely concluded that phytochemicals could be a better future therapeutic option over the currently available therapeutic strategies for the treatment of a variety of diseases.

Acknowledgment AK acknowledges the University Grants Commission and Department of Science and Technology, New Delhi, India, for providing financial support in the form of UGC-Start-Up Research and Early Career Research Award, respectively. AK also acknowledges the University Grants Commission, Government of India, for providing Center of Advanced Studies to

Department of Zoology, Banaras Hindu University, India. SK acknowledges the University Grants Commission, India, for providing financial support in the form of UGC-Start-Up Research Grant [No.F.30-372/2017(BSR); FD Diary No.6755] and the Central University of Punjab for providing necessary infrastructure facilities. DK acknowledges the Central University of Jammu for providing necessary infrastructure facilities along with financial support in the form of Minor Research Grant (No: CUJ/Acad/Proj-PLS/2017/158) and also to SERB, DST for providing financial support. AB acknowledges the Central University of Jammu for providing necessary infrastructure facilities and the financial support in the form of Minor Research Grant (No: CUJ/Acad/Proj-CMB/2017/159).

References

1. Ameh SJ, Obodozie OO, Inyang US, Abubakar MS, Garba M (2010) Current phytotherapy – a perspective on the science and regulation of herbal medicine. J Med Plants Res 4:072–081
2. WHO (2008) Traditional medicine. WHO fact sheet no. 134. Revised. December 2008
3. BBC News (2006) Can herbal medicine combat aids? Wednesday, 15 March, 13:10 GMT. http://newsvote.bbc.co.uk/mpapps/pagetools/print/news.bbc.co.uk/2/hi/Africa/4793106.stm. Accessed 16 Apr 2011
4. van Breemen RB (2015) Development of safe and effective botanical dietary supplements. J Med Chem 58:8360–8372
5. Pferschy-Wenzig EM, Bauer R (2015) The relevance of pharmacognosy in pharmacological research on herbal medicinal products. Epilepsy Behav 52:344–362
6. Betz JM, Brown PN, Roman MC (2011) Accuracy, precision, and reliability of chemical measurements in natural products research. Fitoterapia 82:44–52
7. Pauli GF, Chen SN, Simmler C, Lankin DC, Gödecke T, Jaki BU, Friesen JB, McAlpine JB, Napolitano JG (2014) Importance of purity evaluation and the potential of quantitative ^{1}H NMR as a purity assay. J Med Chem 57:9220–9231
8. Piersen CE, Booth NL, Sun Y, Liang W, Burdette JE, van Breemen RB, Geller SE, Gu C, Banuvar S, Shulman LP, Bolton JL, Farnsworth NR (2004) Chemical and biological characterization and clinical evaluation of botanical dietary supplements: a phase I red clover extract as a model. Curr Med Chem 11:1361–1374
9. Bisson J, McAlpine JB, Friesen JB, Chen SN, Graham J, Pauli GF (2016) Can invalid bioactives undermine natural product-based drug discovery. J Med Chem 59:1671–1690
10. Wagner H (2011) Synergy research: approaching a new generation of phytopharmaceuticals. Fitoterapia 82:34–37
11. Pelkonen O, Xu Q, Fan TP (2014) Why is research on herbal medicinal products important and how can we improve its quality? J Tradit Complement Med 4:1–7
12. Chou TC, Talalay P (1984) Quantitative analysis of dose-effect relationships: the combined effects of multiple drugs or enzyme inhibitors. Adv Enzyme Regul 22:27–55
13. Drews J (2000) Drug discovery: a historical perspective. Science 287:1960–1964
14. Ojima I (2008) Modern molecular approaches to drug design and discovery. Acc Chem Res 41:2–3
15. Wadood A, Ahmed N, Shah L, Ahmad A, Hassan H, Shams S (2013) In-silico drug design: an approach which revolutionarised the drug discovery process. OA Drug Des Deliv 1:4
16. Le Tourneau C, Lee JJ, Siu LL (2009) Dose escalation methods in phase I cancer clinical trials. J Natl Cancer Inst 101:708–720
17. Elsäßer A, Regnstrom J, Vetter T, Koenig F, Hemmings RJ, Greco M, Papaluca-Amati M, Posch M (2014) Adaptive clinical trial designs for European marketing authorization: a survey of scientific advice letters from the European Medicines Agency. Trials 15:383
18. Bernabe RD, Van Thiel GJ, Raaijmakers JA, Van Delden JJ (2014) The fiduciary obligation of the physician-researcher in phase IV trials. BMC Med Ethics 15:1

19. AbouZahr C (2014) Progress and challenges in women's health: an analysis of levels and patterns of mortality and morbidity. Contraception 90:S3–S13
20. Clark IA, Vissel B (2013) Treatment implications of the altered cytokine-insulin axis in neurodegenerative disease. Biochem Pharmacol 86:862–871
21. Ren SY, Xu X (2015) Role of autophagy in metabolic syndrome-associated heart disease. Biochim Biophys Acta 1852:225–231
22. Taylor M (2015) Complementary and alternative approaches to menopause. Endocrinol Metab Clin North Am 44:619–648
23. Imai A, Ichigo S, Matsunami K, Takagi H (2015) Premenstrual syndrome: management and pathophysiology. Clin Exp Obstet Gynecol 42:123–128
24. Nickel JC (2007) Urinary tract infections and resistant bacteria: highlights of a symposium at the Combined Meeting of the 25th International Congress of Chemotherapy (ICC) and the 17th European Congress of Clinical Microbiology and Infectious Diseases (ECCMID), Munich, Germany. Rev Urol 9:78–80
25. Gabay MP (2002) Galactogogues: medications that induce lactation. J Hum Lact 18:274–279
26. Marcellin L, Chantry AA (2015) Breastfeeding (part III): breastfeeding complications–guidelines for clinical practice. J Gynecol Obstet Biol Reprod (Paris) 44:1084–1090
27. Nelson HD (2008) Menopause. Lancet 371:760–770
28. Chlebowski RT, Manson JE, Anderson GL, Cauley JA, Aragaki AK, Stefanick ML, Lane DS, Johnson KC, Wactawski-Wende J, Chen C, Qi L, Yasmeen S, Newcomb PA, Prentice RL (2013) Estrogen plus progestin and breast cancer incidence and mortality in the Women's Health Initiative Observational Study. J Natl Cancer Inst 105:526–535
29. Manson J, Branch H (2014) The women's health initiative: the latest findings from long-term follow-up. Womens Health (Lond Engl) 10:125–128
30. Meier B, Berger D, Hoberg E, Sticher O, Schaffner W (2000) Pharmacological activities of Vitex agnus-castus extracts in vitro. Phytomedicine 7:373–381
31. Tamagno G (2009) Are changes of prolactin levels the effectors of vitex agnus castus beneficial effects on the pre-menstrual syndrome? Maturitas 63:369
32. Chen XP, Li W, Xiao XF, Zhang LL, Liu CX (2013) Phytochemical and pharmacological studies on Radix Angelica sinensis. Chin J Nat Med 11:577–587
33. Jarboe CH, Schmidt CM, Nicholson JA, Zirvi KA (1966) Uterine relaxant properties of Viburnum. Nature 212:837
34. Nicholson JA, Darby TD, Jarboe CH (1972) Viopudial, a hypotensive and smooth muscle antispasmodic from Viburnum opulus. Proc Soc Exp Biol Med 140:457–461
35. Cometa MF, Parisi L, Palmery M, Meneguz A, Tomassini L (2009) In vitro relaxant and spasmolytic effects of constituents from Viburnum prunifolium and HPLC quantification of the bioactive isolated iridoids. J Ethnopharmacol 123:201–207
36. van Andel T, de Boer HJ, Barnes J, Vandebroek I (2014) Medicinal plants used for menstrual disorders in Latin America, the Caribbean, sub-Saharan Africa, South and Southeast Asia and their uterine properties: a review. J Ethnopharmacol 155:992–1000
37. Ahmad B, Rehman MU, Amin I, Arif A, Rasool S, Bhat SA, Afzal I, Hussain I, Bilal S, Mir M (2015) A review on pharmacological properties of zingerone (4-(4- hydroxy-3-methoxyphenyl)-2-butanone). Sci World J 2015:816364
38. Occhiuto F, Pino A, Palumbo DR, Samperi S, De Pasquale R, Sturlese E, Circosta C (2009) Relaxing effects of Valeriana officinalis extracts on isolated human nonpregnant uterine muscle. J Pharm Pharmacol 61:251–256
39. Mirabi P, Dolatian M, Mojab F, Majd HA (2011) Effects of valerian on the severity and systemic manifestations of dysmenorrhea. Int J Gynaecol Obstet 115:285–288
40. Kashani L, Saedi N, Akhondzadeh S (2010) Femicomfort in the treatment of premenstrual syndromes: a double-blind, randomized and placebo controlled trial. Iran J Psychiatry 5:47–50

41. Foo LY, Lu Y, Howell AB, Vorsa N (2000) The structure of cranberry proanthocyanidins which inhibit adherence of uropathogenic P-fimbriated Escherichia coli in vitro. Phytochemistry 54:173–181
42. Guay DR (2009) Cranberry and urinary tract infections. Drugs 69:775–807
43. Vasileiou I, Katsargyris A, Theocharis S, Giaginis C (2013) Current clinical status on the preventive effects of cranberry consumption against urinary tract infections. Nutr Res 33:595–607
44. de Arriba SG, Naser B, Nolte KU (2013) Risk assessment of free hydroquinone derived from Arctostaphylos Uva-ursi folium herbal preparations. Int J Toxicol 32:442–453
45. Micklefield GH, Redeker Y, Meister V, Jung O, Greving I, May B (1999) Effects of ginger on gastroduodenal motility. Int J Clin Pharmacol Ther 37:341–346
46. Wu KL, Rayner CK, Chuah SK, Changchien CS, Lu SN, Chiu YC, Chiu KW, Lee CM (2008) Effects of ginger on gastric emptying and motility in healthy humans. Eur J Gastroenterol Hepatol 20:436–440
47. Greenlee H, Abascal K, Yarnell E, Ladas E (2007) Clinical applications of Silybum marianum in oncology. Integr Cancer Ther 6:158–165
48. Forinash AB, Yancey AM, Barnes KN, Myles TD (2012) The use of galactogogues in the breastfeeding mother. Ann Pharmacother 46:1392–1404
49. Borrelli F, Ernst E (2008) Black cohosh (Cimicifuga racemosa) for menopausal symptoms: a systematic review of its efficacy. Pharmacol Res 58:8–14
50. Powell SL, Gödecke T, Nikolic D, Chen SN, Ahn S, Dietz B, Farnsworth NR, van Breemen RB, Lankin DC, Pauli GF, Bolton JL (2008) In vitro serotonergic activity of black cohosh and identification of N(omega)-methylserotonin as a potential active constituent. J Agric Food Chem 56:11718–11726
51. Wuttke W, Jarry H, Haunschild J, Stecher G, Schuh M, Seidlova-Wuttke D (2014) The non-estrogenic alternative for the treatment of climacteric complaints: black cohosh (Cimicifuga or Actaea racemosa). J Steroid Biochem Mol Biol 139:302–310
52. Dietz BM, Mahady GB, Pauli GF, Farnsworth NR (2005) Valerian extract and valerenic acid are partial agonists of the 5-HT5a receptor in vitro. Brain Res Mol Brain Res 138:191–197
53. Lethaby A, Marjoribanks J, Kronenberg F, Roberts H, Eden J, Brown J (2013) Phytoestrogens for menopausal vasomotor symptoms. Cochrane Database Syst Rev 12:CD001395
54. Aghamiri V, Mirghafourvand M, Mohammad-Alizadeh-Charandabi S, Nazemiyeh H (2016) The effect of Hop (Humulus lupulus L.) on early menopausal symptoms and hot flashes: a randomized placebo-controlled trial. Complement Ther Clin Pract 23:130–135
55. Mersereau JE, Levy N, Staub RE, Baggett S, Zogovic T, Chow S, Ricke WA, Tagliaferri M, Cohen I, Bjeldanes LF, Leitman DC (2008) Liquiritigenin is a plant-derived highly selective estrogen receptor beta agonist. Mol Cell Endocrinol 283:49–57
56. World Health Organization (2016) Global report on diabetes. World Health Organization, Geneva
57. Cefalu WT, Stephens JM, Ribnicky DM (2011) Diabetes and herbal (botanical) medicine. In: Benzie IFF, Wachtel-Galor S, Herbal medicine: biomolecular and clinical aspects, 2nd edn. CRC Press/Taylor & Francis, Boca Raton; Chapter 19
58. Roglic G (2016) WHO Global report on diabetes: a summary. Int J Non-Commun Dis 1:3–8
59. Anderson RA, Zhan Z, Luo R, Guo X, Guo Q, Zhou J, Kong J, Davis PA, Stoecker BJ (2015) Cinnamon extract lowers glucose, insulin and cholesterol in people with elevated serum glucose. J Tradit Complement Med 6:332–336
60. Bailey CJ, Day C (1989) Traditional plant medicines as treatments for diabetes. Diabetes Care 12:553–564
61. Shane-McWhorter L (2009) Dietary supplements for diabetes: an evaluation of commonly used products. Diabetes Spectr 22:206–213
62. Oubré AY, Carlson TJ, King SR, Reaven GM (1997) From plant to patient: an ethnomedical approach to the identification of new drugs for the treatment of NIDDM. Diabetologia 40:614–617

63. Cao H, Polansky MM, Anderson RA (2007) Cinnamon extract and polyphenols affect the expression of tristetraprolin, insulin receptor, and glucose transporter 4 in mouse 3T3-L1 adipocytes. Arch Biochem Biophys 459:214–222
64. Sheng X, Zhang Y, Gong Z, Huang C, Zang YQ (2008) Improved insulin resistance and lipid metabolism by cinnamon extract through activation of peroxisome proliferator-activated receptors. PPAR Res 2008:581348
65. Adisakwattana S, Lerdsuwankij O, Poputtachai U, Minipun A, Suparpprom C (2011) Inhibitory activity of cinnamon bark species and their combination effect with acarbose against intestinal α-glucosidase and pancreatic α-amylase. Plant Foods Hum Nutr 66:143–148
66. Madar Z (1984) Fenugreek (*trigonella foenumgraceum*) as a means of reducing postprandial glucose levels in diabetic rats. Nutr Rep Int 29:1267–1273
67. Raghuram TC, Sharma RD, Sivakumar, Sahay BK (1994) Effect of fenugreek seeds on intravenous glucose disposition in non-insulin dependent diabetic patients. Phytother Res 8:83–86
68. Shanmugasundaram ER, Panneerselvam C, Samudram P, Shanmugasundaram ERB (1983) Enzyme changes and glucose utilisation in diabetic rabbits: the effect of gymnema sylvestre. J Ethnopharmacol 7:205–234
69. Persaud SJ, Al-Majed H, Raman A, Jones PM (1999) Gymnema sylvestre stimulates insulin release in vitro by increased membrane permeability. J Endocrinol 163:207–212
70. Jellin JM, Batz F, Hitchens K (1999) Pharmacist's letter/prescribers letter natural medicines comprehensive database. Therapeutic Research Faculty, Stockton, Available from www.NaturalDatabase.com
71. Joseph B, Jini D (2013) Antidiabetic effects of *Momordica charantia* (bitter melon) and its medicinal potency. Asian Pac J Trop Dis 3:93–102
72. Yuan HD, Kim JT, Kim SH, Chung SH (2012) Ginseng and diabetes: the evidences from *in vitro*, animal and human studies. J Ginseng Res 36:27–39
73. Chandalia M, Garg A (2000) Beneficial effects of high dietary fiber intake in patients with type 2 diabetes mellitus. NEJM 342:1392–1398
74. Alarcon-Aguilar FJ, Valdes-Arzate A (2003) Hypoglycemic activity of two polysaccharides isolated from *Opuntia ficus-indica* and *O. streptacantha*. Proc West Pharmacol Soc 46:139–142
75. Stevinson C, Pittler MH, Ernst E (2000) Garlic for treating hypercholesterolemia: a meta-analysis of randomized clinical trials. Ann Intern Med 133:420–429
76. Reinhart KM, Talati R, White CM, Coleman CI (2009) The impact of garlic on lipid parameters: a systematic review and meta-analysis. Nutr Res Rev 22:39–48
77. Ghosh D, Konishi T (2007) Anthocyanins and anthocyanin-rich extracts: role in diabetes and eye function. Asia Pac J Clin Nutr 16:200–208
78. Guigas B, Naboulsi R, Villanueva GR, Taleux N, Lopez-Novoa JM, Leverve XM, El-Mir LY (2007) The flavonoid silybinin decreases glucose-6-phosphate hydrolysis in perifused rat hepatocytes by an inhibitory effect on glucose-6-phosphatase. Cell Physiol Biochem 20:925–934
79. Cheng B, Gong H, Li X, Sun Y, Zhang X, Chen H, Liu X, Zheng L, Huang Kl (2012) silybinin inhibits the toxic aggregation of human islet amyloid polypeptide. Biochem Biophys Res Commun 419:495–499
80. Khan A, Safdar M, Ali Khan MM, Khattak KN, Anderson RA (2003) Cinnamon improves glucose and lipids of people with type 2 diabetes. Diabetes Care 26:3215–3218
81. Mang B, Wolters M, Schmitt B, Kelb K, Lichtinghagen R, Stichtenoth DO, Hahn A (2006) Effects of a cinnamon extract on plasma glucose, HbA, and serum lipids in diabetes mellitus type 2. Eur J Clin Invest 36:340–344
82. Crawford P (2009) Effectiveness of cinnamon for lowering hemoglobin A1C in patients with type 2 diabetes: a randomized, controlled trial. J Am Board Fam Med 22:507–512
83. Akilen R, Tsiami A, Devendra D, Robinson N (2010) Glycated haemoglobin and blood pressure-lowering effect of cinnamon in multi-ethnic Type 2 diabetic patients in the UK: a randomized, placebo-controlled, double-blind clinical trial. Diabet Med 27:1159–1167

84. Suppapitiporn S, Kanpaksi N, Suppapitiporn S (2006) The effect of cinnamon cassia powder in type 2 diabetes mellitus. J Med Assoc Thai 89(Suppl 3):S200–S205
85. Sahib AS (2016) Anti-diabetic and antioxidant effect of cinnamon in poorly controlled type-2 diabetic Iraqi patients: a randomized, placebo-controlled clinical trial. J Intercultural Ethnopharmacol 5:108–113
86. Neelakantan N, Narayanan M, de Souza RJ, van Dam RM (2014) Effect of fenugreek (Trigonella foenum-graecum L.) intake on glycemia: a meta-analysis of clinical trials. Nutr J 13:7
87. Sharma RD, Raghuram TC, Rao NS (1990) Effect of fenugreek seeds on blood glucose and serum lipids in type I diabetes. Eur J Clin Nutr 44:301–306
88. Lakholia AN (1956) The use of bitter gourd in *diabetes mellitus*. Antiseptic 53:608–610
89. Fuangchana A, Sonthisombata P, Seubnukarnb T, Chanouanc R, Chotchaisuwatd P, Sirigulsatiene V (2011) Hypoglycemic effect of bitter melon compared with metformin in newly diagnosed type 2 diabetes patients. J Ethnopharmacol 134:422–428
90. Madkor HR, Mansour SW, Ramadan G (2011) Modulatory effects of garlic, ginger, turmeric and their mixture on hyperglycaemia, dyslipidaemia and oxidative stress in streptozotocin-nicotinamide diabetic rats. Br J Nutr 105:1210–1217
91. Banerjee SK, Maulik SK (2002) Effect of garlic on cardiovascular disorders: a review. Nutr J 19:4
92. Bunyapraphatsara N, Yongchaiyudha S, Rungpitarangsi V, Chokechaijaroenporn O (1996) Antidiabetic activity of *aloe vera* L. juice. II. Clinical trial in diabetes mellitus patients in combination with glibenclamide. Phytomedicine 3:245–248
93. Yongchaiyudha S, Rungpitarangsi V, Bunyapraphatsara N, Chokechaijaroenporn O (1996) Antidiabetic activity of *aloe vera* L. juice. I. Clinical trial in new cases of diabetes mellitus. Phytomedicine 3:241–243
94. Frati-Munari AC, Gordillo BE, Altamirano P, Ariza CR (1988) Hypoglycemic effect of opuntia streptacantha lemaire in NIDDM. Diabetes Care 11:63–66
95. Frati AC, Gordillo BE, Altamirano P, Ariza CR, Cortes-Franco R, Chavez-Negrete A (1990) Acute hypoglycemic effect of opuntia streptacantha lemaire in NIDDM (Letter). Diabetes Care 13:455–456
96. Velussi M, Cernigoi AM, Viezzoli L, Dapas F, Caffau C, Zilli M (1993) Silymarin reduces hyperinsulinemia, malondialdehyde levels, and daily insulin need in cirrhotic diabetic patients. Curr Therap Res 53:533–545
97. Fallahzadeh MK, Dormanesh B, Sagheb MM, Roozbeh J, Vessal G, Pakfetrat M, Daneshbod Y, Kamali-Sarvestani E, Lankarani KB (2012) Effect of addition of silymarin to renin-angiotensin system inhibitors on proteinuria in type 2 diabetic patients with overt nephropathy: a randomized, double-blind, placebo-controlled trial. Am J Kidney Dis 60:896–903
98. Huseini HF, Larijani B, Heshmat R, Fakhrzadeh H, Radjabipour B, Toliat T, Raza M (2006) The efficacy of *Silybum marianum* (L.) Gaertn. (silymarin) in the treatment of type II diabetes: a randomized, double-blind, placebo-controlled, clinical trial. Phytother Res 20:1036–1039
99. Di Pierro F, Putignano P, Montesi L, Moscatiello S, MarchesiniReggiani G, Villanova N (2013) Preliminary study about the possible glycemic clinical advantage in using a fixed combination of *Berberis aristata* and *Silybum marianum* standardized extracts versus only *Berberis aristata* in patients with type 2 diabetes. Clin Pharmacol 5:167–174
100. Büchner FL, Bueno-de-Mesquita HB, Linseisen J, Boshuizen HC, Kiemeney LA, Ros MM, Overvad K, Hansen L, Tjonneland A, Raaschou-Nielsen O, Clavel-Chapelon F, Boutron-Ruault MC, Touillaud M, Kaaks R, Rohrmann S, Boeing H, Nöthlings U, Trichopoulou A, Zylis D, Dilis V, Palli D, Sieri S, Vineis P, Tumino R, Panico S, Peeters PH, van Gils CH, Lund E, Gram IT, Braaten T, Martinez C, Agudo A, Arriola L, Ardanaz E, Navarro C, Rodríguez L, Manjer J, Wirfält E, Hallmans G, Rasmuson T, Key TJ, Roddam AW, Bingham S, Khaw KT, Slimani N, Bofetta P, Byrnes G, Norat T, Michaud D, Riboli E (2010) Fruits and vegetables consumption and the risk of histological subtypes of lung cancer in the European Prospective Investigation into Cancer and Nutrition (EPIC). Cancer Causes Control 21:357–371

101. Michels KB, Giovannucci E, Joshipura KJ, Rosner BA, Stampfer MJ, Fuchs CS, Colditz GA, Speizer FE, Willett WC (2000) Prospective study of fruit and vegetable consumption and incidence of colon and rectal cancers. J Natl Cancer Inst 92:1740–1752
102. Larsson SC, Holmberg L, Wolk A (2004) Fruit and vegetable consumption in relation to ovarian cancer incidence: the Swedish Mammography Cohort. Br J Cancer 90:2167–2170
103. Gutheil WG, Reed G, Ray A, Dhar A (2012) Crocetin: an agent derived from saffron for prevention and therapy for cancer. Curr Pharm Biotechnol 13:173–179
104. Aggarwal BB, Sung B (2009) Pharmacological basis for the role of curcumin in chronic diseases: an age-old spice with modern targets. Trends Pharmacol Sci 30:85–94
105. Aggarwal BB, Harikumar KB (2009) Potential therapeutic effects of curcumin, the anti-inflammatory agent, against neurodegenerative, cardiovascular, pulmonary, metabolic, autoimmune and neoplastic diseases. Int J Biochem Cell Biol 41:40–59
106. Shankar T, Shantha NV, Ramesh HP, Murthy IA, Murthy VS (1980) Toxicity studies on turmeric (Curcuma longa): acute toxicity studies in rats, guinea pigs and monkeys. Indian J Exp Biol 18:73–75
107. Soni K, Kuttan R (1992) Effect of oral curcumin administration on serum peroxides and cholesterol levels in human volunteers. Indian J Physiol Pharmacol 36:273–275
108. Kuttan R, Sudheeran PC, Josph CD (1987) Turmeric and curcumin as topical agents in cancer therapy. Tumori 73:29–31
109. Cheng AL, Hsu CH, Lin JK, Hsu MM, Ho YF, Shen TS, Ko JY, Lin JT, Lin BR, Ming-Shiang W, Yu HS, Jee SH, Chen GS, Chen TM, Chen CA, Lai MK, Pu YS, Pan MH, Wang YJ, Tsai CC, Hsieh CY (2001) Phase I clinical trial of curcumin, a chemopreventive agent, in patients with high-risk or premalignant lesions. Anticancer Res 21:2895–2900
110. Sharma RA, McLelland HR, Hill KA, Ireson CR, Euden SA, Manson MM, Pirmohamed M, Marnett LJ, Gescher AJ, Steward WP (2001) Pharmacodynamic and pharmacokinetic study of oral curcuma extract in patients with colorectal cancer. Clin Cancer Res 7:1894–1900
111. Sharma RA, Euden SA, Platton SL, Cooke DN, Shafayat A, Hewitt HR, Marczylo TH, Morgan B, Hemingway D, Plummer SM, Pirmohamed M, Gescher AJ, Steward WP (2004) Phase I clinical trial of oral curcumin: biomarkers of systemic activity and compliance. Clin Cancer Res 10:6847–6854
112. Garcea G, Berry DP, Jones DJ, Singh R, Dennison AR, Farmer PB, Sharma RA, Steward WP, Gescher AJ (2005) Consumption of the putative chemopreventive agent curcumin by cancer patients: assessment of curcumin levels in the colorectum and their pharmacodynamic consequences. Cancer Epidemiol Biomarkers Prev 14:120–125
113. He ZY, Shi CB, Wen H, Li FL, Wang BL, Wang J (2011) Upregulation of p53 expression in patients with colorectal cancer by administration of curcumin. Cancer Investig 29:208–213
114. Cruz-Correa M, Shoskes DA, Sanchez P, Zhao R, Hylind LM, Wexner SD, Giardiello FM (2006) Combination treatment with curcumin and quercetin of adenomas in familial adenomatous polyposis. Clin Gastroenterol Hepatol 4:1035–1038
115. Irving GR, Howells LM, Sale S, Kralj-Hans I, Atkin WS, Clark SK, Britton RG, Jones DJ, Scott EN, Berry DP, Hemingway D, Miller AS, Brown K, Gescher AJ, Steward WP (2013) Prolonged biologically active colonic tissue levels of curcumin achieved after oral administration--a clinical pilot study including assessment of patient acceptability. Cancer Prev Res (Phila) 6:119–128
116. Irving GR, Iwuji CO, Morgan B, Berry DP, Steward WP, Thomas A, Brown K, Howells LM (2015) Combining curcumin (C3-complex, Sabinsa) with standard care FOLFOX chemotherapy in patients with inoperable colorectal cancer (CUFOX): study protocol for a randomised control trial. Trials 16:110
117. James MI, Iwuji C, Irving G, Karmokar A, Higgins JA, Griffin-Teal N, Thomas A, Greaves P, Cai H, Patel SR, Morgan B, Dennison A, Metcalfe M, Garcea G, Lloyd DM, Berry DP, Steward WP, Howells LM, Brown K (2015) Curcumin inhibits cancer stem cell phenotypes in ex vivo models of colorectal liver metastases, and is clinically safe and tolerable in combination with FOLFOX chemotherapy. Cancer Lett 364:135–141

118. Dhillon N, Aggarwal BB, Newman RA, Wolff RA, Kunnumakkara AB, Abbruzzese JL, Ng CS, Badmaev V, Kurzrock R (2008) Phase II trial of curcumin in patients with advanced pancreatic cancer. Clin Cancer Res 14:4491–4499
119. Epelbaum R, Schaffer M, Vizel B, Badmaev V, Bar-Sela G (2010) Curcumin and gemcitabine in patients with advanced pancreatic cancer. Nutr Cancer 62:1137–1141
120. Kanai M, Yoshimura K, Asada M, Imaizumi A, Suzuki C, Matsumoto S, Nishimura T, Mori Y, Masui T, Kawaguchi Y, Yanagihara K, Yazumi S, Chiba T, Guha S, Aggarwal BB (2011) A phase I/II study of gemcitabine-based chemotherapy plus curcumin for patients with gemcitabine resistant pancreatic cancer. Cancer Chemother Pharmacol 68:157–164
121. Vadhan-Raj S, Weber D, Wang M, Giralt S, Alexanian R, Thomas SK, Alexanian R, Zhou X, Patel P, Bueso-Ramos CE, Newman RA, Aggarwal BB (2007) Curcumin downregulates NF-KB and related genes in patients with multiple myeloma: results of a phase I/II study. Blood 357a:110
122. Golombick T, Diamond TH, Badmaev V, Manoharan A, Ramakrishna R (2009) The potential role of curcumin in patients with monoclonal gammopathy of undefined significance-its effect on paraproteinemia and the urinary N-telopeptide of type I collagen bone turnover marker. Clin Cancer Res 15:5917–5922
123. Golombick T, Diamond TH, Manoharan A, Ramakrishna R (2012) Monoclonal gammopathy of undetermined significance, smoldering multiple myeloma, and curcumin: a randomized, double-blind placebo-controlled cross-over 4g study and an open-label 8g extension study. Am J Hematol 87:455–460
124. Bayet-Robert M, Kwiatkowski F, Leheurteur M, Gachon F, Planchat E, Abrial C, Mouret-Reynier MA, Durando X, Barthomeuf C, Chollet P (2010) Phase I dose escalation trial of docetaxel plus curcumin in patients with advanced and metastatic breast cancer. Cancer Biol Ther 9:8–14
125. Ryan JL, Heckler CE, Ling M, Katz A, Williams JP, Pentland AP, Morrow GR (2013) Curcumin for radiation dermatitis: a randomized, double-blind, placebo-controlled clinical trial of thirty breast cancer patients. Radiat Res 180:34–43
126. Ide H, Tokiwa S, Sakamaki K, Nishio K, Isotani S, Muto S, Hama T, Masuda H, Horie S (2010) Combined inhibitory effects of soy isoflavones and curcumin on the production of prostate-specific antigen. Prostate 70:1127–1133
127. Mahammedi H, Planchat E, Pouget M, Durando X, Curé H, Guy L, Van-Praagh I, Savareux L, Atger M, Bayet-Robert M, Gadea E, Abrial C, Thivat E, Chollet P, Eymard JC (2016) The new combination docetaxel, prednisone and curcumin in patients with castration-resistant prostate cancer: a pilot phase II study. Oncology 90:69–78
128. Kim SG, Veena MS, Basak SK, Han E, Tajima T, Gjertson DW, Starr J, Eidelman O, Pollard HB, Srivastava M, Srivatsan ES, Wang MB (2011) Curcumin treatment suppresses IKKβ kinase activity of salivary cells of patients with head and neck cancer: a pilot study. Clin Cancer Res 17:5953–5961
129. Ghalaut VS, Sangwan L, Dahiya K, Ghalaut PS, Dhankhar R, Saharan R (2012) Effect of imatinib therapy with and without turmeric powder on nitric oxide levels in chronic myeloid leukemia. J Oncol Pharm Pract 18:186–190
130. Panahi Y, Saadat A, Beiraghdar F, Sahebkar A (2014) Adjuvant therapy with bioavailability-boosted curcuminoids suppresses systemic inflammation and improves quality of life in patients with solid tumors: a randomized double-blind placebo-controlled trial. Phytother Res 28:1461–1467
131. Zhao H, Zhu W, Xie P, Li H, Zhang X, Sun X, Yu J, Xing L (2014) A phase I study of concurrent chemotherapy and thoracic radiotherapy with oral epigallocatechin-3-gallate protection in patients with locally advanced stage III non-small-cell lung cancer. Radiother Oncol 110:132–136
132. Zhao H, Xie P, Li X, Zhu W, Sun X, Sun X, Chen X, Xing L, Yu J (2015a) A prospective phase II trial of EGCG in treatment of acute radiation-induced esophagitis for stage III lung cancer. Radiother Oncol 114:351–356

133. Zhao H, Zhu W, Jia L, Sun X, Chen G, Zhao X, Li X, Meng X, Kong L, Xing L, Yu J (2016) Phase I study of topical epigallocatechin-3-gallate (EGCG) in patients with breast cancer receiving adjuvant radiotherapy. Br J Radiol 89:20150665
134. Alumkal JJ, Slottke R, Schwartzman J, Cherala G, Munar M, Graff JN, Beer TM, Ryan CW, Koop DR, Gibbs A, Gao L, Flamiatos JF, Tucker E, Kleinschmidt R, Mori M (2015) A phase II study of sulforaphane-rich broccoli sprout extracts in men with recurrent prostate cancer. Invest New Drugs 33:480–489
135. Cipolla BG, Mandron E, Lefort JM, Coadou Y, Della Negra E, Corbel L, Le Scodan R, Azzouzi AR, Mottet N (2015) Effect of sulforaphane in men with biochemical recurrence after radical prostatectomy. Cancer Prev Res (Phila) 8:712–719
136. Atwell LL, Zhang Z, Mori M, Farris PE, Vetto JT, Naik AM, Oh KY, Thuillier P, Ho E, Shannon J (2015) Sulforaphane bioavailability and chemopreventive activity in women scheduled for breast biopsy. Cancer Prev Res (Phila) 8:1184–1191
137. Markovits J, Linassier C, Fossé P, Couprie J, Pierre J, Jacquemin-Sablon A, Saucier JM, Le Pecq JB, Larsen AK (1989) Inhibitory effects of the tyrosine kinase inhibitor genistein on mammalian DNA topoisomerase II. Cancer Res 49:5111–5117
138. López-Lazaro M, Willmore E, Austin CA (2007) Cells lacking DNA topoisomerase II beta are resistant to genistein. J Nat Prod 70:763–767
139. Spagnuolo C, Russo GL, Orhan IE, Habtemariam S, Daglia M, Sureda A, Nabavi SF, Devi KP, Loizzo MR, Tundis R, Nabavi SM (2015) Genistein and cancer: current status, challenges, and future directions. Adv Nutr 6:408–419
140. deVere White RW, Hackman RM, Soares SE, Beckett LA, Li Y, Sun B (2004) Effects of a genistein-rich extract on PSA levels in men with a history of prostate cancer. Urology 63:259–263
141. Lazarevic B, Boezelijn G, Diep LM, Kvernrod K, Ogren O, Ramberg H, Moen A, Wessel N, Berg RE, Egge-Jacobsen W, Hammarstrom C, Svindland A, Kucuk O, Saatcioglu F, Taskèn KA, Karlsen SJ (2011) Efficacy and safety of short-term genistein intervention in patients with localized prostate cancer prior to radical prostatectomy: a randomized, placebo-controlled, double-blind phase 2 clinical trial. Nutr Cancer 63:889–898
142. Messing E, Gee JR, Saltzstein DR, Kim K, diSant'Agnese A, Kolesar J, Harris L, Faerber A, Havighurst T, Young JM, Efros M, Getzenberg RH, Wheeler MA, Tangrea J, Parnes H, House M, Busby JE, Hohl R, Bailey H (2012) A phase 2 cancer chemoprevention biomarker trial of isoflavone G-2535 (genistein) in presurgical bladder cancer patients. Cancer Prev Res (Phila) 5:621–630
143. Lazarevic B, Hammarström C, Yang J, Ramberg H, Diep LM, Karlsen SJ, Kucuk O, Saatcioglu F, Taskèn KA, Svindland A (2012) The effects of short-term genistein intervention on prostate biomarker expression in patients with localised prostate cancer before radical prostatectomy. Br J Nutr 108:2138–2147
144. Brown VA, Patel KR, Viskaduraki M, Crowell JA, Perloff M, Booth TD, Vasilinin G, Sen A, Schinas AM, Piccirilli G, Brown K, Steward WP, Gescher AJ, Brenner DE (2010) Repeat dose study of the cancer chemopreventive agent resveratrol in healthy volunteers: safety, pharmacokinetics, and effect on the insulin-like growth factor axis. Cancer Res 70:9003–9011
145. Singh CK, Ndiaye MA, Ahmad N (2015) Resveratrol and cancer: challenges for clinical translation. Biochim Biophys Acta 1852:1178–1185
146. Nguyen AV, Martinez M, Stamos MJ, Moyer MP, Planutis K, Hope C, Holcombe RF (2009) Results of a phase I pilot clinical trial examining the effect of plant-derived resveratrol and grape powder on Wnt pathway target gene expression in colonic mucosa and colon cancer. Cancer Manag Res 1:25–37
147. Patel KR, Brown VA, Jones DJ, Britton RG, Hemingway D, Miller AS, West KP, Booth TD, Perloff M, Crowell JA, Brenner DE, Steward WP, Gescher AJ, Brown K (2010) Clinical pharmacology of resveratrol and its metabolites in colorectal cancer patients. Cancer Res 70:7392–7399
148. Howells LM, Berry DP, Elliott PJ, Jacobson EW, Hoffmann E, Hegarty B, Brown K, Steward WP, Gescher AJ (2011) Phase I randomized, double-blind pilot study of micronized resvera-

trol (SRT501) in patients with hepatic metastases--safety, pharmacokinetics, and pharmacodynamics. Cancer Prev Res (Phila) 4:1419–1425
149. Zhu W, Qin W, Zhang K, Rottinghaus GE, Chen YC, Kliethermes B, Sauter ER (2012) Trans-resveratrol alters mammary promoter hypermethylation in women at increased risk for breast cancer. Nutr Cancer 64:393–400
150. Popat R, Plesner T, Davies F, Cook G, Cook M, Elliott P, Jacobson E, Gumbleton T, Oakervee H, Cavenagh J (2013) A phase 2 study of SRT501 (resveratrol) with bortezomib for patients with relapsed and or refractory multiple myeloma. Br J Haematol 160:714–717
151. Benayache S, Benayache F, Benyahia S (2001) Leaf oils of some eucalyptus species growing in Algeria. J Essent Oil Res 13:210–213
152. Benhassaini H, Enabderrahmane MB, Chikhi K (2003) Contribution à l'évaluation de l'activité antiseptique de l'oléorésine et des huiles essentielles du pistachier de l'Atlas sur certaines sources microbiennes: candida albicans (ATC 20027), candida albicans (ATCC 20032) et saccharomyces cerevisiae: ethnopharmacologie, fév. 30, pp 38–46
153. Nostro A (2006) Activity of plant extracts and plant-derived compounds against drug-resistant microorganisms. Wiley-VCH, Weinheim, pp 199–231
154. Warnke PH, Becker ST, Podschun R, Sivananthan S, Springer IN, Russo PA, Wiltfang J, Fickenscher H, Sherry E (2009) The battle against multi-resistant strains: renaissance of antimicrobial essential oils as a promising force to fight hospital-acquired infections. J CranioMaxillofacial Surg 37:392–397
155. Burt S (2004) Essential oils: their antibacterial properties and potential applications in foods-a review. Int J Food Microbiol 94:223–253
156. Giron LM, Aguilar GA, Caceres A, Arroyo GL (1988) Anticandidal activity of plants used for the treatment of vaginitis in Guatemala and clinical trial of a *Solanum nigrescens* preparation. J Ethnopharmacol 22:307–313
157. Jayaprakasha GK, Negi PS, Jena BS (2006) Antimicrobial activities of pomegranate. In: Seeram NP, Schulman RN, Heber D (eds) Pomegranates: ancient roots to modern medicine. CRC Press, NewYork, p 168
158. Reddy MK, Gupta SK, Jacob MR, Khan SI, Ferreira D (2007) Antioxidant, antimalarial and antimicrobial activities of tannin-rich fractions, ellagitannins and phenolic acids from Punica granatum L. Planta Med 73:461–467
159. Naz S, Siddiqi R, Ahmad S, Rasool SA, Sayeed SA (2007) Antibacterial activity directed isolation of compounds from Punica granatum. J Food Sci 72:341–345
160. Menezes SM, Cordeiro LN, Viana GS (2006) Punica granatum (pomegranate) extract is active against dental plaque. J Herb Pharmacother 6:79–92
161. Somu CA, Ravindra S, Ajith S, Ahamed MG (2012) Efficacy of a herbal extract gel in the treatment of gingivitis: a clinical study. J Ayurveda Integr Med 3:85–90
162. Sastravaha G, Gassmann G, Sangtherapitikul P, Grimm WD (2005) Adjunctive periodontal treatment with Centella asiatica and Punica granatum extracts in supportive periodontal therapy. J Int Acad Periodontol 7:70–79
163. Hajimahmoodi M, Shams-Ardakani M, Saniee P, Siavoshi F, Mehrabani M, Hosseinzadeh H, Foroumadi P, Safavi M, Khanavi M, Akbarzadeh T, Shafiee A, Foroumadi A (2011) In vitro antibacterial activity of some Iranian medicinal plant extracts against Helicobacter pylori. Nat Prod Res 25:1059–1066
164. Jepson RG, Mihaljevic L, Craig J (2002) Cranberries for treating urinary tract infections. (Cochrane Rewiew). In The Cochrane Library, Issue 4, Update Software, Oxford
165. Paranjpe P, Kulkarni PH (1995) Comparative efficacy of four Ayuredic formulations in the treatment of acne vulgaris: a double-blind randomised placebo-controlled clinical evaluation. J Ethnopharmacol 49:127–132
166. Aydin A, Ersoz G, Tekesin O, Akcicek E, Tuncyurek M, Batur Y (1997) Does garlic oil have a role in the treatment of *Helicobacter pylori* infection? Turkish J Gastroenterol 8:181–184
167. Di Mario F, Cavallaro LG, Nouvenne A, Stefani N, Cavestro GM, Iori V, Maino M, Comparato G, Fanigliulo L, Morana E, Pilotto A, Martelli L, Martelli M, Leandro G, Franzè A (2007)

A curcumin-based 1-week triple therapy for eradication of Helicobacter pylori infection: something to learn from failure? Helicobacter 12:238–243
168. Liu JP, Manheimer E, Tsutani K, Gluud C (2001) Medicinal herbs for hepatitis C virus infection. Cochrane Database Syst Rev 4:CD003183
169. James JS (1996) Curcumin: clinical trial finds no antiviral effect. AIDS Treat News 1:1–2
170. Basu P, Dutta S, Begum R, Mittal S, Dutta PD, Bharti AC, Panda CK, Biswas J, Dey B, Talwar GP, Das BC (2013) Clearance of cervical human papillomavirus infection by topical application of curcumin and curcumin containing polyherbal cream: a phase II randomized controlled study. Asian Pac J Cancer Prev 14:5753–5759
171. Ahn WS, Yoo J, Huh SW, Kim CK, Lee JM, Namkoong SE, Bae SM, Lee IP (2003) Protective effects of green tea extracts (polyphenon E and EGCG) on human cervical lesions. Eur J Cancer Prev 12:383–390
172. Halegoua-De Marzio D, Kraft WK, Daskalakis C, Ying X, Hawke RL, Navarro VJ (2012) Limited sampling estimates of epigallocatechin gallate exposures in cirrhotic and noncirrhotic patients with hepatitis C after single oral doses of green tea extract. Clin Ther 34:2279–2285
173. Zhao M, Zheng R, Jiang J, Dickinson D, Fu B, Chu TC, Lee LH, Pearl H, Hsu S (2015b) Topical lipophilic epigallocatechin-3-gallate on herpes labialis: a phase II clinical trial of AverTeaX formula. Oral Surg Oral Med Oral Pathol Oral Radiol 120:717–724
174. Müller L, Meyer M, Bauer RN, Zhou H, Zhang H, Jones S, Robinette C, Noah TL, Jaspers I (2016) Effect of broccoli sprouts and live attenuated influenza virus on peripheral blood natural killer cells: a randomized, double-blind study. PLoS One 11:e0147742

Functional Foods As Personalised Nutrition: Definitions and Genomic Insights

22

Sujata Mohanty and Kopal Singhal

Abstract

Diet has always been the major factor determining the health and well-being of an individual. Technological advancements have lead human beings to fall prey to sedentary lifestyle which in turn has become the root cause of major lifestyle disorders such as type 2 diabetes and cardiovascular diseases. The advent of the genomics era has proven to be a boon in such a scenario. With the human genome sequence in hand and the use of different omics techniques, it is now possible to understand the gene-diet interactions and find significant components in functional foods modulating the disease aetiology. The knowledge of functional foods and their integration with individual requirements can together become the basis of the personalised nutrition. This chapter provides a basic understanding of the functional foods, discusses the different phytochemicals as functional foods and finally highlights the different genomics approaches to personalised nutrition.

Keywords

Diet · Phytochemicals · Functional food · Health · Personalised nutrition

22.1 Introduction

When Watson and Crick proposed the double-helical structure of DNA, they had no idea that somewhere down the lane the whole genome of an individual was going to be sequenced. The idea behind sequencing the whole human genome was to gather

S. Mohanty (✉) · K. Singhal
Department of Biotechnology, Jaypee Institute of Information Technology, Noida, Uttar Pradesh, India
e-mail: sujata.mohanty@jiit.ac.in

V. Rani, U. C. S. Yadav (eds.), *Functional Food and Human Health*,
https://doi.org/10.1007/978-981-13-1123-9_22

insight into the similarities and differences shared among the different individuals, thereby elucidating the mechanism of human evolution and disease aetiology [54, 93]. With the successful completion of the HGP (Human Genome Project) by NIH and US Department of Energy in 2003, researchers all over the world had access to an enormous range of data highlighting these distinctive features of the human genome. One of the major outcomes of the project was the fact that only 99.9% of the human genome was found to be similar among the different individuals. The rest 0.1% of it accounts for the vast majority of the phenotypic variations observed in the different individuals [24]. These differences contribute towards the distinctive response of individuals facing the same situation and therefore shape the individual choices and demands. Though every living being requires food for their growth and development, their choices of food are varied. The traditional way of early man's life believed in simple living. However, with the course of human evolution, man has adopted an advanced way of life altering his food choices, thereby giving rise to such a large number of nutrition-based disorders [40]. This smart technology era has actually made human bodies weaker with its sedentary lifestyle giving rise to a range of diseases which are today defined as lifestyle disorders. These lifestyle disorders include heart diseases, obesity, hypertension, type 2 diabetes, etc. [22]. Since the advent of gene therapy techniques, researchers have been trying to develop therapeutics targeting a particular group of people suffering from a common ailment [56]. This is what marked the beginning of a personalised era with targeted treatment therapies for specific cohorts. As it is always said that "Prevention is better than cure", we need to design alternatives to this unhealthy lifestyle and which is where personalised nutrition steps in.

Food is the basic necessity of living being, which provides energy through various physiological and metabolic activities to fuel up its day-to-day activities. The choice of food varies highly among societies built upon demographic and sociocultural trends and includes the food items mostly depending on the availability in their local market. However, it is seen that there is always a balance maintained with respect to each nutrients among those diets. Since long, the knowledge of diets and nutrition among these societies is cultured and inherited generation wise, popularly called as "Grandmother's recipe". This science of personalised nutrition is nothing new, and varied food choices have always been a part of the civilised society. The dietary preference of vegetarians is different from those of non-vegetarians; children and pregnant women require different set of nutrition than the adult population. However, in the context of genomics, personalised nutrition can be defined as an amalgam of nutrigenetics and nutrigenomics whereby the genetic makeup of an individual is going to decide the dietary plan of the person [14, 39]. Nutrigenetics caters to the individualistic differences in the genome of an organism that govern the metabolism of the food we take [33, 67]. Individuals have been characterised into different groups on the basis of SNPs and their effect on the predisposition of disorders. The frequency of a particular polymorphism has been linked to the increased chances of occurrence of the disease in those individuals, e.g. several SNPs have been identified in the predisposition of a major lifestyle order T2DM (type 2 diabetes mellitus) [11]. Nutrigenomics on the other hand reveals how a certain diet regulates our gene expression. This field incorporates the different transcriptomics, proteomics and metabolomics techniques to analyse the effect of the food we take in controlling the expression of functional genes [66, 74].

Understanding the individualistic differences and their response to dietary regime of the person together with the impact of the diet on the person's genome will become the building blocks for personalised nutrition. We are not far from the day when our genomic identity is going to be linked to our food preferences.

22.2 Functional Foods

22.2.1 History

Like all other technological advancements, Japan is also known to have an edge in the knowledge of functional foods that confers a specific health benefit to the individuals [44, 100]. The Ministry of Health and Welfare in Japan assigned a specific name for these foods, i.e. foods for specified health use (FOSHU) [3]. Since then, several definitions of functional foods have come up from different organisations across the globe, e.g. according to National Academy of Sciences' Food and Nutrition Board, functional foods is "any modified food or food ingredient that may provide a health benefit beyond the traditional nutrients it contains" [28]. However, the International Life Sciences Institute version of it is "foods that, by virtue of the presence of physiologically-active components, provide a health benefit beyond basic nutrition" [88]. The American Dietetic Association definition of functional foods is more or less the same which states that foods that are "whole, fortified, enriched, or enhanced," and consumed "as part of a varied diet on a regular basis, at effective levels" for imparting satisfactory health benefits to consumers [88]. In a nutshell, we can say that functional foods are the enhanced, fortified foods that provide health benefits along with the regular nutritional value when taken in a specified amount regularly. Conventional foods with specific nutritional value have long been known to exist, but these functional foods can be considered analogous to genetically modified range of products wherein they are fortified with an additional nutritional benefit. Functional foods act like preventive measure against several diseases as against the therapeutics which comes into play only after manifestations of disease and are aimed to treat the medical condition [100]. An increasing concern for healthy diet is a promising approach for the development and advancements of these foods with targeted functional roles.

22.2.2 Sources of Functional Foods

These functional foods can be broadly categorised on the basis of their origin, i.e. naturally derived products or industrially synthesised products (Fig. 22.1). The naturally occurring functional foods can be further subdivided on the basis of whether they have been derived from plants or from animal sources. The plant-derived

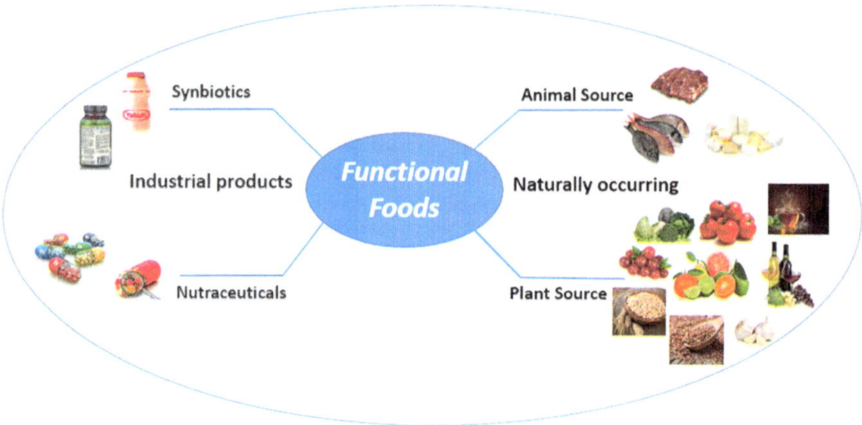

Fig. 22.1 Categories of functional foods on the basis of origin

functional foods include oats, flaxseeds, cruciferous vegetables, citrus fruits, garlic, tea, grapes, wine, etc. and the animal-derived foods include egg, meat, fish, milk, curd, cheese etc.. The industrial products include nutraceuticals and chemically synthesised pro- and prebiotics [89].

22.2.3 Current Scenario of Functional Foods

The awareness that diet has a major role to play for health benefits of human being in prevention and treatment of disease made researchers to discover and study the health benefits of the different diet components. Moreover, research outcomes of various such studies in animal models and human introduced us with the food items, their doses and the beneficiary measures. Some of these studies with respect to different phytochemicals as shown in Fig. 22.2 are being discussed here.

22.2.3.1 Catechins

Catechins are chemically phenols more specifically flavan-3-ol and possess antioxidant properties. Below are some of the major sources of catechin-rich foods.

Beverages

Beverages, mainly the tea and coffee, can play a role of decreasing stress and enhancing focus and concentration in human. *Tea* is known to be the world's second most consumed beverages, made from the leaves of the *Camellia sinensis* [77]. Different categories of tea are prepared, e.g. green tea, black tea and oolong tea depending on their processing protocol, i.e. none, partial or complete oxidation of the tea leaves, and accordingly their colour varies. *Tea* is very popular among all age groups and well accepted by all professionals and especially highly liked among scientific communities. The health benefits of green *tea* have been extensively

Fig. 22.2 Types of phytochemicals used as functional foods

studied in many animal models and also in human. Inhibitory effect of green tea on induced tumorigenesis throughout the digestive tract has been evidenced by experiments done in rats [47, 104, 105]. Green *tea* is rich in antioxidants like polyphenols (most abundant of epigallocatechin-3-gallate) and catechins [103]. Therefore, intake of green tea is known to lower the incidences of cardiovascular diseases, neurological disorder [9, 59] and also cancer; hence, green tea extract is added in many dietary supplements and also to the cosmetics. Green *tea* is also known to improve the blood flow and lower the cholesterol level, evidencing its positive effects on human health, and as an antioxidant remains effective in reducing atherosclerosis risk associated with oxidative stress [102]. Large numbers of in vitro and in vivo studies have been conducted which established a correlation between tea intake and cancer prevention [60].

The fermented *tea*, known as black *tea*, has been found to be very helpful in decreasing the body weight in case of obese patients and can prevent them from having other health problems like dyslipidemia, hyperglycaemia, hypertension and inflammation [68, 71]. Recently, a study was undertaken to evaluate the effect of black *tea* on postprandial blood glucose and insulin response in prediabetic subjects, and the results of this study revealed that black *tea* can decrease postprandial blood glucose after sucrose intake [50]. There are also evidences of the inhibitory effect of black *tea* polyphenols against amylase and glucosidase enzymes and act as a glucose transporter. Global research in *tea* polyphenols, e.g. catechins and gallic acid in green tea and theaflavins and thearubigins with other catechin polymers in black and oolong *teas*, has generated meaningful in vitro and in vivo data which clearly reveals the preventive and therapeutic properties of *tea* polyphenols in human diseases such as hypertension, cancer, cardiovascular and other metabolic diseases [20, 57, 99].

Similar to *tea*, *coffee* is a very popular drink among the intellectual group. The bioactive compound present in it is caffeine, which has stimulatory effect on the central nervous system and a positive effect on long-term memory [13]. Research with volunteer group shows that caffeine can retain memory till 24 h after its consumption, but its strengthening effect on memory is still held a question mark. Instead of its earlier known adverse health effects, recent research indicates its consumption to be beneficial and again introduces as a component in some of the energy drinks especially targeting adolescents. Moreover, caffeine is found to play an important role in weight loss programme by increasing both lipolysis and thermogenesis and sports performance by increasing mental alertness and extending the time to exhaustion in high-intensity endurance exercise and at the same time reducing muscle pain and soreness [7, 16]. Caffeine is found in various plant products, e.g. leaves, beans or fruits, and mostly their source reflects in their brand names.

Moderate consumption of fermented beverages such as *wine* and *beer* (one/two glasses per day) reduces overall mortality, mainly from coronary diseases as evidenced from several epidemiological and clinical studies [58, 73]. Regular and moderate *wine* consumption is also known to be associated with decreased incidence of cardiovascular disease (CVD), hypertension, diabetes and certain types of cancer, including colon, basal cell and ovarian and prostate carcinoma. Experimental evidence of red *wine* reducing cardiovascular disease is observed in western countries especially among French population due to the presence of flavonoids in red *wine* [38]. It is seen that the phenolic content of red *wine* is 20–25 times higher than white wine [48]. In vitro studies with phenolic substances in red *wine* and normal human low-density lipoprotein (LDL), it is observed that red wine inhibits the copper catalysed oxidation of LDL. Surprisingly, a recent study says hot cocoa contains more antioxidant properties than almost all beverages, e.g. green *tea*, black *tea, coffee and wine* [55, 72].

22.2.3.2 Flavonoids

With the advent of various extraction and characterization techniques, the ingredients of food items meant for health benefit of human are being identified and alarmed the public, especially the educated mass to be cautious about their regular diet plan. Addition of functional foods to the regular diet plan may decrease the risk of many lifestyle and neurological diseases. To corroborate the food demand, industries started exploring the possibility of developing nutritional supplements rich in natural antioxidants, vitamins, minerals, etc. from various fruits and vegetables.

Flavonoids are the phytonutrients rich in polyphenols that are responsible for the pigments and colours in fruits and vegetables.

Berries

Berries belong to the soft fruits category and are distributed throughout the world in various forms. Some are cultivated commercially and available both in fresh and dried forms, e.g. *blackberry*, *blueberry*, *cranberry*, *raspberry*, etc., whereas others are restricted to different regions and are included in the traditional diet, for example, *crowberry*, *loganberry*, *gooseberries, lingonberries* and many more. Berries are the potentially important source of vitamins and micronutrients mainly the flavonoids, e.g. anthocyanins, flavonols and proanthocyanidins [80]. Cranberry alone is known to be rich in phenolic phytochemicals such as phenolic acids, flavonoids and ellagic acid, and its consumption lowers incidences of urinary tract infections and has a capacity to inhibit peptic ulcer-associated bacterium, *Helicobacter pylori* [92]. Isolated compounds from cranberry have also been shown to reduce the risk of cardiovascular diseases. Immense research have been carried out on the health benefits of these *berries* and strongly evidenced that the dietary intake of *berry* fruits has a positive effect on human health and disease. In laboratory experiment, it was observed that dietary freeze-dried *berries* inhibit chemically induced cancer of the oesophagus by 30–60% and of the colon by up to 80% in rodents [85]. Berries can reduce carcinogen level by metabolising them and can stop DNA damage. They are also known to inhibit promotion and progression of premalignant cells by reducing the tissue inflammation and inhibiting angiogenesis. These functions of *berries* can be possible by modulating the expression of the genes involved. Antimicrobial effects of *berry* extracts against the intestinal pathogens are also established in some studies [76], but more investigation on the synergistic effects of different *berries* may be carried out in the future.

Citrus Fruits

Various citrus fruits have been consumed traditionally as remedies for chronic diseases, e.g. diabetes, cardiovascular diseases, inflammation, dementia and some form of cancer [17, 70, 97]. Interestingly, both citrus pulps and peels are known to be the promising source of bioactive compounds, e.g. carotenoids, phenolic acids, flavonoids, terpenoids and vitamin C. Epidemiologically, in vitro, animal and clinical studies of citrus juices reported to have a beneficial influence on most of these chronic diseases. Lemons possess the highest antioxidant potential among the studied citrus fruits and are preferable for dietary prevention of cardiovascular and other diseases [41]. The molecular mechanism behind their disease prevention may be clearer in the future with the knowledge of epigenetics.

Pomegranate

Punica granatum, the *pomegranate*, is a small plant belonging to family Punicaceae. *Pomegranate* peel and juice are used as a traditional medicine in treating diabetes which are rich in polyphenols and contain anthocyanin. The *pomegranate* peel more efficiently works than the juice, so the peels can be used as nutrient supplements mainly in prediabetic condition. The burden of diabetes due to the changes in lifestyle of people is increasing day by day, and it also becomes a major concern throughout the world as the number of overweight and obesity cases is getting increased [2, 31]. To this respect, *pomegranate* may act as one of the familiar naturally occurring or industrial fortified functional foods. In addition, due to its rich antioxidant properties, it is found to decrease disease ailments like atherosclerosis and other oxidative stress-related diseases including cardiovascular disease. The *pomegranate* flower extract is also experimentally proved to be successful in protecting against hepatic lipid peroxidation in rats. Though several animal studies were carried out to estimate the effect of *pomegranate* in various biological activities, further investigations on human and clinical studies are yet to be carried out to come up with a strong conclusion and deep understanding about the molecular mechanism behind its potential action as cardioprotective and antidiabetic.

Syzygium cumini, commonly known as *jambolan* or *jamun*, contains a wide variety of secondary compounds, mainly terpenes, and phenolic compounds, such as phenolic acids, flavonoids and tannins, thus found to be a potential source of bioactive molecules which can fight against diabetes and other cardiometabolic disorder [5]. There are a number of in vitro and in vivo studies that have been carried out using its seeds, pulp and roots [12, 79]. In addition, *S. cumini* also show other pharmacological properties like antioxidant, anti-hyperlipidemic, anti-hyperglycaemic and anti-inflammatory [6].

22.2.3.3 Isothiocyanates

Isothiocyanates are the sulphur containing phytochemicals mainly present in cruciferous vegetables and have long been known to fight cancer. Epidemiological studies show *brassica* vegetables including all types of cabbages, cauliflower, broccoli, kale, Brussels sprout, etc. decrease the risk of cancer [95]. This cancer-preventive effect of brassicas may be plausible due to their relatively high content of glucosinolates [45, 94]. Storage and processing of these vegetables influence the bioavailability of glucosinolates and its metabolic product, isothiocyanates [83]. To evaluate the associations between the consumption of brassicas and the risk of lung cancer, 6 cohort studies and 74 case-control studies were conducted, the results of which showed an inverse associations between the consumption of brassicas and risk of lung cancer, stomach cancer and all cancers taken together. Overall, the combination of epidemiological and experimental data provides suggestive evidence for a cancer-preventive effect of a high intake of brassica vegetables [91].

Broccoli, a member of cabbage family Brassicaceae, contains multiple nutrients and vitamins including good amount of vitamins C, E and K and the minerals such as iron, zinc and selenium and the polyphenols, namely, kaempferol, quercetin glucosides and isorhamnetin [10, 18]. It has high anticancer properties due to the presence of antioxidant properties. Diindolylmethane found in broccoli is known to be a potent modulator of the innate immune response system towards antiviral and antibacterial activity.

22.2.3.4 Carotenoids

Do you ever wonder the reason behind the bright colours in fruits and vegetables? Yes, carotenoids are the organic pigments providing this enormous range of colours. But coloration is not the only function of these carotenoids, they also possess antioxidant properties. Increase in lifestyle diseases is one of the major global concerns, which enforces researchers to understand the causes lying behind and come up with some alternative solutions for communities. The diseases like obesity, low-insulin-dependent diabetes, neurological disorder and cancer are occurring in greater frequency which is mainly linked to changes in dietary pattern towards high-calorie foods such as sugar, refined grain flour and sweetened beverages [53]. The need of the hour demands a return to traditional dietary patterns, which may help to reduce these disease problems because of better balance of calories and beneficial nutrients.

22.2.3.5 Corn

Whole-grain foods are valuable sources of nutrients (dietary fibre, B vitamins, vitamin E, selenium, zinc, copper and magnesium) and also contain phytochemicals, such as phenolic compounds that together with vitamins and minerals play important roles in disease prevention [82]. High intake of *dietary fibres* appear to be at significantly lower risk for developing coronary heart disease, stroke, hypertension, diabetes, obesity and certain gastrointestinal diseases [63]. Whole-grain *wheat*, whole-grain *corn* and other carbohydrates and carbohydrate-containing foods are known to have prebiotic effects, and their intake lowers blood pressure and serum cholesterol levels [81]. In vitro inhibitory activities of enzymes belong to the plant foods (pumpkin, beans and maize), e.g. glucosidase, α-amylase and angiotensin I-converting enzyme (ACE) provide a positive health effects and make it one of the strong biochemical rationales for further in vivo studies. Therefore, dietary management may be a future strategy for non-insulin-dependent diabetes mellitus through the control of glucose absorption and reduction of associated hypertension.

Another carotenoid-rich source is tomato which protects the lymphocytes from DNA damage, which generally occurs due to oxidative stress. Lycopene, a potent antioxidant carotenoid, is known to be present in tomatoes, which may save cells from oxidative stress and can also be used in cancer prevention [52]. The synergistic effect of broccoli with tomato on tumour growth suppression has been well evidenced.

22.2.3.6 Saponins: Legumes

Saponins are phytochemicals which constitute of glucosides. They mainly include leguminous pulses. Pulses are an important source of the 15 essential minerals needed by humans. Pulses and cereals may provide one of the best solutions to protein calorie malnutrition, especially in the developing countries [78]. Nutritional survey studies show that people do not have sufficient vegetables and particularly legumes in their daily diet plan despite their well-established health benefits [98]. Beans are low-cost food item, but improve nutrition status. Therefore, through awareness programme, beans can be included in the daily diet of lower-income group, especially the women, and can reduce the risk of chronic diseases [101]. Consumers of chickpeas have been shown to have higher nutrient intakes of dietary fibre, polyunsaturated fatty acids, vitamin A, vitamin E, vitamin C, folate, magnesium, potassium and iron as compared to nonconsumers [98]. Cowpea is a drought- and heat-tolerant grain and is rich in bioactive polyphenols and peptides [8]. Some of these polyphenols are even not commonly found in other pulses. Due to its significant human health beneficiary action against diabetes, cancer and cardiovascular diseases, cowpea may be a component in modern food diet plan. It can be believed that the next dietary strategy will soon replace the energy-dense foods with those that are nutrient dense.

22.2.3.7 Garlic, Onion and Other Allyl Sulphides

Allyl sulphides are sulphur-containing organic compounds. Garlic has traditionally been used to add flavour to the different cuisines. But this plant belonging to the genus *Allium* also possesses several medicinal properties. The allyl sulphur component of garlic makes it a potential candidate for lowering the risk of heart diseases and stroke [61]. It is also known to play a role in reducing cancer events by acting as HDAC (Histone Deactylase) inhibitor. The organo-sulphurs in garlic can be metabolised to allyl mercaptan which are known to induce histone hyperacetylation in colon cancer cells [65]. Along with these, the allyl sulphides also have antimicrobial and antithrombic activities [4]. These allyl sulphides include onion, chives and leeks all of which are known for their anticancerous and cardioprotective properties. The role of these allium sulphides as antineoplastic agent has also been studied in vivo and in vitro; however, no concurrent results have been produced in human beings so far and remain an area for research [51].

22.2.3.8 Some Other Important Functional Foods

The dates of *Phoenix dactylifera* contain at least 15 minerals, both saturated and unsaturated fatty acids, proteins, vitamins and various salts like calcium, potassium, boron, cobalt, copper, fluorine, iron, magnesium, manganese, potassium, phosphorous, sodium and zinc[1]. A high amount of unsaturated fat, oleic acid content (41.1–58.8%) is found in the seeds of date, which suggests that the seeds could be used as a source of oleic acid. There are 23 types of amino acids present in dates, which is much more in comparison to the common fruits like orange or banana. The dates are also enriched with dietary fibres, which make it to consider as an ideal healthy diet. Six types of vitamins, e.g. vitamin C, vitamin $B_{1,}$ thiamine, B_2 riboflavin, nicotinic acid (niacin) and vitamin A are also present, however, in various concentrations in different varieties of dates. Therefore, dates may be a prescribed diet for the malnutrition diseases. However, further research is needed to establish its other health-promoting activities.

Similar to the *berries, flaxseed* is also experimentally proved to be playing a protective role at the early promotion stage of carcinogenesis [87]. 50 g of flaxseed was consumed by young healthy adults for 4 weeks and observed that alpha-linolenate and n-3 polyunsaturates were increased significantly in adipose tissue and plasma lipids, respectively [25]. Plasma LDL cholesterol was also reduced by up to 8%, and total urinary lignan excretion was increased more than fivefold. *Flaxseed* is a rich source of mammalian lignan precursor secoisolariciresinol-diglucoside and α-linolenic acid (ALA).

Olive and *olive oil*, though being a traditional food component of the Mediterranean diet since thousands of years, are largely consumed in the world at present [84, 90]. Health beneficial action of *olive* and *olive oil* is well established as they provide a rich of natural antioxidants [15, 90]. The presence of two phenolic compounds from *olive oil*, namely, oleuropein and hydroxytyrosol, has been assigned for the preclinical treatment of several cardiovascular and metabolic disorders due to their antioxidant activities [42, 75]. Also increase *olive oil* consumption has been shown to modulate immune function, particularly the inflammatory processes associated with the immune system [96]. Though sufficient data are lacking in human study, experimental animal studies reveal that *olive oil* consumption helps to suppress the occurrence of chemically induced mammary tumours and rheumatoid arthritis. Therefore, it may be a food supplements for woman at risk for breast cancer. The virgin *olive oil* is found to be more beneficial to health as it contains abundant antioxidants, namely, lipophilic and hydrophilic phenols [15]. It can be concluded based on research evidences that *olive* and *olive oil* may have a role in the prevention of coronary heart disease and certain cancers because of their high levels of monosaturated fatty acids and phenolic compounds, which draws special attention of nutraceutical companies to supply compounds with high amount of antioxidant activities.

22.2.4 Flipside of Functional Foods

As the famous quote says that "Excess of anything is bad", the same applies to these functional foods. All these foods in addition to their health benefit effects contain several potentially toxic compounds, which compels the scientists to explore more in this direction [9, 21]. Research was done in green tea to evaluate its toxicity in rats and mice. Though no case of mortality was observed in rats, but in both male and female mice, mortality occurred at a dose greater than 1000 mg/kg. The cause of early onset of death was mainly due to liver necrosis. The females are seen to be affected more than males. Other than liver necrosis, lowering on body weight and food intake, thymic atrophy, several eye diseases (conjunctiva, reddening and discharge and chemosis), intestinal dilation and secretion of haemorrhagic fluid are also noticed as posttreatment symptoms [21, 46]. Research in different models provides insight into the probable toxic effect of green *tea* in other non-model system including human. Therefore, the consumption of green tea should also be in a dose-specific manner and desire for more can be a threat to health. Similar to green tea, other functional foods also need to be taken in appropriate amount for effective and harmless results.

22.2.5 Consumer Acceptance of Functional Foods

Globalisation has proven to be a bane for the people for remaining updated with the recent trends and advancements throughout the world. Thus, while the concept of functional foods firstly emerged in Japan, the other countries over the world did not remain behind in experimenting with their food choices and incorporating these functional foods as a part of their regular diet. The Latin phrase "*Mens sana in corpore sano*" clearly states that a sound body produces a sound mind. Therefore, with busy schedules and sedentary lifestyles people nowadays are readily opting fortified food as an easy way to healthy life as has been highlighted by ADA in American population [49]. However, the cost and awareness of these food products are an issue that needs to be addressed in developing nations with majority of economically weaker sections in society. Technology is going to become a major tool in solving this drawback and bringing these functional foods to each and every doorstep all over the world.

22.3 Genomics Approach to Personalised Nutrition

Wouldn't it be great if we knew exactly how and where a specific food component is going to affect our genome? Knowing the answer to the above question can help us to control the different diet-related disorders. Post-genomics era, our understanding of disease aetiology widened to include the impact of genetic makeup on disease pathogenesis. As and when the scope of research in the field of epidemiology broadened, scientists began to consider the role of environment in modulating gene expression and its consequences on one's susceptibility to diseases [23, 67, 74]. A separate body of nutritionist started proposing the role of diet in regulating this gene expression. All this marked the beginning of integration of fields of nutrition and genomics [64]. The following are the definitions of some of the important terms related to this fusion.

22.3.1 Gene–Diet Interactions

Genes are like molecular machines that encode proteins required for the normal functioning of the body. Dietary components interact with several biological entities to regulate the expression of genes [27, 30, 35, 74]. These components have both antagonistic and agonistic effects on the expression of the genes. However, the genome of an organism also regulates the uptake of the nutrients, and the response of each and every individual to this is varied. Therefore, gene-diet interactions form the basis of nutritional genomics. Since a large number of nutritional diseases are linked to deficiency or insensitivity of a particular diet component, understanding this interplay of genome and diet has become imperative to reduce the risk of these diet-related disorders.

22.3.2 Nutrigenetics

Sir Gregor John Mendel observed the variations in the phenotypic characters and linked them to the genotype of the organism. Since then, genetic variations have explained a lot of phenomena including an individual's response to the nutrients. Nutrigenetics is the science explaining the differential response of the individuals to the same diet on account of the variations in the genome [37, 74]. This field of nutrition also provides an explanation to the fact that different person are at different risk of acquiring/developing diseases and the individualistic requirements of specific nutrition. For example, the diet for the lactose-intolerant patients should not include any milk product, whereas for the patients of PKU (phenylketonuria), any source of the amino acid phenylalanine should be avoided.

22.3.3 Nutrigenomics

The advent of lifestyle-related disorders encouraged scientists to look into the effect of diet and nutrition on human metabolism and its influence on the expression of genes in what came to be known as the science of nutrigenomics. Nutrigenomics emphasises on the fact that food and diet preferences should not be limited to taste, culture and traditional beliefs but they should be based on the effect of a certain food component in regulation of gene expression [34, 37, 74]. Researchers have since then identified an array of food products having beneficial or detrimental effect on the genome by controlling the gene expression. Identification of the food component which can activate the expression of a certain gene thereby controlling the production of some metabolites forms the core of nutrigenomics. Therefore, nutrigenomics facilitate our understanding of how nutrition affects the metabolic pathways and its correlation with diet-related diseases. Identification of the transcriptional regulators in diet that can regulate the expression of disease-related genes as well as the polymorphisms in genome will prove to be beneficial for the medical researchers to understand the aetiology of diet-related diseases and provide a solution to the society.

22.3.4 Epigenetics

The term epigenetics was coined by a Greek researcher Conrad H. Waddington in 1942 which describes the changes in the expression of genes without any change in the nucleotide sequence. Thus epigenetics changes are those that modulate the genome architecture by chromatin remodelling or by regulating the bioactive molecules required for the synthesis of the nucleotides or proteins, e.g. histone acetylation or methylation [34, 74]. Methylation of DNA results in tight linkage between DNA and histone proteins thereby turning off the expression of genes, whereas acetylation loosens this DNA-histone linkage and turns on the gene expression, e.g. broccoli and other cruciferous vegetables increase histone acetylation, whereas folic acid and vitamin B group provide methyl group thereby enhancing DNA methylation. Not only dietary components but certain non-coding RNAs such as microRNAs also regulate gene expression and chromatin remodelling. However, such changes are reversible and subject to the availability of the particular molecule that either enhances or loosens the DNA-histone association.

22.3.5 Nutritional Epigenomics

Epigenetics and its correlation with diet have been studied widely. The ancient belief that during pregnancy a mother should consume food for two is very much based on epigenetics of diet. The kind of nutrition taken during pregnancy forms an imprint for the developing foetus and plays a significant role in regulating the gene expression of the individual throughout their lives. The significance of diet in regulating our genome and its role in defining specific phenotypes in the individuals has been the area of nutritional epigenomics [36, 74]. The understanding of the role of each and every component of diet in controlling our gene expression has thus become crucial for the field of nutritional epigenomics. The genomic signature of a person is going to guide the nutritional requirement.

22.3.6 Nutritional Transcriptomics and Proteomics

Nutritional epigenomics is a concept that can be brought into practical lives by incorporating transcriptomics techniques into it [34, 74]. Transcriptomics is the study of a complete RNA transcripts produced by the genome. These transcripts form the building blocks of proteins, and the study of a complete set of biologically active proteins is what is known as proteomics. Both these transcriptomics and proteomics techniques are influenced by the bioactive molecules present in our diet, i.e. some of the nutrients regulate the activity of transcripts or proteins. Through the approach of these techniques, researchers aim to study the specific interaction of a particular nutrient in regulating these mechanisms. In vitro studies done on certain animal models can at times help simulate the in vivo conditions inside the human being and hypothesise about the effect of a component on the genome. Further, the clinical trials can be done to derive a conclusive result.

22.3.7 Biomarkers

WHO defines biomarkers as "almost any measurement reflecting an interaction between a biological system and a potential hazard, which may be chemical, physical, or biological". The measured response may be "functional and physiological, biochemical at the cellular level, or a molecular interaction" [64, 86]. Thus, in simple terms, we can say that biomarkers help us to detect any biological condition. In terms of nutrigenomics, we need to find early markers which can help us to prevent the occurrence of disease, i.e. we need to identify those markers whose level is a direct indication of the onset of a disease [69]. Nutritional deficiency results in modifications in the genome in the form of DNA damage which acts as a biomarker, e.g. change in the length of telomere and micronuclei has been found to be associated with the deficiency of zinc, folate and vitamin D [32].With the use of metabolomics, one can aim to identify the different nutritional biomarkers, the knowledge of which will give us an edge on our understanding of disease pathogenesis. This is a potential field for the researchers to identify significant biomarkers of diet-related diseases.

22.3.8 Genome-Wide Association Studies (GWAS)

Post-genomics era has allowed us to study the genetic variations at intra- and interpersonal. This fusion of nutrition and genetics brings in the need for GWAS in order to know the role of genetic variation in the disease pathogenesis. GWAS tries to find the genes associated with a disease condition by linking the frequency of SNPs (single nucleotide polymorphisms) to the occurrence of disease, i.e. they try to study the genetic variations in the diseased cohorts and then link it with disease occurrence [11, 34]. It helps us to identify the genes, genotypes and bioactive compound associated with a particular disease aetiology. In comparison to the candidate gene association studies, GWAS does not require a prior knowledge of the gene function and will also identify novel gene variants linked to the diseases. However, the cost associated with this technique is quite high and may become a drawback for its successful execution. With the whole genome sequence of individual in hand, this technique is very useful to provide concurrent results.

22.4 Health Aspects of Functional Foods

These wide range of functional foods confer additional health benefits to the individuals. Targeted nutrition rich in a particular bioactive molecule helps in disease prevention or cure. There are numerous ways in which functional foods aid in disease suppression. The changed lifestyle has been accompanied with a number of lifestyle disorders which can be easily prevented by taking proper diet enriched with the nutrients and bioactive molecules. An increased life expectancy is accompanied with a desire to live a healthy life in later years of one's life. The kind of nutrition we take is a direct measure of intestinal health, and therefore functional foods play a significant role in maintaining healthy gut flora [19]. Figure 22.3 gives an overview of the different areas involved in the development of personalised nutrition. Understanding and developing these areas by research are thus the need for the nutritional genomics scientists. Having known the factors involved in personalised nutrition, let us now take a look at some of diseases related to nutritional genomics.

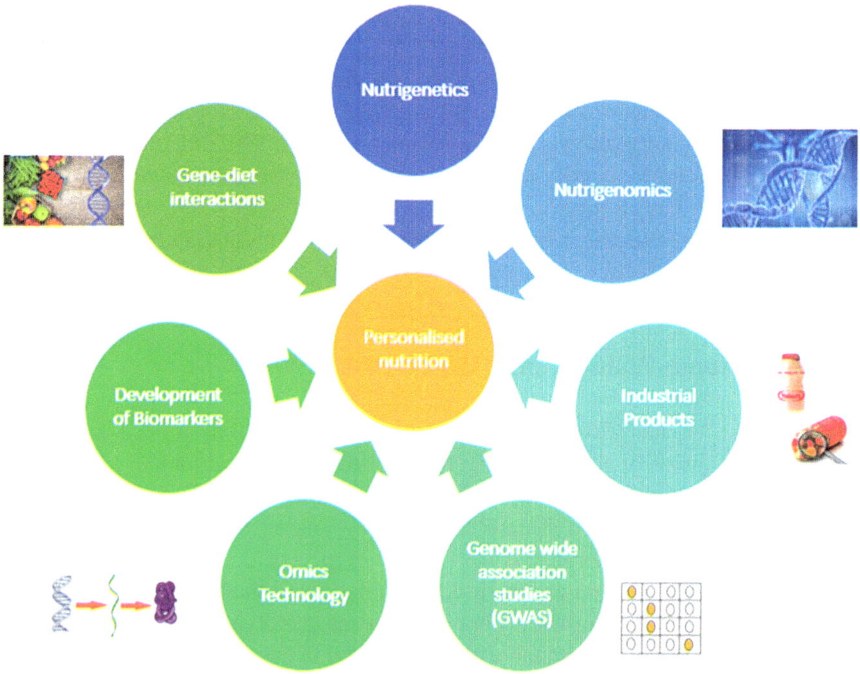

Fig. 22.3 Factors influencing and defining personalised nutrition

22.4.1 Obesity

WHO defines obesity as "abnormal or excessive fat accumulation that may impair health". Technological advancements have made us prey to the sedentary form of lifestyle with most of us choosing easier diet options as processed food and minimising our intake of natural products. An average person's physical work has also reduced a lot thus paving way for this highly prevalent diet-related disorder. However, the repercussions of obesity are not limited to the physical appearance, but it very well affects the metabolism of the body making a person prone to certain diseases such as hypertension, diabetes mellitus, cardiovascular diseases, cancer, etc. [29]. Obesity is a multifactorial disease, but it is mostly influenced by environment and our diet patterns; however, here again, the individual variations and differences result in different susceptibility to developing obesity. Changing our food habits can definitely help control the disease. Restricting ourselves to healthy diet with less calorie intake or reducing our fat intake can provide some solutions. Intake of certain bioactive molecules such as polyphenols, phytosterols, terpenoids and alkaloids which can modulate expression of adipose tissue genes can help reduce the level of obesity [62]. Certain compounds, such as α-tocopherol obtained from the consumption of green tea, function by reducing obesity-induced inflammation. However, diet alone cannot eliminate obesity, and an active lifestyle is a must for prevention of this disease.

22.4.2 Cancer

The role of environment in inducing cancer has been well known. Our food material can contain certain mutagens which can increase one's susceptibility to cancer. Certain hormone-dependent cancers such as ovarian cancer, breast cancer and prostate cancer are affected by the nutrition uptake. Since these cancers are dependent on hormones which in turn are governed by the kind of diet we take, we can conclude that diet plays a major role in the onset of cancer [30]. Certain salts and preservatives are known to be carcinogenic for gastric cancers, deficiency of folate and vitamin B12 has been linked to high risk of cancer. However, the predisposition of cancer is also governed by one's genotype, i.e. people who consume red meat and have NAT fast acetylator genotype have a higher risk of developing colon cancer as NAT is an acetyltransferase which activates the heterocyclic aromatic amines (HAA) which upon acetylation get activated and binds to DNA to cause cancer [37]. Also in one of the studies done in Sudanese population, the GST M-1 genotype with the consumption of aflatoxins composed of peanut butter put the individuals at high risk of developing hepatocellular carcinoma [35]. Other dietary components well known for their role in cancer prevention include long-chain polyunsaturated fatty acids (LC-PUFA), several phytochemicals, dietary fibres, omega-3 fatty acid-rich oils, flavonoids, phenols, isothiocyanates, allyl sulphur compounds, indoles and selenium. All these components work in one way or the other to control and prevent cancer. The crux of the situation is that there is a need to identify, administer and analyse the bioactive molecules that have anticancerous properties through clinical trials so that this knowledge of nutrigenomics can be used for the benefit of mankind.

22.4.3 Cardiovascular Diseases (CVD)

Cardiovascular diseases include a group of several heart-related disorders including coronary artery disease (CAD), cerebrovascular disease (stroke), peripheral artery disease, rheumatic heart disease, congenital heart disease and heart failure. The reasons behind the manifestation of these symptoms lie in the genetic makeup and the environmental conditions of the patients [27]. However, the single most important factor responsible for the occurrence of this disease is atherosclerosis which results in hardening of the blood vessels ultimately leading to blockage and strokes. Hormonal changes or changes in the level of enzymes due to genetic variations are some of the genetic factors contributing towards the occurrence of these conditions. Further, consistently high levels of total cholesterol including LDL cholesterol and triglycerides in the plasma assist in the process of atherosclerosis; however, high levels of HDL prevent its occurrence. A study reports that people who have E4 allele in their apolipoprotein E gene show high levels of HDL as they increase their fat intake as compared to the individuals with E1, E2 or E3 allele consuming the same amount of fat [26]. Owing to the significant role played by nutrition in the pathogenesis of CVD, the advice of doctors has been on adopting a healthy lifestyle. However, in order to better understand the disease prognosis and aetiology, we need to study the interaction of diet with the genome and consider the individual differences.

22.4.4 Type 2 Diabetes Mellitus (T2DM)

Diabetes mellitus is a major metabolic disorder, and a large number of individuals have fallen prey to this disease [36]. Out of the two categories of diabetes mellitus, type 1 is insulin dependent and mainly occurs in juvenile population, whereas type 2 is non-insulin-dependent and mainly starts its onset in the adult stage. Type 2 diabetes mellitus (T2DM) occurs due to a deficiency in insulin secretion or insulin resistance [43]. The onset of this disease is mainly dependent on the combination of genetic and environmental factors, and therefore gene-diet interactions again play a major role in the pathogenesis of this disease [67]. Several nutrients such as naringins, anthocyanins and genistein are involved in reducing the risk to T2DM by regulating the processes of insulin resistance, glucose uptake and insulin secretion. Therefore, intake of dietary components known to prevent diabetes or delay its onset should be the first choice of the vulnerable population.

22.5 Future Perspective

All the above-mentioned diseases are a result of changes in regular gene-diet interactions; the knowledge of significant nutrients involved in the predisposition of these diseases is a must for reducing their occurrence and prevalence. This recent interest in the field of nutritional genomics has resulted in the marketing of a number of bioactive compounds containing products. With each passing day, the variety of such nutrition-based products has been increasing; however, lack of awareness and inertia against change in dietary habits is one of the major challenges in this area. Consumers should be made aware of the benefits of such nutritional products. Also the cost and efficacy of these dietary components need to be evaluated and checked upon before their industrial production and marketing. From the perspective of research, scientists need to identify more such products involved in preventing major metabolic disorders. Also most of the research has been done taking individual components at a time; however, a human diet is an amalgamation of several such compounds so studies should be focussed on the synergistic effects of different bioactive molecules administered together. A major challenge comes up regarding the privacy of the genetic information obtained from an individual for the development of personalised nutrition. Effective measures need to be taken in order to maintain secrecy and prevent wrongful use of any individual's genetic information. The knowledge of the whole genome and the different omics techniques will prove to be a boon for the upcoming advancements in this field of personalised nutrition.

References

1. Al-Shahib W, Marshall RJ (2003) The fruit of the date palm: its possible use as the best food for the futureInt. J Food Sci Nutr 54(4):247–259
2. Althunibat OY, Al-Mustafa AH, Tarawneh K, Khleifat KM, Ridzwan BH, Qaralleh HN (2010) Protective role of Punicagranatum L. peel extract against oxidative damage in experimental diabetic rats. Process Biochem 45(4):581–585
3. America IN (1999) Safety assessment and potential health benefits of food components based on selected scientific criteria. ILSI North America technical committee on food components for health promotion. Crit Rev Food Sci Nutr 39:203–316
4. Ariga T, Seki T (2006) Antithrombotic and anticancer effects of garlic-derived sulfur compounds: a review. Biofactors 26(2):93–103
5. Atale N, Chakraborty M, Mohanty S, Bhattacharya S, Nigam D, Sharma M, Rani V (2013) Cardioprotective role of Syzygiumcumini against glucose-induced oxidative stress in H9C2 cardiac myocytes. Cardiovasc Toxicol 13(3):278–289
6. Atale N, Saxena S, Nirmala JG, Narendhirakannan RT, Mohanty S, Rani V (2016) Synthesis and characterization of Sygyziumcumini nanoparticles for its protective potential in high glucose-induced cardiac stress: a green approach. Biotechnol Appl Biochem:1–15
7. Avois L, Robinson N, Saudan C, Baume N, Mangin P, Saugy M (2006) Central nervous system stimulants and sport practice. Br J Sports Med 40(suppl 1):i16–i20
8. Awika JM, Duodu KG (2016) Bioactive polyphenols and peptides in cowpea (Vignaunguiculata) and their health promoting properties: a review. J Funct Food 38:686. https://doi.org/10.1016/j.jff.2016.12.002
9. Babu A, Pon V, Liu D (2008) Green tea catechins and cardiovascular health: an update. Curr Med Chem 15(18):1840–1850
10. Baenas N, Moreno DA, Garcia-Viguera C (2012) Selecting sprouts of Brassicaceae for optimum phytochemical composition. J Agri Food Chem 60(45):11409–11420
11. Ban HJ, Heo JY, Oh KS, Park KJ (2010) Identification of type 2 diabetes-associated combination of SNPs using support vector machine. BMC Genet 11(1):26
12. Banerjee A, Dasgupta N, De B (2005) In vitro study of antioxidant activity of Syzygiumcumini fruit. Food Chem 90(4):727–733
13. Borota D, Murray E, Keceli G, Chang A, Watabe JM, Ly M, Toscano JP, Yassa MA (2014) Post-study caffeine administration enhances memory consolidation in humans. Nat Neurosci 17(2):201–203
14. Bouwman LI, Koelen MA (2007) Communication on personalised nutrition: individual-environment interaction. Genes Nutr 2(1):81
15. Bulotta S, Celano M, Lepore SM, Montalcini T, Pujia A, Russo D (2014) Beneficial effects of the olive oil phenolic components oleuropein and hydroxytyrosol: focus on protection against cardiovascular and metabolic diseases. J Transl Med 12(1):219
16. Campbell B, Wilborn C, La Bounty P, Taylor L, Nelson MT, Greenwood M, Ziegenfuss TN, Lopez HL, Hoffman JR, Stout JR, Schmitz S (2013) International Society of Sports Nutrition position stand: energy drinks. J Int Soc Sports Nutr 10(1):1
17. Cancalon PF (2016) Citrus juices health benefits. In: Beverage impacts on health and nutrition. Springer, pp 115–127
18. Cartea ME, Francisco M, Soengas P, Velasco P (2010) Phenolic compounds in Brassica vegetables. Molecules 16(1):251–280
19. Cencic A, Chingwaru W (2010) The role of functional foods, nutraceuticals, and food supplements in intestinal health. Nutrients 2(6):611–625
20. Chacko SM, Thambi PT, Kuttan R, Nishigaki I (2010) Beneficial effects of green tea: a literature review. Chin Med 5(1):13
21. Chan PC, Ramot Y, Malarkey DE, Blackshear P, Kissling GE, Travlos G, Nyska A (2010) Fourteen-week toxicity study of green tea extract in rats and mice. Toxicolpathol 38(7):1070–1084

22. Chen R, Snyder M (2012) Systems biology: personalized medicine for the future? Curr Opin Pharmacol 12(5):623–628
23. Choi SW, Claycombe KJ, Martinez JA, Friso S, Schalinske KL (2013) Nutritional epigenomics: a portal to disease prevention. Adv Nutr 4(5):530–532
24. Collins FS, McKusick VA (2001) Implications of the human genome project for medical science. JAMA 285(5):540–544
25. Cunnane SC, Hamadeh MJ, Liede AC, Thompson LU, Wolever TM, Jenkins DJ (1995) Nutritional attributes of traditional flaxseed in healthy young adults. Am J Clin Nutr 61(1):62–68
26. Davignon J, Gregg RE, Sing CF (1988) Apolipoprotein E polymorphism and atherosclerosis. Arterioscler Thromb Vasc Biol 8(1):1–21
27. Dimitriou ME, Dedoussis GV (2012) Gene–diet interactions in cardiovascular disease. Curr Nutr Rep 1(3):153–160
28. Earl R, Thomas PR (eds) (1994) Opportunities in the nutrition and food sciences: research challenges and the next generation of investigators. National Academies Press
29. Elliott RM, Johnson IT (2007) Nutrigenomic approaches for obesity research. Obes Rev 8(s1):77–81
30. Elsamanoudy AZ, Neamat-Allah MAM, Mohammad FAH, Hassanien M, Nada HA (2016) The role of nutrition related genes and nutrigenetics in understanding the pathogenesis of cancer. JMAU 4(3):115–122
31. Esmaillzadeh A, Tahbaz F, Gaieni I, Alavi-Majd H, Azadbakht L (2004) Concentrated pomegranate juice improves lipid profiles in diabetic patients with hyperlipidemia. J Med Food 7(3):305–308
32. Fenech M (2015) Perspectives in nutrigenomics and nutrigenetics. Sight and Life 29(1):64–70
33. Fenech M, El-Sohemy A, Cahill L, Ferguson LR, French TA, Tai ES, Milner J, Koh WP, Xie L, Zucker M, Buckley M (2011) Nutrigenetics and nutrigenomics: viewpoints on the current status and applications in nutrition research and practice. J Nutrigenet Nutrigenomics 4(2):69–89
34. Ferguson LR, De Caterina R, Gorman U, Allayee H, Kohlmeier M, Prasad C, Choi MS, Curi R, De Luis DA, Gill A, Kang JX (2016) Guide and position of the international society of nutrigenetics/nutrigenomics on personalised nutrition: part 1-fields of precision nutrition. J Nutrigenet Nutrigenomics 9(1):12–27
35. Gaboon NE (2011) Nutritional genomics and personalized diet. Egypt J Med Hum Genet 12(1):1–7
36. Gallou-Kabani C, Junien C (2005) Nutritional epigenomics of metabolic syndrome. Diabetes 54(7):1899–1906
37. Garg R, Sharma N, Jain SK (2014) Nutrigenomics and nutrigenetics: concepts and applications in nutrition research and practice. Acta Med Int 1(2):124–130
38. German JB, Walzem RL (2000) The health benefits of wine. Annu Rev Nutr 20(1):561–593
39. Ghosh D (2010) Personalised food: how personal is it? Genes Nutr 5(1):51–53
40. Ginsburg GS, Willard HF (2009) Genomic and personalized medicine: foundations and applications. Transl Res 154(6):277–287
41. Gorinstein S, Martin-Belloso O, Park YS, Haruenkit R, Lojek A, Ciz M, Caspi A, Libman I, Trakhtenberg S (2001) Comparison of some biochemical characteristics of different citrus fruits. Food Chem 74(3):309–315
42. Guasch-Ferre M, Hu FB, Martinez-Gonzalez MA, Fito M, Bullo M, Estruch R, Ros E, Corella D, Recondo J, Gomez-Gracia E, Fiol M (2014) Olive oil intake and risk of cardiovascular disease and mortality in the PREDIMED study. BMC Med 12(1):78
43. Hansen L, Pedersen O (2005) Genetics of type 2 diabetes mellitus: status and perspectives. Diabetes Obes Metab 7(2):122–135
44. Hasler CM (2002) Functional foods: benefits, concerns and challenges—a position paper from the American council on science and health. J Nutr 132(12):3772–3781
45. Jeffery EH (2015) Cancer prevention with Brassica vegetables. Acta Hortic 1106:143–146

46. Jimmy EO, Adelaiye AB, Umoh I, Bassey EI, Ekwere EO (2013) Stomach Histopathologic and Ulcerogenic potentials of tea beverage. J Nat Sci Res 3(8):195–199
47. Ju J, Lu G, Lambert JD, Yang CS (2007) Inhibition of carcinogenesis by tea constituents. Semin Cancer Biol 17(5):395–402
48. Kanner J, Frankel E, Granit R, German B, Kinsella JE (1994) Natural antioxidants in grapes and wines. J Agric Food Chem 42(1):64–69
49. Kapsak WR, Rahavi EB, Childs NM, White C (2011) Functional foods: consumer attitudes, perceptions, and behaviors in a growing market. J Am Diet Assoc 111(6):804–810
50. Kirkham S, Akilen R, Sharma S, Tsiami A (2009) The potential of cinnamon to reduce blood glucose levels in patients with type 2 diabetes and insulin resistance. Diabetes Obes Metab 11(12):1100–1113
51. Knowles LM, Milner JA (2000) Allyl sulfides modify cell growth. Drug Metabol Drug Interact 17(1/4):81–108
52. Kris-Etherton PM, Hecker KD, Bonanome A, Coval SM, Binkoski AE, Hilpert KF, Griel AE, Etherton TD (2002) Bioactive compounds in foods: their role in the prevention of cardiovascular disease and cancer. Am J Med 113(9):71–88
53. Kwon YI, Apostolidis E, Kim YC, Shetty K (2007) Health benefits of traditional corn, beans, and pumpkin: in vitro studies for hyperglycemia and hypertension management. J Med Food 10(2):266–275
54. Lander ES, Linton LM, Birren B, Nusbaum C, Zody MC, Baldwin J, Devon K, Dewar K, Doyle M, FitzHugh W, Funke R (2001) Initial sequencing and analysis of the human genome. Nature 409(6822):860–921
55. Lee KW, Kim YJ, Lee HJ, Lee CY (2003) Cocoa has more phenolic phytochemicals and a higher antioxidant capacity than teas and red wine. J Agric Food Chem 51(25):7292–7295
56. Lesko LJ (2007) Personalized medicine: elusive dream or imminent reality? Clin Pharmacol Ther 81(6):807–816
57. Li S, Lo CY, Pan MH, Lai CS, Ho CT (2013) Black tea: chemical analysis and stability. Food Funct 4(1):10–18
58. Lippi G, Franchini M, Favaloro EJ, Targher G (2010) Moderate red wine consumption and cardiovascular disease risk: beyond the "French paradox". Semin Thromb Hemost 31(1):59–70
59. Mandel SA, Amit T, Kalfon L, Reznichenko L, Youdim MB (2008) Targeting multiple neurodegenerative diseases etiologies with multimodal-acting green tea catechins. J Nutr 138(8):1578S–1583S
60. Maron DJ, Lu GP, Cai NS, Wu ZG, Li YH, Chen H, Zhu JQ, Jin XJ, Wouters BC, Zhao J (2003) Cholesterol-lowering effect of a theaflavin-enriched green tea extract: a randomized controlled trial. Arch Intern Med 163(12):1448–1453
61. Milner JA (2001) Mechanisms by which garlic and allyl sulfur compounds suppress carcinogen bioactivation. In: Nutrition and cancer prevention. Springer, New York, pp 69–81
62. Mohamed GA, Ibrahim SR, Elkhayat ES, El Dine RS (2014) Natural anti-obesity agents. B-FOPCU 52(2):269–284
63. Mozaffarian D (2016) Dietary and policy priorities for cardiovascular disease, diabetes, and obesity. Circulation 133(2):187–225
64. Neeha VS, Kinth P (2013) Nutrigenomics research: a review. JFST 50(3):415–428
65. Nian H, Delage B, Ho E, Dashwood RH (2009) Modulation of histone deacetylase activity by dietary isothiocyanates and allyl sulfides: studies with sulforaphane and garlic organosulfur compounds. Environ Mol Mutagen 50(3):213–221
66. Palou A (2007) From nutrigenomics to personalised nutrition. Genes Nutr 2(1):5–7
67. Phillips CM (2013) Nutrigenetics and metabolic disease: current status and implications for personalised nutrition. Nutrients 5(1):32–57
68. Qin B, Polansky MM, Harry D, Anderson RA (2010) Green tea polyphenols improve cardiac muscle mRNA and protein levels of signal pathways related to insulin and lipid metabolism and inflammation in insulin-resistant rats. Mol Nutr Food Res 54(S1):S14–S23

69. Ramesha C, Kumari SS, Anuradha CM, Lakshmi H, Kumar CS (2010) Nutrigenomic analysis of mulberry silkworm (Bombyxmori L.) strains using polymerase chain reaction-simple sequence repeats (PCR-SSR). Int J Biotechnol Mol Biol Res 1(7):92–100
70. Rampersaud GC, Valim MF (2017) 100% Citrus juice: nutritional contribution, dietary benefits, and association with anthropometric measures. Crit Rev Food Sci Nutr 57(1):129–140
71. Ranilla LG, Kwon YI, Apostolidis E, Shetty K (2010) Phenolic compounds, antioxidant activity and in vitro inhibitory potential against key enzymes relevant for hyperglycemia and hypertension of commonly used medicinal plants, herbs and spices in Latin America. Bioresour Technol 101(12):4676–4689
72. Richelle M, Tavazzi I, Offord E (2001) Comparison of the antioxidant activity of commonly consumed polyphenolic beverages (coffee, cocoa, and tea) prepared per cup serving. J Agric Food Chem 49(7):3438–3442
73. Ruf JC (2002) Overview of epidemiological studies on wine, health and mortality. Drugs Exp Clin Res 29(5–6):173–179
74. Sales NMR, Pelegrini PB, Goersch MC (2014) Nutrigenomics: definitions and advances of this new science. J Nutr Metab 1:202759. https://doi.org/10.1155/2014/202759
75. Schwingshackl L, Hoffmann G (2014) Adherence to Mediterranean diet and risk of cancer: a systematic review and meta-analysis of observational studies. Int J Cancer 135(8):1884–1897
76. Seeram NP (2008) Berry fruits for cancer prevention: current status and future prospects. J Agric Food Chem 56(3):630–635
77. Shukla Y (2007) Tea and cancer chemoprevention: a comprehensive review. Asian Pac J Cancer Prev 8(2):155
78. Singh NP, Pratap A (2016) Food legumes for nutritional security and health benefits. In: Biofortification of food crops. Springer, New Delhi, pp 41–50
79. Singh JP, Kaur A, Singh N, Nim L, Shevkani K, Kaur H, Arora DS (2016) In vitro antioxidant and antimicrobial properties of jambolan (Syzygiumcumini) fruit polyphenols. Food Scitechnol 65:1025–1030
80. Skrovankova S, Sumczynski D, Mlcek J, Jurikova T, Sochor J (2015) Bioactive compounds and antioxidant activity in different types of berries. Int J Mol Sci 16(10):24673–24706
81. Slavin J (2013) Fiber and prebiotics: mechanisms and health benefits. Nutrients 5(4):1417–1435
82. Slavin JL, Jacobs D, Marquart L, Wiemer K (2001) The role of whole grains in disease prevention. J Am Diet Assoc 101(7):780–785
83. Song L, Thornalley PJ (2007) Effect of storage, processing and cooking on glucosinolate content of Brassica vegetables. Food Chem Toxicol 45(2):216–224
84. Stark AH, Kossoy G, Zusman I, Yarden G, Madar Z (2003) Olive oil consumption during pregnancy and lactation in rats influences mammary cancer development in female offspring. Nutr Cancer 46(1):59–65
85. Stoner GD, Wang LS, Zikri N, Chen T, Hecht SS, Huang C, Sardo C, Lechner JF (2007) Cancer prevention with freeze-dried berries and berry components. Semin Cancer Biol 17(5):403–410
86. Strimbu K, Tavel JA (2010) What are biomarkers? Curr Opin HIV AIDS 5(6):463
87. Thomson LU, Rickard SE, Orcheson LJ, Seidl MM (1996) Flaxseed and its lignan and oil components reduce mammary tumor growth at a late stage of carcinogeneis. Carcinogenesis 17:1373–1376
88. Thomson C, Bloch AS, Hasler CM, Kubena K, Earl R, Heins J (1999) Position of the American dietetic association. J Am Diet Assoc 99(10):1278–1285
89. Tufarelli V, Laudadio V (2016) An overview on the functional food concept: prospectives and applied researches in probiotics, prebiotics and synbiotics. J Exp Biol 4:3S
90. Uylaşer V, Yildiz G (2014) The historical development and nutritional importance of olive and olive oil constituted an important part of the Mediterranean diet. Crit Rev Food Sci Nutr 54(8):1092–1101
91. Van Poppel G, Verhoeven DT, Verhagen H, Goldbohm RA (1999) Brassica vegetables and cancer prevention. In: Advances in nutrition and cancer 2. Springer, New York, pp 159–168

92. Vattem DA, Ghaedian R, Shetty K (2005) Enhancing health benefits of berries through phenolic antioxidant enrichment: focus on cranberry. Asia Pac J Clin Nutr 14(2):120
93. Venter JC, Adams MD, Myers EW, Li PW, Mural RJ, Sutton GG, Smith HO, Yandell M, Evans CA, Holt RA, Gocayne JD (2001) The sequence of the human genome. Science 291(5507):1304–1351
94. Verhoeven DT, Goldbohm RA, Van Poppel G, Verhagen H, Van den Brandt PA (1996) Epidemiological studies on brassica vegetables and cancer risk. Cancer Epidemiol Biomarkers Prev 5(9):733–748
95. Verkerk R, Schreiner M, Krumbein A, Ciska E, Holst B, Rowland I, De Schrijver R, Hansen M, Gerhauser C, Mithen R, Dekker M (2009) Glucosinolates in Brassica vegetables: the influence of the food supply chain on intake, bioavailability and human health. Mol Nutr Food Res 53(S2):S219–S219
96. Wahle KW, Caruso D, Ochoa JJ, Quiles JL (2004) Olive oil and modulation of cell signaling in disease prevention. Lipids 39(12):1223–1231
97. Wahyuono RA, Hesse J, Hipler UC, Elsner P, Bohm V (2017) In vitro lipophilic antioxidant capacity, antidiabetic and antibacterial activity of citrus fruits extracts from Aceh, Indonesia. Antioxidants 6(1):11
98. Wallace TC, Murray R, Zelman KM (2016) The nutritional value and health benefits of chickpeas and hummus. Nutrients 8(12):766
99. Weerawatanakorn M, Hung WL, Pan MH, Li S, Li D, Wan X, Ho CT (2015) Chemistry and health beneficial effects of oolong tea and theasinensins. Food Sci Human Wellness 4(4):133–146
100. Weststrate JA, Van Poppel G, Verschuren PM (2002) Functional foods, trends and future. Br J Nutr 88(S2):S233–S235
101. Winham DM, Florian TLA, Thompson SV (2016) Low-income US women under-informed of the specific health benefits of consuming beans. PloS One 11(1):p.e0147592
102. Yang TT, Koo MW (1999) Chinese green tea lowers cholesterol level through an increase in fecal lipid excretion. Life Sci 66(5):411–423
103. Yang CS, Wang H (2016) Cancer preventive activities of tea catechins. Molecules 21(12):1679
104. Yang CS, Ju J, Lu G, Xiao H, Hao X, Sang S, Lambert JD (2008) Cancer prevention by tea and tea polyphenols. Asia Pac J Clin Nutr 17(Suppl 1):245
105. Yang CS, Wang X, Lu G, Picinich SC (2009) Cancer prevention by tea: animal studies, molecular mechanisms and human relevance. Nat Rev Cancer 9(6):429–439

Part IV

Functional Food and Human Health: Miscellaneous Effects of Functional Food and Components

Importance of Probiotics in Human Health

23

Dibyendu Banerjee, Tushar Jain, Sagarika Bose, and Vivek Bhosale

Abstract

Probiotics are live microorganisms like bacteria and yeast that are consumed along with food for various health benefits. These microorganisms play a very important role not only in intestinal microbial balance but also in maintaining health protection of the host organism against harmful microbiota. Probiotic term means "for life." The sale of these products has increased worldwide, and sale figures are estimated to cross $45B by the year 2018 due to its wide range of health profits. Humans have used probiotics for decades and centuries involuntarily. Food additives like fermented milk and yogurt have been a major part in regular meals and diet in several parts of the world. Nowadays, probiotics are the most important prescription by gastrointestinal physicians' because of its potential role in curing certain diseases for which treatment options are limited. The use of probiotics that were initially restricted for antibiotic-associated diarrhea now extends as powerful anti-allergic and anti-inflammatory factors. *Lactobacillus* and *Bifidobacterium* are the most important probiotic groups; however, the probiotic potential of *Pediococcus, Lactococcus, Bacillus,* and yeast is also well known. Probiotics are not only used as food to enhance health

D. Banerjee (✉) · T. Jain
Molecular and Structural Biology Division, CSIR-Central Drug Research Institute, Lucknow, Uttar Pradesh, India

Academy of Scientific and Innovative Research (AcSIR), Chennai, India
e-mail: d.banerjee@cdri.res.in

S. Bose
Molecular and Structural Biology Division, CSIR-Central Drug Research Institute, Lucknow, Uttar Pradesh, India

V. Bhosale
Department of Clinical and Experimental Medicine, CSIR-Central Drug Research Institute, Lucknow, Uttar Pradesh, India

Academy of Scientific and Innovative Research (AcSIR), Chennai, India

V. Rani, U. C. S. Yadav (eds.), *Functional Food and Human Health*,
https://doi.org/10.1007/978-981-13-1123-9_23

but also to cure specific gastrointestinal conditions such as irritable bowel syndrome, inflammatory bowel disease (IBD), and infectious diarrhea (caused by viruses, bacteria, or parasites). They also play an important role in improving skin conditions like eczema. Urinary and vaginal health, allergy prevention, colds, and oral health can be treated with probiotics. This chapter throws light on different probiotics, their health benefits, and regulatory or safety issues linked with their usage.

Keywords
Probiotics · Functional food · Gut microbiome · Probiotic bacteria · Gut health

23.1 Introduction and History

Let food be thy medicine, and medicine be thy food.
Everything in excess is opposed to nature.
All disease begins in the gut. – Hippocrates

Probiotic is the name given to microorganisms that are linked with providing health benefits to both humans and animals. The helpful microbiota contribute to intestinal microbial balance and help in maintaining good health. The application of microbes in food originated decades and centuries ago, when people first discovered the health benefits of eating fermented food [1]. In modern age, probiotic-containing food products are directly derived from the initially discovered fermented food products [2]. The Vedic hymns of India, written before 2000 B.C., showed that the fermented milk products were widely used in the diet by the people living in this part of the world since prehistoric times [3]. Between 2000 and 3000 B.C., several records have been found showing that milk, cheese, and butter were commonly used by a large number of different civic establishments (the Egyptians, Greeks, and Romans) [4]. Millennia ago, a Roman naturalist named Pliny the Elder recommended drinking fermented milk to cure different intestinal problems. About a hundred years ago, Dr. Elie Metchnikoff, the researcher known as the "father of natural immunity," first hypothesized that some intestinal bacteria "produce compounds useful against premature aging." The hypothesis was based on his observations that the consumption of certain strains of *Lactobacillus* in yogurt was associated with survival to old age as some communities have high percentage of long-lived individuals. This hypothesis was in vogue for a period, then floated to the edge of restorative practice, before coming back in favor in the mid-1990s as an idea deserving of standard medicinal consideration. Around the early 1900s, Henry Tissier, a French pediatrician, observed that children with diarrhea had a low number of bacteria characterized by a peculiar Y-shaped morphology in their stool as compared with healthy children. He suggested that these "bifid" bacteria can be of some help to restore a healthy gut flora when administered to patients with diarrhea. The term probiotics was coined by Lilly and Stillwell in 1965 in reference to substances produced by protozoa which triggered the development of various life-forms. However, Metchnikoff and Tissier did not use the term probiotic but they were the first to suggest probiotics as microorganisms that help another life-form, such as a human to grow and thrive [5]. Rosalie Stillwell and Daniel Lilly in the mid-twentieth century took a glance at the impacts of therapeutics medications that kill our gut microflora, such as the use of antibiotics or radiation and their effect on human health.

The Human Microbiome Project running at the National Institute of Health, USA, for the last 15 years led to the identification of microbes that live in our body. The researchers now realize the role of microorganisms that thrive on us and how well they affect how we feel or how frequently we become ill [6]. They are omnipresent and synergistic all through the complex gastrointestinal framework, where they have a crucial and defensive capacity [7]. The human gastrointestinal (GI) tract is a shelter to diverse groups of microbes representing over 400 cultivable species [8]. At first, *Escherichia coli* and *Streptococcus* dominate after birth, but in breast-fed infants, the number of the beneficial *bifidobacteria* increases whiles those of *E. coli*, *Streptococcus*, and *Clostridia* decrease. An adjustment in the grown-up vegetation happens when the disassociation occurs from breastfeeding, and by the second year of life, the intestinal flora resembles that of an adult and remains relatively stable throughout life [9]. The density of microorganisms increases successively from the stomach (102–103 colony-forming units (CFU)/g lumenal contents) to the colon (1011–1012 CFU/g lumenal contents). In a healthy adult, the gastrointestinal tract contains ten times more bacteria (1014 bacteria) as compared to eukaryotic cells in the entire human body [10].

Medical benefits sustained by satisfactory information or promising animal data involve cessation and treatment of diarrheal diseases (acute infantile diarrhea, antibiotic-associated diarrhea, nosocomial infections) [11], interruption of systematic infections [12], immune-modulation, treatment as well as prevention of allergies [13], treatment of cholesterolemia [14], anticancer effects [15], administration of systematic disorders [16], and lactose intolerance [17].

23.2 Definitions

Probiotics: The World Health Organization (WHO) defines probiotics as "live microorganisms which when administered in adequate amounts confer a health benefit on the host."

Prebiotics Prebiotics are a category of functional food, defined as: "Non-digestible food ingredients that beneficially affect the host by selectively stimulating the growth and/or activity of one or a limited number of bacteria in the colon, and thus improve host health."

Table 23.1 The most common species of bacteria in probiotic foods

Lactobacillus species	*Bifidobacterium* species	*Others*
L. acidophilus	*B. adolescentis*	*Streptococcus salivarius* ssp. *thermophilus*
L. plantarum	*B. bifidum*	*Lactococcus lactis* ssp. *lactis*
L. casei subspecies rhamnosus	*B. longum*	*Lactococcus lactis* ssp. *cremoris*
L. brevis	*B. infantis*	*Enterococcus faecium*
L. delbrueckii subspecies bulgaricus	*B. breve*	*Leuconostoc mesenteroides* ssp. *dextranicum*
		Propionibacterium freudenreichii
		Pediococcus acidilactici
		Saccharomyces boulardii

23.3 Synbiotics: Probiotics + Prebiotics

Food containing the amalgamation of probiotics and prebiotics are referred to as synbiotics. Advantages include improved survival in upper GIT (gastrointestinal tract) and more efficient implantation.

23.4 Probiotic Strains and Their Selection Criterion

A probiotic organism requires to be nonpathogenic and nontoxic and have resistance to low pH and bile salts to strengthen its odds of survival in the gastrointestinal tract [18]. According to WHO guidelines, the probiotics need authorization by in vitro tests followed by authorization from in vivo performance [19]. The test includes:

1. Gastric acid resistance
2. Resistance to bile acids
3. Adherence to the bodily fluid as well as human epithelial cells and cell lines
4. Antimicrobial property against harmful/pathogenic bacteria
5. Capability to lower the adhesion of pathogen to the surfaces
6. Hydrolase activity of bile salt
7. Resistance to spermicides (applicable to probiotics for vaginal use)

The most popular strains consist of the following genera: *Lactobacillus*, *Streptococcus*, *and Bifidobacterium* (Table 23.1); other microbes involve *Enterococci* and yeasts that are also used as probiotics. These strains are chosen based on selection criteria that are important for their survival in the gut, for example, the origin of strain, in vitro adherence to intestinal cells, and survival during passage through the gastrointestinal tract.

Bifidobacterium **Genus** *Bifidobacteria* are described as Gram-positive, non-spore-forming, nonmotile and catalase-negative anaerobes. They are different shapes involving short, curved rods, club-shaped rods, and bifurcated Y-shaped rods. Currently, there are 30 species included in the genus *Bifidobacterium* [20] which are microbes of vital importance in the active and complex ecosystem of the intestinal tract and genitourinary tracts of humans and other warm-blooded animals. Age and diet are the major factors for their distribution. In infants, *bifidobacteria* dominate over the gut microbiota, which are established shortly after the birth and later decreases with time.

Lactobacilli **Genus** These can be defined as normal dwellers of the human intestine and vagina. They are Gram-positive facultative anaerobes that are aero-tolerant or anaerobic and are widely used in fermentation processes. In the fermentative case, glucose is converted primarily into lactic acid. *Lactobacilli* are also categorized as lactic acid bacteria (LAB). Gomes and Malcata reported 56 species of the genus *Lactobacillus* that are known to promote health, especially in the human gastrointestinal and genitourinary tracts [21].

23.5 Mechanism of Action

Probiotics probably act by obstructing the adhesion sites and strengthening of gut epithelial barrier. The pathogenic bacterial adhesion sites on intestinal surfaces are competitively inhibited by *Lactobacillus* and *Streptococcus thermophilus,* thereby leaving no space for the pathogenic bacteria to grow at those sites. They are also known to produce inhibitory substances that inhibit the growth of harmful bacteria. *Lactobacillus* produces bacteriocins, hydrogen peroxide, acidofilin, and volatile acids that inhibit metabolism in both Gram-positive and Gram-negative bacteria [22]. *Bifidobacteria* lower the pH of the medium and therefore exert antibacterial effects by producing strong acids as metabolic end products (acetate, lactate). Bifidobacteria directly inhibit a range of Gram-positive and Gram-negative pathogenic bacteria by expelling metabolic end products [23]. In addition to this, they also help the host digest a variety of carbohydrates (Table 23.2) by utilizing a wide range of substrates such as polyols, disaccharides, oligosaccharides, or polysaccharides for their sustenance. Probiotics may also help in modulating the immune system to contribute an advantage to the host. Evidence shows that through pattern recognition receptors such as toll-like receptors and nucleotide-binding oligomer ization domain-containing protein-like receptors, the probiotics communicate with the host, thereby modulating key signaling pathways like nuclear factor-κB and mitogen-activated protein kinase (MAP-K), to intensify or repress activation and additional influence of downstream pathways [24]. This recognition is indispensable for eliciting measured anti-microbial responses with lessened inflammatory tissue damage. A better understanding of these processes and functioning will allow

Table 23.2 Types of probiotic substrates

Polyols (sugar alcohols)	*Oligosaccharides*
Xylitol	*Raffinose*
Sorbitol	*Soybean oligosaccharides (raffinose + stachyose)*
Mannitol (stereoisomer of sorbitol)	*Fructo-oligosaccharides (FOS) – the most commonly used prebiotic*
	Oligofructose (mixture of oligosaccharides)
	Galacto-oligosaccharides (trans-galactosylated oligosaccharides (TOS))
Disaccharides (synthetic derivatives of lactose)	
Lactulose	
Lactitol	
Other nondigestible oligosaccharides	***Polysaccharides***
Palatinose	*Inulin*
Isomaltose – oligosaccharides	*Resistant starch (starch modifications resistant to the host endogenous glycolytic enzymes)*
Lacto-sucrose	
Galactosyl lactose	

the selection of appropriate probiotic strains for particular applications and may unveil novel probiotic functions [24].

23.6 Health Benefits of Probiotics

For achieving desired health benefits from probiotics, a "proper dose" needs to be administered. The factors responsible for effective dose of probiotics involve health endpoint, usage of particular probiotics, delivery vehicle, and the administration route [25]. Table 23.3 lists a number of advantages attributed to the use of probiotics and defined in detail below:

Diarrheal Diseases The patients who are on antibiotic therapy when administered with probiotics showed a reduction in the incidence of antibiotic-associated diarrhea in adults as well as in children. Diarrhea among children is commonly caused by a viral strain called *Rotavirus* and is highly prevalent in developing countries. Several *Lactobacillus* species have been tested and found to be effective against *Rotavirus*-associated diarrhea. Diarrhea might be caused by other changes in intestinal microflora; however, *Clostridium difficile* is the major agent. Different strains examined involve *Lactobacillus bulgaricus*, *Lactobacillus acidophilus*, *Lactobacillus rhamnosus* strain GG, and the yeast *Saccharomyces boulardii*. Two different meta-analyses have concluded that probiotics could be used to prevent antibiotic-associated diarrhea [26]. Probiotics like *Lactobacillus* GG, *L. casei*, *Lactobacillus reuteri*, and *S. boulardii* have proven useful in lessening the duration of acute diarrhea in regulated clinical trials [27]. The dose for *rhamnosus* or *S.*

Table 23.3 Health benefits of probiotics

Gastrointestinal applications of probiotics	Non-gastrointestinal applications of probiotics
Traveler's diarrhea, gastroenteritis	Atopic eczema
Nosocomial diarrhea, clostridium difficile diarrhea	Cystic fibrosis
HIV/AIDS diarrhea, enteral feeding- associated diarrhea	Dental caries
Colon cancer	Various cancers
Irritable bowel syndrome (IBS)	Prevention of vaginitis
Inflammatory bowel disease	Juvenile chronic arthritis
Food allergies and lactose intolerance	Depression
Pancreatitis	Boosting immunity
Diarrhea of various origins	Urogenital infections
Antibiotic-associated diarrhea, Rotavirus diarrhea	Aphthous ulcers
H. pylori infection	Halitosis
Small bowel bacterial overgrowth	Enhancement of oral vaccine administration
Alcoholic liver disease	Hypercholesterolemia
Relieves constipation	Asthma

boulardii is currently recommended at 5 to 40 billion CFUs/day [28]. In a meta-analysis based on 25 studies covering 13 probiotic products by McFarland [29] on antibiotic-associated diarrhea (AAD) and *Clostridium difficile*-associated diarrhea (CDAD), a breakpoint was observed. Doses less than 10^{10} cfu tended to be unsuccessful studies. Probiotics also play a role in necrotizing enterocolitis (NEC). Upon conducting 26 studies with 21 different probiotic preparations, the beneficial effect of probiotics on lowering the risk for NEC was observed [30].

Combating Lactose Intolerance The starter cultures used in yogurt contain microbes that improve lactose absorption such as *S. thermophilus* and *Lactobacillus delbrueckii* subspecies *bulgaricus*. The enhancement of lactose absorption is due to the presence of microbial B-galactosidase (lactase) in the bacteria. A study conducted to compare utilization of fresh yogurt (with live yogurt cultures) and consumption of pasteurized product (with heat-killed bacteria) demonstrated better lactose digestion and absorption in people who consumed yogurt with live cultures. In addition to this, reduction in gastrointestinal symptoms was also observed [31].

Bowel Transit and Better Digestion Intake of probiotic microbes can possibly balance out the immunological barriers in the gut mucosa by lowering the production of local pro-inflammatory cytokines. The term inflammatory bowel disease (IBD) involves diverse states of the gut, of which two fundamental sorts are ulcerative colitis (UC) and Crohn's disease (CD). IBD refers to a group of disorders of unknown etiology that are distinguished by chronic or recurrent mucosal inflammation. The very much characterized, nonpathogenic strain *E. coli* Nissle 1917 demonstrated more success in forestalling backslide in Crohn's illness patients [32]. Through studies it was observed that *S. boulardii* has an important role in both relieving the active Crohn's disease symptoms and in lowering the risk of relapse.

Irritable Bowel Syndrome Irritable bowel syndrome is a chronic and serious condition that involves sporadic pain in the abdomen, GI symptoms like flatulence and bloating, and altered bowel habits (diarrhea and/or constipation). However, there are no noticeable structural changes or abnormalities in the intestine. Drouault-Holowacz et al. found that the patients on receiving a probiotic mix (*B. longum* LA 101, *L. acidophilus* LA 102, *Lactococcus lactis* LA 103, *St. thermophilus* LA 104) showed increased satisfactory relief of overall IBS symptoms and of abdominal discomfort/pain with increase in time [33]. *Bacteria in the intestine not only consume gas but also produce gas. Therefore, probiotics can lower the digestive gases accumulated within the intestine. The probiotics and prebiotics, when used in combination with a bulk laxative like ispaghula (psyllium husk), can be a safe and effective treatment for IBS-related constipation* [34].

Prevention of Systemic Infections It has been noted that the translocation of either viable or dead bacteria in small amounts can provide a physiologically necessary aid to the immune system. Dysfunction of the gut mucosal barrier can lead to large quantities of viable microbes passing into the body, which usually belong to Gram-negative aerobic genera and disperse all through the body, creating sepsis. Probiotics have been reported to prevent sepsis in patients suffering from severe acute pancreatitis. In a randomized double-blind trial, patients treated with either *Lactobacillus plantarum* or placebo showed significantly lower rate of infected pancreatic necrosis and abscesses in the *L. plantarum*-treated group as compared to the placebo-treated group [35].

Allergies Allergies are atopic diseases that are characterized by amplified immune response to the components around in the environment or food. The determining factor in a person's predisposition to allergies is the proper balance between microorganisms (*bifidobacteria* and *clostridia*). Through studies it is observed that in breastfed infants, the chance of occurrence of allergies is relatively lower than formula-fed infants because of higher counts of *bifidobacteria* in their flora; formula-fed infants have more *Clostridia* in their gut [36].

Vaginal Infection Vaginal and urinary tract infections (UTIs) are common clinical entities occurring in a variety of patient groups. The microorganisms responsible to cause infections are *Klebsiella* spp., *Proteus* spp., *Staphylococcus* spp., and *Escherichia coli*, which are also known as intestinal uropathogens [37]. Vaginal flora mainly consists of *Lactobacilli* that help in preventing infection by dominating over pathogenic bacteria.

Cancer Prevention Cancer is yet another disease that can be prevented using probiotic products as suggested by several studies. Hosono et al. were the first to observe that milk fermented using *Lactococcus lactis* subsp. *lactis*, *L. delbrueckii* subsp. *Bulgaricus, or Enterococcus faecalis* possessed an antimutagenic activity. Many studies have suggested that the risk of cancer can be lowered using probiotic products in daily life [38, 39]. The mechanisms of anti-carcinogenicity include:

- Ability to bind to carcinogens
- Reaction on bacterial enzymes
- Generation of anti-tumorigenic or antimutagenic compounds
- Improvement of the host's immune response

Research on human subjects revealed that the risk of colon cancer can be significantly lessened using probiotic treatment as this transforms/converts pro-carcinogens to active carcinogens by various mechanisms like binding/inactivating mutagenic compounds, suppressing the growth of pro-carcinogenic bacteria, reducing the absorption of mutagens from the intestine, producing antimutagenic compounds, and enhancing immune function [40]. The incidence of colon/breast cancer and the consumption of fermented dairy products containing *Lactobacilli* and *Bifidobacterium* are indirectly related, as reported in epidemiological and population-based case-control studies [41].

Boosting Immunity Several studies have shown enhancement of our immune functions using probiotic products. There is a strong indication that specific strains of lactic acid bacteria are able to modulate aspects of not only natural but also acquired immune responses, when consumed in certain numbers. Studies have also reported positive impact of probiotics on overall immunogenicity of mucosal and systemic vaccines. Link-Amster and colleagues observed a fourfold increase in serum IgA antibody response to *Salmonella typhi* immunization in volunteers consuming yogurt that contained *B. bifidum* and *L. acidophilus* [43].

Combating *Helicobacter pylori* Infection With over half the world's population colonized, Gram-negative *Helicobacter pylori* is one of the most prosperous human pathogenic bacterium and is formally categorized as bacterial carcinogen [43]. *H. pylori* has also been examined using probiotics treatment. Experiments on animal model suggested that *H. pylori* attachment and colony formation can be inhibited using some *Lactobacilli* strain [44]. Several processes by which probiotics facilitate anti-infective effects have been reported, with producing antimicrobials, strengthening gut barrier function, stimulating the secretion of mucus, stimulating specific as well as non-specific immune responses, and competing for adhesion sites being the major ones.

Lowering Blood Cholesterol Increased cholesterol levels are accountable for ischemic heart disease that leads to morbidity and mortality in humans. The chance of occurrence or incidence and mortality of ischemic heart disease in healthy individuals can be reduced significantly by primary prevention with lipid lowering drugs or dietary alterations. Probiotic bacteria are reported to reduce cholesterol level by deconjugating bile salts: deconjugated bile acid does not absorb lipid as readily as its conjugated counterpart [45]. The clinical trials of serum lipid modulation use a wide range of probiotic products. The oral administration of *L. johnsonii* and *L. reuteri* in pigs leads to the decrease in serum cholesterol. Administration of certain strains of lactic acid bacteria or fermented milk helps in reducing blood cholesterol levels [46].

Treatment of Depression The prevention and treatment of depression could be one of the most promising uses of probiotics. The main cause of depression is stress, and it is said to change the microflora by altering the composition of gut. Upon experimenting it was found that the rats which showed a depressive behavior due to maternal separation when fed with the probiotic *B. infantis* resulted in normalization of their behavior and a healthier immune system [47]. A human study indicated lower levels of stress in subjects who received active probiotic organisms as compared to others who were given placebo treatment. A similar double-blinded human trial was carried out to compare probiotics and placebos among patients who suffered from severe fatigue. Promising results were generated in the case of patients who consumed probiotics by reporting significant improvements in their symptoms [48]. The term "psychobiotics" has been coined for probiotics which show promise for the prevention and/or treatment of psychiatric illness.

23.7 Regulations and Guidelines for Probiotics

Probiotics are widely accessible in food and dietary supplements, as pharmaceutical formulations (capsules, tablets, and powders) and in some other forms as well, but their claims of health benefits may challenge the traditional border between food and medicine. Introduction of a wide range of probiotics into the international market has already taken place as food supplements, dietary supplements, functional foods, natural health products, and many other forms; because of which, the position of the regulatory system for probiotics within existing categories becomes ambiguous and quite unclear. Probiotics are considered as dietary supplement or drugs in the USA [49]. In India, specific guidelines have been issued by ICMR-DBT for probiotics [50]. Regulatory instructions have also been framed by WHO

Table 23.4 Probiotic products marketed in various countries

Country	Probiotic products
India	Amul probiotic (ice cream, dahi, and lassi), Mother Dairy (bActiv Probiotic Dahi, bActiv Probiotic Lassi, bActiv Curd and Nutrifit), Yakult-Danone probiotic drink and Nestle Nesvita
Japan	Yakult, Meiji Bulgaria yogurt and drinks, Morinaga Bifidus yogurt, Calpis Ameal S120 probiotic drink
Europe	Lifeway Kefir, Culturelle probiotic infant formula, France Bravo Friscus, Yoplait yogurt, Danone Activia, Actimel probiotic yogurt
USA	Activia creamy yogurt, Danone Activia Plus, Blue Bunny-Sedona yogurt ice cream, Yovation-Pierre's probiotic ice cream
UK	Vita-Yo creamy probiotic yogurt, Yeo Valley Bio Live yogurt, Unilever's Flora Proactive
Finland	Valio Gefilus, Valio Kidius Gefilus, and Evolus milk drink, Yosa yogurt oat product, Bioferme Valio Villis
Canada	BioBest Plant Sterols probiotic yogurt, Kraft LiveActive cheese and chocolates, Liberte yogurt, Olympic Natural no fat probiotic yogurt
Brazil	Chamyto probiotic drink Nestle, Actimel *L. casei* Defensis, Danito *L. casei* Danone, Sofyl Yakult
Sweden	Provita yogurt, Yogen Fruz frozen yogurt, BioGaia products, LiveActive probiotic products
Australia	F&N's aLIVE fruit chunk yogurt, Yo-Plus digestive yogurt, Wallaby organic yogurt, Vaalia probiotic yogurt, Ombar probiotic chocolate

for probiotics in the year 2002. Within the functional foods, probiotics are the major player and the fastest growing industry and the most active arena.

In modern age, the demand for probiotic products has exponentially increased to an estimated 32.6 billion US$ in 2014, with a predicted annual increase of 20% [51]. The European market makes for 42% of this, and dairy products makes for 70%. In 2005, the USA made only $764 million by the sales of probiotic-containing products, representing 5% of the dietary supplement market [52].

In contrast, both Japan and the European Union make much larger and more mature probiotic markets than the USA, because of its long history of traditional use, different regulatory frameworks, consumer comfort with microbial ingredients in food, and perceived health benefits from probiotic products in these regions [52]. The European Regulation on Nutrition and Health Claims of 2006 provides common regulation health claims to be made on foods in a uniform manner throughout the member states in the European Union (EU) [53]. Globally, probiotics are expected to be a $45B business in the world by 2018 [51]. In India, the major players of probiotic industry are Amul, Mother Dairy, Yakult, Danone, and Nestle, along with other minor players functioning in different regions of the country. Probiotics can be used in combination with medicines, vitamins, etc. In India, 800 different preparations of *Lactobacillus* are accessible [54]. Table 23.4 lists the probiotic products that are marked across the world as supplements.

Given this large market and increasing demand for probiotics, it is time that NIH, in collaboration with FDA, should develop a regulatory document in order to aid clinical researchers in better maintaining the framework. Certified methods in preclinical studies, such as those of in vitro and in vivo animal models of disease, have started to contribute important new data and surrogate markers of both safety and efficacy of probiotics [55–59]. This may help in increase in the success rate of phase 2 and 3 clinical studies by better designing of phase 1 studies both in individuals and in patients.

23.8 Safety of Probiotics

Probiotics are proven to be remarkably safe as per the record; however, some safety measures should be taken in some patient groups like prematurely born neonates or those with immune deficiency. Though probiotics are safe products to use, still there are certain areas that need to be further investigated like probiotic interactions or their appropriate administrative regimens so as to study and learn the mechanisms through which probiotics work. In addition, different probiotic species retain different properties; these properties are strain specific and variable. Therefore, the effects of one probiotic strain cannot be justifiable to others without proper studies to validate the same. Also, further careful considerations are required to address the issues before clinically suggesting patients to make use of probiotic supplements in their course of treatment. With regard to increasing market value, their quality-, safety-, and efficacy-related affairs are becoming significantly indispensable for both the consumers and manufacturers/regulators. As per current scenario, regulatory guidelines stipulating requirements of "Good Manufacturing Practice (GMP)" are not yet covered properly by any country except Canada, and hence regulatory regimes are at best unsatisfactory [60].

Some clinical trials and meta-analyses claim that probiotics have a brief and little impact on human gut microbiome with little clinical benefits [61]. In the USA, probiotic manufacturers have received warning letters of impending legal actions for using exaggerated claims of health benefits not supported by clinical evidence of efficacy where even food product labeling requires language approval by the FDA. There are even reports associating probiotics to severe side effects like dangerous infections, in people with serious underlying medical problems [62]. The people who are critically ill or those who have had surgery, very sick infants, and people with weakened immune systems are the people who are most at risk of severe side effects. Even for the healthy people, there are uncertainties about the safety of the probiotics, because not much studies have been done to closely look at the safety parameter, thereby providing very little or no information about the safety issues.

23.9 Future Perspective

The intestinal microbiota needs to be further evaluated by the use of state-of-the-art technologies. To study the mechanisms of action of probiotics and biomarkers, in vivo and in vitro animal models are required to assess the biologic effects of probiotics. Further studies are also needed to conclude high-risk populations that should use probiotics only under the supervision of a qualified healthcare professional. Proper guidelines are required or needed to be in place for high-quality preparations to formulate or establish product safety and efficacy in disease states for creating sound biological basis for conducting phase 2 and phase 3 clinical studies of specific probiotic strains.

23.10 Conclusions

Probiotics are live microorganisms that possess certain health benefits when consumed. Probiotics can be taken in combination with dietary supplements as well as with medicines. Probiotic industry is flourishing at a tremendous rate because of its market demand/value. They have been extensively studied and explored commercially in various parts of the world. Their advantages to both animal and human health have been studied and reported in hundreds of scientific articles. *Lactobacillus* and *Bifidobacterium* are the major probiotic groups, and reports on other potential probiotic groups like *Pediococcus*, *Lactococcus*, *Bacillus*, and yeasts are also mentioned. They play an important role in different diseases like diarrhea, inflammatory conditions, depression, and the most important cancer, among others. The commercial business of probiotic products is exponentially rising all over the world, demanding GMP and regulatory requirements to be followed more stringently.

Acknowledgments The authors would like to acknowledge CSIR-CDRI for infrastructure and other support necessary for writing the book chapter. TJ would like to thank CSIR for his junior and senior research fellowships. The manuscript bears the CDRI communication number 9484.

References

1. Ozen M, Dinleyici EC (2015) The history of probiotics: the untold story. Benefic Microbes 6(2):159–165
2. Gogineni VK et al (2013) Probiotics: history and evolution. J Anc Dis Prev Remedies. https://doi.org/10.4172/2329-8731.1000107
3. Muller M (1900) Sacred books of the east. The Colonial Press, New York
4. Breasted J (1906) Ancient records of Egypt. University of Chicago Press, Chicago
5. Langdon A, Crook N, Dantas G (2016) The effects of antibiotics on the microbiome throughout development and alternative approaches for therapeutic modulation. Genome Med 8(1):39
6. Grice EA, Segre JA (2012) The human microbiome: our second genome. Annu Rev Genomics Hum Genet 13:151–170
7. Fijan S (2014) Microorganisms with claimed probiotic properties: an overview of recent literature. Int J Environ Res Public Health 11(5):4745–4767
8. Gill HS, Guarner F (2004) Probiotics and human health: a clinical perspective. Postgrad Med J 80(947):516–526
9. Rodriguez JM et al (2015) The composition of the gut microbiota throughout life, with an emphasis on early life. Microb Ecol Health Dis 26:26050
10. Yang I et al (2016) The infant microbiome: implications for infant health and neurocognitive development. Nurs Res 65(1):76–88

11. Bergonzelli GE et al (2005) Probiotics as a treatment strategy for gastrointestinal diseases? Digestion 72(1):57–68
12. Khailova L et al (2013) Probiotic administration reduces mortality and improves intestinal epithelial homeostasis in experimental sepsis. Anesthesiology 119(1):166–177
13. Prakash S et al (2014) Probiotics for the prevention and treatment of allergies, with an emphasis on mode of delivery and mechanism of action. Curr Pharm Des 20(6):1025–1037
14. Ooi LG, Liong MT (2010) Cholesterol-lowering effects of probiotics and prebiotics: a review of in vivo and in vitro findings. Int J Mol Sci 11(6):2499–2522
15. Kumar R, Dhanda S (2017) Mechanistic insight of probiotics derived anticancer pharmaceuticals: a road forward for cancer therapeutics. Nutr Cancer 69:1–6
16. Sinagra E et al (2013) Probiotics, prebiotics and symbiotics in inflammatory bowel diseases: state-of-the-art and new insights. J Biol Regul Homeost Agents 27(4):919–933
17. Shaukat A et al (2010) Systematic review: effective management strategies for lactose intolerance. Ann Intern Med 152(12):797–803
18. Brown AC, Valiere A (2004) Probiotics and medical nutrition therapy. Nutr Clin Care 7(2):56–68
19. Liaskovskii TM, Podgorskii VS (2005) Assessment of probiotics according to the international organizations (FAO/WHO). Mikrobiol Z 67(6):104–112
20. Soccol CR et al (2010) The potential of probiotics: a review. Food Technol Biotechnol 48(4):413–434
21. Gomes AM, Malcata FX (1999) Bifidobacterium spp. and Lactobacillus acidophilus: biological, biochemical, technological and therapeutical properties relevant for use as probiotics. Trends Food Sci Technol 10(4):139–157
22. Dembele T, Obdrzalek V, Votava M (1998) Inhibition of bacterial pathogens by lactobacilli. Zentralbl Bakteriol 288(3):395–401
23. Gibson GR et al (1995) Selective stimulation of bifidobacteria in the human colon by oligofructose and inulin. Gastroenterology 108(4):975–982
24. Bermudez-Brito M et al (2012) Probiotic mechanisms of action. Ann Nutr Metab 61(2):160–174
25. Ouwehand AC (2016) A review of dose-responses of probiotics in human studies. Benefic Microbes 8:1–10
26. Ritchie ML, Romanuk TN (2012) A meta-analysis of probiotic efficacy for gastrointestinal diseases. PLoS One 7(4):e34938. Epub 2012 Apr 18
27. Cruchet S et al (2015) The use of probiotics in pediatric gastroenterology: a review of the literature and recommendations by Latin-American experts. Paediatr Drugs 17(3):199–216
28. Goldenberg SD (2016) Faecal microbiota transplantation for recurrent Clostridium difficile infection and beyond: risks and regulation. J Hosp Infect 92(2):115–116
29. McFarland LV (2015) Deciphering meta-analytic results: a mini-review of probiotics for the prevention of paediatric antibiotic-associated diarrhoea and Clostridium difficile infections. Benefic Microbes 6(2):189–194
30. Aceti A et al (2015) Probiotics for prevention of necrotizing enterocolitis in preterm infants: systematic review and meta-analysis. Ital J Pediatr 41:89
31. Savaiano DA (2014) Lactose digestion from yogurt: mechanism and relevance. Am J Clin Nutr 99(5 Suppl):1251s–1255s
32. Korada SK et al (2016) Can probiotics cure inflammatory bowel diseases? Curr Pharm Des 22(7):904–917
33. Drouault-Holowacz S et al (2008) A double blind randomized controlled trial of a probiotic combination in 100 patients with irritable bowel syndrome. Gastroenterol Clin Biol 32(2):147–152
34. Faruqui AA, Jamaiwar B, Sharma B (2017) Combination of bulk laxative along with probiotics and prebiotics in the management of irritable bowel syndrome with constipation: a safe & effective option. Asian J Med Sci 8(1):6
35. Vaishnavi C (2013) Translocation of gut flora and its role in sepsis. Indian J Med Microbiol 31(4):334–342

36. Slattery J, MacFabe DF, Frye RE (2016) The significance of the enteric microbiome on the development of childhood disease: a review of prebiotic and probiotic therapies in disorders of childhood. Clin Med Insights Pediatr 10:91–107
37. Lim IS, Lee HS, Kim WY (2009) The effect of lactic acid bacteria isolates on the urinary tract pathogens to infants in vitro. J Korean Med Sci 24(Suppl):S57–S62
38. Saber A et al (2017) Cellular and molecular effects of yeast probiotics on cancer. Crit Rev Microbiol 43(1):96–115
39. Chong ES (2014) A potential role of probiotics in colorectal cancer prevention: review of possible mechanisms of action. World J Microbiol Biotechnol 30(2):351–374
40. Liong MT (2008) Roles of probiotics and prebiotics in colon cancer prevention: postulated mechanisms and in-vivo evidence. Int J Mol Sci 9(5):854–863
41. Pericleous M, Mandair D, Caplin ME (2013) Diet and supplements and their impact on colorectal cancer. J Gastrointest Oncol 4(4):409–423
42. Link-Amster H et al (1994) Modulation of a specific humoral immune response and changes in intestinal flora mediated through fermented milk intake. FEMS Immunol Med Microbiol 10(1):55–63
43. Sgouras DN, Trang TT, Yamaoka Y (2015) Pathogenesis of Helicobacter pylori infection. Helicobacter 20(Suppl 1):8–16
44. Sgouras D et al (2004) In vitro and In vivo inhibition of Helicobacter pylori by Lactobacillus casei strain Shirota. Appl Environ Microbiol 70(1):518–526
45. Kumar M et al (2012) Cholesterol-lowering probiotics as potential biotherapeutics for metabolic diseases. Exp Diabetes Res 2012:902917
46. Lye HS et al (2009) The improvement of hypertension by probiotics: effects on cholesterol, diabetes, renin, and phytoestrogens. Int J Mol Sci 10(9):3755–3775
47. Yarandi SS et al (2016) Modulatory effects of gut microbiota on the central nervous system: how gut could play a role in neuropsychiatric health and diseases. J Neurogastroenterol Motil 22(2):201–212
48. Rao AV et al (2009) A randomized, double-blind, placebo-controlled pilot study of a probiotic in emotional symptoms of chronic fatigue syndrome. Gut Pathog 1(1):6
49. Degnan FH (2008) The US Food and Drug Administration and probiotics: regulatory categorization. Clin Infect Dis 46(Suppl 2): S133–S136; discussion S144-51.
50. ICMR-DBT guidelines for evaluation of probiotics in food (2011) Indian J Med Res 134:22–25
51. Market RA (2016) *Probiotics Market by Ingredient (Bacteria, Yeast), Application (Food & Beverages, Dietary Supplements, Animal Feed), Function (Regular, Preventative Healthcare, Therapeutic) – Global Opportunity Analysis and Industry Forecast, 2014 – 2022*
52. Hoffman FA et al (2008) Executive summary: scientific and regulatory challenges of development of probiotics as foods and drugs. Clin Infect Dis 46(Suppl 2):S53–S57
53. Flynn A (2012) Scientific substantiation of health claims in the EU. Proc Nutr Soc 71(1):120–126
54. Sreeja V, Prajapati JB (2013) Probiotic formulations: application and status as pharmaceuticals—a review. Probiotics and Antimicrobial Proteins 5(2):81–91
55. Redman MG, Ward EJ, Phillips RS (2014) The efficacy and safety of probiotics in people with cancer: a systematic review. Ann Oncol 25(10):1919–1929
56. Goldenberg JZ et al (2015) Probiotics for the prevention of pediatric antibiotic-associated diarrhea. Cochrane Database Syst Rev 12:Cd004827
57. Lv Z et al (2015) Efficacy and safety of probiotics as adjuvant agents for Helicobacter pylori infection: a meta-analysis. Exp Ther Med 9(3):707–716
58. Yan Q, Li X, Feng B (2015) The efficacy and safety of probiotics intervention in preventing conversion of impaired glucose tolerance to diabetes: study protocol for a randomized, double-blinded, placebo controlled trial of the Probiotics Prevention Diabetes Programme (PPDP). BMC Endocr Disord 15:74
59. Xu M et al (2015) The efficacy and safety of the probiotic Bacterium Lactobacillus reuteri DSM 17938 for Infantile Colic: a meta-analysis of randomized controlled trials. PLoS One 10(10):e0141445

60. Sanders ME, Levy DD (2011) The science and regulations of probiotic food and supplement product labeling. Ann N Y Acad Sci 1219(Suppl 1):E1–e23
61. Hemarajata P, Versalovic J (2013) Effects of probiotics on gut microbiota: mechanisms of intestinal immunomodulation and neuromodulation. Therap Adv Gastroenterol 6(1):39–51
62. Didari T et al (2014) A systematic review of the safety of probiotics. Expert Opin Drug Saf 13(2):227–239

Aldose Reductase Inhibitors in the Functional Foods: Regulation of Diabetic Complications

24

Arpita Devi, Aramati B. M. Reddy, and Umesh C. S. Yadav

Abstract

Aldose reductase (AR) is implicated in the precipitation of diabetic complications including the microangiopathy such as nephropathy, neuropathy, retinopathy as well as macroangiopathy such as cardiovascular disease in diabetes patients. A number of synthetic inhibitors of AR have been tested in laboratories as well as in the clinical settings but with limited success due to their low bioavailability and safety concerns. Since the epidemiological studies suggest the enormous increase in incidence of diabetes globally, there is urgent need for potent yet safe AR inhibitors. The molecules derived from plants usually show greater biocompatibility and lack toxicity and other harmful side effects due to their coevolution with human and also because they have been the part of the local diet regimen for the millennia. Several researchers have reported that crude extracts as well as pure molecules derived from natural sources such as plants and various plant parts possess excellent AR inhibitory activity. Such compounds have shown remarkable inhibition of elevated AR activity during diabetic condition. Although the chemical nature of the plant extracts is of mixed type, the pure molecules isolated from plants are of varied chemical nature such as flavonoids, alkaloids, phenolics, tannins, coumarins, terpenoids, etc. In this chapter the different plant-derived extracts and compounds have been discussed that have been shown to have potential AR inhibition activity and may qualify to be considered as the functional food for prevention of diabetes and diabetic complications.

A. Devi · A. B. M. Reddy
Department of Animal Biology, School of Life Sciences, University of Hyderabad, Hyderabad, India

U. C. S. Yadav (✉)
Metabolic Disorder & Inflammatory Pathologies Laboratory, School of Life Sciences, Central University of Gujarat, Gandhinagar, Gujarat, India
e-mail: umeshyadav@cug.ac.in

V. Rani, U. C. S. Yadav (eds.), *Functional Food and Human Health*,
https://doi.org/10.1007/978-981-13-1123-9_24

Keywords
Aldose reductase · Diabetes · Functional food · AR inhibitors · Diabetic complications · Herbal compounds · Phytochemicals

24.1 Introduction

Aldose reductase (AR) (EC 1.1.1.21) is an oxidoreductase enzyme which uses NADPH as cofactor to reduce its substrate, which is mostly aldehydic in nature, e.g. glucose, and thus belongs to the family of aldo-keto reductases. The protein is encoded by AKR1B1 gene located on the long arm (q) of chromosome 7 [1]. AR catalyses the reduction of various aldehydes to their corresponding alcohols. It can also oxidise different alcohols to their corresponding aldehydes. AR is the first enzyme of polyol pathway of glucose metabolism and catalyses the following reaction (Fig. 24.1) [2].

AR is located at various parts of the cells including the exterior of the cells. It may localise in the cytoplasm, extracellular exosome, extracellular space, mast cell granule, nucleoplasm, paranodal junction, perinuclear region of cytoplasm, Schmidt-Lanterman incisure and Schwann cell microvillus (source: UniProt). AR has been shown to be excessively expressed in embryonic epithelial cells in response to osmotic stress [3]. It is also highly expressed in the kidney, lens and Schwann cells of neurons in diabetic condition due to high concentration of glucose (30 mM) [4, 5].

24.1.1 Structure of AR and Mechanism of Action

AR is expressed as a 37 kDa protein with 316 amino acids; however, the initiator residue methionine is removed after translation leaving a functional 315 residues long protein [6]. The second residue alanine undergoes post-translational modification to N-acetylalanine [6]. AR has also been shown to undergo mutagenesis which results in loss of its enzymatic activity. Due to mutagenesis, the residues alanine, tyrosine, lysine and histidine at 43, 48, 77 and 110 positions, respectively, are replaced by arginine, phenylalanine, methionine and arginine, respectively [7].

Structurally, AR protein structure involves β-/α-barrel motif containing eight parallel β strands [8] which are joined by eight α-helical segments which run antiparallel to β sheets [9]. The active site of the enzyme is located in the centre of β-barrel structure [9]. The binding of NADPH occurs at the top of β/α barrel in such a way that nicotinamide projects downwards at the centre of the barrel, whereas the pyrophosphate hangs near the opening of the barrel [10]. Recent X-ray crystallographic structure of AR shows that it has two binding sites, one for glutathione and other for the carbonyl [11]. The glutathione-binding pocket in AR active site is lined by Trp 20, Trp 219, Leu 300, Leu 301 and Ser 302, whereas the pocket that binds carbonyl group is lined by Tyr 48, His 110, Trp 111, Phe 122 and NADPH [11]. Tyr 48 is involved in the catalytic activity as it acts as a proton donor [12].

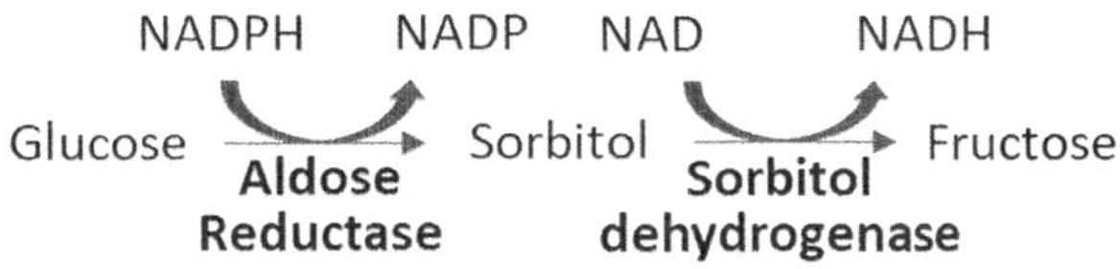

Fig. 24.1 The schematics of polyol pathway where glucose is first converted to sorbitol and finally to fructose

The reduction of aldehyde starts with binding of NADPH after which the substrate binds. As NADPH approaches the active centre, it induces change in the conformation which involves movement of the surface loop formed by residues 213–217 akin to a hinge. Transfer of H^+ from NADPH from C-4 of its nicotinamide ring leads to the reduction of aldehyde to alcohol. Another conformational change precedes with release of alcohol from the enzyme followed by release of $NADP^+$ [13].

24.1.2 Biological Function of AR

AR is involved in a number of physiological events such as carbohydrate catabolism; sorbitol biosynthetic process; steroid hormone and fructose biosynthesis; daunorubicin, doxorubicin, naphthalene and norepinephrine metabolic processes; cellular response to hydrogen peroxide, methylglyoxal and peptides, inner medullary collecting duct development; pregnancy-related events; modulation of JAK-STAT pathway; proliferation of smooth muscle; response to stress, thyroid hormone and water deprivation, stress-activated protein kinase signalling pathways; and tissue homeostasis (source: UniProt). These various metabolic processes appear to be different, although the basic principle behind all of them involves the conversion of toxic aldehyde species into the non-toxic alcohol species which either get excreted from the body or induce varied molecular responses.

24.1.3 Aldose Reductase and Detoxification

Oxidation of unsaturated fatty acids produces short-chain aliphatic aldehydes. Some of these aldehydes are 4-hydroxynonenal (4-HNE), malondialdehyde, methylglyoxal and acrolein. Being highly reactive in nature, they readily react with cellular nucleophiles such as glutathione, histidine, lysine or arginine residues in proteins and guanosine bases in DNA [14]. These conjugates are highly toxic to the cells leading to protein and DNA damages. Thus, reduction of these toxic aldehyde species by aldose reductase is important detoxification event in the cells. Among these aldehydes, 4-HNE is the major component constituting approximately 95% of the aldehydes. Since glutathione conjugate of 4-HNE is an excellent AR substrate, AR readily reduce the conjugate and thus plays a major role in cellular detoxification (Fig. 24.2). AR also readily reduces 4-HNE to 1,4-dihydroxy-2-nonene (4-DHN) which is non-toxic molecule and exported out of the cell by specific exporter. However, there is a thin line between detoxification and induction of cell proliferation. Excessive reduction of 4-HNE leads to the induction of abnormal cell

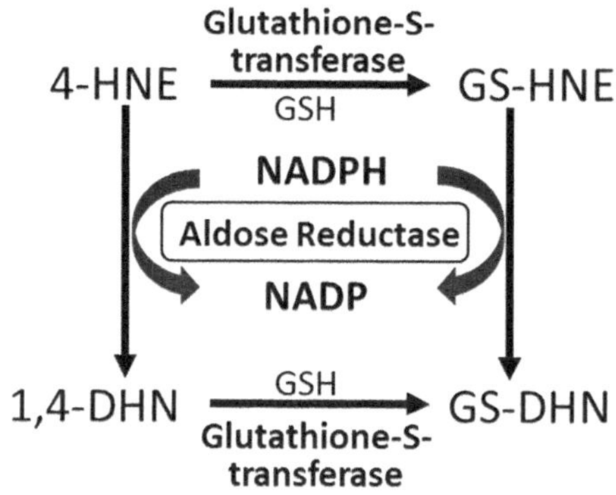

Fig. 24.2 Schematic diagram depicting detoxification of harmful aldehydes by AR. *4-HNE* 4-hydroxynonenal, *GS-HNE* glutathionylated 4-hydroxynonenal, *1,4-DHN* 1,4-dihydroxy-2-nonene, *GS-DHN* glutathionylated 1,4-dihydroxy-2-nonene, *GSH* glutathione, *AR* aldose reductase

proliferation in certain cells, whereas no reduction of 4-HNE would lead to its accumulation and that would be toxic to cells. Therefore, a balance between detoxification and induction of cellular process needs to be maintained for survival of the cell in a healthy condition [15–17].

24.2 Aldose Reductase in Diabetes and Other Inflammatory Pathologies

AR is the first enzyme of polyol pathway, which catalyses the conversion of glucose, an aldehydic sugar, to sorbitol, an alcohol. The expression and activation of AR is dependent upon glucose concentration in the cells, and it becomes active when glucose concentration rises to approximately 30 mM or more. AR has been reported to be involved in many pathological situations. It is mostly involved in oxidative stress which is one of the reasons for the development of many disease conditions such as diabetic complications, cancer, atherosclerosis, ischemia, etc. Overexpression of AR is found to be involved in liver, breast, cervical and colon cancers [18, 19]. However, most of the studies involving AR have been focused on diabetic complications and, recently, a few on cancer.

Glucose is easily diffusible into some cell types such as nephrons, Schwann cells of neurons and lenses; however it requires glucose transporters (GLUTs) in certain other cell types such as muscles and liver cells, which import glucose in response to insulin. Thus, in diabetic condition, high-glucose concentration in the nephrons, Schwann cells and lenses increases the expression and activation of AR in these cells. Increased AR then reduces glucose to sorbitol, which, being impermeable, accumulates in these cells. Since during this conversion, NADPH is used as a cofactor, more and more NADPH is used in this enzymatic reaction, and as a result the ratio of NADPH/NADP is decreased. NADPH, which plays important role in the

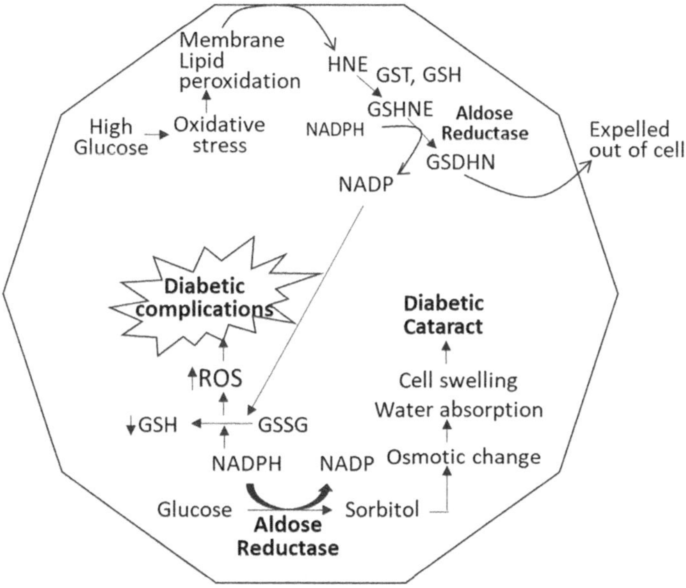

Fig. 24.3 Schematic diagram showing the role of AR in the development of diabetic complications, cataract and detoxification of lipid aldehydes. *ROS* reactive oxygen species, *4-HNE* 4-hydroxynonenal, *GS-HNE* glutathionylated 4-hydroxynonenal, *1,4-DHN* 1,4-dihydroxy-2-nonene, *GS-DHN* glutathionylated 1,4-dihydroxy-2-nonene, *GSSG* oxidised glutathione, *GSH* reduced glutathione, *GST* glutathione-s-transferase, *AR* aldose reductase

regeneration of reduced glutathione, is an important reducing power and indirectly controls oxidative stress in the cells. Thus, decreased NADPH levels cause depletion of reduced glutathione level in the cells, and it becomes difficult to cope with the free radicals formed inside the cells, and the cells undergo oxidative stress. Increased oxidative stress leads to cellular dysfunction as well as apoptosis leading to complications such as neuropathy, nephropathy and retinopathy [5, 20]. In the case of lens cells, the accumulated sorbitol also causes osmotic imbalance in the cells, and to balance the same, more and more water is taken up into the cells. This leads to the opaqueness of lens cells resulting in the formation of cataract (Fig. 24.3) [21].

AR has also been identified to induce inflammatory signals that are initiated by growth factors (GF), cytokines and various carcinogens. Excessive ROS generation in response to cytokines, GF, lipopolysaccharides (LPS) and hyperglycaemia causes lipid peroxidation and formation of lipid-derived aldehydes (LDAs) and also damage to DNA. The LDAs such as 4-HNE and their glutathione (GSH) conjugates are excellent AR substrates [16]. Reduction of glutathione conjugate by AR forms glutathionyl 1,4-dihydroxynonanol (GS-DHN) which is involved in activating NF-κB and AP-1, which then transcribe various inflammatory genes. In this manner, increased synthesis of cytokines and growth factors could lead to enhance cell growth, which may lead to neoplasia [14, 15, 22].

24.3 Therapeutic Potential of Aldose Reductase Inhibitors

AR has been implicated in many pathophysiological processes in addition to diabetic complications such as cardiovascular complications, cancers, ovary dysfunction and liver diseases. Accordingly, AR inhibitors are expected to be important therapeutic molecules in these diseases. So far, epalrestat, the only ARI marketed in Japan, has been clinically studied for diabetic neuropathy in 5000 patients. Patients were treated for 3–12 months with epalrestat which resulted in 75% improvement in pain, coldness, numbness and hypoesthesis and 36% improvement in motor nerve and sensory nerve conduction velocities [23]. Only approximately 2.5% patients showed non-adverse side effects. In a 52-week clinical study with another ARI, tolrestat, given at 200 mg dose daily, showed significant improvement in tibial and peroneal motor nerve conduction velocities. Paraesthetic symptoms and motor nerve conduction velocities were found improved by 24 weeks of treatment [24]. Zopolrestat, yet another ARI, reduced proteinuria in preclinical model of streptozotocin-induced diabetes in mice. Further, administration of zopolrestat (100 mg/kg/day) for 4 months showed reduced amount of protein excretion from 49.97 ± 7.94 mg/day to 15.07 ± 2.17 mg in 24 h [25]. In another study, patients treated with 250 mg sorbinil per day showed a decrease of $41.8 \pm 8.0\%$ sorbitol in nerves and an increase of nearly 3.8-fold in regeneration of myelinated nerve fibres [26]. Ranirestat, another AR inhibitor, when given at 20 mg/day improved sensory nerve function in 12 weeks and motor nerve function in 60 weeks [27].

These preclinical and clinical studies indicate that ARIs could be potential molecules in preventing and treating the diabetic complications. Some of the synthetic drugs that have undergone different trials are given in the table below. However, a number of potential synthetic AR inhibitors as discussed above have failed in clinical trials mainly due to their toxic side effects. Therefore, identification of novel molecules from the natural sources could be helpful in decreasing the side effects associated with the synthetic ARIs (Table 24.1).

24.4 Chemical Structures of Aldose Reductase Inhibitors

Based on their chemical structure, the ARIs have been divided into following four classes:

24.4.1 ARIs Having Carboxylic Acid Group

Tetramethyleneglutaric acid (TMG) was the first ARI developed in the 1960s. Since then, the many more potent carboxylic acid group containing AR inhibitors have been developed. Among them, epalrestat is the only ARI developed in 1983 which is marketed in Japan and recently also launched in China and India. Tolrestat [28], zenarestat, zopolrestat [29], lidorestat and ponalrestat are potent ARI, although they could not pass phase III clinical trial (Table 24.2).

Table 24.1 IC50 values of different ARIs and status of their clinical trials

S.No.	Name	IC50 value	Clinical trials
1.	TMG	1–31.8 μM	
2.	Epalrestat	0.01–1.5 μM	Marketed in Japan, China, India
3.	Tolrestat	0.02–1.2 μM	Withdrawn from market due to side effects
4.	Zenarestat	0.004–0.04 μM	Phase III
5.	Zopolrestat	0.01–0.3 μM	Discontinued after Phase II
6.	Ponalrestat	0.007–2 μM	
7.	Lidorestat	0.005–27 μM	Phase II
8.	Sorbinil	0.01–3.6 μM	Phase III completed
9.	Fidarestat	0.57–17.8 μM	
10.	Imirestat	0.009–0.054 μM	Discontinued after Phase III
11.	Minalrestat	0.01–0.7 μM	Discontinues after Phase III
12.	Ranirestat	0.015 μM	Phase III completed in the USA
			Phase III continuing in Japan

Information collected from NCBI PubChem Compounds, Clinicaltrials.gov.in

Table 24.2 ARIs having carboxylic acid group along with their IUPAC name [Information collected from NCBI PubChem Compound and ZINC database]

S. No.	Common name	IUPAC name
1.	Tetramethylene glutaric acid	2-[1-(carboxylatomethyl) cyclopentyl]acetate
2.	Epalrestat	2-[(5*Z*)-5-[(*E*)-2-methyl-3-phenylprop-2-enylidene]-4-oxo-2-sulfanylidene-1,3-thiazolidin-3-yl]acetic acid
3.	Tolrestat	2-[[6-methoxy-5-(trifluoromethyl)naphthalene-1-carbothioyl]-methylamino]acetic acid
4.	Zenarestat	2-[3-[(4-bromo-2-fluorophenyl)methyl]-7-chloro-2,4-dioxoquinazolin-1-yl]acetic acid
5.	Zopolrestat	2-[4-oxo-3-[[5-(trifluoromethyl)-1,3-benzothiazol-2-yl]methyl] phthalazin-1-yl]acetic acid
6.	Ponalrestat	2-[3-[(4-bromo-2-fluorophenyl)methyl]-4-oxophthalazin-1-yl] acetic acid
7.	Lidorestat	2-[3-[(4,5,7-trifluoro-1,3-benzothiazol-2-yl)methyl]indol-1-yl] acetic acid

24.4.2 ARIs as Spirohydantoin Derivatives and Their Analogues

The first spirosystem was formed by the combination of chroman ring and hydantoin ring. Sorbinil was the first compound developed in this group followed by M79175 and fidarestat [30, 31]. Replacement of the chroman ring with a planar fluorene ring produced imirestat. Replacements of both the chroman and hydantoin rings with isoquinoline-1,3(2H,4H)-dione ring and succinimide ring, respectively, formed minalrestat. Further, replacement of the isoquinoline-1,3(2H,4H)-dione ring of minalrestat with pyrrolo[1,2-a]pyrazine-1,3(2H,4H)-dione led to the development of ranirestat (AS-3201). These drugs, especially sorbinil, have been tested

Table 24.3 ARIs as spirohydantoin derivatives along with their IUPAC name [Information collected from NCBI PubChem Compound and ZINC database]

S. No.	Common name	IUPAC names
1.	Sorbinil	(4S)-6-fluorospiro[2,3-dihydrochromene-4,5′-imidazolidine]-2′,4′-dione
2.	Fidarestat	(2S,4S)-6-fluoro-2′,5′-dioxospiro[2,3-dihydrochromene-4,4′-imidazolidine]-2-carboxamide
3.	Imirestat	2,7-difluorospiro[fluorene-9,5′-imidazolidine]-2′,4′-dione
4.	Minalrestat	2-[(4-bromo-2-fluorophenyl) methyl]-6-fluorospiro[isoquinoline-4,3′-pyrrolidine]-1,2′,3,5′-tetrone
5.	Ranirestat	(3R)-2′-[(4-bromo-2-fluorophenyl)methyl]spiro[pyrrolidine-3,4′-pyrrolo[1,2-a]pyrazine]-1′,2,3′,5-tetrone

Table 24.4 ARI with a flexible sulfonyl group along with their IUPAC name [Information collected from NCBI PubChem Compound and ZINC database]

S. No.	Common name	IUPAC names
1.	Tri-Cl-PSH	1-(2,4,5-trichlorobenzene-1- sulfonyl)imidazolidine-2,4-dione
2.	Di-Cl-PSH	1-(2,5-dichlorobenzene-1- sulfonyl)imidazolidine-2,4-dione
3.	ARI-809	3-[(5-chloro-3-methyl-1-benzofuran-2-yl) sulfonyl]-1H-pyridazin-6-one
4.	M16209	1-[(3-bromo-1-benzofuran-2-yl)sulfonyl]imidazolidine-2,4-dione
5.	M16287	1-[(3-chloro-1-benzofuran-2-yl)sulfonyl]imidazolidine-2,4-dione

clinically, but the outcomes have not been encouraging due to various side effects. Although more studies are required to confirm its effectiveness, fidarestat has also been tested clinically and found to alter the progression of diabetic neuropathy [32]. Some of the ARIs from this group are listed in the table below along with their chemical structures (Table 24.3).

24.4.3 ARI with a Flexible Sulfonyl Group

The first few molecules containing linker sulfonyl group were Tri-CI-PSH and Di-CI-PSH. Then, a group of benzofuransulfonylhydantoins were developed from which two compounds M16209 and M16287 were found to have shown potent AR inhibition activity. A recent inhibitor developed under this category is ARI-809. When administered to diabetic rats at optimum doses, ARI-809 showed improvement in the survival, slowed cataract formation and prevented retinal injury associated with early neuronal, glial and vascular complications [33]. These effects however would be required to be confirmed in the patients and side effects, if any,

Table 24.5 ARIs isolated from microorganisms, fungi, marine algae and sponges

S. No.	Source	Name	Refs.
1.	*Corynebacterium* sp.	YUA001	[34]
2.	*Streptomyces* sp.	1a-[(2E-3,7-dimethylocta-2,6-dien-1-yl]-4,6-dihydroxy-5-methyl-1a,7a-dihydronaptho[2,3-6] oxirene-2,7-dione	[35]
3.	*Thermoactinomyces* sp.	2-hydroxyphenyl acetic acid	[36]
4.	*Thermoactinomyces* sp.	Thermorubin	[36]
5.	*Fusarium proliferatum*	Moniliformin	[37]
6.	*Chaetomella circinoseta*	3-[(2R)-2,6,8-trihydroxy-1-(4-hydroxyphenyl)-5-oxo-benzo[e]benzofuran-2-yl]propanoic acid[R=OH]	[38]
		3-[(2R)-2,6,8-trihydroxy-5-oxo-1-phenyl-benzo[e] benzofuran-2-yl]propanoic acid[R=H]	
7.	*P. citrinum*	Dihydrocitrinin	[39]
8.	*P. citrinum*	Citrinin	[39]
9.	*P. corylophilum*	DHMI (3, 4-dihydro-6-methoxy-3,7-dimethyl-1H-2-benzopyran-8-ol)	[39]
10.	*Crucibulum* sp.	Salfredins	[40]
11.	*Humicola grisea*	Aldostatin, WF-2421	[41, 42]
	Pseudoeurotium zonatum		
12.	*Actinosynnema* sp. *Paecilomyceslilacinus*	Benzothiazole	[43, 44]
13.	*Chaetomella raphigera*	WF-3681	[45]
14.	*Ecklonia cave*	1,(3,5-dihydroxyphenoxy)-2,4,6,8-tetrahydroxyanthracene-9,10-dione	[46]
15.	*Dictyodendrilla* sp.	Phenolic derivatives	[47]
16.	*Dictyodendrilla* sp.	Phenolic derivatives	[47]
17.	*Asparagopsis taxiformis*	Tribromoacetate (1) and dibromoacetate (2)	[48]

need to be assessed. The following table summarises this groups of AR inhibitors and depicts chemical structures of some of the ARIs from this group (Table 24.4).

24.4.4 ARIs Derived from Natural Sources

Numerous ARIs have been isolated from natural sources including microorganisms (bacteria and fungi), marine sponges, algae and land plants. The lists of the ARIs isolated from bacteria, fungi, marine sponges and algae are given in Table 24.5.

Land plants are the major source of ARI till date. A number of plant metabolites have been shown to possess the potential to inhibit AR in vitro. The inhibitors have been isolated from different parts of plants such as root, stem, fruits and leaves of plants including *Phyllanthus niruri*, *Potentilla candicans*, *Corydalis tuber*, *Maesa lanceolata*, *Salvia miltiorrhiza*, *Salvia* sp., *Perilla frutescens*, *Emblica officinalis*, *Anacardium occidentale*, *Salacia oblonga*, *Caesalpinia brevifolia*, *Nuphar*

Table 24.6 Land plant -derived metabolites with AR inhibition potential

S. No.	Source	Name of the compound	Activity found against	Refs.
1.	*Phyllanthus niruri*	Ellagic acid	Rat lens ALR2	[49]
2.	*Phyllanthus niruri*	Brevifolin-carboxylic acid (**R=H**) and ethyl brevifolin carboxylate (**R=Et**)	Rat lens ALR2	[50]
3.	*Corydalis tuber*	Dehydrocorydaline	Rat lens ALR2	[50]
4.	*Maesa lanceolata*	Maesanin	Rat lens ALR2	[51]
5.	*Maesa lanceolata*	Flaviolin	Rat lens ALR2	[51]
6.	*Salvia miltiorrhiza*	Danshenols A and B	Rat lens ALR2	[52]
7.	*Salvia miltiorrhiza*	(−)-danshexinkun A	Rat lens ALR2	[52]
8.	*Salvia miltiorrhiza*	Dihydrotanshinone I	Rat lens ALR2	[52]
9.	*Salvia miltiorrhiza*	Cryptotanshinone	Rat lens ALR2	[52]
10.	*Salvia miltiorrhiza*	Tanshinone I and tanshinone IIA	Rat lens ALR2	[52]
11.	*Salvia miltiorrhiza*	Salvianolic acid A	Rat lens ALR2	[53]
12.	*Salvia deserta*	Lithospermic acid B	Rat lens ALR2	[53]
13.	*Salvia deserta*	Salvianolic acid K	Rat lens ALR2	[53]
14.	*Salvia deserta*	Salviaflaside (**1**) and rosmarinic acid (**2**)	Rat lens ALR2	[53]
15.	*Perilla frutescens*	Perillosides A–D (**1–4**)	Rat lens ALR2	[54]
16.	*Emblica officinalis*	Beta-glucogallin	Rat lens ALR2	[55]
17.	*Emblica officinalis*	Emblicanin A	Rat lens ALR2	[56]
18.	*Emblica officinalis*	Emblicanin B	Rat lens ALR2	[56]
19.	*Emblica officinalis*	Punigluconin	Rat lens ALR2	[56]
20.	*Emblica officinalis*	Pedunculagin	Rat lens ALR2	[56]
21.	*Anacardium occidentale*	Phenol derivatives	Bovine lens ALR2	[57]

(continued)

Table 24.6 (continued)

S. No.	Source	Name of the compound	Activity found against	Refs.
22.	*Salacia oblonga*	Kotalagenin 16-acetate	Rat lens ALR2	[58]
23.	*Salacia oblonga*	19-hydroxyferruginol (**1**) and lambertic acid (**2**)	Rat lens ALR2	[58]
24.	*Salacia oblonga*	Maytenfolic acid (1) and 3β,22α-dihydroxyolean-12-en-29-oic acid (2)	Rat lens ALR2	[58]
25.	(1) *Caesalpinia brevifolia*	(1) 1,3,4,6-tetragalloylglucose	Human placenta ALR2	[59, 60]
	(2) *Nuphar japonicum*	(2) 1,2,3,4,6-pentagalloylglucose		
26.	*Monochasmasa vatieri* and *Plantago asiatica*	Acteoside	Rabbit and bovine lens ALR2	[61, 62]
27.	*Myrcia multiflora*	Myrciaphenone B	Rat lens ALR2	[63]
28.	*Chrysanthemun indicum*	Chlorogenic acid	Rat lens ALR2	[64]
29.	*Chrysanthemun indicum*	Clovanediol	Rat lens ALR2	[64]
30.	*Chrysanthemun indicum*	Caryolane 1, 9-β-diol	Rat lens ALR2	[64]
31.	*Ipomoea batatas*	3,5-dicaffeoylquinic acid	Rat lens ALR2	[65]
32.	*Gossypium*	Gossypol	Human placental ALR2	[66]
33.	*Schisandra chinensis*	Dibenzocyclooctane	Rat lens ALR2	[67]
34.	*Caesalpinia sappan*	Brazilin (R=H) and haematoxylin (R=OH)	Bovine lens ALR2	[68]
	Haematoxylum campechianum			
35.	*Eucalyptus macrocarpa*	Macrocarpals A	Porcine lens ALR2	[69]
36.	*Eucalyptus macrocarpa*	Macrocarpals B	Porcine lens ALR2	[69]
37.	*Eucalyptus macrocarpa*	Macrocarpals D	Porcine lens ALR2	[69]
38.	*Eucalyptus macrocarpa*	Macrocarpals G	Porcine lens ALR2	[69]
39.	*Guaiacum officinale*	Furoguaiaoxidin (R=H) and diethyl ether furoguaiaoxidin (R=Et)	Rat lens ALR2	[70]
40.	*Guaiacum officinale*	Guaiaretic acid	Rat lens ALR2	[70]
41.	*Guaiacum officinale*	Malabaricanol (1), machilin I (2), and tetrahydrofuroguaiacin B (3)	Rat lens ALR2	[70]

japonicum, Monochasmasa vatieri, Plantago asiatica, Myrcia multiflora, Chrysanthemun indicum, Schisandra chinensis, Caesalpinia sappan, Haematoxylum campechianum, Eucalyptus macrocarpa, Guaiacum officinale, etc. The list of compounds with AR inhibition potential and their respective structures is shown in Table 24.6.

24.5 Plant-Derived Functional Food Containing AR Inhibitors

Ever since the revelation of pathological role of AR in the development of diabetic complications such as diabetic retinopathy, nephropathy, neuropathy and cataractogenesis, researchers have attempted to identify molecules that could inhibit this enzyme and minimise the diabetes-associated complications. After initial efforts to use the synthetic inhibitors, and their subsequent failure in clinical trials, the emphasis on the use of natural sources, especially plant-derived AR inhibitors was renewed. As mentioned in Sect. 24.4.4, a number of ARIs have been identified and isolated from the plants and their parts. Some of the plants have been even used in the treatment of diabetic patients for millennia, although only now their key constituents are being identified. Many traditional medicine systems that have been using various plant extracts to revert or minimise the organ damage in diabetic patients were revived, and many plants were identified that possessed AR inhibitory activities.

Saraswat et al. (2008) tested at least 22 plant-derived extract or natural products that are used in Indian culinary practices for their ability to inhibit AR using rat lens or recombinant human AR [71]. The authors have reported that out of 22 samples tested, 10 extracts from common food items including spinach, cumin, cinnamon, fennel, lemon, basil, orange, fenugreek, bitter guard and black pepper showed remarkable AR inhibition with an IC_{50} value of approximately 0.2 mg/mL. The inhibitory potential of these inhibitors was also substantiated in vivo by decreasing the accumulation of AR-catalysed product sorbitol intracellularly [71]. It is noteworthy that these common plants are usually used in many culinary practices as fruit supplements, spices or beverages or as traditional medicine and, thus, are part of everyday diet. Thus, to use them as functional food would help regulate AR activity and manage the diabetes-related complications better. Similarly, Kim et al. (2013) showed *Zea Mays* kernel could have significant AR inhibition capacity [72]. Out of 12 non-anthocyanin phenolic compounds and 5 anthocyanins isolated by ethanol extraction from kernel of purple corn (*Zea mays* L.), a compound identified as hirsutrin showed potent AR inhibitory activity with IC_{50} of 4.78 μM. Further, hirsutrin showed competitive inhibition against recombinant human AR and significantly inhibited galactitol formation in rat lens and erythrocytes samples when incubated with a high concentration of galactose. The purple corn has been grown for centuries in the countries along Andes in South America and indeed is consumed now across the world. A characteristic beverage called 'chichamorada' used as a folk medicine in Peru is produced from purple corn. Similarly, Indian gooseberry (Amla; *Emblica Officinalis*), mangosteen (*Garcinia mangostana*), litchi (*Litchi chinensis*), grapes (*Vitis vinifera*), lemon (*Citrus limettoides*) and organs (*Citrus sinensis*) have also shown AR inhibitory effects in preclinical studies and inhibited diabetic cataractogenesis [73].

Yawadio et al. (2007) showed that two anthocyanins, namely, cyanidin-3-O-b-glucoside and peonidin-3-O-b-glucoside, and other phenolic compound ferulic acid isolated from black and pigmented brown rice (*Oryza sativa L. japonica*) showed significant AR inhibitory activity [74]. This study suggested that consumption of pigmented rice as functional food may be associated with antioxidant potency as well as AR inhibitory property present in them and can promote better health due to the phenolic compounds present in them.

Green tea is consumed across the world as a beverage and has shown many health benefits including in cardiovascular, cancer, obesity and weight loss, diabetes and other associated with general heath. Murata et al. (1994) have shown AR inhibitory activity of various phyto-constituents such as catechins, (-)-epicatechin gallate, (-)-epicatechin and (-)-epigallocatechin gallate, isolated from green tea [75]. Further, Vinson and Zhang (2005) showed that indeed black as well as green teas both could prevent diabetes-induced cataract development in STZ-induced diabetes in rats equally [76]). Both the teas given in drinking water for 3 months to the diabetic rats decreased sorbitol in the lens and red blood cells, thiobarbituric acid-reactive substances and protein glycation in the plasma and lens and inhibited diabetic cataract formation in rats by promoting hypoglycemic effect. Thus, tea, being a simple and relatively low-priced beverage, could be used as part of functional food for preventing complication related with diabetes in humans.

Grapefruit (*Citrus paradisi*) is a subtropical citrus fruit originally cultivated as a cross variety between pomelo (*Citrus maxima*), a native of Southeast Asia, and sweet orange (*Citrus sinensis)* in Barbados, West Indies. The juicy fruit is enriched with vitamins such as vit A and C; polyphenols such as flavonoids, terpenes and limonoids; pigments such as carotene and lycopene in addition to fibres; and minerals such as potassium and magnesium, iron, calcium, copper and phosphorus and known for its health-promoting effects and healing properties. Naringin, derived from grapefruit, has shown lipid-reducing potential as well as insulin-mimetic effects. Sharma et al. [77] showed that naringin-ameliorated insulin resistance improved β-cell function and corrected dyslipidaemia, hepatic steatosis and kidney damage in the rat model of type 2 diabetes [77]. These effects were in part due to reduction in oxidative stress and inflammation and improved adipocytokines levels via modulation of PPARγ, heat shock proteins -27 and -72 by naringin [77].

Curcumin isolated from *Curcuma longa* has been shown to inhibit AR as well as ameliorate the diabetes-related complications including diabetic nephropathy [78], decreased thermal hyperalgesia in a diabetic model of neuropathic pain [79], delayed streptozotocin-induced diabetic cataract in rats [80] and decreased retinal oxidative stress and inflammation in diabetes [81]. These studies suggest multi-pronged effects of curcumin in the diabetic models most likely by inhibiting AR [82]. Since *Curcuma longa* is already used as favourite spice turmeric in the South Asia and is now popular globally as Indian curry, it has potential to be used as functional food to promote general heath and also to treat many ailments.

24.6 Plant-Derived Pure AR Inhibitors

A number of first generation of synthetic AR inhibitors such as sorbinil and tolrestat were tested in preclinical and clinical studies followed by their derivatives fidarestat and zopolrestat which exhibited significant inhibition of AR [83, 84]. However, their clinical use is still marred with relatively low efficacy, poor permeation and safety issues, although the derivative like fidarestat, having IC50 in nanomolar range, is still being studied for various indications, even other than diabetes. In Japan, epalrestat has been marketed for use in diabetic neuropathy patients [85]. Another synthetic AR inhibitor, ranirestat, was tested in the clinical trial in the USA; trial ended in February 2013 that evaluated the effectiveness and safety of orally administered ranirestat (40 and 80 mg) in the persons having mild to moderate sensorimotor polyneuropathy caused by diabetes (https://clinicaltrials.gov/ct2/show/results/NCT00927914 accessed Dec 30, 2016). The results are not yet published though. Therefore, the need for better AR inhibitors is being felt excessively.

The researchers have been looking to the natural sources to augment this need, and increasing evidences of the benefits of numerous dietary supplements taken as food items as well as herbal components used in traditional medicines in different parts of the world encourage the possibilities of finding novel AR inhibitors. It is a widely accepted fact that herbal-derived pharmaceuticals usually lack toxicity and other harmful side effects due to their biocompatibility emanating from coevolution of these sources with human evolution.

Accordingly, a number of purified molecules have been extracted from plants that show AR inhibitory properties (Sect. 24.4.4). It has been found that herbal extracted molecules with AR inhibitory property fall into one of the chemical groups from either flavonoids, alkaloids, phenolics, tannins and coumarins that contain cyclical compounds such as isoquinoline and bisisoquinoline, along with terpenoids or other miscellaneous groups. The structure activity studies have revealed that the presence of isoquinoline ring structure in the plant-derived alkaloids is the primary and necessary condition for it to act as the inhibitor of AR [86]. The cyclical compounds including berberine, palmatine, coptisine and jatrorrhizine belong to this category. Fatmawati S et al. have isolated bioactive metabolites from the root of *Panax ginseng*, commonly identified as ginsenosides, which show AR inhibition activity [87]. One of the several compounds named ginsenoside 20(S)-Rh2 showed potent inhibitory activity (with 147.3 μM concentration as IC50) value against human recombinant aldose reductase probably due to stereo-chemistry of the -OH functional moiety at C-20. The same group of researchers also showed that 3-isomangostin, one of the many prenylated xanthones, was extracted from dichloromethane extract of *Garcinia mangostana* Linn. Root bark possessed potent AR inhibitory property with an IC50 of 3.48 μM [88]. Further, steroidal compounds isolated from the plants also show AR inhibitory activity. A *Balanitesa egyptiaca* fruit, which is used in Egyptian folk medicine for hyperglycaemia-associated problems, was shown to inhibit AR in in vitro assay due to the steroidal saponins found in butanol fraction of the 70% ethanolic extract [89]. The IC50 values of these steroidal saponins were found to be in the range of 1–5 μM as compared to standard AR inhibitor quercetin. At least four flavonoids derived from *Zingiber zerumbet*, viz. kaempferol and kaempferol-3-O-methylether, showed appreciable AR inhibition in vitro in a study by Ajish et al. [90]. An isoflavone, 3,6,7,4′,5′-Pentamethoxy-5,3′-dihydroxyflavone, isolated from *Caesalpinia pulcherrima*, showed antioxidative properties and non-competitive AR inhibition (IC50 value of 16.24 μM) and reversed sorbitol accumulation in retina of STZ-induced diabetic rat in a dose-dependent manner [91].

Additionally, a number of studies have been performed using in vitro and in vivo models of inflammatory pathogenesis including diabetes using the crude extract of plant parts including roots, leaves, fruits and seeds. These studies show AR inhibitory activity of the plant extracts up to various extents and also reversal or amelioration of the hyperglycaemia and other associated complications. Gupta et al. assessed the hydroalcoholic and alkaloidal extracts of *Piper nigrum*, *Murraya koenigii*, *Argemone mexicana* and *Nelumbo nucifera* for their AR inhibitory roles and reported that the alkaloidal extracts of these plants could have potent AR inhibition activity and thus useful in the treatment of diabetic complications [92].

24.7 Conclusion and Future Prospects

Aldose reductase has been implicated in the mediation of diabetic complications including diabetic neuropathy, nephropathy, retinopathy and cataractogenesis. AR is well known to utilise excess glucose during hyperglycemic condition through polyol pathway leading to sorbitol accumulation and increased oxidative stress. Accordingly, a number of AR inhibitors have been tested in clinical trials with success as the synthetic AR inhibitors showed poor response in those trials but showed excessive toxicity. Nonetheless, new inhibitors have been synthesised, and a few such as epalrestat have shown promising results in clinical trials [93] and have been approved for clinical use in Japan for the treatment of diabetic neuropathy [94].

As discussed in this chapter, a number of plants or their extract and molecules are used as food, fruits, spices, beverages, spices and condiments and also as routine traditional medicine due to their antidiabetic properties. In most of the cases, the antidiabetic properties have been attributed to their capability to inhibit elevated AR activity during hyperglycemic condition. The chemical nature of these plant-derived molecules includes flavonoids, alkaloids, phenolics, tannins and coumarins that contain cyclical compounds such as isoquinoline and bisisoquinoline, along with terpenoids or other miscellaneous groups. Most of these compounds acted through inhibition of oxidative stress (antioxidant) or inflammation (anti-inflammatory) by activating the cognate pathways that activate antioxidant and anti-inflammatory players in cells. It appears that since the plant-derived molecules show AR inhibition as well as antioxidant and anti-inflammatory potential, they could be potent antidiabetic agents than the synthetic AR inhibitors which have not been successful in clinical trial and marred with various side effects. Thus, promoting the use of plant-derived AR inhibitors as functional food may be an effective strategy to combat the menace of diabetes which could play more of a preventive role and help decrease the incidence of diabetes.

Acknowledgement Arpita Devi acknowledges financial support from CSIR for her doctoral studies. Dr. ABM Reddy acknowledges UoH-UPE-II, DBT-RGYI and DBT-RNAi, DAE-BRNS for funding. Dr. Umesh C. S. Yadav acknowledges the award of Ramanujan Fellowship and financial support from Department of Science and Technology (DST), SERB Government of India, and GSBTM, Government of Gujarat.

References

1. Graham A et al (1991) The human aldose reductase gene maps to chromosome region 7q35. Hum Genet 86(5):509–514
2. Petrash JM (2004) All in the family: aldose reductase and closely related aldo-keto reductases. Cell Mol Life Sci CMLS 61(7–8):737–749
3. Ferrarett A et al (1993) Aldose reductase is involved in long-term adaptation of EUE cells to hyperosmotic stress. Biochimi Biophys Acta (BBA)-Mol Cell Res 1175(3):283–288
4. Song Z, Fu DT, Chan YS, Leung S, Chung SS, Chung SK (2003) Transgenic mice overexpressing aldose reductase in Schwann cells show more severe nerve conduction velocity deficit and oxidative stress under hyperglycemic stress. Mol Cell Neurosci 23(4):638–647
5. Dunlop M (2000) Aldose reductase and the role of the polyol pathway in diabetic nephropathy. Kidney Int 58:S3–S12
6. Jaquinod M, Potier N, Klarskov K, Jean-Marc R, Sorokine O, Kieffer S, Barth P, Andriantomanga V, Biellmann J-F, Alain D (1993) Sequence of pig lens aldose reductase and electrospray mass spectrometry of non-covalent and covalent complexes. Eur J Biochem 218(3):893–903
7. Tarle I, Borhani DW, Wilson DK, Quiocho FA, Petrash JM (1993) Probing the active site of human aldose reductase. Site-directed mutagenesis of Asp-43, Tyr-48, Lys-77, and His-110. J Biol Chem 268(34):25687–25693
8. Barski OA, Gabbay KH, Bohren KM (1999) Characterization of the human aldehyde reductase gene and promoter. Genomics 60(2):188–198
9. Wilson DK, Bohren KM, Gabbay KH, Quiocho FA (1992) An unlikely sugar substrate site in the 1.65 A structure of the human aldose reductase holoenzyme implicated in diabetic complications. Science 257(5066):81–84
10. Rondeau JM, Tete-Favier F, Podjarny A, Reymann JM, Barth P, Biellmann JF, Moras D (1992) Novel NADPH-binding domain revealed by the crystal structure of aldose reductase. Nature 355(6359):469–472
11. Singh R, White MA, Ramana KV, Petrash JM, Watowich SJ, Bhatnagar A, Srivastava SK (2006) Structure of a glutathione conjugate bound to the active site of aldose reductase. Proteins: Struct, Funct, Bioinf 64(1):101–110
12. Ruiz F, Hazemann I, Mitschler A, Joachimiak A, Schneider T, Karplus M, Podjarny A (2004) The crystallographic structure of the aldose reductase–IDD552 complex shows direct proton donation from tyrosine 48. Acta Crystallogr D Biol Crystallogr 60(8):1347–1354
13. Nakano T, Mark Petrash J (1996) Kinetic and spectroscopic evidence for active site inhibition of human aldose reductase. Biochemistry 35(34):11196–11202
14. Baba SP, Hoetker JD, Merchant M, Klein JB, Cai J, Barski OA, Conklin DJ, Bhatnagar A (2013) Role of aldose reductase in the metabolism and detoxification of carnosine-acrolein conjugates. J Biol Chem 288(39):28163–28179
15. Ramana KV, Bhatnagar A, Srivastava S, Yadav UC, Awasthi S, Awasthi YC, Srivastava SK (2006) Mitogenic responses of vascular smooth muscle cells to lipid peroxidation-derived

aldehyde 4-hydroxy-trans-2-nonenal (HNE) role of aldose reductase-catalyzed reduction of the HNE-glutathione conjugates in regulating cell growth. J Biol Chem 281(26):17652–17660
16. Srivastava SK, Ramana KV, Bhatnagar A (2005) Role of aldose reductase and oxidative damage in diabetes and the consequent potential for therapeutic options. Endocr Rev 26(3):380–392
17. Srivastava S, Chandra A, Bhatnagar A, Srivastava SK, Ansari NH (1995) Lipid peroxidation product, 4-hydroxynonenal and its conjugate with GSH are excellent substrates of bovine lens aldose reductase. Biochem Biophys Res Commun 217(3):741–746
18. Srivastava SK, Ramana, KV Treatment of cancer with aldose reductase inhibitors. U.S. Patent Application No. 12/807,033
19. Alexiou P, Pegklidou K, Chatzopoulou M, Nicolaou I, Demopoulos VJ (2009) Aldose reductase enzyme and its implication to major health problems of the 21st century. Curr Med Chem 16(6):734–752
20. Greene DA, Arezzo JC, Brown MB, Zenarestat Study Group (1999) Effect of aldose reductase inhibition on nerve conduction and morphometry in diabetic neuropathy. Neurology 53(3):580–580
21. Varma SD, Schocket SS, Richards RD (1979) Implications of aldose reductase in cataracts in human diabetes. Invest Ophthalmol Vis Sci 18(3):237–241
22. Yadav U, Ramana KV (2013) Regulation of NF-B-induced inflammatory signaling by lipid peroxidation-derived aldehydes. Oxidative Med Cellular Longev 2013:690545
23. Hotta N, Sakamoto N, Shigeta Y, Kikkawa R, Goto Y, in Japan, D. N. S. G. (1996) Clinical investigation of epalrestat, an aldose reductase inhibitor, on diabetic neuropathy in Japan: multicenter study. J Diabetes Complicat 10(3):168–172
24. Boulton AJM, Levin S, Comstock J (1990) A multicentre trial of the aldose-reductase inhibitor, tolrestat, in patients with symptomatic diabetic neuropathy. Diabetologia 33(7):431–437
25. Beyer-Mears A, Mistry K, Diecke FPJ, Cruz E (1996) Zopolrestat prevention of proteinuria, albuminuria and cataractogenesis in diabetes mellitus. Pharmacology 52(5):292–302
26. Sima AA, Bril V, Nathaniel V, McEwen TA, Brown MB, Lattimer SA, Greene DA (1988) Regeneration and repair of myelinated fibers in sural-nerve biopsy specimens from patients with diabetic neuropathy treated with sorbinil. N Engl J Med 319(9):548–555
27. Bril V, Buchanan RA (2006) Long-term effects of ranirestat (AS-3201) on peripheral nerve function in patients with diabetic sensorimotor polyneuropathy. Diabetes Care 29(1):68–72
28. Showalter HDH, Johnson JL, Werbel LM, Leopold WR, Jackson RC, Elslager EF (1984) Tolrestat, a potent, orally active aldose reductase inhibitor. J Med Chem 27:255–256
29. Mylari BL, Zembrowski J, Aldinger E (1991) iNovel, potent aldose reductase inhibitors: 3, 4 Dihydro 4 oxo 3 [[5 (trifluoromethyl) 2 benzothiazolyl] methyl] 1 phthalazine acetic acid (zopolrestat) and congeners, w+ PVSOBM PG. J Med Chem 34(1):108–122
30. Dirlam NL, Moore BS, Urban FJ (1987) Novel synthesis of the aldose reductase inhibitor sorbinil via amidoalkylation, intramolecular oxazolidin-5-one alkylation and chymotrypsin resolution. J Org Chem 52(16):3587–3591
31. Oka M, Matsumoto Y, Sugiyama S, Tsuruta N, Matsushima M (2000) A potent aldose reductase inhibitor,(2 S, 4 S)-6-fluoro-2′, 5′-dioxospiro [chroman-4, 4′-imidazolidine]-2-carboxamide (Fidarestat): its absolute configuration and interactions with the aldose reductase by X-ray crystallography. J Med Chem 43(12):2479–2483
32. Hotta N, Toyota T, Matsuoka K, Shigeta Y, Kikkawa R, Kaneko T et al (2001) Clinical efficacy of fidarestat, a novel aldose reductase inhibitor, for diabetic peripheral neuropathy. Diabetes Care 24(10).1776–1782
33. Sun W, Oates PJ, Coutcher JB, Gerhardinger C, Lorenzi M (2006) A selective aldose reductase inhibitor of a new structural class prevents or reverses early retinal abnormalities in experimental diabetic retinopathy. Diabetes 55(10):2757–2762
34. Bahn YS, Park JM, Bai DH, Takase S, Yu JH (1998) YUA001, a novel aldose reductase inhibitor isolated from alkalophilic Corynebacterium sp. YUA25. I. Taxonomy, fermentation, isolation and characterization. J Antibiot 51(10):902–907
35. Sankyo Co., Ltd.., JP Pat. 7010857, 1995

36. Hayashi K-i et al (1995) Thermorubin and 2-hydroayphenyl acetic acid, aldose reductase inhibitors. J Antibiot 48(11):1345–1346
37. Deruiter J, Jacyno JM, Cutler HG, Davis RA (1993) Studies on aldose reductase inhibitors from fungi. II. Moniliformin and small ring analogues. J Enzym Inhib 7(4):249–256
38. Shionogi Seiyaku Kabushiki Kaisha, EP Pat. 0557939, 1993
39. Deruiter J et al (1992) Studies on aldose reductase inhibitors from fungi. I. Citrinin and related benzopyran derivatives. J Enzym Inhib 6(3):201–210
40. Matsumoto K et al (1995) Salfredins, new aldose reductase inhibitors produced by Crucibulum sp. RF-3817. I. Fermentation, isolation and structures of salfredins. J Antibiot 48(6):439–446
41. Toyo Jozo Co., Ltd., US Pat. 4749571, 1988
42. Nishikawa M et al (1991) WF-2421, a new aldose reductase inhibitor produced from a fungus, Humicola grisea. J Antibiot 44(2):130–135
43. Yamanouchi Pharm. Co., Ltd., JP Pat. 63030493, 1988
44. Toyo Jozo Co., Ltd., JP Pat. 1181793, 1989
45. Nishikawa M, Tsurumi Y, Namiki T, Yoshida K, Okuhara M (1987) Studies on WF-3681, a novel aldose reductase inhibitor. I. Taxonomy, fermentation, isolation and characterization. J Antibiot 40(10):1394–1399
46. Nakamura H, Yamaguchi S, Hayashi T, Baba M, Okada Y, Tanaka J et al (1997) Studies on the biological activities of marine algae (III) Anti-tumor promoting activity and inhibitory effect on aldose reductase. Nat Med 51:162–169
47. Sato A, Morishita T, Shiraki T, Yoshioka S, Horikoshi H, Kuwano H et al (1993) Aldose reductase inhibitors from a marine sponge, Dictyodendrilla sp. J Org Chem 58(27):7632–7634
48. Sugano M, Sato A, Nagak H, Yoshiok S, Shiraki T, Horikoshi H (1990) Aldose reductase inhibitors from the red alga, Asparagopsis taxiformis. Tetrahedron Lett 31(48):7015–7016
49. Shimizu M et al (1989) Studies on aldose reductase inhibitors from natural products. II.: active components of a Paraguayan crude drug "para-parai mi", Phyllanthus niruri. Chem Pharm Bull 37(9):2531–2532
50. Michinori K et al (1994) Studies of anti-cataract drugs from natural sources. I. Effects of a methanolic extract and the alkaloidal components from Corydalis tuber on in vitro aldose reductase activity. Biol Pharm Bull 17(3):458–459
51. Kubo I, Chaudhuri SK (1994) Structure of maesaquinone. Bioorg Med Chem Lett 4(9):1131–1134
52. Tezuka Y et al (1997) Aodose reductase inhibitory constituents of the root of Salvia miltiorrhiza bunge. Chem Pharm Bull 45(8):1306–1311
53. Kasimu R et al (1998) Comparative study of seventeen Salvia plants: aldose reductase inhibitory activity of water and MeOH extracts and liquid chromatography-mass spectrometry (LC-MS) analysis of water extracts. Chem Pharm Bull 46(3):500–504
54. Fujita T et al (1995) Inhibitory effect of perillosides A and C, and related monoterpene glucosides on aldose reductase and their structure-activity relationships. Chem Pharm Bull 43(6):920–926
55. Puppala M, Ponder J, Suryanarayana P, Reddy GB, Petrash JM, LaBarbera DV (2012) The isolation and characterization of β-glucogallin as a novel aldose reductase inhibitor from Emblica officinalis. PloS One 7(4):e31399
56. Suryanarayana P, Kumar PA, Saraswat M, Petrash JM, Reddy GB (2004) Inhibition of aldose reductase by tannoid principles of Emblica officinalis: implications for the prevention of sugar cataract. Mol Vis 10:148–154
57. Trevisan MTS, Pfundstein B, Haubner R, Würtele G, Spiegelhalder B, Bartsch H, Owen RW (2006) Characterization of alkyl phenols in cashew (Anacardium occidentale) products and assay of their antioxidant capacity. Food Chem Toxicol 44(2):188–197
58. Matsuda H, Murakami T, Yashiro K, Yamahara J, Yoshikawa M (1999) Antidiabetic principles of natural medicines. IV. Aldose reductase and α-glucosidase inhibitors from the roots of Salacia oblonga Wall. (Celastraceae): structure of a new friedelane-type triterpene, kotalagenin 16-acetate. Chem Pharm Bull 47(12):1725–1729

59. Nishizawa M, Yamagishi T, Nonaka GI, Nishioka I (1982) Tannins and related compounds. Part 5. Isolation and characterization of polygalloylglucoses from Chinese gallotannin. J Chem Soc Perkin Trans 1982:2963–2968
60. Nishizawa M, Yamagishi T, Nonaka GI, Nishioka I (1983) Tannins and related compounds. Part 9. Isolation and characterization of polygalloylglucoses from Turkish galls (Quercus infectoria). J Chem Soc Perkin Trans 1983:961–965
61. Kohda H et al (1989) Studies on lens-aldose-reductase inhibitor in medicinal plants II. Active constituents of Monochasma savatierii Franch. et Maxim. Chem Pharm Bull 37(11):3153–3154
62. Ravn H, Nishibe S, Sasahara M, Xuebo LI (1990) Phenolic compounds from Plantago asiatica. Phytochemistry 29(11):3627–3631
63. Matsuda H, Nishida N, Yoshikawa M (2002) Antidiabetic principles of natural medicines. V. Aldose reductase inhibitors from Myrcia multiflora DC.(2): structures of myrciacitrins III, IV, and V. Chem Pharm Bull 50(3):429–431
64. Yoshikawa M et al (1998) Antidiabetic principles of natural medicines. II. aldose reductase and α-glucosidase inhibitors from Brazilian natural medicine, the leaves of Myrcia multiflora DC. (Myrtacae): structures of myrciacitrins I and II and myrciaphenones A and B. Chem Pharm Bull 46(1):113–119
65. Yoshikawa M et al (1999) Medicinal flowers. I. Aldose reductase inhibitors and three new eudesmane-type sesquiterpenes, kikkanols A, B, and C, from the flowers of Chrysanthemum indicum L. Chem Pharm Bull 47(3):340–345
66. Deck LM, Vander Jagt DL, Royer RE (1991) Gossypol and derivatives: a new class of aldose reductase inhibitors. J Med Chem 34(11):3301–3305
67. Tsumura & Co., JP Pat. 01172354, 1989
68. Moon CK, Yun YP, Lee JH, Wagner H, Shin YS (1985) Inhibition of lens-aldose reductase activity by brazilin and haematoxylin. Planta Med 51(01):66–67
69. Murata M, Yamakoshi Y, Homma S, Arai K, Nakamura Y (1992) Macrocarpals, antibacterial compounds from Eucalyptus, inhibit aldose reductase. Biosci, Biotechnol, Biochem 56(12):2062–2063
70. Tsumura & Co., JP Pat. 02180846, 1990.
71. Saraswat M, Muthenna P, Suryanarayana P, Petrash JM, Reddy GB (2008) Dietary sources of aldose reductase inhibitors: prospects for alleviating diabetic complications. Asia Pac J Clin Nutr 17(4):558–565
72. Kim TH, Kim JK, Kang YH, Lee JY, Kang IJ, Lim SS (2013) Aldose reductase inhibitory activity of compounds from Zea mays L. BioMed Res Int 2013:727143
73. Sowmya V, Kalekhan F, Kamath K, Baliga MS Fruits in the prevention of cataractogenesis by targeting the aldose reductase: promise from pre-clinical observations. In: Watson RR (ed) Foods and dietary supplements in the prevention and treatment of disease in older adults. Academic, London, pp 105, 2015–108
74. Yawadio R, Tanimori S, Morita N (2007) Identification of phenolic compounds isolated from pigmented rices and their aldose reductase inhibitory activities. Food Chem 101(4):1616–1625
75. Murata M, Irie J, Homma S (1994) Aldose reductase inhibitors from green tea. LWT-Food Sci Technol 27(5):401–405
76. Vinson JA, Zhang J (2005) Black and green teas equally inhibit diabetic cataracts in a streptozotocin-induced rat model of diabetes. J Agric Food Chem 53:3710–3713
77. Sharma AK, Bharti S, Ojha S et al (2011) Up-regulation of PPAR gamma, heat shock protein-27 and -72 by naringin attenuates insulin resistance, beta-cell dysfunction, hepatic steatosis and kidney damage in a rat model of type 2 diabetes. Br J Nutr 106(11):1713–1723
78. Suresh Babu P, Srinivasan K (1998) Amelioration of renal lesions associated with diabetes by dietary curcumin in streptozotocin diabetic rats. Mol Cell Biochem 181:87–96
79. Sharma S, Kulkarni SK, Agrewala JN, Chopra K (2006) Curcumin attenuates thermal hyperalgesia in a diabetic mouse model of neuropathic pain. Eur J Pharmacol 536(3):256–261
80. Suryanarayana P, Saraswat M, Mrudula T, Krishna TP, Krishnaswamy K, Reddy GB (2005) Curcumin and turmeric delay streptozotocin-induced diabetic cataract in rats. Invest Ophthalmol Vis Sci 46(6):2092–2099

81. Kowluru RA, Kanwar M (2007) Effects of curcumin on retinal oxidative stress and inflammation in diabetes. Nutr Metab 4(1):8
82. Du ZY, Bao YD, Liu Z, Qiao W, Ma L, Huang ZS et al (2006) Curcumin analogs as potent aldose reductase inhibitors. Arch Pharm 339(3):123–128
83. Kawanishi K, Ueda H, Moriyasu M (2003) Aldose reductase inhibitors from nature. Curr Med Chem 10:1353–1374
84. Manzanaro S, Salva J, Angel de la Fuente JA et al (2006) Phenolic marine natural products as aldose reductase inhibitors. J Nat Prod 69:1485–1487
85. Hotta N, Kawamori R, Fukuda M, Shigeta Y (2012) Long-term clinical effects of epalrestat, an aldose reductase inhibitor, on progression of diabetic neuropathy and other microvascular complications: multivariate epidemiological analysis based on patient background factors and severity of diabetic neuropathy. Diabetic Med 29(12):1529–1533
86. Gupta S, Singh N, Jaggi AS (2014) Alkaloids as aldose reductase inhibitors, with special reference to berberine. J Altern Complement Med 20(3):195–205
87. Fatmawati S, Ersam T, Yu H, Zhang C, Jin F, Shimizu K (2014) 20 (S)-Ginsenoside Rh2 as aldose reductase inhibitor from Panax ginseng. Bioorg Med Chem Lett 24(18):4407–4409
88. Fatmawati S, Ersam T, Shimizu K (2015) The inhibitory activity of aldose reductase in vitro by constituents of Garcinia mangostana Linn. Phytomedicine 22(1):49–51
89. Motaal AA, El-Askary H, Crockett S, Kunert O, Sakr B, Shaker S et al (2015) Aldose reductase inhibition of a saponin-rich fraction and new furostanol saponin derivatives from Balanites aegyptiaca. Phytomedicine 22(9):829–836
90. Ajish KR, Antu KA, Riya MP, Preetharani MR, Raghu KG, Dhanya BP, Radhakrishnan KV (2015) Studies on α-glucosidase, aldose reductase and glycation inhibitory properties of sesquiterpenes and flavonoids of Zingiber zerumbet Smith. Nat Prod Res 29(10):947–952
91. Kumar MP, Sankeshi V, Naik RR, Thirupathi P, Das B, Raju TN (2015) The inhibitory effect of Isoflavones isolated from Caesalpinia pulcherrima on aldose reductase in STZ induced diabetic rats. Chem Biol Interact 237:18–24
92. Gupta S, Singh N, Jaggi AS (2014) Evaluation of in vitro aldose reductase inhibitory potential of alkaloidal fractions of Piper nigrum, Murraya koenigii, Argemone mexicana, and Nelumbo nucifera. J Basic Clin Physiol Pharmacol 25(2):255–265
93. Ramirez MA, Borja NL (2008) Epalrestat: an aldose reductase inhibitor for the treatment of diabetic neuropathy. Pharmacotherapy, J Hum Pharmacol Drug Ther 28(5):646–655
94. Sato K, Yama K, Murao Y, Tatsunami R, Tampo Y (2014) Epalrestat increases intracellular glutathione levels in Schwann cells through transcription regulation. Redox Biol 2:15–21

Ferulic Acid: A Natural Antioxidant with Application Towards Neuroprotection Against Alzheimer's Disease

25

Sharanjot Kaur, Monisha Dhiman, and Anil K. Mantha

Abstract

Alzheimer's disease (AD) is a neurodegenerative disorder affecting millions of people worldwide. AD pathology is mainly caused due to the deposition of amyloid beta (Aβ) peptide and upsurge of oxidative stress, the two major factors responsible for AD occurrence and progression. The current treatment strategies available are merely effective and lack long-term benefits against this disease, which necessitates the need for the use of antioxidants for managing/preventing AD pathology. Ferulic acid (FA), a natural antioxidant found in various fruits (such as orange and grapefruit), vegetables (such as broccoli and tomato), coffee beans and barley used in beverages (such as coffee and beer), can be utilized as a natural source of antioxidants against AD. Several studies have pointed towards the protective effects of FA against AD in vitro as well as in vivo. The antioxidant defence system is known to be compromised in AD, whereas the use of FA has been shown to restore the activities of antioxidant enzymes like superoxide dismutase (SOD), catalase (CAT) and heme oxygenase-1 (HO-1). FA has also been shown to destabilize the preformed Aβ fibrils both in vitro and in vivo. Several studies have pointed towards the neuroprotective role of FA through modulating the expression of several key proteins such as p38, Hsp70, ERK1/2, foxo3a and Akt, thus playing a key role towards neuroprotection. In addition, FA is also reported to inhibit the inflammatory responses induced by lipopolysaccharides (LPS) in microglia and also shown to modulate the β-secretase activity and improved AD-like pathology in transgenic mice model study. Inhibition of

S. Kaur · M. Dhiman
Department of Biochemistry and Microbial Sciences, School of Basic and Applied Sciences, Central University of Punjab, Bathinda, Punjab, India

A. K. Mantha (✉)
Department of Animal Sciences, School of Basic and Applied Sciences, Central University of Punjab, Bathinda, Punjab, India
e-mail: anil.mantha@cup.edu.in

V. Rani, U. C. S. Yadav (eds.), *Functional Food and Human Health*, https://doi.org/10.1007/978-981-13-1123-9_25

acetylcholinesterase (AChE) activity and restoration of mitochondrial membrane potential by FA has also been reported. Taken together, it is advocated that the use of FA in routine diet may effectively lower the risk of developing AD and can play an important role in managing the disease. In this book chapter, we will try to document and summarize the structure and functional aspects of FA along with the known benefits of it as a neuroprotective agent against AD.

Keywords
Ferulic acid · Phytochemicals · Amyloid toxicity · Amyloid β · Alzheimer's disease · Neuroprotection

25.1 Introduction

Alzheimer's disease (AD) is the most common cause of dementia, characterized by the loss of memory and impairment of cognitive ability. A Bavarian psychiatrist, Alois Alzheimer, first described AD in his patient in 1907. According to him, his patient had miliary bodies, i.e. plaques and dense bundles of fibrils in the brain [1]. In the late 1960s, a study supported that AD occurs along an age continuum, with most of the cases appearing after the age of 60 [2]. In 1984, George Glenner isolated β-amyloid (Aβ) protein from the brains of patients who died with AD [3, 4]. Later the microtubule-associated protein, Tau, was identified as a constituent of the neurofibrillary tangles (NFTs) found in Alzheimer's brain [5].

The major factor responsible for the development of AD is the accumulation of Aβ peptides, which result from proteolytic processing of amyloid precursor protein (APP) [6]. Aβ enters into the mitochondria and causes reduction in the activities of mitochondrial complexes leading to increased oxidative stress, disturbed Ca^{2+} homeostasis, nitrosylation of Dynamin-related protein-1 (Drp1) leading to fusion/fission imbalance, reduction in ATP and hence decrease in mitochondrial membrane potential causing opening of mitochondrial permeability transition pores (MPTPs), thus leading to the activation of apoptotic cascade through release of cytochrome c (cyt c) [7–9, 65]. The central nervous system (CNS) has very high metabolic rate, as it needs large amounts of ATP for maintaining the ionic gradients across the cell membrane. Most of the energy needed by the neurons is supplied by oxidative metabolism. Therefore, neurons highly depend on mitochondrial function to carry out all the metabolic processes like maintenance of resting membrane potential and neurotransmission. To compensate for such large amounts of energy demands, greater numbers of mitochondria are present in the neurons. Therefore, mitochondrial dysfunction and oxidative stress impair neuronal function [10]. According to a body of evidence, Aβ inhibits the activities of mitochondrial complexes and hence results in neuron vulnerability to oxidative stress [11, 12].

Oxidative stress has been long implicated in the progression of neurodegenerative diseases [13, 14, 69]. Oxidative stress is normally targeted by antioxidant defence system in the cells, the antioxidant enzymes, which take up the reactive oxygen species (ROS) and reactive nitrogen species (RNS) and metabolize them into non-toxic compounds like water (H_2O) and carbon dioxide (CO_2). In neurodegenerative diseases, the antioxidant defence system has been found to be compromised [15, 16]. The activities of superoxide dismutase (SOD), catalase (CAT) and glutathione peroxidase (GSH-Px) were found to be significantly lower in the brain tissue of AD patients as compared to normal (age-matched) controls. However, the expression of SOD1 and SOD2 was found to be increased in the brain tissue, suggesting that increased oxidative stress causes increase in the expression of antioxidant enzymes, but oxidative stress-mediated protein oxidation inactivates the enzymes [17]. Mutations in SOD1 have been associated with familial amyotrophic lateral sclerosis (ALS) [18]. In addition, it has been demonstrated that intracellular Aβ interacts with SOD1 and inhibits its activity, thus leading to increased oxidative stress by decreasing SOD activity in H4 human neuroglioma cells. In addition, it was found that Aβ immunoprecipitated with both wild-type and mutant SOD1 but interacted more with mutant SOD1 [19].

Several studies have pointed towards oxidative stress-mediated protein oxidation, lipid peroxidation and DNA damage in AD. Increased lipid peroxidation has been observed in brains of AD patients and is thought to be associated with generation of Aβ aggregates [20, 21]. The ability of cells to respond to protein damage/oxidation decreases with age. Ageing increases the possibility of aggregate formation by decline in chaperone activity and proteasome activity and decrease in metal-ion homeostasis [70]. ROS inactivate antioxidant enzymes by oxidizing the enzymes and hence impair cellular functions [17]. Induction of oxidative stress by H_2O_2 in vascular smooth muscle cells decrease the activity of SOD, CAT and GSH-Px [22]. Increased oxidative DNA damage has been observed in Aβ(1-42)-treated SH-SY5Y cells [23]. In addition, the expression of DNA repair enzyme, APE1, has been found to be reduced in transgenic AD mouse model, suggesting abnormal regulation of APE1 in response to Aβ pathology [24]. According to an another study, the expression of mitochondrial APE1 was found to be decreased with the treatment of Aβ(25-35) in SH-SY5Y cells [11, 12].

AD has several causes associated with its onset, like Aβ plaques, NFTs, oxidative stress, protein misfolding, mutations, mitochondrial dysfunction and ageing as reviewed recently [11, 12]. Therefore, the treatment strategies that have to be followed should target all these cellular processes so as to be effective against AD pathology. Several studies have indicated the importance of phytochemicals, especially phenolic compounds, in alleviating AD pathology [25, 68].

Ferulic acid (FA) or (E) 3-(-4hydroxy-3methoxyphenyl) prop-2-enoic acid (Fig. 25.1) is an antioxidant, found in vegetables like broccoli, fruits like grapefruit and orange, cereal bran and beverages like coffee and beer [26, 27]. It has been known to protect cells from oxidative stress by an enhancing activity of several antioxidant enzymes and by modulating several cellular functions [28–30]. This book chapter reviews the effects of FA in alleviating AD pathology by modulating several cellular functions.

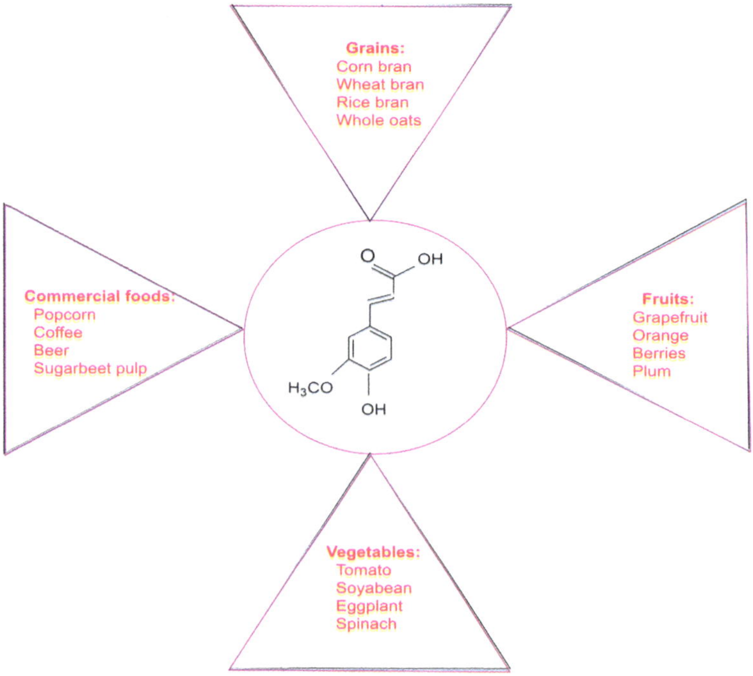

Fig. 25.1 Structure and sources of FA

25.2 Ferulic Acid or (E)-3-(-4hydroxy-3methoxyphenyl) prop-2-enoic Acid

FA is a derivative of caffeic acid. It is found in vegetables like brinjal and broccoli, fruits like orange and tomato, grains, bread and coffee beans and barley used in beverages like coffee and beer [26, 27]. FA, a polyphenol, is one of the intermediates formed during the biosynthesis of lignin from phenylalanine or tyrosine in plants [71]. It is found in both *cis* and *trans* forms in plants [31]. There are several lines of evidence that suggest the protective and neuromodulatory role of plant polyphenols in neurodegenerative diseases like AD and PD, by their ability to cross the blood-brain barrier (BBB) [32, 64].

FA has been known to promote neuronal survival [33], enhance activities of antioxidant enzymes [22], modulate several neuronal signalling pathways [29], exhibit inhibitory role on cholinesterase (ChAT) activity [34] and cause inhibition of Aβ aggregation [35] (Table 25.1). It is also known to scavenge free radicals; this scavenging capacity of FA resides in its three structural motifs, electron-donating benzene ring and carboxylic acid group with adjacent C–C double bond (Fig. 25.1). The benzene ring can form a resonance stabilized phenoxy radical and carboxylic acid group with its adjacent C–C double bond, which can provide additional sites to prevent free radical membrane attack and stabilize free radicals by resonance. The carboxylic acid group can anchor in lipid membrane and provide protection against lipid peroxidation [39, 40].

Table 25.1 Summary of FA hybrids and their functional effects

Name	Structure	Functional effects
Ferulic acid ethyl ester	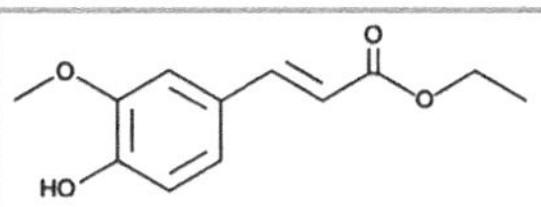	↑ cell viability, Hsp72 and HO-1 ↓ iNOS and ↓ Cyt c release [28, 33]
Tacrine-6-FA	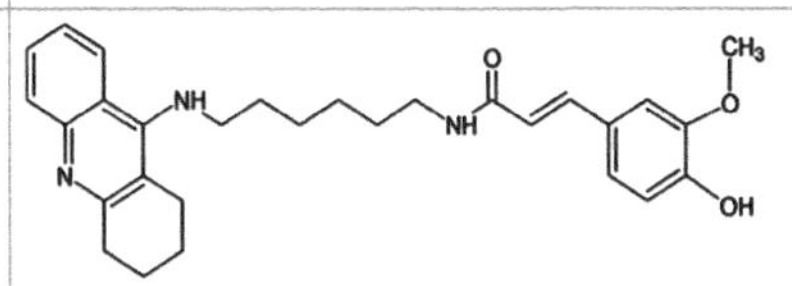	Anti-AChE and anti-amyloidogenic [35]
FA-memoquin hybrid	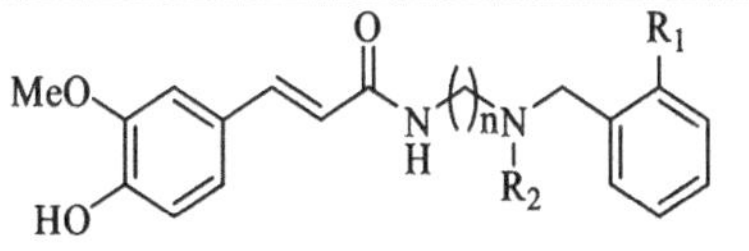	Anti-AChE and anti-amyloidogenic activities [36]
FA-styrylbenzene hybrid	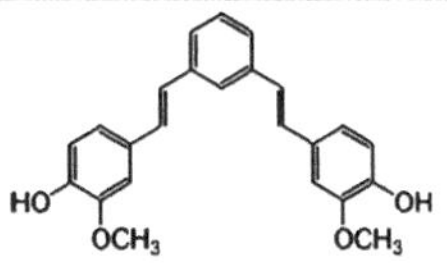	↑ cell viability and inhibit Aβ fibril formation [37]
Rivastigmine-FA hybrid	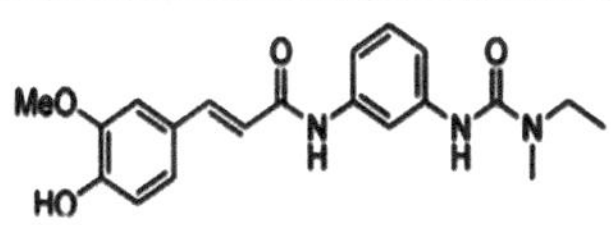	Antioxidant, anti-AChE and anti-amyloidogenic activities [38]

25.3 Antioxidant Potential

Oxidative stress plays a major role in the onset and progression of several neurodegenerative diseases like AD, PD and Huntington's disease, HD [41]. This has led to extensive research on natural antioxidants found in plants that can provide protection/improvement against such pathophysiologies by counteracting oxidative stress. FA, a natural antioxidant, has been extensively studied for its antioxidant properties, like ROS scavenging, and increasing/restoring activities of several antioxidant enzymes against induced oxidative stress conditions [30].

Antioxidant enzymes, SOD and CAT, play a major role in scavenging ROS. SOD converts superoxide anion into H_2O_2, and CAT converts H_2O_2 into O_2 and H_2O [42]. Overproduction of Aβ causes decrease in SOD activity, thus causing more damage to proteins and lipids [43]. According to an another study, Aβ interacts with SOD1 and impairs its activity [19]. Several studies have shown that the restoration of SOD and CAT activities is possible after FA treatment [44, 45]. FA has been reported to restore SOD activity, decrease lipid peroxidation and scavenge free radicals in hypoxia-stressed PC12 cells [29]. Recently, it has been reported that the oxidative stress induction by H_2O_2 decreased SOD, CAT and GSH-PX activities in rat vascular smooth muscle cells; FA increased activities of these enzymes against H_2O_2-induced oxidative stress [22]. Ferulic acid ethyl ester (FAEE) upregulated heme oxygenase-1 (HO-1) and heat shock protein 72 (Hsp-72) and decreased ROS production in cortical rat neuronal cultures against Aβ(1-42)-induced oxidative stress

[33, 66, 67]. In an another study on hepatotoxicity induced by formaldehyde in Wistar albino rats, FA increased SOD, CAT and GSH-Px activities, thus alleviating formaldehyde toxicity [46]. Recently, it has been reported that virgin olive oil (VOO) that contains a mixture of hydroxycinnamic acids, including FA, decreased Aβ(25-35)-induced ROS production in SH-SY5Y cells [47]. According to an another study, FA significantly decreased ROS, protein carbonyl levels and lipid peroxidation in synaptosomes against peroxyl radical generator, 2,2, azobis (2 amidinopropane) dihydrochloride (AAPH), mediated oxidative stress conditions [40]. Recently, it has been found that FA increases SOD expression against monosodium-urate-induced inflammation in rats and alleviates AAPH-induced apoptosis by increasing activities of antioxidant enzymes in PC12 cells [48, 49]. Thus, by modulating the cellular antioxidant system, FA provides neuroprotection via its antioxidant potential.

25.4 Cytoprotective Effects

Accumulating evidences suggest that Aβ leads to mitochondrial dysfunction and decrease in mitochondrial membrane potential, therefore causing opening of MPTPs leading to apoptosis [8]. Several lines of evidence indicate that FA treatment inhibits apoptosis [28, 29]. The pretreatment of FAEE provided protection against Aβ(1-42)-induced oxidative stress by inhibiting loss of phospholipid asymmetry and by reduction in the levels of cytochrome c released in the synaptosomes [28]. According to a study, a hybrid molecule of FA and styrylbenzene showed increase in cell viability of SH-SY5Y cells against Aβ(1-42)-induced oxidative stress and hence alleviated neuronal toxicity induced by Aβ peptides [37]. It has also been reported that FA ethyl ester also increases cell viability of cortical rat neurons against Aβ-induced oxidative stress [33]. According to a recent study, virgin olive oil that contains a mixture of hydroxycinnamic acids (*p*-coumaric acid, FA, vanillic acid and caffeic acid) alleviated Aβ-induced cytotoxicity in SH-SY5Y cells and resulted in increased cell viability. The increase in cell viability by FA against oxidative stress is caused by its ability to modulate several signalling pathways [47].

AD pathology has been associated with the activation of mitogen-activated protein kinase (MAPK) family in response to oxidative stress. Extracellular signal regulated kinase (ERK1/2), a MAPK, activation has been associated with neuronal death in AD [50]. ERK activation and cyclooxygenase (COX-2) and caspase 3 upregulation were found to be induced by oxidative stress in PC12 cells. FA inhibited caspase 3, COX-2 and ERK expression in PC12 cells against hypoxia [29]. The expression of Foxo 3a, a member of Forkhead transcription factor gene family that activates apoptotic genes, was inhibited by FA in sea urchin embryo against Aβ-induced toxicity, thereby providing protection. In addition, it reduced ERK1/2 activation and p53 upregulation induced by Aβ [51]. Phosphorylated ERK1/2, p38, JNK and AKT remarkably decreased in FA+H_2O_2-treated SH-SY5Y cells, indicating protection from oxidative stress induced by H_2O_2 [22]. Recently, it was found that VOO treatment increased the expression of phosphofructokinase (PFKM), hexokinase (HK1) and pyruvate kinase (PKLR) against Aβ-induced lower expression. In addition, the expression levels of phosphorylated ERK1/2 and p38 decreased by VOO and hydroxycinnamic acid treatment against Aβ induced higher expression of pERK1/2 and phosphorylated p38 [47]. In addition, FA alleviated pentylenetetrazole-induced seizures by increasing expression of Hsp70, serotonin and norepinephrine in rats [52]. Taken together, FA exerts cytoprotective effects by modulating various intracellular signalling pathways and also by inhibiting the apoptotic pathways.

25.5 Anti-inflammatory Responses

Brain inflammation is considered as another factor that accelerates the progression of AD [53]. FA is known to exert anti-inflammatory effects. Long-term pretreatment of FA inhibited microglial activation in mice that was induced by intracerebroventricular (ICV) injection of Aβ(1-42) [54]. The pretreatment of FAEE followed by oxidative stress induction by Aβ(1-42) in rat cortical neuronal cultures significantly reduced the type II nitric oxide synthase (iNOS) levels [33]. It also inhibited the neuroinflammation in microglial cells induced by lipopolysaccharides by suppressing the production of nitric oxide (NO), interleukin-1β (IL-1β), prostaglandin E2 (PGE2), iNOS and COX-2 [55]. According to other studies, FA alleviated neuroinflammation in PS/APP mice by decreasing the expression of pro-inflammatory cytokines like TNF-α and IL-1β, thereby improving behavioural impairment in transgenic mice [56, 57]. Recently, it has been reported that FA inhibits LPS-induced upregulation of phosphodiesterase 4 (PDE4) [a regulator of inflammatory processes] and pro-inflammatory cytokines and it upregulated CREB and pCREB [essential for the memory formation] in PC12 cells [58]. Therefore, FA can be ascertained with the functions associated with improving memory impairment via inhibiting the neuroinflammation.

25.6 Mitochondrial Dysfunction and Protection

Impaired mitochondrial dynamics is commonly observed in AD, which leads to impaired mitochondrial function as suggested by several lines of evidence [59]. Antioxidants, like FA, have been shown to improve mitochondrial function [51]. According to a study, the pretreatment of FA along with ascorbic acid followed by induction of cardiotoxicity by isoproterenol (ISO) in rats caused significant improvement in activities of respiratory chain enzymes and mitochondrial antioxidant enzymes [60]. Further, a study indicated that mitochondrial membrane potential was reduced by oligomeric Aβ treatment in sea urchin embryos, whereas treatment of FA along with Aβ restored mitochondrial membrane potential, indicating protection from Aβ toxicity through improved mitochondrial function [51]. In addition, FA was found to restore mitochondrial membrane potential in rat vascular smooth muscle cells against H_2O_2-induced oxidative stress [22].

25.7 AChE Inhibitor and Anti-amyloidogenic Agent

The major factor responsible for AD pathology is the accumulation of Aβ plaques formed by amyloidogenic pathway, which includes cleavage of APP by β-secretase and then by γ-secretase, thus resulting in the formation of Aβ peptides [6]. Inhibition of β-secretase improves behavioural changes in β-site APP cleaving enzyme 1 (BACE1) knockout mice [61]. AChE is mainly involved in the degradation of ACh, but several evidences also point towards its non-cholinergic functions, such as processing and deposition of Aβ peptides [62]. Therefore, inhibiting AChE and β-secretase can help alleviate the AD pathology [56, 62, 63].

Accumulating evidence has suggested anti-amyloidogenic and AChE inhibitory nature of FA [35, 56]. A hybrid molecule of FA and styrylbenzene (SB) inhibited Aβ fibril formation, suggesting its role in interacting with non-fibrous or monomer like Aβ [37]. Hybrid molecules of FA and tacrine were also found to show AChE and butyrylcholinesterase (BChE) inhibitory activities [34]. Later, it was found that tacrine-6-ferulic acid (T6FA) increases choline acetyltransferase (ChAT) activity and inhibits AChE activity in Aβ intra-cerebroventricular AD mouse model study. In addition, T6FA also inhibited auto-aggregation of Aβ, possibly by inhibiting AChE, and thus has been reported for improved memory impairment in mice [35]. Further, it was found that FA reduces cerebral amyloid burden and decreases β-secretase activity in PSAPP mice. This suggested that FA decreases amyloid burden by destabilizing the BACE1 enzyme [56]. Hybrids of FA and memoquin were also found to inhibit AChE activity and Aβ aggregate formation in vitro [36]. Recently, rivastigmine-FA hybrids have been synthesized and found to show antioxidant, anti-AChE and anti-amyloidogenic activities [38]. Taken together, FA shows anti-amyloidogenic effect by inhibiting AChE and by destabilizing the BACE1 enzyme and thus improves behavioural impairment observed in AD.

25.8 Conclusion and Future Perspectives

AD, a neurodegenerative disorder, is mainly characterized by oxidative stress, leading to modulation of several cellular molecules and pathways, which ultimately result in abnormal cellular function and neurodegeneration. Since several factors become associated with AD pathology following oxidative stress induced by Aβ plaques, therapeutic targets against AD should be able to target all the aspects related to its pathology in order to alleviate memory impairment. Plant polyphenols play an important role in alleviating pathologies associated with the neurodegenerative diseases. FA, a natural antioxidant, is a polyphenol that has been known to act as an antioxidant, anti-inflammatory agent and protective agent by modulating various signalling and anti-amyloidogenic pathways. It has been shown to improve memory impairment in AD mouse model studies by its antioxidant, modulatory and anti-amyloidogenic effects. FA may also increase the activities of mitochondrial enzymes, including ETC complexes, and may also modulate the DNA repair enzymes associated with different repair pathways. Hence, new strategies/studies are advocated in this direction. Taken together, FA targets various cellular pathways both in vitro and in vivo. Therefore, including FA in day-to-day diet can help in managing/preventing the disease and can also delay the onset of progressive neurodegenerative disorder, AD.

Acknowledgements This work being supported from the funds recieved from the Department of Science and Technology, New Delhi, India (SR/CSI/288/2012/G), awarded to AKM Junior Research Fellowship (JRF) from University Grants Commission (CSIR-UGC JRF), New Delhi, India, to SK being thankfully acknowledged.

Conflict of Interest Authors declare that no conflict of interest exists.

References

1. Alzheimer A (1907) Ueber eine eigenartige Erkrankung der Hirnrinde. Centralblatt fur Nervenheilkunde and Psychiatrie 30:177–179
2. Selkoe DJ (2001) Alzheimer's disease: genes, proteins, and therapy. Physiol Rev 81:741–766
3. Glenner GG, Wong CW (1984) Alzheimer's disease: initial report of the purification and characterization of a novel cerebrovascular amyloid protein. Biochem Biophys Res Commun 120:885–890
4. Wood JG, Mirra SS, Pollock NL, Binder LI (1986) Neurofibrillary tangles of Alzheimer's disease share antigenic determinants with the axonal microtubule-associated protein tau. Proc Natl Acad Sci 83:4040–4043
5. Brion J, Passareiro E, Nunez J, Flament-Durand J (1985) Mise en evidence immunologique de la protein Tau au niveau des lesions de degenerescence neurofibrillaire de la maladie D' Alzheimer. Arch Biol 95:229–235
6. Haas C, Selkoe DJ (1993) Cellular processing of beta-amyloid precursor protein and the genesis of amyloid beta peptide. Cell 75:1039–1042
7. Aronis A, Melendez JA, Golan O, Shilo S, Dicter N, Tirosh O (2003) Potentiation of Fas-mediated apoptosis by attenuated production of mitochondria-derived reactive oxygen species. Cell Death Differ 10(3):335–344
8. Hroudova J, Singh N, Fisar Z (2014) Mitochondrial dysfunctions in neurodegenerative diseases: relevance to alzheimer's disease. BioMed Res Int 2014:1–9
9. Wang X, Fujioka B, Su H, Zhu X (2008) Dynamin-like protein 1 reduction underlies mitochondrial morphology and distribution abnormalities in fibroblasts from sporadic Alzheimer's disease patients. Am J Pathol 173(2):470–482
10. Kann O, Kovacs R (2007) Mitochondria and neuronal activity. Am J Physiol Cell Physiol 292:C641–C657
11. Kaur N, Dhiman M, Perez-Polo JR, Mantha AK (2015a) Ginkgolide B revamps neuroprotective role of apurinic/apyrimidinic endonuclease 1 and mitochondrial oxidative phosphorylation against Aβ(25-35) induced neurotoxicity in human neuroblastoma cells. J Neurosci Res 93(6):938–947
12. Kaur N, Sarkar B, Mittal S, Dhiman M, Taglialatela G, Perez-Polo RJ, Mantha AK (2015b) Oxidative stress events and neuronal dysfunction in Alzheimer's disease: focus on APE1/Ref-1-mediated survival strategies. In: Free radicals in human health and disease. Springer, Noida, pp 175–207
13. Butterfield DA, Castegna A, Lauderback CM, Drake J (2002) Evidence that amyloid beta-peptide-induced lipid peroxidation and its sequelae in Alzheimer's disease brain contribute to neuronal death. Neurobiol Aging 23(5):655–664

14. Dexter DT, Carter CJ, Wells FR, Javoy Agid F, Agid Y, Lees A, Marsden CD (1989) Basal lipid peroxidation in substantia nigra is increased in Parkinson's disease. J Neurochem 52(2):381–389
15. Bostantjopoulou S, Kyriazis G, Katsarou Z, Kiosseoglou G, Kazis A, Mentenopoulos G (1996) Superoxide dismutase activity in early and advanced Parkinson's disease. Funct Neurol 12(2):63–68
16. Zemlan FP, Thienhaus OJ, Bosmann HB (1989) Superoxide dismutase activity in Alzheimer's disease: possible mechanism for paired helical filament formation. Brain Res 476(1):160–162
17. Omar RA, Chyan YJ, Andorn AC, Poeggeler B, Robakis NK, Pappolla MA (1999) Increased expression but reduced activity of antioxidant enzymes in Alzheimer's disease. J Alzheimer's Dis 1(3):139–145
18. Cudkowicz ME, McKenna YD, Sapp PE, Chin W, Geller B, Hayden DL, Brown RH (1997) Epidemiology of mutations in superoxide dismutase in amyotrophic lateral sclerosis. Ann Neurol 41(2):210–221
19. Yoon EJ, Park HJ, Kim GY, Cho H, Choi J, Park H, Jang J, Rhim H, Kang S (2009) Intracellular amyloid beta interacts with SOD1 and impairs the enzymatic activity of SOD1: implications for the pathogenesis of amyotrophic lateral sclerosis. Exp Mol Med 41(9):611–617
20. Bjorkhem I, Cedazo-Minguez A, Leoni V, Meaney S (2009) Oxysterols and neurodegenerative diseases. Mol Asp Med 30:171–179
21. Reed TT (2011) Lipid peroxidation and neurodegenerative disease. Free Radic Biol Med 51:1302–1319
22. Cao Y, Zhang Y, Qi J, Liu R, Zhang H, He L (2015) Ferulic acid inhibits H2O2 induced oxidative stress and inflammation in rat vascular smooth muscle cells via inhibition of the NADPH oxidase and NF-κB pathway. Int Immunopharmacol 28:1018–1025
23. Gill I, Kaur S, Kaur N, Dhiman M, Mantha AK (2017) Phytochemical ginkgolide B attenuates amyloid-β1-42 induced oxidative damage and altered cellular responses in human neuroblastoma SH-SY5Y cells. J Alzheimer's Dis 2017:1–16
24. Lillenes SM, Stoen M, Gomez-Munoz M, Torp R, Gunther C, Nilsson LNG, Tonjum T (2013) Transient OGG1, APE1, PARP1 and Polβ expression in an Alzheimer's disease mouse model. Mech Ageing Dev 134:466–477
25. Kaur N, Sarkar B, Gill I, Kaur S, Mittal S, Dhiman M, Palada PK, Perez-Polo R, Mantha AK (2016) Indian herbs and their therapeutic potential against Alzheimer's disease: what makes them special. In: Neuroprotective effects of phytochemicals in neurological disorders. Wiley-Blackwell Publisher, New Delhi
26. D'Archivio M, Filesi C, Di Benedetto R, Gargiulo R, Giovannini C, Masella R (2007) Polyphenols, dietary sources and bioavailability. Ann Ist Super Sanita 43:348–361
27. Rechner AR, Pannala AS, Rice-Evans CA (2001) Caffeic acid derivatives in artichoke extract are metabolised to phenolic acids in vivo. Free Radic Res 35:195–202
28. Abdul HM, Butterfield DA (2005) Protection against amyloid beta-peptide(1-42)-induced loss of phospholipid asymmetry in synaptosomal membranes by tricyclodecan-9-xanthogenate (D609) and ferulic acid ethyl ester: implications for Alzheimer's disease. Biochim Biophys Acta 1741:140–148
29. Lin W, Peng Y, Hou C (2015) Ferulic acid protects PC12 neurons against hypoxia by inhibiting the p-MAPKs and COX-2 pathways. Iran J Basic Med Sci 18:478–484
30. Mancuso C, Santangelo R (2013) Ferulic acid: pharmacological and toxicological aspects. Food Chem Toxicol 65:185–195
31. Zhao BL, Li XJ, He RG, Cheng SJ, Xin WJ (1989) Scavenging effect of extracts of green tea and natural antioxidants on active oxygen radicals. Cell Biophys 14:175–185
32. Moosavi F, Hosseini R, Saso L, Firuzi O (2016) Modulation of neurotrophic signaling pathways by polyphenols. Drug Des Dev Ther 10:23–42
33. Sultana R, Ravagna A, Abdul HM, Calabrese V, Butterfield DA (2005) Ferulic acid ethyl ester protects neurons against amyloid β-peptide(1-42)-induced oxidative stress and neurotoxicity: relationship to antioxidant activity. J Neurochem 92:749–758

34. Fang L, Kraus B, Lehmann J, Heilmann J, Zhang Y, Decker M (2008) Design and synthesis of tacrine-ferulic acid hybrids as multi-potent anti-Alzheimer drug candidates. Bioorg Med Chem Lett 18:2905–2909
35. Pi R, Mao X, Chao X, Cheng Z, Liu M, Duan X, Ye M, Chen X, Mei Z, Liu P, Li W, Han Y (2012) Tacrine-6-ferulic acid, a novel multifunctional dimer, inhibits amyloid-β-mediated Alzheimer's disease-associated pathogenesis in vitro and in vivo. PLoS One 7:1–8
36. Pan W, Hu K, Bai P, Yu L, Ma Q, Li T, Sang Z (2016) Design, synthesis and evaluation of novel ferulic acid-memoquin hybrids as potential multifunctional agents for the treatment of Alzheimer's disease. Bioorg Med Chem Lett 26(10):2539–2543
37. Lee KH, Shin BH, Shin KJ, Kim DJ, Yu J (2005) A hybrid molecule that prohibits amyloid fibrils and alleviates neuronal toxicity induced by β-amyloid (1-42). Biochem Biophys Res Commun 328:816–823
38. Chen Z, Digiacomo M, Tu Y, Gu Q, Wang S, Yang X, Nesi G (2017) Discovery of novel rivastigmine-hydroxycinnamic acid hybrids as multi-targeted agents for Alzheimer's disease. Eur J Med Chem 125:784–792
39. Graf E (1992) Antioxidant potential of ferulic acid. Free Radic Biol Med 13(4):435–448
40. Kanski J, Aksenova M, Stoyanova A, Butterfield DA (2002) Ferulic acid antioxidant protection against hydroxyl and peroxyl radical oxidation in synaptosomal and neuronal cell culture systems in vitro: structure-activity studies. J Nutr Biochem 13:273–281
41. Andersen JK (2004) Oxidative stress in neurodegeneration: cause or consequence? Nat Rev Neurosci 5:S18–S25
42. Fridovich I (1999) Fundamental aspects of reactive oxygen species, or what's the matter with oxygen? Ann N Y Acad Sci 893(1):13–18
43. Bayer TA, Schafer S, Breyhan H, Wirths O, Treiber C, Multhaup G (2006) A vicious circle: role of oxidative stress, intraneuronal Abeta and Cu in Alzheimer' disease. Clin Neuropathol 25:163–171
44. Roy S, Metya SK, Sannigrahi S, Rahaman N, Ahmed F (2013) Treatment with ferulic acid to rats with streptozotocin-induced diabetes: effects on oxidative stress, pro-inflammatory cytokines, and apoptosis in the pancreatic β cell. Endocrine 44(2):369–379
45. Xu X, Xiao H, Zhao J, Zhao T (2012) Cardioprotective effect of sodium ferulate in diabetic rats. Int J Med Sci 9(4):291–300
46. Gerin F, Erman H, Erboga M, Sener U, Yilmaz A, Seyhan H, Gurel A (2016) The effects of ferulic acid against oxidative stress and inflammation in formaldehyde-induced hepatotoxicity. Inflammation 39(4):1377–1386
47. Villareal MO, Sasaki K, Margout D, Savry C, Almaksour Z, Larroque M, Isoda H (2016) Neuroprotective effect of Picholine virgin olive oil and its hydroxycinnamic acids component against b-amyloid-induced toxicity in SH-SY5Y neurotypic cells. Cytotechnology 3:1–12
48. Doss HM, Dey C, Sudandiradoss C, Rasool MK (2016) Targeting inflammatory mediators with ferulic acid, a dietary polyphenol, for the suppression of monosodium urate crystal induced inflammation in rats. Life Sci 148:201–210
49. Shen Y, Zhang H, Wang L, Qian H, Qi Y, Miao X, Qi X (2015) Protective effect of ferulic acid against 2, 2′-azobis (2-amidinopropane) dihydrochloride-induced oxidative stress in PC12 cells. Cell Mol Biol (Noisy-le-Grand, France) 62(1):109–116
50. Chong YH, Shin YJ, Lee EO, Kayed R, Glabe CG, Tenner AJ (2006) ERK1/2 activation mediates Aβ oligomer-induced neurotoxicity via caspase-3 activation and tau cleavage in rat organotypic hippocampal slice cultures. J Biol Chem 281(29):20315–20325
51. Picone P, Nuzzo D, Carlo MD (2013) Ferulic acid: a natural antioxidant against oxidative stress induced by oligomeric A-beta on sea urchin embryo. Biol Bull 224:18–28
52. Hussein AM, Abbas KM, Abulseoud OA, Elhussainy EMA (2017) Effects of ferulic acid on oxidative stress, heat shock protein 70, connexin 43 and monoamines in hippocampus of pentylenetetrazole-kindled rats. Can J Physiol Pharmacol, (ja) 95(6):732–742
53. Akiyama H, Barger S, Barnum S, Bradt B, Bauer J, Cole GM, Finch CE (2000) Inflammation and Alzheimer's disease. Neurobiol Aging 21(3):383–421

54. Kim HS, Cho JY, Kim DH, Yan JJ, Le HK, Suh HW, Song DK (2004) Inhibitory effects of long-term administration of ferulic acid on microglial activation induced by intracerebroventricular injection of β-amyloid peptide (1—42) in mice. Biol Pharm Bull 27(1):120–121
55. Huang F, Deng HM, Zhu MM, Xiao F, Yang L, Zhang ZJ, Nie H (2011) Inhibitory effect of ferulic acid on inflammatory response in microglia induced by lipopolysaccharides. Zool Res 32(3):311–316
56. Mori T, Koyama N, Guillot-Sestier MV, Tan J, Town T (2013) Ferulic acid is a nutraceutical β-secretase modulator that improves behavioral impairment and alzheimer-like pathology in transgenic mice. Plos One 8:1–16
57. Yan JJ, Cho JY, Kim KL, Jung JS, Huh SO, Suh HW, Kim YH, Song DK (2001) Protection against β-amyloid peptide toxicity in vivo with long-term administration of ferulic acid. Br J Parmacol 133:89–96
58. Huang H, Hong Q, Tan HL, Xiao CR, Gao Y (2016) Ferulic acid prevents LPS-induced up-regulation of PDE4B and stimulates the cAMP/CREB signaling pathway in PC12 cells. Acta Pharmacol Sin 37:1543–1554
59. Chaturvedi RK, Beal MF (2013) Mitochondrial diseases of the brain. Free Radic Biol Med 63:1–29
60. Yogeeta SK, Raghavendran HRB, Gnanapragasam A, Subhashini R, Devaki T (2006) Ferulic acid with ascorbic acid synergistically extenuates the mitochondrial dysfunction during β-adrenergic catecholamine induced cardiotoxicity in rats. Chem Biol Interact 163(1):160–169
61. Harrison SM, Harper AJ, Hawkins J, Duddy G, Grau E, Pugh PL et al (2003) BACE1 transgenic and knockout mice: identification of neurochemical deficits and behavioral changes. Mol Cell Neurosci 24:646–655
62. Castro A, Martinez A (2006) Targeting beta-amyloid pathogenesis through acetylcholinesterase inhibitors. Curr Pharm Des 12(33):4377–4387
63. Citron M (2004) Beta-secretase inhibition for the treatment of alzheimer disease-promise and challenge. Trends Pharmacol Sci 25:92–97
64. Gilgun-Sherki Y, Melamed E, Offen D (2001) Oxidative stress induced-neurodegenerative diseases: the need for antioxidants that penetrate the blood brain barrier. Neuropharmacology 40(8):959–975
65. Hashimoto M, Rockenstein E, Crews L, Masliah E (2003) Role of protein aggregation in mitochondrial dysfunction and neurodegeneration in Alzheimer's and Parkinson's diseases. NeuroMol Med 4(1–2):21–36
66. Huang WY, Chao XJ, Ouyang Y, Liu A, He X, Chen M, Wang L, Liu J, Yu S, Rapposelli S, Pi R (2012) Tacrine-6-ferulic acid, a novel multifunctional dimer against Alzheimer's disease, prevents oxidative stress-induced neuronal death through activating Nrf2/ARE/HO-1 pathway in HT22 cells. CNS Neurosci Ther 18:950–952
67. Ma ZC, Hong Q, Wang YG, Liang QD, Tan HL, Xiao CR, Tang XL, Shao S, Zhou SS, Gao Y (2011) Ferulic acid induces heme oxygenase-1 via activation of ERK and Nrf2. Drug Discov Ther 5:299–305
68. Ono K, Hasegawa K, Naiki H, Yamada M (2004) Curcumin has potent anti-amyloidogenic effects for Alzheimer's beta-amyloid fibrils in vitro. J Neurosci Res 75:742–750
69. Sultana R, Butterfield DA (2010) Role of oxidative stress in progression of Alzheimer's disease. J Alzheimer Dis 19(1):341–353
70. Cohen E, Bieschke J, Perciavalle RM, Kelly JW, Dillin A (2006) Opposing activities protect against age-onset proteotoxicity. Science 313(5793):1604–1610
71. Fukuda H, Komamine A (1982) Lignin synthesis and its related enzymes as markers of tracheary-element differentiation in single cells isolated from the mesophyll of Zinnia elegans. Planta 155(5):423–430

Flavonoids and Cancer Stem Cells Maintenance and Growth

26

Kushal Kandhari, Hina Agraval, Arpana Sharma, Umesh C. S. Yadav, and Rana P. Singh

Abstract

Normal stem cells are known to possess three important characteristics of self-renewal, restriction on stem cell numbers and ability to divide and differentiate. Compared to normal stem cells, the cancer stem cells (CSCs) have no control on the stem cell numbers. CSCs constitute a miniscule number of cells in the tumour and are responsible for tumour growth, recurrence and progression. CSCs play a vital role in drug resistance, EMT and metastasis, which are responsible for approximately 90% of cancer-related deaths. Thus, targeting CSCs has now gained significant importance in the control and treatment of various cancers. Traditional cancer therapy regimens have not been successful against cancer drug resistance and metastasis. In the recent past, numerous dietary compounds derived from natural sources have been found effective in chemoprevention and treatment of various cancers. Flavonoids are one of such naturally occurring polyphenolic compounds that are found abundantly in fruits, vegetables, tea, seeds, grains, nuts and some traditional medicinal herbs. Various flavonoids have also been shown to have an inhibitory effect on the self-renewal potential and survival of cancer stem cells of different origins. The aim of this chapter is to focus on cancer stem cells and their role in tumour progression and drug resistance and how chemoprevention using flavonoids can become an effective tool to control cancer growth.

Keywords

Cancer stem cells · EMT · Metastasis · Drug resistance · Chemoprevention

K. Kandhari · H. Agraval · A. Sharma · U. C. S. Yadav
School of Life Sciences, Central University of Gujarat, Gandhinagar, Gujarat, India

R. P. Singh (✉)
Cancer Biology Laboratory, School of Life Sciences, Jawaharlal Nehru University, New Delhi, India
e-mail: rana_singh@mail.jnu.ac.in

V. Rani, U. C. S. Yadav (eds.), *Functional Food and Human Health*,
https://doi.org/10.1007/978-981-13-1123-9_26

26.1 Introduction

Cancer is defined as the group of diseases in which cells divide uncontrollably by breaking normal rules of cell division. Normal cells follow a set of instructions that dictate whether and when cells should divide or differentiate, but in cancer cells these instructions are not followed. Cancer cells grow uncontrollably and invade nearby tissues and spread through the body. In the year 2012, International Agency for Research on Cancer (IARC) estimated 14.1 million new cancer cases and 8.2 million cancer deaths worldwide. Cancer is caused by various factors that include environmental factors and genetic and/or epigenetic factors within a single cell. Cancer cells alter important mechanisms of cells such as rate of proliferation, invasion and metastasis ability, replicative immortality and angiogenesis [1]. The cancer cells may comprise CSCs which are a small population of cancer cells having indefinite potential for self-renewal and frequently develop drug resistance [2]. Ever since the identification of CSCs in 1994, they have been a subject of intense study. Their properties, such as the capability to initiate and propagate the tumour growth and develop resistance to the conventional therapies, have garnered focus of the cancer researchers [3]. The fundamental difference between CSCs and normal stem cells is that in the case of the latter, the number of cells generated through cell division is similar to cells that terminally differentiate, thus keeping the number of normal stem cells constant. In contrast, cancer stem cells keep on proliferating and do not differentiate, while the mature cells do not die. Although, in both the cases, some cells do not actively proliferate and function as reserve cell population [4]. Stem cells have been implicated in various important cellular processes, and they are identified and isolated based on various cell surface markers.

CSCs are speculated to have an imperative role in tumour cell proliferation, invasion and metastasis. One of the most important cellular events that are related to heterogeneity and stemness of tumour is epithelial-mesenchymal transition (EMT). In EMT, epithelial cells lose their cell adhesion property and get converted into motile cells having a distinctive mesenchymal morphology, thus allowing the cells to migrate to a different location within or outside the tissue. Apart from these pro-invasive and metastasis-inducing functions, EMT has been shown to promote stemness in tumours in both in vitro and in vivo models [1]. EMT confers metastatic potential to the cancer cells with the help of EMT transcription factors such as Twist and Zeb1 [5]. A few recent studies have shown that CSCs express various EMT markers which promote the generation of the cancer stem cell-like population leading to chemotherapeutic resistance [6, 7]. Thus, CSCs play a pivotal role in EMT and metastasis of cancer cells and vice versa. Indeed, approximately 90% of cancer mortality has been attributed to metastasis and not primary tumours [8]. Therefore, it is pertinent that CSCs are targeted along with the usage of conventional therapies for successful cancer therapy to minimize or negate the chances of relapse (Fig. 26.1).

Prevalent therapies such as chemotherapy, radiation therapy and surgery have not proved significantly successful in reducing the burden of cancer. In addition, chemotherapy and other traditional therapies are confined by toxicity and development

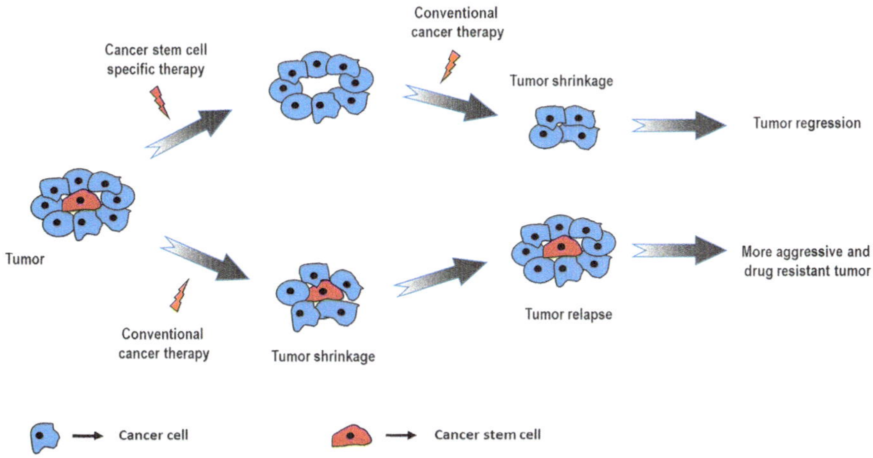

Fig. 26.1 Cancer stem cell therapy versus conventional cancer therapy

of resistance that eventually results in relapse of tumours [9]. Studies have shown that chemoprevention is a much better approach for controlling cancers [10]. Usage of dietary compounds in cancer chemoprevention has been a topic of interest among researchers for the past few years as it has been shown in various studies that there is a clear link between dietary habits and cancer prevention [11]. Flavonoids, available in various food items such as citrus fruits, vegetables, tea and cocoa, are labelled as one of the important components of functional foods, and they exhibit several beneficial properties for human health [12]. These properties include antiallergic, anti-inflammatory, antioxidant, antiviral and anticancer effects [13]. Most flavonoids arise from natural sources and are found to be safe for human consumption [14]. Numerous epidemiological studies support the use of flavonoids for their chemopreventive abilities as well as for their anticancer potential [15]. Several flavonoids have been shown to inhibit CSCs in vitro and in vivo [9, 16]. Flavonoids have been shown to target or intervene signalling pathways that are critical for CSC maintenance and growth such as Hedgehog, Wnt/β-catenin and Notch pathways. Flavonoids have been used in combination with several chemotherapeutic drugs and found to exert synergistic effect to control growth, maintenance, cell survival and stemness of cancer cells and CSCs [14]. Thus, flavonoids can prove to be promising molecules that can target CSCs.

26.2 The Biology of Cancer Stem Cells

26.2.1 Cancer Stem Cells: Origin

It is a well-established fact that cancer arises from a mutation in a single cell, yet tumours eventually become heterogeneous in nature containing undifferentiated and proliferative cells expressing different markers [17]. Heterogeneous nature of

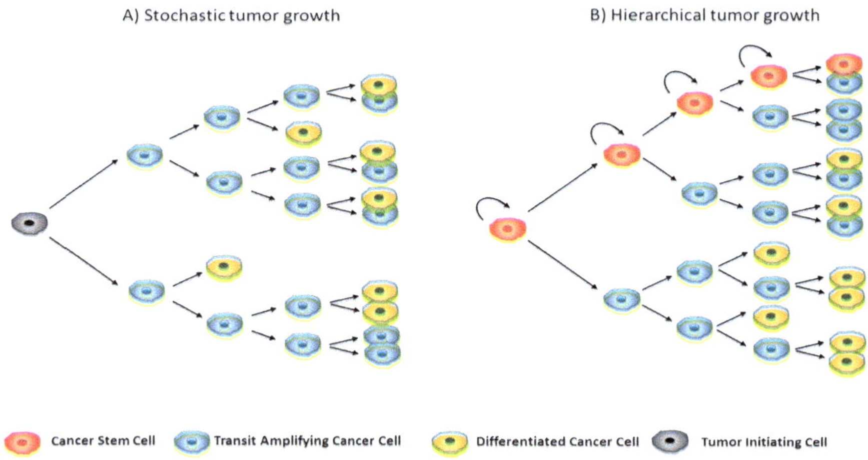

Fig. 26.2 Two models proposed for tumour growth

tumours is believed to be the reason for tumour progression, metastasis, relapse and resistance to therapy. Heterogeneity in tumours is of two types, intra-tumoural and inter-tumoural, and they arise due to mechanisms such as clonal evolution, tumour microenvironment and presence of CSCs [18]. Two models have been proposed for the tumour growth (Fig. 26.2). A stochastic model hypothesizes that tumours arise as a homogenous group of cells and the heterogeneity is the outcome of arbitrary stochastic events. However, hierarchical model propounds that a stem-like precursor cell results in the generation of a heterogeneous group of cells that differentiate with distinct phenotypical as well as biological attributes [1].

Stem cells (SCs) are responsible for various functions such as tissue homoeostasis and regeneration. After division SCs generate transit-amplifying cells, which after some rounds of divisions eventually differentiate terminally and develop the specific tissue. Apart from this, stem cells can also be activated when injuries occur and participate in tissue repair [19]. Another noteworthy feature of stem cells is that they do not express a specific marker for different tissues and are mostly determined by their functional properties such as their scope for long duration self-renewal and their efficiency to differentiate into one or multiple cell lineages [1].

The concept of adult haematopoietic stem cells motivated Dick and colleagues to carry out experiments and demonstrate that in human acute leukaemia, all the leukaemia cells were not able to initiate leukaemia when introduced into immune-deficient mice. Terminally differentiated leukaemia cells were not able to propagate the disease; however, those leukaemia cells which had similar markers as of adult HSCs had more efficacy for initiating leukaemia and thus were termed as leukaemic stem cells (LSCs) [20].

Inspired by the work of Dick and colleagues, many researchers diverted their focus on CSCs and their presence in solid tumours. CSCs in solid tumours were first derived from breast cancer cells when a group of researchers injected the $CD44^{+}CD24^{-/low}$ population of cells into immune-deficient mice [21]. Subsequent studies using the same approach by various research groups led to the discovery of

CSCs in various cancers such as pancreatic cancer, squamous cell carcinomas, lung cancer and melanoma [1].

26.2.2 Cancer Stem Cell Biomarkers in Various Human Malignancies

The first evidence of CSCs in cancer was reported by John Dick in 1994 in AML [20]. Subsequently, CSCs were reported in numerous human malignancies such as brain, colon, ovary, breast, pancreas, prostate and lung [22]. Some of the biomarkers prevalent in CSCs of various malignancies are discussed below.

Specific biomarkers can be used to differentiate CSCs from normal cells and other tumour cells [23]. CSCs express several markers simultaneously, and it is very difficult to define or describe CSCs using one biomarker. Understanding the role of CSC biomarkers may help us to uncover unprecedented facts that will in turn improve current cancer therapy and prognosis [24]. CD133 is the highly investigated biomarker for CSCs. It is a 120 kDa cell surface glycoprotein comprised of two glycosylated extracellular loops and transmembrane domains which are five in number [25]. CD133 and its epitope AC133 have been frequently demonstrated as cell surface marker for CSCs in multiple cancers including central nervous system tumours as well as colon, breast, prostate, ovarian and lung cancers [25]. $CD133^+$ cells were more competent in tumour sphere formation and differentiation than $CD133^-$ cells in lung cancer cell lines like NSCLC and SCLC [26]. Oct-4 and NANOG, which are well-known embryonic stem cell markers, were expressed by $CD133^+$ cells. Upon knockdown of Oct-4 in $CD133^+$ cells, clonogenicity and chemosensitization of cells reduced significantly [27]. Further, $CD44^+/CD49f^{hi}/CD133/2^{hi}$ phenotype cells have been shown to have increased tumorigenicity and self-renewal capacities in vivo and give rise to molecular and functional heterogeneity [28].

CD44 is a multifunctional membrane-bound glycoprotein which binds to hyaluronic acid that is abundant in stem cell niches and performs various functions such as differentiation, migration, cell adhesion, homing and interaction with extracellular matrix [25, 29]. A recent study by Wang et al. suggested an important role of CD44 in identification of lung CSCs. They showed that the expression of CD44 variant exon 6 (CD44v6) in 79 lung cancers was 67.6% (48/71) in NSCLC and only 0% (0/8) in SCLC. This data demonstrates that the expression of CD44v6 is associated with histologic type of tumour [25, 30]. $CD44^+/CD24^{-/low}$ cell surface marker has been most commonly used to characterize breast CSCs. The stemness-inducing properties of $CD44^+/CD24^{-/low}$ cells have been further implicated in colony formation, migration and invasion assays [31].

Aldehyde dehydrogenases (ALDH) are a family of $NAD(P)^+$-dependent enzymes which are involved in differentiation, cellular detoxification and drug resistance by exploiting cellular aldehyde oxidation. Increased activity of ALDH was observed in stem cell populations of different cancers which suggests its role as a common biomarker for both normal and cancer stem cell populations [25, 29]. Overexpression of ALDH1 is observed in lung tumours compared to the normal lung. A study has demonstrated the expression level of ALDH in 12 different human lung cancer cell lines using flow cytometry. A higher expression level of cytosolic forms, ALDH1A1 and ALDH3A1, in some NSCLC cell lines and patients with lung cancer was observed [29]. NSCLC (non-small-cell lung cancer) patients having tumour cells with elevated expression of ALDH1A1 show significant resistance to EGFR tyrosine kinase inhibitors and chemotherapy drugs which is associated with poor clinical outcome [32, 33].

CD338 or ABCG2 (BCRP) is a member of the ATP-binding cassette transporters. The main function of ABCG2 is to pump out wide range of molecules out of cells providing multidrug resistance [34, 35]. ABCG2 could impart subpopulation (SP) phenotype of CSCs and serve as a promising CSC biomarker. Increased expression of ABCG2 has been observed in mammary gland, liver cancer SP cells and lung CSCs. Elevated expression of ABCG2 was also reported in SP cells isolated from different lung cancer cell lines (H460, H23, HTB-58, A549, H441 and H2170). The SP cells isolated from A549 cells were completely vanished after treatment with selective ABCG2 inhibitor suggesting it as an important biomarker in lung cancer [29]. $CD338^+$ cells displayed stemness and tumorigenic potential when compared to $CD338^-$ cells in BRCA1-mutated breast cancer cell line HCC1937 [36].

Alpha-6 integrin has been showed as essential for tumorigenicity of a CSC-like subpopulation within the breast cancer cell line MCF-7 [37]. Mammary progenitor marker $CD61^+$, also known as beta-3 integrin, has been demonstrated to have a CSC-like population having highly enriched tumorigenic capability when compared to $CD61^-$ subset of cells in the mouse model [38]. Expression of Ca^{2+}-dependent cell-cell adhesion glycoprotein P-cadherin has been associated with the expression of breast CSC markers and an onset of transient EMT required for the metastasis [39].

Other prominent stem cell marker is nestin, an intermediate filament protein which is marked with cell signalling, cytoskeletal organization, proliferation and attributing stemness properties to the cells [40]. Musashi-1 is RNA-binding protein which is additional protein focused towards the presence of stem cells in neurosphere culture. Its involvement in tumorigenesis is well accepted [41]. Also, the level is associated with the mitotic activity, determining the grade and aggressiveness of the brain tumour [42]. CD15 is a novel putative marker present on astrocytes; adult neurogenic zones had been shown in the glioma-derived neural sphere [43]. The presence of this trisaccharide 3-fucosyl-N-acteyllactosamine (FAC) group is attributed with the tumour-propagating properties in medulloblastoma [44].

All these above-mentioned markers are useful in cell signalling mechanisms employed between the cells and their microenvironments. Over the period of time, several other cell surface markers were identified in various CSCs; some of them have been enlisted in the table below (Table 26.1).

26.2.3 Cancer Stem Cells: EMT and Metastasis

EMT in cancer cells is orchestrated by the involvement of various environmental factors; signalling pathways; group of pleiotropic transcription factors (TFs) such as Snail1, Twist1 and Zeb1; microRNAs (miRNAs); and other numerous mechanisms

Table 26.1 Cancer stem cell markers in different cancers

Tumour type	Cancer stem cell marker	References
Acute myeloid leukaemia	$CD38^-$, $CD34^+$	[45]
Breast cancer	EPCAM (ESA)$^+$, $CD44^+$, $CD24^-$, ALDH, CD29, CD133	[21]
Colorectal cancer	$CD133^+$, $CD44^+$, $CD26^+$, ALDH	[46]
Liver cancer	$CD133^+$/$CD44^+$, $EpCAM^+$, $CD90^+$	[47]
Glioblastoma	$CD133^+$, $CD15^+$	[48]
Head and neck cancer	$CD44^+$, ALDH1	[49]
Hepatocellular carcinoma	$CD45^-$, $CD90^+$	[50]
Lung cancer	$CD133^+$, ABCG2, CD90, CD117, ALDH1	[51, 52]
Medulloblastoma	$CD133^+$, $CD15^+$	[53]
Melanoma	$CD20^+$, $CD271^+$	[54]
Multiple myeloma	$CD138^+$	[55]
Osteosarcoma	$CD117^+$ (c-Kit), CD133+, Stro-1+	[56]
Ovarian cancer	$CD133^+$, $CD44^+$, $CD117^+$, $CD24^+$	[57]
Pancreatic cancer	$CD44^+$, $CD24^+$	[58]
Prostate cancer	Integrin $\alpha2/\beta1$, BMI-1, CD49f (integrin $\alpha6$), $CD133^+$, $CD44^+$, ABCG2/Hoechst 33342, SCA-1, $CD166^+$, $CD151^+$, $p63^+$	[59, 60]
Renal carcinoma	$CD133^+$	[58]
Bladder cancer	$CD44^+$, $CD47^+$, $CK5^+$	[61]

that foster the loss of epithelial and adhesive attributes and gain of mesenchymal and migratory characteristics. Aberrant expression level of EMT TFs in primary tumours has been linked with poor prognosis, tumour invasion and metastasis. [62]. Activation of EMT through the repression of E-cadherin is carried out by several signalling pathways such as transforming growth factor beta (TGF-β), Notch, Wnt and integrin pathway [63, 64]. E-cadherin is known to maintain epithelial state, and decrease in its expression has been acknowledged to be a hallmark of EMT [65]. Recent study has indicated that EMT cascade helps in the genesis of many traits of CSCs in numerous cancers such as colorectal, breast, hepatocellular and pancreatic carcinoma [66]. The work of various researchers suggests that EMT induction and various EMT markers expressed by CSCs are linked with genesis of CSC-like population that in turn impart resistance to chemotherapy [6, 7, 67, 68].

Overexpression of various EMT TFs is downregulated at the time of metastatic colonization. In a study, it has been shown that transient overexpression of Twist1 facilitated the lung metastasis in squamous cell carcinoma [69]. The change in mesenchymal state to epithelial is imminent in formation of secondary cancer and is called as MET (mesenchymal-epithelial transition). In MET, expression of EMT TFs gets switched off to facilitate the metastatic colonization of cancer cells. Similarly, another EMT TF Prrx1, a member of the homoeobox proteins, is an established inducer of EMT in cancer cells. Nonetheless, Prrx1 downregulation has been implicated in metastasis and induction of tumour stemness in various cancer cell lines [70]. Further, increased expression of a well-known EMT inducer Snail in prostate cancer cells was surprisingly shown to inhibit metastasis and colonization [71]. Collectively, these studies suggest that tumour stemness, EMT and metastasis are regulated by diverse mechanisms and point out that for the completion of metastatic colonization, MET is as important as EMT.

Acquisition of EMT and CSC-like phenotype cannot be achieved only by overexpression and downregulation of EMT TFs but also requires the regulation by tumour microenvironment or niche. Circulating tumour-associated fibroblasts are extensively endowed with stem cell markers and augment the metastatic capability of tumours by boosting their migration and extravasation [72, 73]. CSCs are directly involved in establishing distant metastasis, for example, in breast cancer metastasis to the bone, it was shown that the cells present in the bone metastatic site contained cells with characteristic breast cancer stem cell markers [74]. Another study showed that breast CSCs having putative markers such as $CD44^{+}$ and $CD24^{-/low}$ were able to initiate tumours in distant sites and could induce lung metastasis [75]. In pancreatic cancer it was demonstrated that subpopulation of cells with $CD133^{+}/CXCR4^{+}$ phenotype were localized in the invasive front of the tumour and had more migratory potential than $CD133^{+}/CXCR4^{-}$ cells. Only $CD133^{+}/CXCR4^{+}$ cells were able to metastasize to the liver [76].

Metastasis, in most cases, occurs several years after the successful treatment of the initial tumour, thus displaying the unique features of latency in metastatic cells [1]. Cancer stem cells with metastatic characteristics have unique features such as dormancy and plasticity. For example, the cancer cells present in the bone marrow are dormant in nature and thus are protected from cytotoxic agents. To come out of the dormant phase to proliferate and induce metastasis, they require the activation of several signalling pathways such as BMP and MAPK pathways [71]. Some unique markers such as thrombospondin-1 induce the sustained tumour quiescence during the dormant phase of the tumour, but when the dormant phase is over, these quiescent cells enter the cell cycle with the help of active TGF-β1 and periostin secretion from endothelial tip cells that are present in the tumour niche. These evidences suggest that tumour microenvironment and niche play a very important role to sustain tumour dormancy and late metastasis [77].

26.2.4 Cancer Stem Cells: Drug Resistance

Chemotherapy is one of the traditionally used methods for the treatment of many cancers. However, tumour cells frequently acquire the ability to resist the effects of chemotherapeutic drugs [78, 79]. Recent studies suggest that inhabitancy of cancer cells provides them varying capabilities to both initiate and metastasize tumours [20, 21]. CSCs are believed to be predominantly responsible for the genetic heterogeneity present in tumours. When success of treatment in cancer patients is considered, the main concern is about the relative resistance of CSCs to many standard chemotherapeutic drugs [80]. The cancer cells develop capability of synchronous resistance to several drugs that are not structurally related and have varied mechanism of actions,

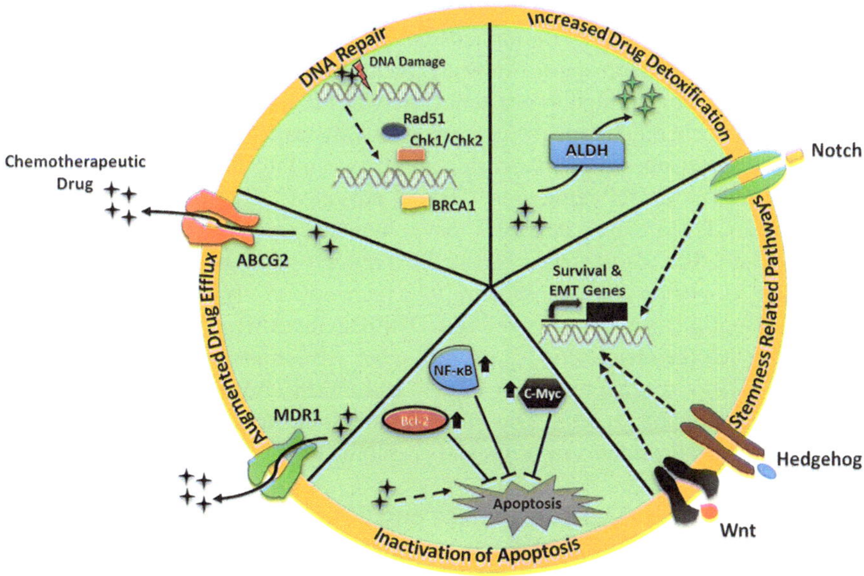

Fig. 26.3 Molecular mechanisms involved in CSC-associated chemoresistance

known as multidrug resistance (MDR). Various strategies used by CSCs for establishing and maintaining drug resistance include the efflux of drugs by ABC transporters such as ABCC1, ABCG2 and ABCB1/P-glycoprotein, secretion of drugs into vesicles and consequent exclusion by exocytosis, reduced uptake of drugs, glutathione system, detoxifying pathways such as cytochrome P-450 pathway and modulations in the apoptotic signalling (Fig. 26.3) [78]. Thus, CSCs are hard to treat because of their abilities to efflux the drugs and their metabolites easily. Therefore, focus should be on tackling the mechanisms involved in induction of MDR.

26.2.4.1 CSCs Chemoresistance and Enhanced Drug Efflux Mechanisms

ABC transporters have been shown to have a vital role in efflux of most chemotherapeutic drugs, and elevated expression of ABC transporters is the chief mechanism of chemotherapeutic resistance in CSCs [81–83]. ABC transporters efflux a wide variety of chemotherapeutic drugs that include vinblastine, doxorubicin, colchicine, etoposide and paclitaxel and are frequently correlated with MDR in cancer [80]. The side population of CSCs is a subgroup of stem cells that have a high capacity for efflux of mitosis-inhibiting drugs. The drug-transporting ability of these SP cells is due to the presence of certain ABC transporters [78]. ABCB1/P--glycoprotein/MDR1 is one of the well-studied ABC transporters and is usually involved in chemotherapeutic resistance in various cancers, including AML and gastrointestinal cancer [84]. Further, meta-analysis studies showed that ABCB1 may be expressed in 40% or as high as 66% of breast cancers [85]. Some recent evidence showed that chemotherapy may elevate expression of ABCB1. This may suggest why at least some acquired resistance in breast cancer is associated with elevated expression of ABCB1 after neoadjuvant chemotherapy treatment [86, 87]. ABCG2, another ABC transporter known as breast cancer resistance protein (BCRP) because it was first identified in resistant variety of MCF-7 breast cancer cells [88], is also linked to chemotherapeutic resistance in AML patients [80, 89].

Increasing evidences show the correlation between the chemotherapeutic resistances of CSCs to MDR1 expression. $CD133^+$ prostate CSCs [80, 90] and $CD117^+$, $CD44^+$ and $CD133^+$ ovarian CSCs have higher expression of MDR1 [91]. Glioblastoma CSCs also showed elevated expression of MDR1 and resistance to carboplatin, etoposide and doxorubicin [92]. Increased levels of MDR1 and resistance to daunorubicin were observed in $CD34^+/CD38^-$ AML CSCs [93]. CSCs or cells with CSC-like characteristics express other ABC transporters. For example, ABCG2/BCRP is overexpressed in $CD34^+$ and $CD38^-$ AML CSCs which show resistance to mitoxantrone [94]. Thus, these studies indicate a clear link between the expression of MDR proteins in CSCs and chemoresistance in these cells.

26.2.4.2 Aldehyde Dehydrogenase and CSCs Chemoresistance

Increased ALDH activity is another mechanism through which CSCs attain the ability for chemoresistance. ALDH1 is an enzyme that is localized in cytosol which catalyses the oxidation of aldehyde into carboxylic acids [95]. The main function of ALDH enzymes is to remove toxic aldehydes produced during metabolic processes [96]. This detoxification capability of ALDH enzymes implies that it has a role in imparting resistance to cells in the case of certain chemotherapeutic drugs [80]. A study by ALDEFLUOR-positive staining showed that on an average, ALDH activity was present in 8% of normal mammary epithelial cells [97]. Breast cancer cells having high ALDH activity compared to normal epithelial cells were able to develop xenograft tumours with as low as 500 cells [98]. The ability of ALDH expression to impart resistance against cyclophosphamide has been showed in tumours of different origin, and its inhibition can sensitize CSCs to cyclophosphamide [99]. Knockdown of ALDH1A1 in pancreatic cancer suggested that ALDH may also be resistant towards gemcitabine [100]. It suggests that ALDH expression imparts resistance in CSCs against chemotherapeutic drugs and its inhibition can sensitize CSCs to these drugs [99].

26.2.4.3 Chemoresistance of CSCs by Apoptosis Inactivation

B-cell lymphoma-2 (BCL-2) family proteins, traditionally identified as the potential oncogene in acute B-cell leukaemia, are involved in chemotherapeutic resistance of CSCs [98]. BCL-2 expression has also been seen in various cancers as well as in haematopoietic lineage cells [101, 102]. Increased BCL-2 expression in myeloid progenitor cells aggravated leukaemia development in transgenic mice. This suggests an overlay between BCL-2 expression and the capacity of cancer-initiating cells to induce tumour formation [80]. In glioblastoma cells, $CD133^+$ CSCs showed higher BCL-2 expression and increased chemoresistance [103, 104]. A study explained that BCL-2 expression was elevated in $CD44^+/CD24^{-/low}$ breast cancer stem cells; however, the mechanism involving the expression of these proteins is still not clear [105]. Similarly, $CD133^+$ hepatocellular CSCs had higher expression of BCL-2 and also showed resistance to doxorubicin and 5-fluorouracil [106]. Further, in $CD133^+$ colon CSCs where IL-4 utilization was found via autocrine system, IL-4 neutralizing antibodies reduced BCL-XL expression and increased sensitivity towards 5-fluorouracil and oxaliplatin [107]. Aurora-A is an oncogenic serine/threonine kinase that regulates the cell cycle, and BCL-2 family members may be induced in CSCs via this protein. Recent studies in $CD133^+CD29^+CD20^-$ colorectal CSCs showed that these cells have an elevated expression of Aurora-A as well as BCL-2, MCL-1 and BCL-XL. Aurora-A knockdown with shRNA showed a strong decrease in MCL-1 and BCL-2 expression along with a moderate reduction in BCL-XL expression [108, 109, 98].

26.2.4.4 CSCs and Dysregulated DNA Damage Response

Another important mechanism that regulates tumour chemoresistance and progression is elevated DNA damage response. Under hypoxia, cancer cells can stimulate a strong DNA damage response via proteins such as HIF transcription factors [110]. Subsequently, ataxia telangiectasia-mutated (ATM) and ataxia telangiectasia and Rad3-related (ATR) protein are activated after the induction of hypoxia and in response to DNA damage [111]. CSCs are protected from DNA-damaging radiation and chemotherapeutic drugs via these signalling pathways [98]. A study showed that $CD133^+$ glioma CSCs were resistant to ionizing radiation than $CD133^-$ cells [112]. Activated DNA damage response factors ATM, CHK1 and CHK2 were highly expressed in $CD133^+$ cells as compared to $CD133^-$ cells after treating with radiation. Further, treatment of debromohymenialdisine, which is a CHK1/CHK2 inhibitor, to $CD133^+$ glioma stem cells reversed radio resistance. This suggests that knockdown of CHK1 and the reduction in DNA damage response may be an effective approach to target chemoresistant CSCs [98].

26.2.4.5 Chemoresistance and CSC-Related Signalling Pathways

Several pathways participate in imparting chemoresistance to CSCs. Wnt/β-catenin pathway is often required for self-renewal of both normal stem cells and CSCs in different cell types [113, 114]. It has been shown that Wnt signalling pathway activation increased renewal of $OV6^+$ HCC hepatic CSCs, whereas knockdown of β-catenin using lentiviral microRNA showed a decrease in self-renewal capacity [115]. In neuroblastoma cells, Wnt pathway activation by FZD1 induced resistance against Dox [116]. Resistance to paclitaxel and cisplatin in C-kit$^+$ ovarian cells was mediated by ABCG2, which was downregulated by knockdown of β-catenin and reversed chemoresistance to cisplatin and paclitaxel [117]. Further, oxaliplatin treatment induced Notch pathway activation in colon cancer cells, which, when knocked down or inhibited, could prevent chemoresistance of colon CSCs towards oxaliplatin [118]. Furthermore, Notch proteins were found to be upregulated in ovarian CSCs. GIS treatment sensitized ovarian CSCs to cisplatin via inhibition of Notch-mediated maintenance of MDR1 expressing CSCs [119, 120]. The redox transcription factor NF-κB, a key inflammatory mediator, is also associated with chemoresistance of different CSCs. In $CD44^+$ ovarian CSCs, constitutive activation

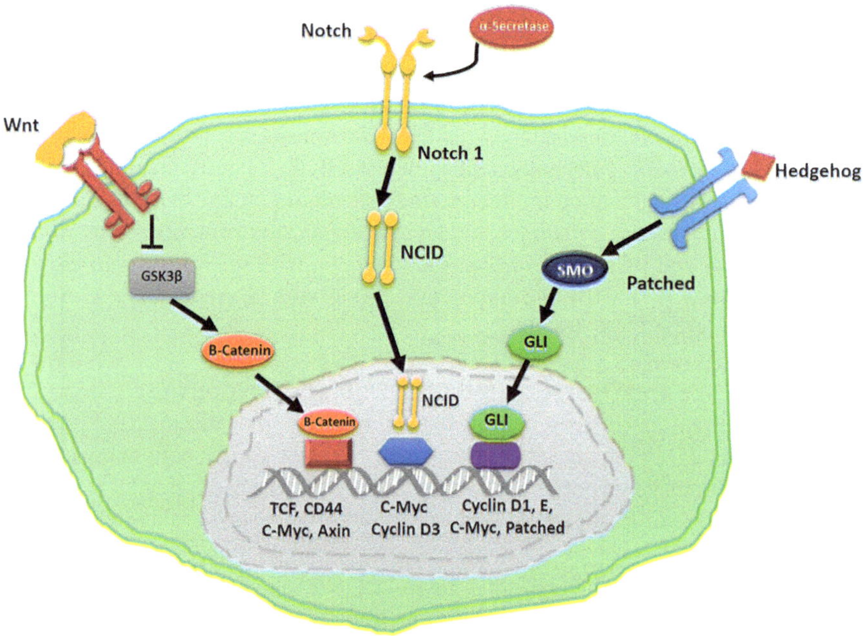

Fig. 26.4 Molecular pathways regulating CSC growth and maintenance

of NF-κB and other pro-inflammatory signals was correlated with resistance towards carboplatin and paclitaxel. Inhibition of NF-κB by eriocalyxin B increased cell death in resistant $CD44^+$ ovarian CSCs [121] (Fig. 26.4).

26.3 Flavonoids in Cancer Chemoprevention and Their Implications in Cancer Stem Cell Biology

26.3.1 Flavonoids: Occurrence and Characteristics

Flavonoids are naturally occurring plant phytochemicals, which are widely distributed and are important constituents of the diet working as functional foods. Flavonoids have been known to mankind from the last 60 years when for the first time they were identified by Szent-Gyorgyi and colleagues. These are the largest group of the phenolic phytonutrients, which have shown beneficial health-promoting effects and pharmacological properties due to which they have inspired various researchers [122]. There are six major important flavonoid classes as summarized in Table 26.2, which shows different classes of the flavonoids, sources and some of their properties.

Table 26.2 Different classes of the flavonoids

Class of the flavonoids	Rich food sources	Examples of the class	Remarks	Colour	References
Flavanones	Chick peas, cumin, berries, peppermint, citrus fruits	Hesperidin Narirutin Naringenin glycosides Liquiritigenin	Flavour of the citrus fruits like bitter taste	Colourless to pale yellow	[124]
Flavones	Grains, herbs like parsley and rosemary	Apigenin Luteolin Nobiletin Sinensetin Tangeretin	Gives plant tissue colour, gives bitter taste to fruits	Pale yellow	[125]
Flavans (monoflavans, biflavans, triflavans) or flavan-3-ol or flavanol	Fruits and teas (green and black). Biflavans: fruits, hops, nuts, beverages (cocoa and tea), sorghum, barley grains	Catechin Epicatechin Luteoforol Procyanidin Theaflavin	Complicated flavonoids and are not glycosylated but esterified with gallic acid. Contribute to the astringent taste of beer and wines	Colourless	[126]
Flavonols	Fruits, vegetables, berries, herbs, legumes, maize and tea	Quercetin Kaempferol Isorhamnetin Myricetin	Found predominantly in the skin of fruits	Pale yellow	[125]
Isoflavones	Legumes, black beans, soya beans, green split peas, clover sprouts	Daidzein Genistein Biochanin A Formononetin	Best known for their oestrogenic activity	Colourless	[127]
Anthocyanins	Cherries, berry fruits, plums, eggplant and radishes	Delphinidin Cyanidin Petunidin Peonidin Malvidin	Red at 3.5 pH becoming colourless and then shifting to blue as the pH increases	Blue and red coloration	[125]

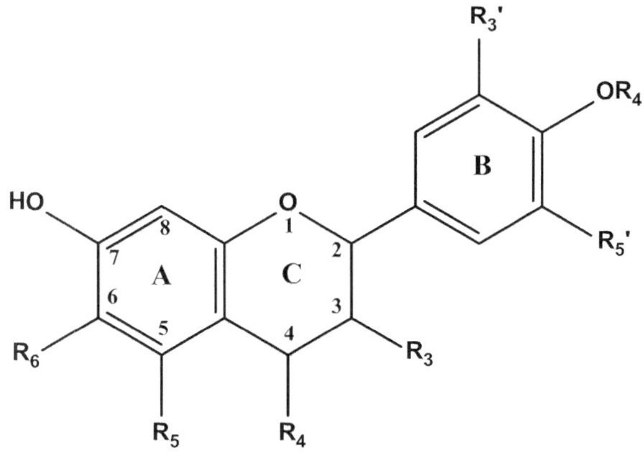

Fig. 26.5 General structure and numbering pattern for common food flavonoids

26.3.1.1 Chemical Structure of Flavonoids

Flavonoids consist of two-benzene ring system (aromatic ring structure) containing aromatic hydroxyl group and are connected by a carbon-carbon bridge. Structurally characterized as C_6-C_3-C_6 carbon skeleton, they occur as both aglycones (without sugar moiety) and glycosides (with sugar moiety). In flavonoids, A ring is joined with a six-member C ring through three-carbon bridge (Fig. 26.5). The different classes of flavonoids formed differ in the substitution at the C ring and level of oxidation. In addition, within a class the variation among the flavonoids is due to different substitution at A and B rings [123].

26.3.2 Flavonoids in Health and Disease

Researchers have shifted their focus on dietary source-derived flavonoids because of the accumulating versatile evidences that implicate their beneficial effects in human health and disease. Fruits and vegetables are the main sources of flavonoids and phenolic compounds that show antioxidant properties both in vitro and in vivo models [128–130]. Numerous epidemiological studies showed the protective roles of flavonoids in cardiovascular diseases, cancer and other age-related diseases [128]. Further, some flavonoids have shown anti-inflammatory, free radical scavenging, antibacterial, antiviral, hepato-protective, anti-allergic and antidiabetic activities [131, 132].

26.3.2.1 Antioxidant Activity

The most widely described property of flavonoids is their potential to act as antioxidants. Due to the potential to scavenge free radicals and ROS, flavonoids have been known as "high-level" natural antioxidants [128, 133]. Antioxidant activity of flavonoids depends on how the functional groups are arranged on the nuclear structure. The number of hydroxyl groups present and their configuration and substitution influence several mechanisms of antioxidant activity including metal ion chelation and radical scavenging ability [134]. Flavonoids exert their antioxidant activity via different mechanisms such as chelation of trace elements that contribute in free radical generation, ROS scavenging and by upregulation of antioxidant enzymes [135, 136]. Flavonoids also inhibit several enzymes that are involved in ROS generation such as glutathione S-transferase, microsomal monooxygenase and NADH oxidase [137]. For example, quercetin exhibits its anti-oxidative property by iron chelation and stabilization [138]; Rutin and epicatechin have demonstrated strong radical scavengers and lipid peroxidation inhibiting ability in vitro [139].

26.3.2.2 Anti-inflammatory Activity

Cyclooxygenase and lipoxygenase are two major enzymes responsible for inflammation and provoking the release of cytokines. Some phenolic compounds are known to inhibit both these enzymes and suppress inflammation [140]. Flavonoids have been shown to inhibit the synthesis of eicosanoids including prostaglandins and leukotrienes [141]. Quercetin inhibits both these pathways [142]. Other anti-inflammatory property of flavonoids includes the capacity to inhibit neutrophil degranulation as it directly hampers with the release of arachidonic acid by neutrophils and other immune cells [143].

26.3.2.3 Protective Effects on Cardiovascular System

Flavonoid consumption can prevent various cardiovascular diseases such as hypertension and atherosclerosis [144]. Endothelial dysfunction is a crucial event for the development of cardiovascular disease as it leads to arterial thrombus formation and eventually atherosclerosis. Consumption of flavonoids has been beneficial in preventing endothelial dysfunction [145]. Quercetin and its glycosides have been shown to protect LDL against oxidative modification [146]. A study has reported an inverse correlation between flavonoid consumption and total plasma cholesterol concentrations [147]. Along with decreasing cholesterol levels, flavonoids are also known to lower blood pressure levels that may be beneficial for heart patients [148]. Aggregation of platelet also plays a pivotal role in cardiovascular diseases as it has been shown that platelets contribute in generation of ROS and inhibit endothelial formation. Tea pigment phenolics have been reported to decrease blood coagulation and prevention of platelet aggregation [149, 150].

26.3.3 Targeting Cancer Stem Cells Using Dietary Flavonoids

26.3.3.1 Flavonoids and Cancer Chemoprevention

Various studies have pointed out the fact that long-term consumption of flavonoid-rich diet can prove to be beneficial and can help in reducing the risk of various chronic diseases including cancers [151]. Epidemiological studies have displayed that flavonoid-rich diet has an inverse relationship with cancers. Some examples include the less incidence of colorectal cancer in Asian population when compared to western population (31 colorectal cancer cases per 100,000 people in the UK and 4 colorectal cancer cases per 100,000 people in India). Apart from this, almost 70%

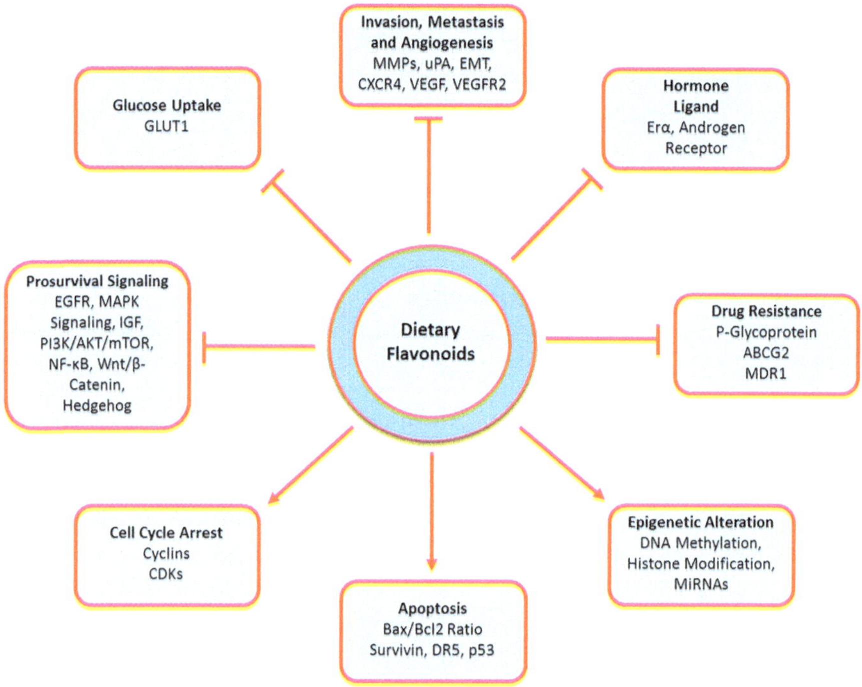

Fig. 26.6 Anticancer properties of dietary flavonoids

of cancers are now linked to the dietary intake [9]. Flavonoids are known as chemopreventers because they impart a cancer-preventive effect, for example, flavonoids such as curcumin, genistein, quercetin, EGCG and luteolin and several others [152]. Cancer chemopreventive efficiency of flavonoids may be because of several factors such as free radical scavenging; induction of apoptosis; downregulation of cell proliferative, adhesion and invasion markers; supressing inflammation; increased cell differentiation; induction of cell cycle arrest; regulation of steroid hormone and oestrogen metabolism; regulating the expressions of oncogenes; tumour suppressor genes; and various growth factor receptors [153]. Flavonoids are indeed shown to modulate the expression level of several proteins that play a role in carcinogenesis. Examples of such proteins affected by flavonoids include p53, Bcl-2, p21, Bax, NF-κB, COX-2, p73, GSH and catalase. EGCG is shown to inhibit cell proliferation and exert apoptosis in several tumour cell lines such as HeLa, CaSki, Hep-2 and SW780, melanoma cells, adrenal NCI-H295 cancer cells and A549 cells [152]. Genistein is shown to scavenge free radicals and exert its chemopreventive effect [154]. Flavonoids such as EGCG and quercetin are also known to exert their anticancer efficacy by modulating epigenetic proteins such as HDAC and DNMT-1 [155]. Several other mechanisms are exploited by flavonoids to exert their chemopreventive effect on cancer as shown in Fig. 26.6.

26.3.3.2 Flavonoids Targeting CSCs

Several recent studies have suggested that flavonoids can also target CSCs and may sensitize them towards anticancer drugs. Although much precise mechanisms are not known, it is hypothesized that flavonoids can target CSCs by either inhibiting their growth, self-renewal and metabolism or by targeting their niche and microenvironment [156].

Flavonoids Targeting CSC Signalling Pathways

Flavonoids can target CSCs by various mechanisms that may include modulating the self-renewal signalling pathways involved in CSC maintenance such as Wnt/β--catenin, Hedgehog and Notch signalling. Flavonoids such as curcumin, genistein and EGCG have been shown to directly or indirectly modulate these signalling pathways and contribute in the reduction of CSC growth and maintenance [157]. EGCG showed alteration in Wnt/β-catenin signalling in breast cancer cells [158]. EGCG also inhibited the downstream molecules of Wnt/β-catenin signalling and inhibited TCF/LEF binding and c-Myc expression [159]. Genistein decreased mammosphere formation and inhibited CSC growth in ER^{+ve} MCF7 and ER^{-ve} MDA-MB-231 breast cancer cell lines [160]. Another well-known flavonoid quercetin was shown to induce apoptosis; downregulate EMT, angiogenesis and stemness proteins; and inhibit CSC-derived xenografts in pancreatic cancer [161]. Curcumin, a flavonoid derived from *Curcuma longa*, has displayed its chemopreventive effect by downregulating CSCs' self-renewal pathways via inhibiting notch signalling in oesophageal cancer [162]. Chemosensitization to 5-FU by curcumin is attributed to its ability to decrease the number of CSC marker positive cells [163]. Further, curcumin treatment induced apoptosis and reduced the TGF-β1-induced cell invasion and proliferation in pancreatic cancer cell line (Panc-1) and also reduced the expression of Shh, GLI1 and vimentin. Further it was also showed that curcumin treatment resulted in the increased E-cadherin expression [164]. Curcumin is also shown to reduce cell growth and downregulate Notch 1, Hes-1 and Bcl-XL expression and induce apoptosis in pancreatic cancer cells. Curcumin pretreatment in combination with Notch-1 downregulation by siRNA synergistically increased growth inhibition and apoptosis [165]. Treatment of curcumin and resveratrol in combination was shown to exert synergistic antitumour activity in vitro in MCF-10A-Tr (cigarette smoke-mediated transformed cells) breast epithelial cells and in tumour xenograft in vivo models. In MCF-10A-Tr cells, Bax to Bcl-xL ratio increased along with the increased curcumin uptake in cells when resveratrol was treated in combination, and this leads to increased apoptosis along with PARP cleavage, release of cytochrome C and cleavage of caspase-3. This combination also modulated Hedgehog pathway and downregulated the expression of Shh, Smo, Gli and c-myc and enhanced the level of p21 in in vitro and in vivo models. Curcumin also inhibited the transactivation of Slug by accumulation of β-catenin and expression of c-Myc and cyclin D1 in azoxymethane-induced colon cancer in rat model [166]. Further, genistein was also shown to decrease cell viability and increase apoptosis in colon cancer cells, through reduction in nuclear β-catenin and increase in mRNA expression of Sfrp2, which is an antagonist of Wnt signalling pathway

[167]. Effect of genistein was also evaluated on breast cancer (MCF-7) cells where it reduced the cell growth and induced apoptosis along with the downregulation of Hedgehog signalling cascade [168]. EGCG inhibited the expression of GSK3-α and GSK3-β along with a reduction in β-catenin phosphorylation in HT29 cells [169]. In pancreatic cancer cells, EGCG downregulated the expression of self-renewal pathway proteins such as Nanog, c-Myc and Oct-4 and also decreased the number of CSCs along with sonic hedgehog pathway proteins such as smoothened, patched, Gli1 and Gli2. EGCG was also reduced EMT by downregulating transcriptional activity of various EMT TFs such as snail, slug and ZEB1. Further, EGCG act synergistically with quercetin and inhibited CSCs' self-renewal potential by inhibiting TCF/LEF and Gli activities [170].

Flavonoids Targeting CSC Markers and Niche

Combination of Src inhibitor dasatinib and curcumin was used against chemoresistant colon cancer cells that were enriched with CSC subpopulations. The combination treatment exhibited an elevated response by reducing cell growth, invasion and colony-forming ability of cancer cells and was also shown to reduce the expression of various CSC markers such as ALDH, CD133, CD166 and CD44 [171]. $ALDH^+/CD133^+$ colon cancer cells were shown to have increased expression of phosphorylated STAT3 in comparison to $ALDH^-/CD133^-$ cells, thus implying that cancer stem cells have more pSTAT3. Curcumin and its analogue GO-Y030 were shown to reduce STAT3 phosphorylation along with a reduction in cell viability, increased apoptosis and decreased sphere formation [172]. Further, curcumin-treated colon cancer cells displayed reduction in levels of CD44, CD166 and EGFR expression suggesting that curcumin can be used as an effective anticancer agent [173]. Difluorinated curcumin (CDF) is a novel curcumin analogue that was shown to have inhibitory effects on colon cancer stem-like cells. Combination of CDF with 5-FU+Ox exerted better anticancer activity than curcumin alone. CDF in combination with 5-FU decreased CD44, CD166 and ABCG2 expression along with apoptosis induction and growth inhibition in colon cancer cells [174].

In GEM-resistant MIAPaCa-2 pancreatic cancer cells, which contain high number of CSC, CDF decreased sphere forming along with reduction in CSC markers such as EpCAM and CD44 [175]. CDF also showed effectiveness in orthotopic xenograft models and decreased the expression of CD44, EpCAM and EZH2 by increasing the expression of let-7, miR-101 and miR-26a [176]. Genistein also displayed anti-CSC properties in pancreatic cancer cells by decreasing the expression of CD44 and EpCAM [177]. Kaempferol, a flavonoid isolated from *Delphinium*, witch hazel, grapefruit and other plant sources, was shown to downregulate ABCG2 expression, thus downregulating CSCs in oesophageal cancer [178]. Luteolin is also investigated for its anticancer properties in various studies, and one such study has shown that it is indeed successful in inhibiting various stem cell markers such as CD44, ALDH1 and many others in breast cancer cells [179]. Many other flavonoids have been investigated for their anti-CSC properties, and most of these flavonoids have shown potential to be very potent anticancer and anti-CSC agents. These flavonoids are summarized in Table 26.3.

Table 26.3 Flavonoids targeting molecular markers of different cancers

Flavonoid	Cancer type	Stem cell target	References
Epigallocatechin gallate (EGCG) [green tea]	Breast cancer	ALDH, c-myc, miR-21, miR-27, Wnt/β-catenin, BCRP	[180–182]
	Head and neck carcinoma	BMI-1, Twist1 and NF-kBp65	[183]
	Prostate cancer	Detoxification enzymes, androgen receptor, Wnt/β-catenin, CD133, CD44, NF-κB	[184, 185]
	Lung cancer	PI3K/AKT, Wnt/β-catenin	[181]
	Pancreatic cancer	Nanog, c-Myc, Oct-4, Hedgehog	[170]
	Liver cancer	Phase II detoxifying enzymes	[186]
	Colon cancer	Wnt/β-catenin	[187]
Curcumin	Breast cancer	ALDH1A1 expression, $CD44^{high}$/$CD24^{low}$ phenotype, Wnt/β-catenin, Hedgehog, STAT-3, NF-κB, Notch-1	[21, 157, 165, 181, 188, 189]
	Lung cancer	miRNA-186	[190]
	Colorectal cancer	DCKL-1, Nanog, ALDH1A, Lgr5, ABCG2, ALDH, CD44, CD133 and CD166 Hedgehog, Wnt/β-catenin	[165, 191, 192, 181]
	Gastric cancer	Wnt/β-catenin	[193]
	Intestinal cancer	Wnt/β-catenin	[194]
	Liver cancer	Side population	[195]
	Pancreatic cancer	NF-κB, CD44, EpCAM, Notch1, Nanog, Hedgehog, miR-21, miR-200, Wnt/β-catenin, let-7, miR-26a, miR-101, Hes-1 and Bcl-XL	[165, 181, 196, 197]
	Prostate cancer	Wnt/β-catenin, androgen receptor, let-7, miR-26a, miR-101, miR-21, miR-200	[181, 197]
	Melanoma	NF-κB	[198]
	Glioma (rat)	SP phenotype	[199]
	Oesophageal	Notch-1	[165]
Quercetin	Pancreatic cancer	CSC population identified using fluorescent tag (Gdeg), $CD44^{high}CD24^{low}$ NF-κB, Wnt/β-catenin	[161, 200, 201]
	Head and neck cancer	ALDH1, Oct-4, Nanog and Nestin	[202]
	Colorectal cancer	CD44,CD133, Wnt/β-catenin	[185, 187, 201]
	Prostate cancer	Wnt/β-catenin	[185]

(continued)

Table 26.3 (continued)

Flavonoid	Cancer type	Stem cell target	References
Genistein	Breast cancer	ALDH, BCRP/K562, NF-κB, Notch-1, Hedgehog, Bcl-2	[168, 203, 204]
	Prostate cancer	miR-221, miR-222, Hedgehog, CD44	[205, 206]
	Pancreatic cancer	CD44, EpCAM, Notch-1, let-7b, c, d, e, miR-26a, miR-146a, miR-200, miR-21	[181, 197, 207]
	Colon cancer	Wnt/β-catenin	[166]
	Renal cell carcinoma	Wnt/β-catenin	[208]
	Gastric cancer	CD44	[209]
	Ovarian cancer	CD133, CD44 and ALDH1	[210]
Silibinin	Breast cancer	Notch-1, NF-κB, Bcl-2, Wnt/β-catenin	[211, 212]
	Colon cancer	NANOG, SOX2, CD44	[213]
	Prostate cancer	ZEB1, NF-κB, Bcl-2, Wnt/β-catenin	[214]
Apigenin	Glioblastoma	CD133, Nanog and Sox2	[215]
	Head and neck cancer	CD44, NANOG, and CD105	[216]
	Prostate cancer stem cells	NF-κB, Oct3/4	[217]
	Breast cancer	Wnt/β-catenin	[187]
Chrysin	Liver cancer	Wnt/β-catenin, CD133 and CD44	[218, 219]
Kaempferol	Breast cancer	BCRP/K562	[203]
	Oesophageal cancer	ABCG2	[178]
Fisetin	Melanoma	Wnt/β-catenin	[220]
Naringenin	Breast cancer	BCRP/K562	[203]
Acacetin	Breast cancer	BCRP/K562	[203]
Luteolin	Breast cancer	ALDH, $CD44^{+}$, Notch-1	[179, 221]
	Colon cancer	Wnt/β-catenin	[222]
	Oral cancer	ALDH1, CD44	[223]

26.3.4 Role of Flavonoids in Cancer Stem Cell Biology and Drug Resistance

The major drawback of using chemotherapy is the presence of the MDR which makes the treatment challenging. There are list of proteins like P-gp, lung resistance protein (LRP), BCRP and ATP-binding cassettes which diminish the bioavailability of the drug to the target causing these drugs to efflux out from the cells [224, 225]. If higher concentration of the drugs is used to combat these problems, it may lead to cytotoxicity and other redundant cellular resistances; hence, there is the need for the development of potential chemosensitizers to overcome the problems of chemoresistance and hazardous toxicity leading to side effects [226, 224].

In this direction, flavonoids containing compounds are useful in targeting the cancer cells. The flavonoids present in honey such as chrysin, genistein, biochanin, quercetin, kaempferol and naringenin have shown induction of apoptosis by generating ROS in both colon cancer and bladder cancer cell lines. These flavonoids have been shown to be effective on various other cancer cell models including CaCo-2 cells and MCF-7 cells where they inhibit P-gp transporters, MRP-2 and BCRP and interact with ATPase's binding proteins reversing MDRs [227–229]. Many of the flavonoids also target drug substrates by increasing their accumulation, hence increasing their efficacy. For example, EGCG present in green tea reduces drug effluxes by targeting P-gp transporters leading to accumulation of rhodamine 123 that reverses MDR [230–232]. A study in MRP1-expressing HEK293 cells showed the inhibitory potential of flavonoids such as quercetin, naringenin, hesperetin, silibinin and daidzein on ABC transporters and MRP1, 4 and 5 and inhibited efflux. Further in L1210/Adr cells (a multidrug-resistant mouse leukaemia cell line), curcumin treatment resulted in decreased expression of *mdr1b* through regulating PI3K, Akt and NF-κB pathways. In addition, curcumin treatment also reversed MDR in ABCG2-expressing HEK cells, which facilitated accumulation of mitoxantrone and doxorubicin and sensitized the cells towards chemotherapeutic drugs [182]. Another flavonoid fisetin decreased the elevated expression of plasma membrane drug transporter P-glycoprotein in various human tumours through increasing cellular glutathione (GSH) content.

Biochanin, a substrate of both P-gp and CYP3A, has been shown to help in paclitaxel accumulation, which alters the bioavailability of the drug in the intestine and helps in raising the efficacy of anticancer drugs [233, 234]. The inhibition of MRP-1-mediated transport by various flavonoids including biochanin A, morin, apigenin, diosmetin, luteolin, baicalein, genistein, robinetin, chrysin, kaempferol, myricetin, naringenin and silymarin has been demonstrated [235].

The major studied mechanisms by which flavonoids augment the multidrug resistance reversal properties include:

1. Inhibiting the overexpression of multidrug resistance gene-1 (MDR1) [236]
2. Binding to the nuclear binding domains (NBD) of P-gp, which is involved in ATP hydrolysis to provide energy for efflux of the drugs out of the cells, thereby enhancing accumulation, bioavailability and efficacy of the drug to efficiently target the cancer cells [237]
3. Suppressing ATPase activity [238]
4. Hindering the ABC transporter functioning by competitively binding to substrate-binding sites [238, 239]

Many evidences have been documented about flavonoid's unique properties of reversing the multidrug resistances in various cancers. Still mechanistic studies are required for a clear understanding in different cross-signalling molecular mechanisms that target the multimodal different transporters, modulate signalling pathways which contributes to drug effluxing and lower the efficacy and concentration of anticancer drugs (Fig. 26.7).

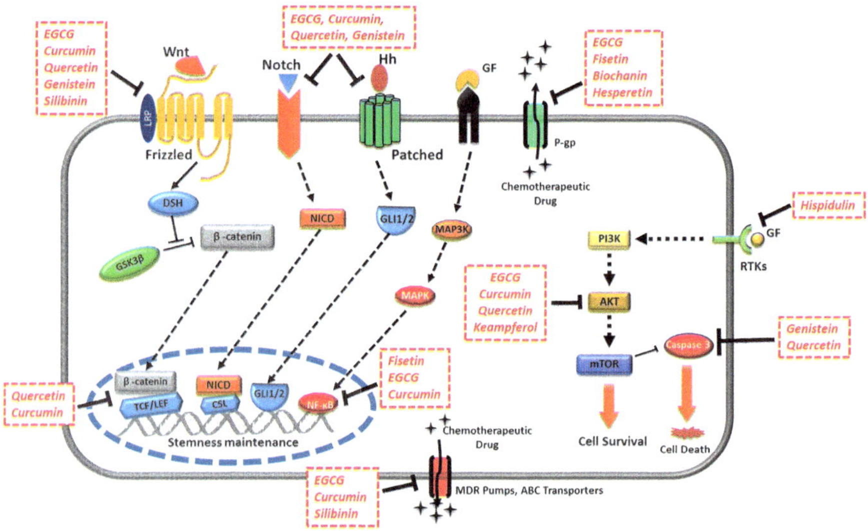

Fig. 26.7 Flavonoids targeting different CSC biomarkers and pathways

26.4 Summary and Future Directions

Cancer remains a major problem due to treatment failure, metastasis and recurrence after surgery, chemotherapy and radiotherapy. Recent reports showed that CSCs comprise a small cell population exhibiting resistance to anticancer drugs and also have been implicated in cancer relapse. A number of studies have identified signalling pathways that are responsible for acquiring and maintaining CSC in tumour mass. Current anticancer chemotherapy has failed to target CSCs. Hence, there is a need of novel compounds, which can target CSCs to achieve complete elimination of cancer. Functional foods and their active constituents, which are consumed regularly as dietary components, have been implicated in inhibiting these signalling pathways and target CSCs in many preclinical and clinical studies. Therefore, a regimen involving functional foods and their components may be helpful in the treatment of cancer in combination with current anticancer therapy to achieve complete remission of cancer. These active constituents will also serve as lead compounds for future anticancer drug discovery targeting CSCs as well as other types of cancers.

Acknowledgements The fellowships from UGC to KK, CSIR to AS and DST INSPIRE to HA and Ramanujan Fellowship Award from DST to UCSY are thankfully acknowledged.

References

1. Nassar D, Blanpain C (2016) Cancer stem cells: basic concepts and therapeutic implications. Annu Rev Pathol 11:47–76. https://doi.org/10.1146/annurev-pathol-012615-044438
2. Reya T, Morrison SJ, Clarke MF, Weissman IL (2001) Stem cells, cancer, and cancer stem cells. Nature 414(6859):105–111. https://doi.org/10.1038/35102167

3. Pattabiraman DR, Weinberg RA (2014) Tackling the cancer stem cells – what challenges do they pose? Nat Rev Drug Discov 13(7):497–512. https://doi.org/10.1038/nrd4253
4. Sell S (2009) History of Cancer Stem Cells. In: Rajasekhar VK, Vemuri MC (eds) Regulatory networks in stem cells. Humana Press, Totowa, pp 495–503. https://doi.org/10.1007/978-1-60327-227-8_37
5. Joosse SA, Gorges TM, Pantel K (2015) Biology, detection, and clinical implications of circulating tumor cells. EMBO Mol Med 7(1):1–11. https://doi.org/10.15252/emmm.201303698
6. Jordan NV, Johnson GL, Abell AN (2011) Tracking the intermediate stages of epithelial-mesenchymal transition in epithelial stem cells and cancer. Cell Cycle 10(17):2865–2873. https://doi.org/10.4161/cc.10.17.17188
7. Krantz SB, Shields MA, Dangi-Garimella S, Munshi HG, Bentrem DJ (2012) Contribution of epithelial-to-mesenchymal transition and cancer stem cells to pancreatic cancer progression. J Surg Res 173(1):105–112. https://doi.org/10.1016/j.jss.2011.09.020
8. Valastyan S, Weinberg RA (2011) Tumor metastasis: molecular insights and evolving paradigms. Cell 147(2):275–292. https://doi.org/10.1016/j.cell.2011.09.024
9. Khan S, Karmokar A, Howells L, Thomas AL, Bayliss R, Gescher A, Brown K (2016) Targeting cancer stem-like cells using dietary-derived agents – Where are we now? Mol Nutr Food Res 60(6):1295–1309. https://doi.org/10.1002/mnfr.201500887
10. Sporn MB, Suh N (2002) Chemoprevention: an essential approach to controlling cancer. Nat Rev Cancer 2(7):537–543. https://doi.org/10.1038/nrc844
11. Lee BM, Park KK (2003) Beneficial and adverse effects of chemopreventive agents. Mutat Res 523–524:265–278
12. Egert S, Rimbach G (2011) Which sources of flavonoids: complex diets or dietary supplements? Adv Nutr 2(1):8–14. https://doi.org/10.3945/an.110.000026
13. Yang CS, Landau JM, Huang MT, Newmark HL (2001) Inhibition of carcinogenesis by dietary polyphenolic compounds. Annu Rev Nutr 21:381–406. https://doi.org/10.1146/annurev.nutr.21.1.381
14. Oh J, Hlatky L, Jeong YS, Kim D (2016) Therapeutic effectiveness of anticancer phytochemicals on cancer stem cells. Toxins (Basel) 8(7):199. https://doi.org/10.3390/toxins8070199
15. Batra P, Sharma AK (2013) Anti-cancer potential of flavonoids: recent trends and future perspectives. 3 Biotech 3(6):439–459. https://doi.org/10.1007/s13205-013-0117-5
16. Sak K, Everaus H (2015) Role of flavonoids in future anticancer therapy by eliminating the cancer stem cells. Curr Stem Cell Res Ther 10(3):271–282
17. Almendro V, Marusyk A, Polyak K (2013) Cellular heterogeneity and molecular evolution in cancer. Annu Rev Pathol 8:277–302. https://doi.org/10.1146/annurev-pathol-020712-163923
18. Kreso A, Dick JE (2014) Evolution of the cancer stem cell model. Cell Stem Cell 14(3):275–291. https://doi.org/10.1016/j.stem.2014.02.006
19. Blanpain C, Fuchs E (2014) Stem cell plasticity. Plasticity of epithelial stem cells in tissue regeneration. Science 344(6189):1242281. https://doi.org/10.1126/science.1242281
20. Bonnet D, Dick JE (1997) Human acute myeloid leukemia is organized as a hierarchy that originates from a primitive hematopoietic cell. Nat Med 3(7):730–737
21. Al-Hajj M, Wicha MS, Benito-Hernandez A, Morrison SJ, Clarke MF (2003) Prospective identification of tumorigenic breast cancer cells. Proc Natl Acad Sci USA 100(7):3983–3988. https://doi.org/10.1073/pnas.0530291100
22. Lobo NA, Shimono Y, Qian D, Clarke MF (2007) The biology of cancer stem cells. Annu Rev Cell Dev Biol 23:675–699. https://doi.org/10.1146/annurev.cellbio.22.010305.104154
23. Chen K, Huang YH, Chen JL (2013) Understanding and targeting cancer stem cells: therapeutic implications and challenges. Acta Pharmacol Sin 34(6):732–740. https://doi.org/10.1038/aps.2013.27
24. Meng X, Li M, Wang X, Wang Y, Ma D (2009) Both CD133+ and CD133- subpopulations of A549 and H446 cells contain cancer-initiating cells. Cancer Sci 100(6):1040–1046. https://doi.org/10.1111/j.1349-7006.2009.01144.x
25. Miyata T, Yoshimatsu T, So T, Oyama T, Uramoto H, Osaki T, Nakanishi R, Tanaka F, Nagaya H, Gotoh A (2015) Cancer stem cell markers in lung cancer. Personal Med Universe 4:40–45. https://doi.org/10.1016/j.pmu.2015.03.007

26. Pine SR, Marshall B, Varticovski L (2008) Lung cancer stem cells. Dis Markers 24(4–5):257–266. https://doi.org/10.1155/2008/396281
27. Alamgeer M, Peacock CD, Matsui W, Ganju V, Watkins DN (2013) Cancer stem cells in lung cancer: evidence and controversies. Respirology 18(5):757–764. https://doi.org/10.1111/resp.12094
28. Meyer MJ, Fleming JM, Lin AF, Hussnain SA, Ginsburg E, Vonderhaar BK (2010) CD44posCD49fhiCD133/2hi defines xenograft-initiating cells in estrogen receptor-negative breast cancer. Cancer Res 70(11):4624–4633. https://doi.org/10.1158/0008-5472.CAN-09-3619
29. Wu X, Chen H, Wang X (2012) Can lung cancer stem cells be targeted for therapies? Cancer Treat Rev 38(6):580–588. https://doi.org/10.1016/j.ctrv.2012.02.013
30. Leung EL, Fiscus RR, Tung JW, Tin VP, Cheng LC, Sihoe AD, Fink LM, Ma Y, Wong MP (2010) Non-small cell lung cancer cells expressing CD44 are enriched for stem cell-like properties. PLoS One 5(11):e14062. https://doi.org/10.1371/journal.pone.0014062
31. Louie E, Nik S, Chen JS, Schmidt M, Song B, Pacson C, Chen XF, Park S, Ju J, Chen EI (2010) Identification of a stem-like cell population by exposing metastatic breast cancer cell lines to repetitive cycles of hypoxia and reoxygenation. Breast Cancer Res 12(6):R94. https://doi.org/10.1186/bcr2773
32. Sullivan JP, Spinola M, Dodge M, Raso MG, Gao B, Schuster K, Shao C, Larsen JE, Laura A, Honorio S, Xie Y, Scaglioni PP, Dimaio JM, Gazdar F, Shay JW, Wistuba II, Minna JD (2011) NIH public access. Cancer 70(23):9937–9948. https://doi.org/10.1158/0008-5472.CAN-10-0881.Aldehyde
33. Huang CP, Tsai MF, Chang TH, Tang WC, Chen SY, Lai HH, Lin TY, Yang JC, Yang PC, Shih JY, Lin SB (2013) ALDH-positive lung cancer stem cells confer resistance to epidermal growth factor receptor tyrosine kinase inhibitors. Cancer Lett 328(1):144–151. https://doi.org/10.1016/j.canlet.2012.08.021
34. Haraguchi N, Utsunomiya T, Inoue H, Tanaka F, Mimori K, Barnard GF, Mori M (2006) Characterization of a side population of cancer cells from human gastrointestinal system. Stem Cells 24(3):506–513. https://doi.org/10.1634/stemcells.2005-0282.
35. Zhou S, Morris JJ, Barnes Y, Lan L, Schuetz JD, Sorrentino BP (2002) Bcrp1 gene expression is required for normal numbers of side population stem cells in mice, and confers relative protection to mitoxantrone in hematopoietic cells in vivo. Proc Natl Acad Sci USA 99(19):12339–12344. https://doi.org/10.1073/pnas.192276999
36. Leccia F, Del Vecchio L, Mariotti E, Di Noto R, Morel AP, Puisieux A, Salvatore F, Ansieau S (2014) ABCG2, a novel antigen to sort luminal progenitors of BRCA1- breast cancer cells. Mol Cancer 13:213. https://doi.org/10.1186/1476-4598-13-213
37. Cariati M, Naderi A, Brown JP, Smalley MJ, Pinder SE, Caldas C, Purushotham AD (2008) Alpha-6 integrin is necessary for the tumourigenicity of a stem cell-like subpopulation within the MCF7 breast cancer cell line. Int J Cancer 122(2):298–304. https://doi.org/10.1002/ijc.23103
38. Vaillant F, Asselin-Labat ML, Shackleton M, Forrest NC, Lindeman GJ, Visvader JE (2008) The mammary progenitor marker CD61/beta3 integrin identifies cancer stem cells in mouse models of mammary tumorigenesis. Cancer Res 68(19):7711–7717. https://doi.org/10.1158/0008-5472.CAN-08-1949
39. Ribeiro AS, Paredes J (2014) P-cadherin linking breast cancer stem cells and invasion: a promising marker to identify an "intermediate/metastable" EMT state. Front Oncol 4:371. https://doi.org/10.3389/fonc.2014.00371
40. Messam CA, Hou J, Major EO (2000) Coexpression of nestin in neural and glial cells in the developing human CNS defined by a human-specific anti-nestin antibody. Exp Neurol 161(2):585–596. https://doi.org/10.1006/exnr.1999.7319
41. Okano H, Imai T, Okabe M (2002) Musashi: a translational regulator of cell fate. J Cell Sci 115(Pt 7):1355–1359

42. Toda M, Iizuka Y, Yu W, Imai T, Ikeda E, Yoshida K, Kawase T, Kawakami Y, Okano H, Uyemura K (2001) Expression of the neural RNA-binding protein Musashi1 in human gliomas. Glia 34(1):1–7
43. Capela A, Temple S (2002) LeX/ssea-1 is expressed by adult mouse CNS stem cells, identifying them as nonependymal. Neuron 35(5):865–875
44. Read T-A, Fogarty MP, Markant SL, McLendon RE, Wei Z, Ellison DW, Febbo PG, Wechsler-Reya RJ (2009) Identification of CD15 as a marker for tumor-propagating cells in a mouse model of medulloblastoma. Cancer Cell 15(2):135–147
45. Soltanian S, Matin MM (2011) Cancer stem cells and cancer therapy. Tumour Biol 32(3):425–440. https://doi.org/10.1007/s13277-011-0155-8
46. Ricci-Vitiani L, Lombardi DG, Pilozzi E, Biffoni M, Todaro M, Peschle C, De Maria R (2007) Identification and expansion of human colon-cancer-initiating cells. Nature 445(7123):111–115. https://doi.org/10.1038/nature05384
47. Vilchez V, Turcios L, Zaytseva Y, Stewart R, Lee EY, Maynard E, Shah MB, Daily MF, Tzeng CW, Davenport D, Castellanos AL, Krohmer S, Hosein PJ, Evers BM, Gedaly R (2016) Cancer stem cell marker expression alone and in combination with microvascular invasion predicts poor prognosis in patients undergoing transplantation for hepatocellular carcinoma. Am J Surg 212(2):238–245. https://doi.org/10.1016/j.amjsurg.2015.12.019
48. Singh SK, Hawkins C, Clarke ID, Squire JA, Bayani J, Hide T, Henkelman RM, Cusimano MD, Dirks PB (2004) Identification of human brain tumour initiating cells. Nature 432(7015):396–401. https://doi.org/10.1038/nature03128
49. Boman BM, Wicha MS (2008) Cancer stem cells: a step toward the cure. J Clin Oncol 26(17):2795–2799. https://doi.org/10.1200/JCO.2008.17.7436
50. Kure S, Matsuda Y, Hagio M, Ueda J, Naito Z, Ishiwata T (2012) Expression of cancer stem cell markers in pancreatic intraepithelial neoplasias and pancreatic ductal adenocarcinomas. Int J Oncol 41(4):1314–1324. https://doi.org/10.3892/ijo.2012.1565
51. Shien K, Toyooka S, Ichimura K, Soh J, Furukawa M, Maki Y, Muraoka T, Tanaka N, Ueno T, Asano H, Tsukuda K, Yamane M, Oto T, Kiura K, Miyoshi S (2012) Prognostic impact of cancer stem cell-related markers in non-small cell lung cancer patients treated with induction chemoradiotherapy. Lung Cancer 77(1):162–167. https://doi.org/10.1016/j.lungcan.2012.02.006
52. Bertolini G, Roz L, Perego P, Tortoreto M, Fontanella E, Gatti L, Pratesi G, Fabbri A, Andriani F, Tinelli S, Roz E, Caserini R, Lo Vullo S, Camerini T, Mariani L, Delia D, Calabro E, Pastorino U, Sozzi G (2009) Highly tumorigenic lung cancer CD133+ cells display stem-like features and are spared by cisplatin treatment. Proc Natl Acad Sci USA 106(38):16281–16286. https://doi.org/10.1073/pnas.0905653106
53. Annabi B, Rojas-Sutterlin S, Laflamme C, Lachambre MP, Rolland Y, Sartelet H, Beliveau R (2008) Tumor environment dictates medulloblastoma cancer stem cell expression and invasive phenotype. Mol Cancer Res 6(6):907–916. https://doi.org/10.1158/1541-7786.MCR-07-2184
54. Fang D, Nguyen TK, Leishear K, Finko R, Kulp AN, Hotz S, Van Belle PA, Xu X, Elder DE, Herlyn M (2005) A tumorigenic subpopulation with stem cell properties in melanomas. Cancer Res 65(20):9328–9337. https://doi.org/10.1158/0008-5472.CAN-05-1343
55. Svachova H, Pour L, Sana J, Kovarova L, Raja KR, Hajek R (2011) Stem cell marker nestin is expressed in plasma cells of multiple myeloma patients. Leuk Res 35(8):1008–1013. https://doi.org/10.1016/j.leukres.2011.03.001
56. Shabahang M, Buras RR, Davoodi F, Schumaker LM, Nauta RJ, Evans SR (1993) 1,25-Dihydroxyvitamin D3 receptor as a marker of human colon carcinoma cell line differentiation and growth inhibition. Cancer Res 53(16):3712–3718
57. Klonisch T, Wiechec E, Hombach-Klonisch S, Ande SR, Wesselborg S, Schulze-Osthoff K, Los M (2008) Cancer stem cell markers in common cancers – therapeutic implications. Trends Mol Med 14(10):450–460. https://doi.org/10.1016/j.molmed.2008.08.003

58. O'Brien CA, Pollett A, Gallinger S, Dick JE (2007) A human colon cancer cell capable of initiating tumour growth in immunodeficient mice. Nature 445(7123):106–110. https://doi.org/10.1038/nature05372
59. Patrawala L, Calhoun T, Schneider-Broussard R, Li H, Bhatia B, Tang S, Reilly JG, Chandra D, Zhou J, Claypool K, Coghlan L, Tang DG (2006) Highly purified CD44+ prostate cancer cells from xenograft human tumors are enriched in tumorigenic and metastatic progenitor cells. Oncogene 25(12):1696–1708. https://doi.org/10.1038/sj.onc.1209327
60. Salnikov AV, Gladkich J, Moldenhauer G, Volm M, Mattern J, Herr I (2010) CD133 is indicative for a resistance phenotype but does not represent a prognostic marker for survival of non-small cell lung cancer patients. Int J Cancer 126(4):950–958. https://doi.org/10.1002/ijc.24822
61. Isfoss BL, Busch C, Hermelin H, Vermedal AT, Kile M, Braathen GJ, Majak B, Berner A (2014) Stem cell marker-positive stellate cells and mast cells are reduced in benign-appearing bladder tissue in patients with urothelial carcinoma. Virchows Arch 464(4):473–488. https://doi.org/10.1007/s00428-014-1561-2
62. Chaffer CL, Weinberg RA (2011) A perspective on cancer cell metastasis. Science 331(6024):1559–1564. https://doi.org/10.1126/science.1203543
63. Polyak K, Weinberg RA (2009) Transitions between epithelial and mesenchymal states: acquisition of malignant and stem cell traits. Nat Rev Cancer 9(4):265–273. https://doi.org/10.1038/nrc2620
64. Fabregat I, Malfettone A, Soukupova J (2016) New insights into the crossroads between EMT and stemness in the context of cancer. J Clin Med 5(3):37
65. Yang JY, Zong CS, Xia W, Wei Y, Ali-Seyed M, Li Z, Broglio K, Berry DA, Hung MC (2006) MDM2 promotes cell motility and invasiveness by regulating E-cadherin degradation. Mol Cell Biol 26(19):7269–7282. https://doi.org/10.1128/MCB.00172-06
66. Scheel C, Weinberg RA (2012) Cancer stem cells and epithelial-mesenchymal transition: concepts and molecular links. Semin Cancer Biol 22(5-6):396–403. https://doi.org/10.1016/j.semcancer.2012.04.001
67. Wu KJ (2011) Direct activation of Bmi1 by Twist1: implications in cancer stemness, epithelial-mesenchymal transition, and clinical significance. Chang Gung Med J 34(3):229–238
68. Yang J, Mani SA, Donaher JL, Ramaswamy S, Itzykson RA, Come C, Savagner P, Gitelman I, Richardson A, Weinberg RA (2004) Twist, a master regulator of morphogenesis, plays an essential role in tumor metastasis. Cell 117(7):927–939. https://doi.org/10.1016/j.cell.2004.06.006
69. Tsai JH, Donaher JL, Murphy DA, Chau S, Yang J (2012) Spatiotemporal regulation of epithelial-mesenchymal transition is essential for squamous cell carcinoma metastasis. Cancer Cell 22(6):725–736. https://doi.org/10.1016/j.ccr.2012.09.022
70. Ocana OH, Corcoles R, Fabra A, Moreno-Bueno G, Acloque H, Vega S, Barrallo-Gimeno A, Cano A, Nieto MA (2012) Metastatic colonization requires the repression of the epithelial-mesenchymal transition inducer Prrx1. Cancer Cell 22(6):709–724. https://doi.org/10.1016/j.ccr.2012.10.012
71. Celia-Terrassa T, Meca-Cortes O, Mateo F, Martinez de Paz A, Rubio N, Arnal-Estape A, Ell BJ, Bermudo R, Diaz A, Guerra-Rebollo M, Lozano JJ, Estaras C, Ulloa C, Alvarez-Simon D, Mila J, Vilella R, Paciucci R, Martinez-Balbas M, de Herreros AG, Gomis RR, Kang Y, Blanco J, Fernandez PL, Thomson TM (2012) Epithelial-mesenchymal transition can suppress major attributes of human epithelial tumor-initiating cells. J Clin Invest 122(5):1849–1868. https://doi.org/10.1172/JCI59218
72. Aktas B, Tewes M, Fehm T, Hauch S, Kimmig R, Kasimir-Bauer S (2009) Stem cell and epithelial-mesenchymal transition markers are frequently overexpressed in circulating tumor cells of metastatic breast cancer patients. Breast Cancer Res 11(4):R46. https://doi.org/10.1186/bcr2333
73. Armstrong AJ, Marengo MS, Oltean S, Kemeny G, Bitting RL, Turnbull JD, Herold CI, Marcom PK, George DJ, Garcia-Blanco MA (2011) Circulating tumor cells from patients

with advanced prostate and breast cancer display both epithelial and mesenchymal markers. Mol Cancer Res 9(8):997–1007. https://doi.org/10.1158/1541-7786.MCR-10-0490
74. Balic M, Lin H, Young L, Hawes D, Giuliano A, McNamara G, Datar RH, Cote RJ (2006) Most early disseminated cancer cells detected in bone marrow of breast cancer patients have a putative breast cancer stem cell phenotype. Clin Cancer Res 12(19):5615–5621. https://doi.org/10.1158/1078-0432.CCR-06-0169
75. Liu H, Patel MR, Prescher JA, Patsialou A, Qian D, Lin J, Wen S, Chang YF, Bachmann MH, Shimono Y, Dalerba P, Adorno M, Lobo N, Bueno J, Dirbas FM, Goswami S, Somlo G, Condeelis J, Contag CH, Gambhir SS, Clarke MF (2010) Cancer stem cells from human breast tumors are involved in spontaneous metastases in orthotopic mouse models. Proc Natl Acad Sci USA 107(42):18115–18120. https://doi.org/10.1073/pnas.1006732107
76. Hermann PC, Huber SL, Herrler T, Aicher A, Ellwart JW, Guba M, Bruns CJ, Heeschen C (2007) Distinct populations of cancer stem cells determine tumor growth and metastatic activity in human pancreatic cancer. Cell Stem Cell 1(3):313–323. https://doi.org/10.1016/j.stem.2007.06.002
77. Ghajar CM, Peinado H, Mori H, Matei IR, Evason KJ, Brazier H, Almeida D, Koller A, Hajjar KA, Stainier DY, Chen EI, Lyden D, Bissell MJ (2013) The perivascular niche regulates breast tumour dormancy. Nature Cell Biol 15(7):807–817. https://doi.org/10.1038/ncb2767
78. Moitra K (2015) Overcoming multidrug resistance in cancer stem cells. Biomed Res Int 2015:635745. https://doi.org/10.1155/2015/635745
79. Hall MD, Handley MD, Gottesman MM (2009) Is resistance useless? Multidrug resistance and collateral sensitivity. Trends Pharmacol Sci 30(10):546–556. https://doi.org/10.1016/j.tips.2009.07.003
80. Thomas ML, Coyle KM, Sultan M, Marcato P (2015) Cancer stem cells and chemoresistance: strategies to overcome therapeutic resistance. In: Babashah S (ed) Cancer stem cells: emerging concepts and future perspectives in translational oncology. Springer, Cham, pp 477–518. https://doi.org/10.1007/978-3-319-21030-8_17
81. Gottesman MM (2002) Mechanisms of cancer drug resistance. Annu Rev Med 53:615–627. https://doi.org/10.1146/annurev.med.53.082901.103929
82. Keshet GI, Goldstein I, Itzhaki O, Cesarkas K, Shenhav L, Yakirevitch A, Treves AJ, Schachter J, Amariglio N, Rechavi G (2008) MDR1 expression identifies human melanoma stem cells. Biochem Biophys Res Commun 368(4):930–936. https://doi.org/10.1016/j.bbrc.2008.02.022
83. Schatton T, Murphy GF, Frank NY, Yamaura K, Waaga-Gasser AM, Gasser M, Zhan Q, Jordan S, Duncan LM, Weishaupt C, Fuhlbrigge RC, Kupper TS, Sayegh MH, Frank MH (2008) Identification of cells initiating human melanomas. Nature 451(7176):345–349. https://doi.org/10.1038/nature06489
84. Shaffer BC, Gillet JP, Patel C, Baer MR, Bates SE, Gottesman MM (2012) Drug resistance: still a daunting challenge to the successful treatment of AML. Drug Resist Updat 15(1–2):62–69. https://doi.org/10.1016/j.drup.2012.02.001
85. Trock BJ, Leonessa F, Clarke R (1997) Multidrug resistance in breast cancer: a meta-analysis of MDR1/gp170 expression and its possible functional significance. J Natl Cancer Inst 89(13):917–931
86. Lizard-Nacol S, Genne P, Coudert B, Riedinger JM, Arnal M, Sancy C, Brunet-Lecomte P, Fargeot P (1999) MDR1 and thymidylate synthase (TS) gene expressions in advanced breast cancer: relationships to drug exposure, p53 mutations, and clinical outcome of the patients. Anticancer Res 19(4C):3575–3581
87. Rudas M, Filipits M, Taucher S, Stranzl T, Steger GG, Jakesz R, Pirker R, Pohl G (2003) Expression of MRP1, LRP and Pgp in breast carcinoma patients treated with preoperative chemotherapy. Breast Cancer Res Treat 81(2):149–157. https://doi.org/10.1023/A:1025751631115
88. Doyle LA, Yang W, Abruzzo LV, Krogmann T, Gao Y, Rishi AK, Ross DD (1998) A multidrug resistance transporter from human MCF-7 breast cancer cells. Proc Natl Acad Sci USA 95(26):15665–15670

89. van den Heuvel-Eibrink MM, Wiemer EA, Prins A, Meijerink JP, Vossebeld PJ, van der Holt B, Pieters R, Sonneveld P (2002) Increased expression of the breast cancer resistance protein (BCRP) in relapsed or refractory acute myeloid leukemia (AML). Leukemia 16(5):833–839. https://doi.org/10.1038/sj.leu.2402496
90. Rentala S, Mangamoori LN (2010) Isolation, characterization and mobilization of prostate cancer tissue derived CD133+ MDR1+ cells. J Stem Cells 5(2):75–81
91. Fong MY, Kakar SS (2010) The role of cancer stem cells and the side population in epithelial ovarian cancer. Histol Histopathol 25(1):113–120
92. Nakai E, Park K, Yawata T, Chihara T, Kumazawa A, Nakabayashi H, Shimizu K (2009) Enhanced MDR1 expression and chemoresistance of cancer stem cells derived from glioblastoma. Cancer Investig 27(9):901–908. https://doi.org/10.3109/07357900801946679
93. Ho MM, Hogge DE, Ling V (2008) MDR1 and BCRP1 expression in leukemic progenitors correlates with chemotherapy response in acute myeloid leukemia. Exp Hematol 36(4):433–442. https://doi.org/10.1016/j.exphem.2007.11.014
94. Raaijmakers MH, de Grouw EP, Heuver LH, van der Reijden BA, Jansen JH, Scheper RJ, Scheffer GL, de Witte TJ, Raymakers RA (2005) Breast cancer resistance protein in drug resistance of primitive CD34+38- cells in acute myeloid leukemia. Clin Cancer Res 11(6):2436–2444. https://doi.org/10.1158/1078-0432.CCR-04-0212
95. Ikawa M, Impraim CC, Wang G, Yoshida A (1983) Isolation and characterization of aldehyde dehydrogenase isozymes from usual and atypical human livers. J Biol Chem 258(10):6282–6287
96. Marchitti SA, Brocker C, Stagos D, Vasiliou V (2008) Non-P450 aldehyde oxidizing enzymes: the aldehyde dehydrogenase superfamily. Expert Opin Drug Metab Toxicol 4(6):697–720. https://doi.org/10.1517/17425255.4.6.697
97. Ginestier C, Hur MH, Charafe-Jauffret E, Monville F, Dutcher J, Brown M, Jacquemier J, Viens P, Kleer CG, Liu S, Schott A, Hayes D, Birnbaum D, Wicha MS, Dontu G (2007) ALDH1 is a marker of normal and malignant human mammary stem cells and a predictor of poor clinical outcome. Cell Stem Cell 1(5):555–567. https://doi.org/10.1016/j.stem.2007.08.014
98. Abdullah LN, Chow EK (2013) Mechanisms of chemoresistance in cancer stem cells. Clin Transl Med 2(1):3. https://doi.org/10.1186/2001-1326-2-3
99. Dylla SJ, Beviglia L, Park IK, Chartier C, Raval J, Ngan L, Pickell K, Aguilar J, Lazetic S, Smith-Berdan S, Clarke MF, Hoey T, Lewicki J, Gurney AL (2008) Colorectal cancer stem cells are enriched in xenogeneic tumors following chemotherapy. PLoS One 3(6):e2428. https://doi.org/10.1371/journal.pone.0002428
100. Duong HQ, Hwang JS, Kim HJ, Kang HJ, Seong YS, Bae I (2012) Aldehyde dehydrogenase 1A1 confers intrinsic and acquired resistance to gemcitabine in human pancreatic adenocarcinoma MIA PaCa-2 cells. Int J Oncol 41(3):855–861. https://doi.org/10.3892/ijo.2012.1516
101. Pegoraro L, Palumbo A, Erikson J, Falda M, Giovanazzo B, Emanuel BS, Rovera G, Nowell PC, Croce CM (1984) A 14;18 and an 8;14 chromosome translocation in a cell line derived from an acute B-cell leukemia. Proc Natl Acad Sci USA 81(22):7166–7170
102. Graninger WB, Seto M, Boutain B, Goldman P, Korsmeyer SJ (1987) Expression of Bcl-2 and Bcl-2-Ig fusion transcripts in normal and neoplastic cells. J Clin Investig 80(5):1512–1515. https://doi.org/10.1172/JCI113235.
103. Liu G, Yuan X, Zeng Z, Tunici P, Ng H, Abdulkadir IR, Lu L, Irvin D, Black KL, Yu JS (2006) Analysis of gene expression and chemoresistance of CD133+ cancer stem cells in glioblastoma. Mol Cancer 5:67. https://doi.org/10.1186/1476-4598-5-67
104. Shervington A, Lu C (2008) Expression of multidrug resistance genes in normal and cancer stem cells. Cancer Investig 26(5):535–542. https://doi.org/10.1080/07357900801904140
105. Madjd Z, Mehrjerdi AZ, Sharifi AM, Molanaei S, Shahzadi SZ, Asadi-Lari M (2009) CD44+ cancer cells express higher levels of the anti-apoptotic protein Bcl-2 in breast tumours. Cancer Immun 9:4

106. Ma S, Lee TK, Zheng BJ, Chan KW, Guan XY (2008) CD133+ HCC cancer stem cells confer chemoresistance by preferential expression of the Akt/PKB survival pathway. Oncogene 27(12):1749–1758. https://doi.org/10.1038/sj.onc.1210811
107. Todaro M, Alea MP, Di Stefano AB, Cammareri P, Vermeulen L, Iovino F, Tripodo C, Russo A, Gulotta G, Medema JP, Stassi G (2007) Colon cancer stem cells dictate tumor growth and resist cell death by production of interleukin-4. Cell Stem Cell 1(4):389–402. https://doi.org/10.1016/j.stem.2007.08.001
108. Cammareri P, Scopelliti A, Todaro M, Eterno V, Francescangeli F, Moyer MP, Agrusa A, Dieli F, Zeuner A, Stassi G (2010) Aurora-a is essential for the tumorigenic capacity and chemoresistance of colorectal cancer stem cells. Cancer Res 70(11):4655–4665. https://doi.org/10.1158/0008-5472.CAN-09-3953
109. Gonzalez C (2002) Aurora-A in cell fate control. Sci STKE: Signal Transduct Knowl Environ 2002(162):pe48. https://doi.org/10.1126/stke.2002.162.pe48
110. Olcina M, Lecane PS, Hammond EM (2010) Targeting hypoxic cells through the DNA damage response. Clin Cancer Res 16(23):5624–5629. https://doi.org/10.1158/1078-0432.CCR-10-0286
111. Hammond EM, Denko NC, Dorie MJ, Abraham RT, Giaccia AJ (2002) Hypoxia links ATR and p53 through replication arrest. Mol Cell Biol 22(6):1834–1843
112. Bao S, Wu Q, McLendon RE, Hao Y, Shi Q, Hjelmeland AB, Dewhirst MW, Bigner DD, Rich JN (2006) Glioma stem cells promote radioresistance by preferential activation of the DNA damage response. Nature 444(7120):756–760. https://doi.org/10.1038/nature05236
113. Reya T, Duncan AW, Ailles L, Domen J, Scherer DC, Willert K, Hintz L, Nusse R, Weissman IL (2003) A role for Wnt signalling in self-renewal of haematopoietic stem cells. Nature 423(6938):409–414. https://doi.org/10.1038/nature01593
114. Bisson I, Prowse DM (2009) WNT signaling regulates self-renewal and differentiation of prostate cancer cells with stem cell characteristics. Cell Res 19(6):683–697. https://doi.org/10.1038/cr.2009.43
115. Yang W, Yan HX, Chen L, Liu Q, He YQ, Yu LX, Zhang SH, Huang DD, Tang L, Kong XN, Chen C, Liu SQ, Wu MC, Wang HY (2008) Wnt/beta-catenin signaling contributes to activation of normal and tumorigenic liver progenitor cells. Cancer Res 68(11):4287–4295. https://doi.org/10.1158/0008-5472.CAN-07-6691
116. Flahaut M, Meier R, Coulon A, Nardou KA, Niggli FK, Martinet D, Beckmann JS, Joseph JM, Muhlethaler-Mottet A, Gross N (2009) The Wnt receptor FZD1 mediates chemoresistance in neuroblastoma through activation of the Wnt/beta-catenin pathway. Oncogene 28(23):2245–2256. https://doi.org/10.1038/onc.2009.80
117. Chau WK, Ip CK, Mak AS, Lai HC, Wong AS (2013) c-Kit mediates chemoresistance and tumor-initiating capacity of ovarian cancer cells through activation of Wnt/beta-catenin-ATP-binding cassette G2 signaling. Oncogene 32(22):2767–2781. https://doi.org/10.1038/onc.2012.290
118. Meng RD, Shelton CC, Li YM, Qin LX, Notterman D, Paty PB, Schwartz GK (2009) gamma-Secretase inhibitors abrogate oxaliplatin-induced activation of the Notch-1 signaling pathway in colon cancer cells resulting in enhanced chemosensitivity. Cancer Res 69(2):573–582. https://doi.org/10.1158/0008-5472.CAN-08-2088
119. McAuliffe SM, Morgan SL, Wyant GA, Tran LT, Muto KW, Chen YS, Chin KT, Partridge JC, Poole BB, Cheng KH, Daggett J Jr, Cullen K, Kantoff E, Hasselbatt K, Berkowitz J, Muto MG, Berkowitz RS, Aster JC, Matulonis UA, Dinulescu DM (2012) Targeting Notch, a key pathway for ovarian cancer stem cells, sensitizes tumors to platinum therapy. Proc Natl Acad Sci USA 109(43):E2939–E2948. https://doi.org/10.1073/pnas.1206400109
120. Zhang S, Balch C, Chan MW, Lai HC, Matei D, Schilder JM, Yan PS, Huang TH, Nephew KP (2008) Identification and characterization of ovarian cancer-initiating cells from primary human tumors. Cancer Res 68(11):4311–4320. https://doi.org/10.1158/0008-5472.CAN-08-0364

121. Leizer AL, Alvero AB, Fu HH, Holmberg JC, Cheng YC, Silasi DA, Rutherford T, Mor G (2011) Regulation of inflammation by the NF-kappaB pathway in ovarian cancer stem cells. Am J Reprod Immunol 65(4):438–447. https://doi.org/10.1111/j.1600-0897.2010.00914.x
122. Mahomoodally MF, Gurib-Fakim A, Subratty AH (2008) Antimicrobial activities and phytochemical profiles of endemic medicinal plants of Mauritius. Pharm Biol
123. Middleton E Jr (1998) Effect of plant flavonoids on immune and inflammatory cell function. In: Manthey JA, Buslig BS (eds) Flavonoids in the living system. Springer, Boston, pp 175–182
124. Kuhnau J (1976) The flavonoids. A class of semi-essential food components: their role in human nutrition. World Rev Nutr Diet 24:117–191
125. Herrmann K (1976) Flavonols and flavones in food plants: a review. Int J Food Sc Technol 11(5):433–448
126. Pierpoint WS (1986) Flavonoids in the human diet. Prog Clin Biol Res 213:125–140
127. Mazur W, Fotsis T, Wahala K, Ojala S, Salakka A, Adlercreutz H (1996) Isotope dilution gas chromatographic-mass spectrometric method for the determination of isoflavonoids, coumestrol, and lignans in food samples. Anal Biochem 233(2):169–180. https://doi.org/10.1006/abio.1996.0025
128. Yao LH, Jiang YM, Shi J, Tomas-Barberan FA, Datta N, Singanusong R, Chen SS (2004) Flavonoids in food and their health benefits. Plant Foods Hum Nutr 59(3):113–122
129. Cook N (1996) Flavonoids?Chemistry, metabolism, cardioprotective effects, and dietary sources. J Eur Ceram Soc 7(2):66–76. https://doi.org/10.1016/s0955-2863(95)00168-9
130. Rice-Evans CA, Miller NJ, Bolwell PG, Bramley PM, Pridham JB (1995) The relative antioxidant activities of plant-derived polyphenolic flavonoids. Free Radic Res 22(4):375–383
131. Nijveldt RJ (2001) Flavonoids: a review of probable mechanism of action and potential applications. Am J Clin Nutr 74:418–425
132. Shashank K, Abhay KP (2013) Review article chemistry and biological activities of flavonoids: an overview. Sci World J 4(2):32–48
133. Fukumoto LR, Mazza G (2000) Assessing antioxidant and prooxidant activities of phenolic compounds. J Agric Food Chem 48(8):3597–3604
134. Heim KE, Tagliaferro AR, Bobilya DJ (2002) Flavonoid antioxidants: chemistry, metabolism and structure-activity relationships. J Nutr Biochem 13(10):572–584
135. Mishra A, Kumar S, Pandey AK (2013) Scientific validation of the medicinal efficacy of Tinospora cordifolia. Sci World J 2013:292934. https://doi.org/10.1155/2013/292934
136. Land ET (2009) Free radicals in biology and medicine. Int J Radiat Biol 58(4):725–725. https://doi.org/10.1080/09553009014552071
137. Brown JE, Khodr H, Hider RC, Rice-Evans CA (1998) Structural dependence of flavonoid interactions with Cu2+ ions: implications for their antioxidant properties. Biochem J 330(Pt 3):1173–1178
138. van Acker SA, van den Berg DJ, Tromp MN, Griffioen DH, van Bennekom WP, van der Vijgh WJ, Bast A (1996) Structural aspects of antioxidant activity of flavonoids. Free Radic Biol Med 20(3):331–342
139. Mishra A, Sharma AK, Kumar S, Saxena AK, Pandey AK (2013) Bauhinia variegata leaf extracts exhibit considerable antibacterial, antioxidant, and anticancer activities. Biomed Res Int 2013:915436. https://doi.org/10.1155/2013/915436
140. Nijveldt RJ, van Nood E, van Hoorn DE, Boelens PG, van Norren K, van Leeuwen PA (2001) Flavonoids: a review of probable mechanisms of action and potential applications. Am J Clin Nutr 74(4):418–425
141. Dames J, Bourdon V, Remacle-Uolon G, Lecomte J (1985) Pro-inflammatory flavonoids which are inhibitors of prostaglandin biosynthesis. Prostaglandins Leukot Med 19(1):11–24. https://doi.org/10.1016/0262-1746(85)90157-x
142. Kim HP, Mani I, Iversen L, Ziboh VA (1998) Effects of naturally-occurring flavonoids and biflavonoids on epidermal cyclooxygenase and lipoxygenase from guinea-pigs. Prostaglandins Leukot Essent Fatty Acids 58(1):17–24

143. Tordera M, Ferrandiz ML, Alcaraz MJ (1994) Influence of anti-inflammatory flavonoids on degranulation and arachidonic acid release in rat neutrophils. Zeitschrift fur Naturforschung C. J Biosci 49(3–4):235–240
144. Hertog MG, Feskens EJ, Hollman PC, Katan MB, Kromhout D (1993) Dietary antioxidant flavonoids and risk of coronary heart disease: the Zutphen Elderly Study. Lancet 342(8878):1007–1011
145. Fuhrman B, Aviram M (2001) Flavonoids protect LDL from oxidation and attenuate atherosclerosis. Curr Opin Lipidol 12(1):41–48
146. Fuhrman B, Lavy A, Aviram M (1995) Consumption of red wine with meals reduces the susceptibility of human plasma and low-density lipoprotein to lipid peroxidation. Am J Clin Nutr 61(3):549–554
147. Arai Y, Watanabe S, Kimira M, Shimoi K, Mochizuki R, Kinae N (2000) Dietary intakes of flavonols, flavones and isoflavones by Japanese women and the inverse correlation between quercetin intake and plasma LDL cholesterol concentration. J Nutr 130(9):2243–2250
148. Hodgson JM, Croft KD (2006) Dietary flavonoids: effects on endothelial function and blood pressure. J Sci Food Agric 86(15):2492–2498. https://doi.org/10.1002/jsfa.2675
149. Lou FQ, Zhang MF, Zhang XG, Liu JM, Yuan WL (1989) A study on tea-pigment in prevention of atherosclerosis. Chin Med J (Engl) 102(8):579–583
150. Rein D, Paglieroni TG, Pearson DA, Wun T, Schmitz HH, Gosselin R, Keen CL (2000) Cocoa and wine polyphenols modulate platelet activation and function. J Nutr 130(8S Suppl):2120S–2126S
151. Xiao ZP, Peng ZY, Peng MJ, Yan WB, Ouyang YZ, Zhu HL (2011) Flavonoids health benefits and their molecular mechanism. Mini Rev Med Chem 11(2):169–177
152. Gibellini L, Pinti M, Nasi M, Montagna JP, De Biasi S, Roat E, Bertoncelli L, Cooper EL, Cossarizza A (2011) Quercetin and cancer chemoprevention. Evid-Based Complement Alternat Med 2011:591356. https://doi.org/10.1093/ecam/neq053
153. Park E-J, Pezzuto JM (2012) Flavonoids in cancer prevention. Anti-Cancer Agents Med Chem 12(8):836–851. https://doi.org/10.2174/187152012802650075
154. Es-Safi NE, Ghidouche S, Ducrot PH (2007) Flavonoids: hemisynthesis, reactivity, characterization and free radical scavenging activity. Molecules 12(9):2228–2258
155. Tan S, Wang C, Lu C, Zhao B, Cui Y, Shi X, Ma X (2009) Quercetin is able to demethylate the p16INK4a gene promoter. Chemotherapy 55(1):6–10. https://doi.org/10.1159/000166383
156. Zhang G, Yang P, Guo P, Miele L, Sarkar FH, Wang Z, Zhou Q (2013) Unraveling the mystery of cancer metabolism in the genesis of tumor-initiating cells and development of cancer. Biochim Biophys Acta 1836(1):49–59. https://doi.org/10.1016/j.bbcan.2013.03.001
157. Li Y, Wicha MS, Schwartz SJ, Sun D (2011) Implications of cancer stem cell theory for cancer chemoprevention by natural dietary compounds. J Nutr Biochem 22(9):799–806. https://doi.org/10.1016/j.jnutbio.2010.11.001
158. Kim J, Zhang X, Rieger-Christ KM, Summerhayes IC, Wazer DE, Paulson KE, Yee AS (2006) Suppression of Wnt signaling by the green tea compound (-)-epigallocatechin 3-gallate (EGCG) in invasive breast cancer cells. Requirement of the transcriptional repressor HBP1. J Biol Chem 281(16):10865–10875. https://doi.org/10.1074/jbc.M513378200
159. Reguart N, He B, Taron M, You L, Jablons DM, Rosell R (2005) The role of Wnt signaling in cancer and stem cells. Future Oncol 1(6):787–797. https://doi.org/10.2217/14796694.1.6.787
160. Montales MT, Rahal OM, Kang J, Rogers TJ, Prior RL, Wu X, Simmen RC (2012) Repression of mammosphere formation of human breast cancer cells by soy isoflavone genistein and blueberry polyphenolic acids suggests diet-mediated targeting of cancer stem-like/progenitor cells. Carcinogenesis 33(3):652–660. https://doi.org/10.1093/carcin/bgr317
161. Zhou W, Kallifatidis G, Baumann B, Rausch V, Mattern J, Gladkich J, Giese N, Moldenhauer G, Wirth T, Buchler MW, Salnikov AV, Herr I (2010) Dietary polyphenol quercetin targets pancreatic cancer stem cells. Int J Oncol 37(3):551–561
162. Subramaniam D, Ponnurangam S, Ramamoorthy P, Standing D, Battafarano RJ, Anant S, Sharma P (2012) Curcumin induces cell death in esophageal cancer cells through modulating Notch signaling. PLoS One 7(2):e30590. https://doi.org/10.1371/journal.pone.0030590

163. Shakibaei M, Buhrmann C, Kraehe P, Shayan P, Lueders C, Goel A (2014) Curcumin chemo-sensitizes 5-fluorouracil resistant MMR-deficient human colon cancer cells in high density cultures. PLoS One 9(1):e85397. https://doi.org/10.1371/journal.pone.0085397
164. Sun XD, Liu XE, Huang DS (2013) Curcumin reverses the epithelial-mesenchymal transition of pancreatic cancer cells by inhibiting the Hedgehog signaling pathway. Oncol Rep 29(6):2401–2407. https://doi.org/10.3892/or.2013.2385
165. Dandawate P, Subramaniam D, Anant S (2016) Targeting cancer stem cells by functional foods and their constituents. In: Food toxicology. CRC Press, Boca Raton, pp 433–460. doi:https://doi.org/10.1201/9781315371443-23
166. Zhang Y, Li Q, Zhou D, Chen H (2013) Genistein, a soya isoflavone, prevents azoxymethane-induced up-regulation of WNT/beta-catenin signalling and reduces colon pre-neoplasia in rats. Br J Nutr 109(1):33–42. https://doi.org/10.1017/S0007114512000876
167. Zhang Y, Chen H (2011) Genistein attenuates WNT signaling by up-regulating sFRP2 in a human colon cancer cell line. Exp Biol Med (Maywood) 236(6):714–722. https://doi.org/10.1258/ebm.2011.010347
168. Fan P, Fan S, Wang H, Mao J, Shi Y, Ibrahim MM, Ma W, Yu X, Hou Z, Wang B, Li L (2013) Genistein decreases the breast cancer stem-like cell population through Hedgehog pathway. Stem Cell Res Ther 4(6):146. https://doi.org/10.1186/scrt357
169. Pahlke G, Ngiewih Y, Kern M, Jakobs S, Marko D, Eisenbrand G (2006) Impact of quercetin and EGCG on key elements of the Wnt pathway in human colon carcinoma cells. J Agric Food Chem 54(19):7075–7082. https://doi.org/10.1021/jf0612530
170. Tang SN, Fu J, Nall D, Rodova M, Shankar S, Srivastava RK (2012) Inhibition of sonic hedgehog pathway and pluripotency maintaining factors regulate human pancreatic cancer stem cell characteristics. Int J Cancer 131(1):30–40. https://doi.org/10.1002/ijc.26323
171. Nautiyal J, Kanwar SS, Yu Y, Majumdar AP (2011) Combination of dasatinib and curcumin eliminates chemo-resistant colon cancer cells. J Mol Signal 6:7. https://doi.org/10.1186/1750-2187-6-7
172. Lin L, Liu Y, Li H, Li PK, Fuchs J, Shibata H, Iwabuchi Y, Lin J (2011) Targeting colon cancer stem cells using a new curcumin analogue, GO-Y030. Br J Cancer 105(2):212–220. https://doi.org/10.1038/bjc.2011.200
173. Yu Y, Kanwar SS, Patel BB, Nautiyal J, Sarkar FH, Majumdar AP (2009) Elimination of colon cancer stem-like cells by the combination of curcumin and FOLFOX. Transl Oncol 2(4):321–328
174. Kanwar SS, Yu Y, Nautiyal J, Patel BB, Padhye S, Sarkar FH, Majumdar AP (2011) Difluorinated-curcumin (CDF): a novel curcumin analog is a potent inhibitor of colon cancer stem-like cells. Pharm Res 28(4):827–838. https://doi.org/10.1007/s11095-010-0336-y
175. Bao B, Ali S, Kong D, Sarkar SH, Wang Z, Banerjee S, Aboukameel A, Padhye S, Philip PA, Sarkar FH (2011) Anti-tumor activity of a novel compound-CDF is mediated by regulating miR-21, miR-200, and PTEN in pancreatic cancer. PLoS One 6(3):e17850. https://doi.org/10.1371/journal.pone.0017850
176. Bao B, Ali S, Banerjee S, Wang Z, Logna F, Azmi AS, Kong D, Ahmad A, Li Y, Padhye S, Sarkar FH (2012) Curcumin analogue CDF inhibits pancreatic tumor growth by switching on suppressor microRNAs and attenuating EZH2 expression. Cancer Res 72(1):335–345. https://doi.org/10.1158/0008-5472.CAN-11-2182
177. Bao B, Wang Z, Ali S, Kong D, Banerjee S, Ahmad A, Li Y, Azmi AS, Miele L, Sarkar FH (2011) Over-expression of FoxM1 leads to epithelial-mesenchymal transition and cancer stem cell phenotype in pancreatic cancer cells. J Cell Biochem 112(9):2296–2306. https://doi.org/10.1002/jcb.23150
178. To KK, Yu L, Liu S, Fu J, Cho CH (2012) Constitutive AhR activation leads to concomitant ABCG2-mediated multidrug resistance in cisplatin-resistant esophageal carcinoma cells. Mol Carcinog 51(6):449–464. https://doi.org/10.1002/mc.20810
179. Cook MT, Liang Y, Besch-Williford C, Goyette S, Mafuvadze B, Hyder SM (2015) Luteolin inhibits progestin-dependent angiogenesis, stem cell-like characteristics, and growth of human breast cancer xenografts. SpringerPlus 4:444. https://doi.org/10.1186/s40064-015-1242-x

180. Mineva ND, Paulson KE, Naber SP, Yee AS, Sonenshein GE (2013) Epigallocatechin-3-gallate inhibits stem-like inflammatory breast cancer cells. PLoS One 8(9):e73464. https://doi.org/10.1371/journal.pone.0073464
181. Dandawate P, Padhye S, Ahmad A, Sarkar FH (2013) Novel strategies targeting cancer stem cells through phytochemicals and their analogs. Drug Deliv Transl Res 3(2):165–182. https://doi.org/10.1007/s13346-012-0079-x
182. Kawasaki BT, Hurt EM, Mistree T, Farrar WL (2008) Targeting cancer stem cells with phytochemicals. Mol Interv 8(4):174–184. https://doi.org/10.1124/mi.8.4.9
183. Li YJ, Wu SL, Lu SM, Chen F, Guo Y, Gan SM, Shi YL, Liu S, Li SL (2015) (-)-Epigallocatechin-3-gallate inhibits nasopharyngeal cancer stem cell self-renewal and migration and reverses the epithelial-mesenchymal transition via NF-kappaB p65 inactivation. Tumour Biol 36(4):2747–2761. https://doi.org/10.1007/s13277-014-2899-4.
184. Johnson JJ, Bailey HH, Mukhtar H (2010) Green tea polyphenols for prostate cancer chemoprevention: a translational perspective. Phytomedicine 17(1):3–13. https://doi.org/10.1016/j.phymed.2009.09.011
185. Tang SN, Singh C, Nall D, Meeker D, Shankar S, Srivastava RK (2010) The dietary bioflavonoid quercetin synergizes with epigallocathechin gallate (EGCG) to inhibit prostate cancer stem cell characteristics, invasion, migration and epithelial-mesenchymal transition. J Mol Signal 5:14. https://doi.org/10.1186/1750-2187-5-14
186. Yu R, Jiao JJ, Duh JL, Gudehithlu K, Tan TH, Kong AN (1997) Activation of mitogen-activated protein kinases by green tea polyphenols: potential signaling pathways in the regulation of antioxidant-responsive element-mediated phase II enzyme gene expression. Carcinogenesis 18(2):451–456
187. Amado NG, Fonseca BF, Cerqueira DM, Neto VM, Abreu JG (2011) Flavonoids: potential Wnt/beta-catenin signaling modulators in cancer. Life Sci 89(15-16):545–554. https://doi.org/10.1016/j.lfs.2011.05.003
188. Zhou Q, Ye M, Lu Y, Zhang H, Chen Q, Huang S, Su S (2015) Curcumin improves the tumoricidal effect of mitomycin C by suppressing ABCG2 expression in stem cell-like breast cancer cells. PLoS One 10(8):e0136694. https://doi.org/10.1371/journal.pone.0136694
189. Chung SS, Vadgama JV (2015) Curcumin and epigallocatechin gallate inhibit the cancer stem cell phenotype via down-regulation of STAT3-NFkappaB signaling. Anticancer Res 35(1):39–46
190. Zhang J, Du Y, Wu C, Ren X, Ti X, Shi J, Zhao F, Yin H (2010) Curcumin promotes apoptosis in human lung adenocarcinoma cells through miR-186* signaling pathway. Oncol Rep 24(5):1217–1223
191. James MI, Iwuji C, Irving G, Karmokar A, Higgins JA, Griffin-Teal N, Thomas A, Greaves P, Cai H, Patel SR, Morgan B, Dennison A, Metcalfe M, Garcea G, Lloyd DM, Berry DP, Steward WP, Howells LM, Brown K (2015) Curcumin inhibits cancer stem cell phenotypes in ex vivo models of colorectal liver metastases, and is clinically safe and tolerable in combination with FOLFOX chemotherapy. Cancer Lett 364(2):135–141. https://doi.org/10.1016/j.canlet.2015.05.005
192. Kantara C, O'Connell M, Sarkar S, Moya S, Ullrich R, Singh P (2014) Curcumin promotes autophagic survival of a subset of colon cancer stem cells, which are ablated by DCLK1-siRNA. Cancer Res 74(9):2487–2498. https://doi.org/10.1158/0008-5472.CAN-13-3536
193. Park CH, Hahm ER, Park S, Kim HK, Yang CH (2005) The inhibitory mechanism of curcumin and its derivative against beta-catenin/Tcf signaling. FEBS Lett 579(13):2965–2971. https://doi.org/10.1016/j.febslet.2005.04.013
194. Jaiswal AS, Marlow BP, Gupta N, Narayan S (2002) Beta-catenin-mediated transactivation and cell-cell adhesion pathways are important in curcumin (diferuylmethane)-induced growth arrest and apoptosis in colon cancer cells. Oncogene 21(55):8414–8427. https://doi.org/10.1038/sj.onc.1205947
195. Marquardt JU, Gomez-Quiroz L, Arreguin Camacho LO, Pinna F, Lee YH, Kitade M, Dominguez MP, Castven D, Breuhahn K, Conner EA, Galle PR, Andersen JB, Factor VM, Thorgeirsson SS (2015) Curcumin effectively inhibits oncogenic NF-kappaB signaling and

restrains stemness features in liver cancer. J Hepatol 63(3):661–669. https://doi.org/10.1016/j.jhep.2015.04.018
196. Wang Z, Zhang Y, Banerjee S, Li Y, Sarkar FH (2006) Notch-1 down-regulation by curcumin is associated with the inhibition of cell growth and the induction of apoptosis in pancreatic cancer cells. Cancer 106(11):2503–2513. https://doi.org/10.1002/cncr.21904
197. Hong M, Tan HY, Li S, Cheung F, Wang N, Nagamatsu T, Feng Y (2016) Cancer stem cells: the potential targets of chinese medicines and their active compounds. Int J Mol Sci 17(6). https://doi.org/10.3390/ijms17060893
198. Surh YJ (2003) Cancer chemoprevention with dietary phytochemicals. Nat Rev Cancer 3(10):768–780. https://doi.org/10.1038/nrc1189
199. Fong D, Yeh A, Naftalovich R, Choi TH, Chan MM (2010) Curcumin inhibits the side population (SP) phenotype of the rat C6 glioma cell line: towards targeting of cancer stem cells with phytochemicals. Cancer Lett 293(1):65–72. https://doi.org/10.1016/j.canlet.2009.12.018
200. Adikrisna R, Tanaka S, Muramatsu S, Aihara A, Ban D, Ochiai T, Irie T, Kudo A, Nakamura N, Yamaoka S, Arii S (2012) Identification of pancreatic cancer stem cells and selective toxicity of chemotherapeutic agents. Gastroenterology 143(1):234–245 e237. https://doi.org/10.1053/j.gastro.2012.03.054
201. Atashpour S, Fouladdel S, Movahhed TK, Barzegar E, Ghahremani MH, Ostad SN, Azizi E (2015) Quercetin induces cell cycle arrest and apoptosis in CD133(+) cancer stem cells of human colorectal HT29 cancer cell line and enhances anticancer effects of doxorubicin. Iran J Basic Med Sci 18(7):635–643
202. Chang WW, Hu FW, Yu CC, Wang HH, Feng HP, Lan C, Tsai LL, Chang YC (2013) Quercetin in elimination of tumor initiating stem-like and mesenchymal transformation property in head and neck cancer. Head Neck 35(3):413–419. https://doi.org/10.1002/hed.22982
203. Imai Y, Tsukahara S, Asada S, Sugimoto Y (2004) Phytoestrogens/flavonoids reverse breast cancer resistance protein/ABCG2-mediated multidrug resistance. Cancer Res 64(12):4346–4352. https://doi.org/10.1158/0008-5472.CAN-04-0078
204. Pan H, Zhou W, He W, Liu X, Ding Q, Ling L, Zha X, Wang S (2012) Genistein inhibits MDA-MB-231 triple-negative breast cancer cell growth by inhibiting NF-kappaB activity via the Notch-1 pathway. Int J Mol Med 30(2):337–343. https://doi.org/10.3892/ijmm.2012.990
205. Chen Y, Zaman MS, Deng G, Majid S, Saini S, Liu J, Tanaka Y, Dahiya R (2011) MicroRNAs 221/222 and genistein-mediated regulation of ARHI tumor suppressor gene in prostate cancer. Cancer Prev Res (Phila) 4(1):76–86. https://doi.org/10.1158/1940-6207.CAPR-10-0167
206. Zhang L, Li L, Jiao M, Wu D, Wu K, Li X, Zhu G, Yang L, Wang X, Hsieh JT, He D (2012) Genistein inhibits the stemness properties of prostate cancer cells through targeting Hedgehog-Gli1 pathway. Cancer Lett 323(1):48–57. https://doi.org/10.1016/j.canlet.2012.03.037
207. Xia J, Duan Q, Ahmad A, Bao B, Banerjee S, Shi Y, Ma J, Geng J, Chen Z, Rahman KM, Miele L, Sarkar FH, Wang Z (2012) Genistein inhibits cell growth and induces apoptosis through up-regulation of miR-34a in pancreatic cancer cells. Curr Drug Targets 13(14):1750–1756
208. Hirata H, Ueno K, Nakajima K, Tabatabai ZL, Hinoda Y, Ishii N, Dahiya R (2013) Genistein downregulates onco-miR-1260b and inhibits Wnt-signalling in renal cancer cells. Br J Cancer 108(10):2070–2078. https://doi.org/10.1038/bjc.2013.173
209. Yu D, Shin HS, Lee YS, Lee D, Kim S, Lee YC (2014) Genistein attenuates cancer stem cell characteristics in gastric cancer through the downregulation of Gli1. Oncol Rep 31(2):673–678. https://doi.org/10.3892/or.2013.2893
210. Ning Y, Luo C, Ren K, Quan M, Cao J (2014) FOXO3a-mediated suppression of the self-renewal capacity of sphere-forming cells derived from the ovarian cancer SKOV3 cell line by 7-difluoromethoxyl-5,4'-di-n-octyl genistein. Mol Med Rep 9(5):1982–1988. https://doi.org/10.3892/mmr.2014.2012
211. Kim TH, Woo JS, Kim YK, Kim KH (2014) Silibinin induces cell death through reactive oxygen species-dependent downregulation of notch-1/ERK/Akt signaling in human breast cancer cells. J Pharmacol Exp Ther 349(2):268–278. https://doi.org/10.1124/jpet.113.207563

212. Lu W, Lin C, King TD, Chen H, Reynolds RC, Li Y (2012) Silibinin inhibits Wnt/beta-catenin signaling by suppressing Wnt co-receptor LRP6 expression in human prostate and breast cancer cells. Cell Signal 24(12):2291–2296. https://doi.org/10.1016/j.cellsig.2012.07.009
213. Kumar S, Raina K, Agarwal C, Agarwal R (2014) Silibinin strongly inhibits the growth kinetics of colon cancer stem cell-enriched spheroids by modulating interleukin 4/6-mediated survival signals. Oncotarget 5(13):4972–4989. https://doi.org/10.18632/oncotarget.2068.
214. Ting H, Deep G, Agarwal R (2013) Molecular mechanisms of silibinin-mediated cancer chemoprevention with major emphasis on prostate cancer. AAPS J 15(3):707–716. https://doi.org/10.1208/s12248-013-9486-2
215. Kim B, Jung N, Lee S, Sohng JK, Jung HJ (2016) Apigenin inhibits cancer stem cell-like phenotypes in human glioblastoma cells via suppression of c-Met signaling. Phytother Res 30(11):1833–1840. https://doi.org/10.1002/ptr.5689
216. Ketkaew Y, Osathanon T, Pavasant P, Sooampon S (2016) Apigenin inhibited hypoxia induced stem cell marker expression in a head and neck squamous cell carcinoma cell line. Arch Oral Biol 74:69–74. https://doi.org/10.1016/j.archoralbio.2016.11.010
217. Erdogan S, Doganlar O, Doganlar ZB, Serttas R, Turkekul K, Dibirdik I, Bilir A (2016) The flavonoid apigenin reduces prostate cancer CD44(+) stem cell survival and migration through PI3K/Akt/NF-kappaB signaling. Life Sci 162:77–86. https://doi.org/10.1016/j.lfs.2016.08.019
218. Quan MF, Xiao LH, Liu ZH, Guo H, Ren KQ, Liu F, Cao JG, Deng XY (2013) 8-bromo-7-methoxychrysin inhibits properties of liver cancer stem cells via downregulation of beta-catenin. World J Gastroenterol 19(43):7680–7695. https://doi.org/10.3748/wjg.v19.i43.7680
219. Ren KQ, Cao XZ, Liu ZH, Guo H, Quan MF, Liu F, Jiang L, Xiang HL, Deng XY, Cao JG (2013) 8-bromo-5-hydroxy-7-methoxychrysin targeting for inhibition of the properties of liver cancer stem cells by modulation of Twist signaling. Int J Oncol 43(5):1719–1729. https://doi.org/10.3892/ijo.2013.2071
220. Syed DN, Afaq F, Maddodi N, Johnson JJ, Sarfaraz S, Ahmad A, Setaluri V, Mukhtar H (2011) Inhibition of human melanoma cell growth by the dietary flavonoid fisetin is associated with disruption of Wnt/beta-catenin signaling and decreased Mitf levels. J Invest Dermatol 131(6):1291–1299. https://doi.org/10.1038/jid.2011.6
221. Sun DW, Zhang HD, Mao L, Mao CF, Chen W, Cui M, Ma R, Cao HX, Jing CW, Wang Z, Wu JZ, Tang JH (2015) Luteolin inhibits breast cancer development and progression in vitro and in vivo by suppressing notch signaling and regulating MiRNAs. Cell Physiol Biochem 37(5):1693–1711. https://doi.org/10.1159/000438535
222. Pandurangan AK, Dharmalingam P, Sadagopan SK, Ganapasam S (2014) Luteolin inhibits matrix metalloproteinase 9 and 2 in azoxymethane-induced colon carcinogenesis. Hum Exp Toxicol 33(11):1176–1185. https://doi.org/10.1177/0960327114522502
223. Tu DG, Lin WT, Yu CC, Lee SS, Peng CY, Lin T, Yu CH (2016) Chemotherapeutic effects of luteolin on radio-sensitivity enhancement and interleukin-6/signal transducer and activator of transcription 3 signaling repression of oral cancer stem cells. J Formos Med Assoc 115:1032. https://doi.org/10.1016/j.jfma.2016.08.009
224. Choi C-H (2005) ABC transporters as multidrug resistance mechanisms and the development of chemosensitizers for their reversal. Cancer Cell Int 5(1):1
225. Tan B, Piwnica-Worms D, Ratner L (2000) Multidrug resistance transporters and modulation. Curr Opin Oncol 12(5):450–458
226. Tan B, Piwnica-Worms D, Ratner L (2000) Multidrug resistance transporters and modulation. Curr Opin Oncol 12(5):450–458
227. Schumacher M, Hautzinger A, Rossmann A, Holzhauser S, Popovic D, Hertrampf A, Kuntz S, Boll M, Wenzel U (2010) Chrysin blocks topotecan-induced apoptosis in Caco-2 cells in spite of inhibition of ABC-transporters. Biochem Pharmacol 80(4):471–479
228. De Castro WV, Mertens-Talcott S, Derendorf H, Butterweck V (2008) Effect of grapefruit juice, naringin, naringenin, and bergamottin on the intestinal carrier-mediated transport of talinolol in rats. J Agric Food Chem 56(12):4840–4845

229. Zhang S, Wang X, Sagawa K, Morris ME (2005) Flavonoids chrysin and benzoflavone, potent breast cancer resistance protein inhibitors, have no significant effect on topotecan pharmacokinetics in rats or mdr1a/1b (–/–) mice. Drug Metab Dispos 33(3):341–348
230. Jodoin J, Demeule M, Béliveau R (2002) Inhibition of the multidrug resistance P-glycoprotein activity by green tea polyphenols. Biochim Biophys Acta 1542(1):149–159
231. Mei Y, Wei D, Liu J (2003) Reversal of cancer multidrug resistance by tea polyphenol in KB cells. J Chemother 15(3):260–265. https://doi.org/10.1179/joc.2003.15.3.260.
232. Zhu A, Wang X, Guo Z (2001) Study of tea polyphenol as a reversal agent for carcinoma cell lines' multidrug resistance (study of TP as a MDR reversal agent). Nucl Med Biol 28(6):735–740. https://doi.org/10.1016/s0969-8051(00)90202-6
233. Ekhart C, Rodenhuis S, Smits PH, Beijnen JH, Huitema AD (2009) An overview of the relations between polymorphisms in drug metabolising enzymes and drug transporters and survival after cancer drug treatment. Cancer Treat Rev 35(1):18–31
234. Peng SX, Ritchie DM, Cousineau M, Danser E, Dewire R, Floden J (2006) Altered oral bioavailability and pharmacokinetics of P-glycoprotein substrates by coadministration of biochanin A. J Pharm Sci 95(9):1984–1993. https://doi.org/10.1002/jps.20664
235. Li Y, Revalde JL, Reid G, Paxton JW (2010) Interactions of dietary phytochemicals with ABC transporters: possible implications for drug disposition and multidrug resistance in cancer. Drug Metab Rev 42(4):590–611. https://doi.org/10.3109/03602531003758690
236. Kioka N, Hosokawa N, Komano T, Hirayoshi K, Nagate K, Ueda K (1992) Quercetin, a bioflavonoid, inhibits the increase of human multidrug resistance gene (MDR1) expression caused by arsenite. FEBS Lett 301(3):307–309
237. Di Pietro A, Dayan G, Conseil G, Steinfels E, Krell T, Trompier D, Baubichon-Cortay H, Jault J-M (1999) P-glycoprotein-mediated resistance to chemotherapy in cancer cells: using recombinant cytosolic domains to establish structure-function relationships. Braz J Med Biol Res 32(8):925–939
238. Shapiro AB, Ling V (1997) Effect of quercetin on Hoechst 33342 transport by purified and reconstituted P-glycoprotein. Biochem Pharmacol 53(4):587–596
239. Wu CP, Calcagno AM, Hladky SB, Ambudkar SV, Barrand MA (2005) Modulatory effects of plant phenols on human multidrug-resistance proteins 1, 4 and 5 (ABCC1, 4 and 5). FEBS J 272(18):4725–4740

Phytochemicals: An Alternate Approach Towards Various Disease Management

27

Vijay Nema, Yogita Dhas, Joyita Banerjee,
and Neetu Mishra

Abstract

With the emergence of allopathic medicine system and the industrial revolution, the era of synthetic chemicals came into being which included medicines also. It reduced the load of cultivating and processing herbal medicines to get a larger amount of medicines with moderate effects in longer durations. It was gradually paralleled and later superseded by the use of purified or synthetic chemicals as drugs for treating various diseases including infections. The use of such molecules was a great success, and a revolution during world wars as the discoveries of antibiotics and their synthetic analogs took place. The use of these molecules continued and at later times became humongous as compared to traditional and herbal formulations. However, as is true with any other material, the overuse of these drugs started showing its negative aspects like side-effects, development of resistance etc. The problem specifically became huge with respect to antimicrobial compounds as the microbes started developing resistance towards all such molecules, while the problem of toxic side effects continued. The problem of drug resistance has been also observed in case of some diseases such as cancer and type 2 diabetes. This made the drug development program rethink if we should reduce the use of the synthetic compounds and start exploring back if there are safer avenues available. Exploration and research on phytochemicals present in medicinal plants and functional foods, thus, came as a safer alternative. This chapter tries to explore the information about knowledge available about phytochemicals and recent developments in this area for finding newer and better antimicrobials, anticancerous and antidiabetics.

Both Vijay Nema and Neetu Mishra have equally contributed.

V. Nema
Division of Molecular Biology, National AIDS Research Institute, Pune, India

Y. Dhas · J. Banerjee · N. Mishra (✉)
Symbiosis School of Biological Sciences, Symbiosis International (Deemed University), Pune, Maharashtra, India

V. Rani, U. C. S. Yadav (eds.), *Functional Food and Human Health*,
https://doi.org/10.1007/978-981-13-1123-9_27

Keywords
Phytochemicals · Antimicrobials · Antiviral · Anticancerous · Antidiabetics

27.1 Introduction

Locally available plants with medicinal potential are a great resource of food as well as for efficacious drugs. Due to their potential of producing diverse range of bioactive molecules, plants are a rich source of different types of medicines and natural products. Hence they have an important role in drug development programmes in the pharmaceutical industry [16]. Exploration of biologically active compounds from natural sources has always been a field of interest for new sources of drugs useful in infectious diseases. A number of studies have been reported antimicrobial screening and antioxidant activity of extracts of medicinal plant with chemical composition, biological and therapeutic activities [149].

The medicinal value of any plant is due to the presence of biologically active components or some physiological components which have some definite action on the human body termed as "phytochemicals". Flavonoids, phenols, alkaloids, saponins, steroids, glycosides, unsaturated lactones, tannins, terpenoids, sulphur compounds, essential oils, glucosinolates, etc. are classified under phytochemicals [42, 166]. These substances are produced by plants in order to counter microorganism, insects or herbivores [32]. Different parts of plants are used for their medicinal property such as stem, leaves, roots, twigs, fruit, flower and modified plant organs [89]. The excessive use of antibiotics causes potential harm to human body, and also resistance occurs against harmful pathogens. So the demand of exploring natural compounds having activity against harmful pathogens is increasing day by day.

Though the phototherapeutics is in trend, still there exist a need of exploration and development of cost-effective, herbal drugs with better bioactive potential and least side effects. Hence, knowing the potential of different indigenous plants and validating the same using scientific methods is a need of the day.

27.2 Antimicrobial Activities of Phytochemicals

27.2.1 Antifungal Activity of Phytochemicals

Fungal infections/diseases are one of the major problems associated with our daily life. Fungal infections in humans are usually associated with the immunocompromised host (AIDS, cancer, transplant patients, etc.). Multidrug resistance (MDR) has emerged, and the number of infections caused by MDR species has increased rapidly [107]. Despite the constant introduction of new and effective synthetic drugs to the market, there is great need to find the novel drug from the medicinal plants with the selective mode of action on the fresh target with lesser side effects. This would be an economical alternative and are of great importance in human health.

27.2.1.1 Phytochemicals in Different Skin Infections Caused by Fungi

Among the various infections caused by fungi (mycoses), the infections of the skin, hair, nail, and mucous membranes are the most common and have a worldwide distribution. The most common infections that affect humans are dermatomycoses (infections of the hair, skin, and nails), pityriasis capitis (occurs on the scalp) and pityriasis versicolor (occurs on the other part of the body such as neck and arms), piedra and tinea nigra palmaris caused due to colonization of the keratinized layers of the body [46, 121]. About 20 species of dermatophyte fungi from the genera *Trichophyton*, *Microsporum* and *Epidermophyton* are responsible for such infections. Most frequent *Trichophyton* spp. responsible for dermatomycoses are *T. mentagrophytes*, *T. rubrum* and *T. tonsurans* [143].

Loizzoet et al. [85] investigated the antifungal activity of different extracts of *Senecio inaequidens* and *Senecio vulgaris*. The hexane extract of *S. vulgaris* showed significant antifungal activity against *T. tonsurans* (IC_{50} of 0.031 mg/ml). Also, *Spilanthes calva* extracts show the antifungal activity against *F. oxysporum* and *T. mentagrophytes* [126].

Infection may also be caused by the members of the genera *Fusarium*, *Scopulariopsis*, *Malassezia* and *Aspergillus*. *Malassezia* species are lipophilic yeast which can cause skin infections and other dermatological disorders, folliculitis and catheter-related invasive fungal infections in neonatal, paediatric and immunocompetent patients. *M. globosa* are responsible for dandruff of the scalp and a rash on the body called tinea versicolor. *Fusarium* species also cause opportunistic infections in immunocompromised individuals, patients of persistent neutropenia, previous fungal infections, etc. [68]. *Fusarium* species generally cause superficial infections (keratitis and onychomycosis), allergic diseases and disseminated diseases. The most frequent *Fusarium* species are *F. solani*, *F. oxysporum*, *F. verticillioides* and *F. moniliforme* [38]. Many phenolic compounds such as salicylic acid, phenol and benzoic acid show antifungal activity against *F. oxysporum* at different concentrations [9]. Fusarium onychomycosis is a common condition that represents up to 50% of all nail problems and 30% ofall cases of dermatophytoses [44].

Houghton et al. [67] evaluated the effects of four *Medicago* species extracts (aerial parts and roots), 3 parent triterpenoids and 19 saponin compounds. All saponin extracts inhibited the growth of three dermatophytic species (*M. gypseum*, *T. interdigitale* and *T. tonsurans*). However, the aglycones showed less antifungal effects than the glycosides, which displayed a range of activities. The monodesmosidic glycosides of medicagenic acid were the most active compounds, particularly the 3-*O*-*β*-glucopyranoside, which displayed a MIC less than 0.09 mM against all three fungi, although those of hederagenin and zanhic acid also showed low levels of activity (MIC 3.3 mM against two fungal species) [67].

Additionally, in human, *Aspergillus* species, such as *A. flavus*, *A. glaucis*, *A. terreus*, *A. parasiticus*, *A. repens*, *A. nidulans*, *A. niger* and *A. tubingensis* can cause aspergillosis [20, 135]. Aspergillosis can cause both upper and lower respiratory tract infection-bronchopulmonary aspergillosis, which can spread to the brain, bone or endocardium [38].

It has been evident that some of the essential oils (oregano, thyme, rosemary and clove) and their main constituents such as eugenol, carvacrol and thymol show efficient anti-*Aspergillus* activity [23]. Phenolic extracts of *Spirulina* (*Arthrospira*) show antifungal activity by inhibiting ergosterol (a component of the fungal cell membrane), glucosamine (present in the fungal cells of some genera as a growth indicator) and some other proteins. The phenolic compounds from *Spirulina* can reverse the effect of drug resistance by inhibiting the function of ABC transporters [9]. Table 27.1 further describes a few plants which have been tested against the various fungal pathogens responsible for skin infections and their mode of action.

27.2.1.2 Polyphenols in Candidiasis

There has been a significant increase in the incidence and prevalence of candidiasis, an opportunistic fungus infection majorly caused by *Candida albicans*. Candida species can mainly cause infections in immunocompromised patients, including those with AIDS, cancer, neutropenia, neutrophil dysfunction, disruption of mucosal barriers, transplant recipients and premature neonates [37, 96, 158]. Seven *Candida* species are classified as having major clinical relevance, namely, *C. albicans*, *C. tropicalis*, *C. glabrata*, *C. parapsilosis*, *C. krusei*, *C. stellatoidea* and *C. kyfer* [15, 58]. *C. albicans*, dimorphic in nature, helps for their survival inside the host. There are various phenolic compounds which affect its dimorphic transition. For instance, many new phenolic compounds isolated from the leaves of *Baseonema acuminatum* and root bark of *Lycium chinense* have been known to show antifungal activity against different strains of *C. albicans* [9]. Furthermore, Amber et al. [6] observed that terpenes (including methyl chavicol and linalool) from *Ocimum sanctum* affected the synthesis of ergosterol and caused cell membrane damage in *Candida* species, suggesting that it may acts as a good antifungal agent.

Many types of extracts derived from *Curcuma zedoaria*, *Psidium guajava*, *Aristolochia cymbifera*, *Lippia alba*, *Hydrocotyle bonariensis*, *Herrerias sapatilha*, *Eleutherine bulbosa*, *Baccharis trimera*, *Calamintha adscendens*, *Albizia inundata*, *Bauhinia forficata* and *Cymbopogon citratus* have been reported to show antifungal activity against *Candida* spp. [107]. Moreover, Polaquini et al. [124] determined that crude extract of *Azadirachta indica* decreased yeast adhesion capacity does not inhibit the growth of *Candida* biofilm. Machumi et al. [88] reported the antifungal activity of nine diterpenoids from *Clerodendrum rio phylum* against *Candida* spp. and *A. fumigates*.

Additionally, curcumin (CUR), an effective phenolic compounds produced by *Curcuma longa*, has been showed to exert toxic effect against *C. albicans* and non-*albicans* species. CUR increases reactive oxygen species (ROS) levels, induces early apoptosis and shows antioxidant, antimutagenic and antitumour activity [144]. CUR acts on farnesol, a quorum sensing molecule, which is responsible for the increase in the level of ROS and inhibits hyphal development by targeting the TUP1 gene (involved in the hyphal development) [76]. CUR induces apoptotic pathway by activating *Ca*MCA1 gene (Metacaspase 1) which is the homolog of mammalian caspases [26].

Table 27.1 Various skin infections caused by fungi and mode of action of various phytochemicals against them

Plant species	Plant part used	Fungal pathogen	Mode of action	Bioactive compound	References
Curcuma longa	Rhizomes	*C. neoformans, Aspergillus* spp.	Increases reactive oxygen species (ROS), induces early apoptosis	Curcumin	[9, 172]
	Rhizomes	*T. mentagrophytes, T. rubrum, M. gypseum*	Antidermatophytic activity	Ar-turmerone, turmeric oil	[72]
Punica granatum	Leaves	*M. furfur, M. restricta*	Antidermatophytic activity	Not documented	[121]
Emblica officinalis	Fruit	*M. furfur, M. globosa*			
Curcuma xanthorrhiza	Not given	*M. furfur, M. pachydermatis*	Anti-*Malassezia* activity	Xanthorrhizol	[129]
Plectranthus barbatus, Plectranthus amboinicus	Leaves, essential oil	*Aspergillus* sp. *Fusarium* sp. *M. furfur*	Acts as a fungitoxic	Caryophyllene, germacrene D, viridiflorol, cymene, terpinene, carvacrol, pinene, α-terpinolene, thymol, cineole	[12, 133]
Aegle marmelos	Leaves	*A. niger, F. oxysporum*	Not documented	γ-cadinene, δ-carene, α-pinene	[70]
Melaleuca alternifolia (tea tree oil)	Essential oil	*Aspergillus* spp., *C. neoformans, Fusarium* spp., *M. furfur, M. canis, Trichophyton* spp.	Antidermatophytic activity	Terpinen-4-ol, 1,8-cineole, γ-terpinene, α-terpinene, terpinolene	[63, 64]
Mentha suaveolens insularis	Arial part	*C. zemplinina*	A delay in lag phase and reduction of growth rate	Piperitone, piperitenone, α-terpineol pulegone	[122]

(continued)

Table 27.1 (continued)

Plant species	Plant part used	Fungal pathogen	Mode of action	Bioactive compound	References
Hypericum perforatum	Adventitious roots	*M. furfur*	Inhibits biofilm formation	Xanthones	[140]
Salvia officinalis, Lavandula angustifolia, *Mentha piperita*	Essential oil	*T. mentagrophytes*, *T. rubrum*, *T. tonsurans*	Effective against experimentally induced dermatomycoses in male Wistar rats	Linalool, 1,8-cineole, menthol	[142]
Croton spp. and *Lippia sidoides*	Essential oil	*T. rubrum*, *T. mentagrophytes*, *Microsporum canis*, *C. neoformans*, *H. anomala*	Antidermatophytic activity	Thymol, eugenol, estragole, anethole	[48]
Eugenia cariophyllata				Essential oil, eugenol	[53]
Juniperus communis				α and β pinene, sabinene, limonene, myrcene	[119]
Syzygium aromaticum	Essential oil	*Aspergillus*, dermatophyte spp.	Inhibits germ tube formation, reduces ergosterol level	Eugenol	[123]

The essential oil from *Ocimum gratissimum* showed fungicidal activity against all *Candida* species with *C. parapsilosis* was the most susceptible and *C. tropicalis* was the least [105]. Essential oils from *Cinnamomum zeylanicum* Blume, *Citrus limon* Risso, *Eucalyptus citriodora* HK., *Eugenia uniflora* L., *Peumus boldus* and *Rosmarinus officinalis* L. have been reported to exert fungistatic action against at least one tested fungal strain of *Candida* species [36]. Table 27.2 describes few plants which have been tested against candidiasis and their mode of action.

27.2.1.3 Phytochemicals in Cryptococcosis

Greater than 50 *Cryptococcus* species live ubiquitously in the environment, but only *Cryptococcus neoformans* and *C. gattii* are significant pathogens of humans. *C. neoformans* accounts for 99% of the cryptococcal meningitis cases, while *C. gattii* is responsible for the remaining 1% of cases, in the immunocompromised people [131]. *C. neoformans* is an encapsulated, opportunistic fungal pathogen that can cause life-threatening cryptococcosis, particularly in immunocompromised individual, those infected with HIV/AIDS [108]. Cryptococcosis results from inhalation of fungal cells with subsequent lung infection and pneumonia. In the immunocompromised state, fungus can disseminate to the brain results in inflammation of the meninges and brain parenchyma, a condition known as meningoencephalitis, which is the most severe and accounts for major mortality in cryptococcosis. *C. neoformans* causes an estimated one million cases of meningoencephalitis globally per year in patients with AIDS, >60% of these cases die within 3 months of diagnosis [79, 131].

Currently, the antifungal drugs used for cryptococcosis treatment are amphotericin B, 5-flucytosine and fluconazole. However, these antifungal drugs have side effects such as chills, fever, headache, nausea, vomiting, hypokalaemia, nephrotoxicity, hypotension and anaemia. Because of these side effects, pharmacological management of the disease is difficult, and, also, the drug resistance has been reported [120].

Phenolic compounds, cudraxanthone S and cudraflavanone B, toxyloxanthone C and wighteone isolated from *Cudrania cochinchinensis* reported to have an antifungal activity against several fungi such as *C. neoformans*, *C. albicans* and *A. fumigates*. It has been determined that the phenolic compounds can also show synergistic effect with present antifungal agents. Thymol shows synergistic effect with azoles such as fluconazole by disrupting the cell wall or membrane integrity or by inhibiting ergosterol biosynthetic pathway. Similarly, the synergistic effect of 2,5,-dihydroxybenzaldehyde has been noted with itraconazole and amphotericin B against *C. neoformans* [15, 60]. Thus phenolic compound can act as antifungal agents and also enhance the activity of present antifungal agents. Machumi et al. [88] studied the antifungal activity of nine diterpenoids from *Clerodendrume rio phylum* against *C. neoformans* and reported that taxodione and 6-hydroxysalvinolone exhibited strong antifungal activity.

Table 27.2 List of plants tested against candidiasis and their mode of action

Plant species	Plant part used	Fungal pathogen	Mode of action	Bioactive compound	References
Liverwort	Not given	*C. albicans*	Inhibit morphogenetic switch and biofilm formation, due to up-regulation of DPP3 gene	Bisbibenzyl	[172]
Dumortiera hirsuta			Inhibits biofilm formation	Riccardin D	
Marchantia polymorpha			Induce apoptosis	Plagiochin E	[60]
Curcuma longa	Rhizomes	*C. albicans* and non-*albicans* species	Increases reactive oxygen species (ROS), induces early apoptosis	Curcumin	[9, 172]
Lythrum salicaria L.	callus	*C. albicans*	Not documented	Gallic acid	[91]
Coriandrum sativum	Leaves	*C. albicans*	Not documented	2E-decenal, 2E-decen-1-ol, n-decanol	[147]
Daucus littoralis	Leaves, stems, roots, flowers, fruits			Germacrene D, acorenone B	
Dracocephalum foetidum	Leaves			Limonene, geranial	
Euphrasia rostkoviana	Essential oil			n-hexadecanoic acid, thymol, myristic acid, linalool	
Artemisia biennis	Aerial parts	*C. albicans*		*β*-Ocimene, β- farnesene, acetylenes	
Lippia sidoides	Leaves	*C. albicans*		Thymol, carvacrol	
Plectranthus barbatus, *Plectranthus amboinicus*	Leaves, essential oil	*Candida* sp.	Acts as a fungitoxic	Caryophyllene, germacrene D, cymene, viridiflorol, terpinene, carvacrol, pinene, α-terpinolene, Thymol, Cineole	[12, 133]
Aegle marmelos	Leaves	*C. albicans*	Not documented	*γ*-Cadinene, *δ*-carene, *α*-pinene	[70]

Cinnamomum zeylancium	Bark, leaves	*C. albicans*, *C. parapsilosis*, *C. krusei*	Induce apoptosis, have strong cytotoxic activity	Cinnamaldehyde	[156]
Melaleuca alternifolia (tea tree oil)	Essential oil	*Candida* spp.	Antidermatophytic activity	Terpinen-4-ol, 1,8-cineole, α and γ-terpinene, terpinolene	[63, 64]
Croton spp., *Lippia sidoides*		*C. albicans*, *C. tropicalis*, *C. krusei*		Thymol, eugenol, estragole, anethole	[48]
Eugenia caryophyllata				Eugenol	[53]
Juniperus communis				α, β pinene, sabinene, limonene, myrcene	[119]
Diospyros melanoxylon	Bark, leaves	*C. albicans*, *C. krusei*, *C. tropicalis*, *C. parapsilosis*	Acts as a fungicidal	Not reported	[113]
Syzygium aromaticum	Essential oil	*Candida* spp.	Inhibits germ tube formation, reduces ergosterol level	Eugenol	[123]

Table 27.3 Cryptococcosis and mode of action of phytochemicals isolated from different plant species

Plant species	Plant part used	Fungal pathogen	Mode of action	Bioactive compound	References
Euphorbia tirucalli	Aerial parts	*C. neoformans*	Fungistatic, fungicidal	Diterpenoids, triterpenoids	[37]
Origanum	Stem and leaves	*C. neoformans*, *R. rubrum*, *C. albicans*	Not documented	Essential oil	[1]
Cannabis sativa	Essential oil	*C. neoformans*, *Candida glabrata* and *C. krusei*		α-Humulene, β-Caryophyllene	[162]
Magnolia obovata Thunb.	Not given	*C. neoformans*	Inhibit calcineurin pathway	Magnoloside A	[82]
Artemisia biennis	Aerial parts	*C. neoformans*	Not documented	β-Ocimene, β-Farnesene, acetylenes	[147]

Essentials oils from cinnamon, geranium "Bourbon", geranium "Palmarosa", savoury, and thyme showed good fungistatic action, but the pure components (terpenoids, citral, geraniol, and citronellol) showed better activity than the essential oils. Table 27.3 describes plants which have been tested against cryptococcosis and their mode of action.

27.2.1.4 Phytochemicals for Other Opportunistic Fungal Infections

Humans are continuously exposed to a huge number of fungal spores. Most of these spores are from avirulent species of fungus, and hence we experience no major harm from them. Also our immune system is capable of taking care of any probable harmful effect on our health. However, the same can causes disseminated fatal infections in immunocompromised patients, and hence are called opportunistic pathogenic fungi.

Geotrichosis

Geotrichum species are a rare cause of invasive fungal infections, mainly cause infection by *Geotrichum candidium* in immunocompromised patients, patients with haematological disease and acute leukaemia. *Geotrichum* species occurs widely as a normal microflora of the skin and the mucosa of the respiratory and digestive tracts [125]. This invasive infection can potentially occur in the bloodstream and also reported in another part of the human body such as pulmonary, CNS, hepatosplenic and urinary tract [96].

Penicilliosis

Penicilliosis is caused by *Penicillium marneffei* which can cause fatal infection in immunocompromised individuals due to HIV infection, specifically in late HIV infection wherein the CD^{4+} count falls below 100/μl [33]. Symptoms include a fever, weight loss, malaise, cough, dyspnoea, haemoptysis, oral lesions and skin manifestation (such as subcutaneous abscesses, papule-like ulcers) [39]. Previous study showed naturally derived phenol thymol as a promising possibility with antifungal activity against *Penicillium*; however, further studies are needed to study its mechanism of action and toxicity [148].

Mucormycosis

Mucormycosis (zygomycosis or phycomycosis) is caused by *Mucorales* as an opportunistic mycotic infection, mainly in patients with diabetes mellitus and diabetic ketoacidosis. Fungi belonging to the genus *Rhizopus*, *Mucor*, *Rhizomucor* and *Abscidia* are the main causative agents of zygomycoses; the most common among all is *Rhizopus*. In diabetic patients, mucormycosis caused by *Rhizopusoryzae* produces enzyme ketoreductase which enables them to feed on patient's ketone bodies [38]. Many antifungal drugs are available in the market such as amphotericin B, itraconazole, fluconazole and miconazole, but most of the *R. oryzae* strains are resistant to these drugs. For instance, thymol, the main constituents of essential oil from *Thymus vulgaris* showed an immense activity against fungal growth. It has been determined that thymol directly interacts with the ergosterol, the main component of fungal cell wall and leads to the disruption of the membrane integrity, fluidity and loss of the intracellular content [34].

Fungaemia

Emerging opportunistic yeast pathogens that can cause fungaemia majorly in immunocompromised patients include *Hansenulaanomala* (*Pichia anomala*), *Saccharomyces cerevisiae* and *Rhodotorula* species. Patients those in intensive care units and patients with central venous catheters, cancer (those undergoing bone marrow transplantations), abdominal surgery, cirrhosis, autoimmune diseases or burns are at high risk for such infection [96]. Outbreaks of the *Hansenulaanomala* (*Pichia anomala*) yeast have been reported in neonatal and paediatric intensive care units and also in immunocompromised patients. This yeast is mainly found in the environment such as plants, soil and fruit juices but also reported as a human colonizer. Among all wide range of invasive infections caused by *Hansenulaanomala,* fungaemia especially in association with a central venous catheter is most common [96].

Rhodotorula infections (*Rhodotorula* fungaemia) occur worldwide but are most frequent in the Asia-Pacific region and overall mortality is 15%. *Rhodotorula* fungaemia is most commonly caused by *Rhodotorula mucilaginosa* (also known as *Rhodotorula rubra*), followed by *Rhodotorula glutinis* and *Rhodotorula minuta* [154]. *Rhodotorula* infections are reported in association with ventriculitis, fungaemia, central venous catheter infection, keratitis, endocarditis and meningitis. The nonsystemic *Rhodotorula* infections were reported in immunocompetent patients such as endophthalmitis and peritonitis (usually associated with continuous ambulatory peritoneal dialysis). *Rhodotorula* infections are characterized by the non-healing oral ulcers and white patches [38, 87].

Table 27.4 Other opportunistic fungal infections and phytochemicals isolated from plants

Plant species	Plant part used	Fungal pathogen	Mode of action	Bioactive compound	References
Melaleuca alternifolia (tea tree oil)	Essential oil	*M. canis*, *Penicillium* spp., *R. rubra*, S. *cerevisiae*	Anti-dermatophytic activity	Terpinen-4-ol, 1,8-cineole, γ-terpinene, α-terpinene, terpinolene	[63, 64]
Plectranthus barbatus, *P. amboinicus*	Leaves, essential oil	*Penicillium* sp., *S. cerevisiae*	Fungitoxic	Caryophyllene, germacrene D, viridiflorol, cymene, terpinene, carvacrol, pinene, thymol, cineole	[12, 133]
Ginkgo biloba	Rhizome	*S. cerevisiae*, *G. candidum*, *Rhodotrula* species	Not documented	Not reported	[19, 69]
Mentha suaveolens spp. *insularis*	Arial part	*S. cerevisiae*	Delay in lag phase	Piperitone, piperitenone, α-Terpineolpulegone	[122]
Camellia sinensis	Leaves	*S. cerevisiae* and yeast-like species	Antimycotic activity	Epicatechin-3-*O*-gallate, epigallocatechin-3-*O*-gallate	[155]

Saccharomyces cerevisiae is a commensal and widespread in nature, inhabiting the gastrointestinal tract of humans and maintaining the normal homeostasis of the lower gastrointestinal tract [38]. Fungaemia from *S. cerevisiae* can be linked to the use of live yeast capsules (called *Saccharomyces boulardii*), which are administered as probiotics for prevention of diarrhoea and adjunctive therapy for diarrhoea associated with *Clostridium difficile* [103]. Major components of oregano extract, includes carvacrol, thymol and eugenol have been shown to exist potent antifungal activities. Carvacrol is responsible for the disruption of both Ca^{2+} and H^{+} homeostasis in yeast and that these disruptions likely lead to loss of cell viability [91]. Plants which have been examined against different opportunistic fungal infections are listed in Table 27.4.

27.2.2 Antiviral Activities of Phytochemicals

Products of medicinal values from plant sources provide unlimited opportunities for new drug leads because of the unmatched availability of chemical diversity and large-scale cultivation potential. As antivirals of chemical origin have often been associated with toxicity to human beings, looking for alternates has always been a prerequisite in antiviral drug discovery. Hence, need to seek therapeutic drugs from edible natural products have grown throughout the world. Traditionally a few countries with rich flora diversity have been practicing the use of plant-based medicines in most of their ailments with different names and mostly undocumented wisdom. India, China, Africa and some sub-Saharan countries are known to possess such wisdom. Chinese medicines and Ayurveda from India are the leading practices with herbal products and well-documented evidence.

With the emergence of new allopathic practice and science of drug discovery, identifying active components from such plant-based treatments has become a mandatory part of the research. The further discussion under this segment would highlight a few such publications. The approach is to identify those plants which are known to possess some antiviral activity against a particular virus and then trying to isolate the individual components to see which one is really working on the virus and what is the mechanism. A few recent studies with respect to different diseases caused by the virus have been discussed below.

27.2.2.1 Phytochemicals in HIV/AIDS

The acquired immunodeficiency syndrome (AIDS), caused by infection with the human immunodeficiency virus (HIV), has become a devastating epidemic worldwide severely hitting countries in Asia and sub-Saharan Africa [93]. Studies conducted in the sub-Saharan Africa have indicated that traditional healers and local communities extensively use medicinal plants to manage the effects of HIV/AIDS [109]. Other studies have reported that the extract of Panax ginseng possesses antiretroviral activity by acting on the reverse transcriptase enzyme of HIV [30]. A study from India has shown that extracts from *Acacia catechu* suppress HIV-1 replication by inhibiting the activities of the viral protease and Tat [110]. Similarly extracts prepared from different parts of *R. parviflora* [98], *P. barbatus* [74], *Hoodia gordonii* [75], *Albizia procera* [114], *Sanguisorba officinalis* [83], etc. have been reported to have anti-HIV activity in laboratory assays. In an interesting study about verifying the traditional medicine practice, it was proved that a three step traditional treatment with *Cassia sieberiana* root, *Vitex doniana* root and *Croton megalobotrys* bark inhibited HIV-1 replication with dose dependence and without concomitant cytotoxicity. Also, extracts did not interfere with antiviral activities of licensed antiretroviral drugs (ARVs) when applied in combination and exhibited comparable efficacies against viruses harbouring mutations to licensed protease, reverse transcriptase or integrase inhibitors [152].

The plant-derived products like polycitone A are reported to be active against the resistant forms of HIV [137]. Oleanolic acid, a well-known triterpene occurring in numerous varieties of plants has shown anti-HIV activities [145]. Another approach used in discovering more drugs for HIV makes use of available libraries of molecules which have been derived from plant sources and are well-characterized in terms of their activities in human cells. Herein, molecules with structural similarity to reported interactors of HIV-1 proteins are screened virtually as well as against the preserved and clinical strains of the virus in the laboratory for their antiviral potential and mechanism of action. A study by Tietjen and co-workers reported the similar work using pan-African natural product library wherein they identified Ixoratannin A-2 and Boldine as novel HIV-1 inhibitors [151].

As HIV makes the host's immune system compromised, a range of drugs which don't really target virus but target the immune system also provide a great help in managing the infection. These are called immunomodulators or immunostimulants which serve as mediators to induce the production of certain enzymes that inhibit viral replication in the cell. Alkaloids, carbohydrates, lectins, polyphenols, stilbenoids and peptides of plant origin are among the many classes of drugs that are used as immunomodulators. Plants of genus *Nigella*, *Tinospora cordifolia*, flowering tops of *Echinacea purpurea*, etc. have shown immunomodulatory activity in HIV and similar viral infections [25, 97, 118, 171]. Table 27.5 describes plants reported for anti-HIV activity and their mechanisms of action.

27.2.2.2 Phytochemicals Against Influenza

Influenza, commonly known as "the flu", is an infectious disease caused by an influenza virus. Three types of influenza viruses, namely, type A, type B and type C, affect people. Influenza A and B viruses circulate and cause outbreaks and epidemics. Due to this reason, relevant strains of influenza A and B viruses are included in seasonal influenza vaccines. Yearly vaccinations against influenza are recommended by the World Health Organization for those at high risk. A vaccine made for 1 year may not be useful in the following year since the virus evolves rapidly. Antiviral drugs for influenza are available in some countries and are often helpful in reducing severe complications and deaths. The drugs range from protein inhibitors, neuraminidase inhibitors (e.g. Zanamivir, Oseltamivir) and ion channel blockers (e.g. Amantadine, Rimantadine). However, resistance to most of these drugs has been reported. Medicinal plants are the important resources for new molecules with least side effects with better efficacy. Different plants from ancient literature and traditional medicine have been picked for the analysis of their activity against influenza virus and identification of active compounds for further drug development. *Allium sativum*, *Azadirachta indica*, *Curcuma longa* L., *Syzygium aromaticum*, *Thymus vulgaris*, *Mentha piperita*, etc. are being used traditionally to treat symptoms of influenza and have been proved very effective. Following this traditional wisdom, many research studies and trials tried to find the active ingredients from traditionally known plants and tested them clinically. Table 27.6 summarizes plants which have been tested for their potential against different influenza virus strains.

27.2.2.3 Phytochemicals Tested Against Other Viruses

Herpes Simplex Virus1 and 2 (HSV-1 and HSV-2)

Both HSV-1 and HSV-2 are ubiquitous and contagious. HSV-1 causes sores around the mouth and lips (cold sores). HSV-2 in most of the cases causes genital herpes. Both the conditions are not treatable. The medicines available and being used till now are for symptomatic relief, and the infection can come back again even after treatment. For instance, Famvir, Zovirax and Valtrex are among the drugs used to

Table 27.5 Some examples wherein plants were explored for their anti-HIV potential

Mode of action	Plant species name	Plant part	References
HIV-1 reverse transcriptase inhibition	*Cinnamomum loureiroi*	Stem bark	[138]
	Quercus infectoria	Fuit	
	Plumbago indica L.	Root	
	Artocarpus heterophyllus Lam.	Seed	
	Ocimum sanctum L.	Leaves	
	Allium sativum L.	Bulb	
	Acoruscalamus L.	Rhizomes	
	Cascabela thevetia (L.)	Roots	[150]
	Lippold		
HIV replication	*Terminalia catappa*	Leaves and bark	[41]
	Syzygium claviflorum	Leaves	[80]
	Xanthoceras sorbifolia	Wood ethanolic extract	[80]
	Scutellaria baicalensis	Flavonoid	[80]
	Tripterygium wilfordii	Triptolide	[161]
	Ancistrocladus congolensis	Root bark	[24]
Immunomodulators	PhytoV7	A complex of phytochemicals	[168]
	Echinacea purpurea	Extract	[99]
	Immunity 1 (Fuzheng 1)- China	Mixture of herbs	[95, 163]
Non-nucleoside specific reverse transcriptase inhibitors (NNRTI)	*Calophyllum cordato-oblongum*	Calanolides (Coumarin)	[80]
Broad anti-HIV potential	*Justicia gendarussa*	Methanol extract of the stems and barks	[173]
Not well documented	*Moringa oleifera*	Dry leaf powder	[100]

treat the symptoms of herpes. An estimated 417 million people aged 14–49 were reported to be infected with HSV-2, while 140 million adults were estimated to have genital infection with HSV-1 in 2015 [56]. Hence, there is an urgent need to find drugs which have antiviral activities against HSVs and also do not have side effects. Efforts are ongoing for having an efficacious vaccine; however, no promising vaccine has come to the market yet. On the other hand, different molecules of synthetic as well as the herbal origin are being tested for activity against HSVs. Extract of *Cornus canadensis,* used in native American traditional medicine has been tested for antiherpes simplex virus type 1 (HSV-1) activity and reported to be active with the need for further work [81]. Another work from Indian subcontinent reports rich antiherpes virus activities of bioactive fraction and isolated pure constituent of

Table 27.6 Plants tested for anti-influenza activity

Mode of action	Plant species name	Plant part	References
Inhibition of viral nucleoprotein synthesis and polymerase activity	*Plumbago indica* and *Allium sativum*	Root	[29]
Inhibition of viral attachment to the host cells	*Eupatorium perfoliatum L.*	Hydroalcoholic extracts from the aerial parts	[40]
Non-specific	*Glycine max*	Cheonggukjang (fermented soybeans) extracts	[167]
Antiviral	*Coptidis rhizoma, Isatidis folium, Lonicerae flos, Scutellarias radix, Cyrtomium rhizome, Houttuynia cordata, Gardeniae fructus and Chrysanthemi indici flos*	Extracts	[65]
Inhibition of viral RNA polymerase activity	*Cryptoporus volvatus*	Cryptoporic acid E	[51]

Mallotus peltatus, an ethnomedicine from Andaman Islands [14]. Taking the lead from traditional medicine wherein, teas made from leaves and bark of *Gallesiagorazema* have been used for various therapeutic purposes including treatment of abscesses, orchitis, gonorrhoea and for rheumatic pain relief. Silva Júnior et al. [139] tested the activity of this plant against HSV-1 and HSV-2 and found it promising for further development. In another study about traditionally known plant *Ficus religiosa* water and chloroform bark extracts have shown good activity against wild-type as well as the acyclovir-resistant strain of HSV-2 [54].

Hepatitis Virus

Viral hepatitis is a major global public health problem affecting hundreds of millions of people. Viral hepatitis is a cause of considerable morbidity and mortality both in acute and chronic form in the human population; this includes, in the case of hepatitis B, C and D, chronic active hepatitis and cirrhosis. Hepatocellular carcinoma has been found to be closely associated with hepatitis B and at least in some regions of the world with hepatitis C virus. Hepatitis C has seen a good success in terms of treatment, while hepatitis B is still a struggle with drugs. Though there is an effective vaccine available for hepatitis B, failure to protection because of vaccine-escape viral mutants in some population is also reported. All the antiviral drugs being used for treating hepatitis B virus (HBV) infection have their limitations. Interferon (IFN-α) has limited efficacy and a high incidence of adverse side effects in a proportion of chronic patients. Nucleos(t)ide analogs like lamivudine, adefovir, tenofovir and entecavir are very effective in treating chronic hepatitis B (CHB) but need long-term therapy which eventually leads to drug resistance. On the other hand, natural or plant products have provided promising therapeutics as they have high chemical diversity and biochemical specificity. A well-practiced herb in Indian medicine is *Phyllanthus urinaria. Various studies have proved that it has an antiviral activity.* Mekha Mohan et al. [94] *using bioinformatics tools proved that Phyllanthus urinaria* phytochemicals have HBV-DNA polymerase inhibition potential. Herbs have not only shown the potential of antiviral activity but also have proven effective against drug-resistant strains of hepatitis B virus. Jung et al. [73] have shown the inhibitory effect of *Phyllanthus urinaria L.* extract on the replication of lamivudine-resistant hepatitis B virus in vitro.

27.2.3 Antibacterial Activities of Phytochemicals

Bacteria are often the cause of many diseases in humans and animals [132]. In India, major threats constitute the food borne and waterborne diseases caused by bacteria [31]. Infections like osteomyelitis, endocarditis, meningitis, pneumonia and many more are caused by *Staphylococcus aureus.* They release enterotoxins in food which results into food poisoning and by releasing super antigens into the blood stream they even cause toxic shock syndrome in humans. They form the common microflora of the skin, nasal passage and axillae in humans [49]. The antibacterial activities of plant extracts have been linked to the presence of some bioactive compounds or secondary metabolites. It has been reported 100 plant species are used regularly as a source of medicine and 2500 species of plants are used by traditional healers in India [104]. Flowers of Chinese bell have antibacterial properties. Flowers, pods and gum resins of Babul tree are used in treating diarrhoea [78].

27.2.3.1 Tuberculosis

Tuberculosis (TB) caused by *Mycobacterium tuberculosis* is the leading cause of death from infectious disease and the treatment of which is globally become a challenge. Regardless of availability of several treatments to treat TB, the causative agent, *M. tuberculosis* has nowadays developed multidrug-resistant and extensively drug-resistant strains. Table 27.7 explained various plant species and their bioactive constituent explored for their antibacterial properties against tuberculosis.

27.2.3.2 Pneumonia

Klebsiella pneumoniae is a Gram-negative bacterium and has been considered a respiratory pathogen that causes pneumonia. *K. pneumonia* is commonly found in the gastrointestinal tract and hands of hospital personnel. The genus *Klebsiella* has been reported to fall under the category of ESBL (extended-spectrum beta-lactamase)-producing bacteria. It confers a high level of resistance to all β-lactams and carbapenems drugs which severely challenged antimicrobial therapy [136, 141]. Lin et al. [84] investigated the effect of flavonoids isolated from natural dietary sources in combination with antibiotics as a strategy against ESBL (extended-spectrum beta-lactamase)-producing *K. pneumonia* isolates. The medicinal plants including *Desmodium gangeticum, Nelumbo nucifera, Cannabis* spp., *Sesame white* and *Sesame black* showed antimicrobial activity towards *K. pneumonia* [141].

Table 27.7 Antimycobacterial activities of some plants and plant parts

Plant name	Plant part used	Active constituent	References
Mallotus philippensis	Leaves	Ursolic acid, β-Usitosterol	[61]
Piper nigrum	Seeds	Piperine	[21]
Alstonia scholaris	Bark, flower, fruit and leaf	Not documented	[10]
Allium sativum	Bulb	Either fats and fixed oils or phenol and aryl amine derivative	[127, 160]
Acalypha indica	Leaves		
Adhatoda vasica	Leaves		
Leucas marrubioides	Roots	Not documented	[57]
Cassia fistula Linn	Roots	Alkaloids and tannins could be responsible	[27]
Glycyrrhiza glabra L.	Rhizomes	Isoliquiritigenin and liquiritigenin	[52]
Celastrus vulcanicola	Dried leaves	Dihydro-β agarofuran sesquiterpenes	[153]
Tiliacora triandra	Roots	Bisbenzylisoquinoline alkaloids, tiliacorinine, 2′-nortiliacorinine and tiliacorine	[146]

Various plants such as *Acacia mearnsii*, *Acacia nilotica*, *Acalypa indica*, *Phyllanthus niruri*, *Allium rotundum* and *Cinnamomum zeylanicum* were observed to possess antibacterial activities against *Klebsiella* spp. [13, 90, 106, 111].

27.3 Anticancerous Potential of Phytochemicals

Cancer is one of the leading causes of death in the developed and developing countries. There are certain factors such as environmental, genetics and microbe induced that can lead to the development of cancer [18, 35]. The current approaches such as chemotherapy, radiotherapy, surgery and immunotherapy, individually or in combination, are widely in use for treating cancer. The localized tumours are often treated using surgery or radiotherapy, while chemotherapy is mainly aimed to cure metastatic cancers. The cytotoxic effects of chemotherapeutic drugs towards cancer cells and also normal cells cause various side effects such as vomiting, nausea, alopecia and myelosuppression. The activities of anticancer drugs are limited by their insolubility, instability and low absorption rate at the tissue level and often develop tumour drug resistance [5].

To overcome the adverse effects of chemotherapeutic drugs, there are considerable number of studies that aim to screen various natural products such as plant species, marine sources, and micro-organisms for potential anticancer sources. There are certain bioactive compounds derived from plants that have been confirmed for their anticancer potentials [62]. These bioactive compounds are exclusively in use to treat cancer along with the current therapeutic modalities. The literature reveals that flavonoids, alkaloids and terpenes of plant-derived products have received considerable attention in recent years due to their diverse pharmacological properties including cytotoxicity and cancer chemopreventive effects [115].

The screening of anticancer agents from plant sources began in early 1950s with the discovery of Vinca alkaloids such as Vinblastine and Vincristine from *Catharanthus roseus*. Earlier these plant-derived alkaloids were specially used as hypoglycaemic agents. Later, the vinca alkaloids were used as anticancer drug in combination with chemotherapy. The anticancer properties of vinca alkaloids are due to their interactions with tubulin leading to disruption of microtubule function and causing metaphase arrest [102]. The vinca alkaloids have been in regular use as medicament against testicular carcinoma and Hodgkin and non-Hodgkin lymphomas [102]. Paclitaxel (Taxol), another phytochemical from bark of Pacific yew tree, showed most potential anticancer activities against breast and ovarian cancer [134, 165]. The polyphenol such as epigallocatechin-3-gallate (EGCG) from the leaves of *Camellia sinensis* was reported to inhibit proliferation of cancerous cells by reducing DNA methylation, inhibition of DNA methyltransferase and reactivation of tumour suppressor genes [8]. Table 27.8 explained various plant species and their bioactive compounds explored for their anticancer properties.

27.4 Antidiabetic Activities of Phytochemicals

Diabetes mellitus is a chronic non communicable endocrine disease which occurs when the body fails to produce enough insulin or fails to utilize insulin effectively. It is common in all populations and all age groups [7, 157]. Diabetes is often characterized by chronic hyperglycaemia which is further associated with long-term damage, dysfunction and failure of vital organs, especially the eyes, kidneys, nerves, heart and blood vessels [7]. Constituted chiefly by type 2 diabetes (T2D), diabetes is a global public health threat. The prevalence among adults (aged 20–70 years) is expected to rise to 438 million by the year 2030. Developing countries have the highest economic burden since more than 80% of cases of diabetes occur in these countries. The prevalence estimates of diabetes and impaired glucose tolerance (IGT) are high for all Asian countries. The present trend indicates that more than 60% of the world's diabetic population will be in Asia [128].

ROS are important mediators of beta cell death during the development of diabetes. High glucose has been postulated to generate ROS and nitrogen species in numerous cell types. Many hypotheses explain the formation of free radicals in diabetes, e.g., autooxidation of glucose and non-enzymatic glycation of proteins and lipid, resulting in increased levels of advanced glycation end products (AGEs) [159].

Alarmingly high costs of synthetic drugs available for diabetes show an urgent requirement for the development of some alternative approaches for the prevention and treatment of diabetes. According to the World Health Organization, 80% of the population of developing countries primarily depends on herbal traditional

Table 27.8 Anticancer properties of some plants and plant-derived bioactive compounds

Plant species name	Bioactive compound	Mode of action	References
Murraya koenigii	Koenimbin	Induction of apoptosis and G0 arrest of cell cycle in breast cancer cells	[2]
Catharanthus roseus	Vinblastine, vincristine	Disruption of microtubule function	[102]
Taxus brevifolia	Paclitaxel	Interference with microtubule stabilization and inhibition of cancer cell propagation	[165]
Camellia sinensis	Epigallocatechin-3-gallate	Inhibit cancer cell proliferation	[8]
Curcuma longa	Curcumin	Inhibit cancer cell proliferation and inhibit NF-kB pathway that triggers the intracellular inflammatory response	[130]
Podophyllum peltatum	Podophyllotoxin	Disruption of chromatin structure	[66]
Moringa peregrina	Apigenin	Induces apoptosis, exhibit cytotoxic activities	[43]
Saffron crocus	Crocetin	Inhibits the growth of cancer cells by inhibiting nucleic acid synthesis, inducing apoptosis and enhancing antioxidative system	[62]
Vitis vinifera *Arachis hypogaea*	Resveratrol	Chemopreventive activity, anti-initiation and anti-progression activities of cancer cells, inhibits cancer metastasis via reducing hypoxia inducible factor-1α and MMP-9 expression	[22, 169]
Glycine max *Flemingia vestita*	Genistein	Exhibits antioxidant, antihelmintic, antiangiogenic effects, blocks the uncontrolled cell growth associated with cancer, inhibits enzymes that regulate cell division and cell survival, functions as a tyrosine kinase inhibitor by inhibiting DNA topoisomerase II	[86, 92]
Zingiber officinale	Gingerol	Exhibits antioxidant, anti-inflammation and antitumor promoting properties, further, decreases iNOS and TNF-alpha expression via suppression of IκBα phosphorylation and NF-κB nuclear translocation	[112]
Solanum lycopersicum	Lycopene	Activates cancer preventive enzymes such as phase II detoxification enzymes and inhibits human cancer cell proliferation	[55]
Rosmarinus officinalis	Rosmarinic acid	Dose dependently inhibits migration, adhesion and invasion of cancer cells, induces apoptosis, suppresses NF-kB pathway through inhibition of phosphorylation and degradation of IκBα	[101, 170]

Table 27.9 Antidiabetic activities of plants and their bioactive compounds

Plant species name	Bioactive compounds	Plant parts used	Mode of action	References
Murraya koenigii	Mahanimbine alkaloid, murrayanol, hydrolysable tannins	Leaves	Increases glycogenesis, decreases glycogenolysis, gluconeogenesis	[71, 77]
Trigonella foenum-graecum	Protein, fat, volatile oil, fixed oil, carbohydrate	Seeds	Reduces blood glucose concentration	[3]
Lippa nodiflora	Sterols, flavonoids, coumarins, quinones, tannins	Whole plant	Significant decrease in blood glucose level, HbA_{1c}	[17]
Morus alba	Quercetin, kaempferol, phenolic acids	Fruits	Increases α-glucosidase inhibitory and free radical scavenging activities	[164]
Hybanthus enneaspermus	Polyphenols	Whole plant	Significant effect on oral glucose tolerance	[116]
Ocimum sanctum	Tetracyclic triterpenoids	Aerial part	Reduces blood glucose, cholesterol, triglyceride levels	[117]
Zingiber officinale	Sesquiterpene	Rhizome	Increases insulin level, reduces fasting blood glucose level	[4]
Vaccinium arctostaphylos	Caffeoylquinic acid	Fruit	Antihyperglycaemic, antioxidant and triglyceride lowering activities	[47]
Momordica charantia	Momordicine alkaloid, ascorbic acid	Fruit	Reduces blood glucose, triglyceride, low-density lipoprotein, increases high-density lipoprotein levels	[28]

medications [50]. Plants like *Aegle marmelos* (L.), *Allium cepa* L., *Allium sativum* L., *Aloe vera* (L), *Artemisia santonicum* L., *Azadirachta indica*, *A.* Juss., *Beta vulgaris* L, *Boerhavia diffusa* L., *Brassica juncea* (L.), *Caesalpinia bonducella* (L.), *Cajanus cajan* (L.), *Camellia sinensis Kuntze*, *Casearia esculenta Roxb.*, *Cassia auriculata* L., *Catharanthus roseus* (L), *Citrullus colocynthis*, *Ficus bengalensis* L, *Eugenia jambolana* Lam, *Enicostemma littorale* Blume, *Coccinia indica*, *Helicteres isora* L., *Hibiscus rosasinensis* L., *Mangifera indica* L., *Momordica charantia*, *Morus alba* L., *Mucuna pruriens* (L.), *Murraya koenigii* (L.) and *Ocimum sanctum* L. have been used in the treatment of diabetes mellitus [11, 59, 159]. Various antidiabetic plants with their bioactive compounds and principle mode of actions are listed in Table 27.9.

27.5 Conclusion

The current situation of indiscriminate use of antibiotics and the consequent MDR tendency of micro-organisms to conventional antibacterial and antifungal treatments has been stimulating researchers to seek alternative natural sources of antimicrobial compounds, deriving from the medicinal plants. Fungal diseases/infections still remain one of the most common as well as the important problem associated with daily life. Till date, several antimycotics synthetic drugs are available in the market, but these drugs are having various side effects, which need the long duration of treatment, high cost and many fungal pathogens developing resistance to these drugs. The unpleasant side effects of the existing antibiotics or therapies include itching, nausea and abdominal pain, and sometimes it can be toxic, in many cases such side effects limit its therapeutic use. Herbal medicines can act as a major alternative to cure these infections, with less toxicity. Various medicinal plants and their active constituents are having antifungal as well as antibacterial activities, and they are widely used in modern medicines today.

As we discussed traditional medicines from various parts of the world being tested for activities against various viruses, we are not being able to translate these findings into a ready to use product. The reason is being the huge cost of clinical trials, product purification and stabilization into a formulation, regulatory processes, the absence of model animals and monitory resources. If we need to see more and more herbal products coming into the market with well-documented antiviral activities, we need to support such studies with all that they need. Different government programmes in a few countries have started supporting such work but not to the extent they should be. The work is continued and hopes that we would be able to see new molecules with good efficacy against viral pathogens soon. In the same line of antimicrobial drugs, the current scenario of the gradual development of resistance of cancer cells towards chemotherapeutic drugs has worsens the cancer treatment procedures. Herbal drugs are also in demand to meet the needs of antidiabetic therapies. Thus, plants and their potential bioactive compounds are the gold mines of future medical world that needs to get more attention and direction.

References

1. Adams A, Kumar S, Clauson M, Sahi S (2011) Anti-yeast activities of Origanum oil against human pathogenic yeasts. Adv Biosci Biotechnol 2:103
2. Ahmadipour F, Noordin MI, Mohan S, Arya A, Paydar M, Looi CY, Keong YS, Siyamak EN, Fani S, Firoozi M, Yong CL (2015) Koenimbin, a natural dietary compound of Murraya koenigii (L) Spreng: inhibition of MCF7 breast cancer cells and targeting of derived MCF7 breast cancer stem cells (CD44+/CD24−/low): an in vitro study. Drug Des Devel Ther 9:1193
3. Ajabnoor MA, Tilmisany AK (1988) Effect of Trigonella foenum graceum on blood glucose levels in normal and alloxan-diabetic mice. J Ethnopharmacol 22:45–49
4. Akhani SP, Vishwakarma SL, Goyal RK (2004) Anti-diabetic activity of Zingiber officinale in streptozotocin-induced type I diabetic rats. J Pharm Pharmacol 56:101–105

5. Akindele AJ, Wani ZA, Sharma S, Mahajan G, Satti NK, Adeyemi OO, Mondhe DM, Saxena AK (2015) In vitro and in vivo anticancer activity of root extracts of Sansevieria liberica Gerome and Labroy (Agavaceae). Evid Based Complement Alternat Med 2015:1
6. Amber K, Aijaz A, Immaculata X, Luqman KA, Nikhat M (2010) Anticandidal effect of Ocimum sanctum essential oil and its synergy with fluconazole and ketoconazole. Phytomedicine 17:921–925
7. American Diabetes Association (2011) Standards of medical care in diabetes—2011. Diabetes Care 34:S11–S61
8. Amin A, Gali-Muhtasib H, Ocker M, Schneider-Stock R (2009) Overview of major classes of plant-derived anticancer drugs. Int J Biomed Sci: IJBS 5:1
9. Ansari M, Anurag A, Fatima Z, Hameed S (2013) Natural phenolic compounds: a potential antifungal agent. Microbiology 2:189–195
10. Antony M, James J, Misra CS, Sagadevan LD, Veettil AT, Thankamani V (2012) Anti mycobacterial activity of the plant extracts of Alstonia scholaris. Int J Curr Pharm Res 4:40–42
11. Arumugam G, Manjula P, Paari N (2013) A review: anti diabetic medicinal plants used for diabetes mellitus. J Acute Dis 2:196–200
12. Arumugam G, Swamy MK, Sinniah UR (2016) *Plectranthus amboinicus* (Lour.) Spreng: botanical, phytochemical, pharmacological and nutritional significance. Molecules 21(4):369
13. Arutselvi R, Balasaravanan T, Ponmurugan P, Suresh P, Ramachandran N (2012) Comparative studies of anti-microbial activity of turmeric and selected medicinal plant leaf extracts used in Indian traditional medicine. J Herbs Spices Med Plants 18:231–239
14. Bag P, Chattopadhyay D, Mukherjee H, Ojha D, Mandal N, Sarkar MC, Chatterjee T, Das G, Chakraborti S (2012) Anti-herpes virus activities of bioactive fraction and isolated pure constituent of Mallotus peltatus: an ethnomedicine from Andaman Islands. Virol J 9:98
15. Bailly S, Maubon D, Fournier P, Pelloux H, Schwebel C, Chapuis C, Foroni L, Cornet M, Timsit JF (2016) Impact of antifungal prescription on relative distribution and susceptibility of Candida spp.—trends over 10 years. J Infect 72:103–111
16. Baker JT, Borris RP, Carté B, Cordell GA, Soejarto DD, Cragg GM, Gupta MP, Iwu MM, Madulid DR, Tyler VE (1995) Natural product drug discovery and development: new perspectives on international collaboration. J Nat Prod 58:1325–1357
17. Balamurugan R, Ignacimuthu S (2011) Antidiabetic and hypolipidemic effect of methanol extract of Lippia nodiflora L. in streptozotocin induced diabetic rats. Asian Pac J Trop Biomed 1:S30–S36
18. Banerjee J, Mishra N, Dhas Y (2015) Metagenomics: a new horizon in cancer research. Meta Gene 5:84–89
19. Bashir SF, Gurumayum S, Kaur S (2015) In Vitro antimicrobial activity and preliminary phytochemical screening of methanol, chloroform, and hot water extracts of ginger (Zingiber officinale). In Vitro 8(1):176–180
20. Bathoorn E, Salazar NE, Sepehrkhouy S, Meijer M, de Cock H, Haas PJ (2013) Involvement of the opportunistic pathogen Aspergillus tubingensis in osteomyelitis of the maxillary bone: a case report. BMC Infect Dis 13:59
21. Birdi T, D'souza D, Tolani M, Daswani P, Nair V, Tetali P, Toro JC, Hoffner S (2012) Assessment of the activity of selected Indian medicinal plants against Mycobacterium tuberculosis: a preliminary screening using the Microplate Alamar Blue Assay. Eur J Med Plant 2:308–323
22. Boocock DJ, Faust GE, Patel KR, Schinas AM, Brown VA, Ducharme MP, Booth TD, Crowell JA, Perloff M, Gescher AJ, Steward WP (2007) Phase I dose escalation pharmacokinetic study in healthy volunteers of resveratrol, a potential cancer chemopreventive agent. Cancer Epidemiol Pre Biomark 16:1246–1252
23. Bouddine L, Louaste B, Achahbar S, Chami N, Chami F, Remmal A (2012) Comparative study of the antifungal activity of some essential oils and their major phenolic components against Aspergillus niger using three different methods. Afr J Biotechnol 11:14083–14087

24. Bringmann G, Steinert C, Feineis D, Mudogo V, Betzin J, Scheller C (2016) HIV-inhibitory michellamine-type dimeric naphthylisoquinoline alkaloids from the Central African liana Ancistrocladus congolensis. Phytochemistry 128:71–81
25. Brüll F, Mensink RP, Plat J (2009) Plant sterols: functional lipids in immune function and inflammation? Clin Lipidol 4:355–365
26. Cao Y, Huang S, Dai B, Zhu Z, Lu H, Dong L, Cao Y, Wang Y, Gao P, Chai Y, Jiang Y (2009) Candidaalbicans cells lacking CaMCA1-encoded metacaspase show resistance to oxidative stress-induced death and change in energy metabolism. Fungal Genet Biol 46:183–189
27. Channabasappa HS, Shrinivas JD, Venkatrao KH (2015) Evaluation of antibacterial and Antitubercular activity of Cassia fistula Linn root. Int J Res Pharm Sci 6:82–84
28. Chaturvedi P, George S, Milinganyo M, Tripathi YB (2004) Effect of Momordica charantia on lipid profile and oral glucose tolerance in diabetic rats. Phytother Res 18:954–956
29. Chavan RD, Shinde P, Girkar K, Madage R, Chowdhary A (2016) Assessment of Anti-Influenza activity and hemagglutination inhibition of Plumbago indica and Allium sativum extracts. Pharm Res 8:105
30. Choi KT (2008) Botanical characteristics, pharmacological effects and medicinal components of Korean Panax ginseng CA Meyer. Acta Pharmacol Sin 29:1109–1118
31. Chugh TD (2008) Emerging and re-emerging bacterial diseases in India. J Biosci 33:549–555
32. Cowan MM (1999) Plant products as antimicrobial agents. Clin Microbiol Rev 12:564–582
33. Dahiya P, Kamal R, Puri A, Saini G, Arora A (2012) Penicillinosis in a HIV-positive individual. Indian J Sex Trans Dis 33:38
34. de Lira Mota KS, de Oliveira Pereira F, de Oliveira WA, Lima IO, de Oliveira Lima E (2012) Antifungal activity of Thymus vulgaris L. essential oil and its constituent phytochemicals against Rhizopus oryzae: interaction with ergosterol. Molecules 17:14418–14433
35. De Martel C, Ferlay J, Franceschi S, Vignat J, Bray F, Forman D, Plummer M (2012) Global burden of cancers attributable to infections in 2008: a review and synthetic analysis. Lancet Oncol 13(6):607–615
36. de Oliveira Lima I, Oliveira RD, de Oliveira Lima E, Farias NM, de Souza EL (2006) Atividade antifúngica de óleos essenciais sobre espécies de Candida. Braz J Pharm 16:197–201
37. de Oliveira RB, Atobe JH, Souza SA, Santos DW (2014) Epidemiology of invasive fungal infections in patients with acquired immunodeficiency syndrome at a reference hospital for infectious diseases in Brazil. Mycopathologia 178:71–78
38. Deepa AG, Nair BJ, Sivakumar TT, Joseph AP (2014) Uncommon opportunistic fungal infections of oral cavity: a review. J Oral Maxillofac Pathol: JOMFP 18:235
39. Deesomchok A, Tanprawate S (2006) A 12-case series of Penicillium marneffei pneumonia. J Med Assoc Thail 89:441
40. Derksen A, Kühn J, Hafezi W, Sendker J, Ehrhardt C, Ludwig S, Hensel A (2016) Antiviral activity of hydroalcoholic extract from Eupatorium perfoliatum L. against the attachment of influenza a virus. J Ethnopharmacol 188:144–152
41. Dwevedi A, Dwivedi R, Sharma YK (2016) Exploration of phytochemicals found in Terminalia sp. and their antiretroviral activities. Pharmacogn Rev 10:73
42. Edeoga HO, Okwu DE, Mbaebie BO (2005) Phytochemical constituents of some Nigerian medicinal plants. Afr J Biotechnol 4:685–688
43. El-Alfy TS, Ezzat SM, Hegazy AK, Amer AM, Kamel GM (2011) Isolation of biologically active constituents from Moringa peregrina (Forssk.) Fiori.(family: Moringaceae) growing in Egypt. Pharmacogn Mag 7:109
44. Faergemann J, Baran R (2003) Epidemiology, clinical presentation and diagnosis of onychomycosis. Br J Dermatol 149:1–4
45. Faria NC, Kim JH, Gonçalves LA, Martins MD, Chan KL, Campbell BC (2011) Enhanced activity of antifungal drugs using natural phenolics against yeast strains of Candida and Cryptococcus. Lett Appl Microbiol 52:506–513
46. Fernández-Torres B, Cabanes FJ, Carrillo-Munoz AJ, Esteban A, Inza I, Abarca L, Guarro J (2002) Collaborative evaluation of optimal antifungal susceptibility testing conditions for dermatophytes. J Clin Microbiol 40:3999–4003

47. Feshani AM, Kouhsari SM, Mohammadi S (2011) Vaccinium arctostaphylos, a common herbal medicine in Iran: molecular and biochemical study of its antidiabetic effects on alloxan-diabetic Wistar rats. J Ethnopharmacol 133:67–74
48. Fontenelle RO, Morais SM, Brito EH, Brilhante RS, Cordeiro RA, Lima YC, Brasil NV, Monteiro AJ, Sidrim JJ, Rocha MF (2011) Alkylphenol activity against Candida spp. and Microsporum canis: a focus on the antifungal activity of thymol, eugenol and O-methyl derivatives. Molecules 16(8):6422–6431
49. Foster T (1996) Chapter 12: staphylococcus. In: Baron S (ed) Medical microbiology, 4th edn. University of Texas Medical Branch at Galveston, Galveston
50. Gahlawat DK, Jakhar S, Dahiya P (2014) Murraya koenigii (L.) Spreng: an ethnobotanical, phytochemical and pharmacological review. J Pharm Phytochem 3:109–119
51. Gao L, Han J, Si J, Wang J, Wang H, Sun Y, Bi Y, Liu J, Cao L (2017) Cryptoporic acid E from Cryptoporus volvatus inhibits influenza virus replication in vitro. Antivir Res 143:106–112
52. Gaur R, Thakur JP, Yadav DK, Kapkoti DS, Verma RK, Gupta N, Khan F, Saikia D, Bhakuni RS (2015) Synthesis, antitubercular activity, and molecular modeling studies of analogues of isoliquiritigenin and liquiritigenin, bioactive components from Glycyrrhiza glabra. Med Chem Res 24:3494–3503
53. Gayoso CW, Lima EO, Oliveira VT, Pereira FO, Souza EL, Lima IO, Navarro DF (2005) Sensitivity of fungi isolated from onychomycosis to Eugenia cariophyllata essential oil and eugenol. Fitoterapia 76:247–249
54. Ghosh M, Civra A, Rittà M, Cagno V, Mavuduru SG, Awasthi P, Lembo D, Donalisio M (2016) Ficus religiosa L. bark extracts inhibit infection by herpes simplex virus type 2 in vitro. Arch Virol 161:3509–3514
55. Giovannucci E, Ascherio A, Rimm EB, Stampfer MJ, Colditz GA, Willett WC (1995) Intake of carotenoids and retino in relation to risk of prostate cancer. JNCI J Nat Cancer Inst 87:1767–1776
56. Gottlieb SL, Johnston C (2017) Future prospects for new vaccines against sexually transmitted infections. Curr Opin Infect Dis 30:77
57. Gowrish A, Vagdevi HM, Rajashekar H (2015) In vitro antioxidant and antitubercular activity of Leucas marrubioides Desf. root extracts. J Appl Pharm Sci 5:137–142
58. Greenberg MS, Burket LW, Glick M (2003) Burket's oral medicine: diagnosis & treatment. BC Decker, Hamilton
59. Grover JK, Yadav S, Vats V (2002) Medicinal plants of India with anti-diabetic potential. J Ethnopharmacol 81:81–100
60. Guo N, Liu J, Wu X, Bi X, Meng R, Wang X, Xiang H, Deng X, Yu L (2009) Antifungal activity of thymol against clinical isolates of fluconazole-sensitive and-resistant Candida albicans. J Med Microbiol 58:1074–1079
61. Gupta VK, Shukla C, Bisht GR, Saikia D, Kumar S, Thakur RL (2011) Detection of anti-tuberculosis activity in some folklore plants by radiometric BACTEC assay. Lett Appl Microbiol 52:33–40
62. Gutheil GW, Reed G, Ray A, Anant S, Dhar A (2012) Crocetin: an agent derived from saffron for prevention and therapy for cancer. Curr Pharm Biotechnol 13:173–179
63. Hammer KA, Carson CF, Riley TV (2002) In vitro activity of Melaleuca alternifolia (tea tree) oil against dermatophytes and other filamentous fungi. J Antimicrob Chemother 50:195–199
64. Hammer KA, Carson CF, Riley TV (2012) Effects of Melaleuca alternifolia (tea tree) essential oil and the major monoterpene component terpinen-4-ol on the development of single-and multistep antibiotic resistance and antimicrobial susceptibility. Antimicrob Agents Chemother 56:909–915
65. Han X, Zhang DK, Guo YM, Feng WW, Dong Q, Zhang CE, Zhou YF, Liu Y, Wang JB, Zhao YL, Xiao XH, Yang M (2016) Screening and evaluation of commonly-used anti-influenza Chinese herbal medicines based on anti-neuraminidase activity. Chin J Nat Med 14:794–800
66. Hartmann JT, Lipp HP (2006) Camptothecin and podophyllotoxin derivatives. Drug Saf 29:209–230

67. Houghton P, Patel N, Jurzysta M, Biely Z, Cheung C (2006) Antidermatophyte activity of medicago extracts and contained saponins and their structure-activity relationships. Phytother Res 20:1061–1066
68. Huprikar S, Shoham S (2013) Emerging fungal infections in solid organ transplantation. Am J Transplant 13:262–271
69. Ibrahim MP, Nuhu AA (2016) Phytochemical screening and antibacterial/antifungal activities of Ginkgo biloba extract EGb 761. J Pharm Biol Sci 11(1):43–49
70. Ibrahim NA, Mohammed M, Farid MA, Abdel-Wahed NA (2015) Chemical composition, antimicrobial and antifungal activities of essential oils of the leaves of Aegle marmelos (L.) Correa growing in Egypt. J Appl Pharm Sci 5:1–5
71. James SA, Omwirhiren RE, Joshua IA, Dutse I (2016) Anti-diabetic properties and phytochemical studies of ethanolic leaf extracts of Murraya koenigii and Telfairia occidentalis on alloxan-induced diabetic albino rats. Ornament 49
72. Jankasem M, Wuthi-udomlert M, Gritsanapan W (2013) Antidermatophytic properties of *Ar*-turmerone, turmeric oil, and *Curcuma longa* preparations. ISRN Dermatol 2013:1
73. Jung J, Kim NK, Park S, Shin HJ, Hwang SG, Kim K (2015) Inhibitory effect of Phyllanthus urinaria L. extract on the replication of lamivudine-resistant hepatitis B virus in vitro. BMC Complement Altern Med 15:255
74. Kapewangolo P, Hussein AA, Meyer D (2013) Inhibition of HIV-1 enzymes, antioxidant and anti-inflammatory activities of Plectranthus barbatus. J Ethnopharmacol 149:184–190
75. Kapewangolo P, Knott M, Shithigona RE, Uusiku SL, Kandawa-Schulz M (2016) In vitro anti-HIV and antioxidant activity of Hoodia gordonii (Apocynaceae), a commercial plant product. BMC Complement Altern Med 16:411
76. Kebaara BW, Langford ML, Navarathna DH, Dumitru R, Nickerson KW, Atkin AL (2008) Candida albicans Tup1 is involved in farnesol-mediated inhibition of filamentous-growth induction. Eukaryot Cell 7:980–987
77. Khan BA, Abraham A, Leelamma S (1995) Hypoglycemic action of Murraya koenigii (curry leaf) and Brassica juncea (mustard): mechanism of action. Indian J Biochem Biophys 32:106–108
78. Khare CP (2008) Indian medicinal plants: an illustrated dictionary. Springer, Berlin
79. Kronstad JW, Attarian R, Cadieux B, Choi J, D'souza CA, Griffiths EJ, Geddes JM, Hu G, Jung WH, Kretschmer M, Saikia S (2011) Expanding fungal pathogenesis: cryptococcus breaks out of the opportunistic box. Nat Rev Microbiol 9:193–203
80. Kurapati KR, Atluri VS, Samikkannu T, Garcia G, Nair MP (2015) Natural products as anti-HIV agents and role in HIV-associated neurocognitive disorders (HAND): a brief overview. Front Microbiol 6:1444
81. Lavoie S, Côté I, Pichette A, Gauthier C, Ouellet M, Nagau-Lavoie F, Mshvildadze V, Legault J (2017) Chemical composition and anti-herpes simplex virus type 1 (HSV-1) activity of extracts from Cornus canadensis. BMC Complement Altern Med 17:123
82. Lee WJ, Moon JS, Kim SI, Bahn YS, Lee H, Kang TH, Shin HM, Kim SU (2015) A phenylpropanoid glycoside as a calcineurin inhibitor isolated from Magnolia obovata Thunb. J Microbiol Biotechnol 25:1429–1432
83. Liang J, Chen J, Tan Z, Peng J, Zheng X, Nishiura K, Ng J, Wang Z, Wang D, Chen Z, Liu L (2013) Extracts of the medicinal herb Sanguisorba officinalis inhibit the entry of human immunodeficiency virus-1. J Food Drug Anal 21:S52–S58
84. Lin RD, Chin YP, Lee MH (2005) Antimicrobial activity of antibiotics in combination with natural flavonoids against clinical extended-spectrum β-lactamase (ESBL)-producing *Klebsiella pneumoniae*. Phytother Res 19:612–617
85. Loizzo MR, Statti GA, Tundis R, Conforti F, Bonesi M, Autelitano G, Houghton PJ, Miljkovic-Brake A, Menichini F (2004) Antibacterial and antifungal activity of Senecio inaequidens DC. and Senecio vulgaris L. Phytother Res 18:777–779
86. López-Lázaro M, Willmore E, Austin CA (2007) Cells lacking DNA topoisomerase IIβ are resistant to genistein. J Nat Prod 70:763–767

87. Lunardi LW, Aquino VR, Zimerman RA et al (2006) Epidemiology and outcome of in a tertiary care hospital. Clin Infect Dis 43:e60–63
88. Machumi F, Samoylenko V, Yenesew A, Derese S, Midiwo JO, Wiggers FT, Jacob MR, Tekwani BL, Khan SI, Walker LA, Muhammad I (2010) Antimicrobial and antiparasitic abietane diterpenoids from the roots of Clerodendrum eriophyllum. Nat Prod Commun 5:853
89. Mahesh B, Satish S (2008) Antimicrobial activity of some important medicinal plant against plant and human pathogens. World J Agric Sci 4:839–843
90. Mahmood A, Mahmood A, Qureshi RA (2012) Antimicrobial activities of three species of family mimosaceae. Pak J Pharm Sci 25:203–206
91. Manayi A, Saeidnia S, Faramarzi MA, Samadi N, Jafari S, Vazirian M, Ghaderi A, Mirnezami T, Hadjiakhoondi A, Ardekani MR, Khanavi M (2013) A comparative study of anti-Candida activity and phenolic contents of the calluses from Lythrum salicaria L. in different treatments. Appl Biochem Biotechnol 170:176–184
92. Markovits J, Linassier C, Fossé P, Couprie J, Pierre J, Jacquemin-Sablon A, Saucier JM, Le Pecq JB, Larsen AK (1989) Inhibitory effects of the tyrosine kinase inhibitor genistein on mammalian DNA topoisomerase II. Cancer Res 49:5111–5117
93. Maroyi A (2014) Alternative medicines for HIV/AIDS in resource-poor settings: insight from traditional medicines use in Sub-Saharan Africa. Trop J Pharm Res 13:1527–1536
94. Mekha Mohan PJ, Valsalan R, Nazeem PA (2015) Molecular docking studies of phytochemicals from Phyllanthus niruri against Hepatitis B DNA Polymerase. Bioinformation 11:426
95. Miao M, Cheng B, Guo L, Shi J (2015) Effects of Fuzheng Paidu tablet on peripheral blood T lymphocytes, intestinal mucosa T lymphocytes, and immune organs in cyclophosphamide-induced immunosuppressed mice. Hum Vaccin Immunother 11:2659–2663
96. Miceli MH, Díaz JA, Lee SA (2011) Emerging opportunistic yeast infections. Lancet Infect Dis 11:142–151
97. Mishima S, Saito K, Maruyama H, Inoue M, Yamashita T, Ishida T, Gu Y (2004) Antioxidant and immuno-enhancing effects of Echinacea purpurea. Biol Pharm Bull 27:1004–1009
98. Modi M, Dezzutti CS, Kulshreshtha S, Rawat AK, Srivastava SK, Malhotra S, Verma A, Ranga U, Gupta SK (2013) Extracts from Acacia catechu suppress HIV-1 replication by inhibiting the activities of the viral protease and Tat. Virol J 10:309
99. Moltó J, Valle M, Miranda C, Cedeño S, Negredo E, Clotet B (2012) Herb-drug interaction between Echinacea purpurea and etravirine in HIV-infected patients. Antimicrob Agents Chemother 56:5328–5331
100. Monera-Penduka TG, Maponga CC, Morse GD, Nhachi CF (2017) Capacity for ethical and regulatory review of herbal trials in developing countries: a case study of Moringa oleifera research in HIV-infected patients. J Pharm Policy Pract 10:9
101. Moon DO, Kim MO, Lee JD, Choi YH, Kim GY (2010) Rosmarinic acid sensitizes cell death through suppression of TNF-α-induced NF-κB activation and ROS generation in human leukemia U937 cells. Cancer Lett 288:183–191
102. Moudi M, Go R, Yien CY, Nazre M (2013) Vinca alkaloids. Int J Pre Med 4:1231
103. Munoz P, Bouza E, Cuenca-Estrella M et al (2005) Saccharomyces cerevisiaefungemia: an emerging infectious disease. Clin Infect Dis 40:1625–34
104. Muthu C, Ayyanar M, Raja N, Ignacimuthu S (2006) Medicinal plants used by traditional healers in Kancheepuram District of Tamil Nadu. India J Ethnobiol Ethnomed 2:43
105. Nakamura CV, Ishida K, Faccin LC, Dias Filho BP, Cortez DA, Rozental S, de Souza W, Ueda-Nakamura T (2004) In vitro activity of essential oil from Ocimum gratissimum L. against four Candida species. Res Microbiol 155:579–586
106. Nazari H, Mohammadi A, Amrollahi H, Dehpour A (2012) Essential oil analysis and antibacterial activities of some medicinal plants. Int J Phytomed 4:212
107. Negri M, Salci TP, Shinobu-Mesquita CS, Capoci IR, Svidzinski TI, Kioshima ES (2014) Early state research on antifungal natural products. Molecules 19:2925–2956
108. Nosanchuk JD, Casadevall A (2003) Budding of melanized Cryptococcus neoformans in the presence or absence of L-dopa. Microbiology 149(7):1945–1951

109. Noumi E, Manga PN (2011) Traditional medicines for HIV/AIDS and opportunistic infections in North-West Cameroon: case of skin infections. Am J Trop Med Hyg 1:44–64
110. Nutan, Modi M, Dezzutti CS, Kulshreshtha S, Rawat AK, Srivastava SK, Malhotra S, Verma A, Ranga U, Gupta SK (2013) Extracts from Acacia catechu suppress HIV-1 replication by inhibiting the activities of the viral protease and Tat. Virology J 10:309
111. Olajuyigbe OO, Afolayan AJ (2012) In vitro antibacterial and time-kill assessment of crude methanolic stem bark extract of Acacia mearnsii De Wild against bacteria in shigellosis. Molecules 17:2103–2118
112. Oyagbemi AA, Saba AB, Azeez OI (2010) Molecular targets of [6]-gingerol: its potential roles in cancer chemoprevention. Biofactors 36:169–178
113. Panda S, Dubey D, Dutta S (2010) Anticandidal activity of Diospyros melanoxylon Roxb. Bark from Similipal Biosphere Reserve, Orissa, India. Int J Green Pharm 4:102
114. Panthong P, Bunluepuech K, Boonnak N, Chaniad P, Pianwanit S, Wattanapiromsakul C, Tewtrakul S (2015) Anti-HIV-1 integrase activity and molecular docking of compounds from Albizia procera bark. Pharm Biol 53:1861–1866
115. Parmar J, Sharma P, Verma P, Goyal PK (2010) Chemopreventive action of Syzygium cumini on DMBA-induced skin papillomagenesis in mice. Asian Pac J Cancer Prev 11:261–265
116. Patel DK, Kumar R, Prasad SK, Sairam K, Hemalatha S (2011) Antidiabetic and in vitro antioxidant potential of Hybanthus enneaspermus (Linn) F. Muell in streptozotocin–induced diabetic rats. Asian Pac J Trop Biomed 1:316–322
117. Patil R, Patil R, Ahirwar B, Ahirwar D (2011) Isolation and characterization of anti-diabetic component (bioactivity—guided fractionation) from Ocimum sanctum L.(Lamiaceae) aerial part. Asian Pac J Trop Med 4:278–282
118. Patwardhan B, Gautam M (2005) Botanical immunodrugs: scope and opportunities. Drug Discov Today 10:495–502
119. Pepeljnjak S, Kosalec I, Kalodera Z, Blazevic NI (2005) Antimicrobial activity of juniper berry essential oil (Juniperus communis L., Cupressaceae). Acta Pharm Zagreb 55:417
120. Pereira CB, de Oliveira DM, Hughes AF, Kohlhoff M, Vieira ML, Vaz AB, Ferreira MC, Carvalho CR, Rosa LH, Rosa CA, Alves TM (2015) Endophytic fungal compounds active against Cryptococcus neoformans and C. Gattii. J Antibiot 68:436–444
121. Perera DF, Fernando KM, Wijendra WA (2015) Efficacy of phytochemicals present in leaves of Punica granatum against Malassezia species. Am J Pharmacol Pharm 2:62–71
122. Petretto GL, Fancello F, Zara S, Foddai M, Mangia NP, Sanna ML, Omer EA, Menghini L, Chessa M, Pintore G (2014) Antimicrobial activity against beneficial microorganisms and chemical composition of essential oil of Mentha suaveolens ssp. insularis grown in Sardinia. J Food Sci 79:369–377
123. Pinto E, Vale-Silva L, Cavaleiro C, Salgueiro L (2009) Antifungal activity of the clove essential oil from Syzygium aromaticum on Candida, Aspergillus and dermatophyte species. J Med Microbiol 58:1454–1462
124. Polaquini SR, Svidzinski TI, Kemmelmeier C, Gasparetto A (2006) Effect of aqueous extract from neem (Azadirachta indica A. Juss) on hydrophobicity, biofilm formation and adhesion in composite resin by Candida albicans. Arch Oral Biol 51:482–490
125. Pottier I, Gente S, Vernoux JP, Guéguen M (2008) Safety assessment of dairy microorganisms: Geotrichum candidum. Int J Food Microbiol 126:327–332
126. Rai MK, Varma A, Pandey AK (2004) Antifungal potential of Spilanthes calva after inoculation of Piriformospora indica. Mycoses 47:479–481
127. Rajiniraja MU, Jayaraman GU (2014) Bioautography guided screening of selected Indian medicinal plants reveals potent Antimycobacterial activity of Allium sativum extracts-implication of non-sulfur compounds in inhibition. Int J Pharm Pharm Sci 6:671–676
128. Ramachandran A, Snehalatha C, Shetty AS, Nanditha A (2012) Trends in prevalence of diabetes in Asian countries. World J Diabetes 3:110
129. Rukayadi Y, Hwang JK (2007) *In vitro* anti-Malassezia activity of xanthorrhizol isolated from Curcuma xanthorrhiza Roxb. Lett Appl Microbiol 44:126–130

130. Sa G, Das T (2008) Anti cancer effects of curcumin: cycle of life and death. Cell Div 3:14
131. Sabiiti W, May RC (2012) Mechanisms of infection by the human fungal pathogen Cryptococcus neoformans. Future Microbiol 7:1297–1313
132. Saini S, Porte SM (2015) Antimicrobial properties of Anand Bhairav Ras and its ingredients: an overview. Int J Res Ayurveda and Pharm 6:80–85
133. Santos NO, Mariane B, Lago JH, Sartorelli P, Rosa W, Soares MG, da Silva AM, Lorenzi H, Vallim MA, Pascon RC (2015) Assessing the chemical composition and antimicrobial activity of essential oils from Brazilian plants—Eremanthus erythropappus (Asteraceae), Plectrantuns barbatus, and P. amboinicus (Lamiaceae). Molecules 20:8440–8452
134. Schiff PB, Fant J, Horwitz SB (1979) Promotion of microtubule assembly in vitro by taxol. Nature 277:665–667
135. Scully C, Almeida OP (1992) Orofacial manifestations of the systemic mycoses. J Oral Pathol Med 21:289–294
136. Shaik G, Sujatha N, Mehar SK (2014) Medicinal plants as source of antibacterial agents to counter *Klebsiella pneumoniae*. J Appl Pharm Sci 4:135–147
137. Shoshana LO, Amira RU, Kashman Y, Amnon HI (1999) Polycitone A, a novel and potent general inhibitor of retroviral reverse transcriptases and cellular DNA polymerases. Biochem J 344:85–92
138. Silprasit K, Seetaha S, Pongsanarakul P, Hannongbua S, Choowongkomon K (2011) Anti-HIV-1 reverse transcriptase activities of hexane extracts from some Asian medicinal plants. J Med Plant Res 5:4899–4906
139. Silva Júnior AJ, de Campos-Buzzi F, Romanos MT, Wagner TM, Guimarães AF, Cechinel Filho V, Batista R (2013) Chemical composition and antinociceptive, anti-inflammatory and antiviral activities of Gallesia gorazema (Phytolaccaceae), a potential candidate for novel anti-herpetic phytomedicines. J Ethnopharmacol 150:595–600
140. Simonetti G, Tocci N, Valletta A, Brasili E, D'Auria FD, Idoux A, Pasqua G (2016) In vitro antifungal activity of extracts obtained from Hypericum perforatum adventitious roots cultured in a mist bioreactor against planktonic cells and biofilm of Malassezia furfur. Nat Prod Res 30:544–550
141. Singh V, Thakur KE, Chauhan PK (2012) Effect of poly herbal formulation against klebsiella pneumonia causing pneumonia in children's. Asian J Pharm Clin Res 5:69–75
142. Soković MD, Glamočlija J, Marin PD, Brkić DD, Vukojević J, Jovanović D, Bulajić N, Kataranovski D (2006) Antifungal activity of the essential oil of Mentha. X piperita. Pharm Biol 44:511–515
143. Sokovic MD, Glamoclija JM, Ciric AD (2013) Natural products from plants and fungi as fungicides. In: Fungicides-showcases of integrated plant disease management from around the world, Chap 9. InTech, New York, pp 185–232
144. Stoldt VR, Sonneborn A, Leuker CE, Ernst JF (1997) Efg1p, an essential regulator of morphogenesis of the human pathogen Candida albicans, is a member of a conserved class of bHLH proteins regulating morphogenetic processes in fungi. EMBO J 16:1982–1991
145. Sultana N, Ata A (2008) Oleanolic acid and related derivatives as medicinally important compounds. J Enzyme Inhib Med Chem 23:739–756
146. Sureram S, Senadeera SP, Hongmanee P, Mahidol C, Ruchirawat S, Kittakoop P (2012) Antimycobacterial activity of bisbenzylisoquinoline alkaloids from Tiliacora triandra against multidrug-resistant isolates of mycobacterium tuberculosis. Bioorg Med Chem Lett 22:2902–2905
147. Swamy MK, Akhtar MS, Sinniah UR (2016) Antimicrobial properties of plant essential oils against human pathogens and their mode of action: an updated review. Evid Based Complement Alternat Med 2016:1
148. Tassiana BD, Ferreira SB, Pinheiro LS, Menezes CP, Guerra FQ, Sousa JP, Lima ED (2015) Antifungal activity of phytochemicals against samples of Penicillium. Int J Trop Dis Health 10:1–9

149. Tepe B, Daferera D, Sokmen A, Sokmen M, Polissiou M (2005) Antimicrobial and antioxidant activities of the essential oil and various extracts of Salvia tomentosa Miller (Lamiaceae). Food Chem 90:333–340
150. Tewtrakul S, Nakamura N, Hattori M, Fujiwara T, Supavita T (2002) Flavanone and flavonol glycosides from the leaves of Thevetia peruviana and their HIV-1 reverse transcriptase and HIV-1 integrase inhibitory activities. Chem Pharm Bull 50:630–635
151. Tietjen I, Ntie-Kang F, Mwimanzi P, Onguéné PA, Scull MA, Idowu TO, Ogundaini AO, Meva'a LM, Abegaz BM, Rice CM, Andrae-Marobela K (2015) Screening of the pan-African natural product library identifies ixoratannin A-2 and boldine as novel HIV-1 inhibitors. PLoS One 10:e0121099
152. Tietjen I, Gatonye T, Ngwenya BN, Namushe A, Simonambanga S, Muzila M, Mwimanzi P, Xiao J, Fedida D, Brumme ZL, Brockman MA (2016) Croton megalobotrys Müll Arg. and Vitex doniana (Sweet): traditional medicinal plants in a three-step treatment regimen that inhibit in vitro replication of HIV-1. J Ethnopharmacol 191:331–340
153. Torres-Romero D, Jiménez IA, Rojas R, Gilman RH, López M, Bazzocchi IL (2011) Dihydro-β-agarofuran sesquiterpenes isolated from Celastrus vulcanicola as potential anti-mycobacterium tuberculosis multidrug-resistant agents. Bioorg Med Chem 19:2182–2189
154. Tuon FF, Costa SF (2008) Rhodotorula infection. A systematic review of 128 cases from literature. Rev Iberoam Micol 2:135–140
155. Turchetti B, Pinelli P, Buzzini P, Romani A, Heimler D, Franconi F, Martini A (2005) In vitro antimycotic activity of some plant extracts towards yeast and yeast-like strains. Phytother Res 19:44–49
156. Unlu M, Ergene E, Unlu GV, Zeytinoglu HS, Vural N (2010) Composition, antimicrobial activity and in vitro cytotoxicity of essential oil from Cinnamomum zeylanicum Blume (Lauraceae). Food Chem Toxicol 48:3274–3280
157. Uttra KM, Devrajani BR, Shah SZ, Devrajani T, Das T, Raza S, Naseem M (2011) Lipid profile of patients with diabetes mellitus (a multidisciplinary study). World Appl Sci J 12:1382–1384
158. Vandeputte P, Ferrari S, Coste AT (2011) Antifungal resistance and new strategies to control fungal infections. Int J Microbiol 2012:1–26
159. Vijay P (2015) Anti-diabetic effects of Eclipta alba on alloxan-induced diabetic mice. Int J Pharm Sci Res 6:308
160. Viswanathan V, Phadatare AG, Mukne A (2014) Antimycobacterial and antibacterial activity of Allium sativum bulbs. Indian J Pharm Sci 76:256
161. Wan Z, Chen X (2014) Triptolide inhibits human immunodeficiency virus type 1 replication by promoting proteasomal degradation of Tat protein. Retrovirology 11:88
162. Wanas AS, Radwan MM, Mehmedic Z, Jacob M, Khan IA, Elsohly MA (2016) Antifungal activity of the volatiles of high potency Cannabis sativa L. against Cryptococcus neoformans. Rec Nat Prod 10:214
163. Wang J (2009) Clinical trials on the effect of immunity 1 (Fuzheng 1) on immune reconstitution of HIV patients. Clinicaltrials.gov
164. Wang Y, Xiang L, Wang C, Tang C, He X (2013) Antidiabetic and antioxidant effects and phytochemicals of mulberry fruit (Morus alba L.) polyphenol enhanced extract. PLoS One 8:e71144
165. Wani MC, Taylor HL, Wall ME, Coggon P, McPhail AT (1971) Plant antitumor agents. VI. Isolation and structure of taxol, a novel antileukemic and antitumor agent from Taxus brevifolia. J Am Chem Soc 93:2325–2327
166. Waza SA, Anthony P, Dar S (2015) Phytochemical analysis, antioxidant and antimicrobial activities of methanolic extract of Datura Stramonium seeds. Int J Pharm Sci Res 6:3021–3026
167. Wei B, Cha SY, Kang M, Kim YJ, Cho CW, Rhee YK, Hong HD, Jang HK (2015 Jun 30) Antiviral activity of Chongkukjang extracts against influenza A virus in vitro and in vivo. J Ethnic Food 2(2):47–51

168. Wernik R, Priore JL, Goldman WF, del Carmen Elias A, Borkow G (2015) Improvement in human immunodeficiency virus-1/acquired immune deficiency syndrome patients' well-being following administration of "Phyto V7". World 2:004
169. Wu H, Liang X, Fang Y, Qin X, Zhang Y, Liu J (2008) Resveratrol inhibits hypoxia-induced metastasis potential enhancement by restricting hypoxia-induced factor-1α expression in colon carcinoma cells. Biomed Pharmacother 62:613–621
170. Xu Y, Xu G, Liu L, Xu D, Liu J (2010) Anti-invasion effect of rosmarinic acid via the extracellular signal-regulated kinase and oxidation–reduction pathway in Ls174-T cells. J Cell Biochem 111:370–379
171. Yang J (2009) Brazil nuts and associated health benefits: a review. LWT-Food Sci Technol 42:1573–1580
172. Zhang L, Chang W, Sun B, Groh M, Speicher A, Lou H (2011) Bisbibenzyls, a new type of antifungal agent, inhibit morphogenesis switch and biofilm formation through upregulation of DPP3 in Candida albicans. PLoS One 6:e28953
173. Zhang HJ, Rumschlag-Booms E, Guan YF, Liu KL, Wang DY, Li WF, Cuong NM, Soejarto DD, Fong HH, Rong L (2017) Anti-HIV diphyllin glycosides from *Justicia gendarussa*. Phytochemistry 136:94–100

Herbal Drugs: To Take It or Not to Take It

28

Gandreddi V. D. Sirisha and Kappala Vijaya Rachel

Abstract

Herbal drugs are the outcome of indigenous medical practices, therapeutic experiences and civilizations accumulated over the years. They are of great demand both in developed and developing countries due to their accessibility, traditional knowledge, safety, efficacy and lesser side effects. They also address age-related disorders like osteoporosis, immune disorders, memory loss, etc. for which there is no modern medicine. Herbal drugs have a noticeable share in the economy of most of the countries. Systematic evaluation, GACP (good agricultural and collection practices), GMP (good manufacturing practices) and quality assurance by regulatory authorities ensure better usage of herbal drugs. In 1991, WHO developed guidelines for assessment of herbal medicine.

Keywords

Herbal drugs · Traditional knowledge · Safety · Evaluation · Quality assurance

28.1 Introduction

Herbal drugs are outcome of therapeutic experiences of practicing physicians over the generations from indigenous systems of medicine. The director general of WHO in his progress report in 1991 defined 'traditional medicine including herbal drugs as comprising therapeutic practices that have been in existence, often for hundreds of years, before the development and spread of modern medicine and are still in use today' [10]. Three-fourths of the world population mostly in the developing countries principally depends on herbal drugs based on the experience and legacy from

G. V. D. Sirisha · K. Vijaya Rachel (✉)
Department of Biochemistry and Bioinformatics, Institute of Science, GITAM University, Visakhapatnam, Andhra Pradesh, India

V. Rani, U. C. S. Yadav (eds.), *Functional Food and Human Health*,
https://doi.org/10.1007/978-981-13-1123-9_28

older generations. They also offer cultural acceptability, lesser side effects and better compatibility. According to WHO report, in India 65% of rural area people use Ayurveda and medicinal plants for the treatment of numerous ailments. Worldwide, the use of folklore medicine has shown a renewed attention even in the developed countries. The percentage of people using these medicines at least once in their life time is 70% in Canada, 49% in France, 48% in Australia, 42% in the United States and 31% in Belgium (Fig. 28.1) [13]. The main source of traditional medicine is allelochemicals, and other secondary metabolites are obtained chiefly from plants.

India is in possession of a huge wealth of authenticated and well-practiced knowledge of traditional herbal drugs. China is far head of India in capitalizing herbal wealth. Strategic and judicious product identification of herbal medicine for different diseases mostly for which there is no treatment or only palliative therapy is available makes India compete in the current market. The herbal drug market of India is about $1billion, and export plant-based crude drugs are around $100 million. Its perpetuation into the developed countries demands scientifically generated data, single-plant medicines, medicinal plants free from pesticides, heavy metals, lesser side effects, i.e. safety, and enhanced shelf life [1, 4].

28.2 Herbal Drugs

Herbal drugs constitute traditional medicines which primarily used medicinal plant preparation for therapy. The herbal drugs are an outcome of rich traditions of ancient societies and scientific legacy. The earliest recorded evidence of their use in Indian, Chinese, Egyptian, Greek, Roman and Syrian texts dates back to about 5000 years [4].

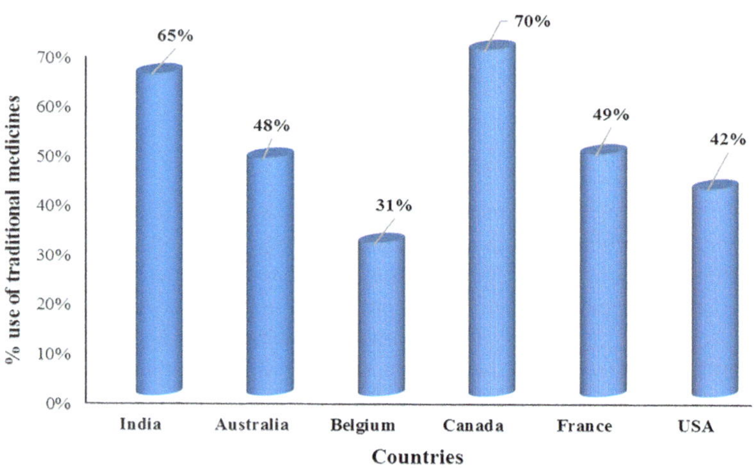

Fig. 28.1 Percentage of use of traditional medicine. Bar graph depicting the use of traditional medicine in different countries. Canada stands first in using traditional medicine, and India stands second position in usage

Herbal medicine is also called botanical medicine or phytomedicine. The Merriam Webster dictionary defines herbal drugs as the art or practice of using herbs and herbal preparations to maintain health and to prevent, alleviate or cure disease. Herbalism has a long tradition of use outside conventional medicine. It is becoming more mainstream as improvements in analysis and quality control, along with advances in clinical research, show the value of herbal medicine in treating and preventing disease. Herbal products with therapeutic indications and/or dosage recommendations are considered to be drugs.

Herbal drugs can be in the form of herbs, herbal materials, herbal preparations and finished herbal products. Herbs include crude plant material such as leaves, flowers, fruit, seed, stems, wood, bark, roots, rhizomes or other plant parts, which can be taken wholly fragmented or in powdered. Herbal materials include fresh juices, gums, fixed oils, essential oils, resins and dry powders of herbs. They are usually processed products obtained after steaming, roasting or stir baking with honey, alcoholic beverages or other materials. Herbal preparations are produced by extraction, fractionation, purification, concentration, or other physical or biological processes. Finished herbal products consist of herbal preparations made from one or more herbs. They also include mixture herbal products [12].

28.3 Herbal Medicine Scenario in India

Indian subcontinent is endowed with 12 mega biodiversity centres with 45,000 different plant species. These include 15,000–18,000 flowering plants, 23,000 fungi, 7244 algae, 1600 lichens, 2504 bryophytes, 1267 pteridophytes, 74 gymnosperms and 17,926 angiosperms. The peninsular geographic topography has ocean on three sides harbouring exclusive varieties of fauna and flora with rich healing properties. According to 2013 statistics, 5662 plants are identified as medicinal plants [7]. About 1500 plants with medicinal uses are mentioned in ancient texts, and around 800 plants have been used in traditional medicine (Government of Telangana [2]) (Table 28.1).

The major pharma companies manufacturing herbal formulations in India are described in Table 28.2. About 7000 firms are in the small-scale sector manufacturing traditional medicines with or without standardization. However, none of the pharma has standardized the bioactivity of herbal medicines in experimental animal models (Kirti [3]).

28.4 Scenario of Herbal Drugs Usage

Plants and their secondary metabolite constituents have a long history of use in modern medicine and in certain systems of traditional medicine and are the sources of important drugs such as atropine, codeine, digoxin, morphine, quinine and vincristine. Main sources of herbal drugs are medicinal plants and plant products, such as fruits, leaves, seeds, etc. [6]. Traditional systems of medication played an

Table 28.1 Ethnomedicinal use of plants as drugs

S. No	Plant family	Plant name	Useful parts	Chemical constitutes	Uses
1	Fabaceae	*Abrus precatorius*	Roots, leaves and seeds	Seeds – abrin, indole alkaloids, anthocyanins	Seeds: as contraceptive and used to treat conjunctivitis
				Roots and leaves –glycyrrhizin, traces of abrin	Leaves and roots: used to treat asthma, bronchitis and fever
					Roots: mixed with cow milk made into a paste are used against diarrhoea and dysentery.
		Clitoria ternatea	Roots, leaves and seeds	Seeds and root bark – tannins, starch	Roots: used to treat eye diseases especially leprosy, tubercular glands, retarded brain development, migraine, leucoderma, ulcers, pulmonary tuberculosis
				Seeds – oligosaccharides	Leaves: used to treat ear and liver diseases and eruptions
		Indigofera tinctoria	Whole plant	Nitrogen, phosphoric acid, potash and lime	Roots, stem and leaves: promote hair growth, used to treat stomach problems, spleen troubles, headache, chronic bronchitis, asthma, ulcers, heart and skin diseases
					Plant extract: used to treat epilepsy and nervous problems
					Root: good for dental problems
		Mucuna pruriens	Seeds and roots	Seeds – proteins, dihydroxyphenylalanine (DOPA), glutathione, calcium, phosphorus, sulphur, manganese, lecithin and gallic acid	Seeds: aphrodisiac, nervine tonic, for scorpion sting
					Pods: antihelminthic
					Roots: purgative, against delirium in fever, cholera

		Psoralea corylifolia	Seeds	Seeds – coumarins (psoralen and isopsoralen)	Seeds: used as antimicrobial, antifeedant and anti-insecticidal
					Used to treat leucoderma, leprosy, psoriasis, amoebiasis, healing of wounds and ulcers
		Trigonella foenum-graecum	Seeds and leaves	Volatile oils, alkaloids, flavonoids, mucilage, protein, fixed oil, vitamin A, B_1, C and minerals	Seeds: carminative, tonic, aphrodisiac and used in dyspepsia with loss of appetite
					Plant and seeds: used as diuretic, emmenagogue, dropsy, cough, diabetes and in enlargement of spleen and liver
					Leaves: leaves applied externally for swellings, burn and used as anti-hair fall
2	Mimosaceae	*Acacia sinuate*	Bark of branches and pods	Pods – saponins (acacinin, oligosaccharides)	Bark: used to treat asthma, leprosy, leucoderma, bone fractures, excess bleeding
				Bark – saponins	Pods: used for skin disorders and also as a hair tonic
		Mimosa pudica	Roots and leaves	Alkaloid mimosine	Root decoction: used in urinary complaints
					Leaves paste: reduces glandular swellings
					Leaf juice: used in dressings for sinus, hydrocel
					Roots and leaves: used in piles and fistula

(continued)

Table 28.1 (continued)

S. No	Plant family	Plant name	Useful parts	Chemical constitutes	Uses
3	Euphorbiaceae	*Acalypha indica*	Whole plant	Kaempferol, β-sitosterol andtriacetonamine	Used to treat tooth- and earaches, bronchitis, pneumonia and asthma
				Leaves and twigs – acalyphine, acalyphamide, amides, quinones, sterols, cyanogenic glycosides	Leaves are useful for scabies and snake bites
					Leaf extracts: used to treat rheumatoid arthritis and skin infections
					Leaf paste: used for burns
		Cicca acida	Fruits, seeds and root bark	Fruits – vitamin C, moisture, protein, fibre, carbohydrate, ash	Fruits: useful for bronchitis, intestinal diseases and jaundice
				Root bark – tannins, saponins and gallic acid	Roots and seeds: used to treat cathartic
		Emblica officinalis	Fruits, leaves, seeds, bark and flowers	Vitamin C, amino acids, fibre, gallic acid, tannic acid and calcium	Fruits: bitter, sweetish, coolant, laxative, diuretic, carminative and used to treat respiratory problems, asthma, jaundice, cough and leprosy
					Dried fruits: used in the preparation of hair oils and chyavanprash
		Euphorbia hirta	Roots and leaves	Flavonoids, terpenoids, alkanes, phenolic acids, shikimic acid and choline	Roots and leaves: used to treat worms in children, bowl complaints, cough and dysentery, intestinal amoebiasis and piles
					Specifically used for bronchial asthma
		Euphorbia tirucalli	Milky latex	Terpenes and sterols alcohol eufol, alfaeuforbol and taraxa sterol e tirucallol	Used to control cough, rheumatism, cold and nervous diseases and also useful for warts, tooth and ear pains and laxative

		Jatropha curcas	Dried seeds and leaves	Dried seeds and leaves yield oils – myristic, palmitic, stearic, oleic and linoleic acid	Leaves: galactagogue, rubefacient, suppurative, have insecticidal properties and used to treat foul ulcers and tumours
					Plant juice: used to treat eczema and scabies
					Oils: purgative, biodiesel is manufactured
		Jatropha gossypifolia	Leaves, seeds and roots	Coumarin, lignoids, terpenoids, flavonoids, phenols, alkaloids, lignins, tannins and cardiac glycosides	Roots: used to treat leprosy and urinary troubles and used as antidote for snake bites
					Leaves and its juices: useful for earaches, eczema and itches and ulcers; seed oils: treat leprosy
		Pedilanthus tithymaloides	Roots and stem latex	Latex – euphorbin, serine, resin and fixed oil	Root: powerful emetic
					Latex: emetic, irritant/caustic properties used in venereal troubles and leucoderma
		Phyllanthus amarus	Roots, fruits and leaves	Phyllanthine, hypophyllanthin and flavonoids	Root extract: controls jaundice and skin diseases
					Leaf extract: controls gonorrhoea, fever and constipation
					Fruit: controls ulcers and wounds

(continued)

Table 28.1 (continued)

S. No	Plant family	Plant name	Useful parts	Chemical constitutes	Uses
4	Amaranthaceae	*Achyranthes aspera*	Leaves, seeds and roots	Triterpenoid, saponins and potassium	Seeds: used as antidote for poisonous insects, snakes and scorpion bites, diuretic
					Flower stalk: used as purgative, pungent digestive, remedy for piles, itch, abdominal enlargements and enlarged cervical glands
					Root powder: cures night blindness
					Leaf juice: mixed with a paste of dry ginger to cure eye injuries of cattle
		Aerva lanata	Whole plant	Amyrin, sitosterol, chrysin	Used to treat piles, kidney stones, intestinal diseases, wounds, headache, scorpion bite, boils, cough, diabetes
5	Araceae	*Acorus calamus*	Rhizomes	Rhizomes yields aromatic volatile oils – asarylaldehydes (a-asarone, b-asarone), acorin, eugenol	Asarone oil is reported to cure gastritis and promote digestion, administered to animals as stomachic
					Dried rhizome: vermifuge and antiseptic, dyspepsia, chronic diarrhoea
6	Acanthaceae	*Adhatoda vasica*	Leaves, flowers, roots and bark	Vasicine, adhatodic acid, aromatic oils	Leaves, roots and bark: chest infections attended with cough and hectic fever, asthma
					Flowers: useful in hectic fever, heat of blood, gonorrhoea
					Oils: possess antispasmodic properties

		Andrographis paniculata	Whole plant	Glucosides, andrographolide, andrographin	Plant: considered as a febrifuge and bitter tonic, infusion of the plant used against malaria
					Fresh juice of leaves is mixed with spices such as cardamom, cloves, dried cinnamon and made into pills for use in household remedy for the minor digestive ailments of children
					Dried stems and roots: used for dyspepsia and influenza
		Barleria prionitis	Whole plant	Phenylethanoid glycoside, iridoid glycosides, alkaloids and potassium	Useful for ulcers, skin diseases, leucoderma, pains, itching, inflammations, bronchitis, cough and for teeth problems
		B.cristata			
		B.strigosa			
7	Rutaceae	*Aegle marmelos*	Root bark, fruits, leaves and roots	Phenols, auroptin, umbelliferon, marmin, lupen, psoralin, xanthotoxin, scopaltin	Unripe fruits: used as a remedy for diarrhoea and dysentery
				Fruit pulp – mucilage, pectin, tannin, volatile oil, marmelosin	Ripe fruits: tonic, astringent, restorative, laxative, good for heart and brain
				Leaves – aegelin, aegelinin	Root bark: used in intermittent fevers and as fish poisons
				Bark – coumarin and marmin	

(continued)

Table 28.1 (continued)

S. No	Plant family	Plant name	Useful parts	Chemical constitutes	Uses
		Feronia elephantum	Fruits and leaves	Fruits – fruit acids, vitamins and minerals	Fruits: used as astringent, stomachic and stimulant
				Leaves – tannins and volatile oil	Mainly used to stimulate the digestive system, to check diarrhoea and dysentery
					Leaves: aromatic and carminative
					Pulp: applied externally as a remedy for bites of venomous insects and reptiles
					Bark: prescribed for biliousness
		Murraya koenigii	Leaves, roots and bark	Glycoside (koenigin), volatile oil and tannins	Increase digestive secretions and relieve nausea, indigestion and vomiting, diarrhoea, dysentery, burning sensation, pruritus, leprosy, skin diseases, leucoderma, helminthiasis, dyspepsia, colic, flatulence, tonic and stomachache
		Ruta chalepensis	Whole plant	Contains rutin	Leaves: used to treat cough and cold in children
					Leaf juice: reduces earache
8	Agavaceae	*Agave Americana*	Leaf sap and agave sap	Oestrogen like isoflavonoids, alkaloids, coumarins and vitamins A, B_1, B_2, C, D and K	Agave sap: used to treat ulcers, syphilis, tuberculosis, jaundice, liver and digestive ailments
					Used as laxative, antiseptic
		Sansevieria roxburghiana	Leaves and rhizomes	Glycosides and saponins	Used to treat ringworm infections and fungal diseases, control leprosy and rheumatism
					Rhizome: treats cough
					Juice of tender shoots: administered to children to clear their throats from phlegm

9	Liliaceae	*Aloe barbadensis*	Leaves	Barbaloin, aloin, anthraquinones, resins, tannins, polysaccharides	Gel contained in the leaf: effective healer of wounds, intestinal ulcers, burns and scalds, cosmetic purposes, painful inflammations, helminthiasis in children and as a purgative
					Aloe: controls leprosy, piles, mental disorders, jaundice, menstrual problems and joint pains
		Asparagus racemosus	Tubers	Saponins, steroidal glycosides, bitter glycosides, asparagine, flavonoids, starch and mucilage	It is a tonic for diuretic and useful as a galactagogue
					Tubers juice mixed with honey is used for dyspepsia
					Roots: used in preparation of medicated oils for nervous and rheumatic complaints
		Gloriosa superba	Tubers and seeds	Alkaloid (colchicine and colchicoside), flavonoids, glycosides, phenols, saponins, steroids, tannin and terpenoids	Useful for chronic ulcers, piles, carcinoma, diarrhoea and skin disorders
10	Zingiberaceae	*Alpinia calcarata*	Rhizomes	Cineol, methyl cinnamate and kaempferol	Useful in treatment of fever, rheumatism, piles, ulcers and constipation
					Rhizome powder: used to treat respiratory and cardiac problems
					It is also used as antibacterial and a digestive stimulant
		Costus speciosus	Rhizomes	Steroidal saponins 'diosgenin'	Controls indigestion, fever, skin diseases, jaundice and is an antidote for snake bite
					Used in contraceptives

(continued)

Table 28.1 (continued)

S. No	Plant family	Plant name	Useful parts	Chemical constitutes	Uses
		Curcuma amada	Rhizomes	d-a-pinene, ocimene, linalyl acetate, safrole, linalool and myrcene	Acts as an appetizer and aphrodisiac, reduces fever, promotes free urine passage, indigestion, chronic ulcers, skin diseases, bronchitis, gout and constipation
		Curcuma angustifolia	Rhizomes	Starch	Useful in treating peptic ulcers and urinary troubles
					Especially used to treat stomachache in children
		Curcuma aromatic	Rhizomes	Curcumin	Useful in the treatment of leprosy, bronchitis, skin eruptions, leucoderma and sprains
					Improves semen and appetite
		Curcuma longa	Rhizomes	Curcumin, volatile oil (zingiberene, turmerone), oleoresin	Used as stomachic, tonic and blood purifier
					Juice: used as an antiparasitic for many skin infections. Externally applied to ulcers, and the paste is made from the powdered rhizome and lime forms a remedy for inflamed joints
					Decoction is used to relieve the pain of purulent ophthalmia

		Elettaria cardamomum	Seeds and pods	Seed and pod oils – borneol, camphor, pinene, cineol, caryophyllene, carvone, terpineol, terpinene	Powerful aromatic stimulant, carminative, stomachic and diuretic, checks nausea and vomiting and also act as cardiac stimulant
				Seeds – palmitic and oleic fatty acids	Powdered seeds boiled with tea water are used as a medicine for scanty urination, diarrhoea,dysentery, palpitation of the heart, exhaustion due to overwork and depression
		Kaempferia rotunda	Tubers yield oil	Cineol, methyl chavicol	Tubers: aromatic, anti-inflammatory, useful in stomach pain, wounds, ulcers, tumours, gastric disorders and swellings
		Zingiber officinale	Rhizomes	Volatile oil, zingiberene, oleoresin	Rhizome: carminative, stimulant and given in dyspepsia and flatulent colic
11	Apocynaceae	*Alstonia scholaris*	Fruit and stem	Echitamine, glycoside of venoterpine, detamine, echitonine and echitine	Ripe fruit: used in syphilis, insanity, epilepsy and as a tonic
					Stem bark decoction: used in fever, respiratory, leprosy and ulcer problems
		Alstonia venenata	Roots, fruits stem bark and root	Alkaloids, alstovenine, venenatine, 3-dehydroalstovenine, reserpine, venoxidine and kopsinine	Roots: useful in skin diseases, erysipelas, leprosy, cobra bite and other venomous bites, epilepsy, fatigue and ear pain
					Fruits: useful in syphilis, insanity and epilepsy
		Catharanthus alba	Whole plant	Vincristine and vinblastine	Roots: highly medicinal for curing cancer
				Roots – anti-fibrillic and hypotensive principles (ajmaline, reserpine)	Plant: used as a folk medicine for diabetes

(continued)

Table 28.1 (continued)

S. No	Plant family	Plant name	Useful parts	Chemical constitutes	Uses
		Rauvolfia serpentine	Roots	Indole alkaloids (reserpine, rescinnamine, ajmaline and yohimbine)	Used in the treatment of high blood pressure, as sedative and specific for insanity, febrifuge and is an antidote for the bites of poisonous reptiles, scorpion stings and also in dysentery
					It cause uterine contraction and promote the expulsion of foetus
		Rauvolfia tetraphylla	Roots	Indole alkaloids (reserpine and ajmaline)	Roots: anthelmintic, useful for the treatment of blood pressure and skin diseases
					Leaf juice: curing the corneal opacity of the eyes
12	Bromeliaceae	*Ananas comosus*	Fruits and leaves	Fruit – bromelian, protein-splitting enzyme	Sour unripe fruit: improves digestion, increases appetite and relieves dyspepsia and constipation
				Vitamin – A and C, fibre	Ripe fruit as a whole cools and soothes used to reduce excessive gastric acid
					Ripe fruit juice: acts as a digestive tonic and is diuretic
					Leaves: useful in encouraging the onset of menstrual periods and easing painful ones
13	Rubiaceae	*Anthocephalus cadamba*	Bark, leaves and fruits	Alkaloids like cadambine, cadamine, a-sitosterol and cadambagenic acid	Bark: tonic, reduces fever
					Leaves decoction: good for ulcers, wounds and metrorrhagia
					Fruits: gastric irritability and fevers

14	Convolvulaceae	*Argyreia nervosa*	Leaves, young shoots, roots and seed	Eragine, isoeragine and penniclavine	Leaves and young shoots: give relief from indigestion and constipation
					Pills: made from leaves used in diabetes
15	Aristolochiaceae	*Aristolochia indica*	Roots and aerial parts	Aristolochic acids, a volatile oil and tannins	Used to treat wounds, sores, snake bite, prevent infections, potent menstruation and is an inducing herb and an abortifacient
					Decoction: treats ulcers, asthma and bronchitis
16	Asteraceae	*Artemisia vulgaris*	Leaves and roots	Volatile oil, a sesquiterpene (lactone, flavonoids, coumarin derivatives and triterpenes)	Used as digestive tonic, improve appetite, encourage the elimination of worms, increase bile flow, induce onset of menstruation and is also used in asthma, nervous and spasmodic affections and as stomachic, antiseptic
		Eclipta alba	Whole plant	Ecliptine, thiophene derivatives and also nicotine	Oil obtained from the plant is good for blackening and strengthening of hair, stops hair fall
					Useful for curing sours on the head and jaundice
					Seeds: good as aphrodisiac

(continued)

Table 28.1 (continued)

S. No	Plant family	Plant name	Useful parts	Chemical constitutes	Uses
		Spilanthes acmella	Leaves, flowers and roots	Spilanthol	Leaves: relieve toothache and infections of throat and gums
					Flower heads: tincture made from flower heads is reported to be used as a substitute for tincture of pyrethrum to treat inflammation of jaw bones
		Tridax procumbens	Leaf	Steroid, saponin, coumarins, alkaloids, amino acids, diterpenes, phenol and flavonoids	Leaf paste is used to stop bleeding of minor wounds or cuts and cures the wounded portion gradually
		Wedelia chinensis	Whole plant	Lactone, wedelolactone, stigmasterol and tannin	Useful in viral hepatitis, wounds, ulcers, seminal weakness, baldness and greying of hair, eye diseases, anti-inflammatory, diuretic, elephantiasis, ear pain and anaemia
17	Moraceae	*Artocarpus heterophyllus*	Leaves, roots, seeds, bark, latex and fruit	Leaves, roots, seeds, bark, latex and fruits – rich source of carbohydrates, amino acids and b-carotene	Jack fruit are cooked when immature, act as tonic, improve sperm content and are useful in free urine passage
				Fruit latex – leucine, isoleucine, tyrosine and valine	Dried bark: mixed with little black pepper is applied for eczema and scab
				Heart wood – morin, artocarpin	
18	Meliaceae	*Azadirachta indica*	Bark, leaves, flowers and seeds	Bark – nimbin, nimbinin, volatile oil, tannins, margosine	Leaves: steeped for malaria, peptic ulcers, intestinal worms, wounds, boils and eczema
				Seeds – fixed oil	Neem oil: antifungal, antiviral and prevents lice, treats leprosy

		Melia azedarach	Bark, leaves, stem bark and root bark	Bark, leave, stem bark and root bark (alkaloids, margosine, nimbin, nimbidin and tannins)	Root bark: used in treating ascariasis
				Seeds – rich in fatty oil consisting of sulphur	Leaves decoction: trichomonas infections and used against skin diseases, fevers, worms and as antitoxic
19	Scrophulariaceae	*Bacopa monnieri*	Whole plant	Alkaloids, brahmine, alkaloid herpestine, saponins (bacoside A and B and β-sitosterol)	Used to treat neuralgia, inflammations, tumours, ulcers, indigestion, constipation, asthma, bronchitis, leprosy, leucoderma, elephantiasis, sterility, fever and general debility
20	Bixaceae	*Bixa orellana*	Roots, stem bark, seeds and leaves	Seeds – bixin	Bark: used to treat gonorrhoea and fever
					Seed pulp and leaves: used to treat dysentery and jaundice
					Seed: mosquito repellent, colouring agent in food material and as a natural dye in textile industry
21	Nyctaginaceae	*Boerhavia diffusa*	Whole plant	Alkaloids, tricontanol, hentriacontane, β-sitosterol, ursolic acid	Used to treat diuretic, ophthalmia, scabies, cardiac disorders, jaundice, bronchitis and reduces swellings
		Mirabilis jalapa	Leaves and tubers	Alkaloids, flavonoids, phenols, steroids, triterpenes, glycosides, tannins, saponins and lignins	Brussel leaves: employed for poulticing abscess and boils
					Leaf juice: external application to wounds and bruises
22	Arecaceae	*Borassus flabellifer*	Leaves, roots and fruits	Minerals like iron, proteins, sucrose, glucose, ash, thiamine, riboflavin, Vitamin – C	Root: useful in leprosy, helps in delivery
					Fruit: stomachic, aphrodisiac
					Fermented juice: diuretic and also used to treat diabetes and chronic gonorrhoea

(continued)

Table 28.1 (continued)

S. No	Plant family	Plant name	Useful parts	Chemical constitutes	Uses
		Phoenix farinifera	Tubers, gum and fruits	Galactose and galacturonic acid	Tuber juice: laxative
					Gum of stem: useful in the treatment of diarrhoea, urinary and venereal diseases
					Fruits: used for consumption, seminal weakness, gastric problems and general debility
23	Caesalpiniaceae	*Bauhinia purpurea*	Roots, bark and seeds	Bark – tannin	Bark: useful in treatment of diarrhoea, pains and ulcers. It is a cooling agent used for constipation, anthelmintic, skin diseases, leprosy, intestinal worms and also for tumours
				Seeds – oil	Root: carminative
		Caesalpinia crista	Seeds, leaves and roots	Seeds – glycoside 'bonducin', fixed oil, proteins and starch	Fruit: it is acrid, causes heat to the body, cures urinary discharges, leucorrhoea, piles and wounds
					Seeds: antiperiodic, antipyretic and tonic, intermittent fevers
					Leaves: anthelmintic, used against elephantiasis and diseases of spleen
		Caesalpinia pulcherrima	Roots, bark, leaves and seeds	Gallic acid,resins,benzoic acid and tannins	Leaves and bark: promote menstruation, used as purgative and tonic
					Bark: used as an abortifacient
					Flowers: cures chest disorders and reduce fevers
		Cassia absus	Leaves and seeds	Chrysophanol and β-sitosterol	Leaves: useful in tumours, anaemia, asthma and hiccups
					Seeds: useful in ophthalmological and skin diseases, laxative

		Cassia alata	Leaves and stems	Anthraglucosides, chrysophanic acid and rhein	Leaves and stem: have antiseptic and laxative properties and are prescribed for constipation, oedema, hepatitis, ringworm infections and scabies
					Leaf juice: applied externally for skin diseases
		Cassia angustifolia	Leaves and pods	Pods – sennosides A and B	It is used for constipation, loss of appetite, disease of liver and spleen, indigestion, malaria, skin diseases, jaundice and anaemia
				Leaves – flavanols, kaempferol, rhein, emodin and sennosides A, B, C and D	
		Cassia fistula	Whole plant	Sennosides A and B, rhein, glucoside, thiocyanogen, tannins, reducing sugars and oxyanthraquinones	Roots: useful in skin diseases, tuberculos glands, syphilis and burning sensation
					Bark: laxative, anthelminthic, diuretic and is useful in boils, pustules, leprosy, ringworm, constipation
					Leaves: laxative
					Flowers: useful in skin diseases, pruritus, dry cough and bronchitis
					Fruits: useful in flatulence, colic, inflammations, rheumatism, intermittent fever, strangury and ophthalmopathy
		Cassia occidentalis	Whole plant	Sennosides, chrysophanol, emodin, glycosides, crude fibre, ash, calcium, iron, niacin	Leaves and seeds: used externally in skin diseases and also for cough, bronchitis, constipation, ulcers, epilepsy and hysteria

(continued)

Table 28.1 (continued)

S. No	Plant family	Plant name	Useful parts	Chemical constitutes	Uses
		Saraca indica	Bark flowers and seeds	Bark – tannins, catechol, sterol and organic calcium compounds	Bark: useful in digestion, fever, menorrhagia, leucorrhoea and pimples
				Methanol fraction –haematoxylene, tannin, saponin and water soluble glycoside (glucose,galactose and mannose)	Flowers: used in syphilis, haemorrhoids and scabies in children
					Dried flowers: used in diabetes, haemorrhagic dysentery
					Seeds: used for treating bone fractures, painful urination and stones in urinary bladder
		Tamarindus indica	Seeds, fruits, leaves and flower	Tartaric acid, citric acid, malic acid, oxalic acid, sugars, pectin, potassium and fats	Fruit: improves digestion, appetite, relieves constipation, dysentery, soothes sore throats, mild laxative, control nausea and vomiting
24	Cannaceae	*Canna indica*	Tubers	Starch	Seeds: powder used as dye
					Root: drug to induce perspiration, acts as diuretic and also controls fever
					Tubers: acts as antidote for poisonous grass in cattle
25	Sapindaceae	*Cardiospermum halicacabum*	Roots, leaves and seeds	Cyanogenic glycosides, alkaloid, L-triacontanol, n-pentacosane and saponins	Root: used to bring on delayed menstruation and to relieve backache and arthritis
					Leaves: stimulate blood circulation and are applied to painful joints to help clearing of toxins
					Seeds: treatment of arthritis

		Sapindus trifoliatus var. *emarginatus*	Roots, bark, seeds and fruits	Roots, bark, seeds and fruits – saponins (sapindoside A and B)	Fruit: bitter bad taste, tonic, stomachic, aphrodisiac, useful in chronic dysentery, diarrhoea, cholera and paralysis
				Nuts contain – kaempferol, quercetin and β-sitosterol	Main usage of soap nuts are like a hair shampoo
					Powder of soap nuts mixed with honey is given to cure tonsillitis
26	Lauraceae	*Cinnamomum camphora*	Leaves, branches, wood, roots and bark	Safrole, d-a-pinene, dipentene, cineol, terpineol, caryophyllene, limonene, sesquiterpine	It is externally applied to relieve arthritic and rheumatic pains, neuralgia cold sores and back pain
					Camphor from trunk wood is used for cardiac problems
		Cinnamomum zeylanicum	Bark, leaves and seeds	Bark, leaves – yield aromatic oils, eugenol and cinnamaldehyde	Used as astringent, stimulant, stops flatulence and also checks nausea and vomiting
				Seeds – fixed oil	Oil: stomachic, can be used against ulcers, relieve toothache and also used as spice
		Litsea chinensis	Roots, stem and bark	Laurotetanine, actinodaphnine, boldine, norboldine, sebiferine, flavonoid, naringenin, naringin, kaempferol, 7-glucoside, quercetin and glycine	Root: used as astringent, tonic, expectorant, aphrodisiac
					Useful for joint pains, thirst, throat troubles, diseases of spleen and paralysis
					Stem bark: useful in the treatment of diarrhoea and dysentery

(continued)

Table 28.1 (continued)

S. No	Plant family	Plant name	Useful parts	Chemical constitutes	Uses
27	Vitaceae	*Cissus quadrangularis*	Whole plant	Tetracyclic triterpenoids and steroidal principles	Cures cough and dyspepsia, scurvy, irregular menstruation and for joining fractured bones
				Stem – protein, fat, wax, fibre, carbohydrates, ash, tartaric, vitamin C, sodium, calcium, potassium, magnesium	
28	Lamiaceae	*Coleus aromaticus*	Leaves	Volatile oil from leaves – carvacrol, thymol, camphor	Leaves: useful for urinary disorders, dyspepsia
					Leaf juice: mixed with sugar is a powerful aromatic carminative
					Leaves decoction: chronic cough and asthma
		Coleus forskohlii	Tuberous roots	Diterpene 'forskolin' (coleonol)	Used against hypertension, glaucoma, asthma, congestive heart failure and certain cancers, obesity
		Leucas aspera	Leaves, flowers and roots	Flowers – aromatic oils	Plant juice: antidote for snake bite, controls cough, skin diseases, epilepsy, headache, rheumatism, jaundice, asthma, bowel complaints, conjunctivitis and as nasal drops in catarrh and cephalalgia
				Seeds – fixed oils	
		Majorana hortensis	Leaves	volatile oil, comprising geraniol, thymol, linalool, carvacrol and other terpenes, flavonoids, rosmarinic acid and triterpenoids	Used against cold, cramps, digestive disorders, swellings, rheumatism, colic and headache
		Ocimum basilicum	Seeds, leaves, roots and flowers	Yields aromatic oils – eugenol, methyl chavicol and linalool	Seeds: used in diarrhoea and chronic dysentery, cooling agents
					Leaf juice: cough
					Root: bowl complaints of children

		Ocimum canum	Leaves and seeds	Volatile oil, linalool, methyl chavicol, eugenol, caryophyllene, methyl cinnamate, cineole and terpenes	Plant decoction: cough
					Leaves: dysentery, cold, influenza, catarrh and bronchitis
					Leaf paste: external application for parasitic skin infections, used against kidney stones in homoeopathy
		Ocimum gratissimum	Whole plant	Oil – eugenol, methyl eugenol	Used in treating sunstroke, headache influenza, rheumatism and paralysis
		Ocimum sanctum	Roots, leaves and seeds	Volatile oils – eugenol, methyl chavicol, methyl eugenol and caryophyllene, flavonoids and triterpene	Leaf juice: diaphoretic, antiperiodic, stimulating and expectorant properties used in catarrh and bronchitis, applied to the skin in ringworm and other cutaneous diseases, earache, stomachic, gastric disorders of the children
					Root decoction: diaphoretic in malarial fevers
					Seeds: demulcent in disorder of urinogenital system
29	Burseraceae	*Commiphora wightii*	Bark	Yields resin and oleo gum resin – gives dark brown oil contains furfural	Resin tapped from the trunk is useful in sciatica, facial paralysis and rheumatic disorders, fixative in perfumery industry, in medicine as a substitute for African bdellium
					Also used as astringent, antiseptic, appetizer, improves digestion and leucocytes in blood

(continued)

Table 28.1 (continued)

S. No	Plant family	Plant name	Useful parts	Chemical constitutes	Uses
30	Hypoxidaceae	*Curculigo orchioides*	Roots	Contain – curculigo (saponins,triterpene alcohol)	Roots: soothing effect on skin and diuretic, reduces effects of hydrophobia
					A thin paste applied in case of spleen troubles
31	Periplocaceae	*Decalepis hamiltonii*	Roots	Quercetin, kaempferol, coumarin and rutin, lupeol, β-amyrin, 2-hydroxy 4-methoxy benzaldehyde and ferulic acid	Roots: used to cure indigestion, act as a cooling agent and blood purifier
					Also used to prepare sharbats and cool drinks
32	Dioscoreaceae	*Dioscorea alata*	Tubers	Sucrose on dry weight basis contains albuminoids, ash, fibre, carbohydrates, phosphorus	Useful in the treatment of piles, leprosy, tiredness, weight loss and lack of appetite
				Leaves – D-fructose, D-glucose and polyols	
		Dioscorea hispida	Tubers	Albuminoids, ash, fibre, carbohydrates, phosphorus	Juice is used as an antidote for snake poison
				Contain poisonous principle – dioscorine	
33	Asclepiadaceae	*Gymnema sylvestre*	Leaves	Gymnemic acid and saponins	Useful in inflammations, liver diseases, indigestion, constipation, jaundice, cardiopathy, conjunctivitis, leucoderma and diabetes
		Hemidesmus indicus	Roots	Tannins, saponins, volatile oil contains 2-hydroxy-4-methoxy benzaldehyde and sterols	Roots: used as substitute for sarsaparilla in loss of appetite, disinclination for food, fever, skin diseases, as blood purifier, in leucorrhoea, syphilis, rheumatism, against scorpion sting and snake bites
					Also used as hair tonic and sharbats

		Holostemma ada-kodien	Tubers	Protein, sugar, starch, fibre and ash	Act as laxative, used for cooling and correcting nutritional deficiencies
					Paste applied in eye problems and orchitis
					Also used in diabetes, gonorrhoea, cough and stomachache
		Tylophora indica	Roots and leaves	Tylophorine, flavonoids and tannins	Roots: treatment of asthma, bronchitis, whooping cough, dysentery and diarrhoea, rheumatic and gouty pains
					Paste of root applied externally for poisonous snake bites
					Roots and leaves: used in hydrophobia
34	Oleaceae	*Jasminum sambac*	Roots, leaves and flowers	Benzyl alcohol, benzyl acetate, linalool and linalyl acetate	Leaves and flowers: useful in the treatment of eye complaints, pruritus, leprosy, ulcers, hiccups, galactorrhoea, fever and skin diseases
		Jasminum grandiflorum	Whole plant	Jasmine volatile oil contains benzyl alcohol, benzyl acetate, linalool and linalyl acetate	Plant: bitter, astringent, acrid, thermogenic, aphrodisiac, antiseptic, anodyne, depurative and tonic
					Flowers: refrigerant and relieve tension
					Used externally to soothe dry or sensitive skin , against facial paralysis, mental debility, sterility, leprosy, skin diseases, fixing loose teeth, ulcers, corns and pruritus

(continued)

Table 28.1 (continued)

S. No	Plant family	Plant name	Useful parts	Chemical constitutes	Uses
35	Crassulaceae	*Kalanchoe lanceolata*	Leaves	Malic acid, isocitric acid and citric acids	Leaves: bitter in taste useful in treatment of heat swells, wounds, discolouration of skin, ulcers, boils, diarrhoea, dysentery and tumours
36	Lythraceae	*Lawsonia inermis*	Roots, leaves, flowers and seeds	Lawsone (colouring matter)	Henna used against athlete's foot disease and fungal infection, burning of soles and palms, local inflammation and jaundice, mental disorders, rheumatism, hair tonic, leucoderma and spermatorrhoea
					Flowers: useful in cephalalgia, burning sensation, insomnia and fever
37	Marantaceae	*Maranta arundinacea*	Rhizomes	Starch, crude protein, fat, crude fibre, ash and moisture	Useful in controlling dysentery, diarrhoea, indigestion, stomachache, respiratory diseases, cough and urinary troubles, acts as an antidote for wounds caused by poisonous arrows
38	Cactaceae	*Opuntia ficus-indica*	Whole plant	Mucilage in fruit, sugars, vitamin C, fruit acids and flavonoids in flowers	Plant: bitter, hot, laxative, carminative, antipyretic, cures biliousness, burning, leucoderma and urinary complaints
					Leaves: stomachic, cure inflammations, tumours and pains
					Flowers: cure bronchitis, asthma, reduce bleeding

39	Bignoniaceae	*Oroxylum indicum*	Roots, leaves, fruits, seeds and stem	Roots and stem – oroxylin A, baicalein, christen, tannic acid	Bark: acrid, bitter, pungent, astringent to the bowels, cooling, aphrodisiac, tonic, increases appetite, useful in fevers, bronchitis, intestinal worms, vomiting, dysentery, leucoderma, asthma, inflammation and anal troubles
				Seeds – fixed oil	Fruit: improves appetite
40	Oxalidaceae	*Oxalis corniculata*	Whole plant	Leaves – vitamin C	Leaves: cooling, refrigerant, stomachic, antiscorbutic effects
				Branches – malic acid	Dried leaves: fevers, dysentery, scurvy, biliousness, for removing corns and warts
				Leaves and branches – tartaric acid, citric acid, oxalic acid and potassium	
41	Piperaceae	*Piper longum*	Roots, spikes, branches and fruits	Piperine alkaloid and piperatin	Fruits: used as spice in pickles and preserves
					Fruits and roots: used in diseases of respiratory tract (cough, bronchitis, asthma) as snuff in coma, liver tonic, sedative in epilepsy, general tonic, haematinic, abortifacient and for miscellaneous purposes such as obstruction of bile duct and gall bladder, an anthelmintic, used in dysentery and leprosy
		Piper nigrum	Berries	Volatile oil, piperine (alkaloids) and proteins	Used against indigestion, chronic rheumatism, asthma, cough, throat complaints, worms, fever, phlegm, stimulant, piles and cutaneous troubles

(continued)

Table 28.1 (continued)

S. No	Plant family	Plant name	Useful parts	Chemical constitutes	Uses
42	Plumbaginaceae	*Plumbago zeylanica*	Roots and leaves	Plumbagin, isoshinanolone, plumbagic acid, beta-sitosterol, 4-hydroxybenzaldehyde, trans-cinnamic acid, vanillic acid, 2, 5-dimethyl-7-hydroxy chromone, indole-3-carboxaldehyde	Root: diuretic, expels phlegmatic secretions and is useful in rheumatism, treatment of dyspepsia, piles, diarrhoea, influenza and skin diseases, curing leprosy, plumbagin and possesses abortifacient and vesicant properties
43	Poaceae	*Saccharum officinarum*	Roots and cane	Fatty acid, alcohol, phytosterols, higher terpenoids, flavonoids, O- and C-glycosides, and phenolic acids	Wild sugarcane juice used as antidote for dog bite in tribal areas
					Roots: diuretic, sweet, astringent and aphrodisiac
44	Hippocrateaceae	*Salacia oblonga*	Stem bark and root bark	Kotalagenin 16-acetate, 26-hydroxy-1,3-friedelanedione, maytenfolic acid,3b,22a -dihydroxyolean-12-en-29-oic acid, 19-hydroxyferruginol, lamberticacid, (2)-49-Omethylepigallocatechin, salacinol, kotalanol, dulcitol, galactinol,3-O-a -D-galactopyranosyl(1→6)-O-b -D-galactopyranosyl- sn-glycerol, raffinose and stachyose	Stem bark: used in treating polyuria, diabetes, excessive thirst, discoloured spreading patches on the skin, diarrhoea and fever
					Root bark: used in gonorrhoea, rheumatism and skin diseases, swellings and ear diseases
45	Anacardiaceae	*Semecarpus anacardium*	Fruits and bark gum	Gum resin – bhilawanol, semecarpol	Used in treatment of leucoderma, indigestion, intestinal and heart diseases, rheumatism, diabetes, leprosy and beriberi
					Bark gum: controls venereal and nervous diseases

46	Solanaceae	*Solanum nigrum*	Whole plant	Leaf – riboflavin, nicotinic acid, vitamin C, β-carotene and citric acid	Leaves: used as poultice for rheumatic and gouty joints and skin diseases, free urine passage
				Fruit – glucose and fructose, vitamin and β-carotene	Decoction of berries and flowers: useful in cough, ratbite, bronchitis, pulmonary tuberculosis, fever, diarrhoea, eye diseases and hydrophobia
				Immature green fruit – four steroidal glycol (alkaloids, solamargine, a and b solanigrine and solasodine)	Seeds: useful in giddiness, inflammations and skin diseases
				Leaves, stem, fruits – diosgenin	Root bark: useful in ear pain, eye and nose diseases, hepatitis and for blood purification
		Solanum surattense	Fruits	Solasodine, solasonine and solanocarpine	Useful in helminthiasis, dental care, inflammation, joint pains, flatulence, colic, constipation, indigestion, loss of appetite, rheumatoid fever, cough, asthma, bronchitis, haemorrhoids and cardiac disorders
		Withania somnifera	Roots	Withanaloids (withaferin A and B)	Root powder: used for improving vitality and general health, against tuberculosis, respiratory diseases, rheumatism and free urine passage
					Fruit: diuretic
					Leaves: used in obesity

(continued)

Table 28.1 (continued)

S. No	Plant family	Plant name	Useful parts	Chemical constitutes	Uses
47	Sterculiaceae	*Sterculia urens*	Leaves and branches	Vitamin A	Leaves and tender branches steeped in water yield a mucilaginous extract useful in pleuropneumonia in cattle
					Gum: useful medicine in throat infections
48	Loganiaceae	*Strychnos nux-vomica*	Leaves, seeds, roots and bark	Alkaloids – strychnine, loganin, brusin, chlorogenic acid and fixed oil	Leaves: prevent body itches, tonic, stimulant and febrifuge, nervous disorders, applied as poultice on sloughing wounds and maggot-infested ulcers
					Root and bark: fevers
					Seeds: fevers, ulcers, dysentery, dyspepsia, piles and excited mental conditions
49	Portulacaceae	*Talinum cuneifolium*	Leaves and roots	Steroids, flavonoids, glycosides, saponins	Roots: cough, gastritis and pulmonary tuberculosis
					Powdered leaf: used in treatment of diabetes and mouth ulcers
50	Combretaceae	*Terminalia arjuna*	Bark and flowers	Bark – arjunolic acid, tomentosic acid, α-sitosterol, arjunetic glycoside, ellagic acid, saponin and leucodelphinidin, tannin, calcium, magnesium, aluminium	Bark: used in fractures, liver cirrhosis and hypertension
					Bark and flowers: used in cardiac troubles in Ayurveda and other Indian systems of medicine

		Terminalia bellirica	Bark and fruit	Fruit – tannins and β-sitosterol, gallic acid, ellagic acid and chebulagic acid	Useful in constipation, removes phlegm, cough, flatulence, used as laxative
				Heart wood and bark – ellagic acid	Mature dry fruit: useful in diarrhoea and dysentery
					Oil obtained from seeds: useful against indigestion, skin diseases, leucoderma and grey hairs
		Terminalia chebula	Fruit	Contains – tannins,chebulic acid, resin,fixed oil and anthraquinone	Controls indigestion,asthma,cough,vomitings,heart diseases and eye diseases
51	Ranunculaceae	*Thalictrum foliolosum*	Rhizomes and roots	Rhizomes – water soluble salts of berberine besides magnoflorine, palmitine and jatrorrhizine	Rhizomes and roots: treatment of conjunctivitis
					Root: antiperiodic, diuretic, mild purgative, tonic for indigestion
					Paste of the plant: used in skin diseases and to suppress boils
52	Menispermaceae	*Tinospora cordifolia*	Stem	Tinosporin, columbin, berberin, giloin and starch	Useful in all kinds of fevers, gout, anaemia, asthma, jaundice, leprosy, piles cardiac problems and swellings
53	Verbenaceae	*Vitex negundo*	Leaves, roots and seeds	Leaves – two alkaloids, nishindine and hydrocotylene	Roots: useful in rheumatic swellings, as a hair tonic, improve eye sight, sprains, malaria, cardiac disorders and diarrhoea
				Seeds – malic acid, organic acid	
54	Celastraceae	*Celastrus paniculatus*	Leaves, seeds and bark	Seeds – reddish yellow aromatic oil and alkaloids like celastrin and paniculatin	Oil is used as brain tonic which improves memory power
					Seeds: used in rheumatism, gout, paralysis and leprosy. Seed is made into a paste with cow's urine and is applied to cure scabies

(continued)

Table 28.1 (continued)

S. No	Plant family	Plant name	Useful parts	Chemical constitutes	Uses
55	Apiaceae	*Centella asiatica*	Whole plant	Asiaticoside, brahmoside, triterpenoid, trisaccharide and vitamin C	Used for the treatment of leprosy
					Leaves: taken as tonic for improving memory, useful both internally and externally in syphilitic skin diseases
		Cuminum cyminum	Seeds	Volatile oil, fixed oil, cumaldehyde, phosphorus, calcium, vitamin C, pinene, alpha-terpineol and flavonoids	Useful in diarrhoea, dyspepsia, strengthening of stomach and stopping of bleeding
					Oil: has a very cooling effect, and it is an ingredient in most of the prescriptions administered in gonorrhoea
		Foeniculum vulgare	Leaves, roots and seeds	Seeds volatile oil – contain anethole, limonene, methyl chavicol, a-pinene	Leaves: diuretic
					Roots: purgative
					Fruits: aromatic, stimulant and carminative
					Used to treat diseases like chest, spleen, kidney
					Fennel oil: good vermicide against hookworm
					Seeds: hot infusion of seeds used in indigenous medicines to increase milk secretion and to stimulate sweating

Table 28.2 Major Companies manufacturing herbal formulations in India

S. No.	Name of the company	City and state
1.	Ansar Drug Laboratories	Surat, Gujarat
2.	Acis Laboratories	Kanpur, Uttar Pradesh
3.	Aimil Pharmaceutical	New Delhi
4.	ALRASIN Marketing	Mumbai, Maharashtra
5.	Allen Laboratories	Kolkata, West Bengal
6.	BhartiRasanagar	Kolkata, West Bengal
7.	Dabur India Ltd.	Ghaziabad, Uttar Pradesh
8.	Dattatraya Krishna Sandu Bros.	Mumbai, Maharashtra
9.	Ayur Herbals Pvt. Ltd.	Vadodara, Gujarat
10.	Herbo-med (P) Ltd.	Kolkata, West Bengal
11.	The Himalaya Drug Co.	Bengaluru, Karnataka
12.	Indian Herb & Research Supply Co.	Saharanpur, Uttar Pradesh
13.	J & J Dechane Laboratories pvt. Ltd.	Hyderabad, Telangana
14.	Madona Pharmaceutical Research Pvt. Ltd.	Patna, Bihar
15.	Kruzer Herbals	New Delhi
16.	Shilpachem	Indore, Madhya Pradesh
17.	Hamdard (Wakf) Laboratories	New Delhi
18.	Zandu Pharmaceutical Works Ltd.	Kolkata, West Bengal
19.	Baidyanath Ayurveda Bhavan	Kolkata, West Bengal
20.	Charak Pharmaceuticals	Mumbai, Maharashtra
21.	Vicco Laboratories	Mumbai, Maharashtra
22.	The Emami Group	Kolkata, West Bengal

Legend: Source adopted from (Subhuti Dharmananda [8])

Table 28.3 Herbs of medicinal use in Canada

S. No.	Product name
1.	Echinacea
2.	Garlic (*Allium sativum*)
3.	Ginseng
4.	Camomile (*Chamomilla recutita*)
5.	*Ginkgo biloba*
6.	Evening primrose (*Oenothera biennis*)
7.	Devil's claw (*Harpagophytum procumbens*)
8.	St John's wort (*Hypericum perforatum*)
9.	Tea tree oil (*Melaleuca alternifolia*)
10.	Valerian (*Valeriana officinalis*)

exceptional role in treating diseases, viz., Ayurveda, Siddha, Unani, Homoeopathy and Western or modern medicine, providing the human race, particularly tribes, with remedies ([3]) (Table 28.3), (Fig. 28.2).

Out of the total medicinal consumption in China, traditional herbal drugs account for 30–50%. In the developing countries like Ghana, Mali, Nigeria and Zambia, 60% of children are treated with herbal drugs for fever resulting from malaria as initial home remedy. It is also noted that developed countries like Europe, North

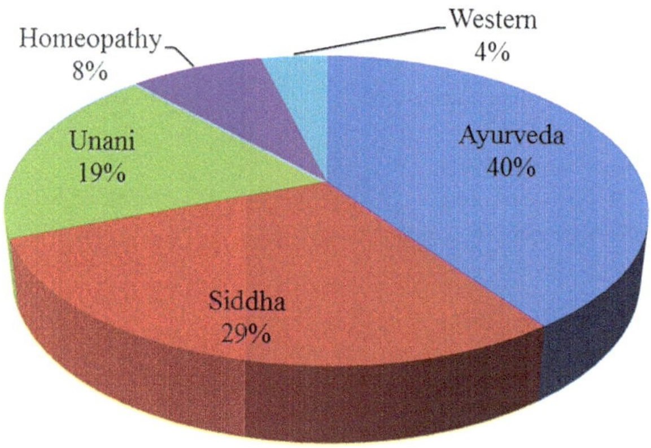

Fig. 28.2 Percentage of different systems of medicine used (Legend: Ayurveda a traditional method of disease treatment occupies 1st place in usage)

America and other industrialized regions, half the population use complementary or alternative medicine at least once. Whereas, in London, San Francisco and South Africa, 75% of individuals with AIDS and HIV use herbal drugs. Even in the most developed country, Canada and Germany, 70% and 90%, respectively, of the people use herbal drugs at least once in their life time. Between 1995 and 2000, the number of doctors who had undergone special training in natural remedy medicine had almost doubled to 10,800 [12, 13] (Table 28.4), (Fig. 28.3).

28.5 Benefits of Herbal Drugs

The yearning to seize the knowledge of traditional healing systems has led to a renaissance of curiosity in herbal drugs [9]. Thereby herbal products are combined with so-called 'complementary, alternative', 'integrative' or 'holistic' therapeutic interventions. Efficiency of herbal drugs is based on randomized, double-blind, placebo-controlled trials which in turn rely on traditional usage, endorsements and clinical trials. Systematic scientific studies are needed to support the traditional usage. In spite of this lacuna, herbal drugs are preferred over modern medicine due to cultural and historical reasons. Owing to their natural origin, herbal drugs are considered to be safe; unnecessary side effects can be avoided and can be part of a healthy lifestyle. Often herbal drugs are last resort to the unconquered diseases. Statistics point to the effectiveness of herbal drugs in therapy. Although modern medicine is developing day by day, herbal drugs on par with allopathy are patronized and used [5].

Nowadays herbal preparations are being marketed as nutritionally and medicinally enhanced foods called health foods. It will be inappropriate to call them as health foods since they are herbal medicines with therapeutic value, for example, genetically modified soybean and canola oil without trans-fatty acids, cereals supplemented with minerals and vitamins and engineered grains. Companies further introduced the term nutraceuticals to pure compounds of organic origin, for

Table 28.4 Popular herbal drugs in developing countries like India

S. No.	Product name	Composition	Uses
1.	*Asparagus racemosus*	Shatavarin ii	Adaptogen
		Shatavarin iv	
2.	Amla (*Embilica officinalis*)	Tannins	Rejuvenating agent
3.	*Andrographis paniculata*	Andrographolides	Liver support
4.	Ashwagandha (*Withania somnifera*)	Withanolides	Adaptogen
5.	*Adhatoda vasica*	Vasicine	Respiratory support
		Total alkaloids	
6.	Boswellin forte, Boswellin (*Boswellia serrata*)	Total organic acid	Anti-inflammatory
		β-Boswellic acid	Arthritis support
7.	Bacopin (*Bacopa monniera*)	Bacosides a and b	Memory support
8.	Bioperine (*Piper nigrum*)	Piperine	Nutrient bioavailability enhancer
9.	*Coleus forskohlii*	Forskolin	Circulation support, sports nutrition,weight management
10.	Calcium sennosides (*Cassia angustifolia*)	Total sennosides	Laxative action
11.	Centellin (*Centella asiatica*)	Total triterpenes	Skin, health, general tonic
12.	Curcumin C[3] complex® (*Curcuma longa*)	Curcuminoids	Antioxidant,anti-inflammatory
13.	Fenusterols® (*Trigonella foenum-graecum*)	Steroidal saponins	Sports nutrition
14.	Gugulipid® (*Commiphora mukul*)	Guggulsterones	Healthy cholesterol
15.	Ginger dry and soft extract (*Zingiber officinale*)	Gingerols	Digestive aid
16.	*Gymnema sylvestre* (*Gymnema sylvestre*)	Gymnemic acids	Healthy blood sugar
17.	Hot sip® (combination of four herbs)	Combination of *Adhatoda vasica*, *Alpinia galanga*, *Glycyrrhiza glabra*, *Piper longum*	Cough and cold aid
18.	*Inula racemosa*	Alantolactone	Circulation support, skin health
19.	Licorice (*Glycyrrhiza glabra*)	Glycyrrhizinic	Expectorant
20.	Momordicin® (*Momordica charantia*)	Bitter principles	Blood sugar support
		Charantin	

(continued)

Table 28.4 (continued)

S. No.	Product name	Composition	Uses
21.	*Mucuna pruriens*	L-dopa	Tonic, energy
22.	Neem leaf extract (*Melia azadirachta*)	Bitter principles	Liver support
23.	*Phyllanthus amarus*	Bitter principles	Liver support
24.	*Piper longum*	Piperine	Respiratory support, thermogenic
25.	*Picroliv®* (*Picrorhiza kurroa*)	Kutkin	Liver support
		Bitter principles	
26.	*Rubia cordifolia*	Total alkaloids	Skin health
27.	*Silbinol* ™ (*Pterocarpus marsupium*)	Pterostilbene	Blood sugar support
28.	*Tinofolin®* (*Tinospora cordifolia*)	Bitter principles	Urinary health support
29.	*Terminalia arjuna*	Arjunolic acid	Circulation support
30.	*Terminalia belerica*	Tannins	Rejuvenating agent
31.	*Terminalia chebula*	Tannins	Rejuvenating agent
32.	Trikatu (Ayurvedic formula) powder	Combination of *Piper longum*, *Piper nigrum*, *Zingiber officinale*	Respiratory support, thermogenic, digestive aid
33.	Triphala (Ayurvedic formula)	Tannins, combination of *Terminalia belerica*, *Terminalia chebula*, *Emblica officinalis*	Digestive aid
34.	Tulsi extract (*Ocimum sanctum*)	Ursolic acid	Blood sugar support, Rejuvenating agent
35.	Tylophora (*Tylophora indica/asthmatica*)	Total alkaloids	Respiratory support

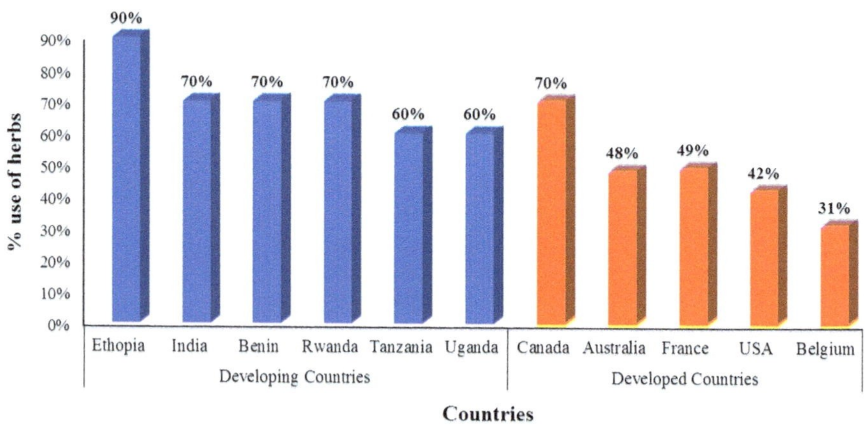

Fig. 28.3 Use of herbal drugs in developed and developing countries (Legend: Canada stands 1st place in developed countries with 70% usage of herbal drugs and in developing countries. Ethiopia stands 1st with 90% usage of herbal drugs)

example, lovastatin (lipid-lowering agent), docosahexaenoic acid (cardiovascular stimulant) and sterols and curcumin (for various health benefits). Misuse of herbal drugs should therefore be checked by regulatory agencies as was done by US-FDA in case of lovastatin as dietary supplement. To conclude since herbal medicines are prepared from raw materials that are renewable in ecofriendly processes, herbal drugs bring economic prosperity to the people who grow these raw materials [4] (Fig. 28.4).

28.6 Disadvantages of Herbal Drugs

Active principles of plant-derived therapeutics are isolated and characterized, and their mode of action is inferred in some cases, for example, ephedrine alkaloids from *Ephedra* species. But most of the herbal drugs marketed do not have such information, or it is partial and incomplete. Often these drugs are a combination of several molecules making it difficult to work out the mode of action, etc. The synergistic action of several active principles is responsible for the therapeutic outcome.

Adverse effects are found to be associated with the use of certain herbal preparations; hence, their use is questioned. This may be due to the interaction of active principles with other drugs which are simultaneously used. Use of ephedra for weight loss and kava-kava to reduce nervous disorders is associated with adverse effects like heart rate irregularities, heart attacks, stroke, seizures, psychoses, liver toxicity and death. So, some countries have issued the regulations banning such products. Some herbs like *Senecio riddellii*, *Rubia tinctorum*, *Aristolochia* and

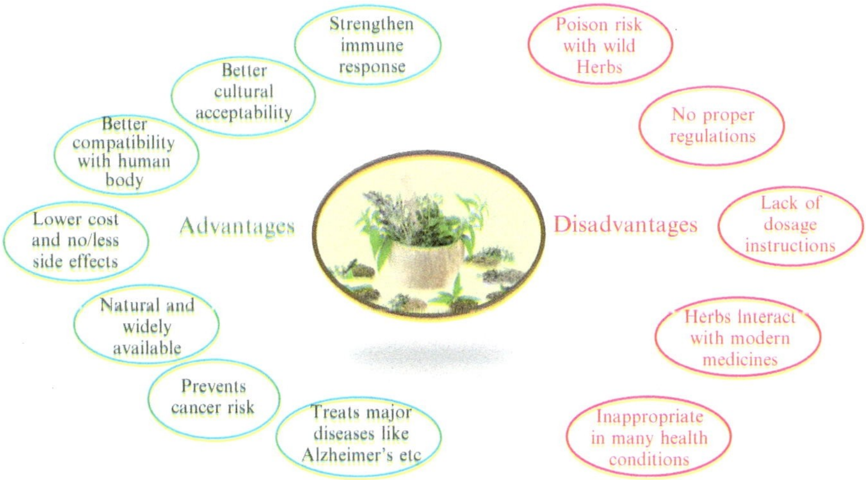

Fig. 28.4 Herbal drugs – a boon or a bane (Legend: Advantages and disadvantages of herbal drugs)

Morinda officinalis are suspected to be carcinogenic. Scientific approach in characterizing and elucidating the mode of action of the herbal products can guarantee the traditional claims as herbal drugs [5].

Secondly, contaminants and residues may occur in herbal medicines which are potentially hazardous. Contaminants can be categorized into biological and chemical contaminants. Various treatments involved during the growth of the plant (herbs) result in contaminants like organic solvents and agrochemicals. GACP (good agricultural and collection practices) and GMP (good manufacturing practices) as well as quality assurance measures can avoid contamination. According to WHO guidelines on GACP, human excreta cannot be used as a fertilizer, and animal manures should be completely composted to avoid microbiological and chemical contamination. Sewage management in agricultural areas should be properly taken care of as toxic elements and chemicals coming from industrial and domestic waste cause serious contamination (Table 28.5).

28.7 WHO Guidelines for Herbal Medicines

A group of experts invited in 1992 by WHO regional office for the Western Pacific recognized the importance of plant-based medicines for the well-being of people worldwide. They said that 'A few herbal medicines have withstood scientific testing, but others are used simply for traditional reasons to protect, restore or improve health. Most herbal medicines still need to be studied scientifically, although the experience obtained from their traditional use over the years should not be ignored. As there is not enough evidence produced by common scientific approaches to answer questions of safety and efficacy about most of the herbal medicines now in use, the rational use and further development of herbal medicines will be supported by further appropriate scientific studies of these products, and thus the development of criteria for such studies'. In 1996, WHO has issued guidelines for the assessment of herbal medicine [5, 11].

28.8 Conclusion

Scientific and systematic approach to the traditional medicine results in safer and cost-effective herbal drugs. A powerful search endorsed with scientific investigation into the Ayurvedic and other Indian systems of medicines will result in better and new herbal drugs. Considerable research on chemistry, pharmacology, pharmacognosy and clinical therapeutics need to be carried out to evaluate the efficacy of unexplored medicinal plants. This not only preserves traditional heritage but also rationalizes the usage of herbal drugs in health care. Constituting and following proper regulatory guidelines will ensure the safe, effective, affordable and quality herbal drugs.

Table 28.5 Major contaminants and residues in herbal drugs

S. No.	Contaminants/ residues	Examples	Sources
1.	Chemical contaminants	Lead, cadmium, mercury, chromium (arsenic, nitrite)	Polluted soil and water, during cultivation/growth, manufacturing process
		Dioxin aldrin, chlordane, DDT, dieldrin, endrin, heptachlor, mirex	Polluted air, soil and water, during cultivation/growth
		Cs-134, Cs-137	Air, soil, water during growth/cultivation
		Mycotoxins	Postharvest processing, storage and transportation
		Bacterial endotoxins	Postharvest processing, storage and transportation
2.	Biological contaminants	*Staphylococcus aureus*, *Pseudomonas aeruginosa*, *Salmonella* species, *Shigella* species, *Escherichia coli*	Soil, postharvest processing, storage and transportation
		Yeast, moulds	Postharvest processing, storage and transportation
		Protozoa – amoebae	Soil, excreta; organic farming/cultivation, manufacturing process
		Helminths – nematoda	
		Cockroach and its parts	Postharvest processing, storage and transportation
		Mouse excreta, earthworms, acarus	Postharvest processing, storage and transportation
3.	Solvents	Acetone, methanol, ethanol, butanol	Soil and water, during cultivation/growth, manufacturing process
4.	Agrochemical residues	Carbamate, chlorinated hydrocarbons, organophosphorus	Air, soil, water, during growth/cultivation, postharvest processing
		2,4-D, 2,4,5-T	Air, soil, water, during growth/cultivation, postharvest processing
		Dithiocarbamate	Air, soil, water, during growth/cultivation
		Ethylene oxide, phosphine, methyl bromide, sulphur dioxide	Postharvest processing
		Thiamethoxam	During cultivation
5.	Residual solvents	Acetone, methanol, ethanol, butanol	Manufacturing process

Legend: Source adopted from WHO guidelines with modifications, 1996

References

1. Brower V (1998) Nutraceuticals: Poised for a healthy slice of the healthcare market? Nat Biotechnol 16:728–730. https://doi.org/10.1038/nbt0898-728
2. Government of Telangana, Telangana State Biodiversity Board. http://www.tsbiodiversity.org/indiabiodiversity.html. Accessed 1 Feb 2017
3. Joshi K (2008) Indian Herbal Sector. India Science and Technology
4. Kamboj V (2000) Herbal medicine. Curr Sci 78:35–39
5. Li L (2000) Opportunity and challenge of traditional Chinese medicine in face of the entrance to WTO (World Trade Organization). Chin J Inf Tradit Chin Med 7:7–8
6. Nanjo F, Goto K, Seto R et al (1996) Scavenging effects of tea catechins and their derivatives on 1,1- diphenyl-2-picrylhydrazyl radical. Free Radic Biol Med 21:895–902. https://doi.org/10.1016/0891-5849(96)00237-7
7. Online project monitoring information system, National Medicinal Plants Board-Ministry of Ayush, 2017. http://www.nmpb.nic.in/
8. Subhuti Dharmananda the ayurvedic medicine industry in India, 2003. http://www.itmonline.org/arts/ayurind.htm
9. Tyler VE (2000) Herbal medicine: from the past to the future. Public Health Nutr 3:447–452
10. WHO (1991) Traditional medicine and modern health care. World Health Organ Rep A 44(20):1–11
11. WHO (1996) Annex II. Guidelines for the assessment of herbal medicines. WHO, Geneva
12. WHO (2000) General guidelines for methodologies on research and evaluation of traditional medicine World Health Organization. World Heal Organ:1–73. doi:WHO/EDM/TRM/2000.1
13. WHO (2003) Traditional medicine. World Health Organ 1:14–17

Printed by Printforce, the Netherlands